500 Electronic Projects

for Inven  *ts*

ARSATH NATHEEM S

Copyright © 2023 by ARSATH NATHEEM S

All rights Reserved. 2023 Edition

ISBN: 9798367025323

Imprint: Independently published

Author: Arsath Natheem S

This electronic project book is geared towards providing exact and reliable data with respect to the subject and issue covered. The publication is marketed with the idea that the publisher is not required to render or otherwise provide any qualified services.

In no way is it valid to reproduce, duplicate, or spread any part of this book in either digital e-format (eBook) or printed format. It is against the law to re-document this textbook, and you can't keep this manuscript unless the publisher gives you written permission to do so. All rights reserved.

The respective authors own all rights not held by the publisher. The attributes that are used are without any consent, and the book of the trademark is without permission or backing by the brand owner. All trademarks and brands within this book are for descriptive purposes only and are the owned by the owners themselves, not associated with this textbook

WHY I WROTE THIS BOOK

When most of us think of the word "Education," the first thing that comes to mind is the kind of formal learning that takes place in elementary, middle, and high schools as well as in colleges. Although this is not the only form of education, tacit and practical knowledge is something that we often cannot learn by reading books; it is obtained through real-world problems and practice, and it relates to experience, intuition, ideals, creative thinking, emotions, values, skills, and attitudes. Project-based learning creates the next generation of STEAM programs and tacit knowledge that give students the power to learn creatively in middle school, high school, and college and help them do well.

Learning through collaborative projects will ultimately result in the student experiencing a more profound impact on their education. It also makes the part about understanding more fun, since students can understand more when they learn in a group. Thus, let's focus on many easy Arduino, esp8266 NodeMCU, and esp32 projects to boost student creativity. Due to the fact that the electronic project is always operational and enhances the current system, students might ultimately acquire a practical, forward-thinking perspective.

"Tell me and I forget, teach me and I may remember, involve me and I learn" — Benjamin Franklin

WHY YOU SHOULD READ THIS BOOK

This book is intended for all of the people who are passionate about innovation. It contains 500 exciting projects that have a detailed functional description and electronic circuits. The target audience for this book is engineering students and electronics hobbyists. It is a massive library of ideas for unique projects and fresh creations. This handbook is for people who want to do creative projects using open-source tools and methods. It has a lot of ideas for how to do creative projects and make something new for society.

This book is divided into four chapters, beginning with basic electronic components and progressing to an introduction to Arduino sensors and boards, esp8266 NodeMCU, esp32, and Raspberry Pi. 300 electronic project ideas with tested circuits are included in the second chapter. The third chapter has 100+ Arduino projects. In the fourth chapter, there are 100 fascinating projects for esp8266 NodeMCU, esp32. Additionally, it contains operational fundamentals for open-source electronics, allowing college students, schoolchildren, and hobbyists to study electronics through useful schematic diagrams at all levels from beginner to expert. I hope that students and researchers will find this project book valuable for their mini-projects and an outstanding project guide for science fair projects and new inventive endeavors.

Dear Aspiring Electronic Lovers!

Congratulations! By choosing this book, you've taken a very significant step toward starting and making your own projects. It is our pleasure to thank you for your purchase and wish you success in your upcoming projects. In addition, these project ideas are not ranked in any particular order. The ideas are yours to choose based on what you feel comfortable with and are passionate about. In addition, it's important to figure out how to execute each of these ideas, since an idea is just the beginning. It is our sincere hope that you will find this book to be an invaluable resource both now and in the future. It's the beginning of a long journey to success.

Hopefully, this book will help you create your story.

Best wishes!

TABLE OF CONTENT

CHAPTER 1: BASIC ELECTRONICS COMPONENTS 1
- Introduction: 1
- Active Components 1
- Passive Components 2
- Electromechanical Components 2
- Resistors 3
- Capacitors 4
- Inductors 5
- Diodes 6
- Transistors 6
- Integrated Circuits 7
- Display Devices 8
- CRT 8
- Power Sources 8
- Relay 9
- LED 10
- Microcontroller 10
- Switches 11
- Seven Segment Display 11
- Test & Measurement Devices 11
- Oscilloscope 11
- Multimeter 12
- Signal or Function Generator 13
- Consumer Electronic Devices 13
- Industrial Electronic Devices 13
- Medical Devices 14
- Aerospace & Defence 14
- Automotive 14

CHAPTER 2: 300 ELECTRONIC PROJECTS 15
- 1. Sound-Operated Switch for Lamps 15
- 2. Battery-Low Indicator 15
- 3. IR sensors (infrared LEDs) to create an object-detection circuit 16
- 4. Automatic Water Pump Controller 17

5. Micro Inverter 18
6. Periodically on /off Mosquito repellents circuit 20
7. Cable Tester 21
8. Automatic 3-Phase Induction Motor Starter 21
9. Multipurpose Listening Device 23
10. Power-on Reminder with LED Lamp 23
11. Quality FM Transmitter 24
12. Medium-Power FM Transmitter 25
13. Simple Shortwave Voice Transmitter 26
14. Four-Stage FM Transmitter 26
15. Whisker for Robots 27
16. Farmhouse Lantern-Cum-Flasher 28
17. Accurate Foot-Switch 29
18. Easy Transistor Tester 30
19. TV Pattern Generator 31
20. Multipurpose Listening Device 31
21. DC-DC Converter 32
22. Remote Control using Wireless Doorbell 33
23. RF Signal Detector 34
24. Audio Mixer with Multiple Controls 35
25. Infrared Interruption Counter 37
26. Clock Tick-Tock Sound Generator and LED Pendulum 38
27. Battery Charger with Automatic Switch-off 38
28. Earth Leakage Tester 39
29. SCR-Controlled EHT Power Supply 40
30. 16-Way Clap-Operated Switch 41
31. Smart Loop Burglar Alarm 42
32. Temperature-Tolerance Checking System 43
33. Radiation Detector Circuit using LM358 IC 44
34. Stereo Headphone Amplifier 45
35. Electronic Thermostat for Fridge 45
36. Affordable Car Protection Unit 46
37. White LED Based Emergency Lamp and Turning Indicator 48
38. Mains-Operated Christmas Star 48

39. Christmas Lights Using LEDs 49
40. DIY: Make Your Own Nifty Night Lamp 50
41. Photometer 51
42. Smart Emergency Light 52
43. Digital Camera Adaptor 53
44. Mock Alarm with Call Bell 54
45. LED Based Reading Lamp 56
46. Multidoor Opening Alarm with Indicator 56
47. Rechargeable Torch Based on White LED 57
48. SMF Battery Guard 59
50. Stress Meter 62
51. Geyser Timer 62
52. Multicell Charger 63
53. Light Dimmer that Doubles as Voltmeter 64
54. 220V Live Wire Scanner 65
55. Smart Switch 66
56. Power Failure and Resumption Alarm 67
57. Zener Value Evaluator 68
58. Doorbell-Cum-Visitor Indicator 70
59. Liquid level alarm 71
60. Electronic Fuse 72
61. Water Tank Overflow Indicator 73
62. Simple smoke detector 74
63. Remote Emergency Alarm for Unmanned Lifts 75
64. Audio Controlled Running Light 76
65. Power Supply Reversal Corrector cum Preventer 76
66. Capacitor Evaluator 77
67. PIN Diode Based Fire Sensor 78
68. Blown Fuse Indicator 80
69. Ding Dong Touch Bell 81
70. Low-cost Stopwatch 82
71. Digitally Adjustable Dancing Lights 83
72. Shock Warning Circuit 84
73. IR Receiver Module Tester 84

- 74. 555 Timer PWM Audio Amplifier 85
- 75. Musical Water Shower 87
- 76. Multipurpose Power Pulser 88
- 77. Continuity Tester with a Chirping Sound 89
- 78. Electronic Combination Lock 91
- 79. Heat Control Unit 92
- 80. Electronic Heart 93
- 81. Ultrasonic Sound Beam Burglar Alarm 94
- 82. Sunset Lamp 95
- 83. Electronic Dice 96
- 84. Solid-state relay 97
- 85. Car Porch Guard 98
- 86. Wire Break Alarm with Delay 99
- 87. Cordless Multi-Door Alarm 100
- 88. School/College Quiz Buzzer 102
- 89. Mobile Phone Detector Using LM358 103
- 90. Anti-Sleep Alarm 104
- 91. Ultrasonic Proximity Detector circuit 106
- 92. Automatic Parking Light for Cars 107
- 93. Peak Hour Timer 108
- 94. Pressure Sensitive Alarm 110
- 95. Crystal Based 50Hz Generator 111
- 96. Electronic Ignition System for Old Cars 112
- 97. Versatile CMOS/ TTL Logic and Clock Probe 114
- 98. PC-based Oscilloscope Using Arduino 115
- 99. Turn Your Old Inverter into an Emergency Power System 116
- 100. DC Motor Speed Controller 117
- 101. AC-Powered Led Lamps Without rectifiers 118
- 102. Easy Transistor Tester 119
- 103. Door Guard 119
- 104. Low-cost Night Lamp 120
- 105. Briefcase Alarm 121
- 106. Low-Cost LPG Leakage Detector 123
- 107. Cupboard light 124

108. Simple Antenna Preamplifier for AM Radios 125
109. Multifunction Power Supply 126
110. Micro-Power Flasher 128
111. Optical Remote Switch 129
112. Infrared Toggle Switch 131
113. Contactless Telephone Ringer 132
114. Automatic Wash Basin Mirror Lamp Controller 133
115. Auto Muting During Telephonic Conversation 134
116. Solar-Powered Pedestal Lighting System 135
117. LED Illumination for Refrigerators 137
118. Electronic Reminder 138
119. Photodiode-Based Fire Detector 140
120. BODMAS Rule Circuit 142
121. Circuit for UPS to Hibernate PC 143
122. Accurate 1Hz Signal Generator 145
123. A Fourth-Order Speech Filter (Based on Texas Instruments Application Note) 145
124. Smart Battery Protector Using a Shunt Regulator 147
125. Microcontroller-Based Tachometer 148
126. Temperature Control & Indicator System 150
127. Stabilized Power Supply for Prototyping 151
128. Infrared Burglar Alarm 152
129. Street Light Controller 153
130. Light Operated Doorbell 154
131. Clock Tick-Tock Sound Generator & LED Pendulum 155
132. Automatic Bike Turning Indicator 157
133. Stabilized Power Supply for Prototyping 159
134. Propeller Message Display with Temperature Indicator 160
135. Low Power Voltage Doubler 162
136. Automatic water pump controller 164
137. Night Lamps 166
138. Continuity Tester with a Chirping Sound 167
139. Circuit for UPS to Hybernate Computer (PC) 169
140. Contactless Telephone Ringer 170
141. Automatic Dimness Controlled Lighting System 171

142. Simple HF Power Amplifier 172
143. Demo Circuit for Over-Voltage Protection 173
144. 3V PC Adaptor 174
145. PC Table Lamp 174
146. Audible Continuity Tester 175
147. Anti-Theft Alarm 177
148. DIAC Controlled Flasher 178
149. Stereo audio Distribution Buffer for headphones 179
150. PIN Diode Based Fire Sensor 180
151. Triple Mode Tone Generator 182
152. Bicycle Guard 184
153. Panic Alarm 185
154. Visual AC Mains Voltage Indicator 186
155. Simple Low-Cost White Noise Generator 187
156. Traffic Light Controller 188
157. Weather Station Using STM32 190
158. Simple Pulse Generator 192
159. Auto Reset Over/Under Voltage Cut-Out 193
160. This Stereo Amplifier Is Simple to Make 195
161. Simple Touch Sensitive Switch 195
162. Multi-way Switch Circuit 196
163. Car Reverse Horn 197
164. Bicycle Indicator 198
165. Fridge Door Alarm Circuit using 555 and LDR 200
166. 3 Phase Motor Programmable on and Off Controller 201
167. Digital Frequency Meter Using Arduino 203
168. Musical AF/IF checker 204
169. Flashing Light with twilight switch 205
170. Automatic Temperature Controlled Fan 207
171. Sound Operated Light 208
172. Electronics Thermometer 209
173. Milli-Ohm Meter with 0.1 To 1-Ohm Range 210
174. 70/40 Watts Hi-Fi amplifier 212
175. Thermistor-Based Fire Alarm System 213

176. Light sensitive switch 215
177. No-Load and Overload Protector for AC Motors 216
178. Microphone Amplifier 218
179. Multi Switch Controlled Relay 219
180. Walky-talky without using inductor or coil 221
181. Ohm Meter 222
182. Electronics Counter 224
183. Clap operated Remote Control for Fans 226
184. Mobile cellphone charger 227
185. Test a Diode | Zener Diode 229
186. Sound Pressure Meter 231
187. Watch Man Watcher 233
188. Under-/Over-Voltage Beep for Manual Stabilizer 234
189. Solar Tracking System 236
190. Automatic Heat Detector 237
191. Automatic Water Pump Controller 239
192. Little Power-Hila Vinegar Battery to power a calculator. 239
193. Night Vision Enhancer 240
194. Emergency Photo Lamp 241
195. 1W LED for Automotive Applications 242
196. Play with Robotic Eye (IR Sensor) 243
197. Faulty Car Indicator Alarm 244
198. Long-range Burglar Alarm Using Laser Torch 245
199. Soldering Iron Temperature Controller 246
200. Make your own Electric Bug Zapper 247
201. Timer for Mosquito Destroyer 247
202. Radiation Sensor 248
203. Handy Tester 249
204. Strain Meter 250
205. Water Pump Controller 251
206. Timer with Musical Alarm 252
207. Simple Key-Hole Lighting Device 252
208. Ball Speed Checker 254
209. Halogen lamp Saver for Bikes 255

210. HDD Selector Switch 256
211. Simple Key-Operated Gate Locking System 257
212. Mains Box Heat Monitor 259
213. Digital Soil Moisture Test 260
214. Over-Heating Indicator for Water Pipe 261
215. Linear Timer for General Use 262
216. Noise Meter 264
217. Mains Failure and Resumption Alarm 265
218. Multipurpose White-LED Light 266
219. IR Based Light Control 268
220. Sequential Device Control using TV Remote Control 270
221. Resistor Calculator 271
222. Triple-Mode Tone Generator 272
223. IR-Controlled Water Supply 273
224. Twilight blinker lamp 274
225. Electronic Street Light Switch 275
226. Standby Power-Loss Preventer 276
227. Touch Sensitive Alarm 277
228. Hum-Sensitive Touch Alarm 278
229. Room Sound Monitor 280
230. Security System Switcher 280
231. Doorbell-controlled Security Switch 282
232. Pencell Charge Indicator 283
233. Power Resumption Alarm and Low Voltage Protector 284
234. Flashing LED Light 285
235. Automatic Soldering Iron Switch 286
236. White LED Light Probe for Inspection 287
237. Calling Bell Using an Intercom 288
238. FM Bug 289
239. Digital Frequency Comparator 290
240. Bhajan and Mantra Chanting Amplifier 292
241. Low-Cost Automation Using PIC16F676 293
242. Mains Box Heat Monitor 295
243. Tachometer 296

244. Timer from Old Quartz Clock 297
245. Keep Away Ni-Cd from Memory effect 298
246. Periodically on off Mosquito ad hoc circuit 299
247. Crystal AM Transmitter 300
248. Programmable Electronic Dice 301
249. PC Based Candle Igniter 302
250. Sound Operated Intruder Alarm 303
251. Versatile LED Display 304
252. Multiutility flash light 305
253. Twi-light using white LEDs 306
254. PC Timer 307
255. Infrared Object Counter 308
256. Pushbutton Control for Single-Phase Appliances 309
257. Soldering Iron Tip preserver 311
258. Overspeed Indicator 312
259. Automatic Washbasin Tap Controller 312
260. 1.5W Power Amplifier 314
261. Wireless Stepper Motor Controller 316
262. Battery Low Indicator 318
263. Speed Checker for Highways 319
264. Simple Stereo Level Indicator 321
265. Manual EPROM Programmer 323
266. Noise Muting FM Receiver 324
267. PC Based Stepper Motor Controller 327
268. Digital Audio/Video Input Selector 329
269. Automatic Bathroom Light with Back-up Lamp 330
270. Simple Low-Power Inverter 331
271. Mains Interruption Counter with Indicator 332
272. FM Adaptor for Car Stereo 333
273. Panic Plate 335
274. Twinkle Twinkle X'mas Star 336
275. Car Fan Speed Controller 337
276. In Car Food and Beverage Warmer 337
277. Flasher 338

278. Home Automation Using Apple HomeKit and ESP8266 340
279. Hot-Water-Ready Alarm 341
280. Optical Smoke Detector 342
281. Capacitance-Multiplier Power Supply 342
282. Wireless PA for Classrooms 344
283. Low-Cost Battery Charger 345
284. Simple Automatic Water-Level Controller 346
285. Touch Based Doorbell 348
286. Electronic Ludo 348
287. Motorbike Alarm 349
288. Dual Motor Control for Robots 351
289. Environment and Weather Monitoring System 352
290. Long-Range IR Transmitter 353
291. Bench Power Supply Using a Computer's Power Supply 354
292. Leakage and Continuity Tester 355
293. 5-Watt Audio Amplifier Using TA7222 357
294. Battery-Powered Night Lamp Using an Old LED Bulb 357
295. Touchscreen and GLCD-Based Home Automation 359
296. GPS Clock using Arduino 362
297. Optical Slave Flash Trigger 363
298. Automatic Water Refilled for Air-Coolers 365
299. Electronic Horn 367
300. LME49710 Based Audio Amplifier 367

CHAPTER 3 370
TOP 100 ARDUINO PROJECTS 370
Introduction 370
Microcontroller 370
Development Board 371
What is the Arduino? 372
Why Arduino Developed? 372
What can Arduino be used to teach? 372
DIFFERENT TYPES OF ARDUINOS 373
What are the benefits of using Arduino UNO? 376
100 ARDUINO PROJECTS IDEAS 378

301. Arduino Based Autonomous Fire Fighting Robot 378
302. Robot Snake based on Arduino controlled by Android 379
303. Intelligent Gas Leakage Detector based on IoT 380
304. Wireless Black Box for Cars 382
305. Smart Charger Monitoring System using Arduino 383
306. Arduino Based Autonomous Fire Fighting Robot 384
307. Automatic Sketching Machine Project 386
308. Arduino based Sun Tracking Solar Panel 387
309. Fire Department Alerting System using Internet of Things and Arduino 388
310. Internet of Things based Irrigation Monitoring & Controller System using Arduino ... 389
311. Internet of Things based Smart Agriculture Monitoring System Project 390
312. Arduino Ultrasonic Sonar/Radar Monitor Project 392
313. Smart Dustbin with IOT Notifications 393
314. IOT Solar Power Monitoring System 395
315. Arduino PID based DC Motor Position Control System 396
316. Open-Source COVID-19 Pulmonary Ventilator 397
317. Arduino based Snake Robot Controlled using Android 399
318. Advanced Automatic Self-Car Parking using Arduino 400
319. IoT Industry Protection System Arduino 401
320. Rotating Solar Panel Using Arduino 402
321. GPS Clock using Arduino 403
322. Touch Free Hand sanitizer dispenser using LDR 404
323. Line Follower Robot with Arduino 407
324. IoT Based Home Automation controlled by smartphone 408
325. Covid-19 Patient Monitoring Device based on LoRa using The Things UNO 409
326. Open-Source Pulse Oximeter for COVID-19 411
327. Touch less doorbells can operate without touching the switch. 412
328. Social Distancing Device (Safety Card) 413
329. Automatic Faucet (Touchless) for COVID-19 Using Arduino 414
330. Automatic Hand Sensing Water and Soap Tank with Tap 415
331. DIY GPS Speedometer using Arduino and OLED 416
332. Automatic Bottle Filling System using Arduino 417
333. Control a Solenoid Valve with Arduino 418
334. An Arduino-based Gesture Controlled Air Mouse that uses Accelerometer 419

335. Arduino Whistle Detector Switch using Sound Sensor 421
336. Obstacle Avoiding Robot using Arduino... 422
337. Speed, Distance and Angle Measurement for Mobile Robots using Arduino and LM393 Sensor (H206).. 423
338. Build a Smart Watch by Interfacing OLED Display with Android Phone using Arduino 424
339. Arduino Bluetooth with MATLAB for Wireless Communication................................ 426
340. Smartphone Controlled Arduino Mood Light with Alarm... 428
341. Interfacing nRF24L01 with Arduino: Controlling Servo Motor 430
342. Build your own self-balancing robot with Arduino 432
343. Automatic Water Dispenser using Arduino 433
344. Interfacing Flame Sensor with Arduino to Build a Fire Alarm System 435
345. IoT Based Electricity Energy Meter using ESP12 and Arduino 436
346. Coronavirus Sterilizer Box | Food Mask Sterilizer 437
347. Play the Space Race Game using the Arduino and Nokia 5110 Graphic Display......... 439
348. Interfacing Tilt Sensor with Arduino.. 441
349. Bluetooth Controlled Servo Motor using Arduino 442
350. Controlling Multiple Servo Motors with Arduino 444
351. Arduino Based Countdown Timer.. 445
352. Automatic Pet Feeder using Arduino 446
353. Arduino Based AC Home Appliances controlling with thermistor and relay 448
354. DIY Arduino Inclinometer using MPU6050 449
355. Smart Blind Stick using Arduino.. 451
356. Home Automation Using Arduino with Bluetooth Control 453
357. Control your Computer with Hand Gestures using Arduino.. 454
358. Floor Cleaning Robot using Ultrasonic Sensor with an Arduino 456
359. Controlling a Stepper Motor using Potentiometer with Arduino 457
360. Arduino Based 3-Way Traffic Light Controller 459
361. Simple Arduino Audio Player and Amplifier with LM386 460
362. Arduino based Bluetooth Biped Bob (Walking & Dancing Robot).............................. 463
363. Arduino Radar System Processing with Ultrasonic Sensor 466
364. Heart Beat Monitoring over Internet using Arduino and ThingSpeak 469
365. IoT based Air Pollution Monitoring System using Arduino 472
366. IOT Based Dumpster Monitoring using Arduino & ESP8266..................................... 473
367. Arduino based Vehicle Tracker using GPS and GSM 475

368. Snake Game on 8x8 Matrix using Arduino 477
369. Prepaid Energy Meter using GSM and Arduino 479
370. Clap Switch using Arduino 482
371. Bluetooth Controlled Toy Car using Arduino 483
372. Automatic Water Level Indicator and Controller using Arduino 485
373. Tachometer using Arduino 487
374. Automatic Room Light Controller with Bidirectional Visitor Counter 489
375. Electronic Voting Machine using Arduino 491
376. Humidity and Temperature Measurement using Arduino 492
377. Automatic Door Opener using Arduino 493
378. LPG Gas Leakage Detector using Arduino 495
379. IR Controlled DC Motor using Arduino 496
380. DC Motor Speed Control using Arduino and Potentiometer 497
381. DIY Smart Vacuum Cleaning Robot using Arduino 499
382. Robot Car controlled by a mobile phone using a G-Sensor and Arduino 500
383. Weight measurement using Arduino, the HX711 Module, and a load cell 503
384. Automated Plant Irrigation System Using Arduino with Message Alerts 505
385. Making calls and sending messages with Arduino and GSM modules 508
386. Fingerprint Based Biometric Attendance System using Arduino 509
387. Generating Tones by Tapping Fingers using Arduino 512
388. The Arduino and Thingsboard are used to create a biometric attendance system based on IoT 515
389. Real Time Face Detection and Tracking Robot using Arduino 516
390. Arduino Touch Screen Calculator using TFT LCD 518
391. Arduino Motion Detector using PIR Sensor 520
392. Interfacing Hall Effect Sensor with Arduino 521
393. Automatic Call answering Machine using Arduino and GSM Module 523
394. Smart Blind Stick using Arduino 526
395. Arduino Metal Detector 528
396. Arduino Based Fire Fighting Robot 531
397. Interfacing Joystick with Arduino 533
398. Arduino RFID Door Lock 535
399. An introduction to Brushless DC Motors (BLDC) and how to control them on an Arduino 537
400. Automatic Medicine Reminder Using Arduino 539

CHAPTER 4: TOP 100 NODEMCU-ESP8266, ESP32 PROJECT IDEAS 542

    Introduction 542

        1. Pinout and description 542

        2. Power Requirement 543

        3. Various Peripherals and I/O 544

        4. On-Board buttons and LED 544

        5. Development Platforms 544

        6. Applications of ESP8266 544

    How are ESP32 and ESP8266 different from each other? 545

    What is Thingspeak? 546

    100 ESP8266, ESP32 PROJECTS 548

    401. ESP 8266 Wifi Controlled Home Automation 548

    402. World Wide Web Control via ESP8266 549

    403. Arduino + ESP8266: How to Make a DIY World Clock and Weather Bot 550

    404.Emergency Button for 7$, Arduino, WIFI and ESP8266 551

    405. DIY Arduino Wi-Fi Shield with ESP8266 for Home Automation That Can Be Controlled by Voice 553

    406. ESP8266 Weather Station with Arduino 554

    407. ESP8266 WiFi Temperature Logger 556

    408. Send Sensor Data (DHT11 & BMP180) to ThingSpeak with an Arduino, Using Cable or WiFi (ESP8266) or Use ESP8266 Alone 557

    409. An Inexpensive IoT Enabler Using ESP8266 562

    410. Firebase: Control ESP8266 NodeMCU GPIOs from Anywhere 564

    411. ESP8266 NodeMCU with Load Cell and HX711 Amplifier (Digital Scale) ... 565

    412. ESP8266 NodeMCU with TDS Sensor (Water Quality Sensor) 567

    413. ESP8266 NodeMCU: K-Type Thermocouple with MAX6675 Amplifier 569

    414. ESP8266 NodeMCU with BH1750 Ambient Light Sensor 571

    415. ESP8266 NodeMCU Web Server: Display Sensor Readings in Gauges 573

    416. ESP8266 NodeMCU Door Status Monitor with Telegram Notifications 575

    417. ESP8266 NodeMCU Web Server: Control Stepper Motor (WebSocket) 578

    418.ESP8266 NodeMCU with HC-SR04 Ultrasonic Sensor with Arduino IDE 580

    420.Multiple Sliders for ESP8266 NodeMCU Web Server (WebSocket): Control the brightness of LEDs (PWM) 583

    421. ESP8266 NodeMCU Plot Sensor Readings in Charts (Multiple Series) 586

422. ESP8266 NodeMCU Integrated with MPU-6050 Accelerometer, Gyroscope, and Temperature Sensor (Arduino) 589

423. The ESP8266 NodeMCU can be used to get Epoch/Unix Time (Arduino) 592

424. ESP8266 NodeMCU MQTT - Publish Temperature, Humidity, Pressure, and Gas Readings from a BME680 Sensor (Arduino IDE) 594

425. Web Server for ESP32/ESP8266 Relay Modules Built with the Arduino IDE (1, 2, 4, 8, 16 Channels) 596

426. DHT Temperature and Humidity Readings Displayed on the on-board ESP8266's OLED Display 597

427. Temperature and Humidity Web Server for ESP8266 DHT11/DHT22 with Arduino IDE 598

428. ESP8266 ADC – Read Analog Values with Arduino IDE, MicroPython and Lua 600

429. Hack a PIR Motion Sensor with an ESP8266 601

430. Power ESP8266 with Mains Voltage using Hi-Link HLK-PM03 Converter 603

431. ESP8266 Multisensor Shield with Node-RED 605

432. Control Sonoff Basic Switch with ESP Easy Firmware and Node-RED 607

433. ESP8266 Daily Task - Publish Temperature Readings to ThingSpeak 609

434. Touchscreen user interface for Node-RED provided by the Nextion Display with ESP8266. 610

435. ESP8266 Voltage Regulator (LiPo and Li-ion Batteries) 613

436. ESP8266 Weather Forecaster 617

437. ESP8266 Publishing DHT22 Readings with MQTT to Raspberry Pi 620

438. ESP8266-Based Do-It-Yourself Wi-Fi RGB LED Mood Light for $10 (Step by Step) 623

439. The ESP8266 is controlled by an Android app (MIT App Inventor) 625

440. ESP8266 - Wireless Weather Station with Data Logging to Excel 629

441. How to Make Two ESP8266 Talk 634

442. ESP8266 in conjunction with Node-RED and MQTT 635

443. Blynk Controlled Automatic Pet Feeder with Timer 637

444. Raspberry Pi Pico Web Server with ESP8266 & MicroPython 639

445. IoT based Smart Agriculture Monitoring System 640

446. Interfacing 5MP SPI Camera with NodeMCU ESP8266 643

447. Real Time GPS Tracker using ESP8266 & Blynk with Maps 645

448. WiFi Controlled Robot using ESP8266 & Android App 647

449. IoT Based Smart Kitchen Automation & Monitoring with ESP8266 649

450. IoT Indoor Air Quality Monitoring with BME680 BSEC & ESP8266 652

451. IoT MQTT Based Heart Rate Monitor using ESP8266 & Arduino 654

452. IoT Bidirectional Visitor Counter using ESPP8266 & MQTT 657

453. IoT IR Thermometer using MLX90614 & ESP8266 on Blynk... 660
454. IoT Temperature Monitor for Industry with MAX6675 and ESP8266 661
455. IoT ESP8266 Lux Meter using BH1750 Light Sensor & Blynk 662
456. IoT Based TDS Meter using ESP8266 for Water Quality Monitoring 664
457. IoT Smart Agriculture & Automatic Irrigation System with ESP8266 665
458. Home Automation using Google Firebase & NodeMCU ESP8266 667
459. IoT Water Flow Meter using ESP8266 & Water Flow Sensor 668
460. IoT Decibel meter with Sound Sensor & ESP8266 670
461. BMP180 Pressure Temperature Monitor on Thingspeak with ESP8266 673
462. IoT ECG Monitoring with AD8232 ECG Sensor & ESP8266 674
463. ESP8266 and Android Home Automation with WiFi and Voice Control 677
464. ESP8266 and DS3231 Based Real Time Clock (RTC) 678
465. MAX30100 Pulse Oximeter with ESP8266 on Blynk IoT App 681
466. IoT Based Air Pollution/Quality Monitoring with ESP8266 683
467. IoT Biometric Fingerprint Attendance System using NodeMCU 685
468. IoT Based RFID Attendance System Using Arduino ESP8266 & Adafruit.io 687
469. Voice Based Home Automation with NodeMCU & Alexa using fauxmoESP 688
470. IoT Based Patient Health Monitoring using ESP8266 & Arduino 691
471. Gas Level Monitoring Using ESP8266 & Gas Sensor Over the Internet 693
472. IoT Live Weather Station Monitoring Using NodeMCU ESP8266 694
473. IoT Based Analog/Digital OLED Clock using NodeMCU 696
474. Guide for TCA9548A I2C Multiplexer: ESP32, ESP8266, Arduino 698
475. ESP32: Guide for MicroSD Card Module using Arduino IDE 700
476. ESP32 IoT Shield PCB with Dashboard for Outputs and Sensors 702
477. ESP32 LoRa Sensor Monitoring with Web Server (Long Range Communication) 704
478. Visualize Your Sensor Readings from Anywhere in the World (ESP32/ESP8266 + MySQL + PHP) 706
479. Power ESP32/ESP8266 with Solar Panels (includes battery level monitoring) 708
480. Alexa (Echo) with ESP32 and ESP8266 - Voice Controlled Relay 710
481. How the HC-SR04 Ultrasonic Range Sensor Can Communicate with the ESP32 713
482. How does a Servo Motor Work and How to Interface it with ESP32? 714
483. Designing a Smartwatch using ESP32 - Magnetometer and Gyroscope 717
484. How Does a NEO-6M GPS Module Work and How to Interface it with ESP32 720
485. DIY ESP32 Oscilloscope 725
486. Smart Wi-Fi Video Doorbell using ESP32 and Camera 727

487. Monitoring the power output of solar panels over the Internet of Things using ESP32 and ThingSpeak 729
488. Audio Player Built with ESP32 for DIY Projects 732
489. ESP32-CAM Face Recognition Door Lock System 733
490. Automatic Hand Sanitizer Dispenser with COVID19 735
491. Bitcoin $ Price Tracker Using ESP32 & OLED Display 736
492. Connecting ESP32 to Amazon AWS IoT Core using MQTT 737
493. Aquarium Water Quality Monitor with TDS Sensor & ESP32 739
494. Monitoring the Indoor Environment Using an ESP32 and an LCD Display 741
495. Measure Wind Speed with Anemometer on ESP32 TFT Display 743
496. UV Index Meter with ESP32 & UV Sensor ML8511 745
497. IoT Based Electricity Energy Meter using ESP32 & Blynk 747
498. IoT Based Soil Nutrient Monitoring with Arduino & ESP32 749
499. Ultrasonic Range Finder with ESP32 TFT Display & HC-SR04 752
500. DIY IoT Water pH Meter using pH Sensor & ESP32 753

Summary 755
Essential Resources 755
Bibliography 758

CHAPTER 1:

BASIC ELECTRONICS COMPONENTS

Introduction:

There are various basic electronic components that are utilized in the construction of electronic circuits. Without these components, circuit designs can never be entirely finished or end up not functioning very well. Resistors, diodes, capacitors, integrated circuits, and a variety of other electronic components are included in this category. A few of these components have two or more terminals, and they are attached to circuit boards by means of solder. It's possible that some of them come in packaged forms, such as integrated circuits, in which a variety of semiconductor devices are combined.

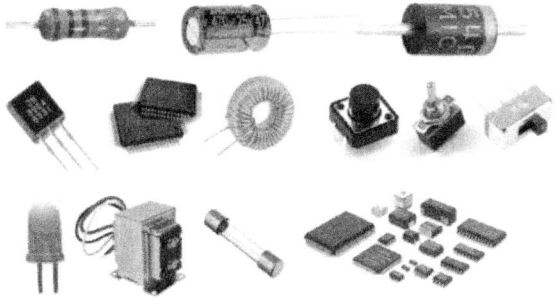

Components are the building blocks of every electronic system, and they can be used in a wide variety of applications outside just electronics. These elements are fundamental to the design of electrical and electronic circuits. These components each have at least two terminals that can be connected to the circuit in order to function properly. The active, passive, and electromechanical applications of electronic components can be used to help categorize the many types of electronic components.

The following are factors to be considered while planning an electronic circuit: Basic electronic components: capacitors, resistors, diodes, transistors, etc. Signal generators and direct current power supply are examples of sources of electricity. Instruments for measurement and analysis, such as Cathode Ray Oscilloscopes (CRO), multimeters, and other similar devices.

Active Components

These parts are crucial in the production of electricity by amplifying electrical signals. Within electronic devices, the operation of these components can be carried out in a manner analogous to that of an AC circuit in order to provide protection against voltage and increased power. Because it is power-driven by a source of electricity, an active component can carry out the activities that are assigned to it.

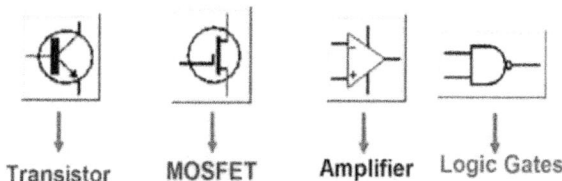

Active Electronic Components

All of these components need access to some sort of energy source, which is typically unavailable within a DC circuit. An oscillator, often known as an integrated circuit, and a transistor are essential parts of any active component of any quality.

Passive Components

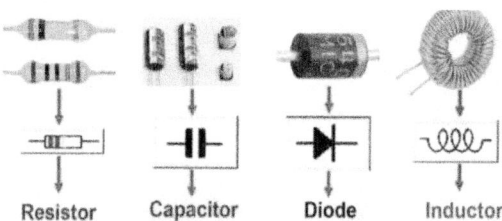

Passive Electronic Components

These kinds of components don't rely on a power source; therefore, they can't use the energy that's available from the AC circuit that they're connected to in an electronic circuit. This means that they can't use the mesh energy that's being introduced into the circuit. As a consequence of this, they are unable to amplify, despite the fact that they are capable of increasing a current, voltage, or both. The majority of these components have two terminals, and they include resistors, inductors, transformers, and capacitors.

Electromechanical Components

These components receive a signal from an electrical source in order to effect various changes in the mechanical world, such as turning a motor. In most cases, these components make use of electrical current in order to create a magnetic field and so make it possible to generate physical movement. These kinds of components can make use of a wide variety of switches and relays in their construction.

Electromechanical devices are those that operate through both an electrical and a mechanical process simultaneously. In order to provide electrical output through the mechanical movement of an electromechanical component, the component must be manipulated manually.

Passive Electronic Components

These components have the ability to either store energy in the form of current or voltage or to maintain that energy. The following discussion will focus on a few of these components.

Resistors

A resistor is a passive electrical device with two terminals that opposes or limits current flow. Ohm's law, which states that "voltage placed across the terminals of a resistor is directly proportionate to the current flowing through it," is the fundamental tenet upon which the operation of a resistor is predicated.

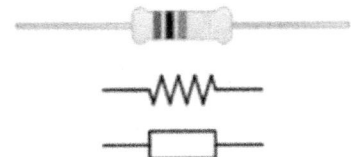

V=IR, Ohms are the units used to measure resistance. Where R is a constant that is referred to as resistance

The following characteristics, such as the power rating, type of material used, and resistance value, are used to further categorize resistors. These many sorts of resistors are utilized for a variety of purposes.

Resistor color codes

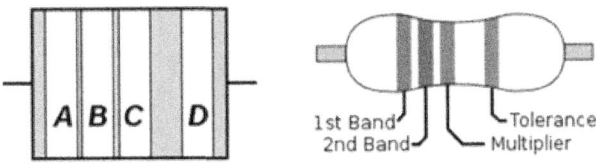

1st band = 1st number
2nd band = 2nd number
3rd band = # of zeros / multiplier
4th band = tolerance

Components and wires are color-coded according to their value and purpose so that they can be easily identified. The use of colored bands in resistor color coding allows for the easy identification of a resistor's resistive value as well as its percentage of tolerance, with the physical size of the resistor showing its wattage rating.

Tolerance: Gold = within 5%

Black:	0
Brown:	1
Red:	2
Orange:	3
Yellow:	4
Green:	5
Blue:	6
Violet:	7
Gray:	8
White:	9

Color Code Number: (BBROYGBVGW)

Fixed Resistors

In an electrical circuit, the proper conditions can be established with the help of this particular sort of resistor. Because the resistance values of fixed resistors are selected during the design phase of the circuit, there is no need to make any adjustments to the circuit as a result of this.

Variable Resistors

A variable resistor is a device that changes the resistance of an electronic circuit according to our needs. These resistors consist of a slider that taps on to a fixed resistor element as well as the resistor element itself. When calibrating a device with three terminals, variable resistors are frequently employed as one of the three-terminal devices. For further information, kindly follow this link. Gain more knowledge about resistors.

Capacitors

Capacitors store electrical energy as an electric field between two conductive plates separated by an insulator. In a timing circuit, a capacitor is used in conjunction with a resistor so that DC signals are blocked while AC signals are allowed through.

The equation for the charge that is being stored is $Q = CV$.

Were, The capacitance of a capacitor is denoted by the letter C.

V denotes the voltage that is being applied.

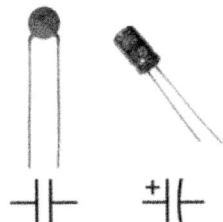

There are various types of capacitors, including film, ceramic, electrolytic, and variable capacitors. Methods based on numbers and color-coding can be utilized to determine its value, and LCR meters can also be utilized to determine the capacitance's value.

Inductors

An inductor is a type of AC resistor that stores electrical energy in the form of magnetic energy. This type of resistor is also known as an inductor. It is measured using the standard unit of inductance, which is the Henry. It is resistant to the fluctuations in the current. Inductance is defined as the capacity to produce magnetic lines from a magnetic field.

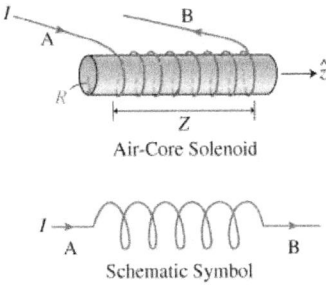

The formula for calculating the inductance of the inductor is as follows: $L = (.K.N2.S)/I$.

Were,

"L" stands for inductance,

The symbol for magnetic permeability is ","

The letter K denotes a magnetic coefficient.

"S" denotes the coil's total area of its cross-section,

"N" denotes the number of times the coils have been wound.

The length of the coil measured in the axial direction is denoted by "I."

A wide variety of passive electronic components, such as sensors, motors, antennas, memristors, and others, are also available. Few of the passive components are mentioned above in order to keep the level of complexity at a minimum throughout this text.

Circuit Symbols of Electronic Components

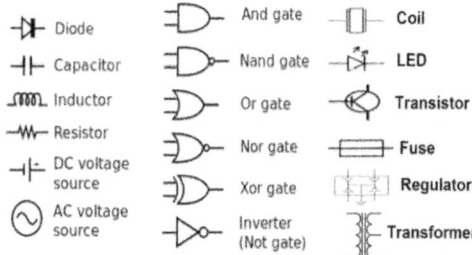

Active Electronic Components

These components are powered by an external source of energy and can control the flow of electrons through them. The term "semiconductors" refers to electronic components such as diodes, transistors, and integrated circuits. Other examples of these components include various types of displays such as LCD, LED, and CRTs, as well as power sources such as batteries, PV cells, and other AC and DC supply sources.

Diodes

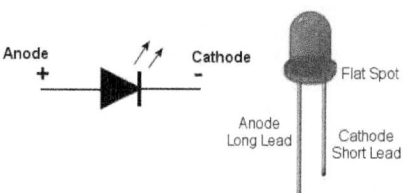

A diode is a device that permits current to flow in just one direction. It is often built of semiconductor material. It features two terminals, one for the anode and one for the cathode. The most common application for these is in the process of converting circuits, such as AC to DC circuits. These can be classified as PN diodes, Zener diodes, LEDs, photodiodes, and a variety of other subtypes.

Transistors

A transistor is a type of semiconductor device that has three terminals. Most of the time, it is used as a switch and sometimes as an amplifier. Control of either voltage or current can be exercised over this switching device. Controlling the voltage that is applied to one terminal allows one to regulate the amount of current that flows through the other two terminals.

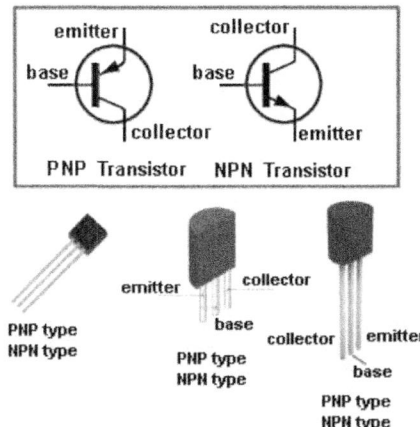

Field-effect transistors (FETs) and bipolar junction transistors (BJTs) are the two varieties of transistors that are available (FET). Additionally, these transistors might be either PNP or NPN types.

Integrated Circuits

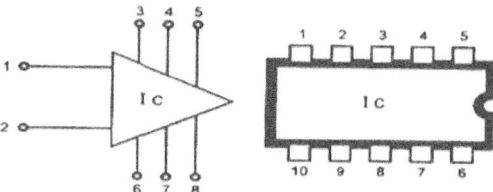

An integrated circuit is a specialized component that is created by fabricating hundreds of transistors, resistors, diodes, and other electronic components on a single tiny silicon chip. Integrated circuits are used in a variety of electronic devices. These are the fundamental components that make up modern electronic gadgets such as mobile phones, laptops, and so on.

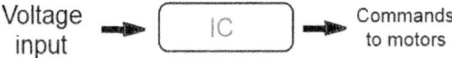

These integrated circuits may be either analog or digital in nature. Op-amps, timers, comparators, switches, and other types of ICs are the most common types of integrated circuits found in electronic circuits. Depending on the task at hand, these might be categorized as either linear or nonlinear integrated circuits (ICs).

Display Devices

LCD stands for liquid crystal display, which refers to a type of flat display technology that is typically implemented in devices such as computer monitors, screens for mobile phones and other electronic devices, calculators, and so on. This innovation makes use of two polarized filters and electrodes to selectively block or let light through from the reflective backing to the eyes of the observer.

In electrical and electronic circuits, the module that is utilized the most of the time is the display, such as a 16X2 LCD. An alphanumeric display is the name given to a particular type of display that has two rows and sixteen columns. The maximum of 32 characters can be displayed with this type of display. For further information regarding 16 x 2 LCD, kindly follow this link.

CRT

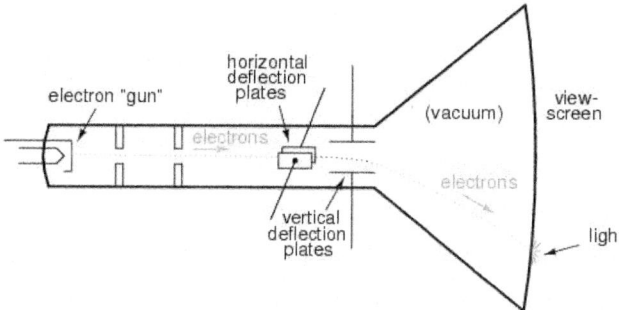

Cathode ray tube display technology is predominantly utilized in televisions and computer screens that operate by moving an electron beam back and forth over the screen's back. An electron gun, an electron beam, and a phosphorescent screen are all exterior components of this tube, which is an extended vacuum tube with a flattened surface.

Power Sources

Batteries and a DC power supply are the two distinct types of power sources that are utilized in the circuits.

DC Power Supply

The direct current power supply, which is just one type of power source, is an extremely important component of electronic circuitry. The primary electronic components are designed to operate with a DC power supply because it is a reliable source of power. Alternating current to direct current (AC to DC), switched mode power supplies (SMPS), linear regulators, and other types of power supplies are utilized in the circuit to give the supply. In some projects that would ordinarily require 12V power, a 5V source is necessary, therefore the DC power supply is replaced with a wall adapter instead.

Batteries

The battery is just one example of a device that can store electrical energy. This device is utilized to convert the chemical energy into electrical energy in order to offer power to a variety of electronic devices. Some examples of these devices include mobile phones, flashlights, computers, and so on.

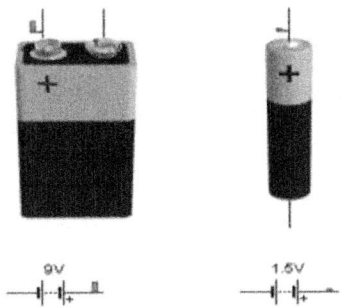

These are made up of one or more cells, and each of these cells has an anode, a cathode, and an electrolyte inside of it. Batteries are sold in a variety of sizes, and these sizes can further be broken down into primary and secondary categories. Primary batteries are only utilized until their power is depleted, after which point they are discarded, whereas secondary batteries can still be put to use even after their power has been depleted. In some of the circuits, 1.5V AA batteries are utilized, whereas in others, 9V PP3 batteries are used.

Relay

When operating the circuits electronically, an electromagnetic switch similar to the relay is used. When operating the circuits electromechanically, a relay is used. Because the operation of a relay requires only a little amount of current, these devices are typically employed in control circuits to effect changes in low-level currents. However, relays can also be used to manage high electric currents when properly configured. A smaller amount of current is all that is required to turn on a new circuit when using a relay switch. These relays are either of the electromechanical or solid-state variety.

An electromechanical relay, also known as an EMR, is comprised of a coil, frame, contacts, armature, and spring. This frame of the relay provides support for the many sections, and the armature is one of the parts that moves. The armature is moved by a magnetic field, which is produced by winding a coil or a wire made of copper around a rod made of metal. The contacts and other conducting elements are what are used to open and close the circuit.

Input, output, and control circuits are the three types of circuits that can be used to construct a solid-state relay, also known as an SSR. The input circuit functions in the same way as a coil, the control circuit acts as a coupling device between the input and output circuits, and finally, the output circuit functions in the same way as the contacts within an electromechanical relay. When compared with electromechanical relays, these relays have a very high level of popularity because to their low cost, high level of dependability, and very high level of speed.

LED

Light-emitting diode is what people mean when they talk about LEDs. It is a device made of semiconductor material that can produce light whenever there is a current source flowing through it. When charge carriers like electrons and holes unite in a semiconductor material, this results in the production of light. These light-emitting diodes (LEDs) are classified as "solid-state devices" when they generate light within a solid semiconductor material.

Indium gallium nitride, also known as InGaN, is the material that is used in the production of LEDs. These LEDs have a high level of brightness and are available in the colours green, blue, and ultraviolet. LEDs that are referred to as AlGaInP (Aluminum Gallium Indium Phosphate) have a high level of brightness and are offered in orange, yellow, and red colour options. GaP, also known as gallium phosphide, can be found in both green and yellow varieties.

LEDs have a wide variety of applications, ranging from magical light bulbs to cell phones to big display boards used for advertising purposes. LEDs are also utilized in some advertising displays. Due of the remarkable capabilities that these gadgets provide, there has been a recent uptick in the rate at which they are being utilized. These gadgets have a very low power consumption and a very small footprint.

Microcontroller

A microcontroller is a type of integrated circuit designed to perform a certain function within an embedded system. Memory, a processor, and input/output peripherals are all integrated onto a single chip in this device. Sometimes people refer to these as MCUs, which stands for microcontroller units; otherwise, embedded controllers are used. Most of the time, these are used in robots, cars, medical devices, office equipment, home appliances, vending machines, mobile radio transceivers, and other things.

The central processing unit (CPU), memory, program memory, data memory, I/O peripherals, and other components are the components that make up the microcontroller. Additionally, it is compatible with components such as ADC, DAC, serial port, and system bus.

Switches

A switch is a type of electrical component that connects or disconnects the conducting path within a circuit, allowing electric current to be supplied or interrupted from one conductor to another. The most common type of switch is an electromechanical device, which consists of one or more electrical contacts that are moveable and coupled to other circuits.

Current will begin to flow in the circuit as soon as the contacts in the circuit have been linked to one another. In a same manner, if the contacts are not linked, then there will not be any passage of current. Switches can be designed in a variety of various designs, and their operation can be carried out manually in the same manner as a button on a keyboard, a light switch, or other similar devices. A switch can also function as a sensor device, namely a thermostat, to determine the location of a machine part, the level of liquid, temperature, pressure, and so on. This can be accomplished in a number of different ways.

The various kinds of switches that may be purchased on the market today include rotary, toggle, pushbutton, mercury relay, and circuit breakers, amongst others. Once the switches have been unlocked, they must use high-powered circuits and have a certain design in order to prevent critical arcing from occurring.

Seven Segment Display

A 7-segment display is a display module that is utilized in a lot of different applications. The primary purpose of this device is to display decimal numbers on a variety of electronic devices, including but not limited to meters, clocks, information systems in public areas, and calculators, among other things.

Test & Measurement Devices

When connecting or designing electrical or electronic circuits, testing and measuring various parameters is very important. These parameters include voltage, frequency, current, resistance, capacitance, and so on. As a result, the testing and measurement tools, such as multimeters, oscilloscopes, signal or function generators, and logic analyzer, are utilized.

Oscilloscope

The most trustworthy piece of testing equipment, such as an oscilloscope, is the one that is used to monitor signals that are continuously changing in some way. When we use this equipment, we are able to observe the changes that take place within an electrical signal, including changes in current, voltage, and the passage of time. Oscilloscopes can be used in a variety of fields, including electronics, industrial medicine, automotive, and telecommunications, among others.

These are built to have CRT screens, which stands for cathode ray tube, although almost all of these devices are digital at the moment, including some more advanced capabilities like memory and storage.

Multimeter

An electronic instrument known as a multimeter is a combination of three separate meters: an ammeter, an ohmmeter, and a voltage meter. The primary function of these devices is to determine various circuit characteristics, such as the voltage, current, and other values, within AC and DC circuits. In the past, meters were of the analog kind, which included a pointing needle. In contrast, modern meters are of the digital type and are therefore referred to as DMs, which stands for "digital multimeters." These instruments are available in a variety of forms, including handheld and bench devices.

FUNCTION OF BASIC ELECTRONIC COMPONENTS	
Terminals and Connectors:	Components necessary for establishing an electrical connection.
Resistors:	Components that resist current flow.
Switches:	Conductive or nonconductive components (open).
Capacitors:	Constituents of an electrical field that are capable of storing electrical charge.
Magnetic or Inductive Components:	These are some examples of electrical components that make use of magnetism.
Network Components:	Components that utilize multiple kinds of Passive Components.
Piezoelectric devices, crystals, resonators:	Components that employ the piezoelectric effect that are passive.
Semiconductors:	Control components that are electronic and do not contain any moving parts.

Diodes:	Elements that are only capable of conducting electricity in a single direction.
Transistors:	A device made of semiconductors that is capable of amplifying signals.
Integrated Circuits or ICs:	A computer circuit that is built into a chip or a semiconductor. It is a whole system, not just one part.

Signal or Function Generator

Signal generators generate various signals to debug and test circuits. Saw tooth, sine, triangle, and square waves are the four types of signals that are produced by the signal generator the majority of the time. In the process of designing electrical circuits, one of the most important tools is a function generator. Other important tools include an oscilloscope and a bench power supply.

Different uses for various electronic components

An electronic circuit that regulates and controls the flow of current in order to carry out several purposes, such as amplifying a signal, transmitting data, and doing computations It is possible to construct it using a wide variety of electronic components including resistors, capacitors, inductors, diodes, and transistors. The following section will cover the various applications of these components.

Consumer Electronic Devices

These components can be found in a wide variety of consumer devices, including but not limited to calculators, personal computers, printers, scanners, and fax machines. Appliances for the home such as air conditioners, refrigerators, washing machines, vacuum cleaners, and microwave ovens, among others.

The components that make up audio and video equipment, such as televisions, DVD players, headphones, VCRs, loudspeakers, and microphones, amongst other things. Modern electronic equipment such as automated teller machines, set-top boxes, smartphones, barcode scanners, DVD players, MP3 players, HDD jukeboxes, and so on.

Industrial Electronic Devices

These parts are used in a variety of applications, including motion control, industrial automation, motor drive control, machine learning, robotics, mechatronics, power conversion technologies, biomechanics, photovoltaic systems, power electronics, and renewable energy applications. Using various forms of communication technologies, the smart grid system collects data in order to respond in a manner that is proportional to the amount of power being consumed.

It is the purpose of computers, intelligence, and carefully organized systems of electricity. These electronic components can be used in a variety of automation applications in industries, including motion control and others. At the moment, machines are gradually replacing people because of their ability to save time and money while simultaneously improving output. In addition, the level of security for works that cannot be controlled is also evaluated.

Medical Devices

In order to record data and conduct physiological research, modern gadgets are currently being put into use. They have been shown to be more effective in diagnosing ailments as well as in the treatment of illnesses. Because of shifts in the patient's pulse, body temperature, blood flow, and respiration, these components can be employed in medical equipment such as respiratory monitors. These monitors are used to determine whether or not the patient's condition has improved.

A defibrillator is a piece of medical equipment that delivers an electric shock to the muscles of the heart in order to restore the organ to its normal functioning state. A glucose meter is what's needed to determine how much sugar is currently present in the blood. The number of heartbeats can be sped up or slowed down with the use of a pacemaker.

Aerospace & Defence

Aerospace and defense use include rocket launchers for space, cockpit controllers, rocket launchers, and boom barriers for military use. Other applications include aircraft systems, radars for the military, missile launching systems, and cockpit controllers.

Automotive

These are the few fundamental electronic components, followed by a condensed description of each of the linked subtopics. In addition to the symbols for electronic components, the reader may also obtain a fundamental understanding of these components. We are the industry leaders when it comes to building electronic projects that combine these fundamental components with sophisticated controllers. As a result, readers are encouraged to leave comments below addressing any assistance they may require for testing these components and practically creating electronic circuits.

CHAPTER 2:

300 ELECTRONIC PROJECTS

1. Sound-Operated Switch for Lamps

This cheap, fully transistorized, sound-operated lamp switch is very sensitive to sound signals and turns on a lamp when you clap within 1.5 meters of the switch. One interesting way it could be used is in discos, where the lights could go on and off in time with the music beats or clapping.

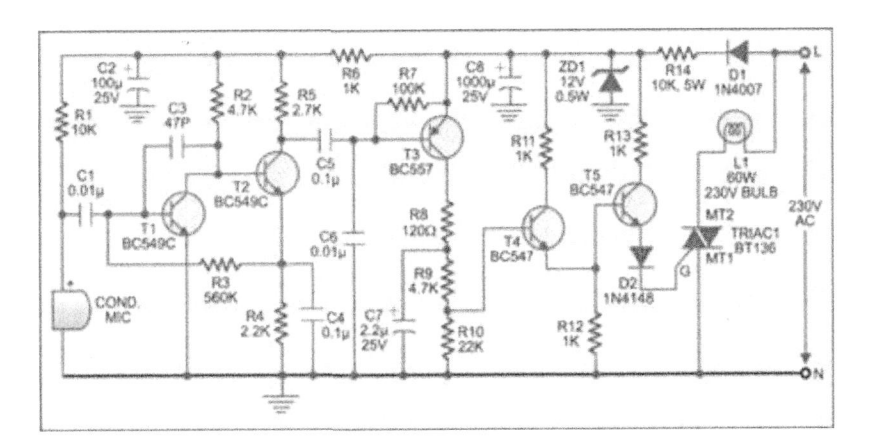

The condenser microphone picks up on sound and turns it into changes in electricity. The two-stage direct-coupled (DC) amplifier, which is made up of transistors T1 and T2, boosts the electrical signals, which are then sent to the switching circuit. The switching circuit is made up of transistors T3, T4, and T5, which only work when sound signals are detected. Transistor T5 gives the triac enough gate voltage to run the 230V lamp. Using resistor R14, diode D1, and zener diode ZD1, the 12V DC power for the circuit is regulated from the AC mains. The circuit can be put together on any kind of PCB.

2. Battery-Low Indicator

Rechargeable batteries should not be depleted below a minimum voltage. This lower limit on voltage is different for each type of battery. This simple battery low indicator circuit can be used with 12V batteries to show when the voltage drops below the value set by the user. A flickering LED is used to show what is going on. Voltage comparator IC LM319 is the heart of the circuit (IC1). It is a dual comparator that has an output that works with TTL. We have only used one comparison in this case. A reference voltage of 1.2 volts is sent to the comparator's non-inverting input (pin 4) from the band-gap reference diode D1 (LM385). A voltage is sent to the comparator's inverting input (pin 5) from the potential divider made up of resistors R2 and R3 and the preset VR1. So, if you are using a 12V battery and want to know when the voltage drops below 10.5V, adjust the voltage at the inverting input using reset VR1 to get a voltage of 1.2 volts (with battery voltage at 10.5V).

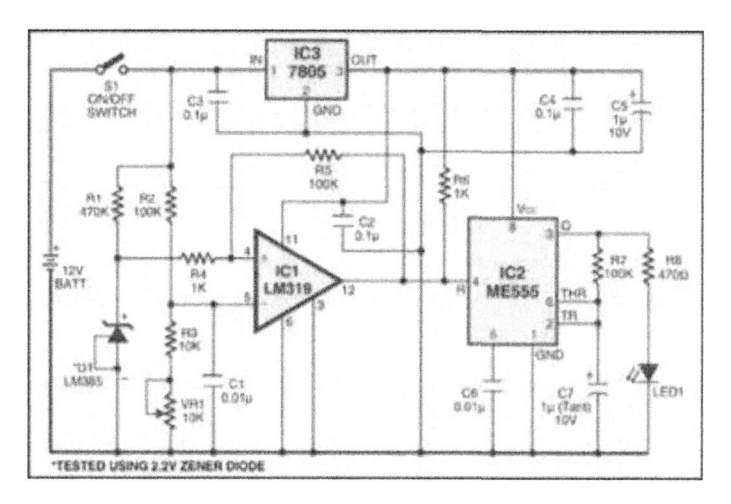

When the battery is first fully charged, the voltage at the inverting input of IC1 is higher than the voltage at the non-inverting input, so IC1's output pin 12 stays low. The reset pin (pin 4) of IC2 is connected to pin 12 of IC1 and stays low. The astable multivibrator built around IC2 does not oscillate. Because of this, LED1 doesn't blink. When the battery voltage drops below 10.5V, the voltage at the inverting input of IC1 becomes lower than the voltage at the non-inverting input, and the output of IC1 goes high. When pin 12 of IC1 goes high, the reset pin of IC2 also goes high, and the astable multivibrator built around IC2 begins to oscillate. When LED1 flickers, it means that the voltage of the battery is low and that the battery needs to be charged before it can be used again. Both IC1 and IC2 run on +5V DC that is controlled by IC 7805's voltage regulator (IC3).

3. IR sensors (infrared LEDs) to create an object-detection circuit

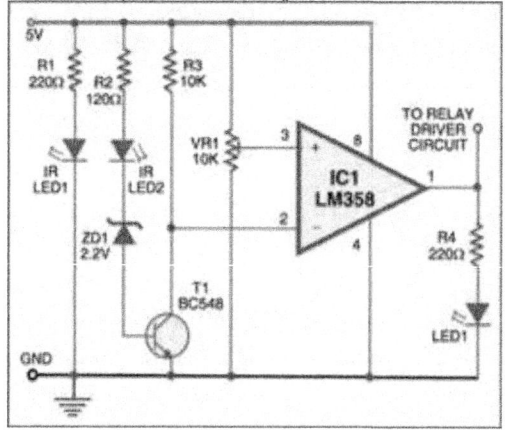

IR sensors are used in many things, like TV remote controls, burglar alarms, and counting how many objects are in a room. Here, we've used infrared LEDs (IR sensors) to make an object-detection circuit and a proximity sensor for robots that follow a path.

The basic idea is to use an IR LED to send out infrared light, which is then reflected by an object in the way and picked up by an LED further down the line. When the light hits the sensing LED, a voltage difference forms between its leads. But this voltage makes an electrical signal that is very weak. So, we used an operational amplifier (op-amp) to find the object with more accuracy.

The circuit for finding objects is up there. The emitter and the receiver are the two parts that make up the circuit. The emitter is made up of an IR LED1 connected in series with a 220-ohm resistor that limits the current going forward. IR LED1 always sends out IR light. The IC LM358 (IC1), the IR LED2, the Zener diode ZD1, the transistor T1, the resistors R2 and R3, and the preset VR1 make up the receiver.

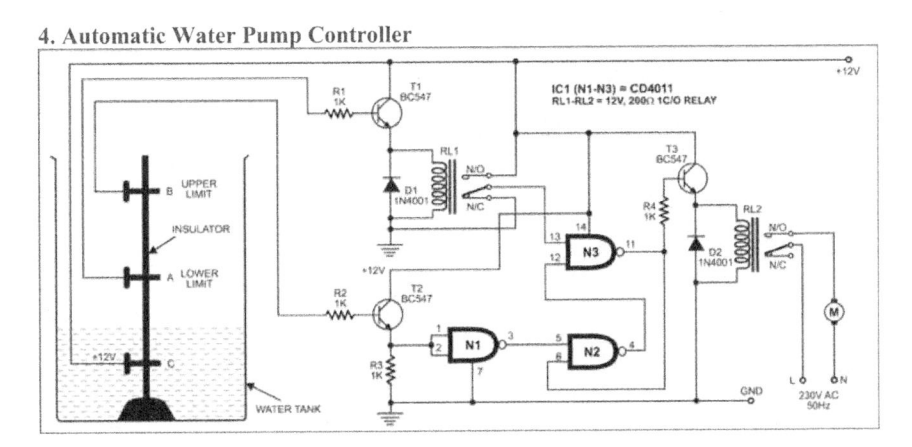

4. Automatic Water Pump Controller

Here is a circuit for an automatic water pump controller that runs the motor of the water pump. When the water level in the overhead tank (OHT) drops below the lower limit, the motor turns on by itself. In the same way, it turns off when the tank is full. The circuit is simple, small, and cheap because it is made up of only one NAND gate IC (CD4011). It gets its power from a 12V DC power supply and doesn't use much power.

The controller circuit is shown in Figure 1. Let's say there are two reference probes, "A" and "B," inside the tank. "A" is the lower-limit probe, and "B" is the upper-limit probe. The 12V DC power supply goes to probe C, which is the minimum amount of water that can always be kept in the tank.

Indicator

Figure 2 shows the circuit for indicators and monitoring. It is made up of five LEDs that light up to show how much water is in the tank above. Since the water at the bottom of the tank gets 12V power, transistors T3 through T7 get base voltage and conduct to light up the LEDs (LED5 down through LED1).

When the water level in the tank drops to level C, the transistor T7 starts to conduct, and LED1 lights up. When the water level in the tank gets to be one-fourth full, transistor T6 conducts, and LED1 and LED2 light up. When the water level in the tank gets to be halfway full, transistor T5 conducts and lights up LED1, LED2, and LED3. When the water level reaches 3/4 of the tank, transistor T4 conducts and lights up LED1 through LED4. When the tank is full, T3 conducts, and all five LEDs light up. So, one can tell how much water is in the tank by how bright the LEDs are. The LEDs can be put anywhere, making it easy to keep an eye on them.

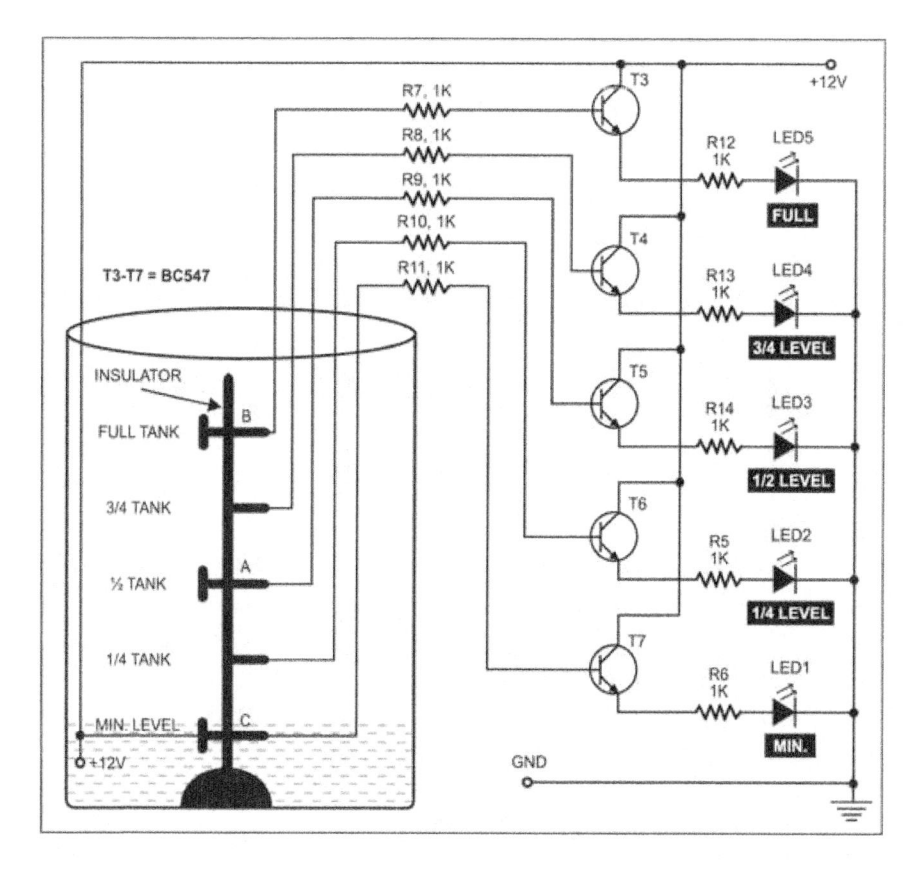

Note: By changing the heights of probes A and B, the user can change how much water needs to go into the tank. The stand and the screws for adjusting it should be insulated so that they don't shor t out.

5. Micro Inverter
This page talks about a simple, low-power inverter circuit that turns 12V DC into 230V AC. It can be used to power very light loads like night lights and cordless phones, but by adding more MOSFETs, it can be changed into a powerful inverter.

Battery charger circuit

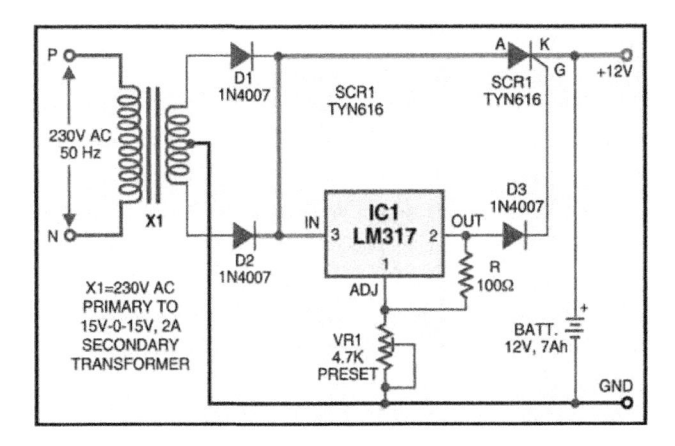

This circuit has an inverter circuit and a two-stage battery charger with a cut-off and a battery level indicator. As shown below, the charging circuit is built around IC1 (LM317). When 230V AC is available from the mains, IC1 sends the gate voltage through diode D3 to SCR1 (TYN616) (1N4007). SCR1 starts the charging process. For setting the voltage at the output, VR1 can be used.

Following is a picture of the circuit for the battery level indicator and inverter. The system for checking the battery level is made up of the transistors T1 and T2 (both BC547) and a few other separate parts.

When the battery is fully charged, say to more than 10.50V, LED1 lights up and the piezo-buzzer PZ1 doesn't make any noise. On the other hand, when the battery voltage drops (say, below 10.50V), LED1 stops glowing and a piezo-buzzer sounds. This means that the battery is empty and needs to be recharged before it can be used again.

Inverter circuit

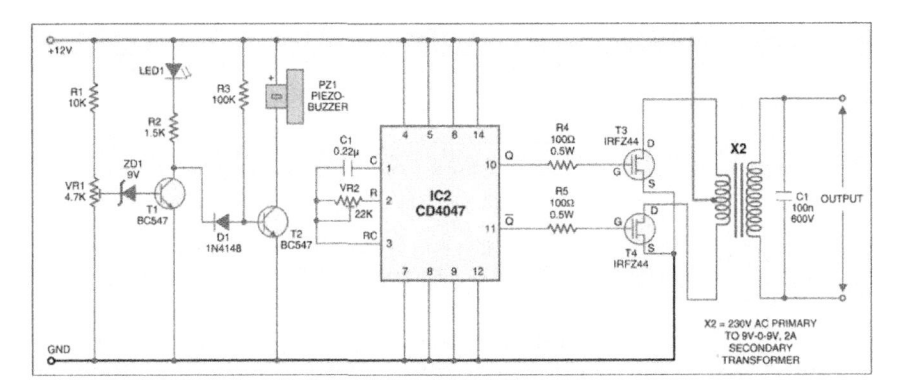

The inverter is built around IC2 (CD4047), which is wired as an astable multivibrator with a frequency of about 50 Hz. Power MOSFETs are driven directly by IC2's Q and Q outputs (T3 and T4). The push-pull configuration uses the two MOSFETs (IRFZ44). The output of the inverter is cleaned up by capacitor C1.

6. Periodically on /off Mosquito repellents circuit

Some mosquito repellents on the market use a poisonous liquid to make poisonous vapors that make mosquitoes leave the room. Due to the constant release of poisonous fumes into the room, the natural balance of the air's components for good health reaches or passes the critical level after midnight. Most of the time, these vapors get into the brain through the lungs. They also have a small or large percentage of anesthetic effects on mosquitoes and other living things. If you are exposed to these toxic vapors for a long time, you may have problems with your brain or nerves.

Here is a circuit that turns the mosquito repellent on and off after a certain amount of time. This lets you control how much toxic gas gets into the room.

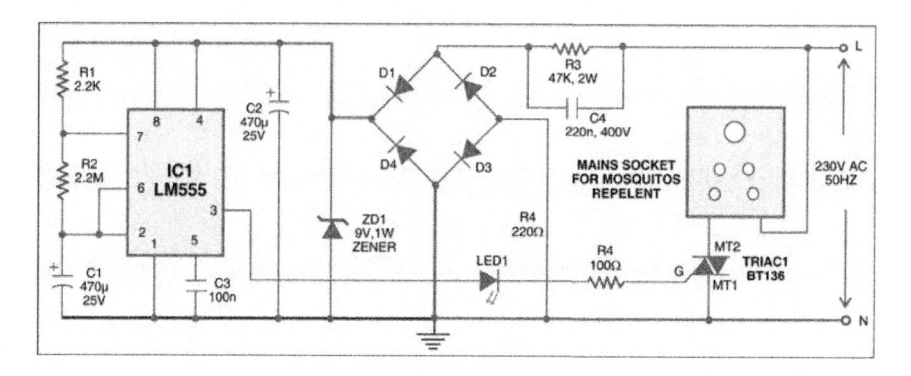

The circuit periodically turns the mosquito repellent "on" and "off" for about 20 minutes each time. So, if you leave the mosquito repellent on from 10 pm to 6 am (eight hours), it will be "on" for four hours and "off" for four hours of the total time. During "off" time, the air in the room tries to get back to its natural state. Another important thing about the circuit is that when it goes from "on" to "off," it doesn't make any noise or click like a relay does, so it doesn't wake you up when you're sleeping.

Circuit Operation

The circuit is made up of IC 555 (IC1), which is a timer, and triac BT136, which is an automatic switch (TRIAC1). The power for the circuit comes from the AC mains. This is done by stepping it down and reversing it. By getting rid of the transformer, space and money are saved. Timer IC1 gets a steady 9V DC power supply from Zener diode ZD1 and capacitor C2.

The resistors R1 and R2 and the capacitor C1 make up the timer section. Through LED1 and R4, the output of timer IC 555 is sent to the gate terminal of BT136. When the timer output goes high, it turns on TRIAC1's gate, and LED1 shows the "on" time. During "off" time, IC1's output is low, so TRIAC1 isn't turned on and LED1 doesn't light up.

7. Cable Tester

Have you ever wondered if a certain cable was giving your device power from the wall? Here is a way to test the continuity of a cable without touching it directly. The circuit can detect AC signal frequencies, and if the cable is conducting, an LED will light up. The circuit is very sensitive and can pick up signals from the cable's surface, so there's no need to touch the cable directly. Other cables, like modem, audio/video, and dish antenna cables, can be tested with the circuit.

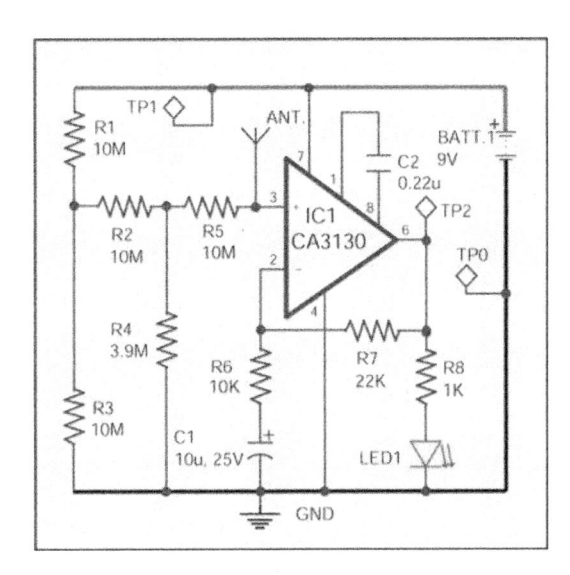

Shows how the cable tester works. The op-amp CA3130 is the center of the circuit (IC1). The operational amplifier IC1 has MOSFET inputs and a CMOS output. It has PMOS transistors on the inputs, which give it a very high input impedance and very good speed performance. Even weak signals are easily picked up by the input.

The feedback resistors R6 and R7 set the closed-loop voltage gain for IC1, which is wired as a non-inverting amplifier. By adding capacitor C1, a gradual roll-off is made possible. The resistors R1 through R5 are used to set the level of the non-inverting input of IC1. For phase compensation, a capacitor C2 is added between pin 1 (offset null) and pin 8 (strobe). How the circuit works is easy to understand. Bring the circuit's antenna close to the wire you want to test. If the cable is doing its job well, the LED will light up.

8. Automatic 3-Phase Induction Motor Starter

Star-to-delta converters are often used to start a 3-phase induction motor. When the motor is turned on, the stator coils are connected in a "star" shape. This is changed to a "delta" shape when the motor reaches 3/4 of its full speed and the stator coils have built up enough "back electromagnetic force" (emf).

The 3-phase induction motor starter circuit shown here has two main benefits: it stops a single phase from happening and it automatically changes from star to delta. It can only be used with motors that are rated for connection in a delta configuration at the given line voltage and have both ends of each of the three stator windings available separately.

How induction motors work

The AC induction motor, which is also known as the squirrel cage motor, has a simple rotor that looks like a cage and a stator with three windings. The AC line current in the stator creates a changing field that causes a current to flow in the rotor. This current interacts with the field and makes the motor turn. The number of poles in the stator windings and the frequency of the AC input voltage determine the AC motor's base speed. When you put weight on a motor, it slips the same amount as the weight.

Circuit Description

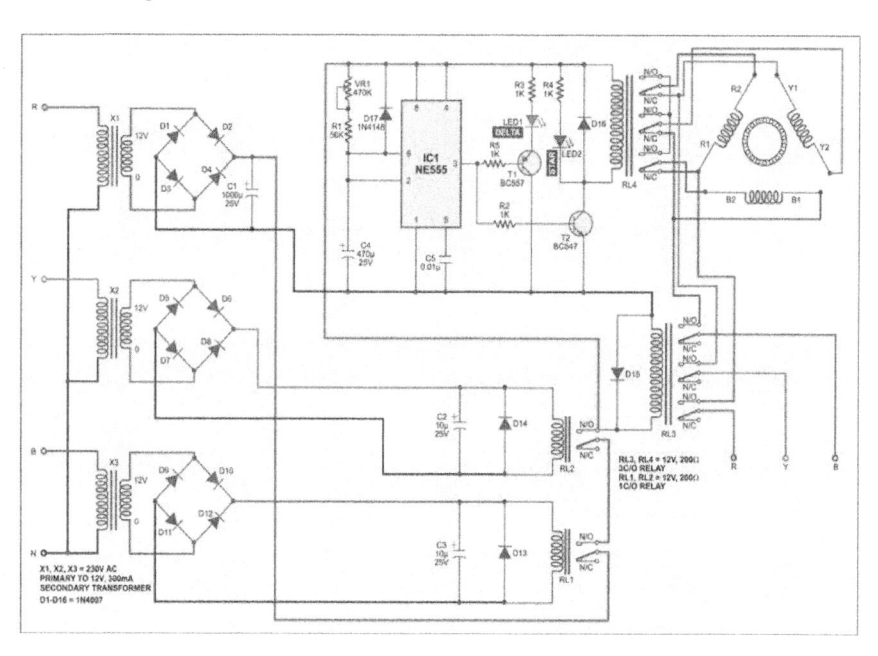

Figure: Starter circuit for a 3-phase induction motor

To step down the 3-phase supply, three single-phase transformers are used. Transformers X1, X2, and X3 take the phases R, Y, and B and step them down. This gives the secondary output of 12V at 300 mA. A full-wave rectifier and a capacitor are used to change the direction of the current coming out of the transformer. Relays RL1, RL2, and RL3 are controlled by the three 12V DC supplies. When all three phases are present, the 12V DC supply from the R phase is sent through the contacts of relays RL1 and RL2 to the coil of relay RL3 and the timer circuit. As a result, relay RL3 energizes.

Circuit shows the automatic star-to-delta converter's circuit, which is made up of a single-phase preventer and a timer. The timer NE555 (IC1), which is set up as a monostable multivibrator, is also set off at the same time. Its length of time is set by capacitor C4, resistor R1, and setting VR1. With VR1, you can set how long it takes for the motor to reach 3/4 of its full speed. The negative pulse that triggers IC1 comes from the combination of resistors VR1, R1, and C4.

Through resistor R2, the output of pin 3 of the timer is linked to the base of transistor T2. So, transistor T2 is driven to saturation, and relay RL4 is turned on (indicated by glowing of LED2). So, when the power is turned on, relays RL3 and RL4 turn on (if all three phases are present) to connect the stator windings in a star shape. If you follow the connections, you'll see that the R phase is connected to the R1 end of the R windings, the Y phase is connected to the Y1 end of the Y windings, and the B phase is connected to the B1 terminal of the B stator windings. To make a "star connection," the other ends of all the stator windings (R2, Y2, and B2) are connected.

9. Multipurpose Listening Device

This listening device can pick up distant sounds that are very quiet and clear. It can be used in large meeting rooms, auditoriums, movie theatres, college lecture halls, etc. The circuit can be kept in a shirt pocket in a small plastic box. It's especially good for watching TV at a low volume so as not to wake up other people in the house.

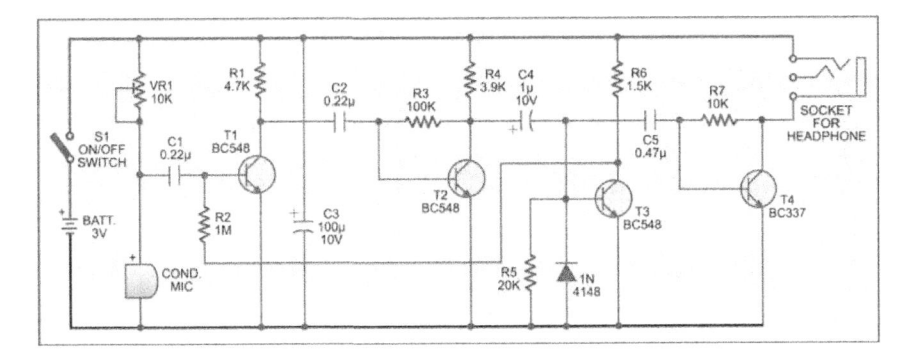

A microphone uses electricity to pick up sounds. It's good to use because it has a wide range of frequencies and is sensitive. To make the circuit, you need four transistors and a few resistors and capacitors. Two 1.5V AA-size batteries power the circuit. Transistors T1 and T2 boost the sounds that the microphone picks up. With the 1-mega-ohm resistor R2 in the feedback path, transistor T3 keeps the level steady. This is important because the device needs to make weak sounds louder and weaken loud sounds to a safe level.

10. Power-on Reminder with LED Lamp

The majority of the time, the equipment at workstations is left on without anyone noticing it. In this situation, these might get too hot and break. Here is an add-on for the workbench power supply that sounds a buzzer for about 20 seconds every hour or so to let you know if the devices you have plugged in are on or off. It also has a white LED that gives off enough light to find things when the power goes out.

In Fig. 1, the circuit of a power-on reminder with an LED light is shown. Here, IC NE555 (IC1) is wired as an astable multivibrator. With resistors R1 and R2, preset VR1, and capacitor C1, the time period is set to about six minutes so that the buzzer goes off every hour. IC1's output is connected to IC CD4017's clock input (IC2). IC2 gets a power-on reset pulse from capacitor C3 and resistor R3.

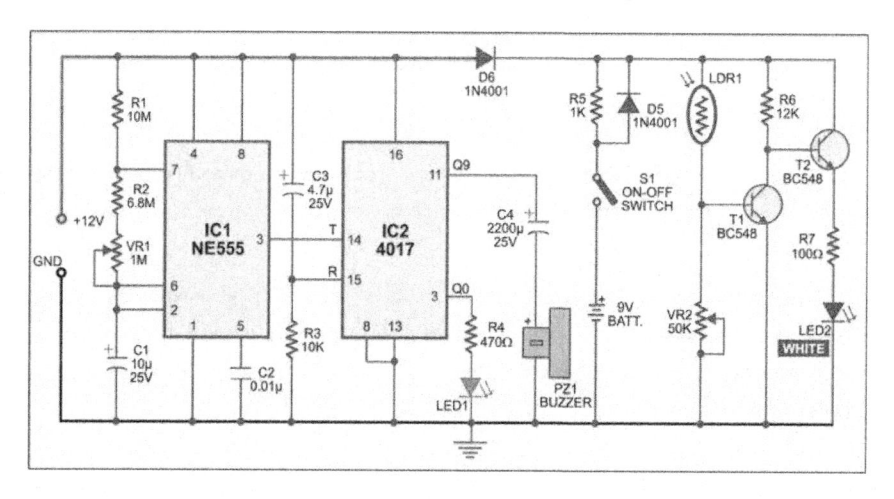

When the power is turned on to the circuit, pin 3 of IC2 goes high. After about an hour, pin 11 (Q9) goes high, and the buzzer goes off. This cycle keeps going until the two npn transistors are reached. In the dark, when no light hits it, the LDR has a very high resistance. So, when the power goes out, the white LED (LED2) lights up because transistor T1 is driven in the wrong direction. The circuit for the lamp is powered by a 9V rechargeable battery, which is charged when mains is present by the resistor R5. So, when it's dark, the LED stays off because the power to the circuit is turned off.

11. Quality FM Transmitter
This is one of the easiest ways to make an FM transmitter. It doesn't need to be tuned, and the 90MHz transmit frequency is locked in with a crystal. It gets rid of the problem of frequency drift that happens in LC-based FM transmitters. Depending on the antenna used, this design works best for sending voice or music up to fifty meters away. The low-power audio amplifier LM386 (IC1), transistor PN2222A (T1), 30MHz crystal (XTAL1), varactor diode 1SV149 (D1), and a few other parts make up the simple FM transmitter. Inductor L1 is a three-turn coil made of 20SWG wire that is 8mm in diameter and is taped every half turn.

The inductor L1 and capacitor C5 in the tank circuit at the collector of T1 are tuned to three times the frequency of the crystal, or 90MHz. The third harmonic of 90MHz is amplified and connected to a wire antenna via capacitor C4. An FM receiver can pick up this signal from 30 to 50 meters away.

Circuit and how it works

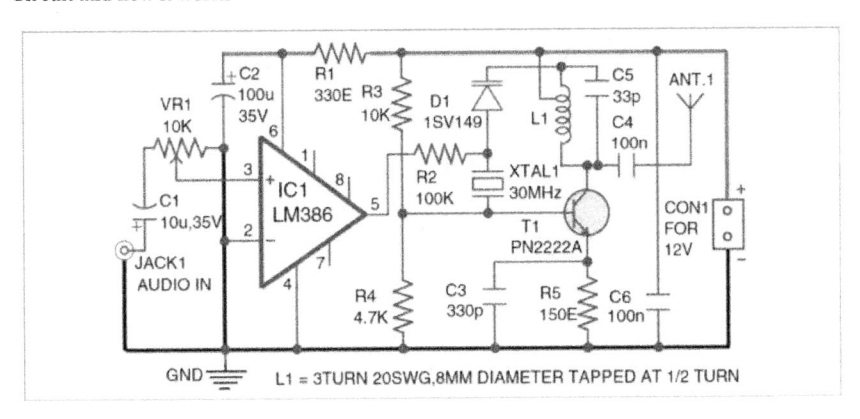

The gain of the amplifier built around IC1 is twenty. Its frequency changes the oscillator by changing the capacitance of varactor diode D1, which is connected to XTAL1 through resistor R2. The frequency deviation is set by the potmeter VR1, which should be adjusted for the best sound and least distortion. The transmitter can take audio from things like a computer, a music player, or a cellphone. Use a piece of wire that is 83 cm long for the antenna.

12. Medium-Power FM Transmitter
At 9V DC, this FM transmitter has a range of about 100 meters.

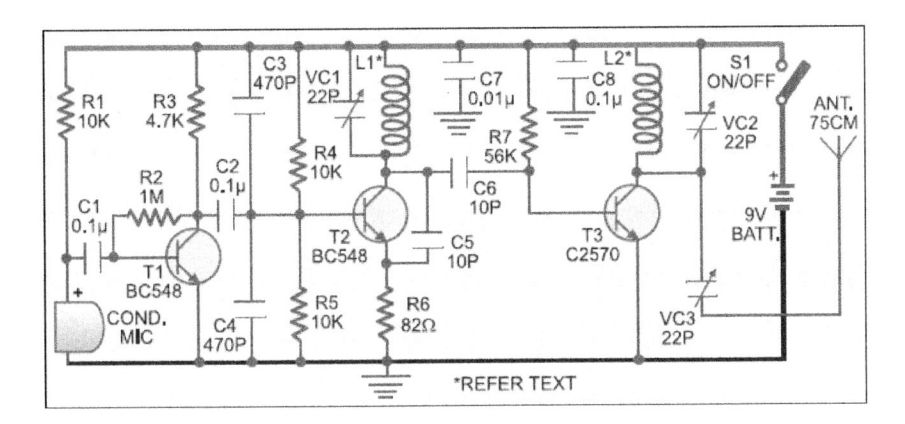

There are three stages in the circuit. The first stage is a BC548 transistor-based microphone preamplifier. The next step is a VHF oscillator, which is just another BC548 with wires attached. (BC series transistors are usually used in stages with low frequency. But these also work well as oscillators in RF stages.) The oscillator signals are amplified in the third stage by a class-A tuned amplifier. When the extra RF amplifier is used, the transmitter can reach farther. Coil L1 is made up of four turns of enamelled 20SWG copper wire wrapped around an air core that is 4mm in diameter and 1.5cm long. Coil L2 is made up of six turns of enamelled 20SWG copper wire wrapped around a 4mm air core.

13. Simple Shortwave Voice Transmitter

This simple shortwave transmitter is easy to build and can send speech on the lower shortwave (SW) band from 5MHz to 12MHz if the right crystals are used. It would be a good first project.

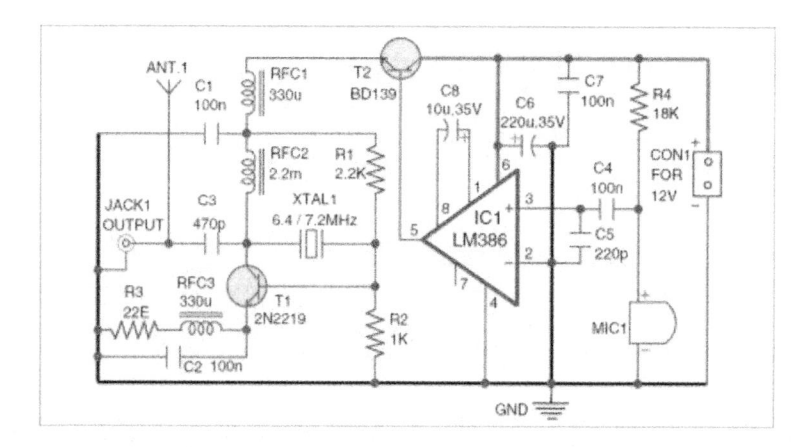

Figure shows the circuit diagram of a simple shortwave voice transmitter. It is made up of low-power audio amplifier LM386 and power transistors 2N2219 (T1) and BD139 (T2) (IC1).

Transistor T1 is set up as a pierce oscillator, and XTAL1 sets the frequency. The resistors R1, R2, and R3 set the bias, and the crystal gives feedback. At the collector of T1, a 6.4- or 7.2-MHz carrier wave is made, which is sent to the antenna via capacitor C3. The RF choke RFC2 lets DC current through but has a high resistance to RF. RF choke RFC1 and capacitor C1 help filter out more noise. The low-power audio amplifier chip LM386 is used to make the modulator (IC1). It has a high audio gain of 200, which is set by capacitor C8.

A microphone called MIC1 sends an audio signal that is amplified and sent to a transistor called T2. Transistor T2 makes sure that IC1 doesn't get too much power, which keeps audio distortion to a minimum. Also, the audio signal changes the supply voltage of transistor T2, which is connected in series with the oscillator. The carrier wave is changed by 80 percent because of this. The output power is only 40mW, but with 80 percent modulation and a 10-meter-long wire antenna, it can go up to 120mW.

14. Four-Stage FM Transmitter

This four stage FM transmitter circuit employs four radio frequency stages: a VHF oscillator based on transistor BF494 (T1), a preamplifier based on transistor BF200 (T2), a driver based on transistor 2N2219 (T3), and a power amplifier based on transistor 2N3866 (T4) (T4). The input of the oscillator is hooked up to a condenser microphone.

Circuit Operation

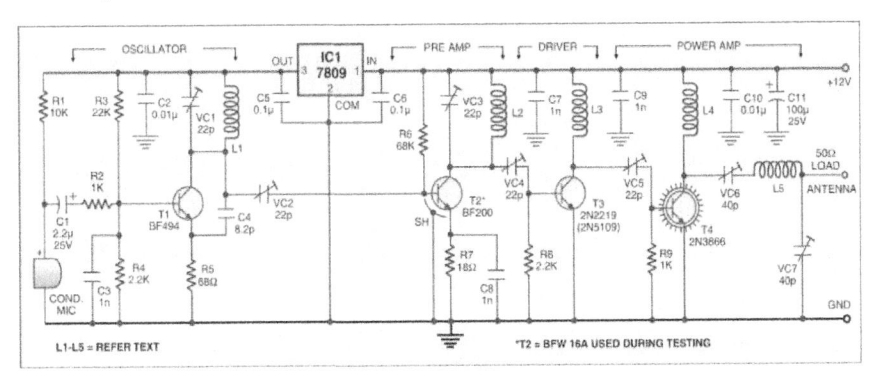

How the circuit works is easy to understand. When you talk close to the microphone, the oscillator transistor T1 picks up frequency-modulated signals at its collector. The VHF preamplifier and the pre-driver stage make the FM signals stronger. The transistor 2N5109 can also be used instead of the 2N2219. The driver is a class-C amplifier, and the preamplifier is a tuned class-A RF amplifier. Signals are then sent to the class-C RF power amplifier, which sends RF power to a horizontal dipole or ground plane antenna with 50-ohm impedance. Use a heat sink with the 2N3866 transistor to get rid of heat. Adjust the trimmer VC1 across L1 carefully to get a frequency between 88 and 108 MHz. Also, change the trimmers VC2 through VC7 so that you get the most power at the most range. Regulator IC 78C09 gives the oscillator a stable 9V supply, so changes in the supply voltage won't change the frequency. You can also power the circuit with a 12V battery.

15. Whisker for Robots

Whiskers on robots are simple sensors that work like an animal's whiskers, picking up on things in the environment that are close by. When the sensor is moved, it sends a pulse to the robot to let it know there is an obstacle.

Steel guitar strings can be used to make whiskers that are sensitive and inexpensive for general use. These strings are very flexible, easy to use, and conduct electricity. Take a copper tube that is 2 cm long and 4 mm wide. Cut a 10-centimeter piece of guitar string and slide it into the copper tube. Use a rubber bush insert at the tube's end to keep the string in the middle of the tube. Figure 1 shows how the last mechanical whisker should look. Solder a wire to one end of the copper tube and another wire to the end of the steel string.

Only one of the four channels that can be made with one LM339 is shown in the diagram (IC1). The IC LM339 series is made up of four independent precision voltage comparators. The offset voltage for all four comparators can be as low as 2 mV. These are made so that they can work on a wide range of voltages from a single power source. Here, we have only used the first one. Pin 5 of the IC LM339 is a non-inverting input, and resistors R2 and R3 connect it to a fixed voltage. As shown in Fig. 2, the inverting input (pin 4) is pulled high by R1 and connected to the mechanical whisker. Because the LM339 only has open-collector outputs, the pull-up resistor R4 is used. The way it works is easy.

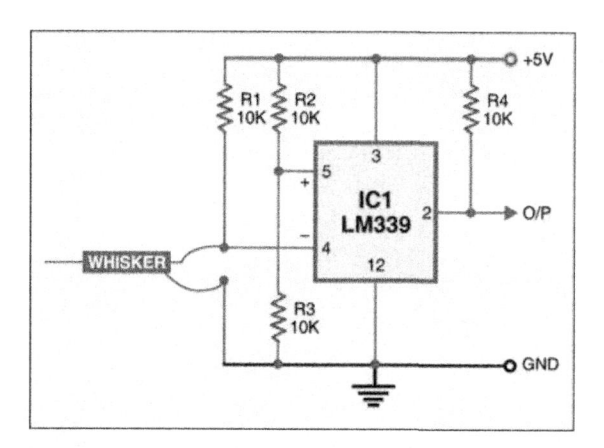

At first, when there are no interruptions, the comparator's output stays low. When something touches the guitar string of the whisker switch, pin 4 of IC1 goes low and pin 2 of IC1's first comparator goes high for a short time. We've only talked about one comparator, but the other three can be used in the same way. Through this comparator output, the robot's brain, or microcontroller, can now find out what the whisker switch is doing.

16. Farmhouse Lantern-Cum-Flasher

This circuit is made up of two transistors and a dual op-amp IC called LM358. It can be powered by either a 6V battery that doesn't need to be maintained or a lead-acid accumulator battery. It can be used in two different ways: as a flasher or as a dimmer. The dimmer mode saves battery power, and the flasher mode lets you use the lantern as a signal.

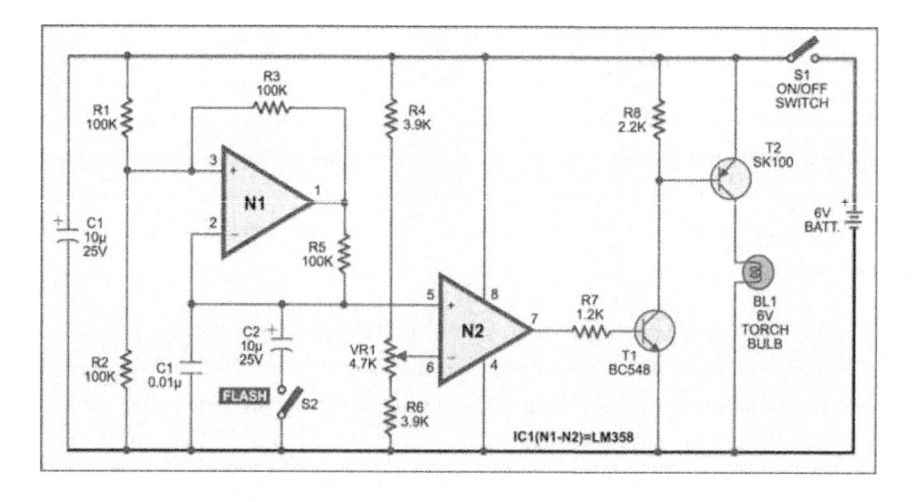

The first op-amp (N1) is set up so that it always sends out pulses at the same rate. The second operational amplifier (N2) is a comparator, and the lamp is driven by a pair of complementary transistors (T1 and T2). When switch S2 is open, the circuit is in dim mode.

With the help of potmeter VR1, which controls the reference voltage at the inverting pin of comparator N2 and, by extension, the pulsewidth at the comparator's output, the intensity of the lamp can be changed by changing the pulsewidth.

When switch S2 is shut, the circuit works as a flasher or beacon. With the help of potmeter VR1, you can change how often the bulb blinks. Assemble the circuit on a general-purpose PCB, then connect the 6V battery and the power "on"/"off" switch S1. Now, put this thing somewhere in your house that makes sense. If both switches S1 and S2 are off, the light bulb flashes at a rate set by potmeter VR1. If switch S2 is open, the light bulb stays on all the time.

17. Accurate Foot-Switch

Some industrial controls need to be switched in a precise way. For example, even a small mistake when using a foot-switch for precise drilling work can cause a lot of loss. This low-cost switch that you operate with your foot can stop that from happening. IC NE555 is wired in a mode called "one-shot." Only when both switches S1 and S2 are pressed at the same time does pin 3 go high. You can turn off any switch without changing the state of the output. When both switches are let go, the output goes low.

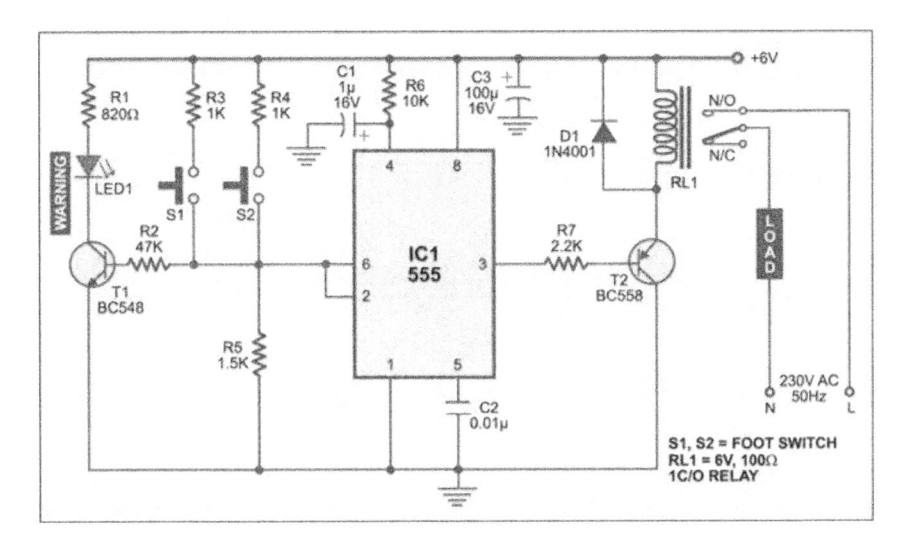

As shown in Fig. 2, the switches are put under a foot paddle. LED1 is used to show that something is wrong. If S1 or S2 are pressed by accident, LED1 will blink to let the operator know. Then, if the operator makes a mistake, he can pull his foot back or press the other switch to start the circuit. LED1 is going to be put on the desk of the operator.

The circuit is easy to understand. The three resistors R2, R3, and R4 divide the voltage. It has two comparators, a flip-flop, and a section for putting out power. By pressing either S1 or S2, the input voltage is put between the upper comparator (2/3Vcc) and the lower comparator (1/3Vcc). So, it doesn't change the state of the flip-flop inside IC NE555. When you press both switches at the same time, the flip-flop is set and the output of the NE555 goes high. Relay RL1 drives the load when transistor T2 turns it on. When you let go of any of the switches, the comparator voltage in the NE555 goes back to where it started, but this doesn't change the state of the flip-flop. When both switches are turned off, the input level relative to ground goes below the low trigger level, which resets the output. When the voltage divider is used, the circuit works well over the full range of voltages that are allowed. The power-on reset is done by the RC circuit at pin 4.

When only S1 is pressed, R3 (1 kilo-ohm) is less than R5 (1.5 kilo-ohm), so IC1 is not set off. But transistor T1 (BC548) gets pushed forward, and LED1 lights up. When both S1 and S2 are pressed, the resistance between +Vcc and pin 2 of IC1 is about 500 ohms. This is less than R5, which is 1.5 kilo-ohms, so IC NE555 is triggered.

18. Easy Transistor Tester

Before soldering a transistor, you can use this simple transistor tester circuit to find out if it works or not. You can also easily tell what kind of npn or pnp it is. The tester has LED lights that show where the pins are and how well the transistors are working.

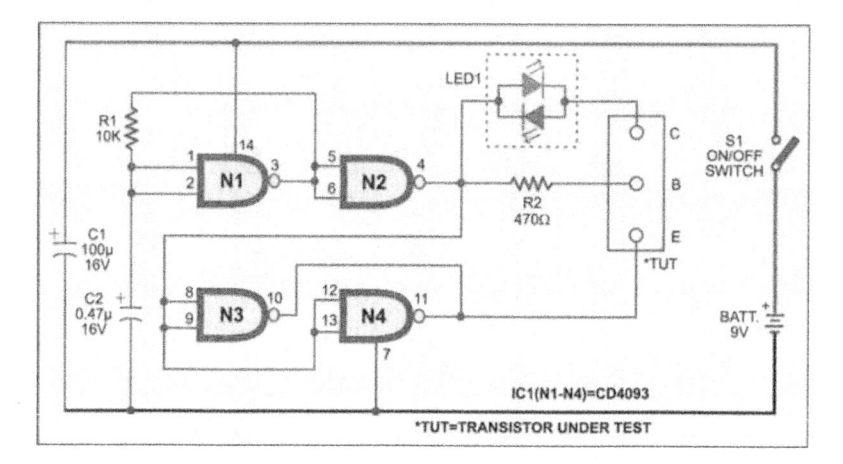

The gates of the low power IC 4093 are used in the circuit. It is a four-gate NAND IC that can be made in different ways. With resistor R1 and capacitor C2, the first gate (N1) is set up as a simple oscillator. The rest of the gates are used as buffers and inverters. The output of gate N1 (pin 3) goes to gate N2 (pin 4), which then drives the green half of bicolor LED1. The collector current for the transistor being tested can be found at the cathode of the green LED1 (TUT). The signal is inverted when the output of gate N2 is sent to the base of the TUT. The outputs of gates N3 and N4 are connected to the TUT's emitter.

19. TV Pattern Generator

This single-IC TV pattern generator can help find problems with TVs. With the help of this circuit, you can fix the alignment of the TV's timing circuits. The pattern generator makes the TV screen show vertical stripes (bars) that help you line up the vertical scanning synchronization circuit of the receiver. To test the TV, you have to connect the video and audio outputs of the circuit to the TV's video and audio inputs one at a time. If the TV's video section is working, the circuit shows white lines going up and down the TV screen, and if the audio section is working, you can hear sound coming from the TV's speakers. You can also change the width of the lines going up and down.

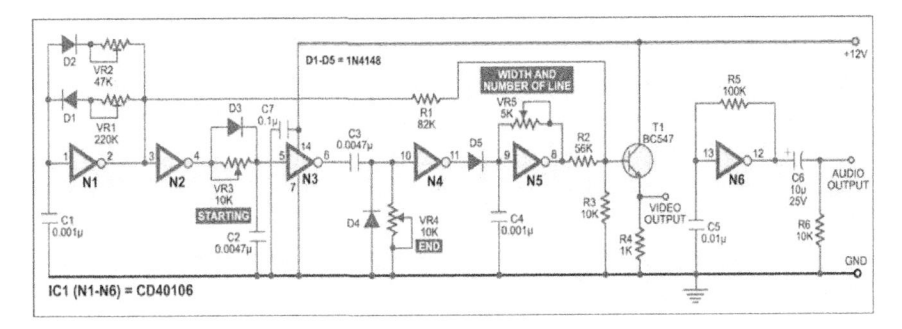

The hex Schmitt inverter IC CD40106 is used in the circuit (IC1). Pulses that are horizontally synchronized (Hsync) for the PAL video signal are made by NOT gate N1. With the presets VR1 and VR2, you can control how long the oscillator is on and how long it is off. For PAL, you need to set VR2 so that it is "off" for about 4.7 s and set VR1 so that it is "on" for about 60 s.

When you connect the video output of the circuit to the video input of the TV, if you see vertical lines on the TV screen, the video part of the TV is working. Potmeter VR3 lets you change where the lines start. Potmeter VR4 lets you change where the lines end. Potmeter VR5 lets you change the line width and the number of lines.

If you don't have an oscilloscope, set VR1 and VR2 to 150k and 22k, respectively, to get the required "on" and "off" times for the oscillator and see the vertical line pattern on the TV. The NOT gate N6 is the main part of the audio frequency oscillator. The frequency of its oscillation is set by resistor R6 and capacitor C5. Connect the sound from the circuit to the TV's sound input. If you can hear sound coming out of the TV's speakers, the audio part is working.

20. Multipurpose Listening Device

This listening device can pick up distant sounds that are very quiet and clear. It can be used in large meeting rooms, auditoriums, movie theatres, college lecture halls, etc. The circuit can be kept in a shirt pocket in a small plastic box. It's especially good for watching TV at a low volume so as not to wake up other people in the house.

Sounds are picked up by an electric microphone. It's good to use because it has a wide range of frequencies and is sensitive. To make the circuit, you need four transistors and a few resistors and capacitors.

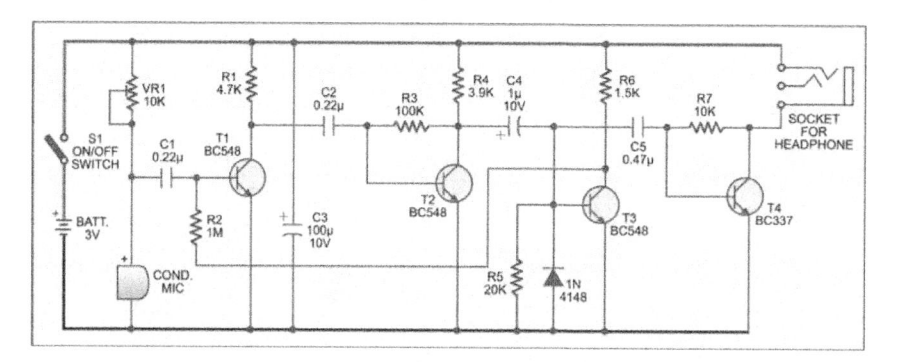

Two 1.5V AA-size batteries power the circuit. Transistors T1 and T2 boost the sounds that the microphone picks up. With the 1-mega-ohm resistor R2 in the feedback path, transistor T3 keeps the level steady. This is important because the device needs to make weak sounds louder and weaken loud sounds to a safe level.

21. DC-DC Converter

Here is a simple DC-to-DC converter circuit that turns 6V DC into 12V DC at a low cost. It is easy to put together and doesn't need a transformer. The circuit is built around IC 555, which makes the 2 to 10 kHz frequency needed to drive the power transistor BD139 (T2). The IC's output frequency can be changed with a 47k potmeter (VR1), and the base of transistor T2 is given that frequency through resistor R3. Transistor T2 is put on a heat sink made of aluminum. Both the inductor L1 and the capacitor C5 (2200F, 35V) store energy. The 12V Zener diode makes sure that the voltage across the circuit's output is always the same. The inductor is made up of 100 turns of enamelled 24SWG copper wire wrapped around a toroidal ferrite core with a 40mm diameter. The more turns there are on the core, the more current the circuit can send to the load at the output.

With the help of resistors R4 and R5, the current going out is controlled by the transistor BC549 (T1). The Zener diode controls the output voltage, and capacitor C5 smooths it out. At the end of this circuit, you can get regulated 12V DC at 120 mA. When there are more loads than 100 ohms, the circuit might not work as well or deliver as much current. For higher voltages and higher currents, use a large capacitor (C5) and an inductor. Zener diodes with different ratings can be used to get different output voltages.

22. Remote Control using Wireless Doorbell

Using a wireless doorbell and this circuit, you will be able to wirelessly control an appliance that is located in a different location. The appliance turns on when the wireless doorbell's transmitter sends a signal, and it turns off by itself after a set amount of time.

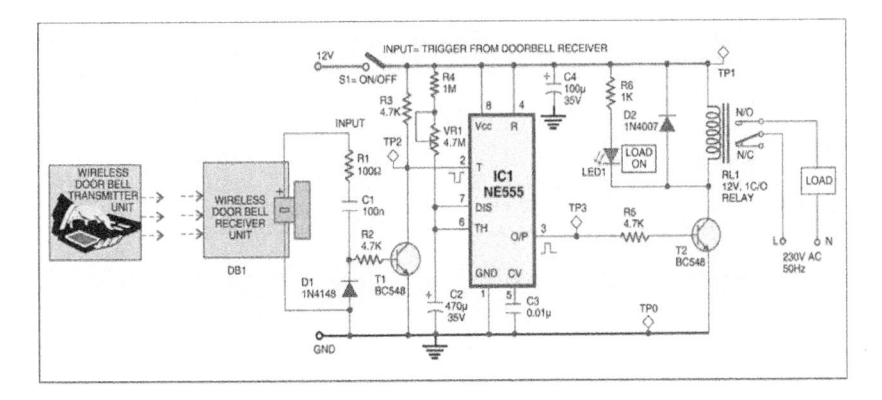

Circuit and working

The circuit has a doorbell (DB1), a NE555 timer (IC1), and a few other separate parts. The sound from the battery-powered doorbell receiver is connected to series resistor R1. Then, capacitor C1 and resistor R2 are used to AC couple the sound to the base of npn transistor BC548 (T1). To protect transistor T1, which is set up as a common-emitter amplifier with a 4.7-kilo-ohm collector load, diode D1 limits the signal's negative swings to about 0.6V below ground. When the doorbell rings, it sends out an AC signal, which diode D1 turns into a DC signal. Transistor T2 works, so its collector is pulled down. This change from high to low turns on timer IC1, which is set up to work in monostable mode. The output of IC1 is high for a time that can be set with trimpot VR1 between 8 and 50 minutes. This high output turns on the appliance by making relay RL1 close and open. How the circuit works is easy to understand. Press the button on the wireless transmitter to turn on the appliance for the set amount of time. When LED1 lights up, it means that the appliance is turned on. After the set amount of time, the appliance will turn off by itself.

23. RF Signal Detector

This easy-to-build RF signal detector circuit can be utilized to detect the existence of radio frequency (RF) signals as well as electromagnetic noise in your home, place of business, or retail establishment. It can be helpful when testing RF circuits or making them. It can also be used to find out if your building has any electrical noise.

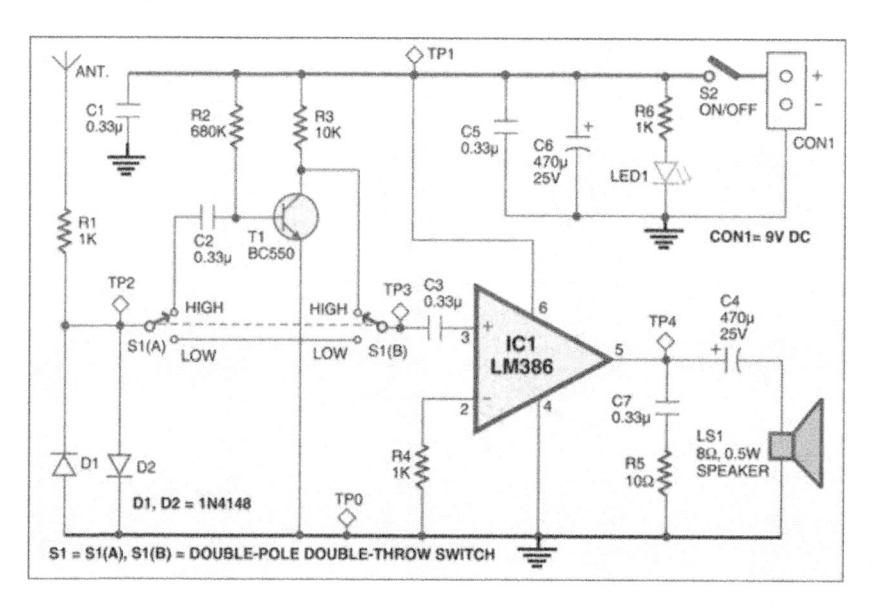

Electromagnetic noises can come from electrical installations that spark or from other places. You need a wide-band receiver that can pick up RF signals and noise in order to find out if they are there or not.

Circuit and working

Fig.1 shows how the RF signal detector works. It has a telescopic antenna, an input protection resistor (R1), two diodes (D1 and D2), a selector switch (S1), a pre-amplifier with a low-noise and high-gain transistor (T1), and an audio amplifier IC (LM386) (IC1). Switch S1 is used to choose between the circuit's high and low sensitivities. When high sensitivity is chosen, the gain of IC LM386 is added to the gain of the transistor stage with T1. When low sensitivity is chosen, only the around 20-gain IC LM386 is used. Because the circuit needs to be simple and small, there is no volume control. When there is a strong electromagnetic signal near the antenna, like from a cell phone, telephone, or electrical motor, you can hear sound from the speaker (LS1). The sound from the speaker gets louder as you move this detector closer to the source of the RF or noise. So, you can find the exact place where the source is. You can improve the range of reception and sensitivity by using an antenna with the right shape and length.

24. Audio Mixer with Multiple Controls

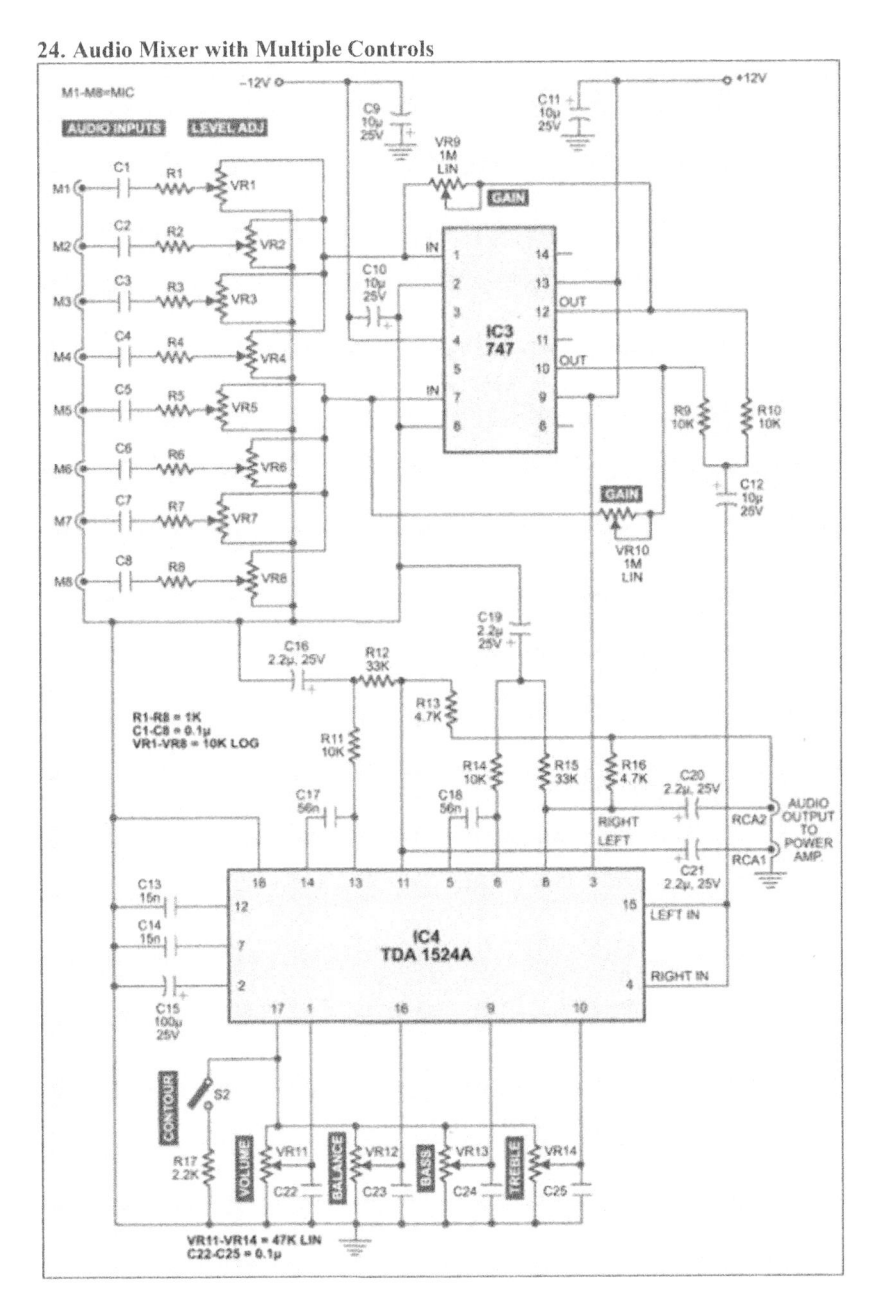

When utilizing a single microphone to record sound from many symphonic instruments being played by various players, the only option to vary the sound balance is to change the position of the musicians relative to the microphone. Before you hit the record button on a direct-to-stereo-master tape, you need to make sure that all the voices and instruments sound good. Here is an audio mixer circuit with eight inputs and controls for bass, treble, volume, and balance. You can use these controls to balance the sounds from each source until you get the mix you want. The audio mixer uses up to eight microphones to pick up sounds from different places.

Circuit of audio mixer with bass, treble, volume and balance control

In this case, the dual operational amplifier IC 747 (IC3) is used to mix several inputs that don't affect each other. The bias network and power source for the two internal amplifiers are the same. The IC is protected against short circuits and has a wide range of common-mode and differential voltages. In this application, IC 747 is powered by +12V and –12V regulated DC supplies. After each signal's level has been changed, the microphone output signals M1 through M4 are mixed together and sent to the differential input terminals (pins 1 and 2). In the same way, the outputs of microphones M5 through M8 are connected across the differential input terminals (pins 7 and 6) of the second amplifier in op-amp IC 747 after each level adjustment.

Audio mixer circuit

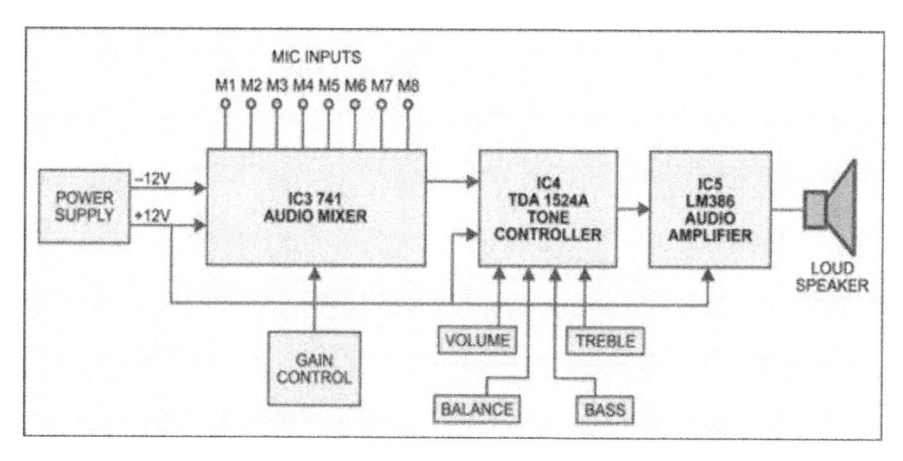

Fig. 1: Block diagram of the audio mixer with bass, treble, volume and balance controls

The block diagram of the audio mixing system and the audio power amplifier are shown in Fig. 1. The circuit of the audio mixer and a tone controller are shown in Fig. 2. Logarithmic variable resistors VR1 through VR4 and VR5 through VR8 are used to adjust the level when the outputs of each microphone are fed into the inputs of the two amplifiers inside IC 747. The outputs of the two amplifiers, which come from pins 12 and 10, are combined at the intersection of resistors R9 and R10 before being sent to the next stage (the tone controller) through the 10 F capacitor C12. With the help of potentiometers VR9 and VR10, the overall gain of each amplifier can be changed.

The amplified mixed signal from IC 747 is sent to the stereo tone controller IC TDA1524A's input pins 15 and 4. (IC4). The TDA1524A is made to be an active stereo tone and volume control for car radios, TV receivers, and other devices that get their power from the wall. It has controls for the bass and treble, the balance, and the volume with a built-in contour that can be turned off. All of these functions can be controlled by single linear potentiometers or by DC voltages. This IC does a good job of controlling the tone. Even though it might work okay with a 9V DC supply, a 12V supply can improve the bass response. For the IC to last longer and work better, it needs a good heat sink.

25. Infrared Interruption Counter

As the sensor, most optical interruption counters use a light bulb with a light-dependent resistor (LDR) or a regular phototransistor. The interruption counter only works well when it's dark out. It can't be used outside because light from the sun, a light bulb, or other sources could cause it to count wrong.

Infrared interruption counter

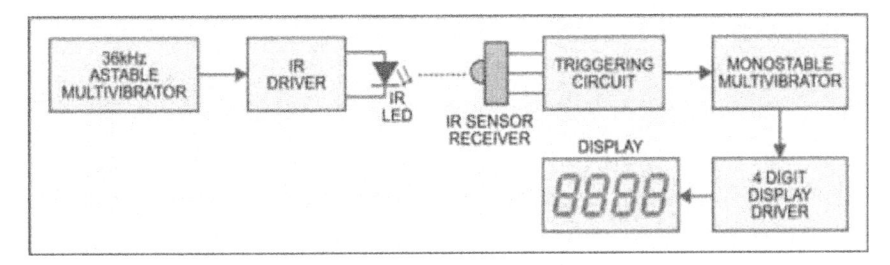

Fig. 1: Block diagram of infrared interruption counter

The interruption counter described here uses an infrared (IR) sensor that can pick up a modulated infrared beam at a certain frequency. To send out modulated IR signals, a small circuit with an IR LED is used as a transmitter. Fig. 1 is a block diagram of the infrared interruption counter that shows how the system works as a whole. The astable multivibrator makes a 36kHz frequency, and the npn transistor BC547 drives the IR LED to send the modulated infrared signal. The IR signal that is being sent keeps falling on the IR sensor (receiver).

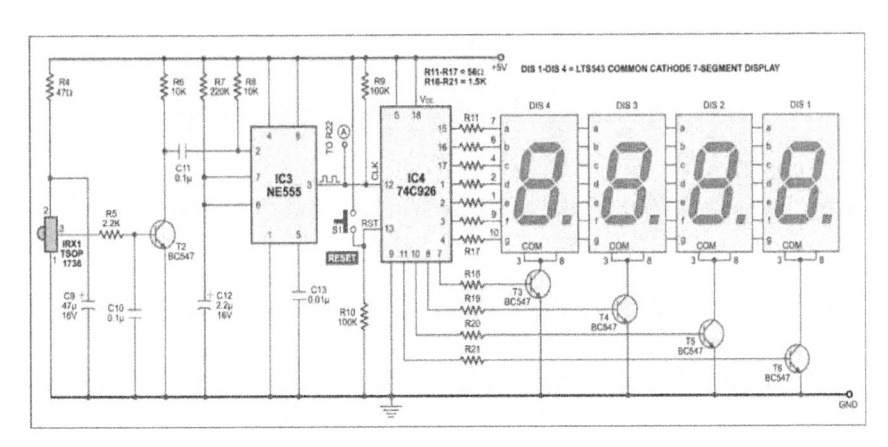

Fig. 2: Circuit of infrared interruption counter

When a person walks in front of the IR beam hitting the sensor, the triggering circuit turns on, which sets off the monostable multivibrator. The monostable's output moves the count of the 4-digit counter-cum-display driver forward so that the count can be shown on 7-segment, common-cathode displays.

26. Clock Tick-Tock Sound Generator and LED Pendulum

Wooden-case, battery-operated wall clocks with pendulums are available in the market. Some even have chimes. What is missing is the tick-tock sound of old mechanical pendulum clocks.

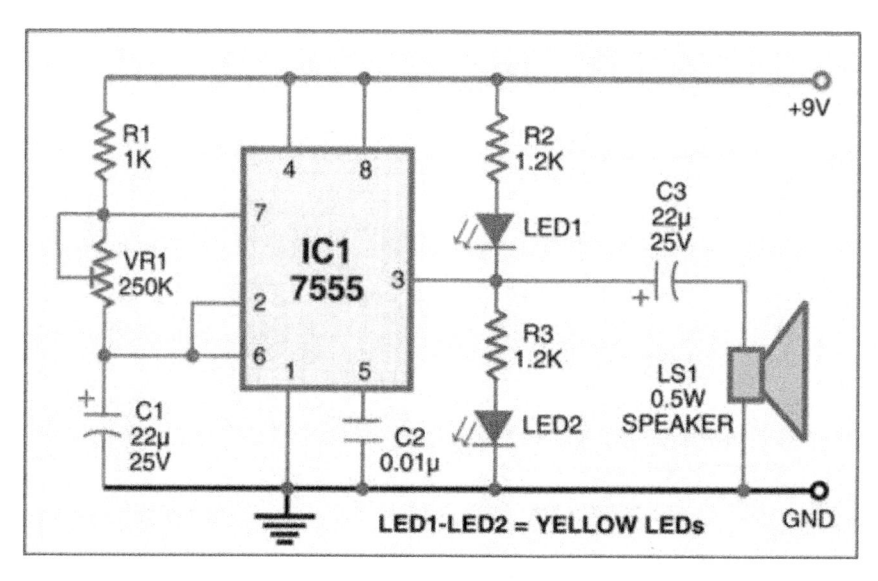

Fig.1 Circuit of tick-tock sound generator

Shown above is the circuit of a tick-tock sound generator. It is built around timer IC 7555 (IC1) wired as an astable multivibrator. Resistor R1, preset VR1 and capacitor C1 generate frequencies that can be varied with the help of preset VR1. At output pin 3 of IC1, a small speaker is connected via capacitor C3 that produces the tick-tock sound. At pin 3 of IC1, two yellow LEDs are also connected on the clock face to flash in time with sound (see Fig. 2). Adjust preset VR1 such that ticking sound closely resembles the tick-tock of old pendulum clocks. Yellow LEDs (LED1 and LED2) match well with the wooden clock.

27. Battery Charger with Automatic Switch-off

Here is a 555 timer IC-based battery charger that turns off by itself. When your rechargeable batteries are fully charged, this smart charger turns off by itself. A bistable multivibrator is wired around a timer IC 555 to make up the circuit. The output of the bistable is sent to an ammeter (via diode D1) and a potmeter (VR1) before it goes to charge three Ni-Cd batteries.

Circuit Diagram:

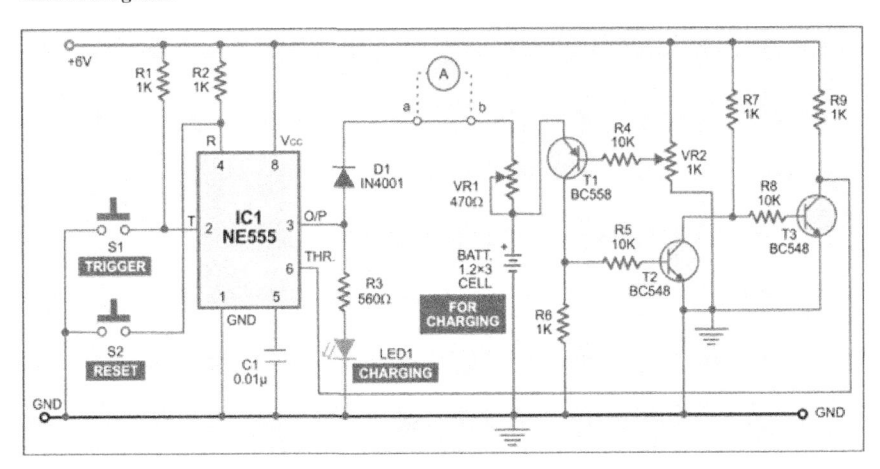

Most Ni-Cd cells have a full charge potential of 1.2V. Press the switch S1 to turn on the bistable, and adjust the potentiometer VR1 so that the ammeter reads 60mA. Now, take out the ammeter and connect a jumper wire between its points "a" and "b." Connect the batteries' positive output to the emitter of pnp transistor T1. Potmeter VR2 is used to keep the base of transistor T1 at 2.9V. The npn transistors T2 and T3 switch the output of transistor T1 around twice.

So, when the batteries are fully charged to 31.2V = 3.6V, a voltage higher than this makes transistor T1 conduct. Also, transistor T2 works, and transistor T3 turns off. The timer 555 turns off when the threshold level reaches 6V, which is more than 2/3VCC = 2/ 36 = 4V. During charging, the timer's threshold is kept low. When the batteries are being charged, the green LED (LED1) lights up. When the batteries are fully charged, the green LED turns off.

Note that this circuit can only be used with 1.2V, 600mAH, Ni-Cd rechargeable batteries that take 15 hours to fully charge with 60 mA of current.

28. Earth Leakage Tester

During the rainy season, earth leakage from electrical wiring is a big problem. Pipelines are less safe from earth leakage and may give you an electric shock when you least expect it. Leakage current through a wet wall could come from electrical machines or faulty wiring. Normal air conditioning testers can't tell if there is an earth leak if the current isn't high enough to turn on a neon light.

Parts of mains-powered equipment that are made of metal are connected to an earth wire so that users don't get shocked if the electrical insulation breaks. Connections to ground through earth connection also keep static electricity from building up when handling devices that are sensitive to static electricity. In a mains electrical wiring system, earth is a conductor that provides a low-impedance path to the earth so that dangerous voltages don't show up on equipment.

Circuit for earth fault indicator

Above figure shows how the earth fault indicator works. A BC547 transistor (T1), two LEDs (LED1 and LED2), two 1N4007 diodes (D1 and D2), and five resistors (R1 through R5) make up the circuit (R1 through R5).

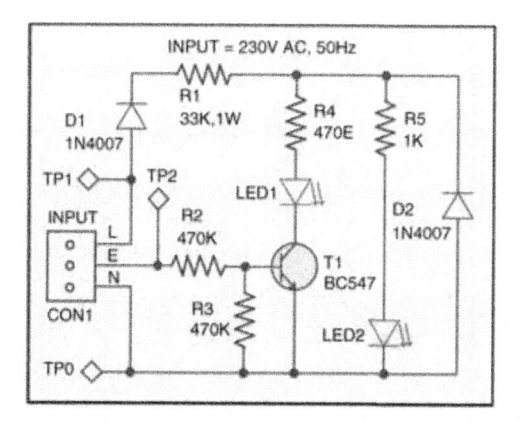

Fig: Circuit of earth fault indicator

The voltage that appears between the earth and neutral terminals is used by the circuit. In a single-phase circuit, the neutral-to-earth voltage measured at the load is a function of the load current and the impedance of the neutral wire. For a reasonable level of efficiency, different standards limit this voltage drop in a branch circuit to 3% (or 5% for both the feeder and the branch circuit). Based on this, the neutral-to-earth voltage limit for a single-phase 120V AC circuit is 3.6V AC and for a single-phase 230V AC circuit it is 6.6V AC.

This circuit doesn't need any other power source to work. The 230V AC mains supply is the only source of power for the circuit. The 230V AC mains are turned into a low voltage for the circuit by diode D1 and resistor R1. If the earth connection is correct, a transistor switch will turn on the green LED (LED1). As shown in Fig. 2, the base of T1 is connected to the earth pin of the mains supply by a network of resistors R2 and R3.

29. SCR-Controlled EHT Power Supply
An extra-high-tension (EHT) power supply is shown here; it is able to deliver variable DC voltage ranging from around 100 V to 5 kV, and it can handle load current of up to 500 mA. One way to get such high DC voltages that can deliver load currents of up to a few amperes is to use back-to-back high-power SCRs to control the AC voltage that goes into the primary of a power transformer. This power supply is made with the same idea in mind.

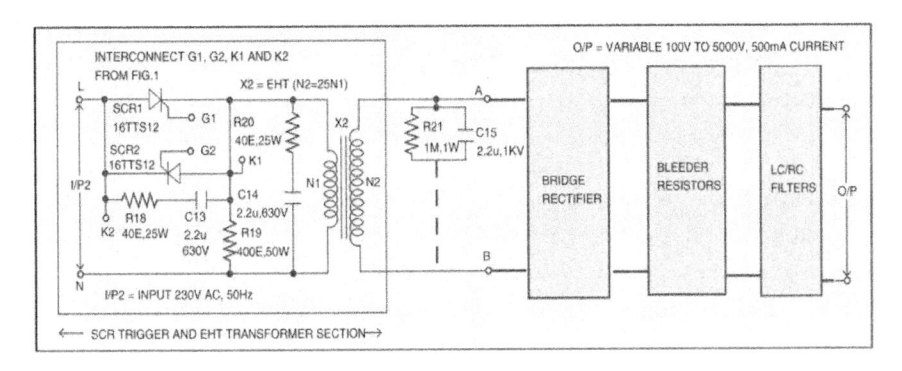

Fig: EHT generator

The most important part of this circuit's design is that the SCR trigger pulses are timed to match the sine wave that comes in. This makes sure that the SCRs used to control the positive and negative parts of the sine wave can be turned on at the same angle, which is set by the intersection of a synchronized sawtooth wave and DC control voltage (explained later). Using SCRs and power transformers with the right ratings, this basic circuit can be made to work with higher voltages and load currents.

30. 16-Way Clap-Operated Switch

You don't even have to get out of bed to use this switch. All you have to do is clap. You just have to clap near the microphone, which you can keep by your bed. You have 16 ways to turn on or off up to four different pieces of electrical equipment, like a TV, fan, light, etc.

Clap Operated Switch Circuit

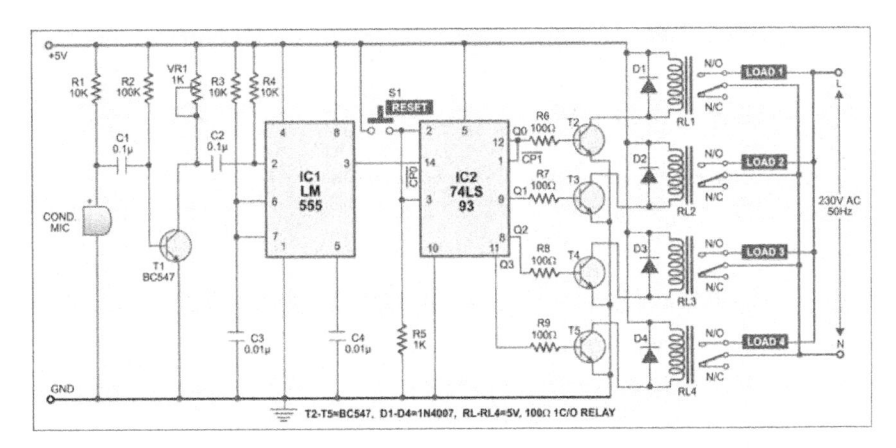

This circuit is made up of the 555 timer IC (IC1), the 74LS93 CMOS IC (IC2), and five BC547 npn transistors (T1, T2, T3, T4 and T5). The pre-amplifier is the transistor T1, and the rest are used to drive the relays.

At the base of transistor T1, which is biased by resistor R1, there is a small condenser microphone (10 kilo-ohms). The microphone turns the sound of clapping into electricity, which transistor T1 amplifies. The output of the transistor goes to the monostable circuit, which is made up of wires around IC 555. The timer's output pin 3 is hooked up to the clock input of the 74LS93 divide-by-16 IC.

Through 100ohm resistors, the outputs of IC2 are sent to npn transistors T2, T3, T4, and T5 to drive relays RL1, RL2, RL3, and RL4 that are connected to appliances 1 through 4. The back electromagnetic field (EMF) made by the relays can damage the transistors. Freewheeling diodes D1 through D4 connected across the relays protect the transistors from this EMF.

31. Smart Loop Burglar Alarm

When the loop breaks, simple loop burglar alarms go off. What if a smart thief figures out how this alarm works? He could just use another conductor to short out the loop, and then cut the part of the loop that was shorted out. Here is the circuit for a smart loop burglar alarm that gets around this problem by putting a sensing resistor (R) in the loop. The sensing resistor has to be kept in the area to be protected (say, a room).

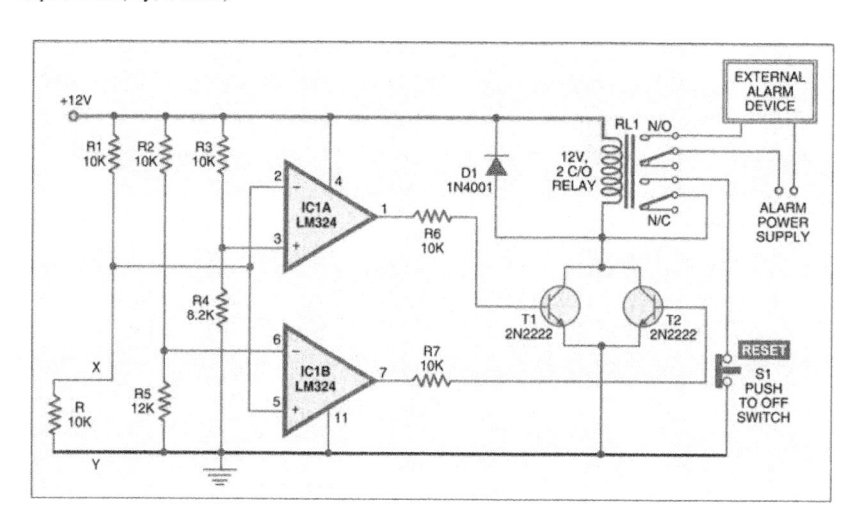

Fig. 1: Circuit of smart loop burglar alarm

In Figure 1, you can see how the smart loop-type burglar alarm works. This circuit is made up of four op-amps (LM324) and a few other separate parts. Two of the LM324's op-amps are set up to compare voltages. The 12V DC power supply goes to both resistors R and R1. The result, 6V, is sent to the inverting input of the first op-amp (IC1A) and the non-inverting input of the second op-amp (IC1B). The input of the first operational amplifier (IC1A) is kept below 6V, while the input of the second operational amplifier (IC1B) is kept above 6V. So, when everything is normal, the outputs of both op-amps stay low.

The alarm will go off as soon as the thief shorts out the loop that is outside the protected region. This will cause the output of op-amp IC1A to go high, which will cause transistor T1 to conduct, which will cause relay RL1 to energies, and the alarm will go off. Additionally, when the burglar breaks the loop, the output of op-amp IC1B swings high, which causes transistor T2 to conduct, and relay RL1 to energies, which causes the alarm to ring. So, the alarm goes off when the loop is cut or cut short.

Put the circuit together on a PCB that can be used for many things and put it in a suitable cabinet. The relay should be a 12V, 2C/O type with contacts that can handle the right amount of current. To power the circuit, use a regulated 12V DC power adapter or a 12V battery. Before connecting the circuit to the power supply, make sure that the components on the assembled PCB are properly soldered. Connect the first C/O contacts of the relay to the 230V AC alarm bell and the second C/O contacts to the reset switch S1. Put switch S1 on the cabinet's front panel. Use a thin wire to connect the loops.

32. Temperature-Tolerance Checking System

Most electrical components' properties fluctuate with temperature; therefore, they're chosen based on the equipment's predicted operational temperature range. So, it's important that the equipment stays within the range of temperatures it was made for.

Circuit and Working

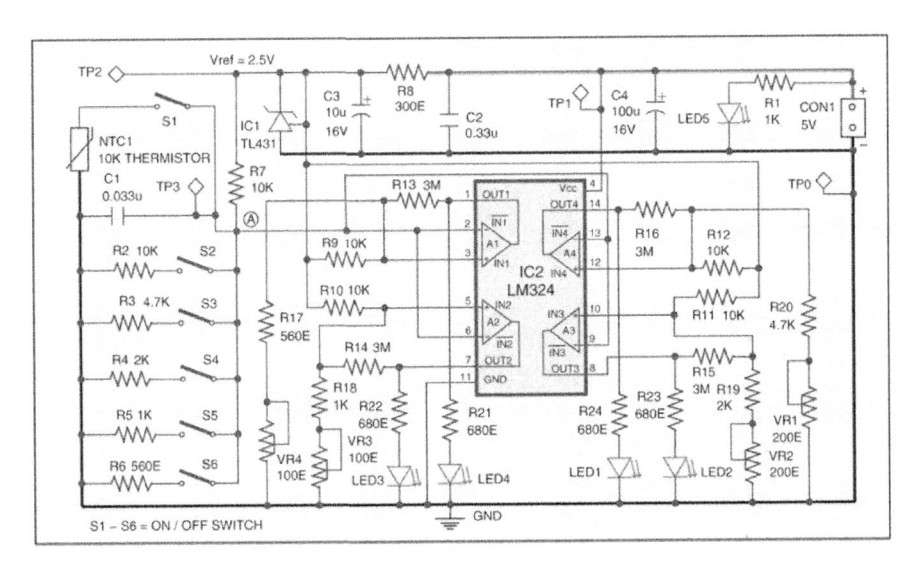

This circuit sounds an alarm when the temperature gets high enough to be dangerous. Here, the temperature levels are already set to 45°C, 65°C, 85°C, and 105°C, but they can be changed to fit the equipment being used.

Figure 1 shows the circuit for keeping track of the temperature. It is made up of a 10k NTC thermistor (NTC1), a shunt regulator (TL431), a common comparator (LM324), and a few other parts. When the temperature of thermistor NTC1 reaches a certain level, IC2 turns on the corresponding LEDs (LED1 through LED4) (refer Table I). NTC1 measures the temperature, and the voltage it makes at point "A" is sent to the "inverting" end of all four op-amps (A1 through A4). For this to work, switch S1 should be turned off. The reference voltage at the non-inverting terminals of each op-amp is compared to this voltage level. Voltage dividers use the 2.5V reference source, which is made by IC1's shunt regulator, to get the reference voltage. The presets VR1 through VR4 can be used to set the reference voltage for each comparator.

Calibration is done with the switches S2 through S6 and the resistors R2 through R6. For example, op-amp A4 of IC2 compares the voltage levels at point A (which change as the temperature changes) with the reference voltage at its pin 12. VR1 is used to change the voltage level at pin 12 of A4. Switches S2 through S6 are used to simulate voltage levels that correspond to different temperatures.

To tune A4 for 45°C, you keep switch S3 closed and adjust preset VR1 until LED1 lights up. During calibration, S1 should be open. The same thing is done to calibrate comparators A3, A2, and A1 for 65°C, 85°C, and 105°C, respectively. Table I lists the presets and LEDs for each level of temperature. Open switches S2 through S6 after the calibration is done.

33. Radiation Detector Circuit using LM358 IC

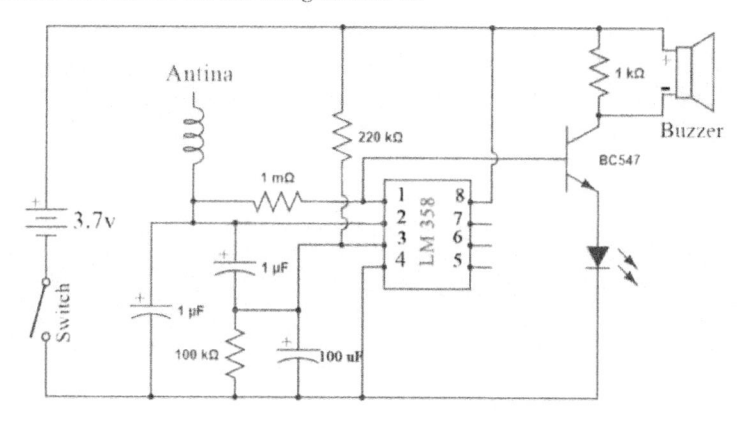

Fig. Radiation Detection Circuits

There are a great number of locations in which a mobile phone of any kind might not work. Take, for instance, our many examination rooms. Therefore, it is essential to enforce a ban on the use of cellular phones in certain settings. And it should go without saying that in order to accomplish this objective, administrators require a detector that is easily capable of detecting the existence of mobile radiation within a certain range. As a result, they are able to utilize a tool that is known as a phone call detector. It makes it possible to detect the existence of mobile phones within a certain range of distance. [Case in point:] Therefore, the purpose of this tutorial is to demonstrate how to construct a circuit for a radiation detector. An operational amplifier, sometimes known as an op-amp, is a DC-coupled high gain voltage amplifier. We are making this circuit with the help of an LM358 integrated circuit. Signal conditioning, filtering, and amplification were the intended functions of this integrated circuit that they built. Between the output terminals and the input terminals of the IC, we are able to conveniently use components such as capacitors and resistors.

34. Stereo Headphone Amplifier

This is a simple and affordable circuit for a stereo amplifier that can be used to power a headphone set with a low impedance. The circuit consists primarily of passive components such as resistors, diodes, and capacitors along with a couple of inexpensive transistors (BC547 and BC557). The headphones are driven by a single stage of preamplifier and a stage of npn-pnp push-pull operation. The left and right inputs, respectively, are handled by the transistors T1 and T6, which are important to the construction of the preamplifier stage. The signal from the amplified left input is sent to the push-pull stage, which is constructed using the transistors T2 and T3, and it is this stage that powers the left headphone. In a similar manner, the signal from the amplified right input is sent to the push-pull stage that is constructed using transistors T4 and T5 and is responsible for driving the right headphone.

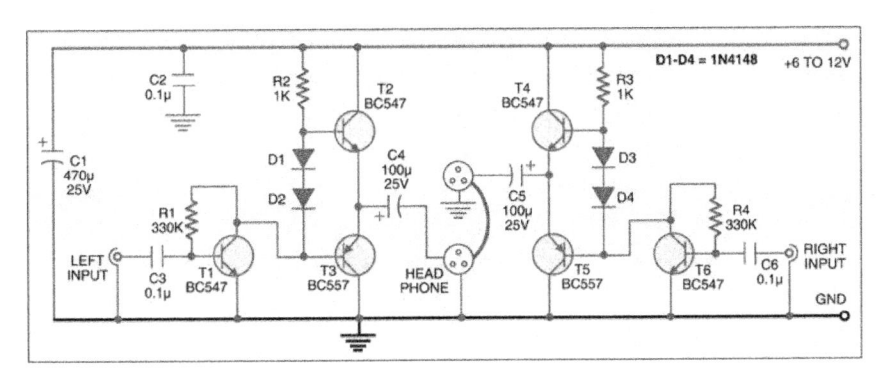

The circuit has the capability of producing 100-200 mW with a power source of 6V-12V. Due to the extremely low amount of electricity that it draws, it can also be powered by a single 9V PP3 battery. Construct the circuit on a PCB that can be used for a variety of purposes, and then place it in an appropriate cabinet along with the headphone connector. You should be able to purchase the headphones at any local market. Solder the components on the PCB with caution to prevent dry soldering from occurring. After the circuit has been put together, it should be powered by a 9V PP3 battery. The circuit can be used immediately at this point. In the event that the device must be powered by the mains, any power supply that is based on a standard voltage regulator (6V-12V, 100mA) may be used to do so.

35. Electronic Thermostat for Fridge

The electronic thermostat is a piece of standard equipment that is used for monitoring the temperature of a system and ensuring that it is kept at a specific level. It is a component that is frequently found in refrigerators, space heaters, and air conditioners. Although some of the commercially available refrigerators include thermostats as part of their design, the supplied circuit is simple to construct and can be programmed to turn the appliance on or off at a range of temperatures that the user specifies. The temperature in my refrigerator's deep freezer stays between -5 degrees Celsius and -15 degrees Celsius (former temperature switches the fridge on and the latter turns it off).

Electronic thermostat circuit

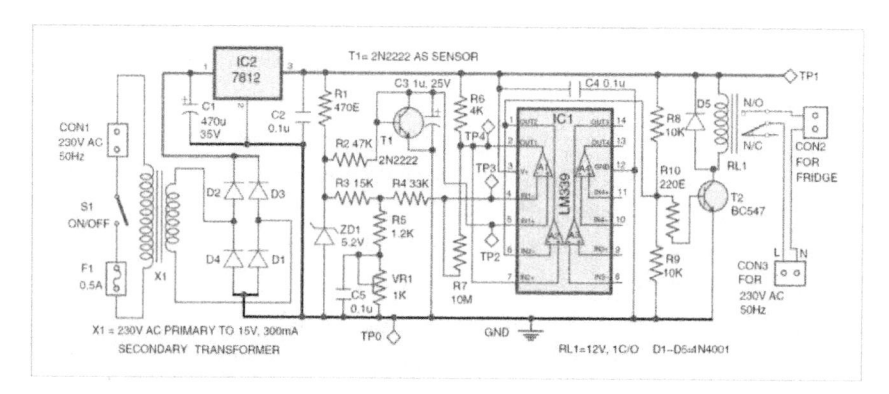

Instead of travelling straight to the refrigerator (or the input of the automatic voltage stabilizer, if one is being used in conjunction with the refrigerator), the 230V AC mains power is routed through the normally open (N/O) contact of relay RL1. The relay contact is closed and the appliance receives 230V AC power when the relay is energized because of the conduction of T2 (BC547). This causes the appliance to begin cooling once it has received power. The LM339 in IC1 is a quad comparator, and it is being put to use as a voltage comparator with hysteresis. The inverting input of integrated circuit 1 is connected to a reference voltage that is derived from Zener diode ZD1 (5.2V), which is then further split by the potential divider arrangement that is given by R3, R5, and VR1. If there is any AC noise, it will be bypassed by the decoupling capacitor C5.

Heart of the circuit

The temperature sensor consists of a npn silicon transistor with its collector shorted to its base and is located between the non-inverting input of integrated circuit 1 (IC1) and the ground. The temperature-dependent characteristic of the base-emitter junction of the transistor is utilized here as the temperature-sensing property of the transistor. The junction voltage drops by 2 millivolts for every degree as the temperature rises. The reference voltage is set by adjusting trimpot VR1 to a value that is equal to the sensor's junction voltage corresponding to the lower negative temperature, which in this case is -5 degrees Celsius.

36. Affordable Car Protection Unit

There are specialized protection systems available to purchase, but they come at a high price. The following is a description of a car protection unit that is affordable, does not require any modifications to be made within the vehicle, and can secure your car audio and other valuables from being stolen.

Put the circuit next to your bed, then take the two wires from the unit to the car (which should be parked outside your house), and attach one wire end to the cover and the other to the ground, while shorting the other wire end with something heavy like a brick. Therefore, the mechanism is hidden from view from the outside.

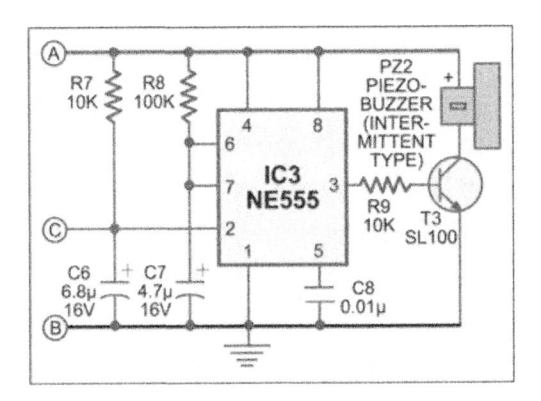

In the event that the cover is removed or attempted to be removed, the alarm on the circuit will sound to notify you. The alarm can be silenced by using switch S1 to perform a reset on the device. The protection circuit for the automobile is made up of two-timed integrated circuits: one for the alarm circuit (see IC2 in Fig. 1) and another to indicate that the battery has taken over as the primary source of power (see IC3 in Fig. 2). The protector is powered by AC mains by default, and the backup battery is only activated in the event that the mains supply is interrupted. Because the current from the battery is not particularly high, the battery will endure for a very long time.

Circuit operation

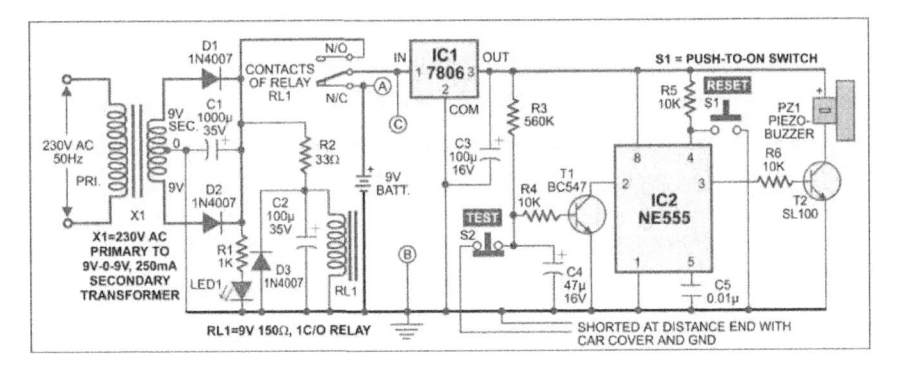

As long as the connection between the two wires is maintained, the transistor T1 will continue to be disabled. When the shorting is removed, transistor T1 becomes forward biased, which causes its collector voltage to drop, which in turn triggers integrated circuit IC2, which causes the piezo buzzer to begin emitting a sound. In the event that the power from the mains supply is lost, the battery-takeover indicator that is depicted in Figure 2 and wired to the points A, B, and C shown in Figure 1 will be immediately activated at pin 2 of IC3. Because of its high output, the battery-operation warning will sound for a brief period of time. For the purpose of activating the protection unit, IC1 gets power from the battery. After ensuring that the device is properly configured and connecting both wires together, activate the test switch labelled S2. Even if there is nothing wrong with the circuit, the alarm will still go off. Now, deactivate the alarm by removing your finger from the test switch S2 and briefly pressing the reset switch S1.

37. White LED Based Emergency Lamp and Turning Indicator

White light emitting diodes (LEDs) are gradually becoming more popular as a replacement for traditional incandescent and fluorescent light bulbs as a result of their great power efficiency and low operating voltage. These have the potential to be utilized to their full potential as an emergency lamp as well as an indicator for turning vehicles. Emergency lamp and turning indicator circuits based on white LEDs are presented in this article for the stated purpose.

Circuit for an Emergency Lamp Based on LEDs

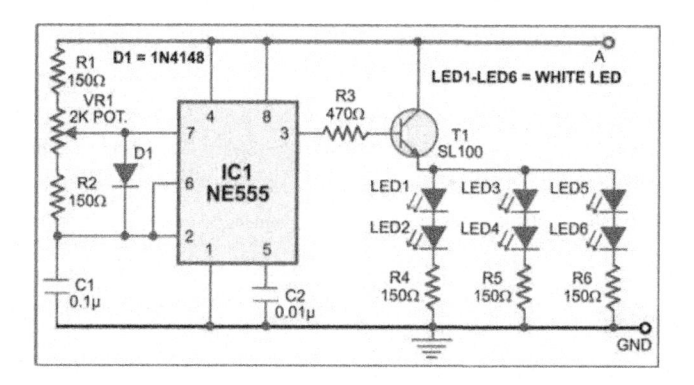

The schematic for the circuit of an emergency lamp with white LEDs is shown in figure 1. In addition, arrays of white LEDs can be utilized in vehicles to serve the purpose of daytime running lamps. The power supply for the emergency lamp is comprised of seven Ni-Cd cells measuring AA size and generating a total of 8.4 volts. The duty-cycle variation of an astable multivibrator that is operating at 1 kHz is what controls the brightness of the light. IC1 serves as the focal point of construction for the astable multivibrator. Its output is wired to LED-driver transistor T1, which is located nearby. White LEDs can be connected in parallel in a maximum of six different branches, with each branch having a maximum of two LEDs connected in series (only three branches are used here). There are a variety of alternative combinations of battery voltages and the number of LEDs connected in series that can be produced, depending on the application, in order to maintain the resistive losses to a minimum.

38 Mains-Operated Christmas Star

Presented for your perusal is a fundamental circuit of a Christmas star that even a novice will have no trouble putting together. SANI, the most important MAINS-OPERATED CHRISTMAS STAR the benefit of using this circuit is that it does not require a step-down transformer or integrated circuits (ICs). THEO PRINCE PHILLIPS

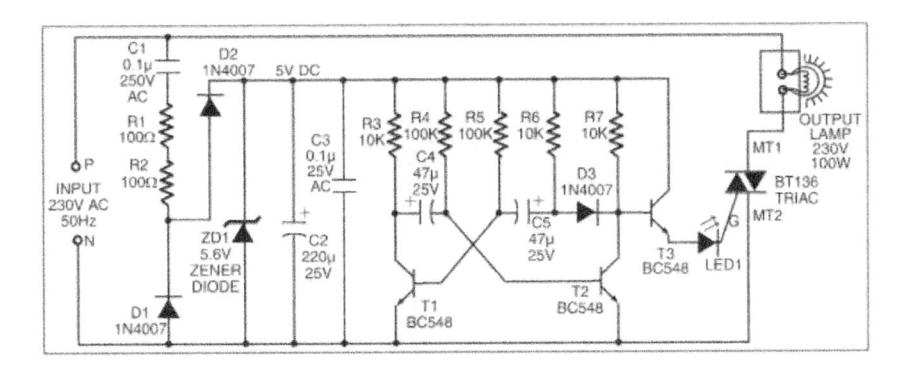

Components such as resistors R1 and R2, capacitors C1, C2, and C3, diodes D1 and D2, and zener ZD1 are used to create a rather consistent 5V DC supply voltage. This voltage is utilized to provide the requisite current to operate the multivibrator circuit and activate the triac BT136 through LED1

T1 and T2 are the two BC548 transistors that are used in the construction of the multivibrator circuit, along with a few passive components. Capacitors C4 and C5 as well as resistors R3 through R7 in the multivibrator circuit are what determine the frequency of the oscillations produced by the circuit.

The output of the multivibrator circuit is connected to the transistor T3, which in turn drives the triac by way of the LED1. The bulb shines as a result of the transistor T3 energizing the triac BT136 during the positive half cycles of the output from the multivibrator.

39. Christmas Lights Using LEDs

When celebrating special occasions, it is common practice to decorate with various light effects. Electronic circuits are becoming increasingly diverse as a result of the creativity and innovation of product designers. As illustrated in Figure 1, here is a Christmas light circuit that is simple to put together and may be used for decorating. It is made up of a few resistors, a couple capacitors, and four transistors, and it has eighteen LEDs. Because T1 and T2 are wired up to function as an astable multivibrator, it is guaranteed that one of the transistors will be conducting at all times. Therefore, clock pulses are produced by the combination.

Christmas lights using LEDs

The values of the time constants produced by the R6-C2 and R8-C1 pair combinations have been carefully chosen in order to provide a low-frequency clock that is discernible by the naked eye. There is a connection made between the collectors of transistors T1 and T2 and driver transistors T3 and T4. In order to illuminate two rows of LEDs that are connected in parallel with alternating clock pulses, these are utilized. Approximately 2 hertz is the frequency at which the LEDs LED1 through LED9 and LED10 through LED18 light up in a randomized order. Simply by altering the values of capacitors C2 and C1, you are able to make simple adjustments to this frequency.

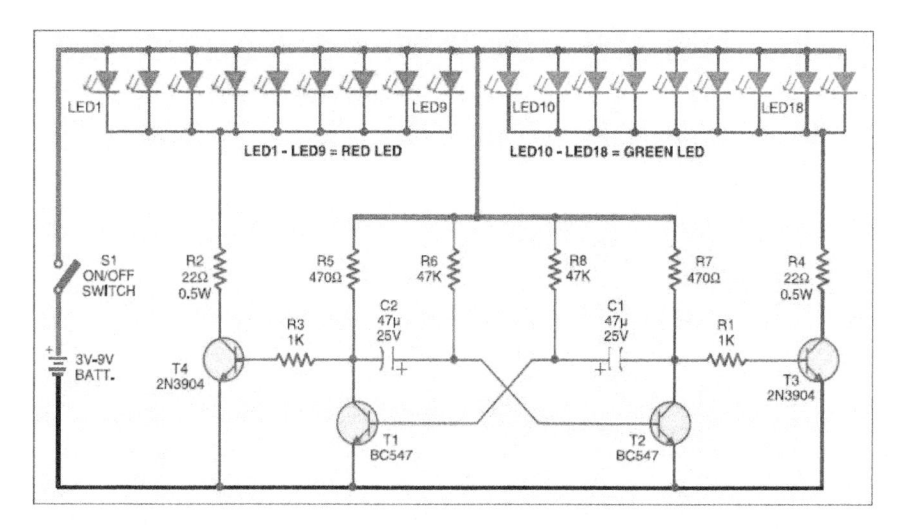

The amount of current that flows through the LEDs can be adjusted by adjusting the values of resistors R2 and R4. For the purpose of mimicking the effects of Christmas decorations, red LEDs (LED1 through LED9) and green LEDs (LED10 through LED18) are utilized. You can adjust the values of the resistors R2 and R4 to achieve the desired degree of brightness variation.

40. DIY: Make Your Own Nifty Night Lamp

During the evening, you may illuminate a small area with the help of this straightforward circuit. In addition to its utility as a snazzy bedroom night lamp, it may also be put to use in a variety of different contexts. In spite of its adaptability and complete automation, the circuit only draws a negligible amount of electric electricity from the grid.

Figure displays the wiring diagram for the cool table lamp's internal components. It is constructed of a rectifier diode 1N4007 (D1), a 0.47F, 400V capacitor (C2), a transistor SS9014 (T1), a linear variable resistor (LDR) (LDR1), a blue LED (LED1), and a few additional components. R1 and R2 both have a resistance of 47 ohms. The 230-volt-alternating-current mains supply is input through the CON1 connector.

This light operates on a power supply of 230 volts AC mains and does not require a transformer. The components that are utilized here would assist in keeping the general efficiency and safety in check. An alternative that is more cost-effective to a conventional power supply is a capacitive transformer-less power supply.

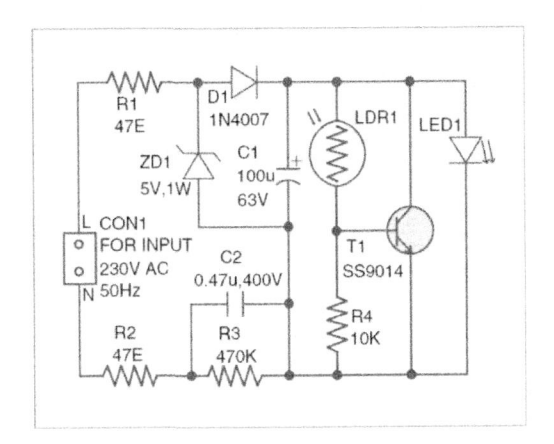

Fig.Circuit diagram of the nifty night lamp

The constant-current arrangement that is utilized in this setting generates approximately 30mA at 5V DC. While resistors R1 and R2 are responsible for limiting inrush current, capacitor C2 is the most important component in this system. The values of resistors R1 and R2 are selected in such a way that they do not lose an excessive amount of power but are still large enough to limit the amount of current that flows during the inrush phase. When connected in parallel with capacitor C2, 470-kilohm resistor R3 forms a filter that reduces the amount of EMI that is transmitted back to the grid. When the device is unplugged from the AC mains, it also performs the function of a bleeder resistor, which empties the capacitor (wall plug).

41. Photometer

To read and do close work without putting stress on your eyes, you need 100 to 1000 lux of light. Bright sunlight or specular illumination gives off 50,000 lux, while twilight or dim light only gives off 10 lux. The light from a fluorescent lamp is better for reading and close work than the light from a tungsten lamp, which is only 1600 lumens. If the eyes are exposed to dim light for a long time, they will get used to it and become very tired. Here is a circuit for measuring the amount of light. It can show a linear scale of illuminance that compares different levels of light, such as minimum, moderate, and maximum.

Circuit for light intensity measurement

The light sensor in the circuit is a p-n photodiode, and the display driver IC LM3915 (IC1) gives a logarithmic current scale that is proportional to the amount of light hitting it. Compared to cadmium-sulphide (CdS) photodetectors, the photodiode has a fast switching speed of one nanosecond and can respond at a high frequency. The photodiode's "n" material is made of pure silicon that has been doped with phosphor to give off free electrons. The 'p' material is made of pure silicon with boron as the doping material. It gives away free holes.

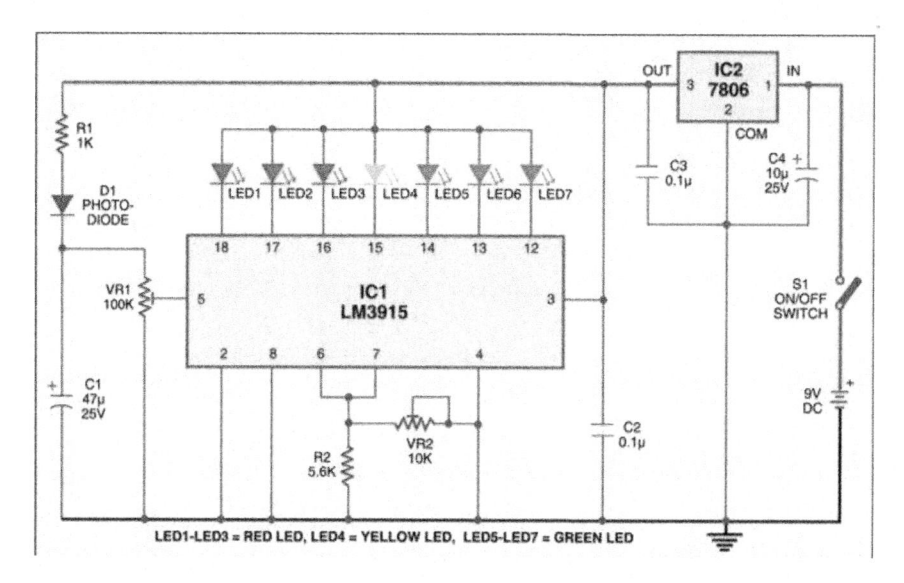

Circuit Operation

When it is dark, a reverse saturation current builds up in the photodiode, and when it is light, free electrons move. In a photodiode, the amount of light is proportional to how straight the flow of current is. Usually, the light current at 5 volts is between 10 and 90 A, and a p-n photodiode's radiation sensitivity is between 0.4 and 4 A/mW/cm2. To get the light current, the photodiode is connected to the circuit in forward-biased mode through the current-limiting resistor R1. When light hits the photodiode, it sends a current to the input pin 5 of IC1 through the preset VR1.

The IC LM3915 has a complete voltmeter circuit built in so that the LED scale can show how the input voltage changes. When 1.25 volts are added to the input pin 5 of IC1, each LED from pin 18 on down lights up. The IC has ten comparators with a current source output and a reference and ladder voltage divider that provides the necessary reference voltage. Pin 7 of the IC is kept at a reference voltage of 1.25 volts, and resistor R2 controls the amount of current going through the LEDs, which controls how bright they are.

By changing the position of preset VR2, you can set the lowest light level at which the first LED (LED1) should light up. By adjusting preset VR1, you can set the maximum voltage at which LED7 should light up. This is done by adjusting the voltage that goes into the IC. LED1 shows the least amount of light, LED4 (yellow) shows medium light, and LED7 shows normal light. Each LED takes a little longer to turn on or off because of capacitor C1.

42. Smart Emergency Light

An emergency light is useful in a variety of situations, but it is particularly useful in the event that the power goes out. Now you don't have to worry about dark nights when the power goes out. Here is a smart emergency light with white LEDs that turns on by itself when the main power goes out.

Emergency light working

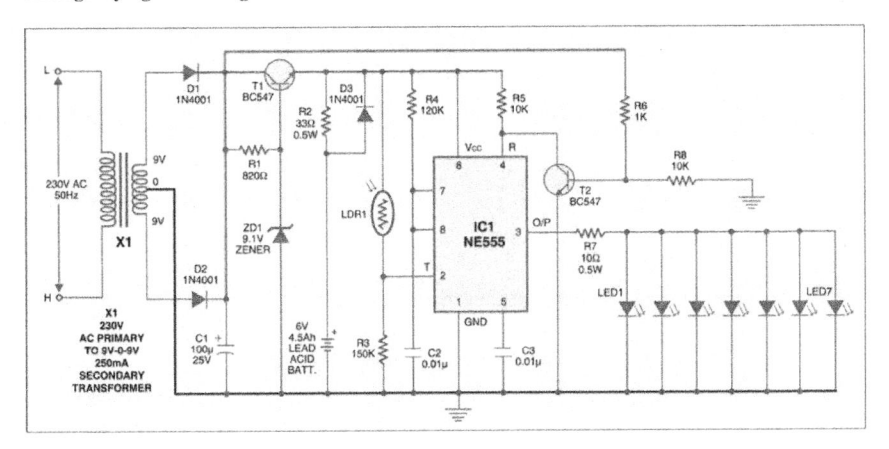

The circuit has sections for getting power, charging the battery, and switching. Transformer X1, diodes D1 and D2, transistor T1, resistors R1 and R2, and zener diode ZD1 make up the power supply and charger. The 9V-0-9V, 250mA step-down transformer X1 takes power from the AC mains to power the circuit. AC voltage is turned into DC voltage by diodes D1 and D2. Filter capacitor C1 smooths out the DC voltage. The DC voltage that is not controlled is controlled by transistor T1, resistor R1, and zener diode ZD1. With the help of resistor R2, the lead-acid battery is charged by the regulated DC voltage. When there is no mains power, diode D3 links the battery power supply to the switching circuit.

The switching circuit is made up of a NE555 timer (IC1) that is wired in a way that makes it only work in one mode. When a low voltage is put on pin 2 of IC1's trigger, the timer turns on and pin 3 of its output goes high. It stays in that state until pin 2 of IC1 is pressed again. Between the positive end of the battery and pin 2 of IC1's trigger, there is a light-sensitive resistor called LDR1. Between pin 2 of IC1 and ground is a resistor called R3. The resistance of LDR1 stays high when it's dark (at night) and low when there's a lot of light around (in daytime). The switching circuit is controlled by this effect.

43. Digital Camera Adaptor
You do not have to be disheartened the next time the low battery indicator appears on your digital camera while you are on a picnic outing. You can easily connect your digital camera to this adaptor by simply plugging it into the cigarette lighter socket in your vehicle and then connecting the camera to it. The converter will serve as an interface between the DC source and the camera battery, and it will supply enough charging current to completely re-energize the battery in one hour. A current of 300 milliamps to one ampere and a voltage of 5 volts will swiftly charge either a lithium-ion or a nickel metal hydride battery in a digital camera.

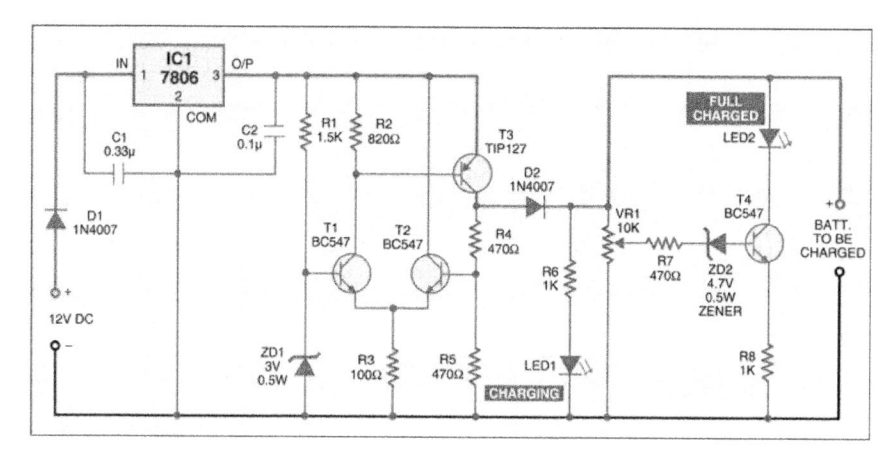

The internal workings of the digital camera adaptor are depicted in Figure 1. The circuit makes use of a differential transistor amplifier in addition to a current amplifier in order to generate a consistent voltage as well as the necessary current. In differential amplifier architecture, NPN transistors T1 and T2 form a differential amplifier that is analogous to the inputs of an operational amplifier (op-amp). Zener diode ZD1 maintains a constant voltage of 3 volts across the base of T1. The base voltage is obtained by T2 from a potential divider that consists of R4 and R5 and is connected to the output rail. The voltage at the base-emitter junctions of transistors T1 and T2 is compared, and the base current of current amplifier T3 is adjusted accordingly.

The current amplifier is a PNP transistor called TIP127 (T3), and it is utilized to supply a suitable amount of current for charging. This device is a Darlington transistor with a medium power rating and a current gain of 1000. In other words, if a current of 1 mA is supplied to the base of T3, the collector current will be at its absolute maximum of 1 A. Due to the fact that T3 operates in inverter mode, a negative pulse will be driven by T1 when it conducts when a positive voltage is provided to the base of T1. Current of approximately 1mA is supplied by the 820-ohm resistor R2 to the base of the transistor T3, allowing the transistor to output current of 1A from its collector.

44 Mock Alarm with Call Bell

This is a completely functional fake alarm that can be used to scare away any potential home invaders. The alarm is programmed to go off at sundown and will remain "on" till the next morning. The machine gives off the impression that it is some kind of high-tech alarm by flashing light-emitting diodes (LEDs) and making intermittent beeping noises. In addition, the circuit keeps a lamp illuminated during the night by switching it on and off at regular intervals of thirty minutes. In addition to that, there is a facility for call bells.

Mock alarm circuit

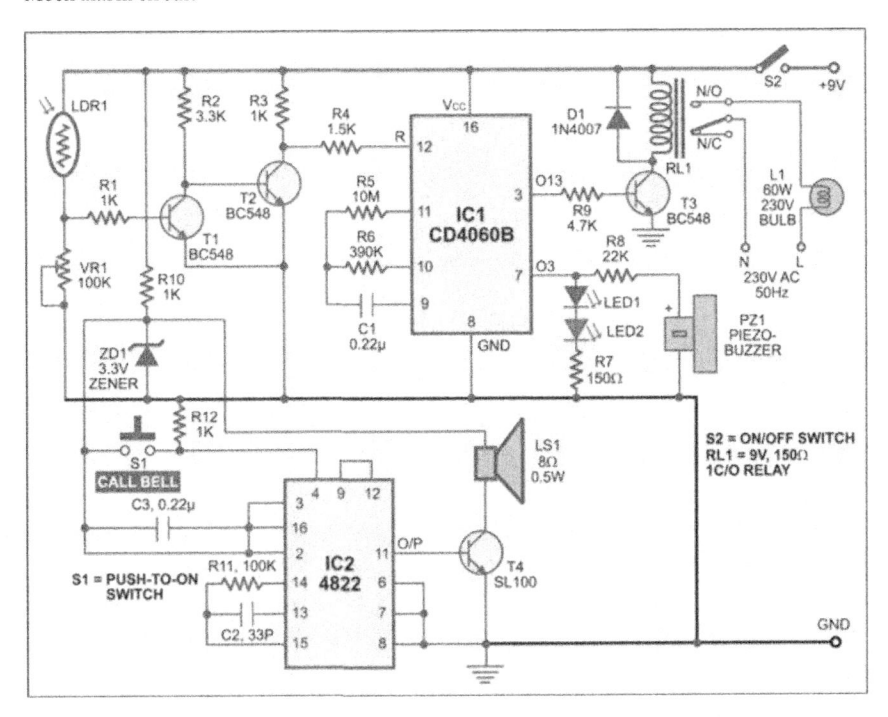

The CMOS integrated circuit CD4060B (IC1) is the central component of this circuit. This IC contains an internal oscillator as well as a 14-stage binary divider, which allows it to offer a long delay without the need for a resistor and capacitor with a large value.

To connect the 9V power source to the circuit, press the switch labelled S2. The light-dependent resistor LDR1 has a low resistance during the day, allowing transistor T1 to be able to conduct. This causes transistor T2 to enter the cut-off mode because its base is dragged to ground by transistor T1, which is a consequence of this situation. As long as transistor T2 is disabled, the reset pin 12 of IC1 will maintain its high state. Because of this, the oscillator in IC1 (which is made up of the resistors R5 and R6 as well as the capacitor C1) stays disabled, and its outputs do not change. Through the use of preset VR1, one is able to modify the sensitivity of LDR1.

The circuit operates from a power supply that is regulated to 9 V. Assemble the circuit on any PCB that can be used for a variety of purposes, and then encase it in a plastic box that is watertight and has holes in the top and back for mounting the LDR and LEDs, respectively. Position the light-dependent resistor (LDR) such that sunlight can directly hit it. Install the device on the pillar that is located at the entrance gate. Install lamp L1 in the porch of the home, which is a sufficient distance from the LDR, to prevent illuminating the device unnecessarily. Maintain the speaker's presence within the room.

45. LED Based Reading Lamp

This LED-based reading lamp circuit utilizes ultra-bright white LEDs to supply adequate light for reading while only consuming about 3 watts of power overall. In the event that there is a problem with the AC mains, the battery backup circuit will immediately turn the LEDs on. When the power is restored, the lamp circuit will immediately switch back to operating off of AC mains, after which the battery supply will be automatically terminated.

LED based reading lamp

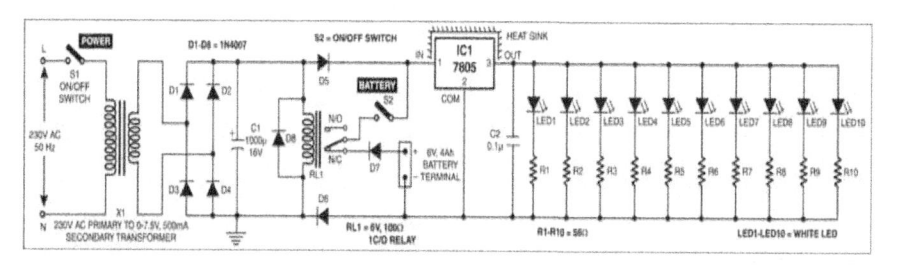

The power supply circuit is made up of a step-down transformer with 0-7.5V and 500mA current capacity called X1, rectifier diodes numbered 1 through 4, and a filter capacitor called C1. Regulator IC 7805 (IC1) supplies regulated 5V to LEDs, ensuring that there is no variation in the intensity of the lamp light even if the mains power supply is unstable. This is accomplished by ensuring that LEDs are able to operate at a constant voltage. In total, ten white LEDs are linked in parallel across the 5V power supply. These LEDs are numbered from LED1 to LED10. To control the amount of current flowing through the white LEDs, resistors R1 through R10, each with a value of 56 ohms, are connected in series with the lights. It is possible to increase the brightness of the lamp light by adding additional LEDs in the same fashion; however, the lamp can only accommodate a maximum of 15 LEDs.

When the power switch S1 is turned off, the relay RL1 becomes energized, and if the battery switch S2 is turned off, it will disconnect the 6V, 4Ah battery from the input to the regulator IC1. The battery is connected across the N/C contact of the relay RL1. When the power switch S1 is in the open position, relay RL1 is de-energized, and its N/C contacts connect the battery to the input of IC1 in the appropriate manner. Diodes D5 and D6 are reverse-current protection diodes, and their job is to prevent current from flowing from the battery towards the power supply section. Diode D7 is there to protect the battery from being discharged in the wrong direction. Check that the battery has been fully charged before you attempt to connect it.

46. Multidoor Opening Alarm with Indicator

This door opening alarm will notify you if an unauthorized person enters your home. You have the ability to use it for up to three different doors. Simply install a small device on each door frame that contains a reed switch, and then attach a magnet to the moving door in such a way that it is aligned with the reed switch when the door is closed. This will complete the installation. You are responsible for maintaining a separate unit within your room that contains the power supply, three LEDs, and a buzzer. This unit is to be kept by your side. Each door unit receives a connection from this unit through a ribbon cable with three cores. The positive terminal is connected to one of the cores, the negative terminal is connected to the second core, and the third core is connected to the output of the unit. Reed switch terminals are connected in a short circuit whenever the door is shut, which prevents the alarm from going off.

Door opening alarm circuit

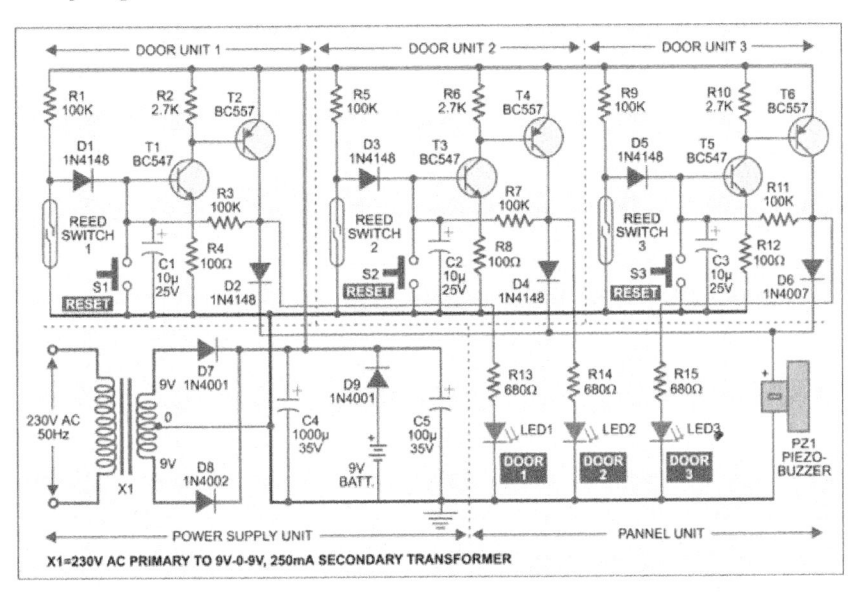

When someone opens door 1, the transistors in the door unit that corresponds to that door conduct electricity, which causes the buzzer to sound. When door 1 is opened, LED1 lights up to show that it's been opened. Even after you have finished closing the door, the alarm will continue to go off because of the diode latching action. Only by hitting the reset switch on the door unit will it be possible to turn it off. The three units that are installed on the doors receive regulated power ranging from 9 to 12 volts DC, which is derived from the AC mains and used to operate the circuit. In addition to it, a battery backup is supplied. When one, two, or all three of the doors are opened at the same time, one, two, or all three of the LEDs will light up. By expanding the number of door units that are connected to the audio-visual indicator unit, this setup can be modified to accommodate a greater number of entrances.

47. Rechargeable Torch Based on White LED

There are a few drawbacks associated with rechargeable flashlights. You are going to need to charge the batteries and change the bulbs at regular intervals. For example, the typical incandescent light-emitting diode (LED) torch uses approximately 2 watts of power. This white LED-based rechargeable torch has a service life that is 60 percent longer than the typical incandescent torch and only consumes 300 milliwatts of power throughout its operation.

Circuit for a rechargeable flashlight with LEDs

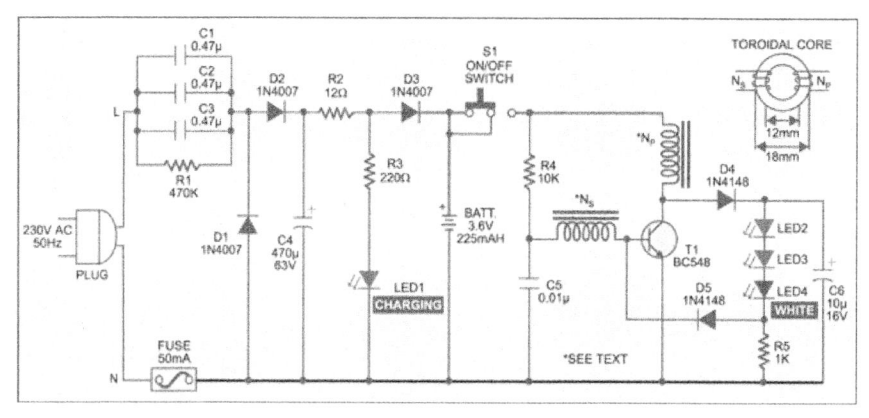

Fig. 1: Circuit diagram of the LED based rechargeable torch

The schematic for the torch's rechargeable white LED-based circuit is shown in Figure 1. The current that can flow through the charger circuit is governed by the reactive impedance of capacitors C1 through C3, which are rated for 250V AC. After the battery has been charged, the capacitors have a channel to discharge, which is provided by the resistor that is placed across the capacitors. When the red LED1 lights up, it means that the charging circuit is now operational. The flashlight is powered by three NiMH rechargeable button cells, each of which has a voltage of 1.2V and a capacity of 225 mAH. The minimum amount of time required for a regular recharge is twelve hours. Following a complete recharge, the battery will have a time of approximately 2.5 hours in which it can be used continuously. Recharging the battery to its maximum capacity as soon as possible after use will ensure its continued dependability and longevity. The current during charging is approximately 25 mA.

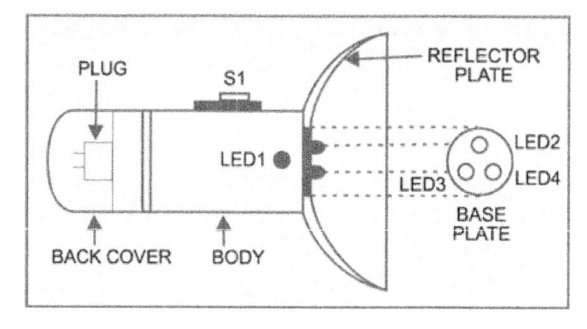

Fig. 2: Suggested enclosure for the torch

To provide power to the white LEDs, you will need a circuit that boosts the voltage (LED2 through LED4). Increasing the voltage requires the utilization of an inverter circuit. The schematic provides information regarding the winding of the inverter transformer, which makes use of an insulated ferrite toroidal core. In the primary coil (NP) and secondary coil (NS), there are a total of 30 and 3, respectively, turns of wire with a gauge of 35 standard gauge. In the event that the inverter does not oscillate, you will need to switch the polarity of either the primary or the secondary winding, but not both. The output is maintained in a consistent and regulated state thanks to a reference voltage that comes from resistor R5, which also provides reflected biasing to the transistor. Fig.2 shows the case that is suggested for the torch.

48. SMF Battery Guard

In the event that the electricity from the mains supply is interrupted, the emergency light will immediately switch over to a light source that is powered by a rechargeable battery. The lamp will turn off once the power from the mains supply is restored. Portable emergency lights almost always make use of batteries that are rated for 6 volts and are maintenance-free and sealed. Because many inexpensive lights do not provide any notice when the battery is running low, there is a possibility that the battery will not work. The following diagram shows a battery guard circuit that can be used with emergency lights that do not have deep-discharge prevention circuitry. In the event that the voltage of the battery drops below the predetermined threshold, it will turn off the bulb (5V).

Block diagram

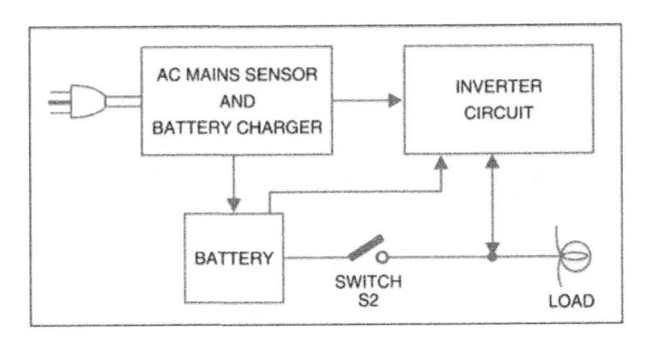

Block diagram of an emergency light

The schematic representation of an emergency light may be seen up above. It is primarily composed of a power supply, a battery pack that can be recharged, an autonomous sensing circuit that can detect the presence (or absence) of a mains supply, and a charging circuit. In the event that there is a disruption in the power supply from the mains, the sensing circuit will connect the lamp (or an inverter for a compact fluorescent tube) to the battery. It will then break this connection once the power from the mains is restored. The battery is charged via the charger circuit whenever there is access to the AC mains supply.

The battery protection circuit is depicted in the following image. Two transistors, one Zener diode, and a few passive components are utilized in its construction. LED1 displays the current state of the battery voltage. When LED1 begins to glow, this signals that the battery voltage is lower than the safe voltage and that the connection between the bulb and the battery has been severed.

Battery guard circuit

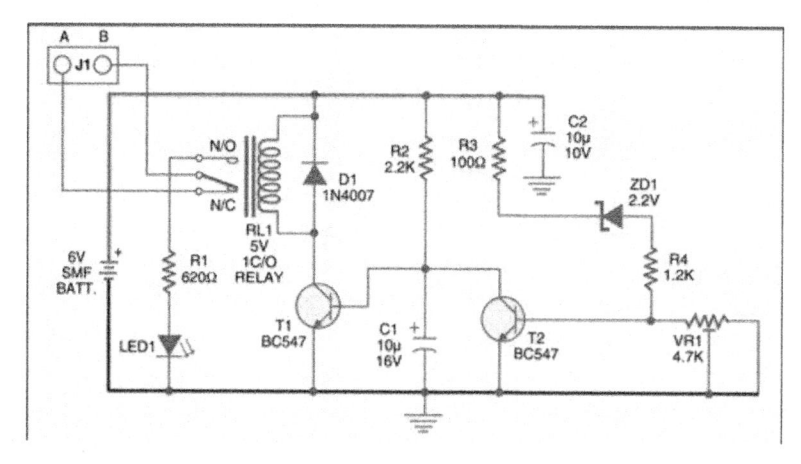

Battery guard circuit

Circuit operation

When the battery has been charged to its maximum capacity, the voltage from the battery is transmitted to the base of transistor T2 through resistors R3, ZD1, and R4. Once transistor T2 begins conducting, the previous transistor, T1, is forced to turn off. The relay is left in its de-energized state, and the lamp is wired to the battery jumper J1 that is used to connect points A and B.

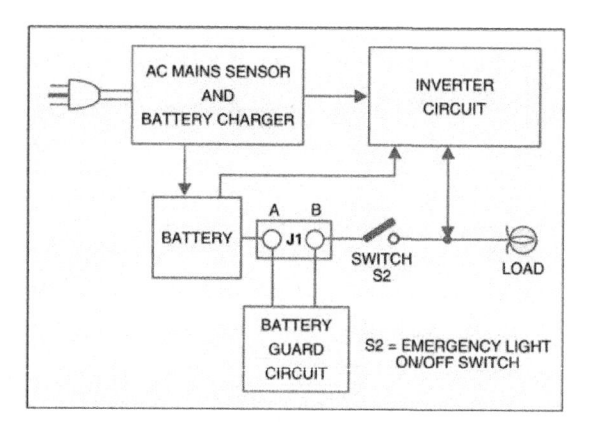

Fig.3 wiring of battery guard circuit to emergency light circuit

When the voltage of the battery drops below the secure threshold, the transistor T2 is turned off. As a consequence of this, T1 becomes conducting, and the relay that was used to disconnect the load from the battery becomes energized.

Set the conduction voltage for transistor T2 using preset VR1 so that it conducts a voltage of 5V or higher. When the voltage is less than 5V, the transistor T2 will stop conducting, which will cause the transistor T1 to start conducting. As a consequence of this, the relay becomes energized, which disconnects the bulb from the battery. As a consequence of this, the battery no longer undergoes deep-discharging.

Brake failure indicator circuit

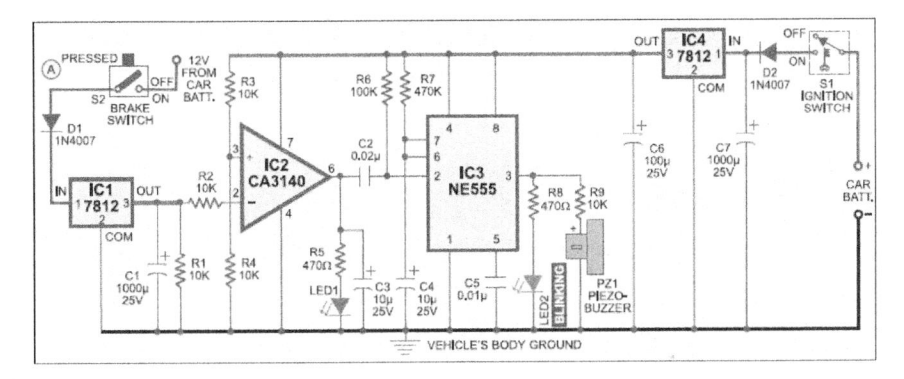

A voltage comparator is provided by an op-amp integrated circuit with the part number CA3140 (IC2), and an alarm is generated by a timer with the part number NE555 (IC3) in a monostable mode. The voltage level that is present across the brake switch is measured by the voltage comparator IC2. Its non-inverting input, pin 3, is connected to a potential divider consisting of resistors R3 and R4 that each have a resistance value of 10 kilo-ohms. Through diode D1, IC 7812 (IC1), and resistor R2, the inverting input (pin 2) of IC2 is connected to the brake switch. When the brake is applied, a larger voltage is transmitted to it. When the brake is not being applied, the normal behavior is for the output of IC2 to remain high, causing LED1 to emit a red light. Through the use of the coupling capacitor C2, the output of IC2 is connected to the trigger pin 2 of the monostable. In order to maintain the input stability of IC2, resistor R1 is utilized. A ripple-free, regulated supply is provided to the inverting input of IC2 by IC1 and C1, respectively.

In order to generate a pulse with a duration of one second, IC3 has been wired as a monostable. In order to turn on the buzzer and LED2, the timing components R7 and C4 bring the output high for a period of one second. Because of R6, the trigger pin of IC3 is normally high, which means that both the buzzer and LED2 are in their "off" states. When the brake pedal is depressed, pin 2 of IC2 receives a higher voltage from the brake switch. As a result, the output of IC2 drops to a lower value, which turns off the red LED. In order to trigger the monostable, the low output of IC2 sends a brief negative pulse to it through the capacitor C2. This causes the buzzer and LED2 to light up, indicating that the braking system is functioning properly. When there is a drop in pressure in the brake system owing to leakage, LED1 continues to show that it is "on," but the buzzer does not sound when the brakes are applied.

50. Stress Meter

The stress meter provides us with an electronic representation of the emotional suffering that we are experiencing. One is able to determine the amount of stress that the body is experiencing with the assistance of this equipment. A warning beep will sound in addition to the stress meter providing a visual indicator through a Light Emitting Diode (LED) display in the event that the stress level is extremely high. The device is little enough to be worn on the wrist like a watch.

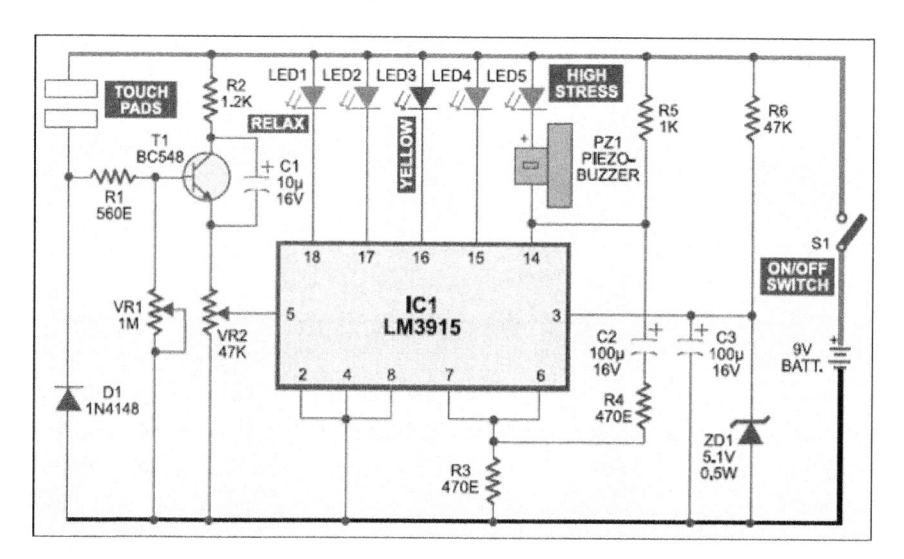

The operation of the device is predicated on the hypothesis that changes in emotional state are accompanied by shifts in the resistance of the skin. When there is a high amount of stress, the skin's resistance is decreased, and when the body is relaxed, the skin's resistance is increased. The increase in blood supply to the skin at times of extreme stress is the cause of the decreased resistance that the skin exhibits. Because of this, the permeability of the skin is increased, which in turn enhances its ability to conduct electric current.

This characteristic of the skin is utilized here in order to determine the level of stress. The stress meter's touch pads are responsible for sensing the voltage fluctuations that occur across the touch pads and communicating this information to the circuit. The circuit has a high degree of sensitivity and can detect even a slight change in voltage when applied to the touch pads.

51. Geyser Timer

After the predetermined time of 22 minutes, this geyser timer circuit will emit an alert signaling that the water has reached the desired temperature. The component in question is a timer IC 555 that has been hooked up as an astable multivibrator and has an adjustable time period of 15 seconds. The astable output, which has been inverted by an inverter, then drives the decade counters IC3 and IC4, both of which are IC 7490s and are connected in cascade. The output of the decade counter is connected, respectively, to the decoders IC5 and IC6 (both of which are IC 7442). The outputs of the decoders, which are the Q8 outputs of IC5 and IC6, are connected to inverters, and the outputs of the inverters are connected to an AND gate. In order to activate the alarm, the AND output is wired up to the reset pin of the astable multivibrator that was constructed using an additional timer IC 555. You are free to switch off the geyser at this time.

Geyser timer circuit

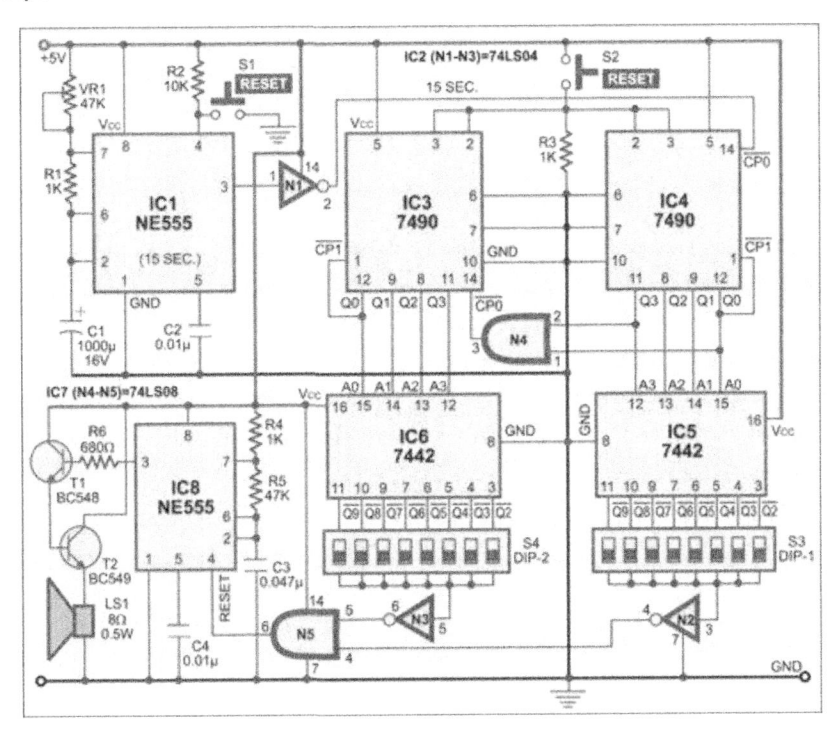

As the indicator for the power supply, a green LED (LED1) has been implemented. Both the timer and the geyser should be turned on at the same time. When the alarm goes off, it indicates that the water in the geyser has reached the temperature at which it may be utilized and has been heated.

52. Multicell Charger

You are able to charge up to two Ni-Cd cells or Ni-MH cells in a secure manner with the assistance of this multicell charger. The circuit is convenient in size, not overly expensive, and simple to use. The 230V AC mains are reduced to 12V AC (at 500 mA) by the step-down transformer X1, transformed into pulsing DC voltage by the diodes D1 and D2, and then sent to the battery charger terminals via the current-limiting resistor R1 and the silicon-controlled rectifier SCR1. SCR1 is the most important component of the charger. It conducts normally because of the gate biasing voltage that is made accessible through resistor R2 and diode D3, and the battery is in charging mode, as indicated by LED1. The charging current is kept within safe parameters by the resistance provided by R2. This circuit has a charging current of around 250 milliamperes (mA).

Multicell Charger Circuit

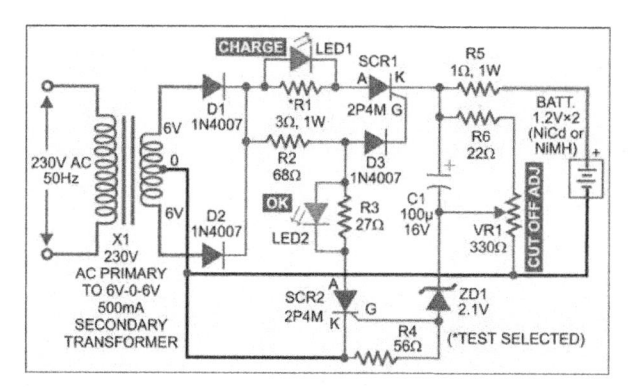

When the battery has reached its maximum capacity, the SCR2 conducts, which causes the gate of the SCR1 to be pulled down. LED2 will illuminate when it is in this state. Now take the batteries out of the charging device. A Ni-Cd cell with a rating of 500 mAH will typically require approximately 2.5 hours to achieve full charge, while a Ni-MH cell with a rating of 1500 mAH will typically require approximately 7 hours to reach full charge. The amount of time it takes to charge can change based on the settings of the charger and the circumstances of the input supply line.

53. Light Dimmer that Doubles as Voltmeter
Measure the alternating current mains voltage without a multimeter. To convert the light dimmer that is normally found at the bottom of a table lamp into a voltmeter all that is required of you is a minor adjustment.

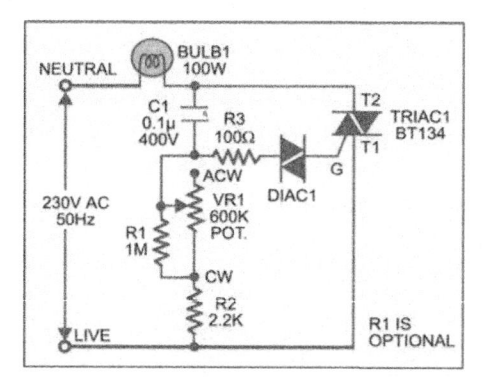

When the dimmer is turned counterclockwise to a position where the illumination of the filament can just be seen, the point at which the dimmer is turned can be used as a reference point for measuring the voltage.

Circuit operation

Turn on the lamp using a variac, and provide it with 50 volts of power. Turn the potmeter knob counterclockwise until the glow of the filament can just be seen, and then put a mark next to the cursor indicating that the voltage is 50V. Continue to use the variac to get the voltage up to 100, 150, 180, 200, and 220, and make sure the scale is calibrated for each of these voltages as you go. At this point, a voltage scale is developed. The only catch is that the voltage is growing in a counterclockwise manner, which in and of itself should not be a problem at all. On the other hand, contrary to what is depicted in the sketch, the scale will not be linear. The accuracy of the measurement will be determined by the calibration standard that was applied, and the tolerance will be on the order of 1 percent 5 volts. When trying to achieve higher precision and tolerance, the diameter of the knob on the potmeter and the fineness of the cursor can be of assistance.

A regular fan regulator can be used with a lamp that has 40, 60, or 100 watts, and the regulator can be calibrated to match the wattage of the lamp. The lowest voltage that can be measured is inherently constrained to the level necessary to achieve the "just visible" requirement. In this configuration, the maximum scale voltage will be close to 220 volts. R1 will be open circuited.

54. 220V Live Wire Scanner
This is a live wire scanner capable of detecting 220V. The wire that serves as the sensor is attached to the clock input of the integrated circuit (IC). As the sensor, we have utilized a 10-centimeter length of 22-gauge wire here.

200V Live Wire Scanner Circuit

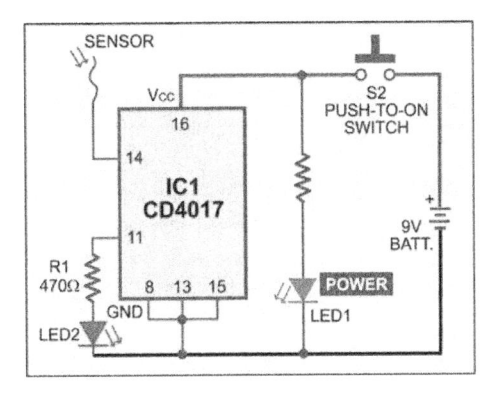

The circuit is activated by the electric field generated by the mains when the sensor, which can be a metallic conductor or a copper wire, is brought into close proximity to the live wire. The CMOS integrated circuit has a high input impedance, which means that the electric field created in the sensor is sufficient to clock it. The LED is powered by the output that is acquired at pin 11 of the CD4017. The existence of mains may be determined by the LED (LED2) blinking, while the active status of the scanner can be determined by the LED1. The circuit can be utilized to locate stray leakage that originates from electrical appliances such as fans, mixers, refrigerators, and so on. It is possible to assemble it quickly on any general-purpose board, or the discrete components may be soldered directly into the integrated circuit. The circuit receives power from a 9V PP3 battery. If you use a mains adaptor, check to see that it is properly regulated and isolated; if not, even a stray electric field from the mains transformer could cause the circuit to clock.

Caution: If you want to eliminate the risk of being exposed to live AC mains, use cable that has been insulated.

55. Smart Switch

A mechanical switch or a relay provides a straightforward option for turning on the mains voltage. However, due to the large amount of space that is required for the relay and the other components that are associated with it, they cannot be installed in a typical switch box. A superior option is provided by this smart switch circuit, which is depicted here.

Smart switch circuit

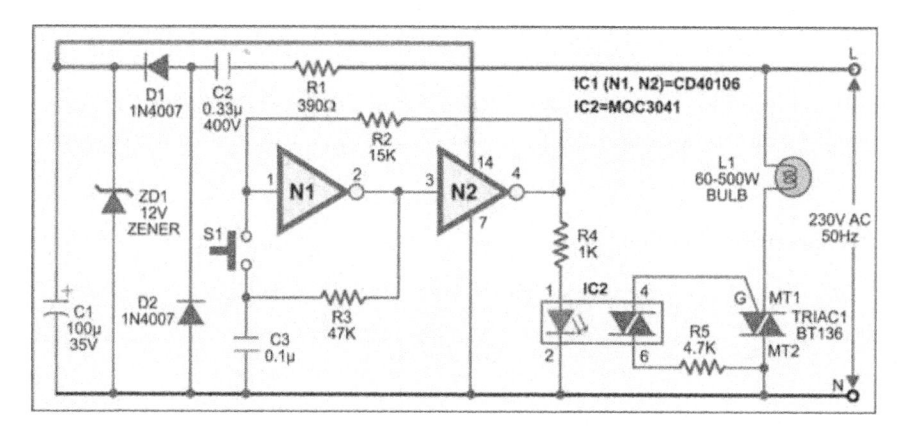

It is nothing more than a simple on/off controller that makes use of an electronic circuit that operates in the same manner as a standard switch. Your switch panel can get a more aesthetically pleasing appearance by using a flat push button control. The switching circuit consists of an optocoupler circuit that gets input from a bistable switch made by a couple of Schmitt trigger gates that control a triac. The input from the bistable switch is then passed on to the optocoupler circuit. To turn the load on or off, one need only push the push button switch for a moment at a time.

The optocoupler controls the triac in such a way that it is toggled whenever the switch is pressed. In contrast to the arbitrary phase switching, the optocoupler contains a zero-crossing detector that acts as a shield against radio interference. Because the mains is not isolated, the only way to protect yourself from a potentially fatal shock is to utilize a push button switch of sufficient quality and insulation. Verify that the triac has the capacity to withstand the amount of current that will be passed through it.

In the event that it is necessary, multiple push buttons can be wired in parallel to provide the ability to toggle the triac from a variety of locations.

56. Power Failure and Resumption Alarm

This circuit provides an audio-visual indicator if there is a disruption in the mains power supply or when it is restored. The dual timer IC LM556 serves as the primary component of the circuit. When there is power from the mains supply, the bicolor LED gives off a green glow; however, if there is no power, it turns red.

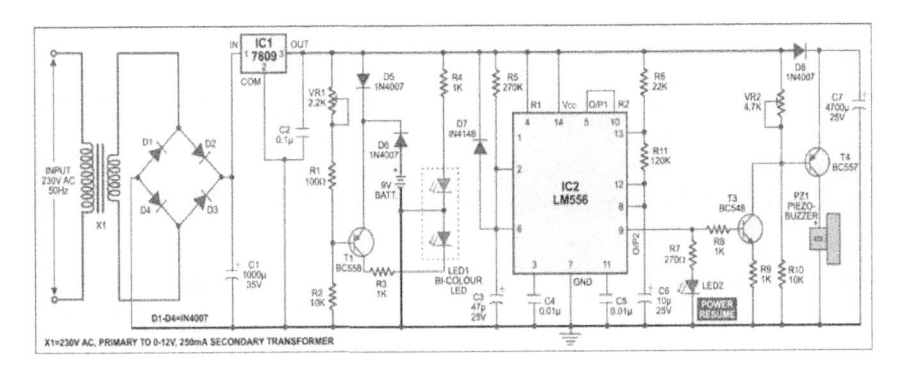

The secondary output of the transformer X1 is 12 volts with 250 milliamperes, and it is stepped down from the mains voltage of alternating current. The output of the transformer is fed into a full-wave bridge rectifier that is made up of the diodes D1 through D4, then it is filtered by capacitor C1 and last it is regulated by IC 7809 (IC1) to provide regulated 9V DC that is used to run the circuit. The power source for the red light that indicates there is no power has been determined to be a 9V battery in conjunction with a pnp transistor designated as T1. Simply adjusting the preset value of VR1, the transistor T1 can be readily made to conduct or switched off.

When the power supply is first turned on, the pnp transistor T1 is in its cut-off state. As a result, the green light emitted by the bicolor LED1 is visible. When the power goes out, the pnp transistor T1 begins conducting, and the red light from the bicolor LED1 begins to shine. Because IC2's pin 14 does not have access to the Vcc voltage, the output of that chip's pin 9 stays low, and the transistor T3 does not conduct. However, capacitor C7 (4700 F) maintains an appropriate charge, and as a consequence, transistor T4 conducts, and piezobuzzer PZ1 emits sound continuously for around eleven seconds until capacitor C7 empties completely.

When power is restored, the bicolor LED1 returns to its normal green glow, and the buzzer continues to sound for approximately 14 seconds. Monostable and astable operation modes of the dual timer IC LM556 (IC2) portions have been implemented here, respectively. The position of the external timing capacitor in the monostable section is what decides whether or not a positive or negative output pulse is generated. Diode D7 assures that even a brief interruption in power supply will result in the generation of a pulse as soon as the power is restored. A positive output pulse is created when the capacitor C3 is connected to ground, and the following relationship is used to explain how this occurs:

$T = 1.1 \times R5 \times C3$.

This positive output is available at pin 5 of IC2, where it can be found. Due to the fact that IC2 is a dual-timer IC, the first output of this component is supplied straight to the reset pin 10 of the second section. As a result, the second timer contained within IC2 begins to oscillate. Its frequency of oscillations, shown by the symbol F0, is controlled by the resistors R6 and R11, as well as the capacitor C6.

$F0 = 1.4/(R6+2R11) \times C6$.

At its pin 9, the integrated circuit LM556 sends out frequencies in the form of pulses. These pulses are linked to a npn transistor, which, depending on the output at pin 9 of IC2, either conducts or cuts off. In order to show that power has been restored, a red LED2 that has a current-limiting resistor of 270 ohm connected to pin 9 is used. Because the collector output of transistor T3 is supplied directly into the base of pnp transistor T4, the base biasing of T4 changes, which results in the buzzer sounding off for around 14 seconds.

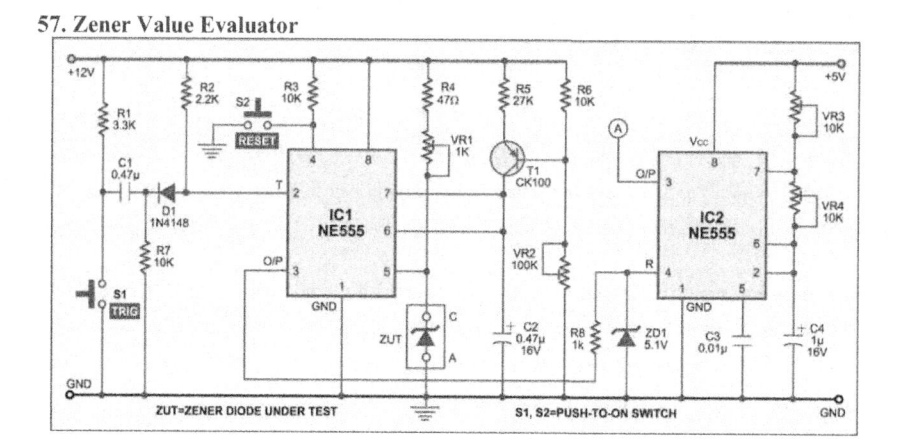

57. Zener Value Evaluator

You may determine the breakdown voltage value of any zener diode by using this straightforward Zener value evaluator circuit in conjunction with a different zener diode whose value is already known. The zener evaluator and the display unit are the two components that make up the rest of the circuit. While only 5V is required to power the display part, the zener evaluator section requires regulated voltages of 12V and 5V to function properly. Establish connections between the corresponding terminals of the display section and the +5V, point A, and ground outputs of the zener evaluator section.

A linear ramp generator that is based on a NE555 timer and an astable multivibrator that is based on a different NE555 make up the components of the zener evaluator circuit. The resistor that was previously present in the monostable has been swapped out for a source of constant current that is formed by the transistor T1. The linear charging of capacitor C2 is accomplished by the source of constant current provided by transistor T1.

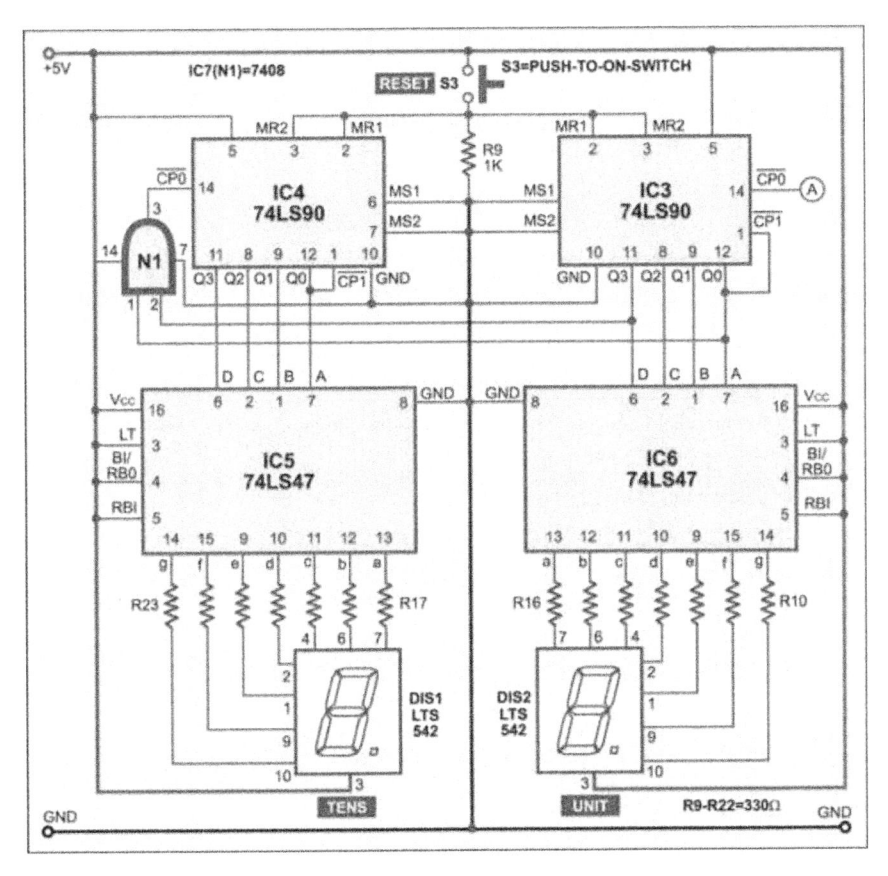

Display Unit

The following equation can be used to determine the time period T of the linear ramp that is created by IC1 at its pin 6 across capacitor C2:

When you plug in the values that are displayed in Figure 1, you get:

T equals 0.1500 seconds (approx.)

This value corresponds to the Ton of the monostable when the zener is not connected to the control voltage on pin 5.

Now connect the zener to the terminal for the control voltage, and activate the monostable (IC1) by briefly depressing switch S1. The astable multivibrator receives the output pulse width from IC1, which it then uses (IC2). The astable multivibrator has an oscillation time period of around 7 milliseconds (ms), and it continues to do so for as long as the ramp output of IC1 remains high. Figure 2 depicts the display unit that consists of decade counter ICs 74LS90, decoder/driver ICs 74LS47, and 7-segment common-anode displays LTS542. Both IC3 and IC4 are decade counters, and they both count the frequency that is applied to the clock pin 14 of IC3, which comes from pin 3 of IC2. IC 74LS90 is a 4-bit ripple decade counter. When the output of IC3 is '10' (1001), it supplies a clock for additional counting at pin 14 of IC4 (through AND gate N1). Simply holding down the reset switch, S3, for a few moments will reset both IC3 and IC4.

Common-anode displays DIS1 and DIS2 are connected to 7-segment decoders/drivers IC6 and IC5, respectively. These components, in turn, are connected to common-anode displays for the purpose of displaying the frequency of astable multivibrator IC2, which is used to evaluate the unknown value of the Zener diode. The outputs of decade counters IC3 and IC4 are connected to 7-segment

58. Doorbell-Cum-Visitor Indicator

This doorbell cum visitor indicator circuit also has the capability of identifying guests who come to your house when you are not there to greet them. When you are at home, it can function as a standard doorbell for your convenience.

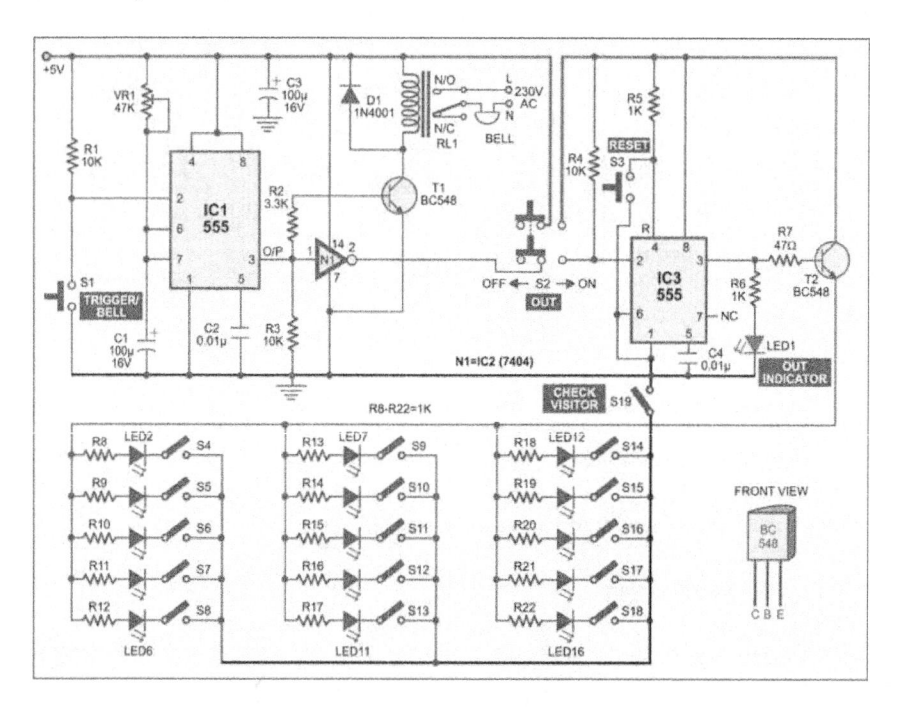

The circuit (see Fig. 1) includes the following components: a monostable constructed around the timer IC 555 (IC1), a relay driver transistor BC548 (T1), an inverter section constructed around the IC 7404 (IC2), a latching section constructed around the IC 555 (IC3), and an LED display driver transistor BC548 (T2).

The relay is controlled by the monostable output, which is sent into transistor amplifier T1. As shown in Figure 1, the normally-opened (N/O) contact of the relay establishes a connection between an electric bell and the mains supply. Inverter N1, which in turn enables the latching circuit designed around IC3 in conjunction with DPDT slide switch S2, receives the output of the monostable, which also passes to inverter N1. Through LED driver transistor T2, the output of IC3 at its pin 3 is supplied to LED1 (the visitor-out indicator), as well as LED2 all the way through LED16.

Before you leave the house, make sure that the switch labelled "Out" is in the "on" position and that reset switch S3 has been pressed once. When a guest at your house hits the doorbell switch, S1, it will set off the monostable (IC1) and also energies the relay, which will cause the bell to ring. As soon as the monostable output passes through inverter N1, the latching circuit will be enabled, and LED1 will continue to shine constantly to let others know that you have left the house.

59. Liquid level alarm

A direct current is run through the metallic probes that have been installed in the water tank so that the water level may be sensed by the water-level controllers for the tanks. This results in electrolysis and corrosion of the probes, which in turn inhibits the conduction of current and reduces the performance of the device. As a direct result of this, it is necessary to regularly replace the probes in order to guarantee an adequate flow of current. This issue can now be resolved thanks to the liquid level alarm that has been provided. By sending an alternating current pulse at 1 kilohertz via the probes, electrolysis can be prevented; as a result, the probes have a longer lifespan.

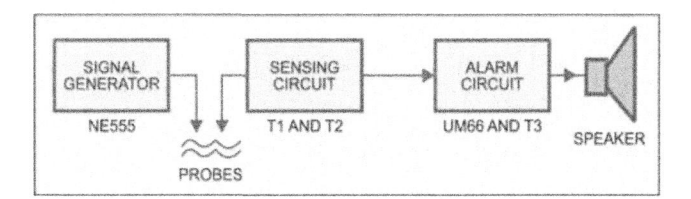

Figure 1 displays the block diagram for the liquid level alarm that was just installed. The generated signal is sent from the signal generator to the first metallic probe in the system. The second metallic probe is plugged into the detecting circuit, and then the alarm circuit is attached to it. Figure 2 presents an illustration of the whole circuit that makes up the liquid level alarm. The astable multibrator that was built around IC 555 (IC1) generates a 1kHz square wave signal, and a DC blocking capacitor is used to send this signal to one of the probes. When there is no water in the tank, the pnp transistor known as T1 does not get a base bias that is negative. However, when water continues to be added to the tank, it begins to receive a 1 kHz signal from IC1 through the probes that are submerged in the water. This signal causes it to conduct during the negative half cycle of the 1 kHz signals. npn transistor T2 maintains base bias and conducts thanks to the presence of capacitor C7 (2.2 microfarads), which enables it to supply 3.3 volts direct current to melody generator IC UM66 (IC2).

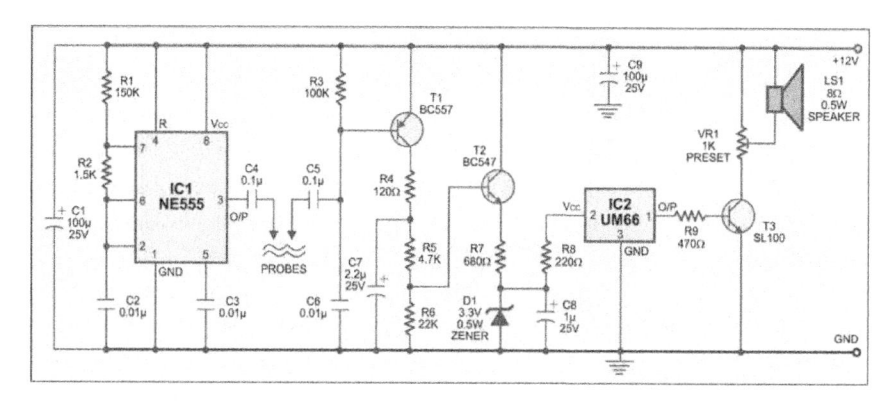

The output loudness can be adjusted by the use of Preset VR1. It is able to be adjusted to set the volume of the alarm sound coming from the speaker to the level that is desired. The circuit is capable of detecting any conductive liquid and operates of power that is unregulated at 12 volts.

60. Electronic Fuse

A workbench power supply is an item that ought to be included in each and every electronics workshop. The power supply needs to have a regulator installed, as well as some kind of protection against short circuits. For current detection, the vast majority of power supply protection circuits make use of a resistor with a low value and a high wattage that is linked in series with the load. When determining whether or not to activate the protective circuit, the voltage drop that occurs across the sensor resistor is taken into consideration. A poly-fuse application, also known as a re-settable fuse by itself, serves as the foundation for the electronic fuse circuit that has been described here.

Circuit operation

When the electricity is first applied to the circuit, the silicon-controlled rectifier SCR1 is in its "off" position. The load is connected to the normally opened (N/O) contact of relay RL1, which is energized by the polyfuse. Relay RL1 receives its power from the polyfuse. When the load's current draw exceeds a certain threshold value (which is dependent on the number of turns in the winding on the reed relay; see Figure 2 and the accompanying table for further information), the contacts of the reed relay RL2 will close, which will cause the SCR1 to be triggered. As a direct consequence of this, relay RL1 becomes de-energized, and the load is severed. Up to the point where SCR1 is deactivated, the polyfuse will maintain its high-resistance state. Either turning off the power supply or pressing the S1 reset switch will reset the circuit. Both of these options are available. That the power supply is operating normally can be seen by the LED2 indicator. The user is being alerted through the use of a buzzer when the power supply unit enters the protection mode, which is indicated by the LED1 light.

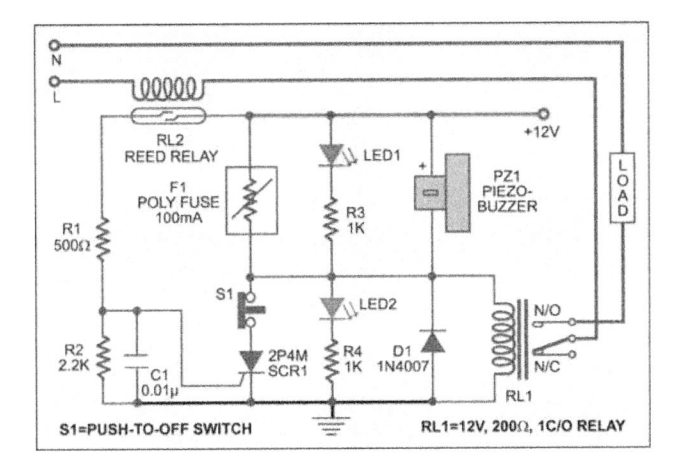

Because the turns of the reed relay winding are determined by the current that flows through the load, you should consult the table for winding specifics that correspond to the requirements you have for the load current. At EFY, testing was carried out for an AC load current of approximately 1.85A at 230V AC mains. As a result, 16 turns of 22SWG copper-enameled wire were wound on the reed relay.

61. Water Tank Overflow Indicator

Water is an essential natural resource, yet there is a limited supply of it. This overflow indicator for the water tank comes in helpful for reducing the amount of water that is wasted. Whenever the water tank becomes too full, it triggers an audible as well as a visual alert.

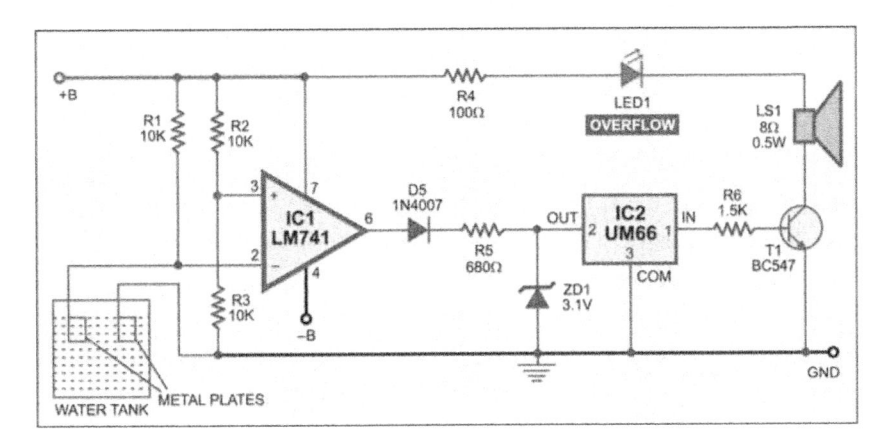

The secondary output of the power supply unit is 9V-0-9V AC at 300 milliamperes, and it is generated by stepping down the mains AC using transformer X1. In order to deliver +9V at the '+B' point and −9V at the '−B' point, the output of the transformer is first filtered by capacitors C1 and C2 before being rectified by a full-wave bridge rectifier that contains diodes D1 through D4. Connect the '+B,' '−B,' and 'GND' terminals of the power supply unit to the corresponding terminals of the overflow indicator circuit for the water tank.

Circuit operation

When the water in the tank comes into contact with the metal plate sensors, a ground connection is made with pin 2 of IC1. At this time, the voltage at pin 3 of IC1 is greater than that at pin 2. Across Zener diode ZD1, the high output of the operational amplifier (op-amp) generates 3.1 volts. The melody that is produced by melody generator IC2 is what drives the transistor, which in turn causes LED1 to light up and the alert to sound from the loudspeaker. In order to prevent a negative polarity from being applied to the cathode of the Zener diode, rectifier diode D5 is utilized.

62. Simple smoke detector

This basic smoke detector has a great level of sensitivity while remaining quite affordable. As a sensor, it makes use of a Darlington-pair amplifier, which is comprised of two npn transistors, in conjunction with an infrared photo-interrupter module. The circuit triggers an audible and a visual warning whenever there is heavy smoke in the surrounding air.

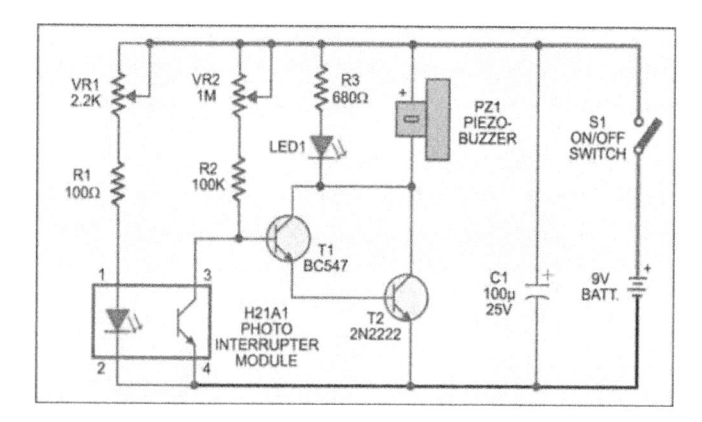

Simple smoke detector circuit

Within a casing made of plastic sits a gallium-arsenide infrared LED that is linked to a silicon phototransistor. This component makes up the photo-interrupter module (H21A1). It is possible to interrupt the signal with smoke by using the slot (gap) that is located between the infrared diode and the transistor. This causes the module output to flip from a "on" state to a "off" state. When smoke passes through the opening, it blocks the infrared photons that are directed at the phototransistor. As a consequence of this, the phototransistor is unable to conduct, and the Darlington pair transistors are able to do so, which causes the buzzer to sound and LED1 to light up. After the smoke in the gap has been cleared, light from the IR LED will fall on the phototransistor, which will cause it to begin conducting electrical current. As a consequence of this, the Darlington pair transistors cease conducting current, which causes the buzzer and LED1 to go off. Adjust the presets VR1 and VR2 to achieve the highest possible sensitivity. The sensitivity of the photo-interrupter module is controlled by the VR1 potentiometer, whereas the sensitivity of the Darlington-pair transistors is controlled by the VR2 potentiometer.

63. Remote Emergency Alarm for Unmanned Lifts

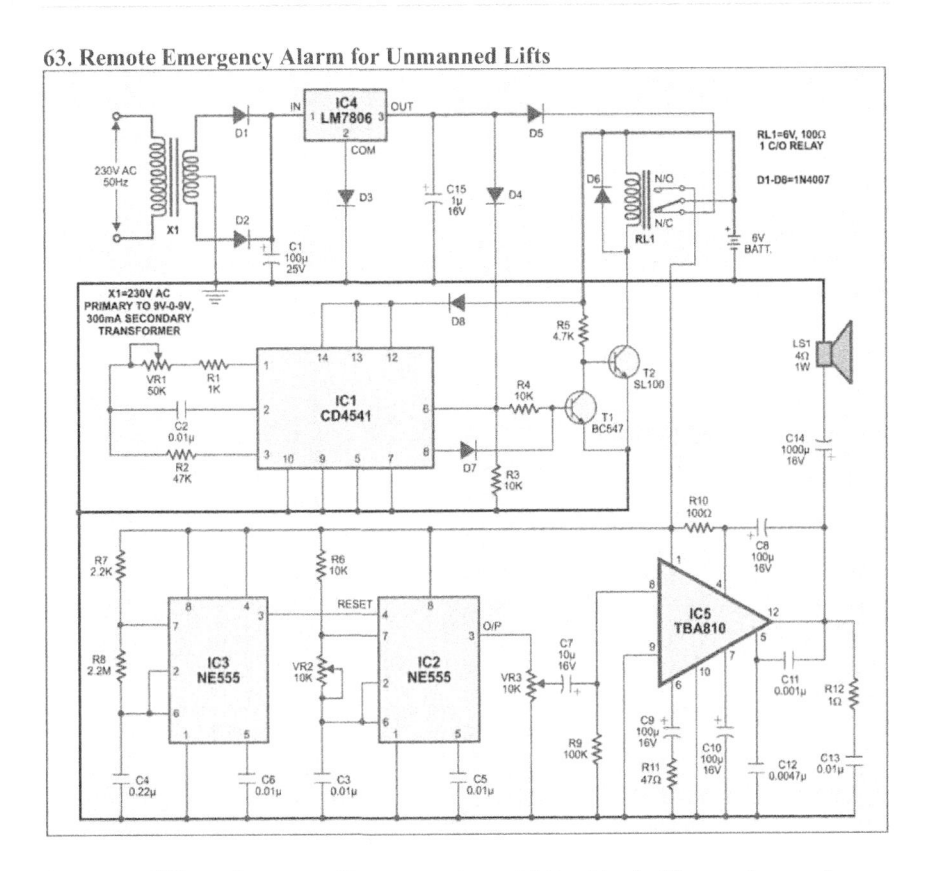

In unmanned lifts or elevators, sudden power outage might be risky for lift users. In case of power outage, this basic remote emergency alarm circuit sounds in the lift/elevator control room. 6V DC powers the circuit. The buzzer will drain the battery if the power outage continues. The buzzer stops at a predetermined time to save battery. As soon as power returns, the alarm circuit disconnects from the battery and it starts charging. Remote ESA Diodes D1 and D2 rectify 230V AC mains for battery charging. Diode D5 and relay RL1's N/C contacts charge the battery. Diode D4 conducts, driving transistor T1 to conduct, which cuts off transistor T2. Relay RL1 de-energizes and the battery charges.

When power quits, transistor T1 turns off to turn on transistor T2 and relay RL1. The battery disconnects from charging and connects to relay RL1's N/O contacts. Timers IC2 and IC3 are powered (6V) through relay RL1's N/O connections.

Timer IC3 creates fixed-interrupt pulses for IC2's reset pin (pin 4), which generates a 1kHz audio pulse for IC5. IC3 controls the alarm's interruption. Pre-programmable IC CD4541 (IC1) stops the buzzer after a certain time. After (R1+ VR1).C2 seconds, IC1's pin 8 becomes high, driving transistor T1. Thus, T2 turns off and RL1 de-energizes. The loudspeaker stops playing when IC2 and IC3 disengage from the battery. IC1 begins counting when pin 6 goes low, when power fails. Connect IC1 pin 6 to the power supply with diode D4. IC1 doesn't fluctuate in mains. When power fails, IC1's master reset pin goes low and it starts counting time. Pins 12 and 13 of IC1 are set high for maximum time count. 50 second's passes. That means the alarm goes off 50 times. IC5 is a low-power audio amplifier with low harmonic and crossover distortion. It amplifies IC2's alarm tone.

64. Audio Controlled Running Light

Discotheques are able to make use of this audio-controlled running light that is powered by the mains. The lamps illuminate in a rhythmic pattern in response to the music being played. Only one of the ten AC lamps will be on continuously even if there is no background noise. Light appears to be "flowing" through the lamps while music is being played.

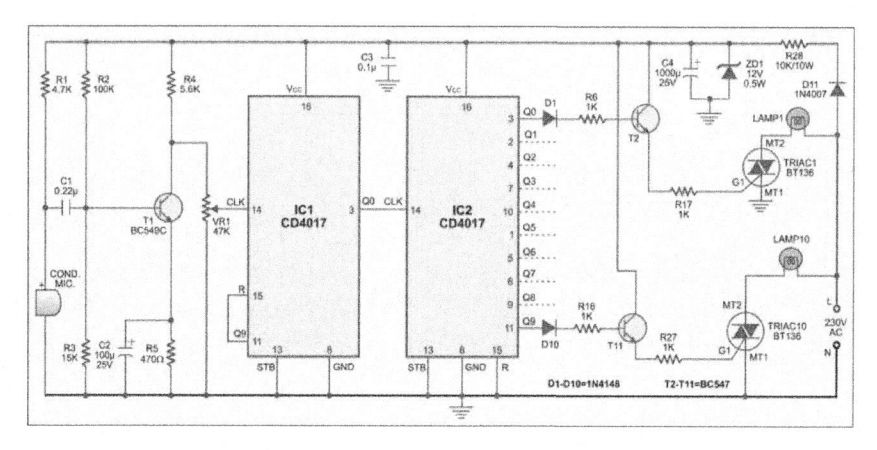

A Running Light Circuit That Is Audio-Controlled

The circuit for the audio-controlled running light is depicted in Figure, while in the meantime. The condenser microphone is responsible for converting sound waves into electric impulses. The signals from the microphone are amplified by the transistor T1, which in turn provides clock pulses to the decade counter IC CD4017 (IC1). Adjusting the signal level can be done with the Preset VR1 knob. The reset signal is transmitted to pin 15 of IC1 through the Q9 output. Signals that have been divided by ten are supplied to the clock pin 14 of IC2, which is yet another decade counter. Triacs receive their triggering pulses from transistors, which are driven by the outputs of IC2. In turn, the triacs are what drive the AC lighting. You can now see the lights synchronizing themselves to the music that is playing. The audio has been brought down to a frequency that is audibly lower thanks to the application of frequency-divider ICs model CD4017. To power the circuit, alternating current (AC) is first rectified by diode D11, then regulated by Zener diode ZD1, and finally filtered by capacitor C4.

65. Power Supply Reversal Corrector cum Preventer

When the power supply polarities of an electronic device get switched by accident, the device could get broken. By adding this tiny circuit to the power supply part of the device, the danger can be stopped. This power supply reversal corrector circuit will immediately fix the power supply if the poles are switched. It will also alert you to the mistake by sounding an alarm and showing a light.

Power Supply Reversal Corrector cum Preventer Circuit

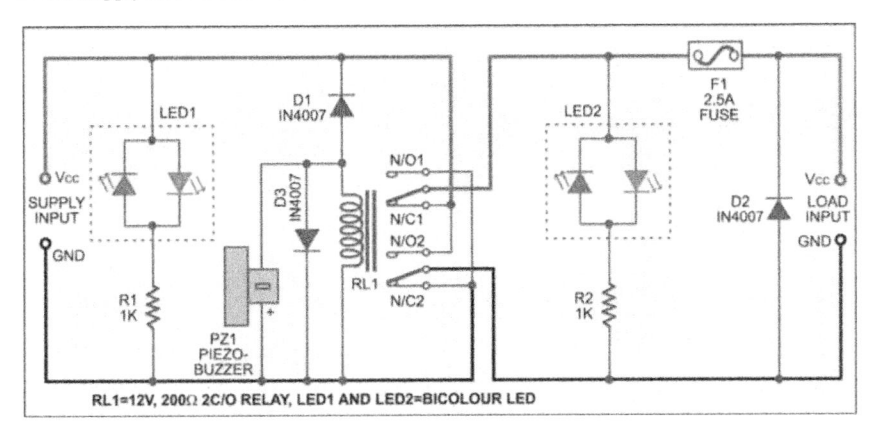

The part that controls things is just a relay (RL1). A diode connected in series makes its power supply one-way (D1). So, the relay is normally not doing anything. This setup is hooked up backwards across the output of the power supply. The load gets power from the add-on circuit's output, which goes through the normally closed contacts of the relay and fuse F1. This keeps the relay from making the power supply work harder than it needs to. The normally closed (N/C) contacts of relay RL1 are connected to the input terminals of the power supply in the correct polarity, while the normally open (N/O) contacts are connected to the reverse-polarity terminals. Both LED1 and LED2 are bicolor LEDs, but they only have two pins instead of the usual three. You can also make a 2-lead bicolor LED by soldering the leads of a green LED and a red LED together in reverse polarity.

When the input polarity is right, the LEDs light up green. If something is wrong, the LEDs light up red. LED1 shows what's going on with the DC input, and LED2 shows what's going on with the output. LED2 is mostly used to let people know that RL1's contacts are stuck.

66. Capacitor Evaluator

You can determine the value of any capacitor by using this circuit in conjunction with a capacitor whose value is already known. This circuit can be broken down into portions that evaluate the capacitor and display the results. While the display unit operates off of 5V, the capacitor evaluator circuit operates off of regulated voltages of 5V and 12V. Connect the 5V supply, the point Y reference point, and the ground reference point of the capacitor evaluator section to the appropriate terminals of the display section.

Capacitor Evaluator Circuit

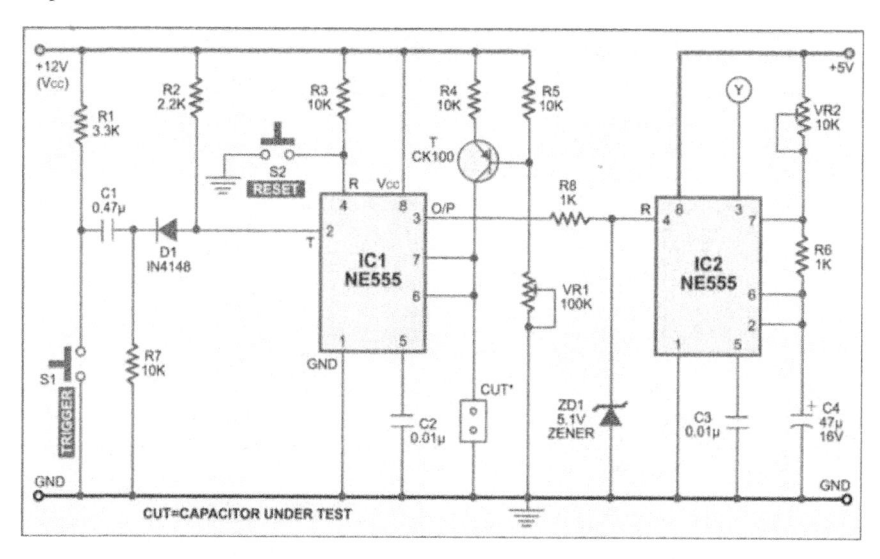

You can determine the value of any capacitor by using this circuit in conjunction with a capacitor whose value is already known. This circuit can be broken down into portions that evaluate the capacitor and display the results. While the display unit operates off of 5V, the capacitor evaluator circuit operates off of regulated voltages of 5V and 12V. Connect the 5V supply, the point Y reference point, and the ground reference point of the capacitor evaluator section to the appropriate terminals of the display section.

The circuit that consists of the astable multivibrator built around timer 555 (IC2), followed by decade counter ICs 74LS90 (IC3 and IC4), which are 4-bit ripple decade counters, counts the duration of the output pulse at pin 3 of IC1. The outputs of IC3 and IC4 are connected to decoder/driver ICs 74LS47 (IC5 and IC6), respectively. These decoder/driver ICs are then connected to 7-segment common-anode displays LTS542 (DIS1 and DIS2) so that the frequency of the astable (IC2) that is used to evaluate the value of unknown capacitors can be displayed on these displays.

67. PIN Diode Based Fire Sensor

Here's a PIN diode-based fire sensor that sounds an alarm. Thermistor-based fire alarms only activate if a nearby thermistor is heated. In this design, a sensitive PIN diode is employed for long-range fire detection. It detects visible and infrared light (430-1100nm). So visible light and IR from the flames can trigger the alert. It detects mains wiring sparks and sounds an alert if they persist. it is an ideal protective device for showrooms, lockers, record rooms and so on. Author's prototype is shown in Fig. 1.

Light and IR sensor is PIN diode BPW34 (Fig. 2). BPW34 has an anode (A) and cathode (C) (K). The photodiode's top flat surface shows the anode end. A slender wire connecting to a small solder point is the anode. When exposed to 900nm light, BPW34's radiant sensitive surface creates 350mV DC open-circuit voltage. Sunlight and firelight affect it. It's a good light sensor.

BPW34 photodiode can be zero-biased or reverse-biased. Light lowers its resistance. Figure 2 shows the fire sensor's PIN diode circuit. It uses a 9V battery, PIN diode BPW34 (D1), op-amp CA3140 (IC1), counter CD4060 (IC2), transistors BC547 (T1 and T2), a piezo buzzer (PZ1), and other parts. In reverse-biased mode, PIN photodiode BPW34 feeds light current into op-amp IC1's input. CA3140 is a MOSFET-input, 4.5MHz BiMOs op-amp.

Input gate-protected MOSFET (PMOS) transistors have a 1.5T ohm resistance. The IC requires 10pA to adjust output high or low. IC1 acts as a current-to-voltage converter as a transimpedance amplifier. IC1 amplifies and transforms PIN diode photocurrent to output voltage. Non-inverting input is connected to photodiode's ground and anode; inverting input gets photocurrent from PIN diode.

Circuitry

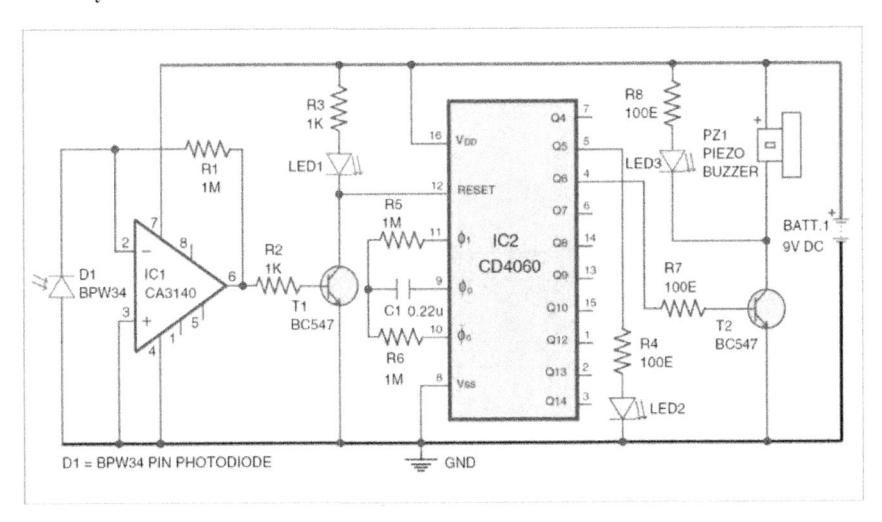

Fig. 3: Circuit diagram of the PIN diode-based fire sensor

Inverting transimpedance amplifier R1's large feedback resistor sets its gain. Low impedance load from non-inverting input to ground keeps photodiode voltage low. The photodiode functions without an external bias. The op-feedback amp's keeps the photodiode current equal to R1's. In self-biased photovoltaic mode, photodiode input offset voltage is minimal. This allows a big gain without an offset voltage. This setup maximizes low-light gain. In ambient light, PIN diode photocurrent is modest, keeping IC1's output low. When the PIN diode detects visible light or IR from fire, its photo current increases and IC1 transforms it to output voltage. LED1 illuminates when IC1's output is high. The circuit detected fire. When T1 conducts, it grounds IC2's reset pin 12 and CD4060 oscillates. IC2 is a ten-output binary counter that oscillates due to C1 and R6. LED2 blinks when IC2 oscillates. After 15 seconds, T2 conducts and activates piezo buzzer PZ1 and LED3. Continuing fire triggers a 15-second alarm. Replace PZ1 with a relay to activate a loud AC alarm (not shown here). The relay contacts activate the AC alarm.

68. Blown Fuse Indicator

A fuse is a type of electrical safety device that can safeguard an electrical circuit from being overloaded with current when it is in operation. Its primary component is a metal wire or strip that, when subjected to an excessive amount of electricity, would melt, causing the flow of current to be temporarily disrupted. It is a device that is meant to be sacrificed; once it has been blown, it creates an open circuit and, depending on the type, either needs to be replaced or rewired. When a piece of machinery indicates that it has no power, the problem may simply be that its fuse has blown. This is a circuit for an indicator that a fuse has blown, which uses LEDs to display the status of the fuse. This simple circuit has a lot of applications and is dependable. Because it contains so few individual parts, it is also very reasonably priced.

The operation of the blown fuse indication

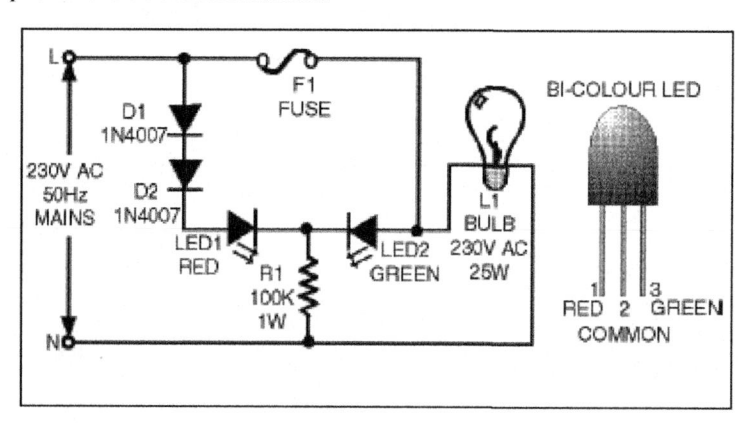

Fig.1 Blow- fuse indicator

When everything is operating normally (when the fuse is not blown), the voltage drop in the first arm is 2V + (2 x 0.7V) = 3.4V, whereas the voltage loss in the second arm is only 2V. Therefore, current is flowing via the second arm, which is the green LED; this causes the green LED to shine, while the red LED continues to be turned off.

The supply to the green LED is cut off when the fuse blows, and because there is only room for one LED in the circuit, the red LED lights up instead. In the event that there is a loss of power, both LEDs will remain in the "off" position.

In the event that the fuse blows, this circuit can quickly and inexpensively be adjusted to emit a siren (see Fig. 2). In order to activate the siren, an optocoupler is utilized. The red LED illuminates when the fuse is blown. Simultaneously, it toggles the siren to the "ON" position.

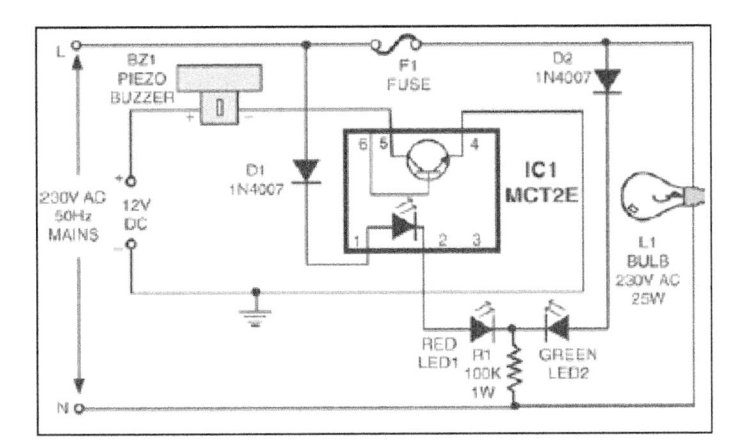

Fig.2 Blow fuse indicator with alarm

It is possible to utilize two LEDs of a single color, such as red and green, in place of a bicolor LED. In the same vein, there can be only one diode used in place of D1 and D2 in the circuit. Due to the possibility that the two LEDs would create different voltage drops, it is necessary to employ two diodes in order to increase the voltage drop.

69. Ding Dong Touch Bell
A doorbell is a signaling device that is often installed next to a door that leads into the entry of a building. When an occupant of the building touches a button on the exterior of the structure, a bell will ring inside the building to notify them of the presence of a visitor. Based on the IC 8021-2 manufactured by Formox Semiconductors, the doorbell circuit that we demonstrate here is both straightforward and economical. It is a DIP integrated circuit with eight pins, but the circuit only makes use of four of those pins. In the event that it's necessary, you can quickly and easily set it up in your home.

When the value of pin 3 is driven low, the integrated circuit (IC) includes a built-in circuitry that makes a ding-dong sound. The IC acts similar to a ROM in that it stores the sound as bits. However, the IC's sound output cannot be used to directly drive a speaker because doing so would place too much strain on the device.

In order to bring the sound up to a level that is sufficiently audible, an amplifier consisting of a complimentary pair and two transistors is utilized. At the output, you can either use a piezo tweeter or a speaker with 8 ohms and 500 milliwatts of power.

Doorbell circuit

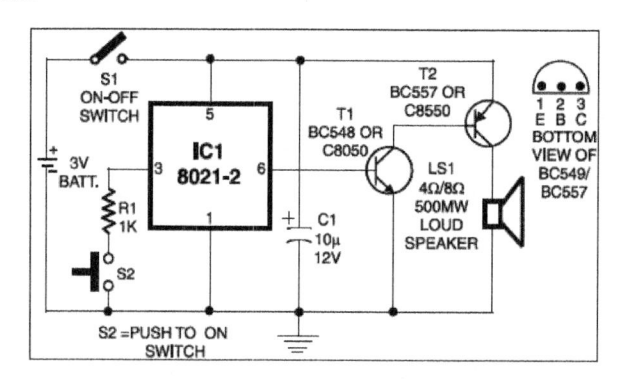

Only a few microamperes of nominal current are drawn from the supply by the integrated circuit while it is in the standby mode. Because of this, the switch S1 can remain in the closed position. When switch S2 is activated, a ding dong sound is made twice at regular intervals. If you try to hit switch S2 a second time while the first ding dong sound is still being created, it has absolutely no effect, and the two ding-dong bell noises will be made regardless of whether or not you press the switch.

70. Low-cost Stopwatch

This is a straightforward circuit for a stopwatch that is inexpensive and has a maximum count of 99 seconds. This circuit makes use of additional discrete components in addition to the LTS543, CD4060, and CD4013. Pulses at a frequency of 1 Hz are produced using a crystal-controlled oscillator. A tiny crystal operating at a frequency of 32.768 kHz is utilized for this purpose.

A low-cost stopwatch circuits

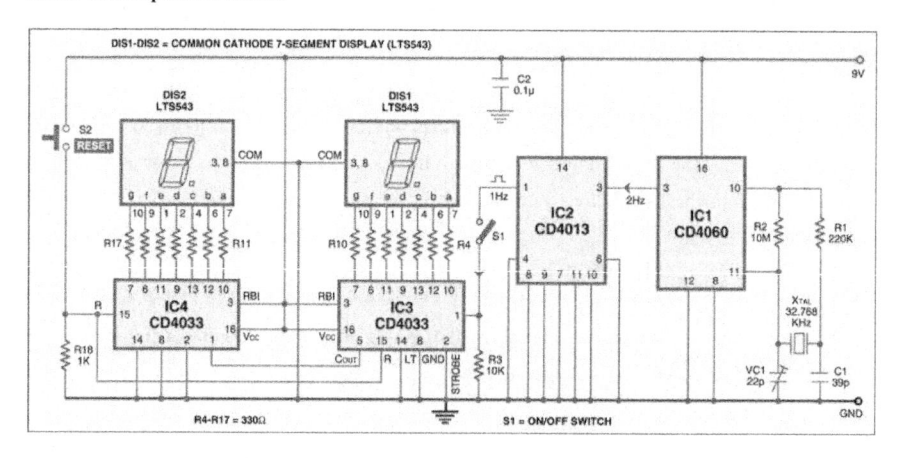

The integrated circuit CD4060 (IC1) has both a divider stage with 14 bits and an oscillator portion. The output of IC1, which is 2Hz, is supplied to the flip-flop IC CD4013 via its pin 3. (IC2). The frequency is reduced from 2 Hz to 1 Hz via IC2. Through switch S1, the output of IC2 is connected to

IC CD4033 (IC3). This connection is made at pin 1. In order to carry out further counting, carry-out pin 5 of IC3 is linked to pin 1 of a different CD4033 (IC4). The time is displayed on the common-cathode displays DIS1 and DIS2 by IC3 and IC4, respectively (maximum 99 seconds). The watch is started and stopped using the switch labeled S1. You are able to return the counter to zero by hitting the reset switch, which is located at S2.

First, ensure that the circuit is connected to a source of 9V power, and then reset the display by pressing and holding the S2 switch for a brief period of time. The number 00 is displayed on the screen. To initiate the counting process, move the switch S1 to the "on" position. At the point in time when an event is over, the switch S1 should be moved to the "off" position. The time that is displayed on the display is the time that will be used for the event.

71. Digitally Adjustable Dancing Lights

There are likely many different kinds of movable dance lights that you have come across (flickering LEDs). The majority of them alter the switching pace through the use of presets, which are variable resistors. Due to the fact that it is a mechanical component, the preset is prone to wear and tear over time and also contributes noise to the circuit when it does so. The adjustable dancing lights circuit provided here uses timer IC 555 to select various values of resistors in order to control the frequency of an astable multivibrator. These resistors are used to control the frequency of the multivibrator.

Adjustable Dancing Lights Circuit

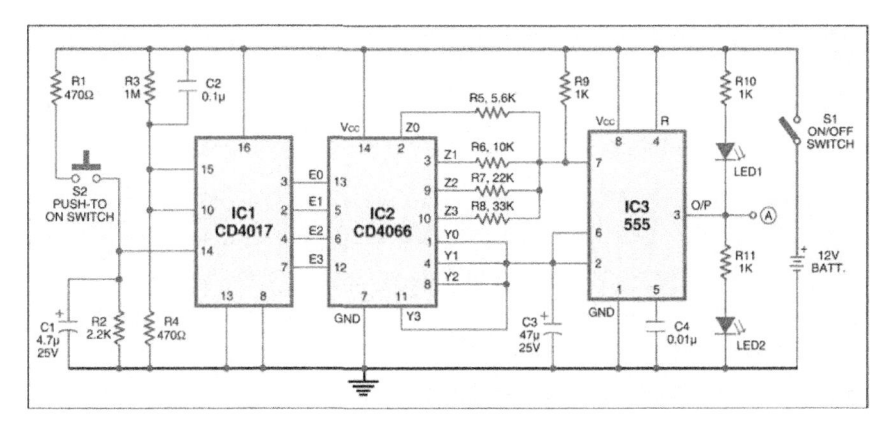

The circuit is constructed around a decade counter IC 4017, a timer 555, and a quad bilateral switch IC 4066, which is used to determine the necessary resistance. The output of the decade counter is used to select which resistor is connected to the output of IC 4066. The chosen resistor modifies the time period of the 555-timer circuit, the timings of who's "on" and "off" states are determined by the following: In the current configuration, R9 is equal to one kilo-ohm, C3 is equal to forty-seven microfarads, and Rx fluctuates depending on the output of IC1 (CD4017). The output of IC1 can be altered by repeatedly pressing switch S2.

The output of the decade counter advances whenever you briefly press switch S2. This causes IC 4066 to select a higher value for resistor Rx, which in turn modifies the switching time of the astable multivibrator. When LED1 and LED2 light up, it means that the multivibrator has been switched "off" and "on," respectively.

72. Shock Warning Circuit

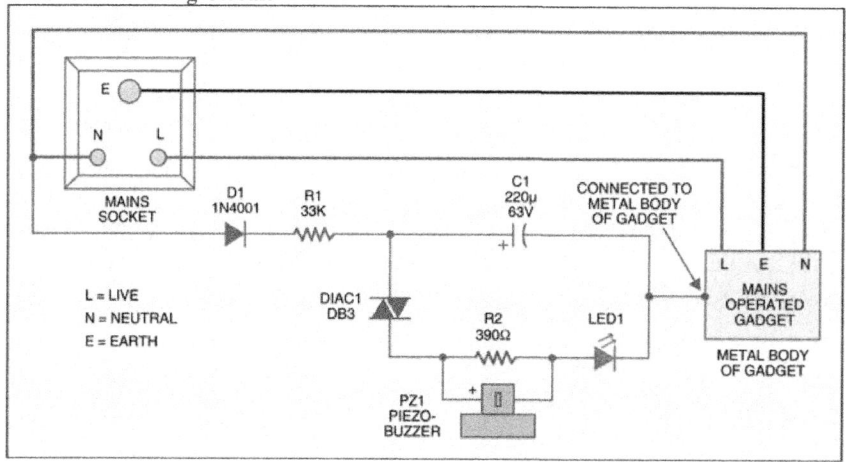

The leaking of electrical current can result in potentially fatal shocks. On the other hand, with this shock warning system, an undesirable circumstance like this one can be averted. It requires an extremely low component count and does not require a separate power source of any kind.

Circuit for the shock detection system

The leakage from an electronic device is continuously monitored by the circuit, which does so by sensing the potential within the device's body. When there is a leakage coming from the body, there is a potential difference between the neutral end and the body of the device, which charges the capacitor C1 via the diode D1 and the resistor R1. When the voltage across capacitor C1 reaches a level that is higher than the breakdown potential of DIAC1 (DB3), the latter begins conducting, which causes the capacitor to swiftly discharge through the buzzer as well as LED1. Because of the energy that is being discharged, LED1 will briefly flash, and a piezo buzzer will sound. The buzzer and the LED will continue to give an intermittent warning as long as there is leakage coming from the body of the device.

73. IR Receiver Module Tester

This is an IR receiver module tester, which can be used to do on-board testing of IR modules used for remote control of electronic devices such as televisions and video disc players. The circuit is quite straightforward, and it also has the capability of acting as a remote tester.

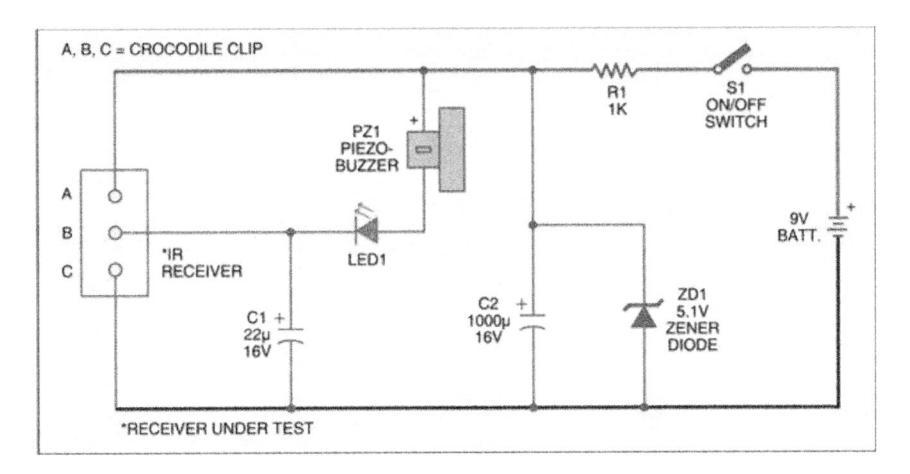

Miniature IR receivers that are sensitive to pulsed infrared radiation are what make up IR receiver modules. These have a pin photo-diode and a preamplifier stage that are both contained in an epoxy casing that also functions as an IR filter. The module is equipped on the inside with an automatic gain controller (AGC), band-pass filter, demodulator, and control circuit. Its output consists of a bipolar transistor with a resistor ranging from 80 to 100 kilobohm in the collector of the transistor. The collector output of the transistor is high and produces 5V at 5 mA when it is functioning normally. When the pin photo-diode detects the existence of pulsed IR photons, the output of the module is active-low; as a result, it causes a sinking current to occur.

The IR receiver module is capable of continuous data transfer at up to 2400 bps or higher, and it is designed with a high level of immunity to light from the surrounding environment. Band-pass filtering and automatic gain control (AGC) work together to eliminate unwanted noise and prevent false triggers. The module will only react to the IR beam if the carrier frequency of the beam is somewhat near to the frequency that is in the center of the band pass.

74. 555 Timer PWM Audio Amplifier

The integrated circuit known as "555" is a multipurpose timer that can be utilized for a wide range of applications. We investigate the use of the 555 timers as an astable multivibrator in this 555 audio PWM project. The common 555 timer integrated circuit (IC) processes audio signals in its own unique method of pulse-width modulation (PWM). The 555 integrated circuit operates in astable mode in this situation.

The switching frequency can be adjusted anywhere between 65 and 188 kilohertz. The amplitude of the input signal and the load impedance are taken into consideration when selecting the PWM frequency to be used. You can make the listening experience more comfortable while maintaining a minimal level of audio distortion by setting VR1.

Working explanation

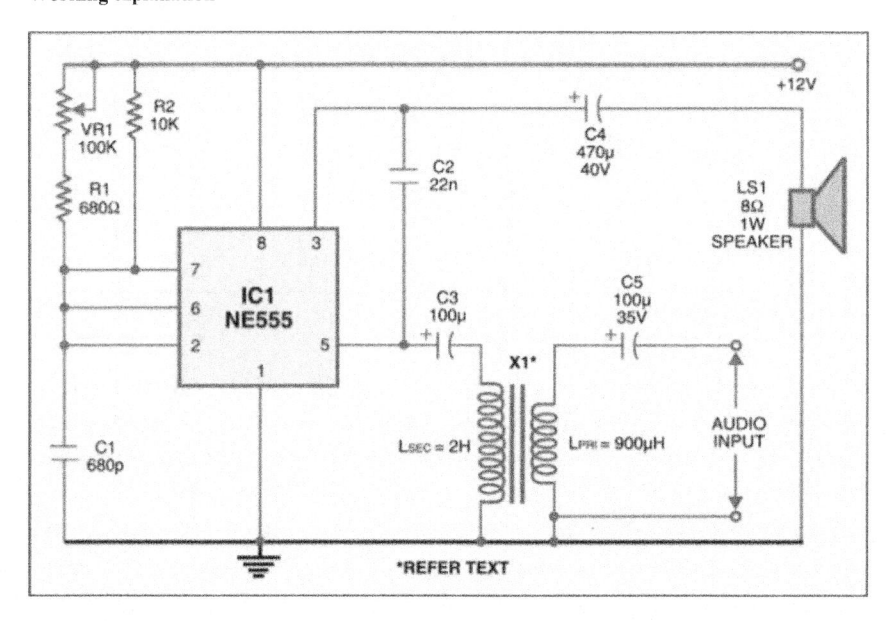

When using pulse-width modulation, the pulse width of the carrier frequency varies according to the amplitude of the audio signal that is being fed into the system. The audio signal is reproduced accurately thanks to feedback capacitor C2, which is located in the circuit. The most typical method for achieving a satisfactory level of carrier frequency rejection is the application of an output L-C filter. It has been left out for the sake of simplicity. In addition, the speakers are incapable of reacting to the signal at the high frequency. They react to the average DC level that is modulated with the audio signal that is sent into the input by us from the source. It goes without saying that the sound quality is not on par with that of a professional system; yet, it would undoubtedly be an unforgettable adventure to listen to audio played through a 555 chip at volume levels that fill a whole room.

555 Audio PWM circuit

Importance should be placed on impedance matching at both the input and the output. Therefore, in order to match the headphone output of a normal CD player to the input of the 555 amp, an input-impedance matching transformer known as X1 is utilized. The load consisted of speakers with an ohm rating of 8 and a watt rating. PWM amplifiers, if they are constructed correctly, have the potential to offer performances that are comparable to those of traditional amplifiers. It is possible to get even higher efficiency and effortless bass.

75. Musical Water Shower

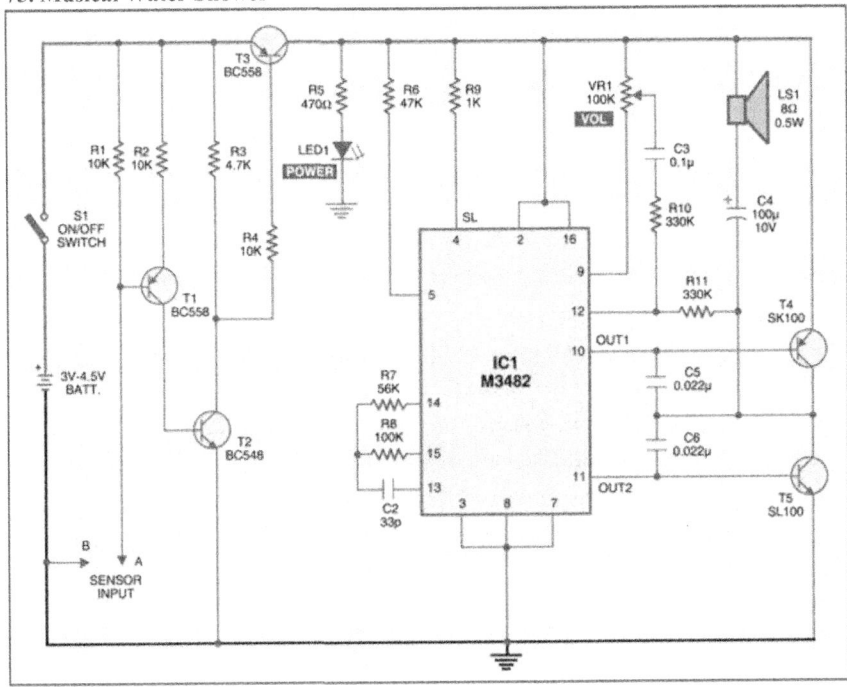

When you take a shower, wouldn't it be nice to have music playing in the background the whole time? The same is true of this simple circuit. As long as your shower is on, it will play different songs over and over again. As soon as the water comes out of the shower, the music starts. When you turn the shower "off" and the water stops coming out of it, the music stops.

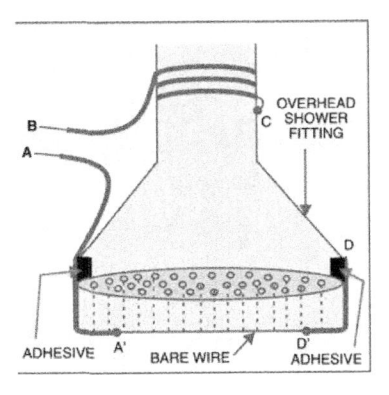

Fig. 2: Sensor arrangement

In Figure 1, you can see how the musical water shower works. It has a pair of complementary amplifiers made up of transistors T1 and T2, a switch made up of transistor T3, and a 12-tone melody generator called IC M3482. The M3482 is a mask-ROM-programmed IC that plays melodies based on the data that was put into it. The built-in preamplifier makes it easy to connect to the driver circuit, which is made up of transistors T4 and T5. The IC can be changed out with another one from the UM348xx series, the WR630173, or the WE4822.

When the shower is turned on, power is sent to the melody section by way of the T3 transistor. As shown in Fig. 2, the overhead shower unit has two insulated copper cables labeled AD and BC. The insulation is taken off of a part of the cable AD (marked A'D'). The cable AD is fixed firmly and runs along the body of the shower so that the bare part doesn't touch the body and cause a short. The part that isn't covered should be long enough so that the shower water falls on it as it runs. At C, the body of the shower is soldered to the end of the copper cable BC. It is tightened by wrapping it around the neck. So, the cable ends at A and B can be plugged into the circuit of Fig. 1 as sensor inputs. Now, when you turn on the shower, water falls on the bare copper wire A'D', which makes an electrical connection with the cable BC that runs through the shower and is also in contact with running water. The cable points A and B on the shower unit are connected to the circuit of Fig. 1 as sensor inputs. This lets the power reach the base of the transistor T1, which is now conducting. This makes the T2 and T3 transistors work. So, the power source is there for the melody circuit (Fig. 1).

The melody generator IC is set up to play twelve different tunes over and over again until the shower water is running and transistors T1, T2, and T3 are conducting. The preset VR1 can be used to change how loud the melody generator tunes are.

When the shower is turned off, both transistors T1 and T2 turn off. At the base of pnp transistor T3, a high voltage builds up. So, transistor T3 stops making electricity. This causes the power to be cut off to the melody generator circuit, which stops making music. When the shower is turned off, the circuit doesn't use much power, so the battery lasts a long time. Put the circuit together on a general-purpose printed circuit board (PCB) and put it in a plastic case with LED1 and switch S1 on the outside. Connect the A and B ends of the sensor wires to the wires on the shower.

76. Multipurpose Power Pulser
This versatile power pulser was designed with a very straightforward concept in mind. A low-frequency oscillator is what drives a voltage regulator, as can be seen in the circuit (Fig. 1), which shows how it works.

Multipurpose power pulser circuit

The LM555 timer chip (IC1) is configured to operate as an astable multivibrator. The frequency at which the circuit is allowed to run freely is generated by the components R1 and R2, VR1, and C1. You may make various adjustments to it by adjusting the potentiometer labelled VR1. The switching on and off of the adjustable voltage regulator LM317T (IC2) is controlled by the output of IC1 at pin 3, which is connected to the NPN transistor SL100B. (T1). You are able to make the output anywhere from 1.25V-15V and 1.5A while using an input power supply of 5V-18V and 1.5A. Incandescent lamps, DC motors, electromagnetic relays, and LEDs are all suitable applications for this pulsed output.

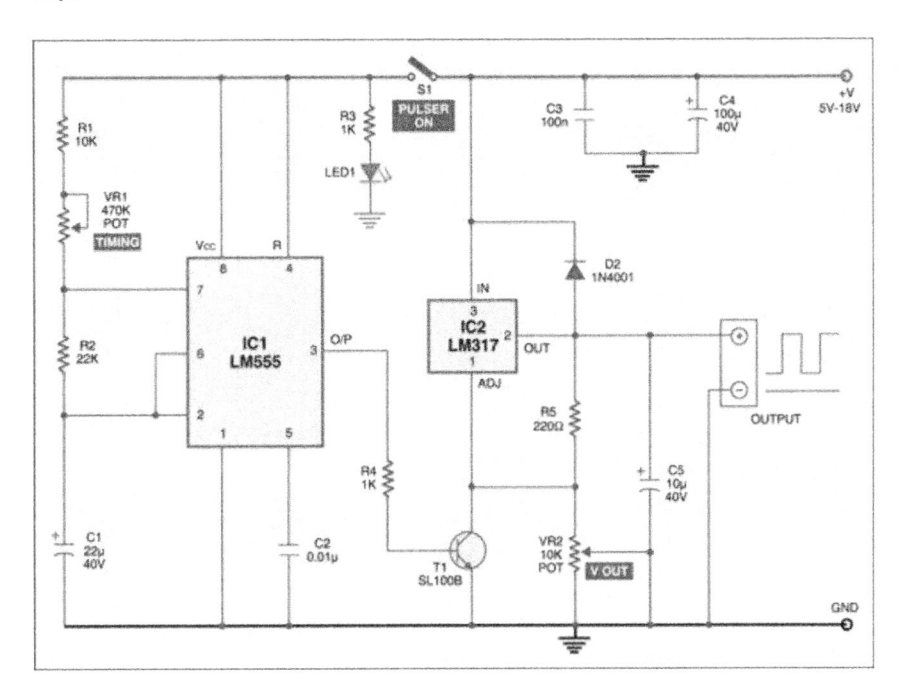

After deciding the load, you want, turn the device on by putting the switch for it, S1, in the "on" position. Now connect a digital multimeter across IC2's output terminal (pin 2), and use the potmeter VR2 to select the desired output voltage. If switch S1 is in the "on" position, then the frequency of IC1 can be adjusted using VR1. It is important to take note that the highest output voltage is just about 15V, even when the DC input is 18V. You are free to choose the frequency of the astable multivibrator by adjusting the values of the components R1, VR1, R2, and C1 in accordance with your preferences.

77. Continuity Tester with a Chirping Sound

The celebration is taking place at your house today. And at this very moment, your attention is focused on locating any malfunctions in the ornamental lights that are strung along the exterior of your home. It is important that the task be completed before the evening. However, daylight makes it more difficult for you to detect whether the neon bulb included within the tester is glowing or not, which adds to your sense of irritation. A neon bulb is typically utilized in traditional continuity testers for live-voltage testing in order to show the presence of live voltage. However, their brilliance is only noticeable when used inside. If it were an extremely bright LED or buzzer, it would be possible to see it in the daytime even when you were outside the house. The live-wire scanner that is being discussed in this article not only makes use of an LED, but also a piezobuzzer that makes a chirping sound like a bird when it locates the place of break in a wire that is located inside the sheath.

The circuit is made up of a few discrete components in addition to a piezobuzzer that is wired all the way around a CMOS Johnson decade counter CD4017 (IC1). This decade counter/divider contains ten decoded outputs (Q0 through Q9), a divided (divide-by-10) output that is accessible at its pins 3, 2, 4, 7, 10, 1, 5, 6, 9 and 11, respectively, and a carryout bit at pin 12. At the positive edge of each input pulse that is received at pin 14, the count of the CD4017 advances by one. After Q9 at pin 11 gets high, this decade sequence will continue to repeat.

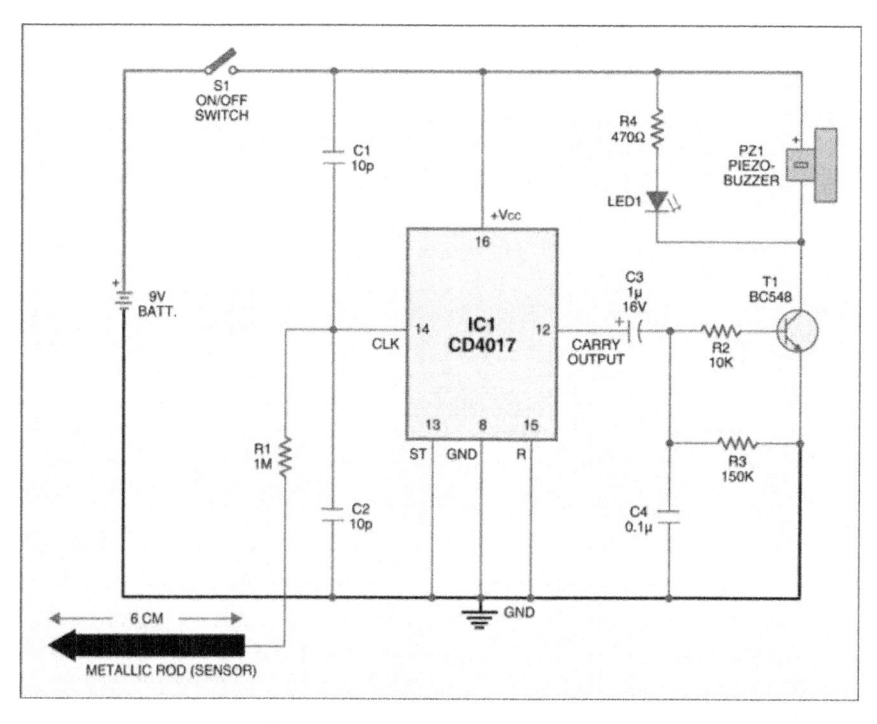

Fig. Continuity tester circuit

Because of the extraordinarily high input impedance of CMOS integrated circuits and the fact that they are voltage-controlled devices, it is simple to set off a CMOS IC with even relatively weak stray signals, such as the electric field of a 220V live-wire. The same fundamental idea underlies the operation of this circuit. Even if a live wire is 20 centimeters distant from a metallic strip, the potential across clock input pin 14 of IC1 can still swing to high and low logic levels when it is attached to a small metallic strip. This is because the potential is connected to a small metallic strip. This causes the counter to start counting, and a squarewave output with a frequency of 50 tenths of a hertz, or 5 Hz, is generated at pin 12 of IC1.

Now, this extremely high sensitivity of the clock input can be decreased to the required extent by connecting a voltage divider composed of passive components of the proper value. This, in turn, decreases the input impedance, which, in turn, lessens the sensitivity. The device is able to respond from up to 10 centimeters distant from the live wire in the event of an emergency. The wire can be scanned with this configuration to look for any breaks or discontinuities inside the sheath where it is housed.

78. Electronic Combination Lock

This electronic lock with a combination of seven digits may be readily hardwired to work with any combination that you want. The circuit consists of ten push button switches, an NPN transistor, and a Johnson counter with 4 bits and a divide-by-8 operation (IC1). When the power is turned "on," an 820-kilohm resistor causes the capacitor C2 that is attached to pin 15 of IC1 to charge to a high level. This maintains the counter in the reset state. Under these circumstances, only the output on pin 2 of the counter IC1 is high, while the other pins' outputs are all low.

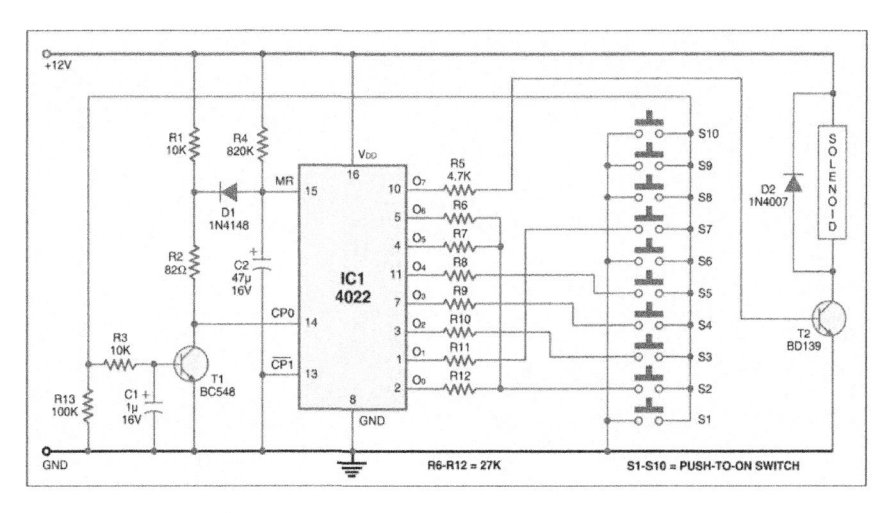

Fig. Electronic combination lock circuit

Circuit operation

When the switch S2 is pressed, transistor T1 begins to current, and capacitor C2 begins to drain by the action of diode D1 and resistor R2. This frees up the reset input for the counter. T1 is turned off and its collector is pulled high when S2 is allowed to be released; this causes a rising edge to be produced on the input clock pin 14 of the counter. Switch contact bounce can cause numerous clock pulses to be generated on pin 14 of IC1, but a simple filter comprised of capacitor C1 and resistor R3 in the base circuit of transistor T1 can prevent this from happening.

The clock pulse increments IC1's count by one, so O0 goes low and O1 goes high. Therefore, the next step is to press switch S7, as it is connected to output O1. The greatest amount of time that can pass in between switches being pressed is equal to the amount of time required for capacitor C1 to charge to a logic high level. In that case, the counter will be started over. After all of the switches have been pressed in the appropriate order (S2-S7-S3-S4-S5-S2-S2 as illustrated), the output of the counter, pin

10, will be high for approximately ten seconds. This output is sent to the driver transistor T2, which then drives the solenoid valve, so unlocking the door.

79. Heat Control Unit

This circuit will tell the heater to "turn on" when the temperature of the water drops below the lower limit that you specify, and it will tell the heater to "turn off" when the temperature rises over the higher limit that you specified.

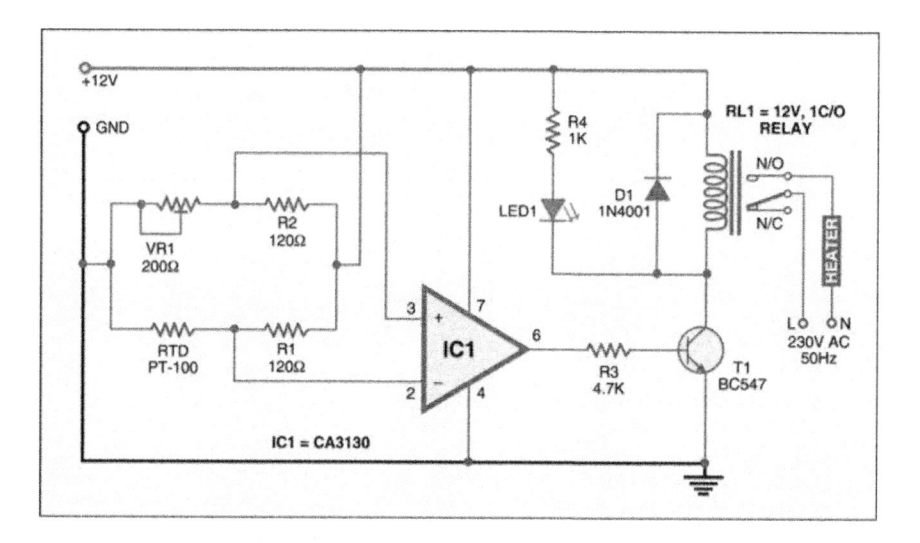

The circuit is comprised of both bridge and op-amp parts, and it is powered by a controlled 12 V supply of power. Resistors R1 and R2, a preset VR1, and a resistance temperature detector are the primary components of the bridge section (RTD). The operational amplifier component is constructed around IC1. Pin 3 of IC1 has a voltage that is greater than pin 2's voltage when water is at its typical temperature, which is approximately 30 degrees Celsius. As a result, the output of the operational amplifier is high, and the relay driver transistor T1 conducts, which causes the relay RL1 to become energized. The N/O contacts of RL1 are what make the connection between the heater and the power supply. As a direct consequence of this, the heater will now begin to heat the water.

The voltage at pin 2 of IC1 rises to a higher level than the voltage at pin 3 as soon as the temperature of the water reaches a certain threshold, say 60 degrees Celsius. As a result, the output of the operational amplifier IC1 becomes low. The conductivity of the relay driver transistor T1 is interrupted, which results in the de-energization of the relay RL1. Because of the pole contact with the N/C of RL1, the power supply to the heater is cut off, which results in an end to any further heating of the water. Now, once the water temperature drops below the level that was previously established, the relay will become energized, which will reconnect the heater to the power source so that the water can be heated. The loop keeps going around. Choose the resistance values for resistors R1 and R2 in accordance with the temperature requirements you have, but make sure that R1 and R2 have values that are proportionate to one another. The bridge can, in theory, be balanced by determining the resistance of the PT-100 (RTD) at the temperature that is required and then adjusting the resistance of the preset. A positive temperature coefficient is exhibited by PT-100. In other words, its resistance improves as the temperature rises.

In order to calibrate, let's assume that the temperature of the water needs to be maintained at 60 degrees Celsius after it has been heated to that level. First, heat the water until it reaches a temperature of 60 degrees Celsius (measure the temperature using a thermometer). Keep sensor PT-100 (RTD) inside water. Adjust the setting in such a way that the potential difference between the inverting and non-inverting inputs of IC1 becomes negative while maintaining RTD at a temperature of 60 degrees Celsius (as determined by the thermometer). That is, the output pin 6 of IC1 should be low when the temperature of the water is 60 degrees Celsius.

80. Electronic Heart

If you have this electrical heart that glows on and off, you are sure to win the affection of that one person you have your eye on. In this particular circuit, a NE555 timer is set up to function as an astable multivibrator. The values of its resistors R1 and R2 as well as its capacitor C2 decide the frequency of its oscillations. In this case, the frequency is somewhere about 0.2 Hz. Each phase is slightly longer than 4 seconds in total duration.

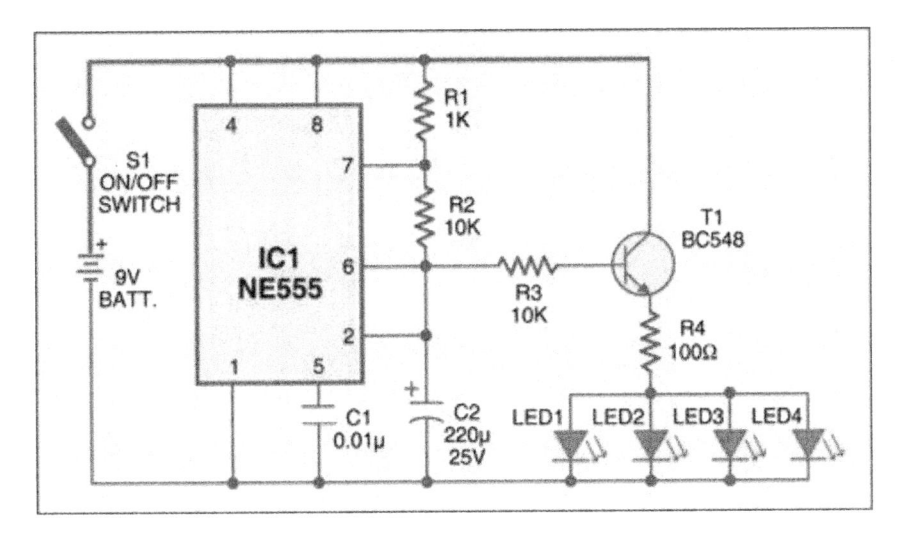

The capacitor C2 charges and discharges at a rate that is exponentially increasing as time passes. As a consequence of this, a sawtooth waveform looks more like a ramp. No matter what the voltage is across the capacitor, it cannot be used directly because doing so will only result in the capacitor being discharged or drained of its charge. A buffer is created by the transistor BC548 when it is set to operate in the common-emitter mode. This prevents the capacitor C2 from becoming loaded, which keeps the frequency of operation of the NE555 unaffected. Its emitter is connected to the four red LEDs that are wired in parallel with a resistor that is 100 ohms. A sawtooth waveform is generated here using the output from IC1's pin 6, which is used as the source. As a consequence, the LEDs will gradually and smoothly decrease to full brightness or return to their previous state. You can achieve the maximum voltage across the LED by making adjustments to the emitter resistance. If the value of the resistance is too low, the light output of the LEDs may have the appearance of a sawtooth that has been clipped. In the event that it is excessively huge, the LEDs will switch off for a portion of the allotted time. The LEDs provide the impression of seamlessly following the voltage when the resistance is kept at 100 ohms. Build the circuit on a PCB that may be used for a variety of purposes, and then insert it within a styrofoam heart. Install the switch S1 on the left side of the heart. When you press it, the heart will begin to glow alternately on and off. When the lights go off, the results will be easier to see.

81. Ultrasonic Sound Beam Burglar Alarm

This one-of-a-kind ultrasonic burglar alarm makes use of an inaudible and invisible sound beam produced by ultrasonic waves to detect any movement. When pushed at a frequency of 40 kHz, ultrasonic transducers are able to function at their most effective level. Therefore, the buzzer and the relay are both controlled by an ultrasonic transmitter and receiver pair that operates at a frequency of 40 kHz.

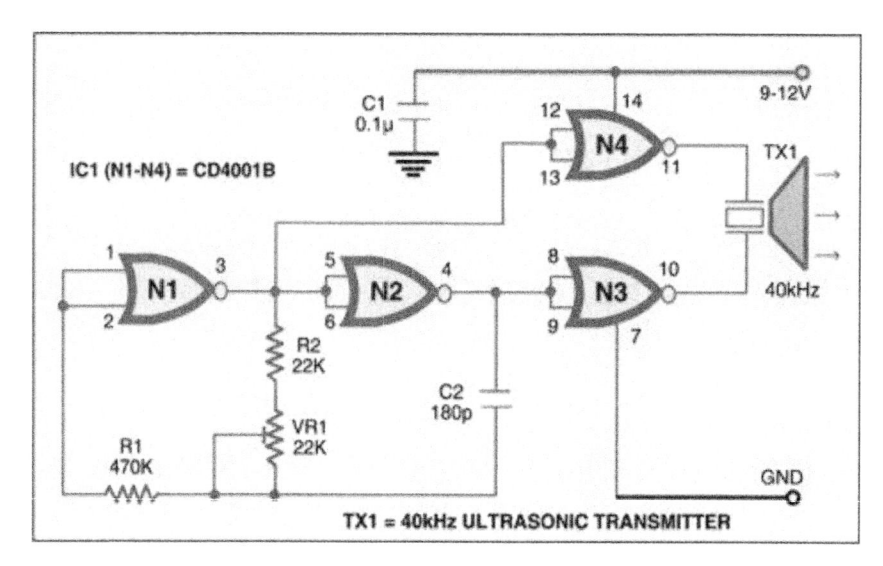

Fig. 1: Ultrasonic Burglar Alarm: Transmitter unit

The CMOS IC CD4001 serves as the focal point of the transmitter unit's construction (see Fig. 1). The oscillator frequency can be changed to roughly 40 kHz by setting the 22-kilo-ohm preset (VR1) on the variable resistor 1. The current drain of the ultrasonic transmitter transducer is extremely low, and it is driven by two complimentary buffer stages of CD4001 components.

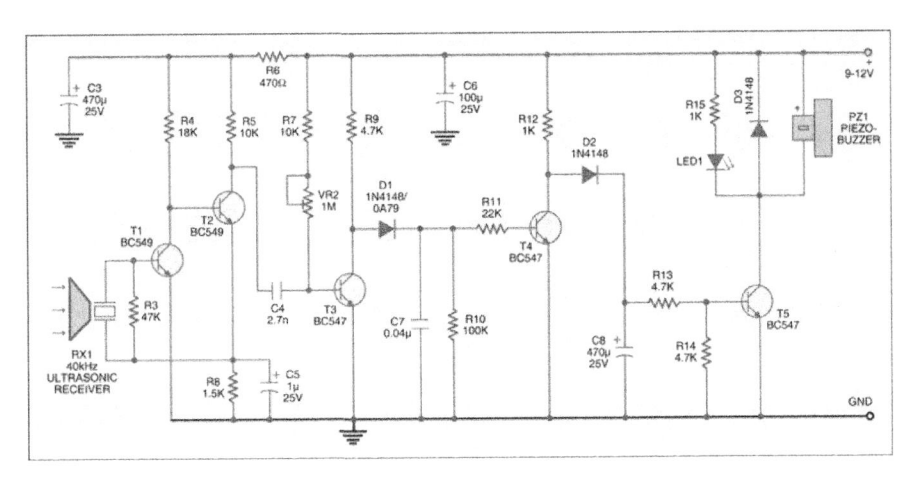

Fig2. Ultrasonic Burglar Alarm: Receiver circuit

A two-stage 40 kHz preamplifier is utilized by the receiver unit, which is then followed by a switching circuit, buzzer driver, and other components (see Fig. 2). Make the necessary adjustments to the value of VR2 inside the receiver's circuit so that the buzzer will no longer ring while the receiver is picking up ultrasonic sound from the transmitter. When the ultrasonic beam is disrupted, either by the movement of a person or any other object that crosses its path, the npn transistor T4 stops conducting, and the transistor T5 begins conducting instead. It turns on the buzzer as well as LED1 for approximately three to four seconds. This duration is dependent on the amount of time it takes for the 470 F capacitor (C8) that is connected at the base of transistor T5 to discharge completely through the resistor R13. By adjusting the values of C8 and R13, the appropriate length of time for the burglar alarm to sound can be selected.

Construction & Testing

A power supply that is controlled between 9-12 volts is used to power the alarm circuit, which consists of a transmitter, a receiver, and other connected circuits. The circuit for the receiver needs a controlled 12V supply. You are also free to use a power supply that is not controlled here. On separate PCBs designed for general use, assemble the circuits for the transmitter and the receiver. Adjust the positioning of the ultrasonic transducer so that the signal will reach the ultrasonic receiver unobstructed. A relay can take the place of the piezobuzzer (PZ1) as the controller of the alarm that is powered by the mains.

82. Sunset Lamp

Automatic lights that use LDR technology flicker because of the shift in light intensity that occurs between dawn and dusk. Therefore, compact fluorescent lamps, often known as CFLs, should not be used in such circuits since the flickering light could potentially damage the electronic circuits included within the lamps. The sunset lamp circuit that is described in this article has the ability to solve the problem and promptly turn on the lamp when the light intensity drops below a predetermined threshold.

Sunset Lamp circuit description

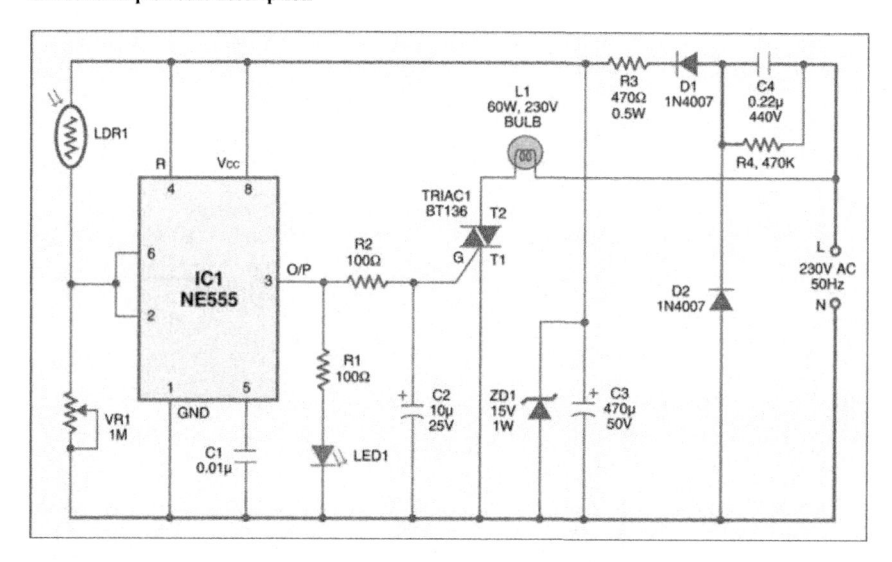

The bistable action is provided by a popular timer IC called NE555 (IC1), which is used in this sunset lamp circuit as a Schmitt trigger. In order to achieve the immediate action, the set and reset operations of the comparators contained within the NE555 are utilized. The upper threshold comparator of IC1 trips when Vcc is 2/3 of its full value, whereas the lower trigger comparator trips when Vcc is 1/3 of its full value. The inputs of the threshold comparator and the trigger comparator of the NE555 are coupled together, and they are connected to the voltage divider that is produced by the LDR1 and the VR1 components. The voltage that is measured across LDR1 is influenced by the amount of light that is present. When the sun is shining, LDR1 has a low resistance, which causes the input voltage to the threshold comparator to rise above 2/3Vcc. As a result, the threshold comparators output drops to zero, which in turn resets the flip-flop that is included within IC1. However, the input to the trigger comparator is still greater than 1/3Vcc, which maintains the low state of the output pin 3 of IC1. Because there is not enough current for it to be fired, the triac BT136 that is connected to the output pin 3 of IC1 stays in its quiescent state. As a result, lamp L1 is in the "off" position during the day.

83. Electronic Dice

This electronic dice eliminates the possibility of wear and tear while maintaining all of the functions of a traditional wooden or plastic die used in the game of Ludo. The time period (T) of the astable multivibrator in this circuit is roughly 0.02 seconds, and it is built around the timer IC NE555 (IC1) in this circuit. Pushbutton switch S1 is the device that it uses to drive the decade counter IC 74LS90 (IC2). The outputs of the decade counter are supplied to the decoder/driver IC 7447 (IC3), which, in turn, is linked to the common-cathode, 7-segment display LTS542 through a connection (DIS-1).

Additionally, the outputs of the decade counter, Q0 through Q2, are sent to the 3-input AND gate N1 (IC4).

Electronic dice circuit

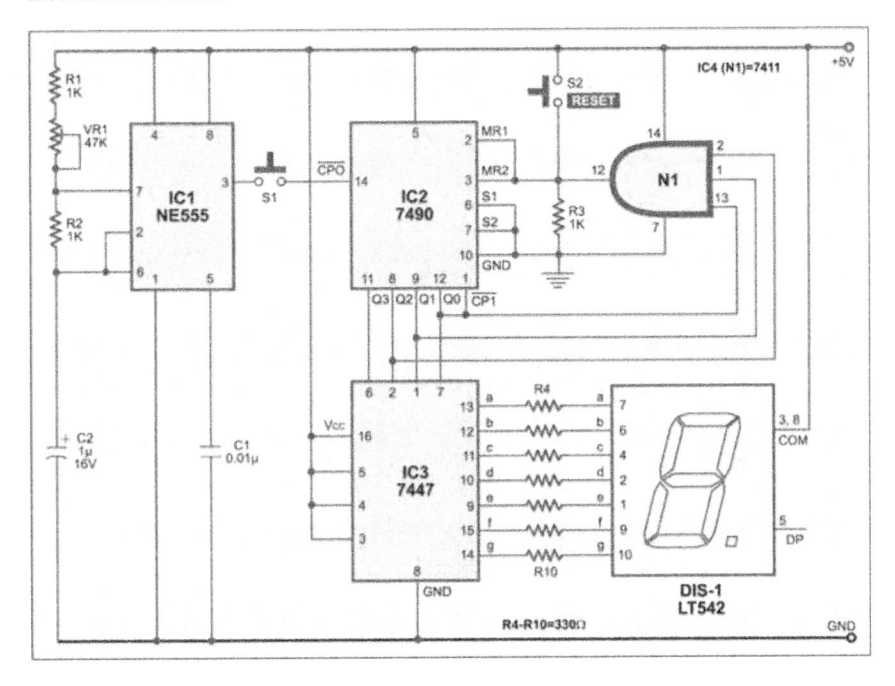

When the decade counter reaches the count of binary 7 (that is, when Q0, Q1, and Q2 all get high), the output of the AND gate (N1) goes high, which in turn resets the counter IC1. A participant must briefly depress the switch S1 in order to participate in the game. The display will show a count somewhere between '1' and '6' in increments of one. Before a player is allowed to play, the push-to-on switch S2 is used to reset the counter in IC1. This ensures that the display will always show the number zero.

84. Solid-state relay

In its most fundamental form, a relay is a switch that, when activated by an electrically separated, low-power control signal, turns on power to a load. Up to this point, the most reliable components for carrying out this duty have been electromechanical relays. The development of a solid-state relay can be attributed to the progress that has been made in the field of semiconductor technology. Solid-state relays have many benefits, including almost infinite switching, bounce-free operation, immunity to electromagnetic interference (EMI), higher operating speeds, low-voltage control, small package size, and multifunction integration. Solid-state relays also have the advantage of being able to integrate multiple functions. In this project, we will explain a DC-operated solid-state relay that makes use of components that are easily accessible (see Fig. 1). It can run off of a TTL compatible PWM input or a battery with a voltage of 3V DC. In Figure 2, you can see the several pin configurations for the MOSFET IRF540, the transistor BC547/BC557, and the opto-coupler.

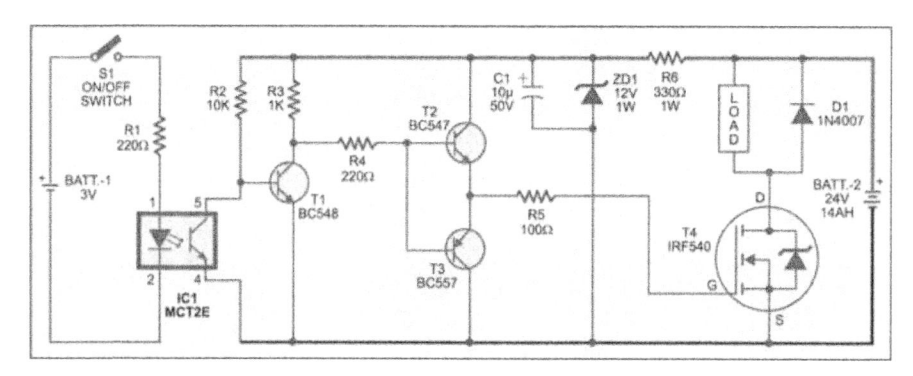

Fig. 1: Solid-state relay circuit

For a current of up to 10 amps to be driven through the load, the load voltage can be supplied by a tubular battery or a car battery with a voltage ranging anywhere from 24V to 96V. A battery operating at 24 volts DC has been utilized here. The value of the series resistor R6, which is currently set at 330 ohms, will alter based on the voltage of the battery in order to produce a current of 30-35mA. Calculating the value of resistor R6 can be done as follows: R6 equals 1000 times (Bat.2 V - 12V) divided by 35, and its power dissipation equals 35 divided by 1000. 2 x R = 0.001225R

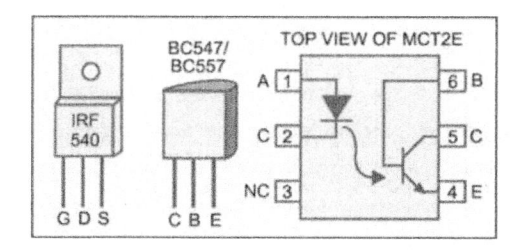

Fig. 2: Pin configurations of MOSFET IRF540, transistors BC547/BC557 and MCT2E

Note: It is recommended that R6's wattage be at least twice as high as its power dissipation. When the S1 button is pressed, the logic 1 state of the opto-coupler input is set, which causes the MOSFET to trigger and turn on the load. The input of the opto-coupler is low (logic 0) while S1 is open, which prevents the MOSFET from being triggered. As a direct consequence of this, the load has not been activated.

85. Car Porch Guard

Using this electronic car porch guard system, you can prevent theft from occurring to your expensive vehicle. As soon as the system detects any potential attempt at theft, it promptly activates the porch light and gives off a piercing alarm. Before moving the vehicle out of the garage at night, a shrewd burglar will disable the security alarm that is installed in the vehicle before making any attempt to start the vehicle. Therefore, it would be incredibly challenging to discover who was responsible for the crime, and the loss wouldn't be noticed until the following morning. In order to foil any effort by the burglar, this circuit sounds an alarm whenever the vehicle is moved further away from the designated parking spot.

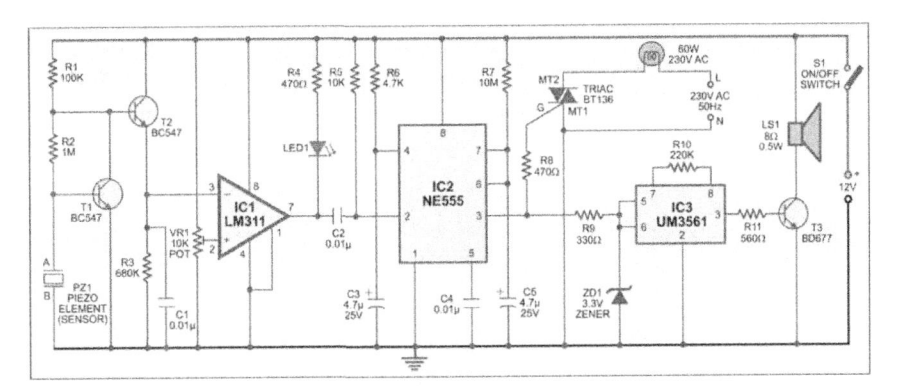

The automobile porch guard's electrical circuit is depicted in Figure 1. A piezo element serves as the sensor for what is fundamentally an alarm that detects vibrations in the environment. When a voltage is put across the piezo element's terminals, the element's capacitance is normally in the range of a few tens of nanofarads, and it will quickly charge up as a result. The charge that is placed on the piezo element by the resistors R1 and R2 will remain on the element until it is shaken by some kind of mechanical motion. When the piezo element detects any vibrations, an alternating current voltage is generated across its terminals. This voltage is then amplified by the transistors T1 and T2 in the circuit. When transistor T2 starts conducting, capacitor C1 starts charging, and a voltage is delivered to the inverting input of integrated circuit LM311 (pin 3). (IC1).

The pin connections of the IC LM311's voltage comparator are distinct from those of other typical op-amps due to the fact that it is not a standard op-amp. In order to maximize the device's versatility, the output stage has both positive (pin 7) and negative (pin 1) outputs. Typically, the negative output is connected to ground. A npn transistor may be found at the output, and its collector is located at pin 7, while its emitter is located at pin 1. Therefore, the integrated circuit will act as a current sink while the output transistor is conducting.

86. Wire Break Alarm with Delay

Here is a simple alarm circuit for a wire that breaks and goes off after 15 to 30 seconds. In the event that the thin wire loop over the door is broken, the alarm will go off after a delay of 15 to 30 seconds, which may be adjusted with VR1. As a result, the residents have enough time to secure the room from the outside and apprehend the burglar.

Wire break alarm circuit

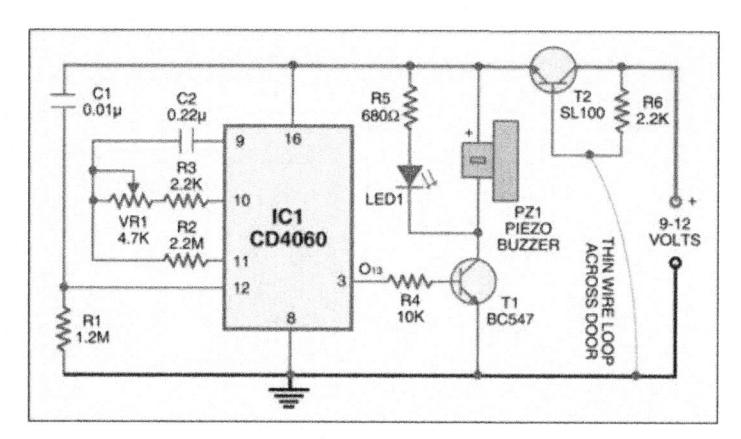

The CD4060, which is a 14-stage ripple-carry binary counter/divider and oscillator, is utilized within the circuit. Since it is hardwired to function as a timer in this instance, an input pulse is not required in order to trigger it. Once the power supply is turned on, the CD4060 is immediately put into operation. After the expiration of the preset delay that was specified through VR1, the output O13 of the CD4060 becomes high. In order to power the timer section that is designed around CD4060, the transistor SL100 (T2) is set up as a switch. The conductivity of the transistor T2 is disrupted when the wire loop is closed. So, power to the timer circuit is not accessible and the piezobuzzer does not sound.

When the wire loop is interrupted by an intruder, however, transistor T2 begins to conduct, providing power to the circuit. After fifteen to thirty seconds, the piezobuzzer begins to emit a sound. Resetting IC1 can be accomplished by either reconnecting the wire loop or turning off the power source.

87. Cordless Multi-Door Alarm
Stop a burglary from happening by setting off this alarm circuit when someone breaks in. Each door is protected by its own circuit, which is made up of a 555 timer IC and a reed switch magnet. All three units get their power from the same source. The buzzer can be plugged into the earth line of a socket in any room of the same building that has a proper earth line connection. There's no need for wires to be run from different rooms to the buzzer unit.

Connect the reed switch S1 near the magnet of gate 1 to the door-1 alarm unit. Connect IC1 as a frequency oscillator, and use VR1 to set the door-1 alarm unit to the frequency you want, say between 1 Hz and 3 Hz. Connect reed switch S2 near gate 2's magnet to the door-2 alarm unit. Connect IC2 as a frequency oscillator as well, and use VR2 to set the frequency of the door-2 alarm between 5 and 7 Hz. Connect the reed switch S3 near the magnet of gate 3 to the door-3 alarm unit. You can also use IC3 as a frequency oscillator. Use VR3 to set the frequency of the door-3 alarm between 10 and 12 Hz.

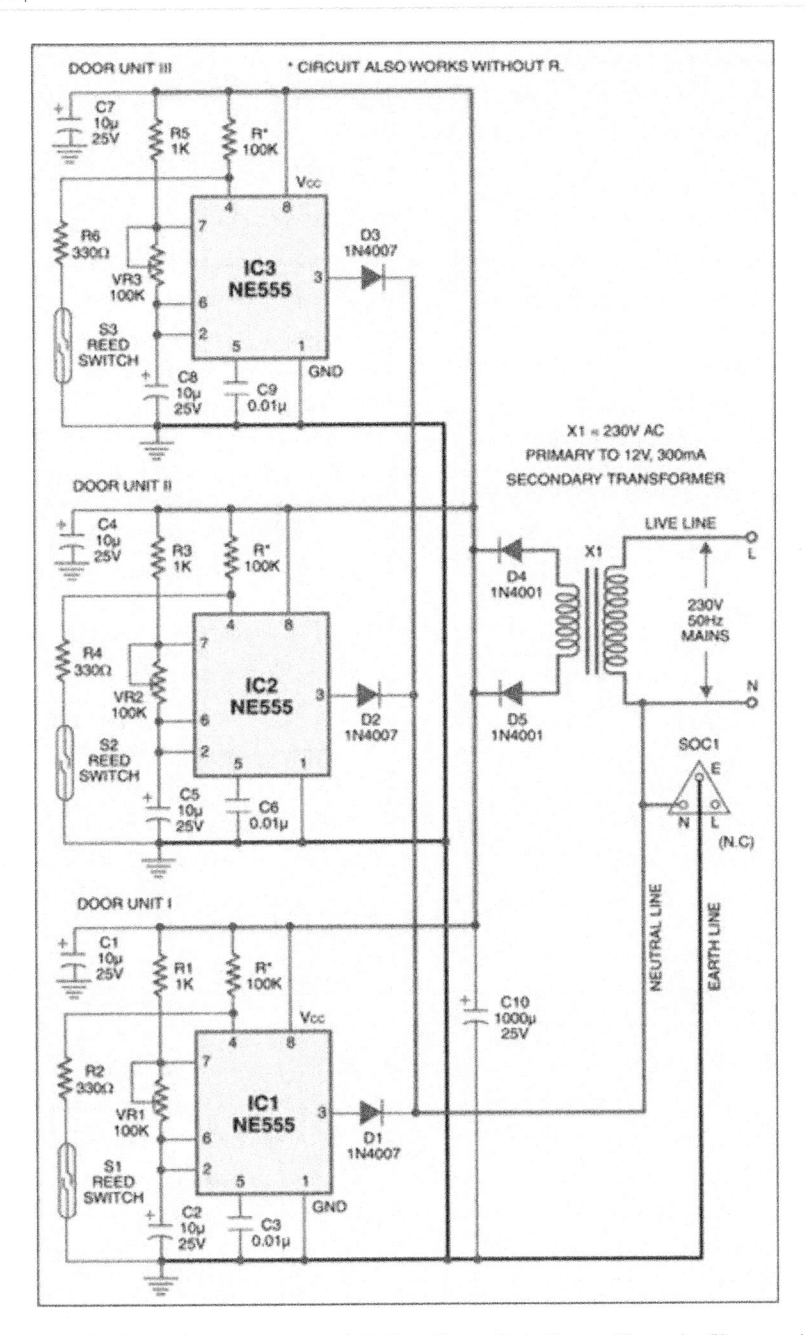

The alarm circuits need a power supply, which is made up of a bridge rectifier and a filter capacitor C10. When all the doors are closed, which is normal, reset pin 4 of IC1 through IC3 stays low. So, these don't move back and forth, and the piezobuzzer PZ1 stays quiet. When door 1 is opened, the magnet moves away from the reed switch S1. When door 1 is opened, IC1 sends out signals at 1-3Hz, and the piezobuzzer beeps. When door 2 is opened, IC2 sends out signals between 5 and 7 Hz, and the piezobuzzer beeps to show that door 2 is open. In the same way, when door 3 is opened, the piezobuzzer beeps at a rate of 10 to 12 Hz to let you know.

88. School/College Quiz Buzzer

When employed in academic competitions, manual buzzers like those found in schools and colleges can lead to a great deal of misunderstanding over the identification of the first reply. Although there exist circuits using PCs and discrete ICs, they are either too expensive or confined to only a tiny number of players. The maximum number of participants allowed in any one quiz tournament is eight people, therefore this circuit for a quiz buzzer can accommodate that many people. The integrated circuit (IC) 74LS373 and a few passive components are used in the circuit. These components are easily obtainable on the market.

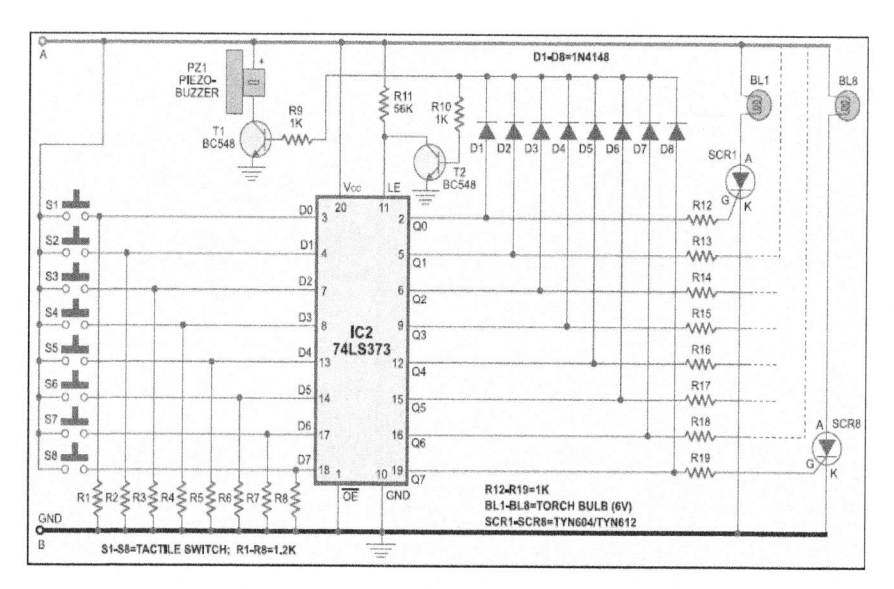

There are two distinct parts of the circuit that may be separated from one another: the power supply and the quiz buzzer. The power supply part is depicted in Figure 1. The AC power lines are the source of the regulated 5V power supply that is used for the quiz buzzer part. The 230V AC mains are reduced to 7.5V AC by transformer X1, which is also responsible for rectifying the current using bridge rectifier BR1, filtering it using capacitor C1, and regulating it using regulator IC1. Capacitor C2 bypasses waves in the regulator output.

The quiz buzzer section is depicted in figure 2. The IC 74LS373, an octal latch, is the most important component of this section. Its job is to convert the logic state present at the data input pins D0 through D7 to the outputs Q0 through Q7 that correspond to those pins. Resistors R1 through R8 are responsible for bringing low the data pins D0 through D7 in their typical configuration.

The push-to-on switches S1 through S8 each have one of their terminals connected to +5V, while the other terminals are each linked to their corresponding data input pins. The cable wire is going to be used to run the switches all the way up to the players. It is possible to store the torch bulbs BL1 through BL8 in boxes, and the front of the boxes could have a piece of white paper with the contestant's name or number printed on it. This would make it simple to identify who owned each box. Position the boxes such that they are visible from above the heads of the audience members so that everyone can see what is happening.

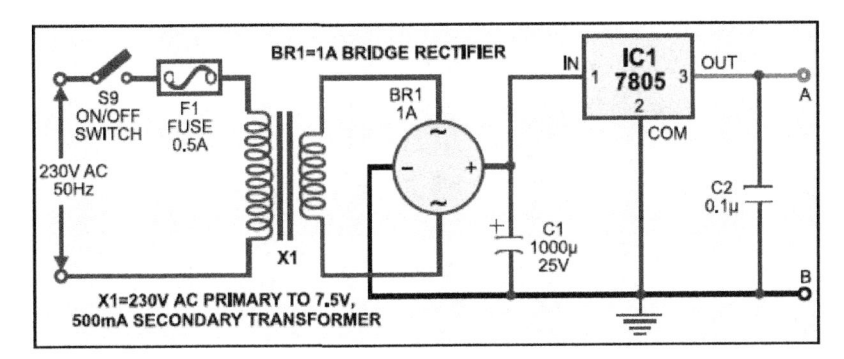

Fig. 2: Circuit of school/college quiz buzzer

The circuit is ready for usage after the power is turned on by utilizing switch S9 (assuming that the terminals 'A' and 'B' of both the power supply and quiz buzzer sections are coupled). At this time, all of the switches, from S1 all the way through S8, are open, and the outputs of IC 74LS373 from Q0 all the way through Q7 are low. As a result, the gates of silicon-controlled rectifiers SCR1 through SCR8 are also low.

As soon as a competitor momentarily hits his respective switch, the matching output data pin turns high. This causes the associated SCR to be triggered, and as a result, the respective bulb lights. The piezobuzzer (PZ1) emits a sound at the precise moment when the transistor T1 begins to conduct.

Concurrently, a high voltage is applied to the base of the transistor T2 in order to make it conduct. In order to latch all of the Q0 through Q7 outputs, pin 11 of IC2's Latch-enable (LE) component is connected to ground. This limits further change in the output state due to any change in the state of switches S1 through S8 by any other participant. Until the on/off switch S9 is used to reset the circuit, just one of the eight torches' bulbs will emit light.

89. Mobile Phone Detector Using LM358

At a distance of between four and five meters, this mobile phone detector is able to identify the presence of a mobile phone that is turned on. Therefore, it can come in handy at a place where mobile phones are not allowed, such as an examination room or a meeting. Even when a mobile phone is set to silent mode, the circuit is able to decipher incoming and outgoing calls, as well as text messages, Internet traffic, and video broadcasts. Its LED will begin blinking as soon as it picks up an

RF signal coming from a mobile phone that is actively being used, and it will continue doing so until the signal is no longer present.

Circuit and working

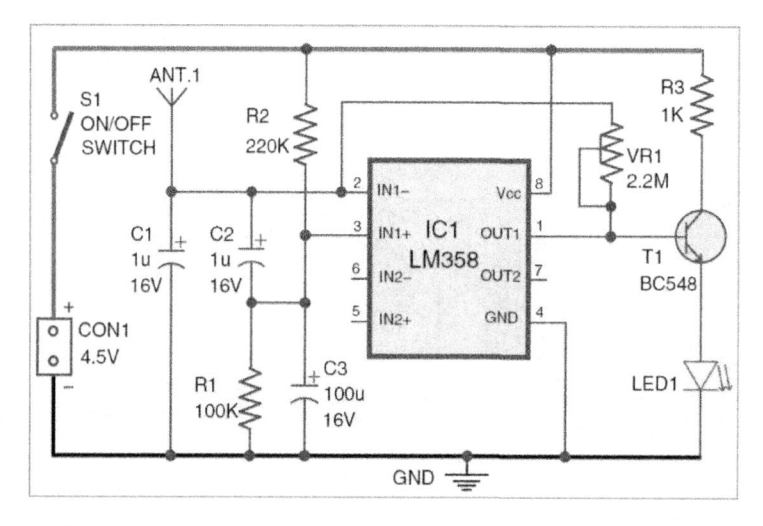

The schematic for the mobile phone detector that makes use of an LM358 may be found in Figure. The LM358 (IC1) and the BC548 npn transistor serve as its primary building blocks (T1). When a mobile phone is turned on, it emits radio frequency (RF) signal that travels through the air in the immediate vicinity. The transmission carries electromagnetic and radiofrequency (RF) radiation that was emitted by the phone.

In order to detect the radio frequency (RF) signal sent by the mobile phone, the circuit makes use of capacitor C1. C1 is responsible for the absorption of any radio frequency (RF) signals that the mobile phone may emit; these signals are then sent to the inputs of IC1. The blinking of LED1 serves as an indication of this fact. The range of the circuit can be adjusted by adjusting the value of Preset VR1 (2.2M). The signal that is obtained at pin 1 of IC1 is fed into transistor T1, which serves to amplify the signal. The circuit can be used for 2G networks as well as GPRS and network search, both manually and automatically. The 3G, WCDMA, and HSDPA network signals are not detected nearly as well by it.

90. Anti-Sleep Alarm

Most accidents on highways at night happen because drivers can't see well because they are constantly looking into the bright headlights of cars coming toward them. The poor vision is caused by the visual pigment in the eyes running out, which makes the person want to sleep to get the pigment back. This alarm doesn't let you sleep. This circuit keeps you alert by making short beeps and flashing lights to remind you that you are not on a bed but instead driving a car. It only works at night because of a switch that is controlled by a light-dependent resistor (LDR).

Anti-sleep alarm circuit

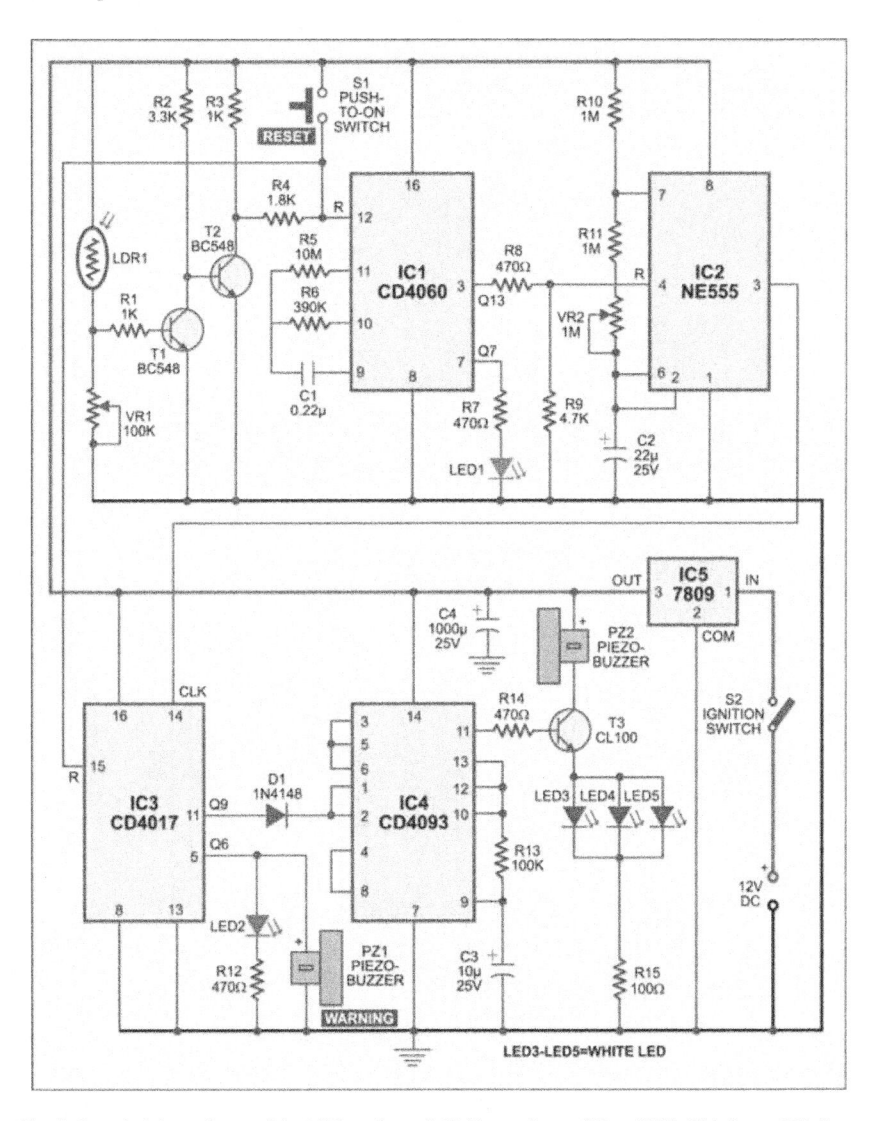

The light switch is made up of the LDR and two BC548 transistors (T1 and T2). This keeps IC1 from oscillating during the day. During the day, when the LDR is exposed to light, T1 conducts to stop T2 from conducting. This makes pin 12 of IC1 high, which stops it from oscillating. So the rest of the circuit stays in a state called "standby."

At night, T1 doesn't work because the LDR is dark. Pin 12 of IC1's reset pin is pulled to ground when T2 conducts. This makes IC1 start to oscillate, which is shown by the flashing of LED1. The values of R5, R6, and C1 determine how fast the oscillator inside the binary counter IC CD4060 goes around and around. Its Q13 output goes from high to low and back again every 15 minutes. With the potmeter VR1, you can change how sensitive the LDR is.

91. Ultrasonic Proximity Detector circuit

Only sounds with a frequency of up to 20 kilohertz can be heard by humans. The frequency of operation for this proximity detector is forty kilohertz.

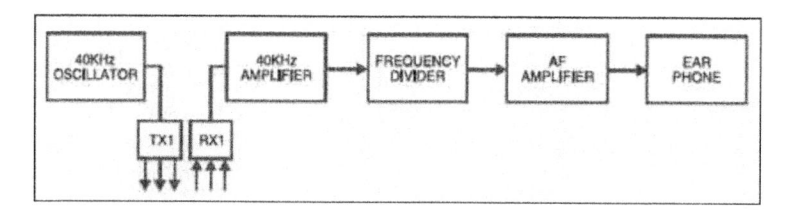

It makes use of two ultrasonic transducers that have been custom-made: one transducer produces sound at a frequency of 40 kilohertz, while the other transducer receives sound at this frequency and turns it into an electrical variation of the same frequency.

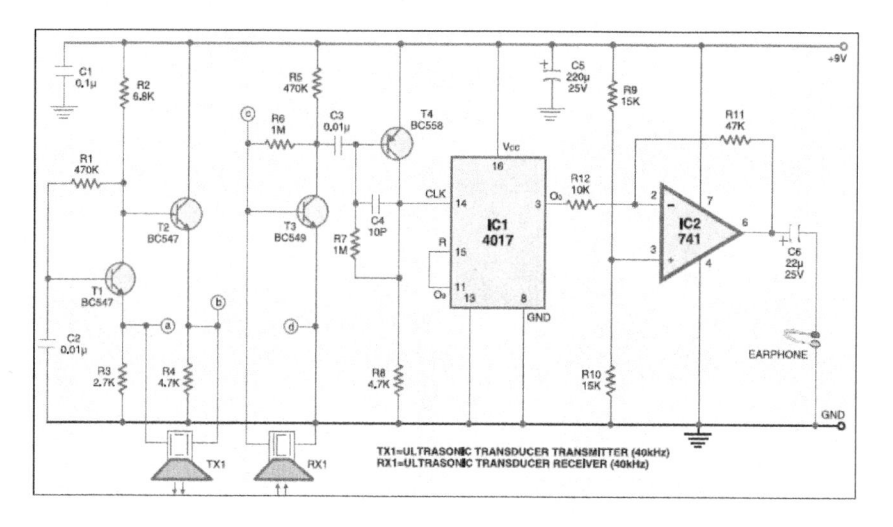

The ultrasonic proximity detector's block design can be seen in Fig. 1, while the detector's circuit is depicted in Fig. 2. Mount the transducers (both the transmitter and the receiver) approximately 5 centimeters apart on a piece of general-purpose printed circuit board (PCB), as indicated in Figure 3, and connect to identical points ('a' through 'd') of the detector circuit (Figure 2) using external wires.

The 40 kHz oscillator is constructed using transistors T1 and T2 as its core components. Some of the transmitted ultrasonic waves will be reflected back and picked up by the reception transducer if there is a solid object in front of the ultrasonic transmitter module (TX1) (RX1). The receiver takes the ultrasonic signals at 40 kHz and converts them into electric impulses at the same frequency. These electric signals are then amplified by the transistors T3 and T4.

```
          ←— 5 CM —→
           TX1   RX1
       a ┌─────┬─────┐ c
       b │  ◯  │  ◯  │ d
         └─────┴─────┘
         EMITTER RECEIVER
            ULTRASONIC
           TRANSDUCERS
```

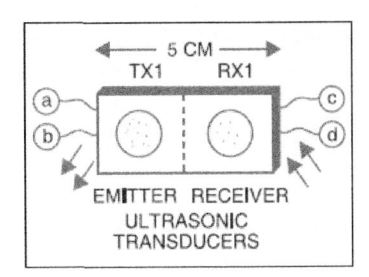

Fig. 3: Transducers mounted on the PCB

The enhanced signals are still within the range of frequencies that are inaudible, which means that they cannot be heard. At the output of the amplifier, there is therefore a frequency-divider stage that makes use of a CMOS decade counter IC4017 (IC1). Because IC1 reduces the frequency of the input signal by a factor of 10, the signal that was originally 40 kHz is now only 4 kHz, which is audible. Op-amp IC 741 (IC2), which has been hooked up to function as an earphone amplifier, receives the 4kHz signals as input.

This circuit has the potential to serve as an electronic watchdog for the visually impaired. Maintain it in their pocket, together with the 9V battery, and ensure that the earphone is inserted into their ear. It is recommended that the transducer modules be aimed in the direction of the walking path. They will be able to adjust their course appropriately if they hear a sound at a frequency of 4 kilohertz through the earphone in the event that an item appears in front of them or nearby. It is important to keep in mind that while you are using this device, you should not have any pets in the room with you. The reason for this is that animals are able to hear ultrasonic sound, which causes them discomfort and causes them to bark unnecessarily.

92. Automatic Parking Light for Cars

At night, parking lights make your parked vehicle more apparent to passing cars, reducing the risk that they will collide with it. However, these lights will draw a significant amount of electricity from the battery in your vehicle. The following description is of a straightforward automatic parking light system that requires no standby current to function. When the light of an oncoming car is detected from either the back or front side, the circuit is programmed to switch on the parking lights automatically for a period of thirty seconds. This automatic feature offers increased protection to a parked car during the night.

The circuit is constructed using discrete components along with the transistors T1 through T3, the MOSFET T4 (BS170), and other transistors. The front and rear sensors are represented by the Darlington phototransistors T1 and T2 (both of which are L14F1) When the light from an incoming vehicle shines on the relevant photo-transistor, it causes transistor T3 to current, which in turn triggers the gate of MOSFET T4 via resistance R3.

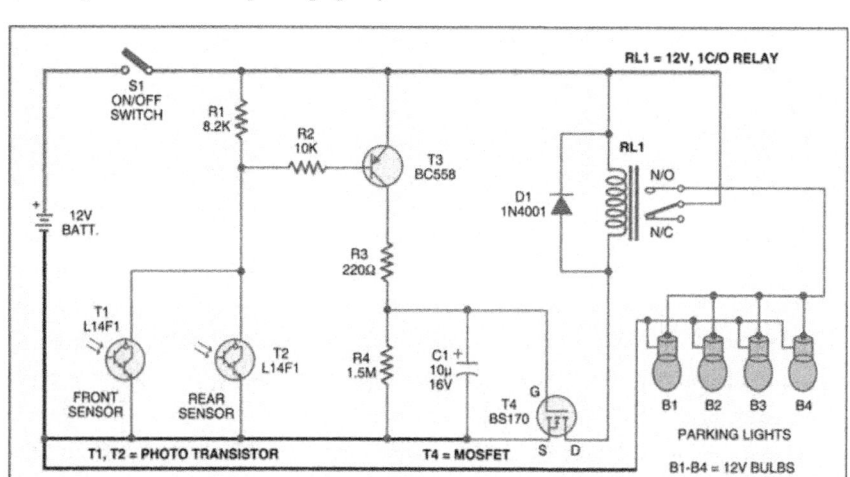

Circuitry for an automatic parking light system

When resistor R4 and capacitor C1 are energized for a certain amount of time, relay RL1 becomes energized. This turns on the parking lights (B1 through B4) that are connected to its N/O contacts. When the allotted amount of time has passed, the relay labeled RL1 will become deactivated, which will result in the parking lights being turned off.

Construct the circuit using a PCB designed for general use and enclose it in an appropriate housing. Mount the front and rear phototransistors (T1 and T2) in such a way that they receive light directly from an approaching vehicle rather than receiving light from an ambient source such as a street light. Install the 12V bulbs marked B1 through B4 in the appropriate locations inside your car or outside in the parking lot.

93. Peak Hour Timer

If the line voltage drops between 6 pm and 9 pm, when most people use electricity, appliances like refrigerators and air conditioners use a lot of power. If these appliances don't have a low-voltage cut-off, it will waste electricity and cause the appliances to heat up. In turn, this could make the compressors in these appliances work less well. This circuit is helpful because it turns off an appliance automatically during the busiest times. The appliance will turn back on after three hours.

An LDR-based switch is used to turn off the power to a circuit automatically. During the day, LDR1 has a low resistance, so transistor T1 pulls its collector voltage to ground by forward biasing. This keeps T2 from also forward biasing. In this state, the rest of the circuit is turned off, and the appliance gets power from the normally-closed (N/C) contacts of relay RL1. When the sun goes down around 6 p.m., LDR1's resistance goes up, and transistor T1 stops working. Then, the circuit gets power from the transistor T2. Using preset VR1, you can change how sensitive LDR1 is to the level of light you want.

The 14-stage ripple counter CD4060 (IC1) is used to make the timer circuit. Taking pin 1 (Q11) as the output gives a timing of 3 hours. When transistor T2 gives power to IC1, it resets through C1 and R3 and begins to oscillate. The voltage to IC1 is kept stable by the reservoir capacitor C2. This means that small changes in the power supply don't affect the oscillation. The oscillation of IC1 is kept going by resistor R4 and capacitor C3. This is shown by the blinking of LED1 connected to its output pin 7. (Q3). The timer's timeout period can be calculated as follows: T = 2n/fosc in seconds, where n is the number of outputs chosen and fosc = 1/2.3 (R4.C3) Pin 10 of IC1 is connected to the resistor R4, and pin 9 is connected to the capacitor C3. When Q11's output goes high, diode D1 stops IC1 from continuing to oscillate. Q11's output stays high until morning, when IC1 is reset.

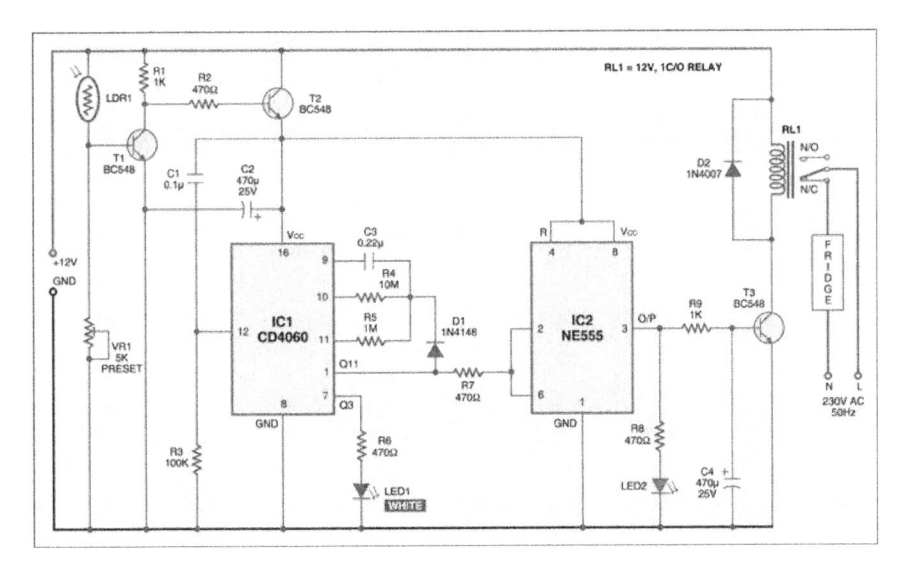

The timer NE555 is used to make IC2 a bistable latch. Its trigger input pin 2 and threshold input pin 6 are connected so that when the power is turned on, the output of IC2 goes high and latches because its trigger and threshold inputs are both floating. When relay driver transistor T3 conducts, relay RL1 is turned on, which disconnects the appliance, like a refrigerator, from the mains. During the busiest times, this problem lasts for three hours. After IC1's timeout period, its Q11 pin goes high to send a positive pulse through resistor R7 to the threshold input of IC2. This pulse that goes high resets IC2 and makes its output go low. Through the N/C contacts of relay RL1, the appliance is turned on when transistor T3 is turned off and relay RL1 is de-energized. Capacitor C4 keeps the base current of transistor T3 steady so that switching is clean and the relay doesn't chatter. The back emf of the relay is taken away by the freewheeling diode D2. Put the circuit together on any type of PCB and put it in a suitable cabinet. The current rating of relay contacts should be enough to handle the load. Connect the relay contacts to the AC mains wire using the right-sized wires. Earth the unit correctly and use a three-pin plug to get AC from it. Put the unit where the LDR will get enough light from the sun.

94. Pressure Sensitive Alarm

Here is a pressure-sensitive alarm that won't break the bank. The alarm makes use of a pressure sensor that was constructed at home. This pressure sensor functions as a variable capacitor and is made up of two copper-clad boards with a piece of sponge sandwiched in between them.

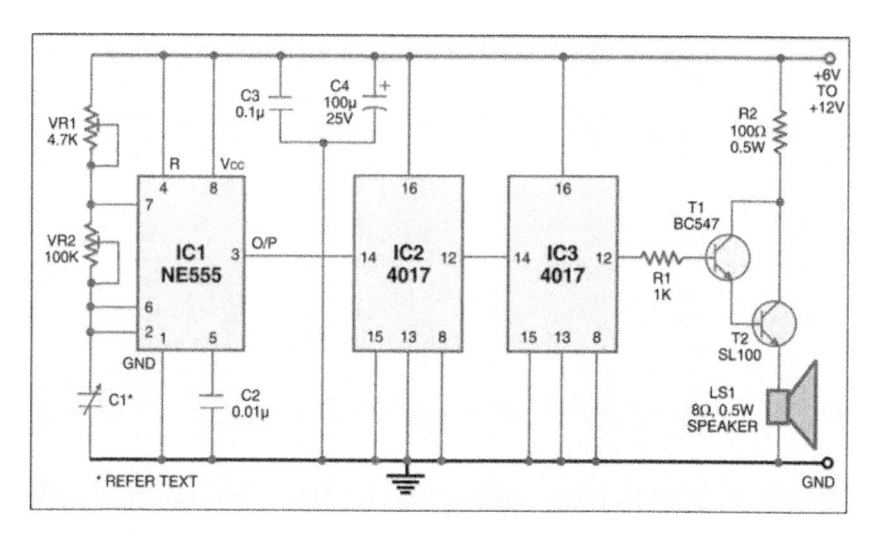

Take two conventional copper-clad boards measuring 6.5 by 6.5 centimeters each and use them as the basis for the pressure sensor (variable capacitor C1). On the substance made of copper, apply varnish or a green mask. To insulate the space between the copper-clad boards, place a layer of soft sponge approximately 2.5 centimeters thick and secure it with insulating tape, as illustrated in figure 1. Ordinary insulated flexible wires should be soldered to the copper plates on both of the cladded boards. These wires are what are required to link the pressure sensor to the alarm circuit's input device.

The pressure sensitive alarm's circuit is depicted in Figure 2, When there is no pressure applied to the sensor plates, the capacitance of the home-made sensor drops to less than 10 pF. This causes the IC 555, which is connected as an astable multivibrator, to become inoperable. When there is pressure applied to the sensor, the gap between the copper-clad boards gets closer together. This causes the capacitance of the sensor to increase, and it might reach up to 50 pF, depending on how close together the copper-clad boards were before. Therefore, the sensor functions within an IC 555-based astable multivibrator as a variable capacitor.

The IC 555 has a frequency of oscillation that is approximately 350 kHz, which is inaudible. This output signal is sent to two 4017-decade counters in the system. The decade counter IC2 changes the signal from 350 kHz to 35 kHz, which is also inaudible to the human ear. The output of IC2, which is operating at 35 kHz, is sent to the decade counter IC3, which alters the frequency such that it may be heard as 3.5 kHz. These frequencies can be altered by adjusting either the preset controls VR1 and VR2 or the pressure applied to the sponge sandwiched between the copper-clad boards. The audio frequency signals coming from IC3 at 3.5 kHz are amplified by the transistors T1 and T2 before they are sent to the loudspeaker. This results in the loudspeaker emitting a high-pitched audible tone.

95. Crystal Based 50Hz Generator

The following diagram illustrates a straightforward circuit for generating 50Hz frequency utilizing a crystal as the active component. Inverter circuits are able to make advantage of the pulses that it produces, which have an alternating frequency of 50 Hz and a duty cycle of 50%. It consists of a dual J-K flip-flop (CD4027), an operational amplifier (LM324), and a few discrete components. The CD4060 is its counter and oscillator, and it has 14 stages.

Crystal based 50Hz generator circuit

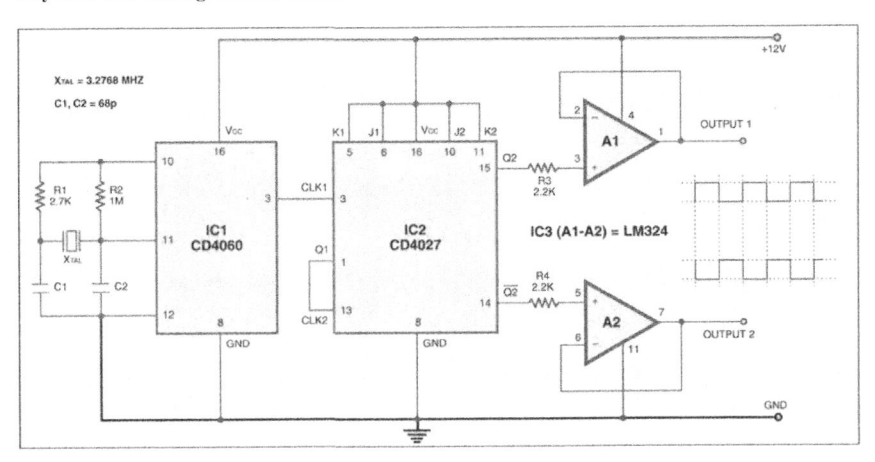

The oscillator is constructed with an IC CD4060 (IC1) as its center and a crystal operating at 3.2768 MHz. Power can be controlled by the use of the resistor R1. The biasing resistor is denoted by R2. R1 supplies the requisite minimum transconductance, which is required for beginning oscillations and keeping them going once they are established. In order to obtain a frequency of 200 Hz at pin 3 of the CD4060, the oscillator frequency is first divided by 14. With the assistance of a dual JK flip-flop CD4027, the frequency of 200Hz is further divided by the number 4. Therefore, the frequency of 50 Hz is generated at pin 15 of CD4027, while the complement of this frequency is generated at pin 14. The frequency that is being created has a duty cycle of fifty percent.

The operational amplifiers A1 and A2 of the LM324 act as a buffer for the frequency of 50 hertz. Inverter circuits make use of the square wave outputs of the operational amplifiers that are available at pins 1 and 7. These outputs have the opposite phase of the square wave. Assemble the circuit on a PCB that can be used for a variety of purposes, and then enclose it in a cabinet that has sufficient room for the battery and the switches. To provide power to the circuit, use either a 12V AC converter or a 12V battery.

96. Electronic Ignition System for Old Cars

In older cars with carburetors, the spark plugs are lit by a system called a contact breaker (CB) point. Using transistorized switching, you can change the CB-point-cum-condenser type of ignition system in your car to an electronic one. Figure 1 shows the electronic ignition coil's circuit, and Figure 2 shows how its wires connect to each other.

Electronic ignition coil

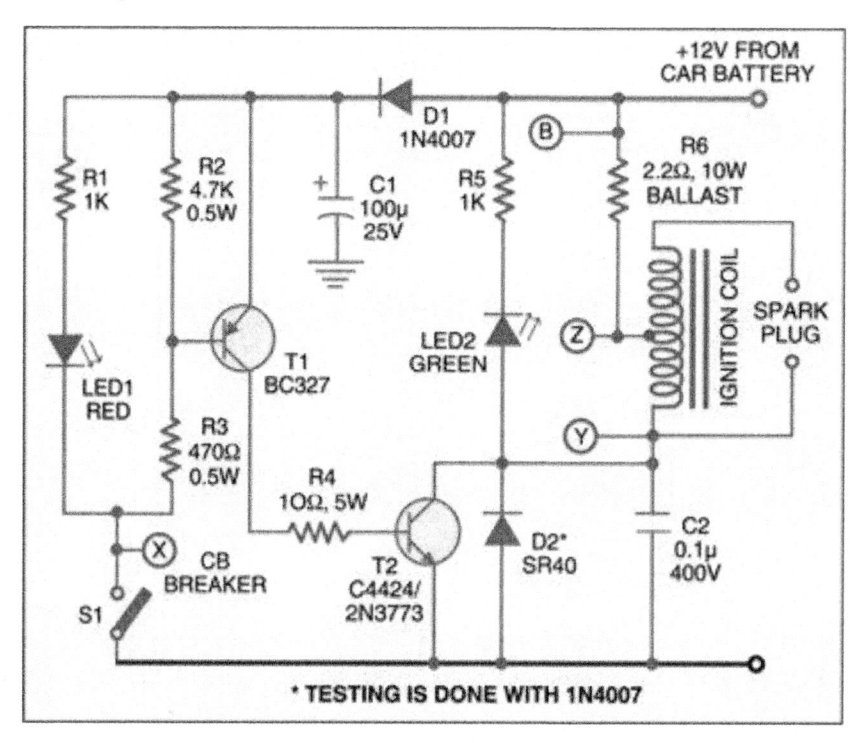

Fig. 1: Circuit diagram of electronic ignition for old cars

In Fig. 1, when the CB point is open, both transistors T1 and T2 are turned off, and the ignition coil doesn't get any power. When the CB point is closed, the T1 and T2 transistors conduct, and a very high current flows through the ignition coil. As a result, the ignition coil builds up a very high voltage that drives the spark plug. A red LED (LED1) lights up to show that the CB point is moving back and forth. The current going through the ignition coil is limited by the 2.2-ohm, 10W ballast resistor R6. Back emf makes the green LED (LED2) in the output section light up, so it needs to be wired with reverse polarity, as shown in Fig. 1. Power transistor C4424 needs a good heat sink (T2). In Fig. 2, the ignition coil, a condenser, and the connection from the EHT to the spark plugs and distributor are shown. For electronic ignition, you don't need the condenser. One of the CB point's pins is already connected to the ground, and the other pin, which goes to the ignition coil, needs to be disconnected. This pin is marked with a "X" on the diagram, and it tells the electronic ignition what input to use.

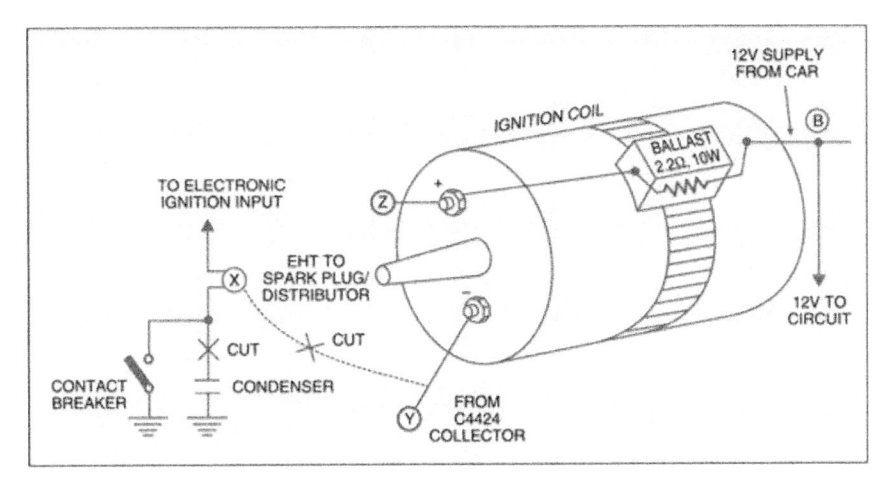

Fig. 2: Wiring diagram of ignition coil

Note: Don't take out the CB-point screw because you don't need to change the switch gap, which will change how the engine is tuned. You just have to take the capacitor off the distributor and cut the wire that goes to the negative point on the ignition coil. During the wiring process, you must be careful not to touch the pins on the ignition coil. At point "B," where the wiring of the car battery is already set up, the circuit gets its power (positive). Make sure you don't short out the wires because that could start a fire.

Circuit operation

Ground (negative) can be taken from any screw on the car's chassis for the ignition system. Put the circuit together on a general-use PCB and put it in a metal box with the right insulation. You can also use a switch box made of plastic. But care must be taken to keep the transistor C4424's heat sink from touching the box because it will be very hot. As we've already talked about, the distributor has a condenser. You might need to open the two clips to take off the distributor cap. Then cut the condenser wire on one side. Take your time. You can take the condenser out of the distributor if you want to. After covering the other end of the condenser with insulation tape, connect one end to ground and leave the other end open. In case the electronic ignition fails, you can just switch back to the old system to start the car. Don't get rid of the condenser.

Note:

Don't short out any wire that goes to ground. The car battery can put out up to 100A of current, which is dangerous. So, you have to be careful when you work with the wiring in the ignition circuit. For the circuit to work well, use an ignition coil from Autolec/Lucas, spark plugs from Champion, and any type of ignition cable that doesn't have a resistor (when you measure the resistance of the ignition cable, it should be nearly zero). Don't use any other transistor to replace C4424, because only C4424 can handle the back-emf from the ignition coil. You don't need to replace the spark plugs. Also, never take the distributor or its alignment nut out of the engine. When the old system is replaced with electronic ignition, the timing of the spark stays the same.

97. Versatile CMOS/ TTL Logic and Clock Probe

You will need a probe that can test the logic level or determine whether or not clock activity is present in order to perform fault diagnosis on any logic circuit. This circuit can be used to test CMOS and TTL logic circuits for logic states and also for the existence of clock activity at any point in the logic circuit, ranging from a few hertz to more than 10 MHz. The circuit is shown here. Through the use of alligator clips, power for the probe circuit is drawn from the circuit that is being tested. A window detector is provided by an LM319 dual-comparator that is connected in the circuit. When switch S1 is in the TTL position, the non-inverting pin of comparator N1 is biased to approximately 2V, and when switch S1 is in the CMOS position, it is biased to 80 percent of Vcc. Only in the event that the logic input at the probe tip has a value that is greater than the biasing voltage will the output of N1 go low. As a consequence of this, the red LED will light up to signify a logic 1 condition at the probe tip.

Comparatively, the inverting pin of comparator N2 is biased at roughly 0.8V (when switch S1 is in the TTL position), which is equivalent to 20% of Vcc (in CMOS position of switch S1). It is only when the input voltage at the probe tip is lower than the biasing voltage that the output of the device will drop low and cause the green LED to light up, indicating that the logic state is 0.

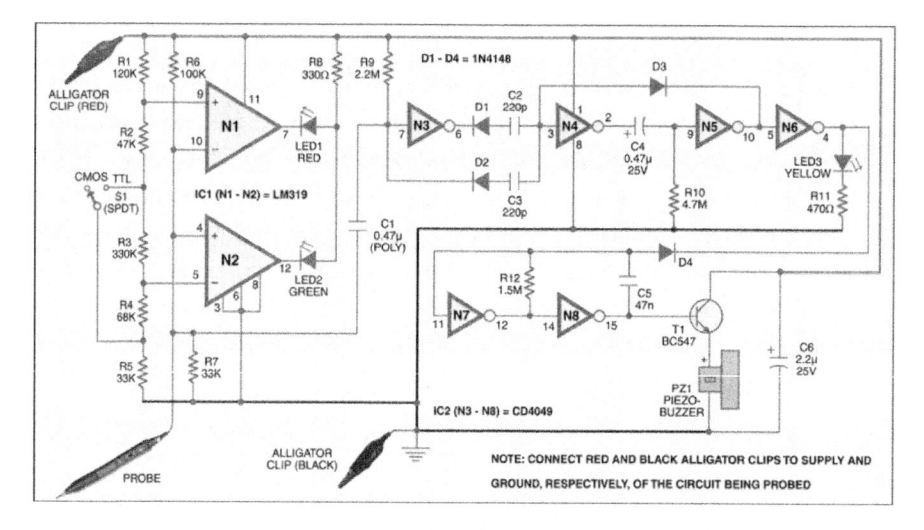

To enable the transmission of AC and clock signals, the probe tip is additionally linked, through capacitor C1, to the input of the CD4049 (N3). It merely performs the function of a buffer and links only the high-to-low going signals that are present at the input and output of the gate to the input of the subsequent gate N4. Additionally, the output of gate N4 is coupled to gate N5, which is configured to operate as a monostable. Positive feedback from the output of gate N5 to the input of gate N4 ensures that future clock pulses at the input of N4 will have no effect unless the capacitor C4 (with a value of 0.47 microfarads) empties sufficiently through the use of a resistor with a value of 4.7 megaohms.

A brief duration of illumination will be provided by a yellow LED that is driven by Gate N6. This LED serves the purpose of indicating the presence of oscillatory input at the probe tip. The output of gate N6 is then utilized in a subsequent step to either disable or activate the oscillator that is formed by gates N7 and N8. It causes the buzzer to beep momentarily during the mono period to indicate that oscillatory input is being received at the probe tip. As a result, we have an audio-visual indicator available at the probe tip during the clock or oscillatory input.

98. PC-based Oscilloscope Using Arduino

Electronics hobbyists and professionals need oscilloscopes to make sure that their designs will work as planned. PC-based oscilloscopes are better than oscilloscopes that stand on their own because they are small, cheap, and can-do offline analysis.

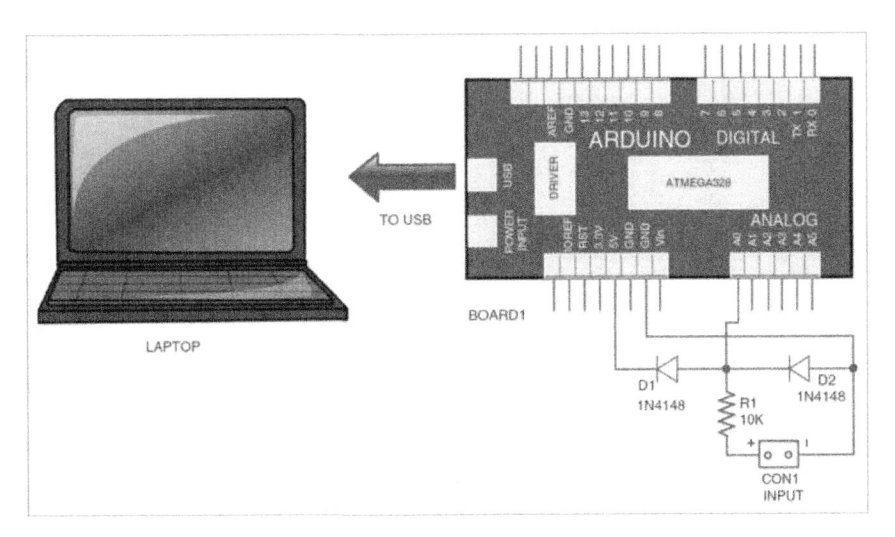

Fig. 1: Circuit of the PC-based oscilloscope using Arduino

Here, we'll show you how to make your own oscilloscope for a very low price by using your PC and an Arduino board as the signal-grabbing hardware. This oscilloscope can pick up signals with a frequency of up to 5kHz. The heart of the oscilloscope is the Arduino board, which reads the values from its built-in analog-to-digital converter (ADC) and sends them to the PC through the USB port. Here is an Arduino sketch that you can compile and then load straight onto the Arduino. You also need to install an executable file or application on your Windows PC. This application acts as the front end and shows waveforms of the input signals on your computer screen.

The Arduino board has an Atmel AVR microcontroller, which, depending on the type of board, can have 8, 16, or 32 bits. You can use any version of the Arduino for this project. The ADC is built into the AVR microcontroller. For the project, we use pin A0 to get the signal from the input. The Arduino's UART-USB converter sends the captured input signal to the PC's UART. When the Arduino connects to the PC, Windows makes a virtual COM port. A Windows app made with NI LabWindows opens the virtual COM port and starts plotting signals visually with Graph libraries. The UART's baud rate limits how fast the oscilloscope can sample. The Arduino sketch is written so that it uses ISR to read the ADC. The UART baud rate is set to 115200, which sends data every 85s. This means that the actual sampling rate is 12kSa/s.

99. Turn Your Old Inverter into an Emergency Power System

An inverter that has been converted into an emergency power system. This system only activates when the electricity from the mains supply is interrupted; more critically, it does not activate when power from the mains supply is available.

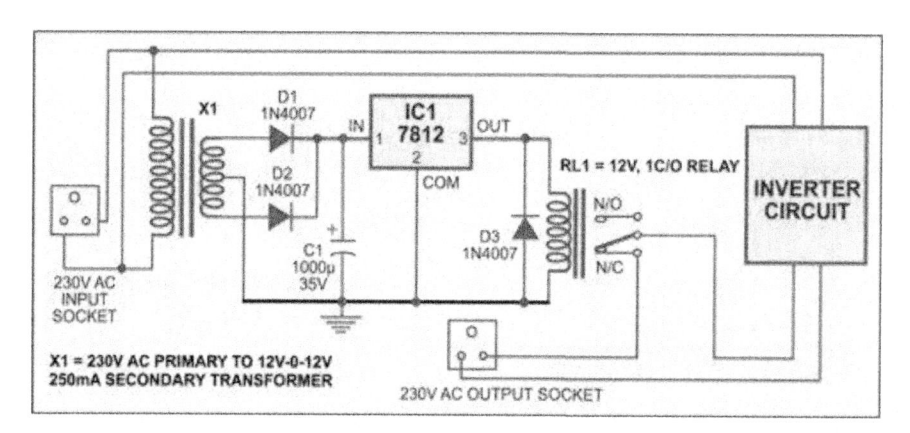

Fig. 1: Auto on/off circuit for inverter

Can you provide me with a configuration (emergency lighting system) in which only a small number of my lights will turn on when the power from the mains is cut off, and more significantly, in which they will not turn on when the supply from the mains is available? This is the question that a reader who wanted to install an emergency lighting system in his factory posed to us when he contacted us. The first thing that came to mind was, "Why doesn't he just use an inverter?" Then, however, we came to the conclusion that a typical inverter would cause the emergency light to turn on even when there is electricity coming from the mains, which is something the reader did not require. We had to act quickly, so we tossed an old inverter and made a few alterations. We connected our specialized inverter to the line that supplied power to the emergency light he had. We were able to modify the inverter in little than two hours, and as a result, the reader was more satisfied. Now that we've made some adjustments, we'd want to go ahead and share them with you in the event that it makes our readers even happier. This straightforward circuit can be implemented as an auto-on/auto-off switch in any residential inverter, provided that the inverter's output load is known. Utilize a relay that has a higher current rating if you need to handle a heavier load.

100. DC Motor Speed Controller

This is a straightforward example of a circuit for a variable speed controller for a DC electric motor, which may be set up to regulate the pace at which windshield wipers go back and forth. In addition to a few additional discrete components, the circuit includes a timer designated as NE555 (IC1), a motor driver transistor designated as BD239 (T1), and a switching transistor designated as BD249 (T2). The negative terminal of the power supply is linked to the ground so that it can be used in automobiles. This configuration is optimized for their use.

Motor speed controller circuit

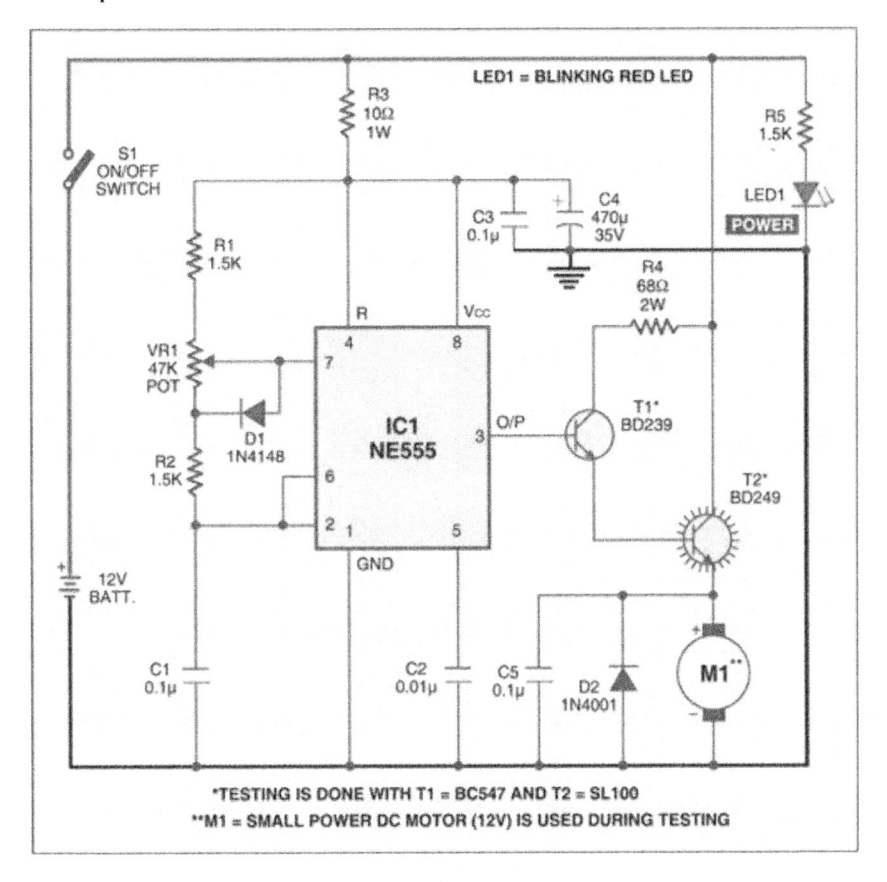

As a low-frequency, free-running astable multivibrator with pulse-width modulation (PWM) dc motor speed controller output, IC1 has been set up in this configuration. The frequency of the oscillations is determined by the R-C components such as R1, VR1, R2, and C1. The potentiometer designated as VR1 serves as the motor's primary speed regulator. Decoupling the oscillator's direct current supply is accomplished via the components R3, C3, and C4. The visual indicator that the electricity is "on" is provided by a red LED (LED1) of the blinking variety.

When the wiper arm of potmeter VR1 is in the top position, a pulse train is produced at the output of IC1 that has long negative and short positive pulse lengths. This pulse train is caused by capacitor C1 charging through resistors R1, R2, and D1. Because of this, the speed of the motor is low. C1 charges through R1, R2, and VR1 when the wiper arm of VR1 is in the bottom position, and it discharges through R2 when VR1 is in the bottom position. The resulting pulse train features prolonged positive pulse widths and condensed negative pulse widths. The motor is currently rotating at a rapid rate. The output from IC1's pin 3 is routed into transistor T1, which then powers the DC motor through high power switching transistor T2 at the speed that the user specifies. The base current of transistor T2 is controlled by resistor R4, which has a limited capacity. The back e.m.f. that is created by the rotation of the DC motor is constrained by the diode D2, which is linked in antiparallel with the motor. Capacitor C5 inhibits parasitic oscillations.

101. AC-Powered Led Lamps Without rectifiers

When connected to an AC mains power supply, most LED lamps need a rectifier. Most of the time, the LED lights also need to be electrically separated from the mains. But rectifiers make noise when they switch and add to the price. Here is a simple circuit for an AC-powered LED lamp that doesn't need rectifiers. Low-power mains transformers come with one or more secondary voltages, such as 2.5 Vrms (3.5 Vpeak), 3 Vrms (4.2 Vpeak), 3.3 Vrms (4.7 Vpeak), 4.5 Vrms (6.4 Vpeak), 5 Vrms (7.1 Vpeak), 6 Vrms (8.5 Vpeak), and 6.3 Vrms (8.9 Vpeak), and power ratings from 1.3 VA to 12 VA. The primary current of these transformers is usually between 15 and 80 mA, which is low enough that resistors can be used to limit it without making a lot of heat. Most of these transformers are easy to use because they are mounted on PCBs. On the other hand, LED lamps with a power rating of 1 W to 10 W are enough to light your workstation and a small room. Depending on the technology and color, most LEDs today have a forward drop voltage of between 1.6 V and 3.7 V. Most of the time, these

LEDs need between 20 mA and 100 mA of current to work, and sometimes they need more. The LEDs' specs are very close to those of the small 2.5Vrms-6.3Vrms transformers.

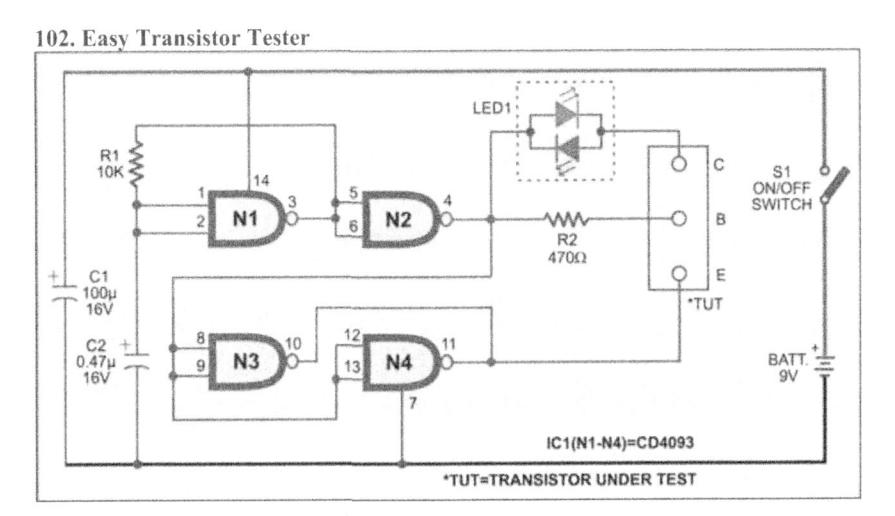

102. Easy Transistor Tester

Before you solder a specific transistor, use this straightforward transistor tester circuit to determine whether or not it is functioning properly. You are also adept at recognizing npn and pnp kinds with ease. The LED on the tester can indicate both the pin-outs of the component and whether or not the transistors are functioning properly. The circuit makes use of the characteristics of the gates of the low power integrated circuit 4093. It is a NAND integrated circuit that has four gates and can be constructed in a variety of different ways. With a resistor R1 and a capacitor C2, this initial gate, N1, is constructed to function as a straightforward oscillator. The inverters and buffers are constructed out of the remaining gates. The output of gate N1 (pin 3) is supplied to gate N2 (pin 4), which is then used to drive the green half of the bicolor LED1 that is being displayed.

The collector current for the transistor that is being tested can be read off of the cathode of the green LED1 (TUT). The signal is inverted by applying a biasing voltage, which comes from the output of gate N2 to the base of the TUT. The outputs of gates N3 and N4 are connected to the emitter of the TUT through a connection.

103. Door Guard

The operational amplifier A741 and a light-dependent resistor are what this door guard is comprised of (LDR). The operational amplifier A741 is put to use as a voltage comparator that is very sensitive. The non-inverting terminal (pin 3) of the A741 is supplied with reference voltage via the preset VR1 terminal. Both the light-detecting resistor LDR1 and the resistor R1 are connected to the inverting pin 2 of IC1. Both the LED1 and the LDR1 are placed on the exterior of the building, on opposing sides of the entryway, so that the light from the LED1 illuminates the LDR1.

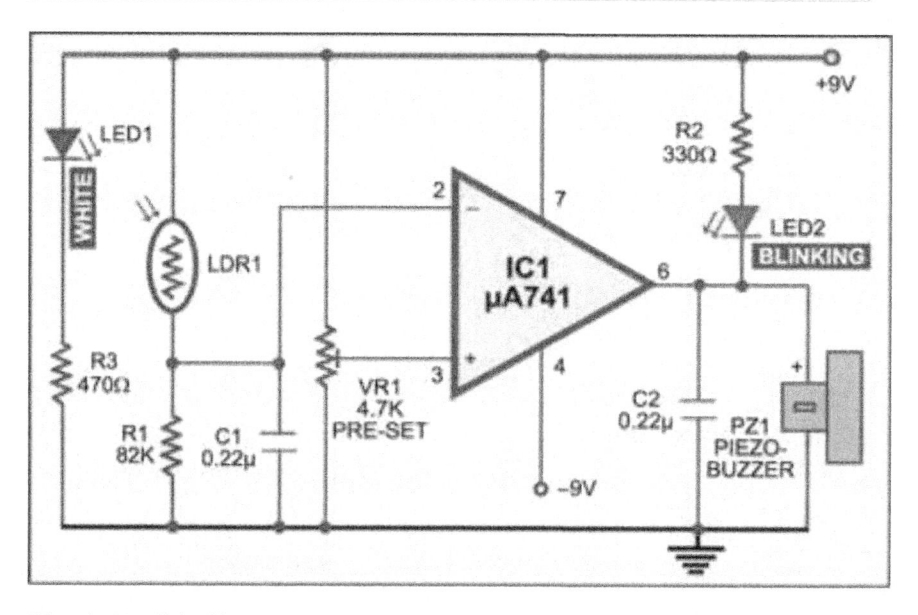

When the LED light shines on LDR1, its resistance drops in comparison to R1, which causes pin 2 of IC1 to become high as a result. As a direct result of this, the output pin 6 of IC1 drops low, which causes LED2 to flash while simultaneously silencing the piezo buzzer PZ1. This denotes that the gate has been shut securely. When the gate is opened by a person to let them in or out, the light from LED1 that is falling on LDR 1 is blocked, which causes the resistance to go up to an extremely high level. As a direct consequence of this, the output on pin 2 of IC1 goes low, while the output on pin 6 of IC1 goes high. During the time when the piezo buzzer is active, LED2 will not blink. This suggests that the gate can now be entered.

104. Low-cost Night Lamp

This straightforward and inexpensive night light turns on by itself as it gets dark and turns off when the sun comes up. It has a battery charging circuit that offers safety against overcharging as well as deep discharging the battery. The circuit can be broken down into three distinct parts: the power supply, the control for overcharging and undercharging, and the lamp control.

The power supply section consists of the bridge rectifier BR1, regulator IC3, and regulator IC4 as well as the transformer X1. The mains voltage of 230 volts AC is reduced to 12 volts AC by transformer X1, which is then rectified by bridge rectifier BR1 and filtered by capacitor C4. Regulators 7809 and 7812 are responsible for bringing the output of the rectifier down to 9V and 12V, respectively. While 9V is utilized to charge the battery and power the lamp control circuit, 12V comes from the supply and is used for the operation of the relay RL1. The overcharging/undercharging control section is constructed using IC2 and a few discrete components as its primary building blocks.

A low-cost night lamp circuit

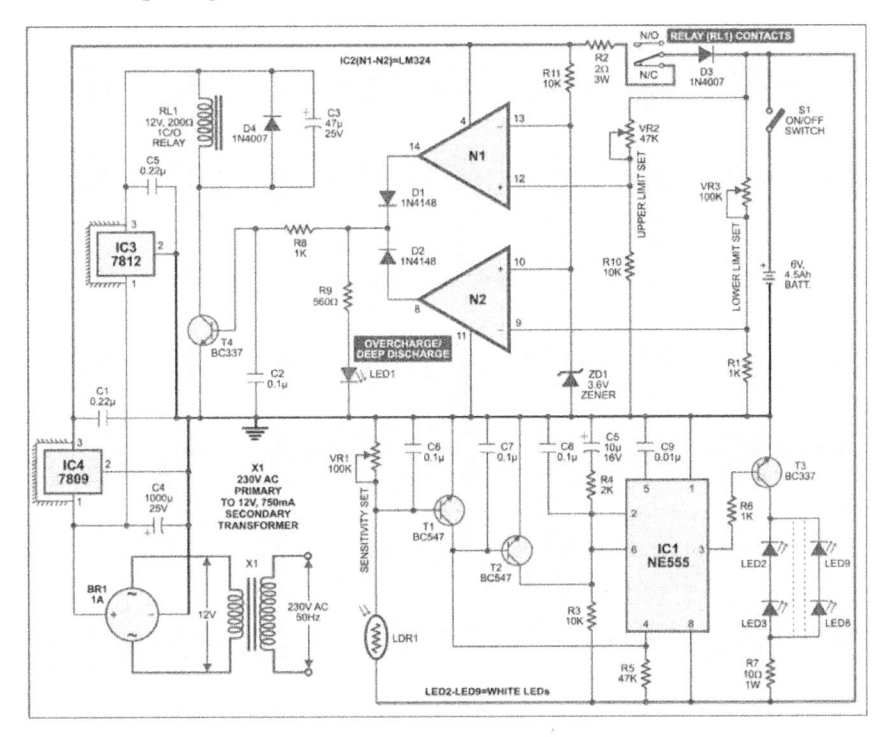

Circuit operation

Even when there is power from the mains supply and the battery voltage is between 5.5V and 6.5V, the relay will not become energized. Under these circumstances, the battery is charged by relay RL1's normally-closed contacts, which are denoted with a N/C designation. When the voltage of the battery drops below 5.5 volts, the voltage at the inverting pin 9 of the circuit drops below 3.6 volts. As a consequence of this, the output of the operational amplifier N2 climbs high, and the transistor T4 conducts, which activates the relay RL1 to stop the battery from going into a deep discharge. In a similar manner, if the voltage of the battery rises above 6.5V, then the voltage at the non-inverting pin 12 of the battery rises above 3.6V. As a consequence of this, the output of the operational amplifier N1 swings high, and the transistor T4 conducts, which activates the relay RL1 to stop the battery from being overcharged. When one of these conditions is present—overcharging or deep discharging—LED1 illuminates to alert the user.

105. Briefcase Alarm

This compact alarm system alerts you with an audible warning if someone approaches your briefcase with the intention of taking your valuables out of it. It is a battery-operated device that may be tucked away discretely in a nook or cranny inside the briefcase. The circuit has a low number of components and is straightforward to put together.

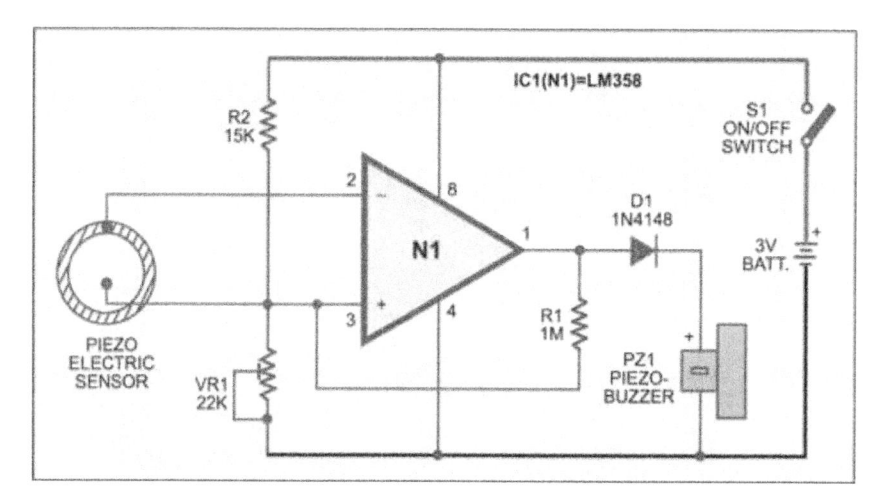

A piezo-element that is typically found in buzzers is the primary component of the circuit. In reaction to either pressure or vibration, the piezo-element can be used to generate electric signals. A couple of tens of nanofarads constitute its capacitance (nF). When a voltage is supplied, the piezo-element rapidly charges and stores the charge until it is mechanically disturbed. This behavior is analogous to that of a capacitor. Any kind of mechanical vibration will cause the charge on the piezoelectric material to reorganize and, as a result, energy will be released. The voltage that is generated across the piezo-sensor is what causes the IC LM358 to be triggered. This component is a low-power transducer amplifier. The integrated circuit LM358 requires 9 volts to function and features two independent high-gain op-amps that together have a massive DC gain of 100 dB.

In this case, the integrated circuit is set up to function as an inverted Schmitt trigger so that it can transform incoming signals into a shaped output waveform. A threshold voltage is established by presetting VR1, which occurs at the non-inverting input threshold of IC1. While the piezo-sensor is linked between the inverting and non-inverting pins, one end of feedback resistor R1 is connected to the output of the IC and the other end is attached to the non-inverting input (pin 3). The output of IC1 is low when it is in the standby mode, which occurs when the signal from the piezo-sensor is low. The feedback resistor R1 raises the non-inverting input voltage to a level higher than the upper threshold voltage (UTV), which is around 1.8 volts.

The charge that is stored in the piezo-sensor is released when it is briefly tapped, causing the inverting input voltage to rise to a level that is higher than the positive input and causing the output to drop. When anything like this takes place, the voltage at the positive input drops through VR1 and reaches the lower threshold voltage (LTV). Due to the absence of the input signal, the voltage that is present at the inverting input at this same instant is quite low. This results in a rise in the output, as well as a beep from the buzzer. The output of the IC will decrease whenever the input value is greater than the UTV, but it will increase whenever the input value falls below the LTV. The hysteresis of the Schimdt trigger is what differentiates the LTV and the UTV in terms of their functionality. The buzzer will continue to sound for a few seconds after being touched by the piezo-sensor, even if the hand is removed from the sensor. This is due to the fact that once the output swings high, even a little reduction in voltage at the negative input has no influence on the circuit. When it has been started, the change in status cannot be easily undone. Construct the circuit on a general-use printed circuit board that is as little as feasible, and then house it in a compact housing. The diameter of the piezo-sensor needs to be between 10 and 15 millimeters at most. You can attach it to the main unit by sticking a tiny wire that is insulated and some glue to the underside of the handle of the briefcase. A battery with 3 V is all that is needed to power the circuit.

106. Low-Cost LPG Leakage Detector

The circuit for an LPG leakage detector is easily accessible on the market; nevertheless, it is rather pricey and is often based on a microcontroller (MCU). This article will provide you with a simple, low-cost, and straightforward circuit for an LPG detector that you may build. The primary purpose of the circuit is to identify any location where there is an LPG leak.

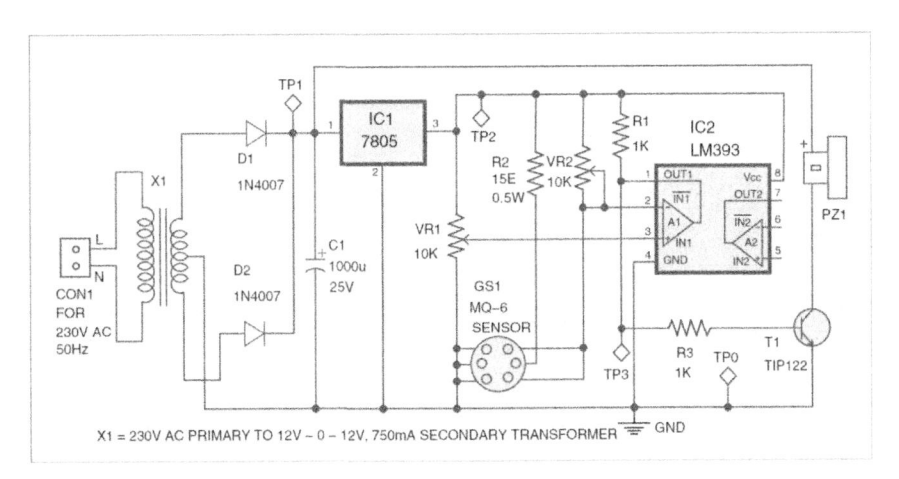

It is constructed of a step-down transformer known as X1, two rectifier diodes known as 1N4007 (D1 and D2), a 1000F capacitor known as C1, a dual comparator known as LM393 (IC2), a darlington transistor known as TIP122 (T2), a 12V high-gain siren/buzzer known as PZ1, and a few additional components. In order to maintain a constant 5V DC output, which is delivered to the circuit, the mains supply is stepped down by transformer X1, rectified by a full-wave rectifier consisting of diodes D1 and D2, filtered by capacitor C1, and then fed to regulator 7805 (IC1).

The dual comparator IC LM393 is the most important component of the circuit (IC2). It is utilized for the purpose of comparing two distinct voltages, specifically the reference voltage and the output voltage of the MQ-6 gas sensor. The potmeter VR1 is used to establish the reference voltage at the non-inverting pin 3 of IC2 in order to modify the voltage levels according to the sensitivity requirements. The output voltage from the LPG sensor (MQ-6) is connected to the inverting pin 2 of

IC2. In the event when the reference voltage at pin 3 of IC2 is lower than the sensor voltage at pin 2 of IC2, the output will drop, indicating that there is no LPG leakage. As long as there is a low output, T1 will remain turned off, and there will be no passage of current through the buzzer. As a result, the buzzer will not sound and will remain in silence mode. In the event that the reference voltage is higher than the sensor voltage, the output will go high, which indicates that there is a loss of LPG. The high output activates transistor T1, and a loud buzzer begins to sound in order to draw the attention of those who are nearby. This circuit, which combines low-cost components and an interactive manner to alter different sensitivity levels based on client needs with the use of a potmeter called VR1, makes it very simple to locate gas leakages. The low-cost components used in this circuit include: After the circuit has been assembled on a PCB, it should be encased in a box that has a hole in it so that the gas may enter. Within a distance of one meter, position the unit so that it is adjacent to the LPG cylinder or the gas burner. Adjusting the sensitivity of the sensor can be done by adjusting the setting VR1.

Before testing the project, you need to make sure the voltages are where they should be according to the test points table. Now, spray the gas from the bottle toward the MQ-6 gas sensor (as illustrated on the left side of the author's prototype), and measure the voltage at TP3; it should be high. If you do not have a gas bottle with gas in it, you can position the LPG leakage detector close to the gas stove burner and turn it on for a few seconds without it igniting. This will detect any leaks. After that, flip the switch to turn off the burner, then adjust VR1 until the buzzer sounds.

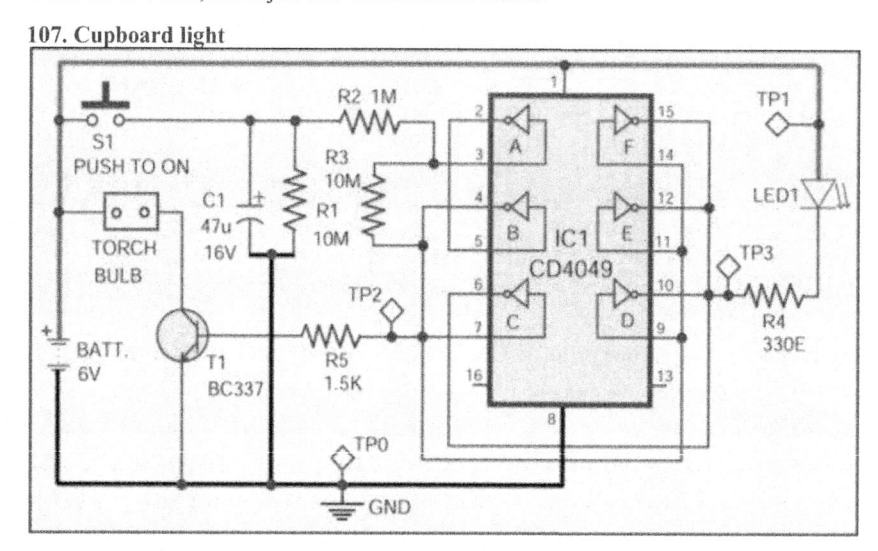

107. Cupboard light

The following is an explanation of a simple circuit that can provide momentary illumination to your cabinet or other similar typically dark locations where a connection to the mains electricity supply is either not possible or not beneficial. The circuit is nothing more than a light that is powered by a battery and has a built-in auto-off feature.

Circuit and working

The lighting system for the cupboard is depicted in Figure 1. The hex inverter buffer CD4049 is the essential component in its design (IC1). The Schmitt trigger is made up of the inverter's "A" and "B," as well as the resistors "R2" and "R3." In order to get a higher current-sinking capacity, the remaining four inverters, denoted by the letters C through E, are connected in parallel. When the switch S1 is held down for just a second, the voltage across capacitor C1 quickly increases to 6 V. This rise in voltage occurs nearly instantly. When the Schmitt trigger output becomes high, it causes the output of four inverters that are connected in parallel to go low, which turns on LED1. At the same time, the flashlight bulb is illuminating the room thanks to npn transistor T1 (BC337).

In conclusion, when the switch S1 is activated, both LED1 and the torch bulb become lit. Through the resistance of R1, the capacitor C1 gradually discharges. The LED and torch bulb will automatically turn off the minute the voltage across it drops to a low condition. With the values that have been provided for the components, the lamp will remain "on" for around two minutes. A rough estimate of the length of time can be found by using the formula (R1C1)/4.

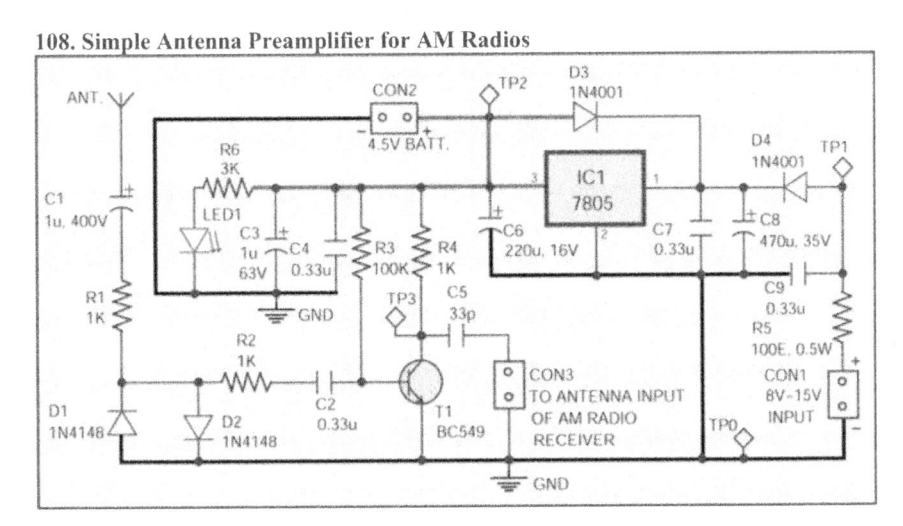

108. Simple Antenna Preamplifier for AM Radios

AM radios are still in use today for the purpose of receiving communications from quite far away (DX reception). However, AM radios often require lengthy external antennae that are between 10 and 30 meters in length. If you reside in a town or a village, doing so is simple, but in more urban settings, it is not always possible. A signal received from an antenna with a length between 2 and 5 meters may be relatively weak, but installing and maintaining such an antenna is simple. If you have no choice but to utilize a somewhat short antenna, you can improve the AM reception by adding the proposed low-noise antenna preamplifier for AM signals. This is because the preamplifier will reduce the amount of noise produced by the antenna. Long-wave (LW) and medium-wave (MW) spectrums are both covered by its 150-1700 kHz frequency operation range that it offers.

Circuit working

The circuit of the antenna preamplifier used in AM radios is seen in Figure. The capacitor C1, the resistor R1, and the diodes D1 and D2 are what provide protection for the input of the preamplifier. In order to be suitable for this application, the Transistor T1 must have a gain that is either moderate or high and a low noise level. Transistors such as the BC550, BC547, and others like them will perform admirably in the LW and MW frequencies. It is essential to maintain a collector current of at least 2 mA at all times. The majority of the time, a current of 2-5 mA is sufficient, and this value can be altered using resistor R3. It's possible that increasing the current to 10 mA will make the volume better, however it will also shorten the battery's life. The preamplifier and the AM radio are physically separated by capacitor C2. The regulator 7805 provides the power supply, which is +5V. (IC1). Connector CON1 is able to accept either an AC or DC power supply that falls within the range of 8-15V.

The operation of the circuit is straightforward. In the low- and medium-frequency ranges, the gain offered by the preamplifier circuit is greater than 10. In order to achieve superior reception, the amplified signals are sent through the AM radio receiver's antenna input. Depending on how well the practical implementation is carried out, the circuit may also function for the short-wave (SW) range.

If you want to see a noticeable improvement in reception, the antenna needs to be at least 2 meters long, and it should be positioned either outside the building or inside the building at least half a meter away from the walls. If you store the antenna close to walls or installations of electrical cables, there may be reception problems, and the antenna may pick up an excessive amount of background noise. Disconnecting the antenna from the preamplifier is a necessary precaution to take if there is a chance of electrical storms in the vicinity.

The preamplifier has the capability of operating off of a 4.5V dry battery that is connected to connection CON2. Because the mains power source could potentially produce considerable noise, it is recommended that you make use of batteries with voltages ranging from 3V to 15V.

109. Multifunction Power Supply

These days, many embedded systems need a +5V power supply with special features like power-fail detection, zero-crossing signals for mains power supply, and the ability to keep built-in batteries charged. Here is the circuit for a power supply for embedded systems like this one. It has +5V, a charger for the battery, zero-crossing signals, and a power-fail signal. The circuit is made up of common, cheap parts. To get started right away, you just need to make a few simple adjustments with potmeter POT1. The circuit in Fig. 1 is made up of a step-down transformer (X1), two bridge rectifiers (BR1 and BR2), an adjustable voltage regulator (IC1), a 5V voltage regulator (IC2), a hex inverter Schmitt trigger (IC3), and a few discrete components.

Connector CON1 is given 230V AC, 50Hz power from the mains. Fuse F1 keeps the input from getting too much power. Some of the noise from the mains power supply is blocked by resistor R1 and capacitor C1. Power transformer X1 has a secondary that is 12V and 2.5A. The standard 5V regulator IC2 is used to power CON3 with 5V and 1A. The 2.5A rectifying bridge is BR1. The main filtering cap (C4) should have a value of at least 4700 F. IC2 gives out a steady +5V with a current of up to 1A. If you need more current, you can use regulators like the 78T05 (3A, 5V) or the 78S05 (2A, 5V). In real life, it is best to keep the load on 78XX to between 0.7 and 0.8A. The voltage from CON4's unregulated output ranges from 10V to 18V, depending on the transformer used and how much current is drawn from the power supply. This voltage isn't controlled, but it can be used for other things. The output is protected by the fuse F2.

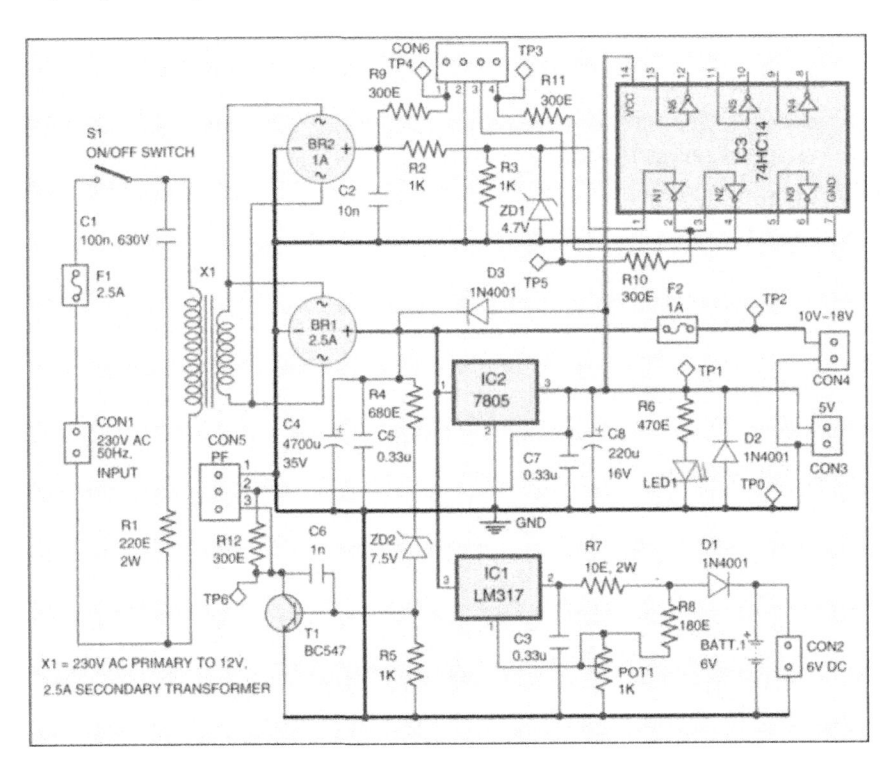

The power supply has a battery charger with a regulator IC1 that can be changed. Some parts of the system are powered by a 6V rechargeable battery when the mains power supply isn't working right. The potmeter POT1 is used to change the maximum voltage level across the rechargeable battery. Resistor R7 limits the amount of charging current that can flow. Regulator IC1 has an output voltage range of +1.25V to +8.2V, which can be changed with potmeter POT1. The regulators IC1 and IC2 are protected by diodes D1, D2, and D3. Bridge rectifier BR2 is only used to send signals at zero-crossing points. The value of capacitor C2 should be low. Its purpose is to cut only the very high frequency, not to filter the power from the mains. IC3 and the parameters of the pulses that are made can change the values of R2 and R3.

When the voltage at test point TP2 drops below about 8V, CON5 sends a power-fail signal. Transistor T1 stops doing its job, and the control unit knows this because pin 3 of CON5 goes high. The power-fail signal's threshold voltage is controlled by zener diode ZD2 (7.5V) and resistors R4 and R5. Switching transistors are best for T1, but most npn silicon transistors with a high gain will also work. Pins 2, 3, and 4 of connector CON6 send out different signals at a rate that is twice as fast as the mains power supply (100 Hz). These signals are active near the zero crossings of the mains power supply and can be used for many things, such as:

- ❖ They can be used to measure the frequency of the mains power supply by the control unit.
- ❖ You can use these to make the control unit work at the same time as the zero crossings of the mains power supply.
- ❖ The strength of signal TP4 is related to the voltage on the transformer's secondary side. It can be measured by the control unit, which can then figure out what X1's secondary voltage is.
- ❖ The outputs TP5 and TP3 work with either TTL or CMOS, depending on IC3. IC3 can be either CMOS or TTL. Some examples are 74HC14, 74HCT14, 74LS14, etc. It should have a Schmitt trigger built in.

110. Micro-Power Flasher

The intruders that are attempting to break into your home can be confused by using this micro-power flasher that you have. The device gives the idea that there are people present within the house at all times of the day and night by continuously emitting flashing light from its front-facing LEDs. The circuit may be powered by four AA-size batteries rated at 1.5V continuously for a significant amount of time.

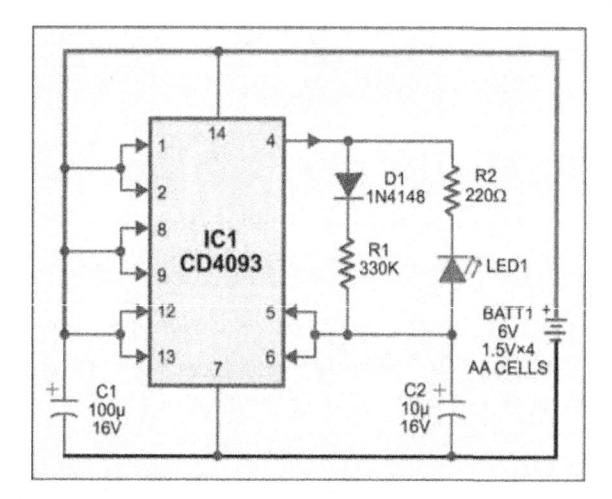

Both the human eye and the human brain have a rapid response to fast shifts in the brightness of the light, particularly flashing light. For full brightness to be perceived, at least 10 milliseconds of illumination from a flashlight is required. The capacity of the brain or retina known as "persistence of vision" maintains the memory of the vision for a period of twenty milliseconds (ms), after which it gradually disappears. The human eye is only able to distinguish individual flashes of the flashing light if there is more than 20 milliseconds of time in between each flash. A flashing light with a period of repetition that ranges from half a second to five seconds is particularly appealing to it. A high-intensity red LED is utilized in conjunction with a low-power CMOS integrated circuit called CD4093 to produce crisp flashes. The CD4093 integrated circuit is a quad NAND gate, however only one of its gates functions as an oscillator here. To prevent floating and the subsequent harm caused by statics, all of the unused inputs have been tied to the positive rail in order to maintain their logic state as "1."

LED1 is wired in such a way that it draws current from the storage capacitor C2, which is attached to it. Since LED1 is not directly driven from the output of the IC, this helps to reduce the amount of power that is drawn from the battery. The amount of current that is drawn from the battery to charge capacitor C2 is determined by the value of the resistor R1. The figure of 330 kilo-ohms indicates that the current drain will be around 18 micro-amperes. (Volts per Resistance = 18 microamperes) The amount of current that can flow through LED1 is controlled by resistor R2.

A 1.5V AA cell with a rating of 1.6 Ah has a typical shelf life of five years provided it is properly stored. A high-quality and low-leakage battery has the potential to power the flasher for close to three years. The circuit can be built on a more compact perforated board, which reduces the associated expenses. Enclose it in a compact case that has a front panel that looks like it came from a burglar alarm. This will ensure that the blinking LED1 will appear to be the indicator that the alarm is active. For a more arresting appearance, use an LED that has a high brightness and is transparent. In the event that the power goes out, the flashlight can also be utilized as a keyhole finder and as a way to navigate through hallways and exit doors.

111. Optical Remote Switch

You are able to control the on/off functionality of any electrical or electronic load by using this optical remote switch. The tiny transmitter unit (shown in Figure 1) and the reception unit (shown in Figure 2) that are used to trigger the relay are standard components of any remote-control system.

An astable multivibrator based on IC 555 is utilized as the optical transmitter in this system. It operates at a frequency of 1 kHz. Two high-intensity red LEDs are utilized, each of which is operated by a pnp transistor, and they are placed at the output of the transmitter.

As soon as the optical transmitter is turned on, the red LEDs begin to illuminate. The 1kHz coding signal is contained within the red light that is emitted. A 9V PP3 battery is required for the optical transmitter to function. Utilize a reflector in your torch so that you can improve its performance and increase its range.

Optical remote switch: transmitter

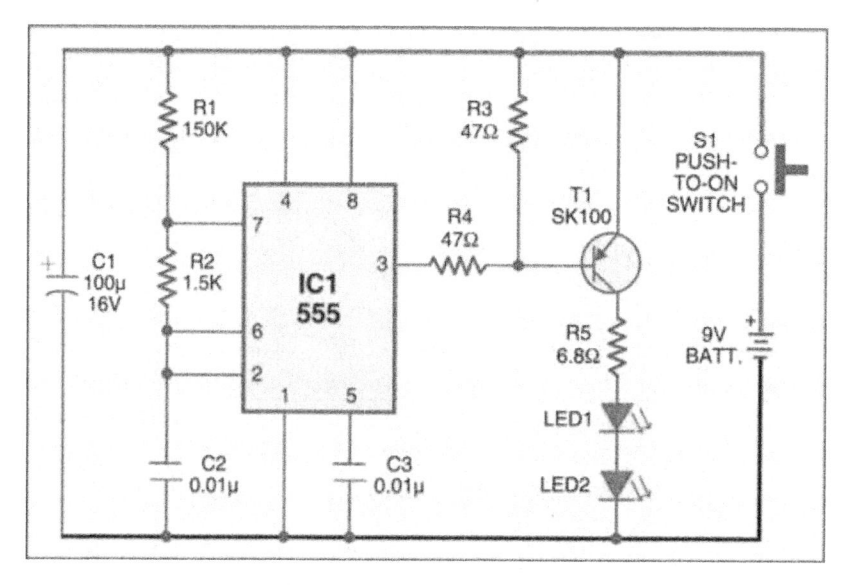

Fig1. Optical remote switch: Transmitter circuit

Optical remote switch: receiver

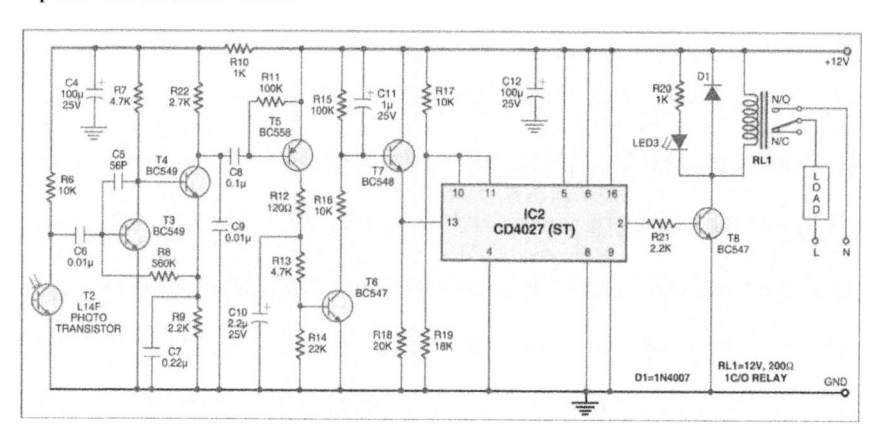

The signal amplifier, switching circuit, flip-flop, and relay are all components that make up the receiver unit. Signals (tones) at an audio frequency of 1 kHz are amplified by the signal amplifier. The switching circuit sends a pulse to the flip-flop circuit, which is designed around a CMOS integrated circuit manufactured by STMicroelectronics. The output of the CD4027 alternates between being high and low in response to each pulse that is entered. A npn transistor amplifies the high-to-low changes of the flip-flop, which then energizes the relay, which, in turn, regulates the load that is applied to the external circuit. For powering the receiver unit, use a controlled 12 V supply. After the 12V DC supply has been connected to the receiver unit, the LEDs that are used in the transmitter should be oriented so that they face the sensor on the receiver (L14F). Maintain a momentary press on the S1 switch on the transmitter. This will cause the relay to become active. The relay can be turned off by repeatedly pressing the S1 switch. This optical remote control performs admirably in environments with moderate to low levels of illumination.

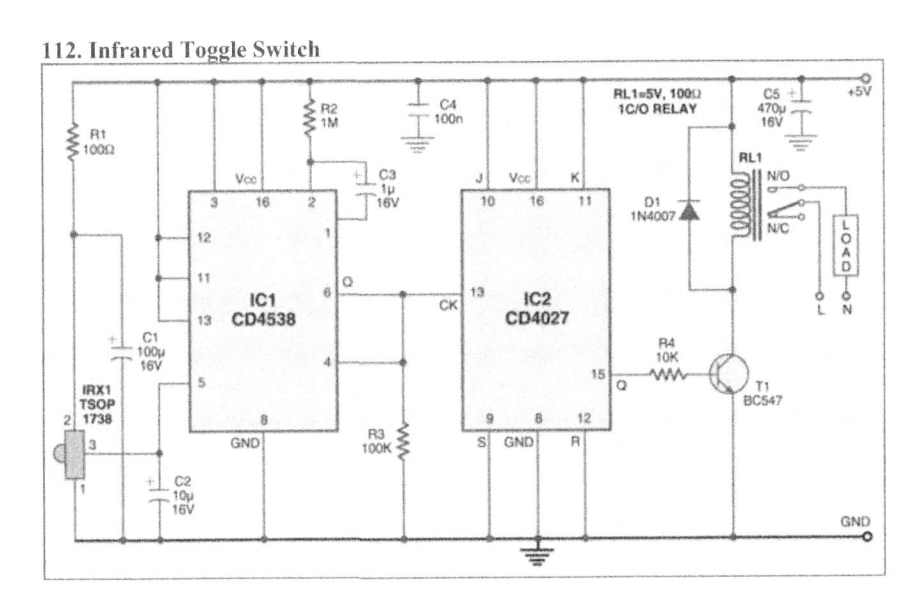

112. Infrared Toggle Switch

Any TV or VCR remote that operates at a frequency of 38 kHz can be programmed to activate this infrared toggle switch. The circuit can be constructed on a veroboard that is quite tiny, and it makes use of components that are not hard to find or expensive.

For the purpose of signal detection, a TSOP1738 (IRX1) integrated infrared receiver module is being utilized here. This minuscule device's power source is decoupled by the components R1 and C1, respectively (receiver module). The output pin 3 of IRX1 is at a high level while it is in the idle state, which is when there are no infrared signals present. This turns off the monostable that is connected around one portion of the famous dual-precision monostable CD4538 (IC1). In order to prevent retriggering while the active state is being used, its Q output (pin 6) is connected to input pin 4. The length of time that monostable oscillations last is determined by the components R2 and C3 (around one second).

Circuit operation

When a genuine infrared pulse train is received by module TSOP1738, the output of the module drops low for a brief period of time. As a direct consequence of this, the monostable (IC1) is instantaneously triggered by the negative-going pulse at input pin 5. Now that the flip-flop (IC2) has been set, its Q output (pin 15) will be high since it is being clocked by the output of IC1. Through the use of the transistor T1, this energizes the relay. Freewheeling diode D1 protects the relay from back e.m.f. while resistor R4 controls the amount of current that flows through the base of transistor T1. As previously discussed, when IC2 receives another pulse train from the remote handset, it is once again timed by IC1, and its output toggles in order to de-energize the relay and disconnect the load from the 230V AC mains. Utilize regulated 5V DC power to supply energy to the device. Noise is reduced thanks to capacitor C4, and a buffering function is provided by capacitor C5.

113. Contactless Telephone Ringer

This fully transistorized, basic circuit built as a contactless telephone ringer gives a signal of an incoming telephone call from a remote location within the building, such as a kitchen or bedroom. In order to put it into action, you will need to wind at least five turns of a short connection insulated wire around one of the wires of a twin telephone cable.

When your phone rings, an alternating current signal with a frequency of approximately 60 hertz is produced in the telephone connection. Resulting in a fluctuating electric field around the telephone wire and the development of a slight induced voltage on the connection cable. When this induced voltage is linked to the base of the transistor T1, it creates a forward bias in the transistor, causing it to conduct current as the ring moves through the telephone cable. As a direct result of this, transistors T2 through T4 conduct as well. As a direct consequence of this, the piezo buzzer emits a sound, and LED1 illuminates, demonstrating that the telephone is now ringing.

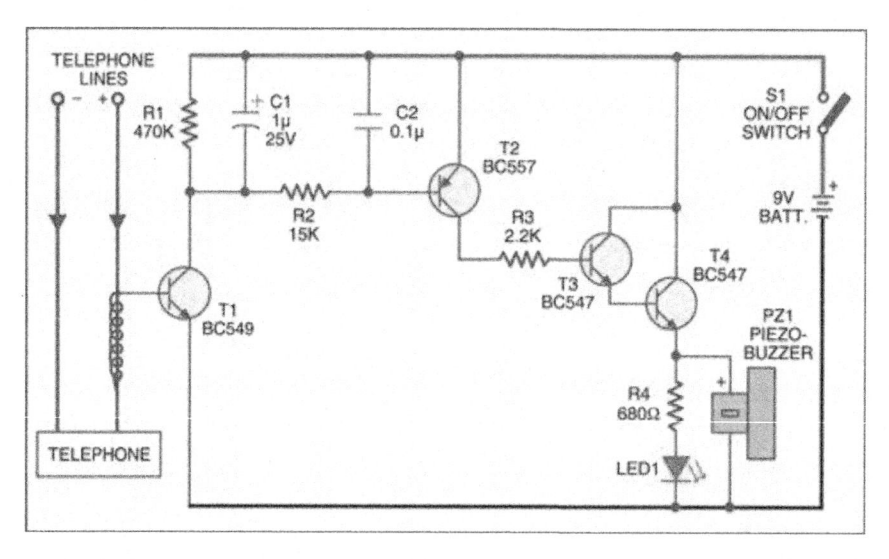

Put together the circuit on a PCB that may be used for a variety of purposes, and then wind an insulated electric cable around one of the telephone wires. A separate connection for the buzzer and LED can be made outside of the area occupied by the telephone receiver in order to provide both audible and visible indication. To power the circuit, you will need a 9-volt battery.

114. Automatic Wash Basin Mirror Lamp Controller

After using the wash basin mirror lamp, it is easy to forget to turn off the light after being in a public space, an auditorium, or even in your own home. When you stand in front of the wash basin mirror lamp, the circuit shown below will turn it on automatically, and it will turn off again when you move away from it, helping you save money on your electric bill.

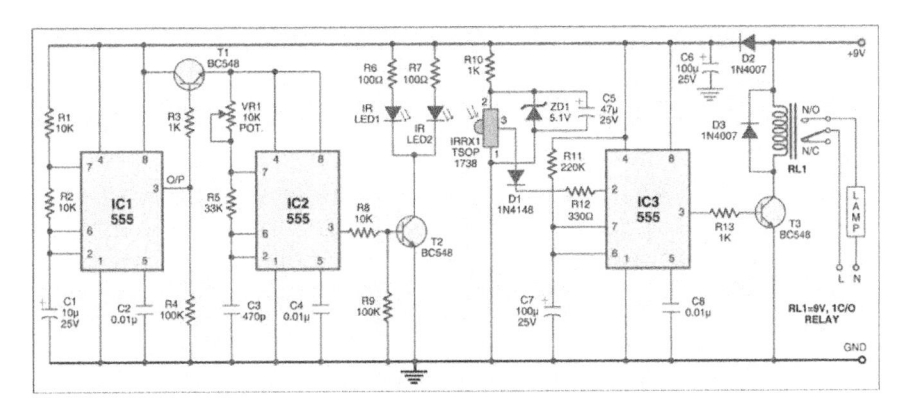

The individual standing in front of the mirror will cause the modulated 38kHz square wave pulses emitted by the infrared LEDs to be reflected back to the sensor that is located behind the mirror. When the receiver sensor detects infrared rays, it energizes the relay for a predetermined delay period, which in turn causes the washbasin lamp to turn on for the duration of that delay. The wiring diagram for the automatic washbasin lamp controller may be found in Figure 1. It is made up of two parts, namely, a receiver and a transmitter for infrared radiation. While IC1 and IC2 serve as the primary building blocks for the transmitter part, IC3 serves as the primary building block for the receiver section.

In the circuit for the transmitter, IC1 performs the role of an astable multivibrator and generates pulsed output at a frequency of 5 Hz. The pulsed output at 5 Hz allows IC2 to function by way of transistor T1. In addition to this, IC2 is set up to function as an astable multivibrator that generates a square wave at 38 kHz. The changeable 10-kilo-ohm potmeter can be used to alter the output frequency of IC2, which can be done at any time. Through driver transistor T2, the output of IC2 is sent to two IR LEDs. The TSOP1738 infrared (IR) sensor, 1N4148 switching diode (D1), IC555 timer (IC3), BC548 relay-driver transistor (T3), and other components make up the receiver circuit. When an IR radiation is detected, the sensor will activate IC3, which is configured to operate as a monostable multivibrator and has a period of around 24 seconds. Simply by adjusting the numbers in R11 and C7, the time period can be altered to take on any value that is required. It is possible to link the output of IC3 to the relay-driver transistor T3.

When the sensor receives modulated 38kHz IR pulses from the IR transmitter, conductivity is established in transistor T3 for a period of 24 seconds. This causes relay RL1 to become energized, and the normally-open (N/O) contact of relay RL1 becomes connected to the mains AC terminal of the mirror light, which then turns on the wash basin mirror lamp for a period of 24 seconds. As a result, the lamp in the wash bowl mirror turns on automatically whenever there is someone standing in front of it. In any other case, it will remain off. The entire circuit is powered by a regulated 9-volt supply of electricity. Put it together on any kind of general-purpose PCB, encase it in an appropriate cabinet, and then mount it behind the mirror of the wash basin.

115. Auto Muting During Telephonic Conversation

Using this simple circuit, you can talk on the phone without being interrupted. As soon as you pick up the phone to talk, the TV, music system, or any other device that might be making noise turns off. When you put the phone back on the cradle, it turns on. This is done with the help of a polarity-guard circuit with a bridge rectifier BR1, an optocoupler MCT2E (IC1), and a timer 555 (IC2), as shown in Fig. 1. Usually, there is about 48V across telephone lines. But it drops to about 9V when you take the phone receiver out of its cradle.

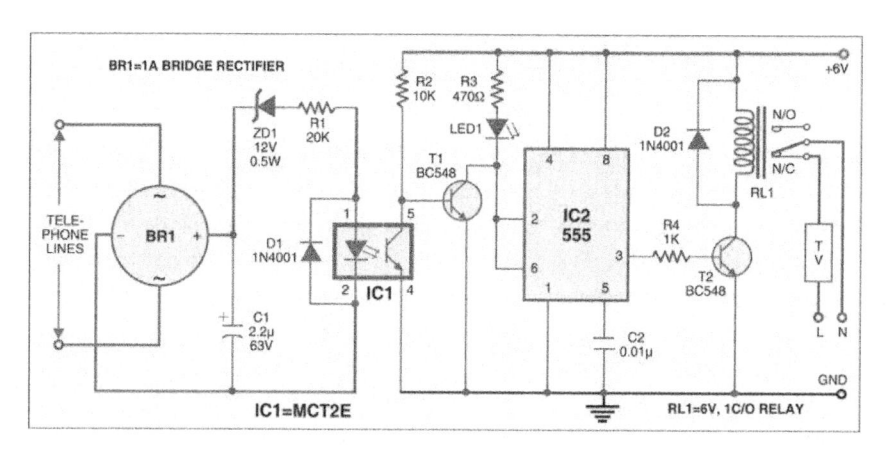

Fig. 1: Auto-muting circuit

When the phone receiver is in the cradle, current flows through the zener diode and the LED inside optocoupler IC1. So, its internal transistor conducts and transistor T1 stops conducting. This makes IC2's pins 2 and 6 high, so its pin 3 output becomes low. Because IC2's output is low, relay-driver transistor T2 doesn't work, and relay RL1 stays off. Because of this, the normally-closed (N/C) contacts of the relay keep the TV or music system on.

When the phone is taken out of the cradle, however, the Zener diode stops working and no current flows through the LED inside optocoupler IC1. So, its internal transistor stops working, and transistor T1 starts working. This makes IC2's pins 2 and 6 low, so its output at pin 3 goes high. Because IC2 is putting out a lot of power, the relay-driver transistor T2 starts to work. So, relay RL1 gets turned on to turn off the power to the TV or music system. At the same time, LED1 lights up to show that someone is talking.

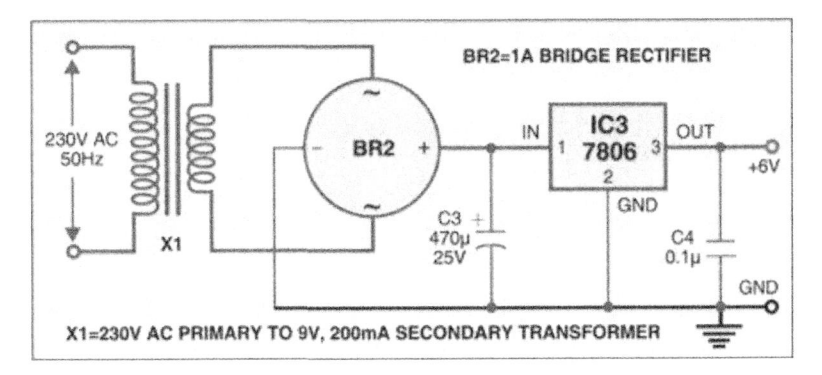

Fig. 2: Power supply for auto-muting circuit

The circuit runs on a 6V power supply that has been regulated. Transformer X1 takes 230V AC mains power and turns it into 9V at 200 mA power. Full-wave bridge rectifier BR2, capacitor C3, and regulator IC3 are used to fix the output of the transformer (7806).

116. Solar-Powered Pedestal Lighting System

The LED lights in this solar-powered pedestal lighting system are powered by the sun. A solar photovoltaic cell first turns solar power into DC electricity, which is then used to charge a storage battery. At night, power LEDs are used to light up the pedestals with the solar energy stored in the battery. The power LEDs on the pedestal are connected to a control unit with a logic circuit that turns them on at night and off during the day. The solar panel that charges the battery is chosen based on how much lighting is needed and how long it needs to be on.

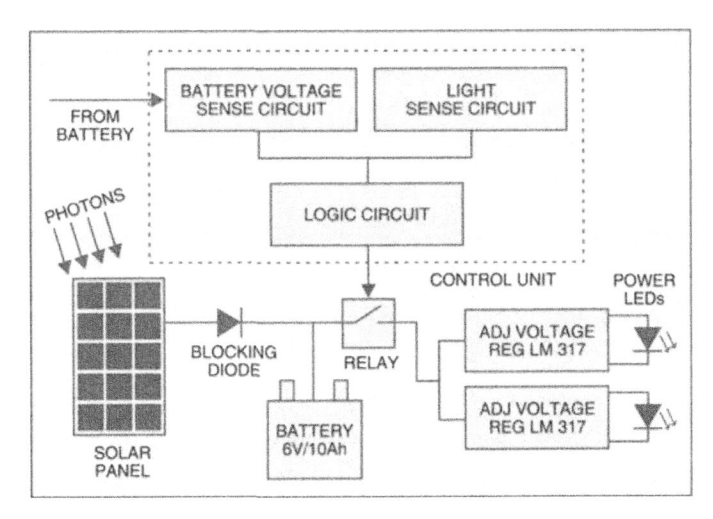

Fig. 1: Block diagram of solar powered pedestal lighting system

Figure 1 is a block diagram of the solar-powered pedestal lighting system that uses power LEDs. Here are the details about the solar panel that is being used:

- Maximum power: 10.0W
- Maximum power voltage: 17.0V
- Current power limit: 0.6A
- Short-circuit current: 0.7A
- Open-circuit voltage: 21.8V

Circuit connections

Through a blocking diode, the output of the solar panel is linked to the battery (D1). Diode D1 keeps the panel from getting power in the wrong direction. During daytime, sunlight is directly converted into DC electricity by solar cells, hence the power flow is from the solar panel to the storage battery. At night, there isn't any sunlight, so the solar panel isn't making power, and there may be a flow of electricity in the opposite direction, from the battery to the solar panel. Diode D1 (1N5408) stops the current from the battery from going into the solar panel. This keeps the solar panel from getting hurt. Lead-acid batteries with a 6V, 4.5Ah rating are used here. The output of the solar panel is directly connected to the battery through a switch so that it can charge. The battery voltage sensing circuit, the light sensing circuit, and the logic circuit are all parts of the control unit for the pedestal lighting system.

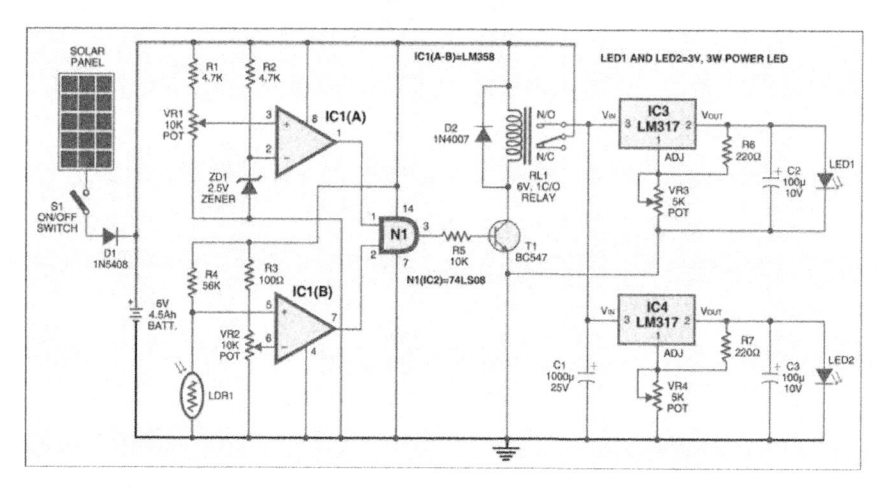

Fig. 2: Circuit of solar powered pedestal lighting system

The battery voltage sensing circuit keeps the battery from dying too quickly. The potential divider is what sets the voltage level of the battery. A comparator circuit built around op-amp IC LM358 compares the voltage of the potential divider with the voltage of the reference (IC1). Zener diode ZD1 sets the reference voltage at pin 2 of IC1 (A). When the battery voltage drops below the set voltage, the comparator output will go low. Potmeter VR1 is used to change the range of the battery cut-off. The logic circuit gets the signal from op-amp IC1 (A).

Circuit operation

The light-sensing circuit tells the power LED circuit to turn on at night and off during the day. Light is sensed by LDR1, which is a light-sensitive resistor. LDR1's resistance value changes based on how bright the light is. When the light is strong, the resistance is low, but when the light is weak, the resistance is high. In the potential divider circuit, LDR1 is hooked up. The voltage at pin 5 of IC1 is compared to the voltage controlled by LDR1. The potential divider at pin 6 of IC1 (B) makes the reference voltage, and potmeter VR2 is used to change it. The output of IC1 (B) is sent to IC2's logic circuit, which is built around its AND gate N1. The battery voltage sensing circuit and the light sensing circuit send their results to the logic circuit (two-input quad AND gate 74LS08).

When both of the AND gate's inputs are high, its output is high as well. The high output from pin 3 of the AND gate is sent to the relay-driver transistor T1 to turn on the relay, which controls the power supply to the LED circuit. When relay RL1 is turned on, the power-LED-driver circuit made up of IC3 and IC4 can work. The power-LED-driver is basically an adjustable voltage regulator circuit (LM317) that can drive up to 3W loads.

117. LED Illumination for Refrigerators

When we open the door of the fridge, the incandescent lamp inside lights up. It suffers from several disadvantages like:

- Because there is only one light source in the top corner, the light doesn't spread out in the same way. Only the top shelves get good light, while the shadows of the food on the lower shelves make them dark.
- In a strange way, the lamp makes the space we are trying to cool hotter, so the compressor has to work for longer.
- When the power goes out, there is no light inside the fridge, which is when you need it the most.

All of the above problems could be solved by using a distributed array of LEDs with a battery backup, which gives off light without shadows and runs cool. White PVC channels are used to make two rows of six white LEDs each. This is called a 26 array of white LEDs. The length of the channel is the same as the height of the fridge's cooling area. The LEDs are set up so that each shelf's top corner has two of them. As shown in, these channels are put in the left and right corners of the inside of the fridge. You can put the wiring behind the shelf or in the place of the bulb holder. The channels should be set up so that the light from the LEDs doesn't go straight into the user's eyes. The forward voltage of a white LED, which is how it works, is 3.5 volts. At this voltage, each branch of the LED circuit gets about 15mA of current. The total amount of voltage needed is about 7 volts. About 8 volts come from the full-wave rectifier. The voltage drops by about 1 volt when the current-limiting series resistor R1 is added. This makes sure that the current in each branch of the circuit is the same, so that the lights are all the same brightness. Use a 63ohm resistor in each of the six branches if you don't get LEDs that match. If so, R1 can be taken away.

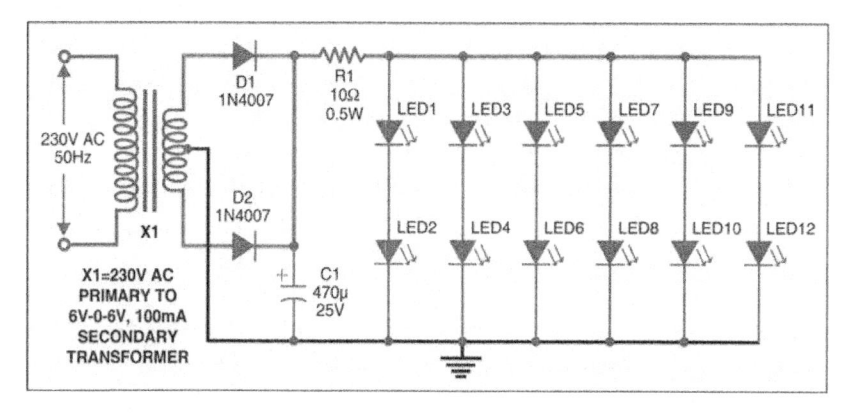

Fig. 1: Illumination LEDs with DC power supply circuit

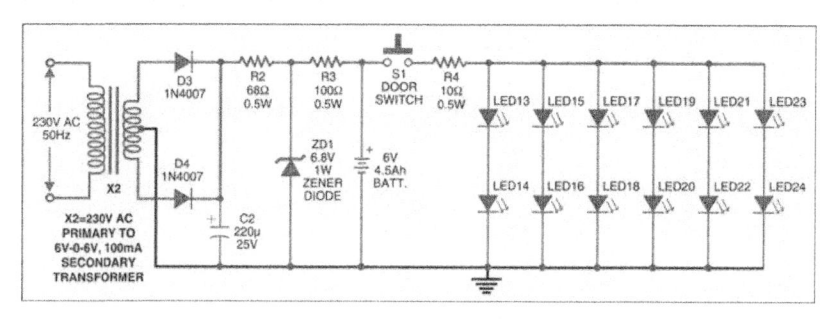

Fig.2: Illumination LEDs with DC power supply circuit

When the power goes out, it's important to keep the door open for as little time as possible. In places where power goes out often, the battery-powered circuit can be used to light up the fridge. The 6V, 4.5Ah sealed lead-acid battery is kept on a slow charge all the time. The battery powers the LEDs, and the switch on the door makes the connection. (In this case, we need to take the door-operated switch's original wiring apart. You might need the help of a professional technician.)

The battery voltage can't go higher than 6.8V because of Zener diode ZD1. So, switching to LED lighting not only makes the inside of the fridge brighter but also uses less electricity. It looks better and has no shadows. Because LEDs produce less heat than incandescent bulbs, the fridge cools better because the compressor needs to run for less time.

118. Electronic Reminder

This easy-to-make electronic alarm will let you know when it's time to do something important. It is especially helpful for housewives and people who work a lot. All you have to do is use the two thumbwheel switches (S3 and S4) to set the time in minutes and press and release the start switch. Exactly when the time you set is up, you will hear and see a signal to let you know that the time you set has passed. The device can be taken with you and works with a 9V battery.

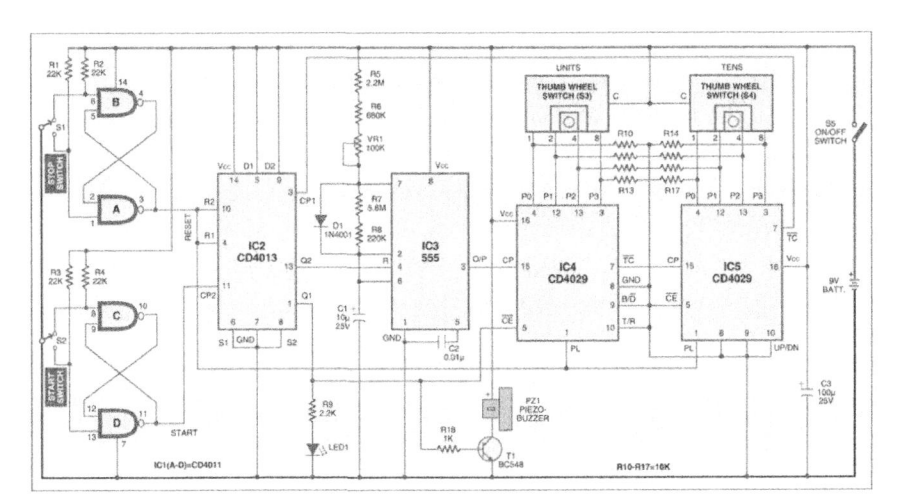

Two counter ICs CD4029 are the heart of this circuit (IC4 and IC5). These are 4 bit binary/decade counters that can be programmed up or down. They are part of the CMOS family of digital integrated circuits. The information on them is fed in parallel to inputs P0 through P3. When the PL input is high, regardless of the clock pulse input, it is loaded into the counter. When the up/down input is high or low, IC4 and IC5 count in up/down mode. These have been set up as 4-bit binary counters in countdown mode with the B/D input low. Every time the clock pulse goes from low to high, the counter moves forward by one.

Most of the time, the output terminal TC, which is used for counting, is high. It goes low when the counter reaches its maximum count (if wired in "up" mode) or its minimum count (if wired in "down" mode), and it goes high again as soon as the clock changes. As shown by the bar, the clock-enable (CE) input is an active-low input. This means that the clock pluses will only work when this input is low. IC2 (CD4013) is a dual D-type flip-flop. Each D flip-flop inside it has its own data, clock, set, and reset inputs. When the clock input goes from low to high, the data bit (which is either low or high) on the D-input is sent to the output. Set and reset are separate inputs that are turned on when these lines go high. This means that when the set input is high, the Q output is also high, no matter what the logic state of the D input is or how the clock changes. In the same way, when reset is high, it overrides everything else and forces the Q output to a low state.

The IC3 (timer 555) is set up as an astable multivibrator with a waveform that repeats every minute. For accurate and stable timing, the resistors R5 and R6 should be metal-film resistors, and the capacitor C1 should be made of tantalum. IC1 (CD4011B) is a quad NAND gate with two inputs. Two of these four NAND gates, A and B, along with pull-up resistors R1 and R2, make up the de-bouncing circuit for micro-switch S1, which sends out the master reset pulse every time it is pressed and released. The other debouncing circuit for microswitch S2 is made up of the last two NAND gates (C and D) and resistors R3 and R4. When microswitch S2 is pressed and then let go, it sends out the start pulse. Here's how the circuit works: At first, microswitches S1 and S2 are in a place where both the reset and start outputs are low. First, use the two BCD switches, also called thumbwheel switches, S3 and S4, to set the time in minutes. If you choose "5" with thumbwheel switch S3 and "6" with thumbwheel switch S4, the time delay is set to 65 minutes. The second step is to press microswitch S1 and then let go of it. A pulse that goes in the right direction resets flip-flop IC2 and loads the information about the time delay into counters IC4 and IC5. Since IC2's Q2 output is low at first, when we press and let go of S2's start switch, IC2 is "clocked," and Q2's output goes high. Reset pin 4 of timer IC3 is given power to turn it on. At this point, the time delay will start to count down.

To sum up, to use the reminder device, you need to set the time after which you want to be reminded, press and release switch S1 and then do the same thing with switch S2. IC5's TC output is normally high, but when the start switch S2 is pressed, it goes low. If the time set by thumbwheel switches S4 and S5 is 65 minutes, it would only go high after 65 minutes. This is a 65-clock cycle, in essence. Keep in mind that the time period of IC3 is about one minute. This pulse at the output (TC) of IC5 clocks at CP1 of IC2, whose output goes from low to high and turns on both LED1 and piezobuzzer PZ1. The clock is turned off when the clock-enable (CE) input of IC4 goes high, which is caused by the Q1 output of IC2. LED1 and piezobuzzer PZ1 stay "on" until the system is reset through switch S5. After the system is reset, you can set the time on the device again.

The counters are wired in count-down mode because that is the only way for the counter IC to finish its count cycle in the number of clock cycles set by the thumbwheel switches. The time can be changed in one-minute increments. If the clock period is changed to, say, 2 minutes, the time resolution will also be 2 minutes, but the most time delay that can be set will go from 99 minutes to 198 minutes. With a time, resolution of one clock cycle, this circuit can only delay time by 99 clock cycles at most.

119. Photodiode-Based Fire Detector

Your computer and television set, along with other electronic appliances, can be safeguarded by a fire alarm that is based on an ultrasensitive photodiode. When it detects a spark or fire in the power supply part of the instrument, it quickly triggers an alarm and cuts off the power supply. The fire sensor in this device is a photodiode, and it promptly sounds the alert. In order to detect a fire, the circuit makes use of the photovoltaic characteristic of the photodiodes. Infrared detectors often make use of photodiodes, which generate a photo voltage that is proportional to the amount of incident light or infrared rays that fall on them. Typically, 1V is created in the photodiode when it is forward biased by accepting the photons. Here the passive infrared rays from the spark or fire are used to activate the photodiode to generate the photo voltage Since the photo voltage is relatively modest, a very sensitive voltage amplifier is needed to activate the remaining half of the circuit. Here dual op-amp IC 741 (IC1) is used as a Schmitt trigger with hysteresis. It does so by transforming the input voltage signals coming from the photodiode into a signal that has been shaped.

In order to provide IC1's inverting input (pin 2) with a voltage that is half of the supply voltage, which is 4.5 volts, resistors R1 and R2 combine to produce a potential divider. Resistor R3 connects the output of IC1 to its inverting input.

Photodiode Based Fire Detector Circuit

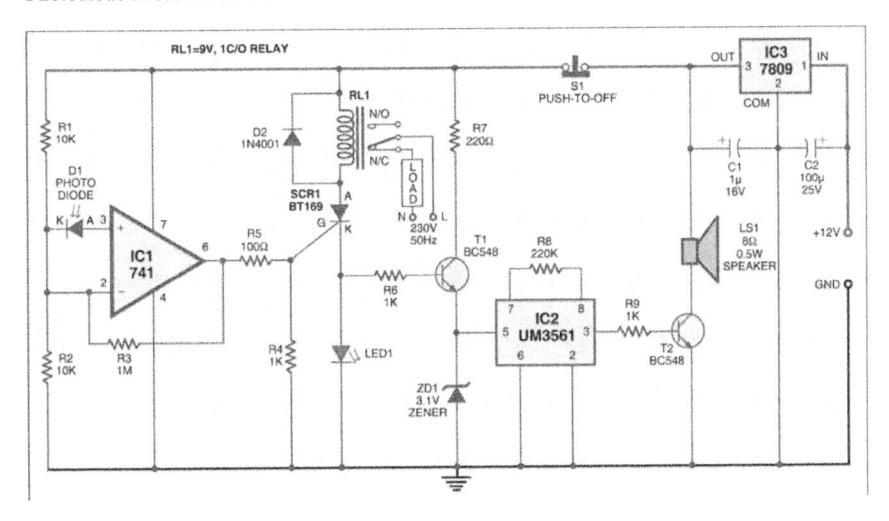

Fig. 1: Circuit of the photodiode-based fire detector

The photodiode is wired so that it connects to the junction of the R1-R2 divider as well as pin 3 of IC1, which is the non-inverting input. In the standby mode, the voltage across the photodiode will be zero, and the output of IC1 will remain low. This occurs when the photodiode is exposed to darkness. The photodiode begins to conduct current and the output of IC1 rises to a high level whenever there is a spark or a fire. Because of the hysteresis of the Schmitt trigger, the amplifier is capable of high levels of sensitivity. Even after the photodiode loses its ability to conduct electricity, the output of IC1 maintains its high state for a few seconds. When the output begins to rise, even a minute shift in voltage at the inverting input has no impact on the state of the circuit. This is incredibly beneficial to activate the alarm even with a single spark.

Circuit operation

SCR BT169 (also known as SCR1) is triggered into action by the high output from IC1, which in turn activates the relay. As soon as the relay becomes active, the power supply to the gadget will be promptly switched off. The SCR1 push-to-off switch will not release its latched state unless it is pressed.

When SCR1 is closed, base bias is applied to T1, which causes it to conduct and activate alarm generator UM3561 (IC2). Zener diode ZD1 keeps the supply voltage for IC2 at a safer level of 3.1 volts. By connecting pin 6 of IC2 to ground, an alert for the fire department will be activated once a fire is detected. The signal that is sent to the speaker comes from the output of IC2, which is amplified by the transistor T2.

120. BODMAS Rule Circuit

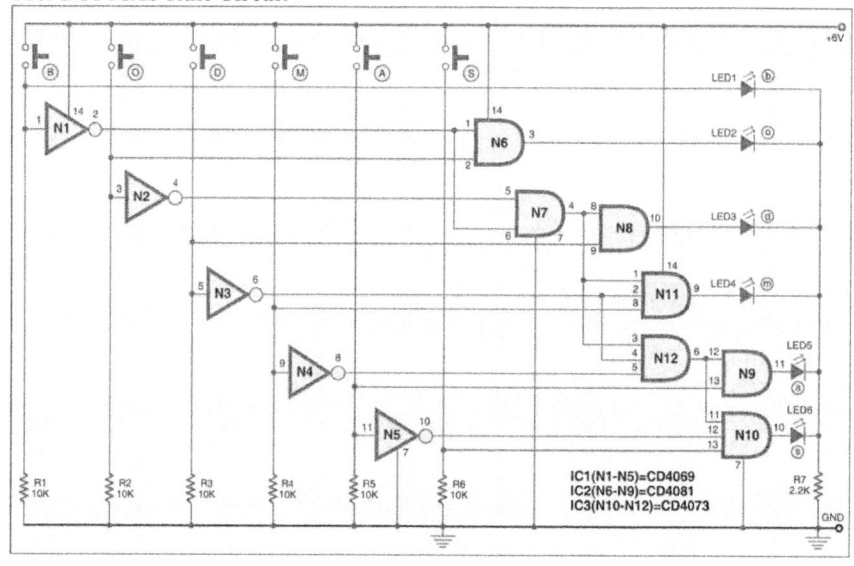

Bodmas rule circuit

The BODMAS rule is an acronym that aids children in remembering the order of mathematical operations. The BODMAS rule is a mnemonic that can be used to remember the order of the several arithmetical operators. The BODMAS acronym stands for bracket, of, division, multiplication, and addition and subtraction. A mathematical expression may involve multiple operators, but only one of those operators must be carried out first in order for the expression to be valid. The following is the order of priority, which we are all aware with from our time spent in school: "Bracket" comes first, followed by "of," "division," "multiplication," "addition," and "subtraction" in that same order.

This circuit would be very helpful to novices who are just starting out in the field of mathematics as a learning aid. This circuit looks for any incorrect answers as well as any incorrect orders. The truth table of the priority checker that was described is displayed here. In the table, the letters written in uppercase (B, O, D, etc.) represent the inputs, while the letters written in lowercase (b, o, d, etc.) represent the outputs.

If an input variable is set to logic "0," it indicates that the relevant mathematical operation is not included in the expression. On the other hand, if the variable is set to logic "1," the relevant mathematical operation is included in the expression. The expression may contain any number of operations (any number of inputs may be at logic '1'), but only one output will be at logic '1.' This indicates that a specific task has the highest priority and must be completed before any others can be done. (The 'don't care' inputs are denoted by an X in the truth table; this means that they might be at logic 0 or logic 1, respectively.) You may simply derive the following Boolean equations based on the truth table, which are as follows:

Truth Table

Input						Output					
B	O	D	M	A	S	b	o	d	m	a	s
0	0	0	0	0	1	0	0	0	0	0	1
0	0	0	0	1	X	0	0	0	0	1	0
0	0	0	1	X	X	0	0	0	1	0	0
0	0	1	X	X	X	0	0	1	0	0	0
0	1	X	X	X	X	0	1	0	0	0	0
1	X	X	X	X	X	1	0	0	0	0	0

$$b = B$$

$$o = \overline{B}. O$$

$$d = \overline{B}. \overline{O}. D$$

$$m = \overline{B}. \overline{O}. \overline{D}. M$$

$$a = \overline{B}. \overline{O}. \overline{D}. \overline{M}. A$$

$$s = \overline{B}. \overline{O}. \overline{D}. \overline{M}. \overline{A}. S$$

Circuit operation

The logic circuit of the priority checker comprises NOT and AND gates, LEDs and a few resistors. The working of the circuit is as simple as its structure. When one of the six microswitches, labeled 'B,' 'O,'..., 'S,' is pressed, the value of the variable that it controls is set to high (logic 1). The flashing of a particular LED, holding the same name as the output variable, will signify its top priority.

For example, pushing 'B,' 'M' and 'A' simultaneously will cause only LED 'b' to turn 'on,' indicating that the bracket action has the highest priority. If 'M' and 'A' are pressed, LED 'm' will glow, indicating that 'multiplication' is to be carried out prior to 'addition.'

121. Circuit for UPS to Hibernate PC

The vast majority of the low-power UPS systems that are currently on the market do not include the functionality necessary to power down the computer before the system powers down by itself due to low battery. Some of them have the capability, but in order to use it they need the appropriate software. In this project, we will discuss a "add-on" circuit for uninterruptible power supplies (UPS) that will hybernate a computer automatically before the UPS shuts down due to low battery voltage. There is no need to install any enabling software.

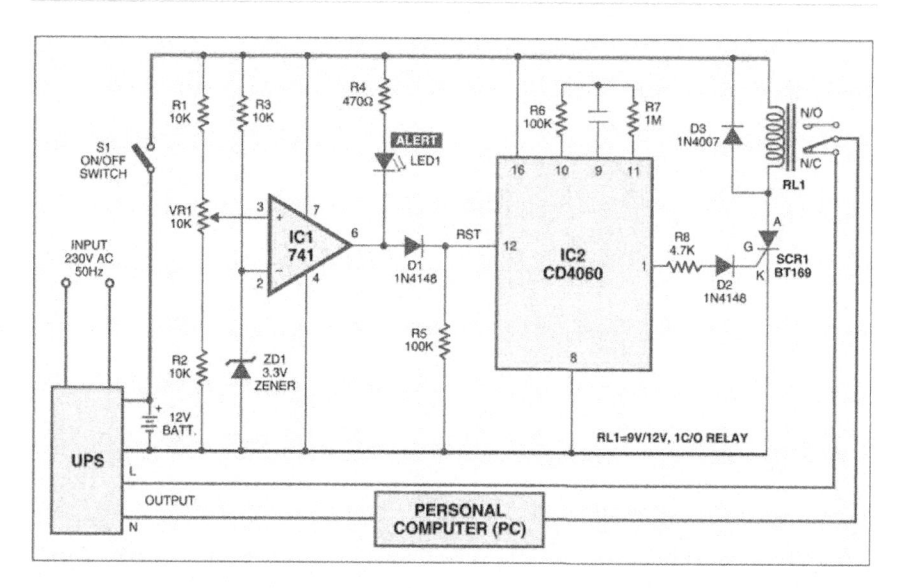

Following the comparator IC 741 (IC1) comes a short-duration positive pulse generator that makes use of a 14-stage ripple-carry binary counter/divider and oscillator IC2. Finally, the circuit concludes with a logic level shifter IC2. Because the voltage at the non-inverting (positive) input terminal of IC1 is typically higher than the voltage at the inverting (negative) input, the output of the comparator is also typically higher than the voltage at the inverting (negative) input. This high output from IC1 is put to use in the process of resetting IC CD4060 (IC2), which is a ripple counter with 14 stages.

When the voltage of the UPS battery drops below the value that has been established, the non-inverting terminal voltage of comparator IC1 at pin 3 begins to decrease. In this circuit, when the voltage of the battery drops below 9.5 volts, the voltage at pin 3 of IC1 drops below 3.3 volts, and the output of the comparator at pin 6 goes low. This causes LED1 to light up, which notifies the user that his computer is going to hybernate within the allotted amount of time (around 3 minutes). After three minutes, a high output will be produced at pin 1 of IC2 because of the low voltage level that is present at pin 12 of that component, which enables it to oscillate and causes the counter to begin counting. The high output is utilized to trigger the SCR1, which in turn activates the RL1 relay. Only when the power source for the circuit is manually turned off through S1 will the relay become de-energized.

The circuit is constructed on a small printed circuit board and then wired to the terminals of the UPS battery. The voltage needed to activate hibernation in the circuit has been configured to be higher than the cut-off voltage of the UPS. (If the cut-off voltage of the UPS is 9.5 volts, set the voltage level at which hybernation activation is activated to 10 volts.) The circuit is placed within the UPS cabinet. Switch S1 and LED1 are wired up to be on the front panel of the cabinet.

122. Accurate 1Hz Signal Generator

In order for stopwatches and other digital circuits to function properly, they need to receive accurate squarewave pulses at 1 Hz. Without the need for a costly crystal oscillator, the following signal generator has a 1Hz frequency and is designed for general use.

1Hz signal generating circuit

A secondary output of 9V-0-9V at 100 mA is delivered by a center-tapped transformer X1, which steps down the primary input of 230V, 50Hz, single-phase AC mains. The output of the transformer is fed into a full-wave rectifier, which consists of the diodes D1 and D2, and it is then filtered by the capacitor C1. This supplies the electrical circuit with the necessary DC voltages for it to function properly.

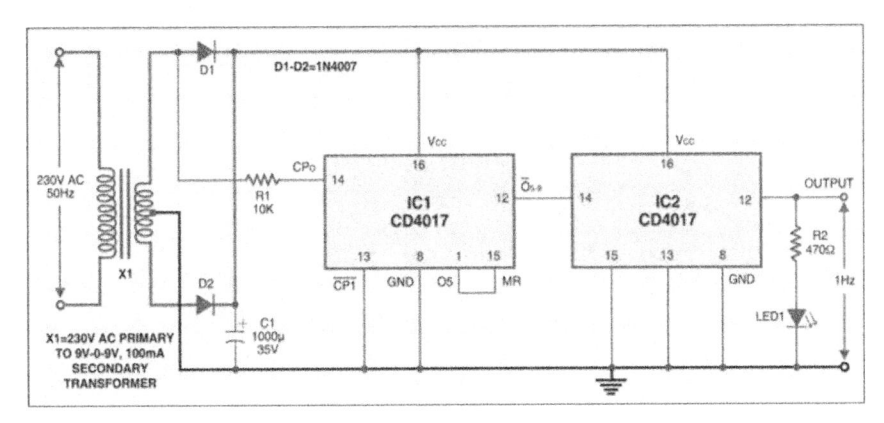

Additionally, the 50Hz input signal is extracted from the secondary winding of transformer X1 and supplied to the clock pin 14 of decade counter CD4017 (IC1), which is configured as a divide-by-5 counter by means of resistor R1. IC1 is now producing a 10Hz output at its pin 12, which is then given to the clock pin of another IC called CD4017 (IC2), which is wired as a divide-by-10-decade counter. On the screen of the oscilloscope, the output of IC2 appears as a clock pulse with a frequency of 1 Hz. Additionally, the output of IC2 causes LED1 to blink once per second when it is connected in series with the output load (resistor R2).

123. A Fourth-Order Speech Filter (Based on Texas Instruments Application Note)

The frequency range of 300 to 3400 Hz is often where human speech can be found in an audio spectrum. It is essential, particularly in telephone lines, to restrict the frequency response to fall within this range as it is required. Another project that utilizes a filter like this is "Digital Speech Security System," which was featured in EFY Electronics Projects Vol. 19. On the other hand, it relies on dedicated filter ICs, which, in addition to being expensive, are difficult to come by. Not just for the project that was specified, but also for a variety of other speech circuits, this circuit will prove to be highly beneficial.

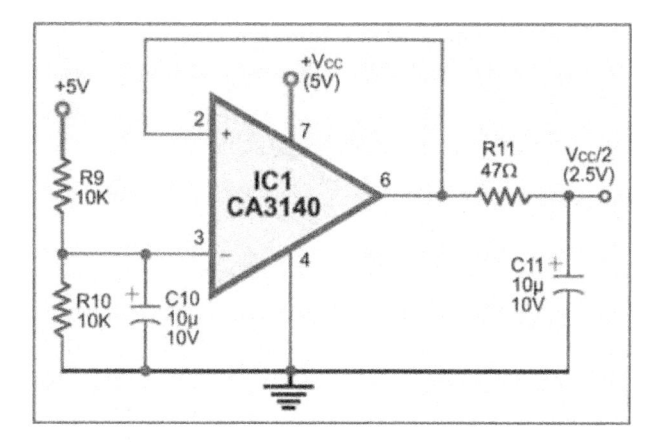

146 | 500 Electronic Project Ideas for Inventors

Fig. 1: Generating stable Vcc/2 from Vcc

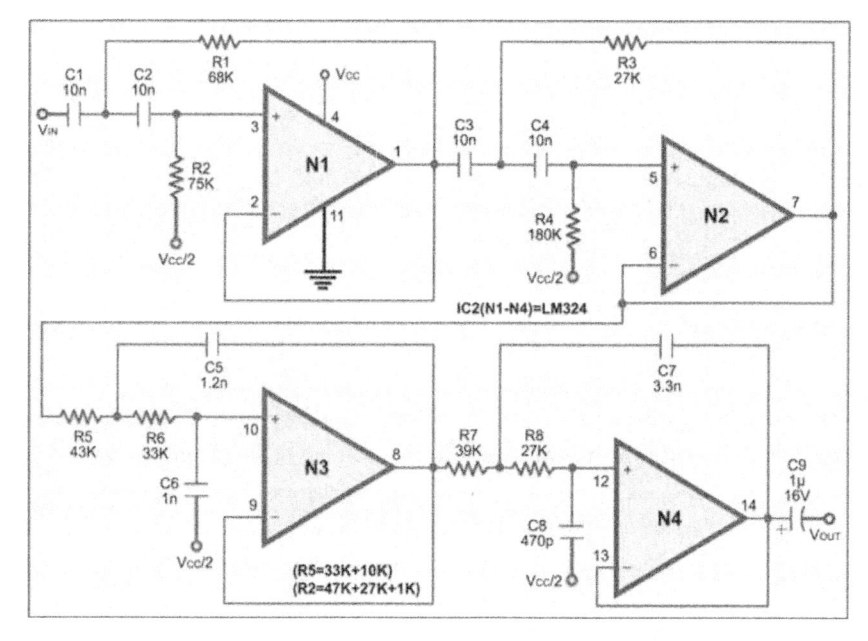

Fig. 2: Fourth-order bandpass speech filter for 300 Hz to 3400 Hz

A fourth-order filter that is able to accept an audio input signal with an amplitude of up to 2.5V can be built using two integrated circuits (ICs). One of these ICs would be a quad op-amp such as LM324, and the other could be a single op-amp CA3130/CA3140 or dual op-amp LM358 operating off a single supply of 5V as Vcc. This filter would be able to Using a unity-gain voltage-follower circuit like the one illustrated in Figure 1, which is biased at half the source voltage developed across R2, one may generate a stable voltage of Vcc/2, which is equal to 2.5V. Bypassing ripples in the supply voltage is the responsibility of the capacitor C1, while the output of the operational amplifier is the responsibility of the capacitor C2. The voltage Vcc/2 is connected to the places that are corresponding to it. The filter circuits that are found at the input of op-amps make use of conventional capacitor values and resistor values with a 5% tolerance. Even while the differences in the components do have a very little impact on how the circuit works, those differences will be nearly impossible to detect.

The response curve of the filter is shown in Figure 3, and it can be seen that the rejection of 50/60Hz by the fourth-order filter is larger than 40 dB, but the rejection by the second-order filter is approximately 15 dB. The filter has been developed to have a roll-off of 0.5 dB at 300 Hz and 3 kHz respectively.

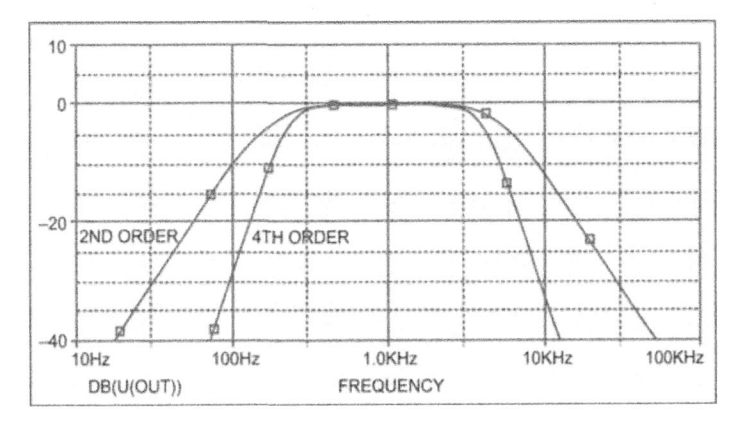

Fig. 3: Frequency response of 2nd and 4th order bandpass speech filter

The response of the filter has been tested and found to be good.

124. Smart Battery Protector Using a Shunt Regulator

When a battery is depleted to a level lower than the minimum voltage that is suggested for it, the battery's expected lifespan is significantly shortened. Before the discharge is completely finished, you are required to disconnect the load. In that case, it has the potential to either damage the battery or reduce its lifespan. In order to stop the chattering in the disconnect switch, this function of the disconnect requires a battery voltage monitor equipped with a significant amount of hysteresis. When the voltage of the battery drops below the threshold and then bounces back when the load is removed, chattering will occur.

A circuit that safeguards batteries

The conventional battery protectors are no match for this uncomplicated circuit. It employs the shunt regulator IC TL431 for both its ease of use and its outstanding performance. Using a power supply for the workbench that is regulated makes it simple to adjust the operating point of the circuit. The technique for making the adjustment is as follows: To begin, use a jumper to short out the Zener diode ZD1. The circuit requires a supply of 12V DC. Relay RL1 energizes.

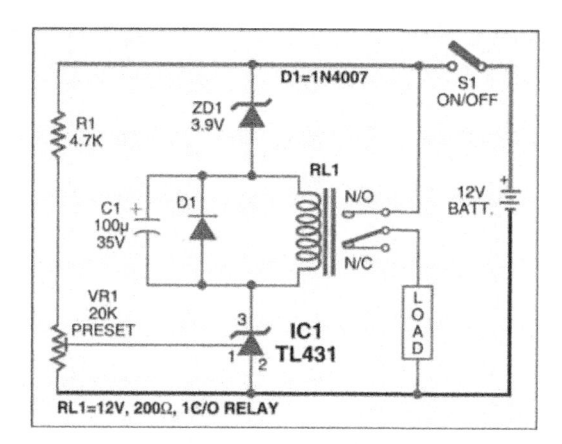

Now turn the power supply down to 10.5 volts, and modify the preset VR1 so that the relay will close when the voltage reaches 10.5 volts. (Refer to the voltage that is advised by the company that makes the battery.) Proceed with the process once more if required.

Now take the jumper out of the circuit and connect the wires. At a voltage of 10.5V, the relay should not operate. Increasing the power supply should be done gradually. It is determined by the relay pull-in voltage as well as the Zener breakdown potential whether or not the relay will activate when it reaches 12 volts or not. Bring the power source for the workbench down to 10.5 volts, and observe how the relay trips perfectly at 10.5 volts. Using a Zener diode with a little bit higher voltage is another way to achieve a huge hysteresis in your circuit. We have demonstrated that it is possible to prevent relay chattering by testing the circuit with a Zener diode rated at 3.9 volts and a relay rated for 12 volts and 200 ohms.

125. Microcontroller-Based Tachometer

A speedometer is just a simple digital electronic transducer. Usually, it is used to measure how fast a shaft is moving. The number of rotations per minute (rpm) is important to know in order to understand any system that moves. For example, there is an ideal speed for drilling a hole of a certain size in a piece of metal. There is also an ideal speed for sanding, which depends on the material being finished. You might also want to measure how fast your fans spin. This microcontroller-based tachometer is easy to make and can measure the rpm of most shop tools and many household machines without a mechanical or electrical connection. Just point the light-sensitive probe tip on top of the spinning shaft toward the spinning blade, disk, or chuck and read the rpm. The only rule is that you must first put down a mask of a different color. On the thing that is spinning, a strip of white adhesive tape would work well. Set it up so that as the object turns, the amount of light that is reflected from its surface changes. When the tape spins past the probe, the phototransistor picks up the momentary rise in light that comes back from the tape. The 4-digit, 7-segment display shows the rpm, which is calculated by the signal processor and microcontroller circuit. It does this by counting the increase in the number of these light reflections that it senses.

Working of a microcontroller-based tachometer

Fig. 1: Circuit of microcontroller-based tachometer

The phototransistor is kept in a plastic tube with a convex lens on one end. A common part used by watch repairmen and in cine film viewer toys is a convex lens with a diameter of about 1 cm and a focal length of 8 to 10 cm. They can give it to you so you can set up the experiment. The phototransistor is attached to a piece of cardboard so that it is about 8 cm away from the lens. The phototransistor's leads are taken off and put into the circuit shown in Fig. 1. Figure 2 shows how a phototransistor should be set up.

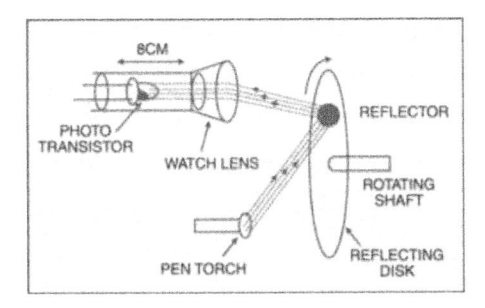

Fig. 2: Suitable arrangement of phototransistor

The signal that is picked up is amplified by the transistor 2N2222 (T5) and then amplified even more by the operational amplifier CA3140 (IC3). The operational amplifier's reference voltage point is found by a resistor divider network made up of R2 and R3. Pin 6 of IC3 is connected to pin 12 of AT89C2051, which is a microcontroller. Note that the inputs (+ and -) of the microcontroller AT89C2051's internal analog comparator are pins 12 and 13. A potential divider made up of resistor R7 and preset VR1 is used across the supply to set Pin 13 to almost half the supply voltage. The AT89C2051's internal comparator picks up the pulses picked up by the phototransistor. Each pulse represents one rotation of the object, so each pulse is seen as a rotation. By counting how many of these pulses there are on average per minute, the RPM can be found. It is shown by a software program that tells the 4-digit, 7-segment display's LEDs to light up.

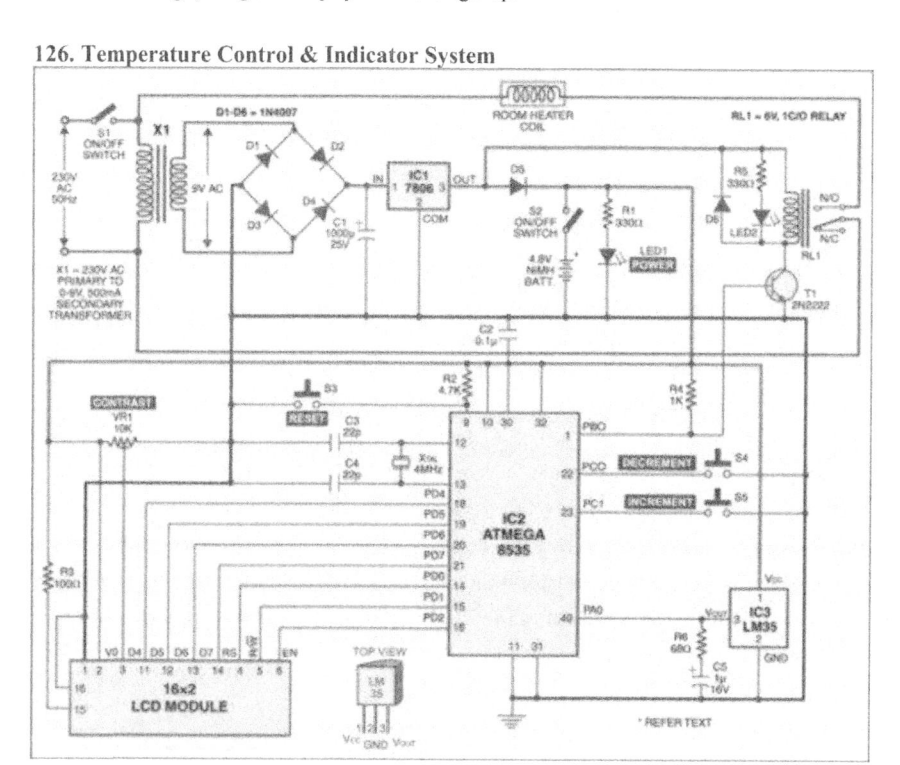

126. Temperature Control & Indicator System

Temperature control system circuit

Here is a temperature indicator-cum-controller that is easy to build and can be connected to a heater's coil to keep the room at a constant temperature. The temperature control system is based on an Atmega8535 microcontroller, which makes it dynamic and faster. It uses an LCD module to show the set values and two buttons to increase or decrease them. In Fig., you can see how the temperature control system works. It has an Atmega8535 microcontroller, an LM35 temperature sensor, a 7806 regulator, an LCD module, and a few other parts.

The 230V, 50Hz AC mains power goes into transformer X1, which steps it down to 9V, 500 mA. The output of the transformer is fixed by a full-wave bridge rectifier made up of diodes D1 through D4. The output is filtered by capacitor C1 and controlled by IC 7806. (IC1). LED1 shows that there is DC power. Resistor R1 stops the current from going too fast. A backup battery is a 4.8V battery that can be charged. Based on the AVR enhanced RISC architecture, the ATmega8535 is an 8-bit CMOS microcontroller that uses little power.

127. Stabilized Power Supply for Prototyping

This stabilized power supply circuit can be connected directly to 230V AC mains to get output voltages of 3V to 12V DC to connect to the prototyping board.

Stabilized power supply circuit

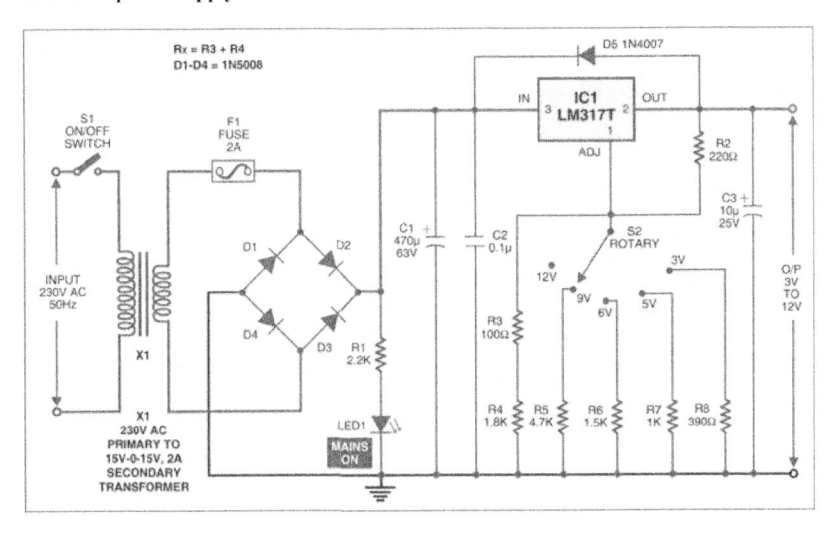

Step-down transformer X1 takes 230V AC from the mains and turns it into 15V AC. Its secondary winding can handle a current of 2 amperes. The AC is changed into pulsating DC with a peak voltage of 21V (151.4142). This is done by a bridge rectifier. When LED1 lights up, it means that the rectifier is ready to send out power. Resistor R1 (2.2 k) keeps the current through LED1 below 10 mA, which is safe. The 470F capacitor C1 smooths out the output of the bridge rectifier. High frequency ripple is blocked by capacitor C2. At the end of the rectifier section, a 3-terminal, positive-voltage regulator from the LM317T series is used to control the voltage. It can give out more than 1.5A over a range of output voltages from 1.2V to 37V. But in this case, it is being used to provide discrete voltages in steps of 3V, 5V, 6V, 9V, and 12V. This is done with the help of a 5-way rotary switch S2, which connects different resistor values between the regulator's Adj pin and ground. R2, on the other hand, is a fixed 220-ohm resistor between the Adj pin and the output pin. The output voltage (Vo) is given by: where "Rx" is the resistance between the regulator's "Adj" pin and "ground."

In the 12V position (the "off" position of the switch), the value of Rx is R3+R4=1900 ohms. In other positions, it is the series equivalent of 1900 ohms in shunt with another resistance chosen by the rotary switch. In different places on the rotary switch, the table shows the equivalent series resistance.

Note; In the circuit diagram, X1 rating is written wrong. That is, you should read 15V-0-15V as 0-15V. Switching with a discrete resistor (with 1% tolerance) is better than using a variable resistor because the wiper contact gets messed up after some use, and a variable resistor's tolerance (change with temperature) is also much higher.

128. Infrared Burglar Alarm

The latching operation of this infrared burglar alarm is something that makes it stand out. The circuit is very sensitive as well. The circuit has parts that send and receive signals. When the IR beam between the transmitter and the receiver breaks, the alarm circuit is set off and the buzzer keeps going off and on. The only way to reset it is to press the reset button.

Infrared burglar alarm circuit

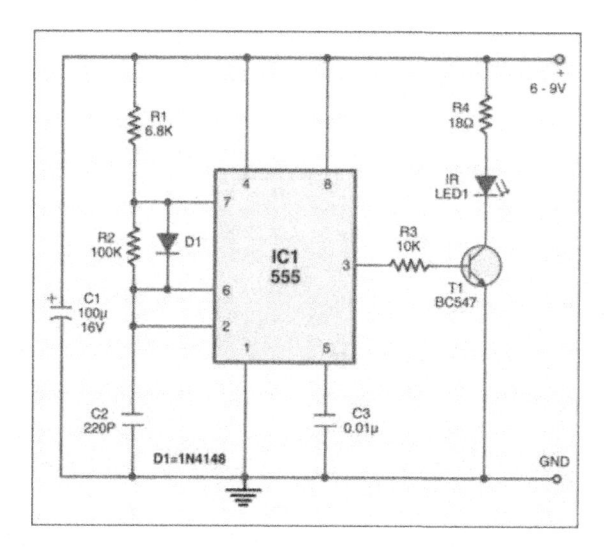

The transmitter section is built around IC 555 (IC1), which is wired as an astable multivibrator and makes a 38kHz frequency. IR LED1 sends out 38 kHz modulated signals up to 4.6 meters away (15 feet). The modulated IR beam from the transmitter keeps hitting the receiver section, which is made up of an IR sensor called TSOP1738, an IC called CD4011, and a few other parts. The TSOP1738 receiver sensor works with a 38kHz frequency. When a modulated IR beam hits the TSOP1738 receiver sensor, its output goes low. When someone crosses the path of the IR rays, the sensor senses this and its output goes high for a short time, which sets off the flip-flop.

IC CD4011 is used in the flip-flop circuit. Here, two IC2 NAND gates are used. Most of the time, IC2's input pins 2 and 13 are high. When the IR beam hits TSOP1738, IC2 keeps its output low. When transistor T2 is turned off, the alarm circuit stops working, and the buzzer stops going off. The IR transmitter and receiver are placed on opposite sides of the entry door or gate. When someone walks through the invisible beam between the transmitter and receiver, the output of the receiver goes low for a moment. This turns on the flip-flop. Pin 3 of IC2's output line goes high and stays there. When transistor T2 conducts, the alarm circuit turns on and the buzzer sounds until switch S1 resets the receiver circuit.

Circuit operation

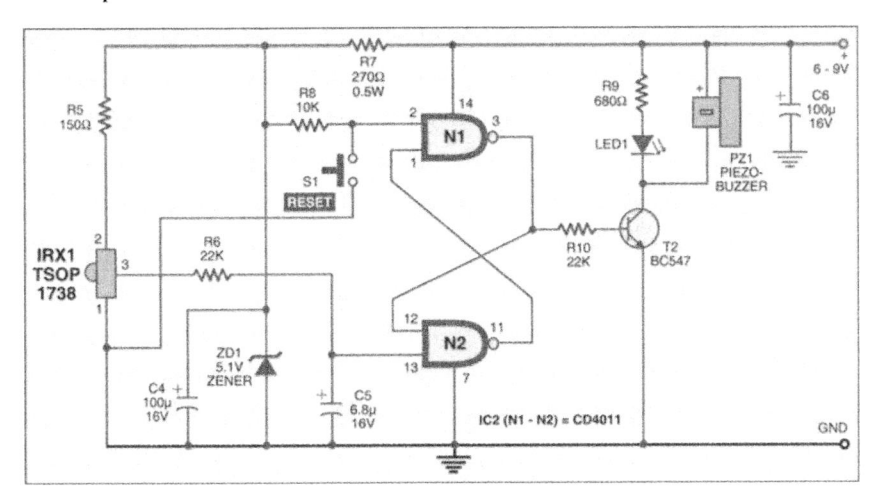

Put together the transmitter and receiver on separate PCBs about the size of a matchbox. For alarm purpose, use a piezobuzzer having internal oscillator. The circuit needs 6 to 9V DC to work. Even supplies with no rules can be used. When the power supply is connected, the alarm may go off. Press the reset button to get the circuit ready to find thieves.

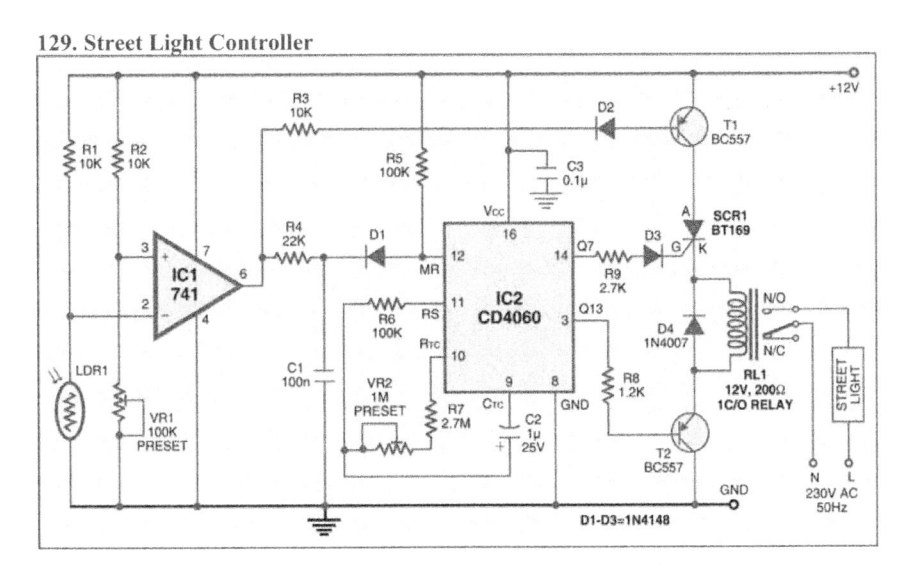

129. Street Light Controller

False triggering as a result of modest variations in the intensity of ambient light and a lack of control over switching action are two issues that are frequently encountered in conjunction with street lights. These issues can be avoided by using this straightforward controller for street lights, which features a switching circuit. The well-known op-amp IC 741 serves as the basis for the circuit's design (IC1), while the 14-bit ripple counter CD4060 serves as the basis for IC2 along with other components. IC1 and LDR1 work together to enable IC2, which then causes transistor T1 to conduct electricity. A second function of IC2 is to act as a trigger for SCR1, which then turns on the street light. When the trigger is pulled, the light is turned "off." Around its pins 9, 10, and 11, the integrated circuit CD4060 features a built-in oscillator. The master reset (MR) control is located at pin 12. When pin 12 is high, the oscillator is inoperable, but when it is low, it becomes operational again.

Circuit operation

During the day, when there is light falling on LDR1, its resistance drops, and as a result, the high output at pin 6 of IC1 turns off the pnp transistor T1 and disables IC2. At this point, the SCR1 relay that turns off the street light has not yet been triggered. During the night, when there is no light falling on the LDR1, its resistance rises, and the low output pin 6 of IC1 causes the pnp transistor T1 to conduct. IC2 is then enabled, and its internal oscillator immediately begins oscillating after this.

Pin 14 (Q7) of IC2 rises high at the predetermined time, which causes SCR BT169 to be triggered via resistor R9 and diode D3, respectively. Because of this, RL1 receives power, and the street light turns on. This time interval can be altered by connecting the gate of SCR1 to other pins of IC CD4060, such as pins 6, 13, etc (not shown in Fig. 1). When the output pin 3 (Q13) of IC2 gets high, the normally conducting transistor T2 is forced into non-conduction. This results in the de-energization of relay RL1, which then turns off the street light. This time can be altered by the use of the variable preset resistor VR2. To put it another way, the street light turns "on" when Q7 of IC2 gets high, and it turns "off" when Q13 goes high, provided that pin 12 of IC2 stays low.

130. Light Operated Doorbell

Automatic hand dryers and toilet flushers typically have light-sensitive switches installed in them. This is a straightforward light switch that can also function in regular lighting conditions. You can install it on the primary entrance to your home so that it serves as both an automatic doorbell and a burglar alarm. The moment that someone's shadow is detected by the sensor of this apparatus, the bell starts to ring.

The light-dependent resistor (LDR) sensor is constructed out of a CdS semiconductor and has a resistance that varies depending on the amount of light present. The presence of light causes its resistance to decrease, while the absence of light causes it to increase to a higher level. Timing circuit IC 555 and melody generator IC UM66 are used in this circuit. The UM66 may operate at a maximum working voltage of 3.3V. In order to supply the maximum voltage that is necessary, a Zener diode is utilized. The input for the melody IC UM66 comes from the timer NE555, and the output of the melody IC is sent, through resistor R2, to the base of the transistor T1. In order to alter the operation of the circuit's threshold, a 470-kilo-ohm preset, denoted by VR1, is utilized.

Circuit with a light-operated switch.

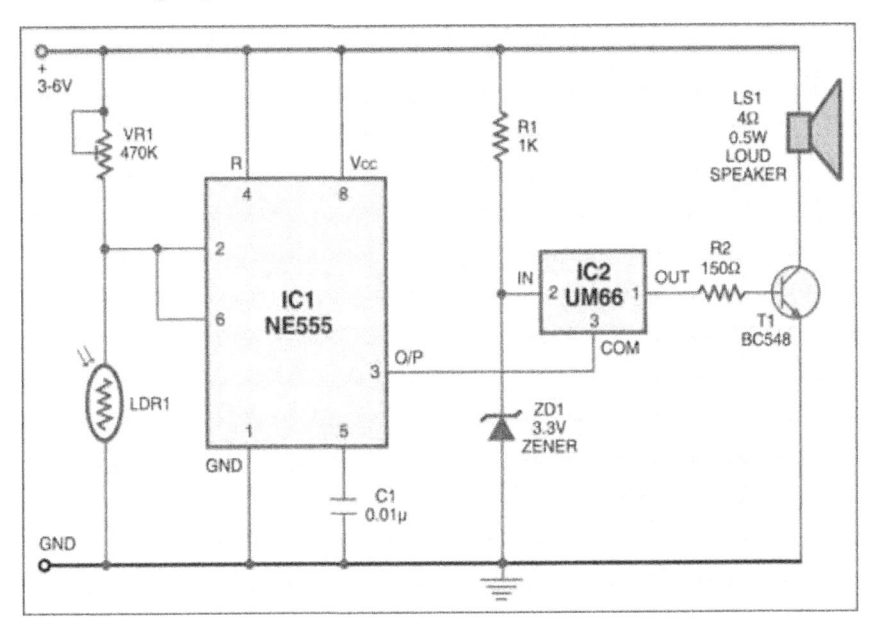

Fig. 1: Light operated doorbell

Circuit operation

When light hits LDR1, its resistance drops, and the potential gap between pin 2 of IC1 and the ground is smaller as a result. Under these circumstances, the output of the timer continues to be high. Due to the high voltage that has been applied to the ground terminal of the UM66, its operation has been halted, and you will not hear any melodies. When someone casts a shadow over the LDR sensor, the resistance of the sensor goes up, which in turn causes an increase in the potential difference between pin 2 of IC1 and ground. The value of the output from the timer drops. The operation of the UM66 is kicked off by this low voltage at the ground terminal. The signal that is amplified by transistor T1 and sent to speaker LS1 is produced by UM66, which generates the pulse. You will be able to hear a pleasant tone produced by the LDR once light no longer falls on it.

131. Clock Tick-Tock Sound Generator & LED Pendulum

Wall clocks with pendulums that have a wooden case and are powered by batteries can be purchased on the market. Some of these clocks feature a tick-tock sound generator. Some of them even have bells. The sound of old-fashioned mechanical pendulum clocks ticking away is something that is lacking. The schematic for a sound generator that goes "tick-tock" is displayed up there. It is constructed around a timer IC 7555 that has been wired as an astable multivibrator (IC1). The preset VR1, resistor R1, and capacitor C1 work together to generate frequencies, and those frequencies can be adjusted with the help of the preset VR1.

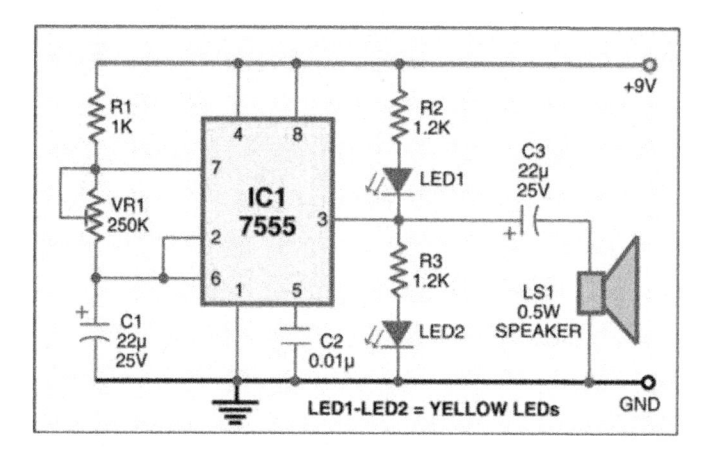

Fig.1 Circuit of tick-tock sound generator

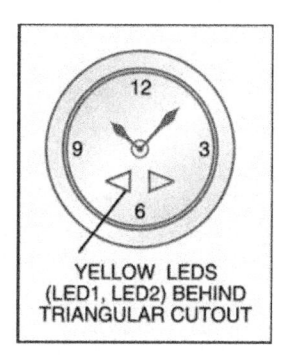

Wooden clock design

The sound of a tick-tock is generated by a small speaker that is connected to the output pin 3 of IC1 by means of the capacitor C3. Additionally, two yellow LEDs are attached to pin 3 of IC1 on the face of the clock so that they flash in sync with the sound (see Fig. 2). Adjust the preset VR1 so that the ticking sound is quite similar to the tick-tock that is produced by traditional pendulum clocks. The wooden clock looks great with the yellow LEDs (LED1 and LED2) that are on it.

The circuit for the LED pendulum may be seen in the image below. It makes use of a different timer 7555 wired in astable mode, and the clock pulse from that timer's pin 3 is supplied to the clock pin 14 of the decade counter 4017. (IC3). The outputs of IC3 are connected with the assistance of the switching diode 1N4148 in such a way that the six LEDs (LED3 through LED8) flash in succession in one direction at first, and subsequently in reverse sequence. With the use of a 10-kilohm setting on VR2, one is able to alter the rate at which the LEDs flash. For the purpose of resetting the decade counter, capacitor C6 has a value of 6.8 nanofarads, and resistor R11 has a value of 100 kilobohms.

LED pendulum circuit

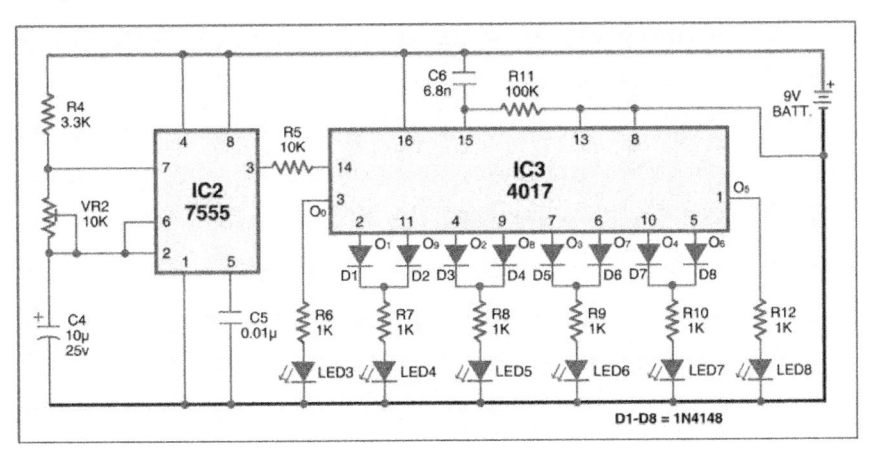

132. Automatic Bike Turning Indicator

The microcontroller (MCU)-based automatic turn signal system for bikes that is on the market is expensive and hard to program. Here is a circuit that is easy to build and doesn't cost much. The circuit is used to tell a bike or two-wheeler to turn left or right. There must be two identical circuits, one for the left and one for the right.

The circuit is made up of an ADXL335 accelerometer sensor, a voltage regulator 7805 (IC1), an LM393 comparator IC (IC2), two NE555 timer ICs (IC3 and IC4), and a few other parts. ADXL335 is a small, thin, low-power, 3-axis (X, Y, and Z) accelerometer with signal-conditioned voltage outputs. It can measure acceleration in the X, Y, and Z directions. In this project, only the Y direction of the ADXL335 is used. The device has a full-scale range of at least 3 volts and can measure acceleration. It can measure both static accelerations, like the force of gravity, and dynamic acceleration, like the force of motion, shock, or vibration.

Dual comparator LM393 is an IC with 8 pins. Pins 1, 2, and 3 make up one comparator, and pins 5, 6, and 7 make up another. The two comparators are used to watch for signals from the left turn and right turn indicators. We use two NE555 timer ICs (IC3 and IC4) set up as a monostable multivibrator, one for the left signal and one for the right signal. IC voltage regulator is used to turn a 9V-12V battery into +5V DC (IC1).

Left signal

When the bike handle is turned to the left, a voltage between 1.2V and 2.6V is sent as a tilt angle output. The pin 2 of IC2 is connected to the Y signal from the ADXL335 sensor, and the pin 3 is connected to the preset signal (VR1). The left tilt angle signal comes out of pin 1 of IC2. Use preset to set the reference voltage to 2.2V at pin 3. (VR1). At first, when the bike handle is turned 90 degrees to the right, the voltage at pin 3 of the comparator will be 2.2V and the voltage at pin 2 will be about

2V. So, the output of the comparator will be high (5V). Pin 2 of IC3 gets this high output signal. Because of this, IC3 won't be able to do much.

When the bike handle is turned to the left, IC2's pin 2 will have 2.6V. This means that pin 1 has a low output. The low output signal is linked to IC3's pin 2, which is a trigger. The high signal at pin 3 of IC3 is caused by the low signal at pin 2. When transistor T1 works, LED1 lights up. The width of a pulse made by monostable IC3 is given by the following equation: t=1.1×R2×C4 seconds

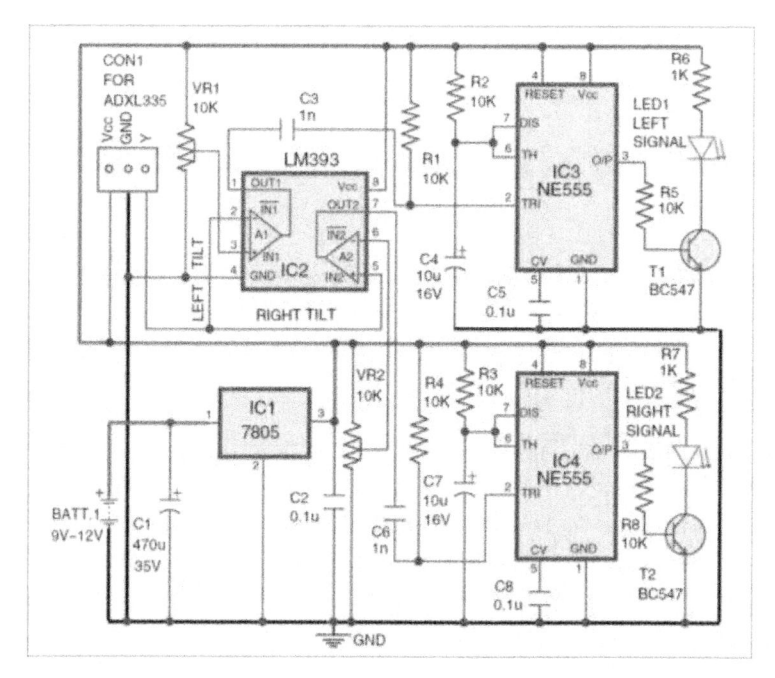

Circuit diagram of bike turning signal system

Right signal

When the bike handle moves to the right, it sends a voltage of 2.6V to 1.2V as a tilt angle output (decreases from high to low). Pin 6 of IC2 is connected to the preset (VR2), and pin 5 is connected to the Y signal from the ADXL335 sensor. VR2 is used to set the voltage reference at the inverting terminal to 1.6V. At first, when the bike handle turns 90 degrees, there will be 1.6V at pin 6 and 2V at pin 5. This means that the output of the comparator will be high (5V). This output goes to IC4's pin 2, which makes its pin 3 output low.

The following equation describes a monostable output pulse:

t=1.1×R3×C7 seconds

When the bike handle is turned to the right, 1.2V will be at pin 5 of IC2. This is less than pin 6's reference voltage of 1.6V. This turns pin 7 of IC3 into a low output. When the output is low, monostable multi-vibrator IC4 starts to vibrate. This causes pin 3 of IC4's output to be high, transistor T2 to conduct, and LED2 to light up.

133. Stabilized Power Supply for Prototyping

This stabilized power supply circuit may be directly linked to 230V AC mains to generate output voltages of 3V to 12V DC for connection to the prototyping board. Step-down transformer X1 takes an input of 230 volts alternating current from the mains and converts it to 15 volts alternating current. The secondary winding of this transformer can carry a current of 2 amperes. The alternating current (AC) is changed into pulsating direct current (DC) by a bridge rectifier, which results in a peak voltage level of 21 volts (151.4142). When there is output available from the rectifier, LED1 will light up to signify this fact. The current that flows through LED1 is kept at a safe level below 10 milliamperes by the 2.2-kilohm resistor known as R1. Capacitor C1 with a value of 470 F helps to smooth out the output of the bridge rectifier. Bypassing the high frequency ripple is accomplished using capacitor C2.

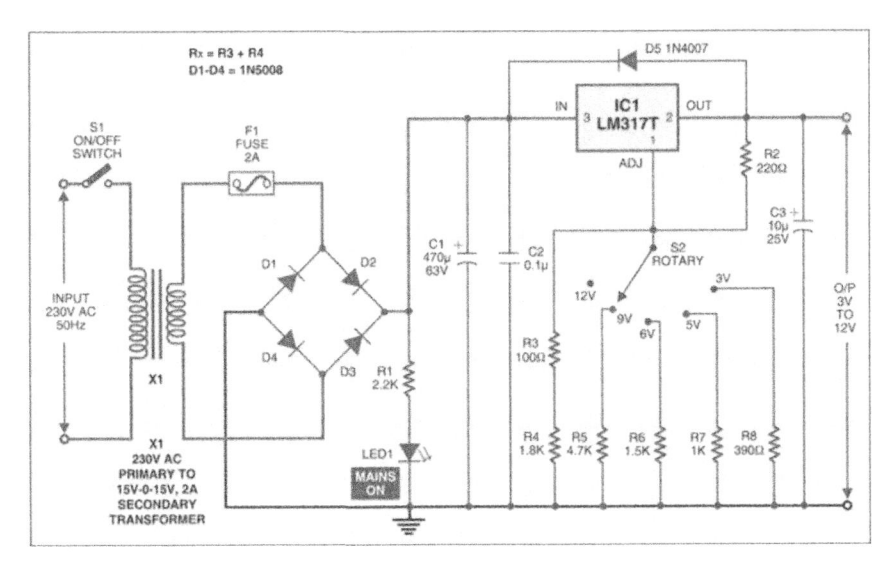

At the output of the rectifier section, a positive-voltage regulator from the LM317T series with three terminals is used for regulation. This regulator can be adjusted. It has the capability of supplying more than 1.5A across an output voltage range that goes from 1.2V to 37V. However, in this case it has been used to supply discrete voltages in steps of 3V, 5V, 6V, 9V, and 12V with the assistance of a 5-way rotary switch S2. This switch brings in different resistor values between the Adj pin of the regulator and ground, while R2 is a fixed resistor that has a value of 220 ohms and is located between the Adj pin and the output pin. The following equation can be used to get the voltage at the output (Vo): And 'Rx' denotes the resistance that may be found connected between the Adj pin of the regulator and ground. Rx has a value of 1900 ohms when it is set to the 12V position (the 'off' position of the switch), but when it is set to any of the other locations, it has a value that is the series equivalent of 1900 ohms when it is shunt with another resistance that is selected by the rotary switch. The rotary switch can be placed in any one of the locations shown in the table, which displays the equivalent series resistance in each position.

Note The X1 rating in the circuit diagram has been printed in the wrong place. Therefore, 0-15V should be understood as the expression 15V-0-15V. Switching with a discrete resistor that has a tolerance of 1% is preferable to employing a variable resistor because, after some use, the wiper contact becomes erratic, and the tolerance (variation with temperature) of a variable resistor is also significantly higher. Discrete resistor switching is preferred.

134. Propeller Message Display with Temperature Indicator

In this project, we will present a propeller display that is controlled by a microcontroller and can show any message that is provided to it through the hyper-terminal of a personal computer. In addition, a temperature detecting integrated circuit, model number TMP125, is installed into the propeller display so that the temperature can be displayed in real time.

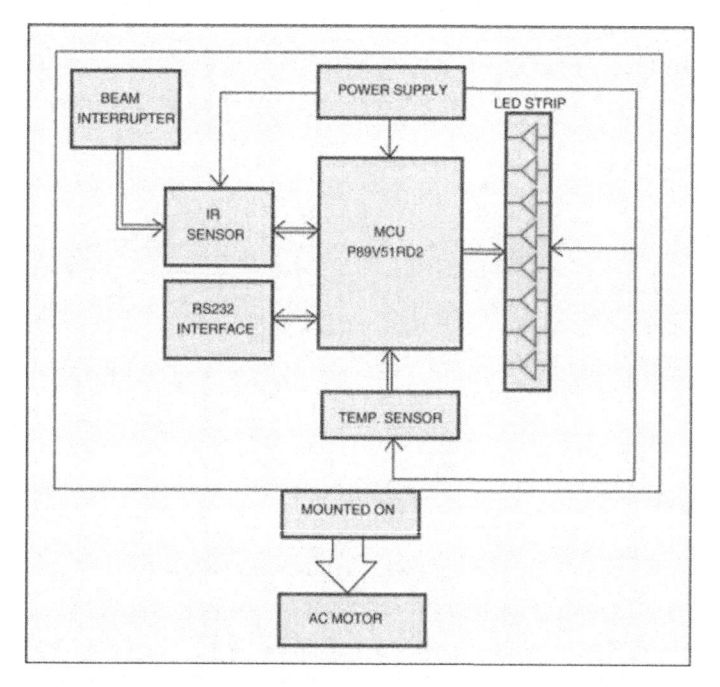

Fig. 1: Block diagram of the propeller message display with temperature indicator

Using only eight LEDs, it displays many characters in a rotating circular route and provides a view that encompasses the whole 360 degrees. It is a dynamic method of display that is both cost-effective and beautiful, and it decreases the complexity of the system overall while simultaneously improving its energy efficiency. A display that faces vertically and has a viewing angle of a full 360 degrees is the primary selling point of this idea. The idea that "persistence of vision" should be the driving force behind this endeavor was inspired by that phrase. This phenomenon is connected to the visual capabilities of the human eye, where it is believed that an afterimage can last for about one-quarter of a second. Therefore, if an image is observed at a rate of 25 times per second, it will appear as though it is one continuous picture to the observer.

Because of the rapid movement of the LED strip in this display, it is possible to make out a matrix of LEDs. In order to display a variety of characters, the duration of a single revolution is broken up into multiple shorter time periods, each of which is dedicated to controlling whether or not a specific LED is lit. The SPI bus is used to communicate the temperature readings from the temperature IC to the controller. The temperature indication is depicted as a block diagram in figure 1, which illustrates the propeller display. The LED strip is positioned vertically in such a way that when the motor rotates, the strip likewise revolves in a circular form, so providing a display that is visible from all 360 degrees. The RS232 interface, which is being controlled by hyper-terminal, is used to send the message that will be displayed. The message appears on the display for a period of thirty seconds, after which it is replaced by the current temperature, which is taken from the sensor that measures temperature. The IR sensor-beam interrupter assembly is responsible for the generation of interruptions.

Circuit and working

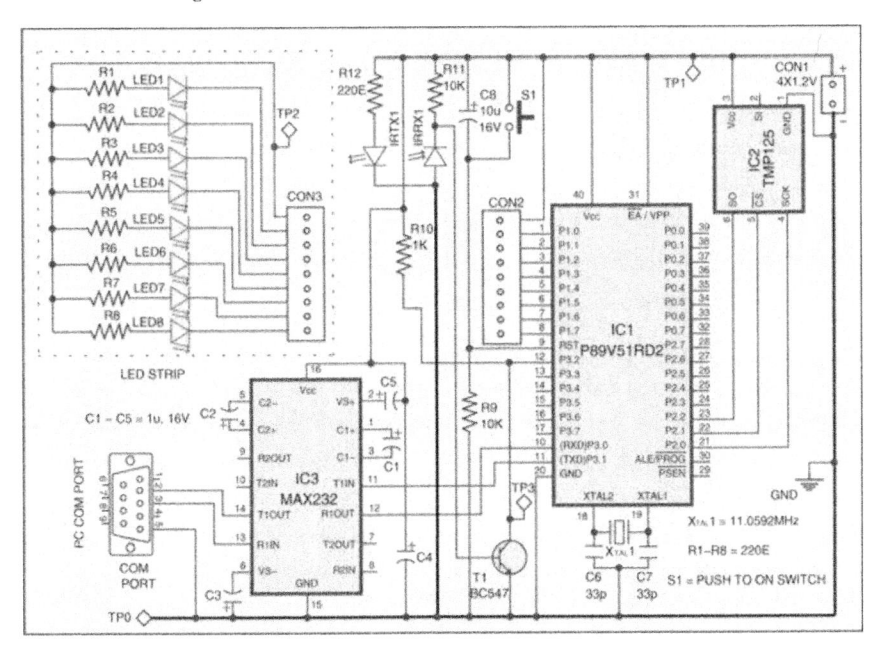

Fig. 2: Circuit of the propeller message display with temperature indicator

The circuit for the propeller display is illustrated in figure 2. The microcontroller P89V51RD2 (IC1), the temperature sensor TMP125 (IC2), the MAX232 (IC3), and a few discrete components are the building blocks for this design. The port pins P1.0 through P1.7 are linked to CON2, which also needs to have a connection made to CON3 of the LED strip, which contains LED1 through LED8. When turned at a high speed, these LEDs come together to form a circular display. Microcontroller. The microcontroller P89V51RD2 is the most important component of the system. The P89V51RD2 is an 8-bit 80C51 microcontroller with 64kB Flash and 1024 bytes of data RAM. It has four 8-bit input/output (I/O) ports, three 16-bit timers and counters, a programmable watchdog timer, eight interrupt sources with four priority levels, enhanced UART and serial peripheral interface (SPI), programmable counter array with PWM and capture/compare functions, and enhanced serial peripheral interface (SPI). It operates up to a 40 MHz crystal and features an oscillator and clock circuits integrated directly onto the device. It is possible to switch between 12 clocks every machine cycle, which is the default setting, and 6 clocks per machine cycle, which can be done through the software. Programming in parallel and in serial fashion can both be done in the Flash program memory.

The combination of the resistor R9 and the capacitor C8 is what provides the power-on reset function. The manual reset procedure is accessed through switch S1. The microcontroller receives its fundamental clock frequency from a combination of an 11.0592MHz crystal (XTAL1) and two capacitors with a value of 33pF each (C6 and C7). A sensor of temperature that has a digital output. The TMP125 is a temperature sensor that is compatible with the SPI protocol and is available in a very small SOT23-6 packaging. It is able to measure temperatures with an accuracy of 2 degrees Celsius over a temperature range of 25 degrees Celsius to +85 degrees Celsius and with an accuracy of 2.5 degrees Celsius over a temperature range of 40 degrees Celsius to +125 degrees Celsius. It does not require any external components. TMP125 is a suitable contender for low-power applications due to its low supply current and wide supply voltage range, which ranges from 2.7 V to 5.5 V. TMP125 is an excellent choice for extended thermal monitoring in a wide variety of applications, including those involving communication, computers, consumers, the environment, industrial processes, and instrumentation. In order to interface the temperature sensor, port pins P2.0 through P2.2 of IC1 are utilized.

135. Low Power Voltage Doubler

Batteries are required for operation of any and all small electronic devices. In order to function properly, several of them require battery voltages that are higher than the typical level. In the event that the battery with that particular voltage is not accessible, we will have no choice but to connect additional cells in series in order to raise the DC voltage. As a result, the true significance of the term "miniaturization" is lost. If the device in question is able to function with a low amount of current, using a voltage doubler is a straightforward method that can be utilized to solve this issue.

In this project, we will discuss a low power voltage doubler circuit that is easily adaptable for use with electronic devices that require a voltage that is higher than that of a conventional battery but only requires a little amount of operational current to function. The circuit is not very complicated because there are not many components in it. Despite this, the efficiency of the output is between 75% and 85% across the whole operating voltage range. At the end of the circuit, the voltage supplied by the battery has almost been increased by a factor of two.

Voltage doubler circuit

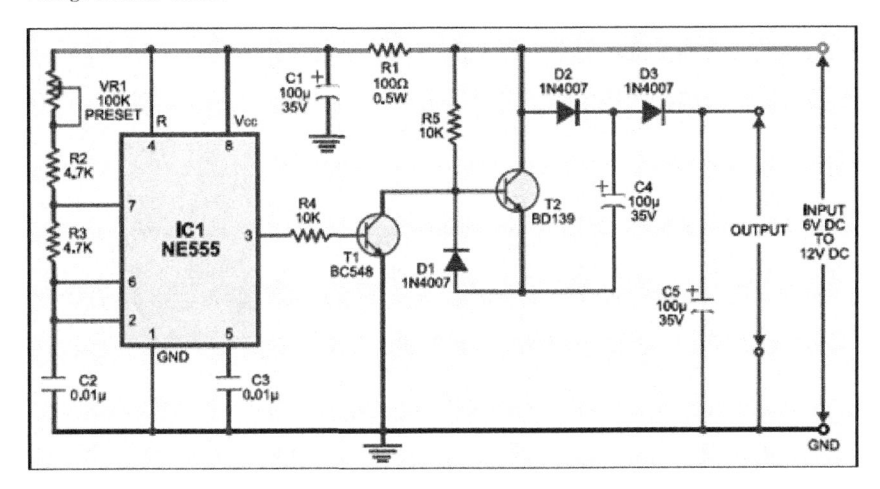

In this configuration, IC1 is wired as an astable multivibrator in order to produce rectangular pulses at a frequency of around 10 kHz. This frequency of the pulses, as well as their duty cycle, can be altered by utilizing preset VR1. The pulses are fed into the switching transistors T1 and T2 in order to drive the output section, which is set up as a circuit that doubles the voltage. The voltage across capacitor C5 has been increased by a factor of two.

Circuit operation

The high-level causes transistor T1 to reach its saturation throughout each cycle of the pulse's occurrence, which prevents transistor T2 from turning on. Therefore, transistor T1 charges capacitor C4 to a voltage level that is somewhat lower than the supply voltage via the path provided by diodes D2 and D1 in the circuit. Nevertheless, when the pulse is at its lowest point, the transistor T1 is turned off while the transistor T2 is forced into saturation. Now, transistor T2 adds one more increment of charge to the capacitor C4's negative pole, bringing it up to a level that is equal to the supply voltage.

As a result, capacitor C5 receives an equal amount of charging through diode D3, which causes it to charge up. Because of this operation, the total voltage across capacitor C5 is increased to almost exactly twice as much as the input voltage. If the output of the pulse generator is maintained with an amplitude and frequency that are sufficiently high, then the output voltage and current will remain constant and will be tailored to meet the requirements of the load. This circuit almost completely lacks ripple voltage, and that is despite the presence of the half-wave function. The efficiency can be estimated to be in the upper 90 percent levels if the connected load does not require a large current.

Because the input voltage is increased by a factor of two, the amount of current drawn from the input power source is also increased by a factor of two at the input but decreased by a factor of two at the output.

136. Automatic water pump controller

Here is a circuit for an automatic water pump controller that runs the motor of the water pump. When the water level in the overhead tank (OHT) drops below the lower limit, the motor turns on by itself. In the same way, it turns off when the tank is full. The circuit is simple, small, and cheap because it is made up of only one NAND gate IC (CD4011). It gets its power from a 12V DC power supply and doesn't use much power.

There are two parts to the circuit: the controller circuit and the indicator circuit. The controller circuit is shown in Figure 1. Let's say there are two reference probes, "A" and "B," inside the tank. "A" is the lower-limit probe, and "B" is the upper-limit probe. The 12V DC power supply goes to probe C, which is the minimum amount of water that can always be kept in the tank.

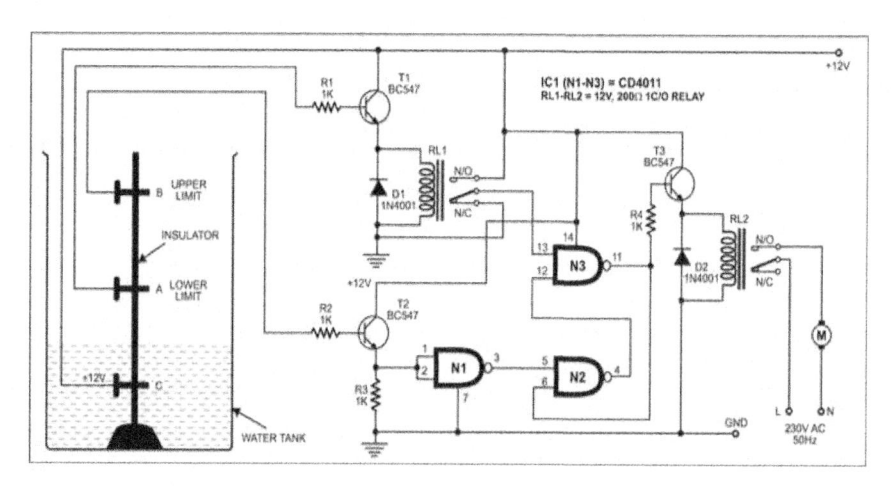

The lower limit "A" is connected to the base of transistor T1 (BC547), which is connected to the 12V power supply through its collector, and to relay RL1 through its emitter. Pin 13 of NAND gate N3 is hooked up to relay RL1.

In the same way, the upper-limit probe "B" is connected to the base of transistor T2 (BC547), whose collector is connected to the 12V power supply and whose emitter is connected to pins 1 and 2 of NAND gate N1 and ground via resistor R3. NAND gate N2's output pin 4 is connected to NAND gate N3's pin 12. Resistor R4 connects the output of N3 to the input pin 6 of N2 and the base of transistor T3. The motor is turned on and off by connecting relay RL2 to the emitter of transistor T3.

Circuit operation

If the tank is filled below probe A, transistors T1 and T2 don't work, and the output of N3 goes high. This high output turns on relay RL2, which drives the motor to pump water into the tank. When the water level in the tank is above probe A but below probe B, the base voltage of transistor T1 is provided by the water in the tank, and relay RL1 turns on to make pin 13 of gate N3 high. But the water in the tank doesn't give transistor T2 a base voltage, so it doesn't work, and the logic built around NAND gates N1 and N2 sends a low signal to pin 12 of gate N3. The result is that the amount of N3 produced stays high and the motor keeps pumping water into the tank.

Automatic water pump controller: Indicator/monitoring circuit

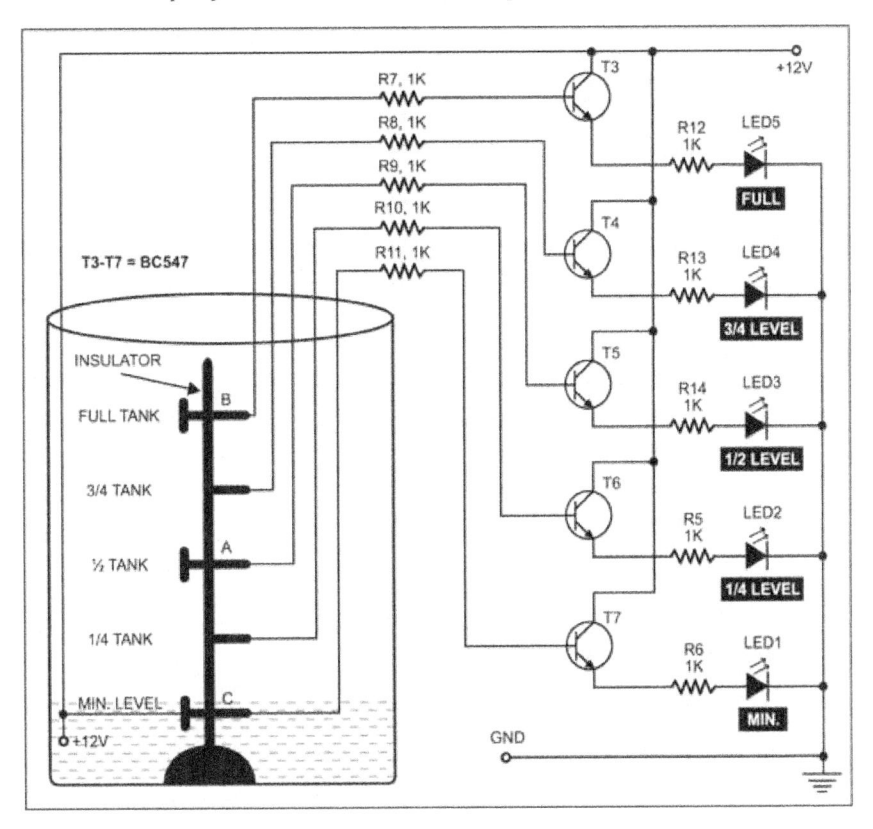

When the tank is full to the level of probe B, the water in the tank still gives base voltage to transistor T1 and turns on relay RL1, which makes pin 13 of gate N3 high. At the same time, water in the tank gives transistor T2 its base voltage, and the logic built around NAND gates N1 and N2 sends a high signal to pin 12 of gate N3. The output at pin 11 of N3 goes low, which stops the motor from pumping water into the tank. When the water level in the tank drops below probe B but stays above probe A, the water in the tank still gives base voltage to transistor T1 and keeps relay RL1 on, which makes pin

13 of gate N3 high. But transistor T2 doesn't work, and the logic made up of NAND gates N1 and N2 sends a high signal to pin 12 of N3. So, the output of N3 stays low, and the motor doesn't move. When the water level goes below probe A, neither T1 nor T2 will work. When the output of NAND gate N3 is high, relay RL2 is turned on, and the motor starts pumping water into the tank again. Figure 2 shows the circuit for indicators and monitoring. It is made up of five LEDs that light up to show how much water is in the tank above. Since the water at the bottom of the tank gets 12V power, transistors T3 through T7 get base voltage and conduct to light up the LEDs (LED5 down through LED1).

When the water level in the tank drops to level C, the transistor T7 starts to conduct, and LED1 lights up. When the water level in the tank gets to be one-fourth full, transistor T6 conducts, and LED1 and LED2 light up. When the water level in the tank gets to be halfway full, transistor T5 conducts and lights up LED1, LED2, and LED3. When the water level reaches 3/4 of the tank, transistor T4 conducts and lights up LED1 through LED4. When the tank is full, T3 conducts, and all five LEDs light up. So, one can tell how much water is in the tank by how bright the LEDs are (see the table). The LEDs can be put anywhere, making it easy to keep an eye on them.

Note: The heights of probes A and B can be changed by the user to change how much water needs to be put in the tank. The stand and the screws for adjusting it should be insulated so that they don't short out.

137. Night Lamps

LEDs are used in the following two different night lamp circuits. Both of them might be utilized in different ways; one as a night-vision clock, and the other as a TV lamp. Both of the circuits are AC-operated and draw an extremely low amount of electricity. Additionally, these are guarded against oscillations in the mains. The night-vision lamp has twelve LEDs laid out in a circular design similar to that of a wall clock, whilst the TV lamp has twenty-four LEDs placed in a prism-like pattern.

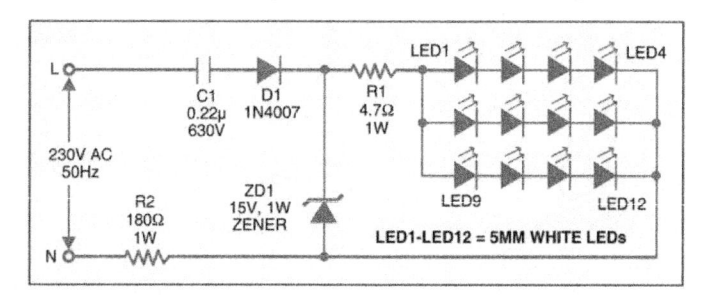

Fig. 1: Night-vision clock circuit

The circuit of the night-vision clock is depicted in Figure 1. The voltage is kept at a safe level by the 15V Zener diode, the current is lowered to a safe limit by capacitor C1 (0.22uF), which is responsible for reducing the current, and diode D1 is responsible for providing rectified DC, which brings out the characteristic color of white LEDs. The circuit is guarded against a high inrush current by the resistor

R2, and it has additional LED protection provided by the resistor R1, which is employed.

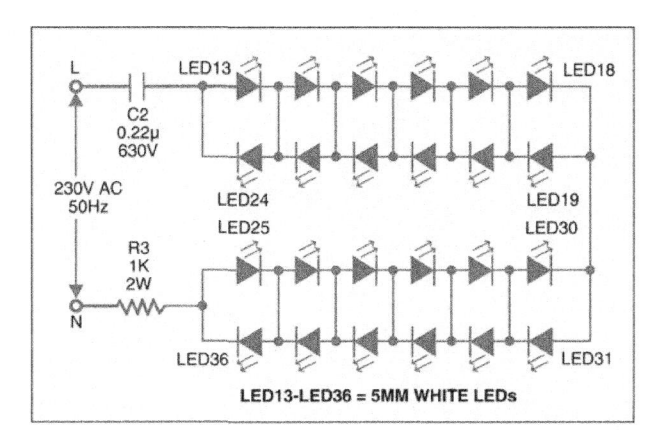

Fig. 2: TV lamp circuit

Assemble the circuit on a PCB that serves a general function, and then mount the white LEDs (LED1 through LED12) in the circular format of the clock dial, using drilled holes and the appropriate spacing between each LED. In a similar manner, it is possible to place LEDs on another dial of the clock. Figure 2 depicts the circuit for the TV lamp. Watching television in a dimly lit room causes pressure on your eyes. As a result, a lighting of 10-15W is advised. The Ken Schultz circuit is what people commonly refer to as this particular circuit. It is perfectly balanced against surges and voltage spikes and uses a total of 24 white LEDs (LED13 to LED36). Almost no heat is generated by the bulb.

Install the white LEDs (LED13 through LED36) in a triangular arrangement before putting the circuit together on a PCB that can be used for a variety of purposes. Next, place it within the TV lamp fitting that was suggested.

138. Continuity Tester with a Chirping Sound

The celebration is taking place at your house today. And at the same time, you are diligently searching for flaws in the ornamental lights that are strung along the exterior of your home. It is important that the task be completed before the evening. However, daylight makes it more difficult for you to detect whether the neon bulb included within the tester is glowing or not, which adds to your sense of irritation. The use of a neon bulb, which glows brightly when live voltage is present, is typical of continuity testers designed for use with live voltage. However, their brilliance is only noticeable when used inside. If it were an extremely bright LED or buzzer, it would be possible to see it in the daytime even when you were outside the house.

The live-wire scanner that is being discussed in this article not only makes use of an LED, but also a piezobuzzer that makes a chirping sound like a bird when it locates the place of break in a wire that is located inside the sheath.

The circuit is made up of a few discrete components in addition to a piezobuzzer that is wired all the way around a CMOS Johnson decade counter CD4017 (IC1). This decade counter/divider contains ten decoded outputs (Q0 through Q9), a divided (divide-by-10) output that is accessible at its pins 3, 2, 4, 7, 10, 1, 5, 6, 9 and 11, respectively, and a carryout bit at pin 12. At the positive edge of each input pulse that is received at pin 14, the count of the CD4017 advances by one. After Q9 at pin 11 gets high, this decade sequence will continue to repeat.

Continuity tester circuit

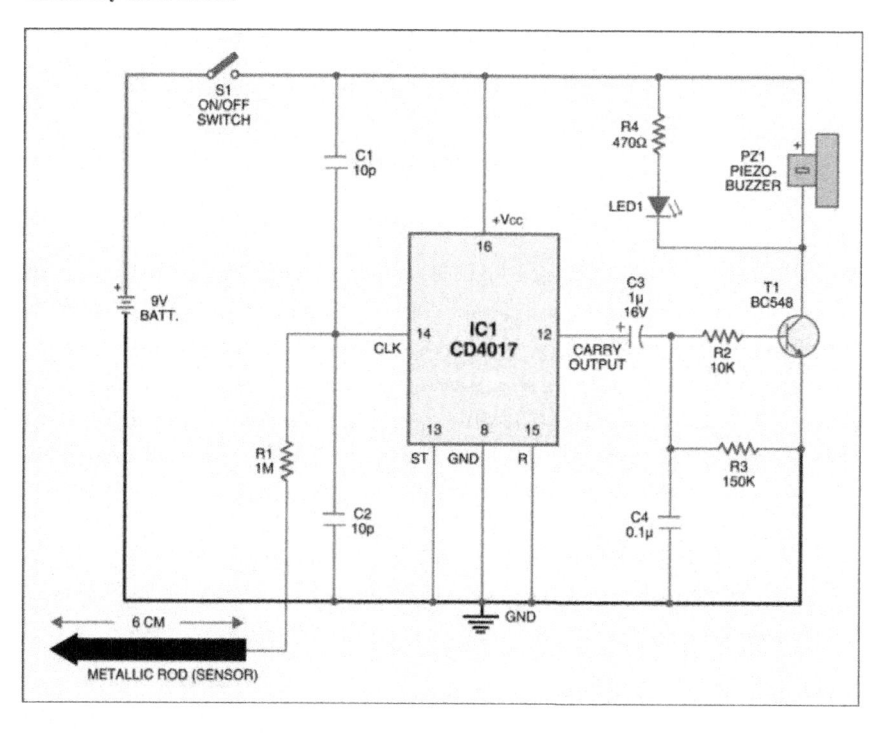

Because of the extraordinarily high input impedance of CMOS integrated circuits and the fact that they are voltage-controlled devices, it is simple to set off a CMOS IC with even relatively weak stray signals, such as the electric field of a 220V live-wire. The same fundamental idea underlies the operation of this circuit. Even if a live wire is 20 centimeters distant from a metallic strip, the potential across clock input pin 14 of IC1 can still swing to high and low logic levels when it is attached to a small metallic strip. This is because the potential is connected to a small metallic strip. This causes the counter to start counting, and a squarewave output with a frequency of 50 tenths of a hertz, or 5 Hz, is generated at pin 12 of IC1.

Now, this extremely high sensitivity of the clock input can be decreased to the required extent by connecting a voltage divider composed of passive components of the proper value. This, in turn, decreases the input impedance, which, in turn, lessens the sensitivity. The device is able to respond from up to 10 centimeters distant from the live wire in the event of an emergency. The wire can be scanned with this configuration to look for any breaks or discontinuities inside the sheath where it is housed.

Functioning of the circuit

When IC1's output pin 12 is driven high, a pulse is sent through capacitor C3 to make transistor T1 conduct, and it also charges capacitor C4 at the same time. Because of the charged capacitor C4, which is discharged through the resistor R3, transistor T1 continues to conduct for a little period of time after pin 12 is brought low. When capacitor C4 is discharged, there is an exponential movement of the transistor T1 from the saturation region to the active region and subsequently to the cut-off zone. Variations in the voltage across the buzzer cause corresponding shifts in the pitch of the sound produced by the buzzer. The chime of the buzzer is silenced at last. A chirping sound is produced as a result of the repetition of this operation at a frequency of 5 Hz. During this operation, LED1 will also blink. This continuity tester circuit does not require any ground terminal to be touched with a finger, as is required by traditional continuity testers. Therefore, it is quite safe in the event that a contact within the circuit becomes dysfunctional. Install all of the components, including a lithium battery, on a PCB designed for general use, then enclose the whole thing in a plastic cabinet.

139. Circuit for UPS to Hybernate Computer (PC)

The vast majority of the low-power UPS systems that are currently on the market do not include the functionality necessary to power down the computer before the system powers down by itself due to low battery. Some of them have the capability, but in order to use it they need the appropriate software.

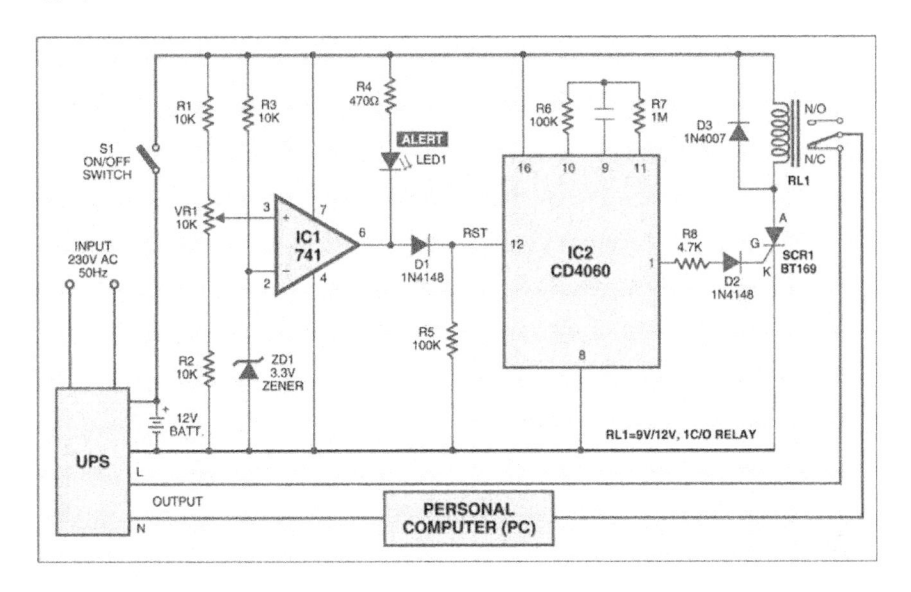

In this article, we will discuss an "add-on" circuit for uninterruptible power supplies (UPS) that will hybernate a computer automatically before the UPS shuts down due to low battery voltage. There is no need to install any enabling software. Following the comparator IC 741 (IC1) comes a short-duration positive pulse generator that makes use of a 14-stage ripple-carry binary counter/divider and oscillator IC2. Finally, the circuit concludes with a logic level shifter IC2. Because the voltage at the non-inverting (positive) input terminal of IC1 is typically higher than the voltage at the inverting (negative) input, the output of the comparator is also typically higher than the voltage at the inverting (negative) input. This high output from IC1 is put to use in the process of resetting IC CD4060 (IC2), which is a ripple counter with 14 stages.

When the voltage of the UPS battery drops below the value that has been established, the non-inverting terminal voltage of comparator IC1 at pin 3 begins to decrease. In this circuit, when the voltage of the battery drops below 9.5 volts, the voltage at pin 3 of IC1 drops below 3.3 volts, and the output of the comparator at pin 6 goes low. This causes LED1 to light up, which notifies the user that his computer is going to hybernate within the allotted amount of time (around 3 minutes).

After three minutes, a high output will be produced at pin 1 of IC2 because of the low voltage level that is present at pin 12 of that component, which enables it to oscillate and causes the counter to begin counting. The high output is utilized to trigger the SCR1, which in turn activates the RL1 relay. It is only when the power supply to the circuit is manually cut off through S1 that the relay will lose its ability to conduct electricity. The circuit is constructed on a small printed circuit board and then wired to the terminals of the UPS battery. The voltage needed to activate hibernation in the circuit has been configured to be higher than the cut-off voltage of the UPS. (If the cut-off voltage of the UPS is 9.5 volts, set the voltage level at which hybernation activation is activated to 10 volts.) Within the UPS cabinet, the circuit has been permanently installed. Both the switch S1 and the LED1 are going to be located on the front panel of the cabinet thanks to some wiring.

140. Contactless Telephone Ringer

This fully transistorized, basic circuit built as a contactless telephone ringer gives a signal of an incoming telephone call from a remote location within the building, such as a kitchen or bedroom. In order to put it into action, you will need to wind at least five turns of a short connection insulated wire around one of the wires of a twin telephone cable. When your phone rings, an alternating current signal with a frequency of approximately 60 hertz is produced in the telephone connection. Resulting in a fluctuating electric field around the telephone wire and the development of a slight induced voltage on the connection cable. When this induced voltage is linked to the base of the transistor T1, it creates a forward bias in the transistor, causing it to conduct current as the ring moves through the telephone cable. As a direct result of this, transistors T2 through T4 conduct as well. As a direct consequence of this, the piezobuzzer emits a sound, and LED1 illuminates, demonstrating that the telephone is now ringing.

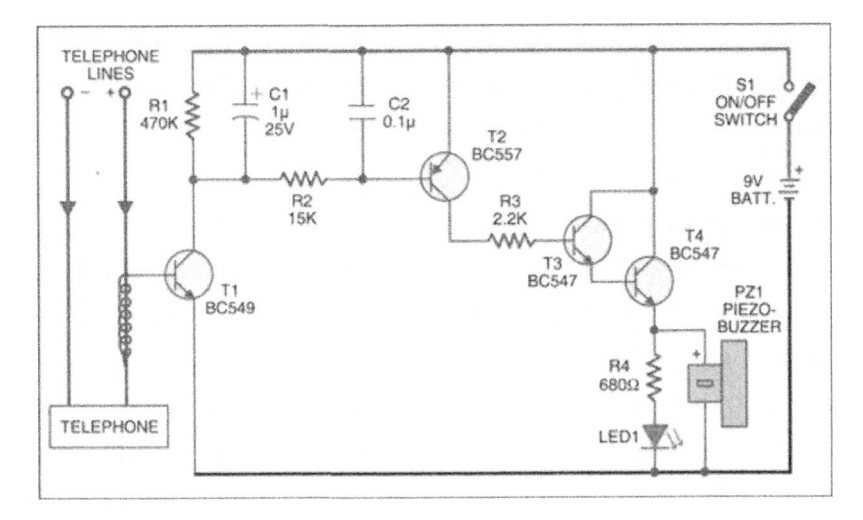

Put together the circuit on a PCB that may be used for a variety of purposes, and then wind an insulated electric cable around one of the telephone wires. A separate connection for the buzzer and LED can be made outside of the area occupied by the telephone receiver in order to provide both audible and visible indication. To power the circuit, you will need a 9-volt battery.

141. Automatic Dimness Controlled Lighting System

Using this automatic control system, you may program the tube lights to turn on automatically when night falls and turn off when the sun comes up. A lighting system that is controlled automatically by the presence or absence of darkness will cause a light source, such as a bulb or tubelight, to glow whenever darkness is detected. The circuit is powered by 5V that has been regulated, and it consists of a triac BT136, a NOT gate 7404, and a light-dependent resistor (LDR).

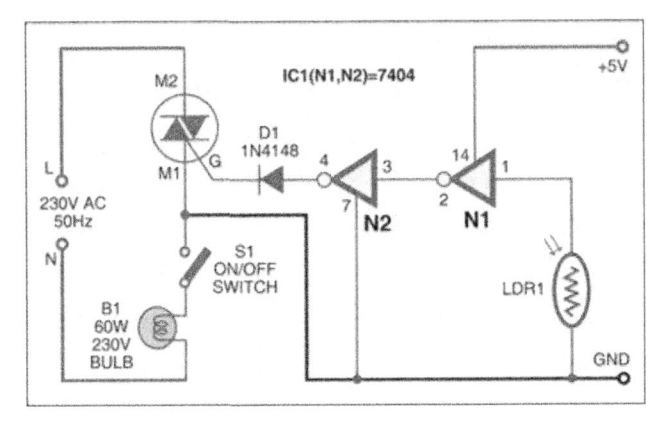

Figure: Circuit of automatic darkness-controlled lighting system

The operation of the circuit is really straightforward. The low resistance of LDR1 causes pin 1 of gate N1 to be low and for pin 2 of its output to be high when it is daytime. This high output is connected to the third input pin of the gate N2. As a direct consequence of this, the output of gate N2 will now be low. Since triac BT136 (triac 1) does not receive any gate signal, it behaves as an open circuit; hence, the light bulb does not glow. When it is nighttime, the high resistance of LDR1 causes pin 1 of gate N1 to become high, while the output pin 2 of gate N1 becomes low. This low output is connected to the third input pin of the gate N2. As a consequence of this, the output of gate N2 becomes high, which is then applied to the gate of triac BT136 (triac 1), where it functions as a short circuit, causing the bulb to begin illuminating.

Construct the circuit on a printed circuit board (PCB) designed for general use, then store it in an appropriate location. Maintain LDR1 in a position where it receives an adequate amount of light during the day. You can also use this circuit to control the lights at the street intersection.

142. Simple HF Power Amplifier

The following item is a high frequency power amplifier for the 40m (7MHz) band that is both affordable and powerful. The circuit can only receive between 20 and 30 milliwatts of RF power before amplifying it to the wattage level. Therefore, its output can be routed to an antenna, and its input can be directly connected to a VFO.

A circuit for a high frequency power amplifier

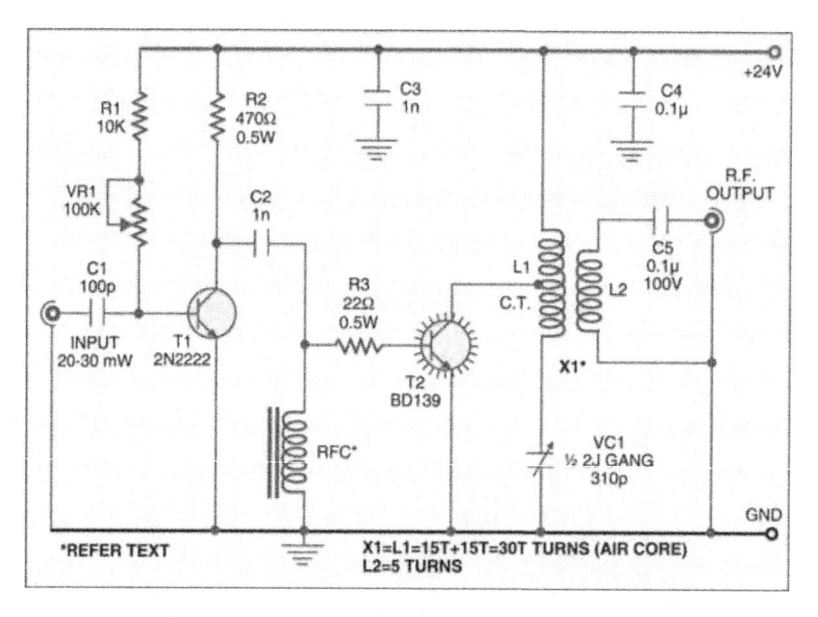

The amplifier is powered by a 24V DC supply. Make the necessary adjustments to potmeter VR1 to achieve maximum power output. The RFC coil is made up of fifteen turns of 36-gauge wire that are wound around a ferrite balun core for a TV. Coil L1 is made up of 30 turns (15T+15T) of 24SWG copper-enameled wire that are coiled over a plastic tube that is 1.5 centimeters in diameter. It is center-tapped, and the collector of transistor T1 is linked to the air core. Five loops of wire with a gauge of 24 gauge were used to wind Coil L2 across the center of L1. It is recommended that the transistor T2 be mounted atop a heat sink made of aluminum with a thickness of 3 millimeters.

143. Demo Circuit for Over-Voltage Protection

Circuits designed to protect voltage-sensitive loads from excessive voltage are called over-voltage protection circuits. Transients in voltage can be caused by a variety of factors, including the switching of loads and transformers, as well as short circuits and open circuits in the rectifier circuit and the regulator circuit. These kinds of transients have the potential to disrupt the normal operation of an electrical circuit or perhaps cause it to become damaged. As a result, the use of an over-voltage protection circuit is obligatory for the purpose of shielding expensive loads from any and all sources of voltage transients. Students who are studying electronics engineering and are required to complete an experiment on over-voltage protection as part of their coursework can use the circuit that is currently being discussed to perform a highly convincing demonstration of the effect.

Circuit and the working

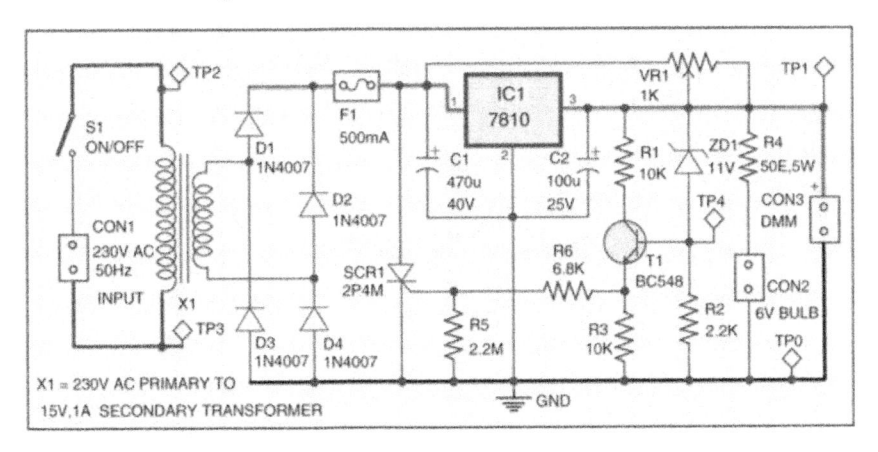

The demonstration circuit for the over-voltage protection is shown in the figure. A rectifier consisting of four 1N4007 diodes (D1 through D4), a 10V voltage regulator IC 7810 (IC1), an SCR 2P4M (SCR1), a transistor BC548 (T1), and a few additional components are used to construct it. In order to provide protection, SCR1 is utilized.

If the voltage is higher than what the device that needs to be protected can withstand (a 6V bulb in this case), the circuit will cut the gadget off from its supply. In order to demonstrate this, a potentiometer named VR1 is linked across regulator IC1, and this connection is utilized to raise the voltage that is produced by regulator IC1. When the voltage at the output of IC1 rises, the voltage at the base of transistor T1 likewise rises. This causes SCR1 to become activated through the medium of resistor R6. When SCR1 is activated, the fuse will blow, which will cut off the device's connection to the power source.

Turn on the circuit once you have set VR1 so that it is at its highest possible setting (let's say, 1k). Take a reading of the output with a digital multimeter at the CON3 terminal. It needs to be somewhere around 10.3 V. Now, gradually lessen the resistance of VR1. The multimeter displays 10.9 V when it is read at around 800. Continue to lower the resistance until the SCR1 relay opens. Following the delivery of a triggering pulse, it was discovered through experimentation that the SCR activates at a resistance of approximately 680, causing a substantial current to flow through the fuse wire. This causes the fuse wire to blow, which in turn causes the load to become disconnected from the supply.

144. 3V PC Adaptor

Connect this PC adaptor circuit to your computer's available USB output port to receive 50mA and 3V DC. Therefore, it can be utilized to do tasks such as recharging two NiCd cells (1.2V times 2) found in a portable music player system.

PC Adaptor Circuit

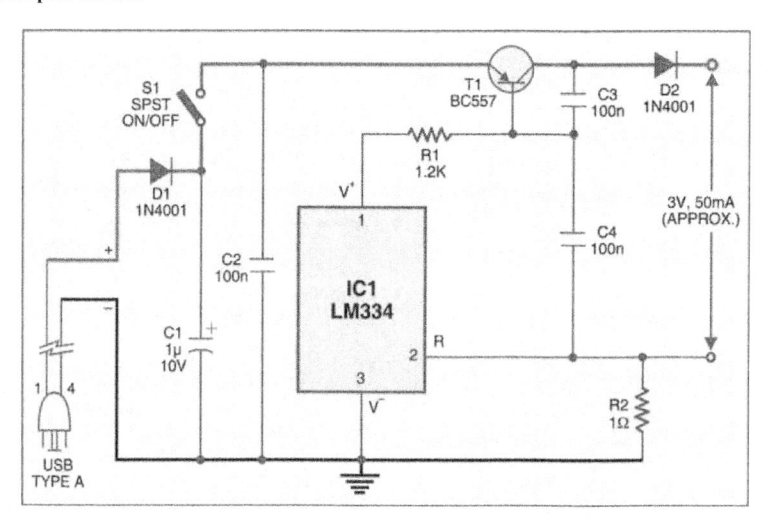

The input of 5V is converted by the circuit into an output of around 50mA at 3V. The circuit is based around the widely used 3-pin adjustable current source LM334 (IC1). A TO92-style pack, which is readily available, is utilized here. Through the use of resistor R2, the output current level is kept constant. The function of the transistor T1 is to enhance the current. For the input connection, all that is required is a regular USB cable measuring one meter in length and fitted with a type-A connector at one end. In order to make the output connection, you can either use a regular DC socket or flying leads equipped with crocodile clips, depending on what you need.

145. PC Table Lamp

This practical circuit for a table lamp is based on the well-known timer integrated circuit (IC) TLC555. When your computer is turned "on," it triggers the activation of an electric bulb. When you turn off the computer, the bulb will likewise go into an "off" state by itself.

Table lamp circuit

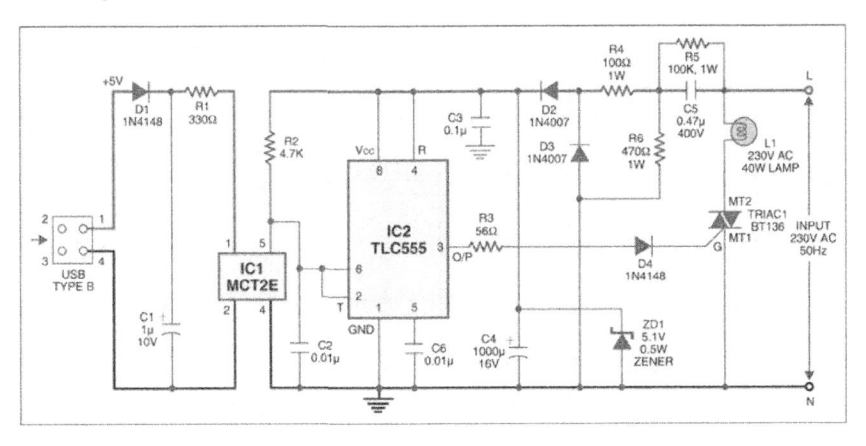

The usage of a step-down transformer in the power part of the circuit has been purposefully eliminated to eliminate the risk of electromagnetic radiation. This was done in order to avoid the danger of electromagnetic radiation. Instead, a capacitive potential divider is utilized to lower the supply voltage before sending it to the rectifier circuit. This results in the control circuit receiving the necessary operating voltage of 5V DC. The most important component of this arrangement is the capacitor C5. Even though there is AC mains supply available, the output pin 3 of IC2 is low when the PC is in the "off" state; as a result, TRIAC1 does not conduct. In order to separate the PC from the control circuit of the table lamp, optocoupler IC1 is being utilized in this situation. Bypassing the USB port's fluctuations is made possible by capacitor C1. The input current is controlled by the resistor labeled R1.

When the personal computer is turned "on," 5V DC is drawn from the USB port and applied to optocoupler MCT2E (IC1). This causes the component to conduct. This causes the input to pins 2 and 6 of IC2 to be pulled down. As a direct consequence of this, the output of IC2 becomes high, which causes TRIAC1 to be activated via resistor R3 and diode D4. With this, the supply to lamp L1 is finished, and the lamp now lights up. As soon as the personal computer is turned off, the input to the optocoupler is turned off as well. This causes the control circuit to stop working, as the input pins 2 and 6 of the timer IC go high while the output pin 3 goes low. The light in the room goes out.

146. Audible Continuity Tester

This inexpensive and easy-to-use auditory continuity tester just requires one LM339 quad comparator IC, a few resistors, and a piezo buzzer. One 9V battery is all that's needed to power the entire circuit, including the comparator. Only one of the quad comparators is put to use in its intended capacity, while the other three comparators, which are connected in parallel, are employed for the purpose of driving directly a medium power piezo buzzer. A voltage divider composed of the resistors R1 and R2 places the inverting pin 8 of comparator 'A' at a level of about 190 mV, while the resistor pair composed of R3 and R4 keeps the non-inverting pin 9 of the same comparator 'A' at a level of around 30 mV at all times. The necessary feedback is provided by resistor R5, which is linked between the output and the input that does not invert the signal.

Audible Continuity Tester Circuit

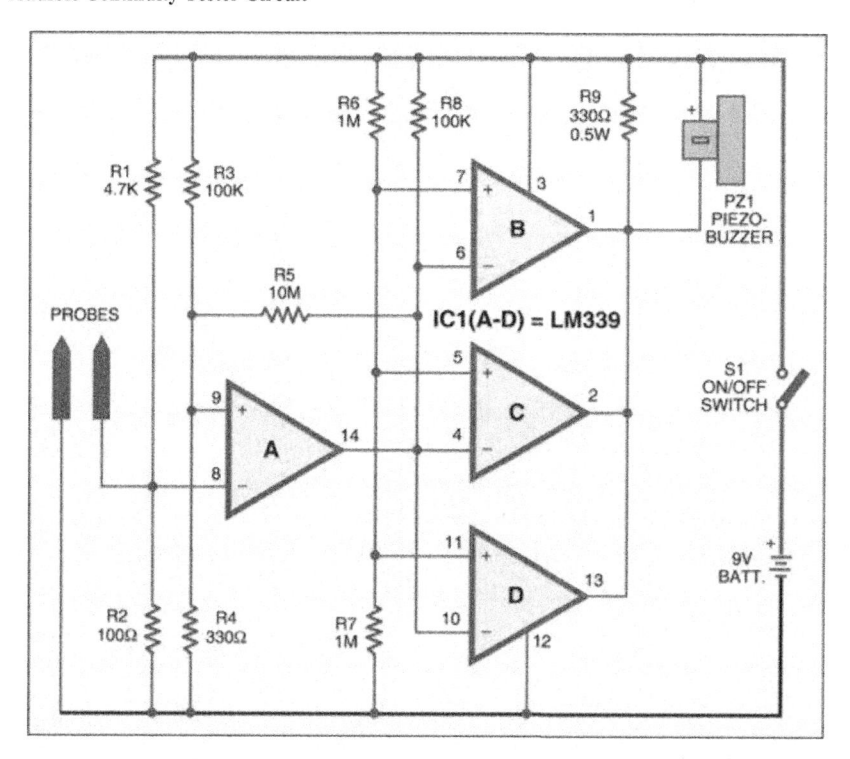

The use of positive feedback to introduce a little amount of hysteresis can be an efficient way to solve the problem. This has the effect of separating the switching points for going up and going down, such that once a transition has begun, the input needs to go through a large reversal before the next transition can take place.

Circuit operation

The output is at a low level while the device is in its quiescent state because the potential at the device's inverting pin is 190 millivolts (mV) higher than the potential at the device's non-inverting pin (30 mV). When the voltage is low at the inverting terminals of the comparators labeled 'B,' 'C,' and 'D,' which are connected in parallel, their outputs transition to the high state (closer to the supply voltage). Because of this, the buzzer's two input terminals are almost at the same voltage, which prevents it from emitting a sound. If a resistance of less than around 15 ohms is connected between the probe tips, then and only then will the inverting terminal of comparator 'A' fall below 30 mV. This will cause the polarity of the output pin 14 of comparator 'A' to switch. Because of this, the aggregate output of comparators 'B,' 'C,' and 'D' will drop, which will result in the buzzer being activated. In order to directly drive a buzzer with a medium amount of power, the output node of the three comparators is capable of sinking more than 50 mA of current. The buzzer will be able to be activated with a resistance of less than 15 ohms if you use the circuit that has been provided here. If you desire an even lower or higher resistance for turning on the buzzer, you can accomplish the same thing by adjusting the values of the resistors R1 through R4.

147. Anti-Theft Alarm

You can prevent a break-in by using this anti-theft alarm circuit that you set up. When someone tries to force their way into your house or place of business by banging, pushing, or hitting the door, the alarm goes off. A condenser microphone serves as the element of the sensor, and it is mounted inside the home somewhere on the front door or, more ideally, on the door frame. The door can be made to make noise by hitting it, pushing it, or knocking on it. This is picked up by the microphone, and the signal is sent on to the preamplifier component of the circuit. The flip-flop is what makes the connection between that section and the buzzer. As a consequence, the buzzer goes off whenever someone strikes or knocks on the front door.

Anti-theft alarm circuit

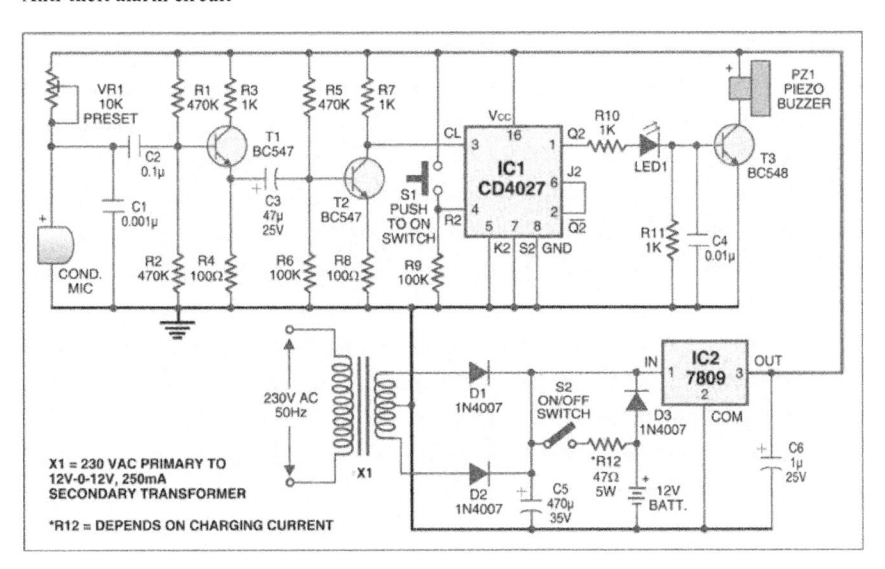

Anti-theft alarm circuit

Preamplifier, flip-flop, and power supply are the three components that can be separated from one another within the circuit. The signal that is created during knocking is amplified in the preamplifier stage by the transistors T1 and T2. This signal was obtained from the condenser microphone. The JK flip-flop CD4027 receives its clock signal from the output of the preamplifier. The JK flip-flop portion makes use of a twin JK flip-flop CD4027, however only one of the two flip-flops is active at any given time. When both the 'J' and the 'K' inputs are low, the flip-flop is stopped from working. The 'K' input of the IC CD4027 (pin 5) is connected to ground, whereas the 'J' input (pin 6) is connected to the Q output (pin 2). The output of the preamplifier is wired to IC1's clock input (pin 3), which allows the IC to function properly. The signal that has been amplified is then sent to the JK flip-flop IC1. The output of this flip-flop gets high as soon as it receives the input from the condenser mic, and it continues to be high until it is reset by the switch S1. The buzzer PZ1 is driven by the output of IC1 through the npn transistor T3. When there is a visitor, the buzzer will alert you to their presence. It will continue to be in the "on" position until you activate the reset push-to-on switch located on pin 4 of IC1.

In the section that deals with the power supply, the secondary winding of transformer X1 is wired up to the rectifier. The rectified output is fed to regulator IC2, which in turn provides a 9V regulated power supply for the circuit to operate. Additionally, the rectified output is used to charge the 12V, 4Ah battery through switch S2. To charge the battery, move switch S2 to the "on" position, and once it is fully charged, move it back to the "off" position. In the event that there is a disruption in the main source of power, the battery will provide backup power. In a nutshell, the sound that is picked up by the condenser microphone is amplified and then used as the input to trigger IC1. As a result, the buzzer will continue to ring until it is reset. At the same moment, the visual indication provided by LED1 is illuminated.

148. DIAC Controlled Flasher

This simple DIAC-controlled flasher is put to use in a variety of industrial contexts, such as in the capacity of a high-voltage indicator or a "on" indicator for machines. It gives a warning indicator by flashing once every second, and the design of it is quite straightforward. Without employing a printed circuit board (PCB), the project can be wired lead-to-lead. It is possible to encapsulate it within the mains box because it is directly powered from 220V AC.

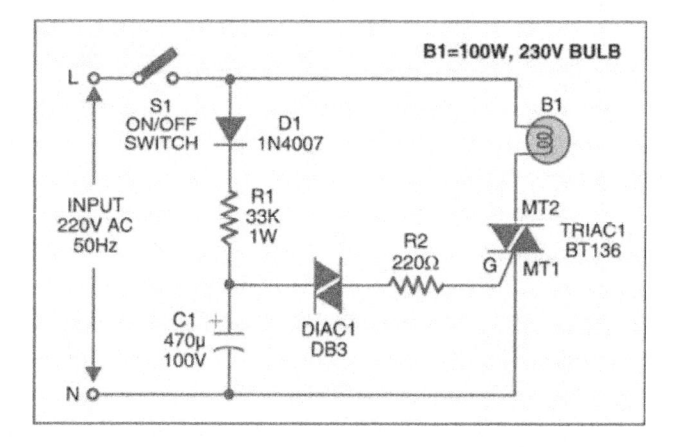

The 230V AC from the mains is rectified by diode 1, and then the 33-kilo-ohm resistor R1 brings the voltage down to a safer level. Because of the reduced flow of current to the diac as a result of the action of resistor R1 and diode D1, the triac continues to be non-conducting. Capacitor C1 has a rating of 470 microfarads and 100 volts. After the capacitor C1 has been completely charged, the voltage across the diac will grow, and it will begin to conduct electricity. This supplies gate current to the triac through R2 in the circuit. When the triac is activated, it completes the electrical circuit of the lamp, causing the lamp to light up.

The silicon-controlled rectifiers (SCRs) that make up the triac (model number BT136) are coupled in a reverse parallel configuration and share the same gate circuit with one another. Both of the SCR parts become conducting when the triac starts conducting. It is possible for DC, rectified AC, AC, or pulse sources like neon lamps or switching diodes like diacs to set off the triac's gate, which is a highly sophisticated component. In the event that the diac is conducting, the capacitor C1 will discharge via the diac and R2. This results in a decrease in the gate current of the triac, which causes it to shut off. When the C1 battery is fully charged again, the lamp will turn back on. Because of this, the flash rate of the lamp is determined by the charging-discharging cycle of capacitor C1.

149. Stereo audio Distribution Buffer for headphones

Most audio signal sources only have one stereo output, which means they can only drive a single pair of headphones with a resistance of about 32 ohms or a single line with a resistance of 600 ohms. But sometimes, like for entertainment, e-learning and training, or at home, more than one person needs to connect their headphones to the same sound source. When these things happen, it's not a good idea to use loudspeakers because they will bother other people in the room.

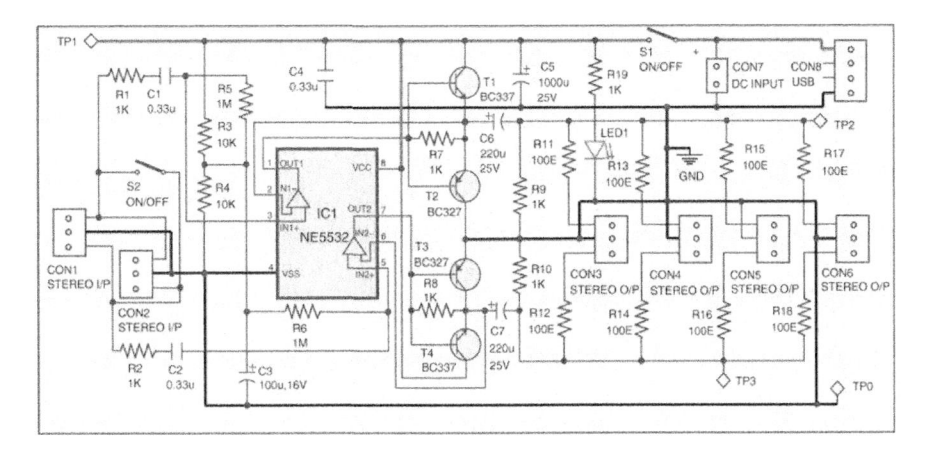

Also, sometimes the signal from one audio output needs to be sent to several power amplifiers that are many meters apart. In this case, the distribution device should be able to drive long audio lines with resistance between 100 and 600 ohms. A distribution buffer (DB) is added between the audio output and the headphones to meet this kind of need. Most headphones have a way to change the volume. So, to keep things simple, there is no volume control added to the distribution buffer.

Circuit and working

Figure shows the circuit of the stereo distribution buffer, which was built to solve practical problems that come up when the signal from a single stereo source needs to be buffered so that it can drive up to four headphones or small high-impedance loudspeakers with resistances higher than 32 ohms. The circuit has two channels that are the same. The buffer has one input for each channel and four outputs for each channel. The stereo signal comes in through CON1 connector. Connector CON2 has the same signal, so the second distribution buffer can be hooked up to that connector. R1 and R2 are input protection resistors with values that are usually between 100 ohms and 2 kilo-ohms. Their exact value depends on the signal source, the cables, the inputs, and other factors. The DC input component is cut with C1 and C2. Most of the time, the amplifier won't work right if DC voltage from the source is present. Also, C1 and C2 work as high-pass filters with the input resistance and should be calculated as follows: F3dB = 1/(6.28RinputC) The input resistance is 1 megaohm, and it mostly depends on what R5 and R6 are set to. Each channel has a gain of 1. Also, there are no potentiometers for controlling the volume because this keeps the circuit simple.

When S2 is closed, both inputs come in at the same time. Connectors CON3, CON4, CON5, and CON6 all have outputs that can be used for headphones or audio interfaces. The outputs can be used in any way that makes sense. The buffer can send out a maximum total current of about 100 mA. The buffer is made up of a dual operational amplifier (IC1) and a few other discrete parts. Other operational amplifiers like the RC4560, OPA2132, TL072, TL082, and even the LM358 can be used. However, for audio, it is best to use amplifiers that can drive loads of 600 ohms or less and have a high input impedance.

Depending on the operational amplifier that is used, the buffer can work with a wide range of power sources. When low-voltage rail-to-rail operational amplifiers are used, a 5V USB power supply can be used to power the buffer. The circuit will also work with 5V/12V DC wall adapters that provide at least 200mA of current.

150. PIN Diode Based Fire Sensor

Here is a PIN diode-based fire sensor that sounds an alarm when it senses fire. Thermistor-based fire alarms have a flaw: the alarm only goes off if the fire is close enough to heat the thermistor. For longer-range fire detection, this circuit uses a sensitive PIN diode as a fire sensor. It can pick up visible light and infrared (IR) light with wavelengths between 430nm and 1100nm. So, it's easy for both visible light and infrared light from the fire to set off the sensor and set off the alarm. It also looks for sparks in the mains wiring and sounds an alarm if they don't go away. It is a great way to protect stores, lockers, and record rooms.

PIN diode in the circuit, BPW34 is used as a light and infrared (IR) sensor. BPW34 is a two-pin photodiode with an anode (A) and a cathode (C) (K). The flat top of the photodiode makes it easy to find the end that is the anode. The anode is a small solder point to which a thin wire is attached. The cathode is the other small solder point.

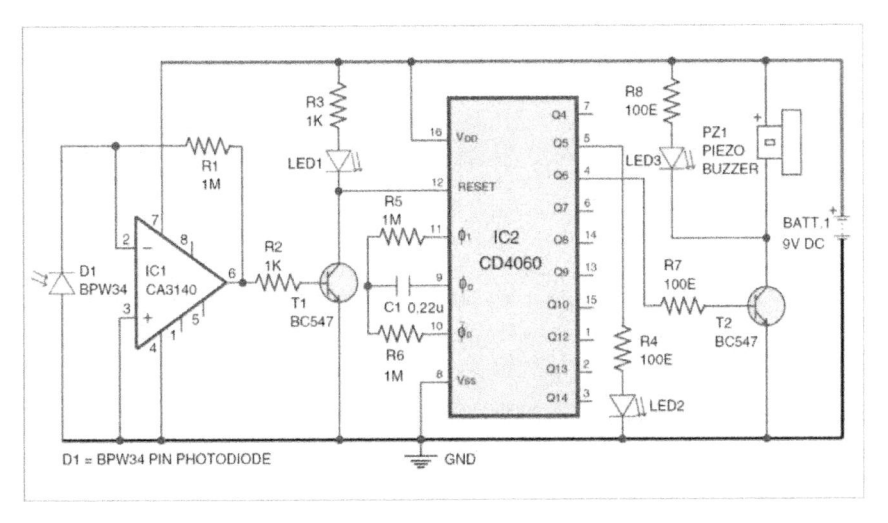

BPW34 is a small PIN photodiode or mini solar cell with a light-sensitive surface that makes 350mV DC when exposed to 900nm light. It is sensitive to both natural sunlight and firelight. So, it works perfectly as a light sensor. The BPW34 photodiode can be used both with no bias and with reverse bias. When light shines on it, its resistance goes down. In Figure shows a diagram of how the PIN diode-based fire sensor works. It is made up of a 9V battery, a PIN diode (D1), an op-amp (IC1), a counter (IC2), two transistors (T1 and T2), a piezo buzzer (PZ1), and a few other parts. In the circuit, the PIN photodiode BPW34 is connected in reverse-biased mode to the inverting and non-inverting inputs of op-amp IC1 to feed photocurrent into the op-input. amp's The CA3140 is a 4.5MHz BiMOs op-amp with MOSFET inputs and a bipolar output.

In the input circuit, gate-protected MOSFET (PMOS) transistors give a very high input impedance, usually around 1.5T ohms. The IC only needs a very small amount of current, as little as 10pA, to change the status of the output from high to low. In the circuit, IC1 is a transimpedance amplifier that converts current to voltage. The photocurrent made by the PIN diode is amplified and turned into the voltage that goes with it by IC1. The non-inverting input is connected to ground and the anode of the photodiode. The photocurrent from the PIN diode goes to the inverting input.

Circuit operation

Since the transimpedance amplifier is in inverting configuration, the gain is set by R1, which has a large value. When the non-inverting input is connected to ground, it gives the photodiode a low impedance load. This keeps the voltage of the photodiode low. The photodiode works in the photovoltaic mode when there is no outside bias. Feedback from the op-amp makes sure that the current through the photodiode is the same as the current through R1. In this self-biased photovoltaic mode, the photodiode's input offset voltage is very low. This lets a big gain happen without a big offset voltage at the output. This configuration is chosen to get a big gain when there isn't much light. In normal light conditions, the PIN diode's photocurrent is very low. This keeps the output of IC1 low. When the PIN diode picks up visible light or IR from a fire, its photo current goes up, and the transimpedance amplifier IC1 turns this current into a voltage that matches the output. When IC1's output is high, transistor T1 turns on, which makes LED1 light up. This means the circuit has picked up on fire. When T1 is open, it connects reset pin 12 of IC2 to ground. This makes CD4060 start to oscillate.

When C1 and R6 cause IC2 to oscillate, each of its ten outputs goes high one at a time. The blinking of LED2 shows that IC2 is moving back and forth. After 15 seconds, when the output Q6 (pin 4) of IC2 goes high, T2 conducts. This turns on the piezo buzzer PZ1 and makes LED3 light up. If there is still fire after 15 seconds, the alarm will sound again. By replacing PZ1 with a relay circuit, you can also turn on an AC alarm that makes a loud noise (not shown here). The contacts of the relay that is used for this purpose are what turn on the AC alarm.

151. Triple Mode Tone Generator

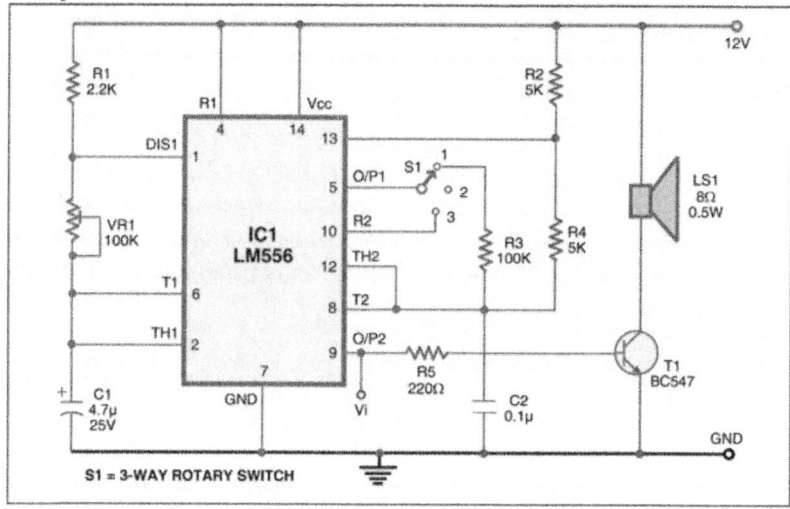

Fig. 1: Triple mode tone generator circuit

The following diagram depicts a straightforward triple mode tone generator circuit that can produce three distinct tones. You can put it to use as a burglar alarm, a call bell, or any other type of security alarm. The circuit of the triple mode tone generator is depicted in Figure 1, and the expansion of that circuit for a car horn is depicted in Figure 2. The circuit is powered by a battery with a voltage of 12 V. An LM556 dual timer IC, which contains two distinct LM555 timers built into it, serves as the primary component of the circuit. The initial timer is set up to function as an oscillator and has a somewhat broad frequency range. The output from the first timer serves as the trigger for the second timer. The location of the rotary switch determines which of three noises are produced by the circuit (S1). A single-pole, three-way switch is denoted by the symbol S1.

Functioning of the circuit

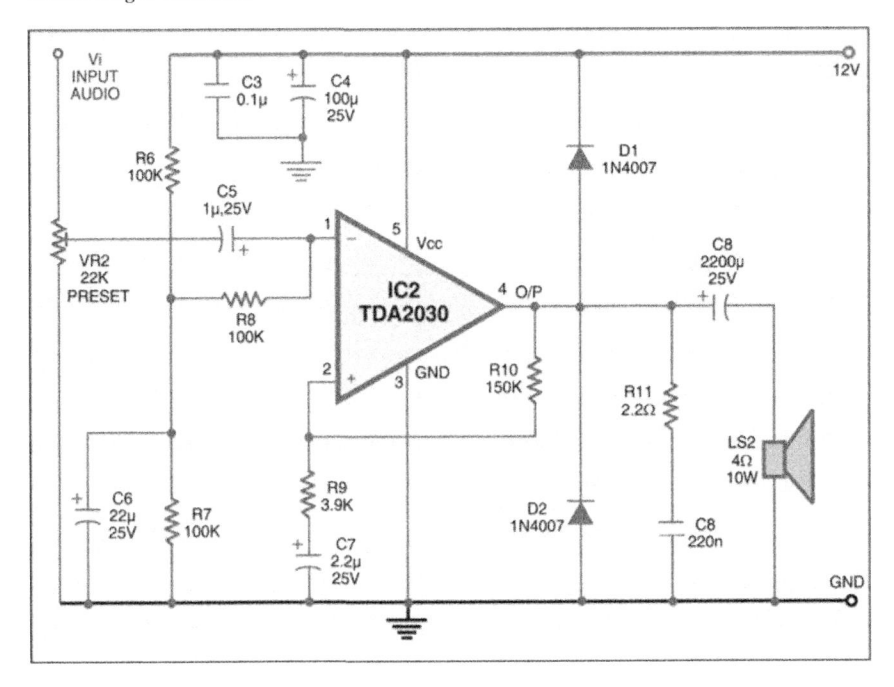

Fig. 2: TDA 2030A amplifier for automobile horn

A two-tone tone can be heard emanating from the output of the second timer located at pin 9 when switch S1 is set to position 1. When the switch S1 is moved to position 2, the output of the second timer, which is located at pin 9, produces a tone that is continuous. A tone burst is produced by the output of the second timer at pin 9 when switch S1 is moved to position 3, which is the default setting. Because the LM556 can only sink a current of 200mA, the output of the second timer is amplified by the transistor T1 which, in turn, causes the speaker to play in accordance with the tone.

After disengaging from the resistor R5, connect the output of the second timer at pin 9 to the audio signal Vi. This will make the sound louder and can also be used to produce a vehicle horn (Fig. 2). The TDA 2030 amplifier delivers a significant output current while simultaneously exhibiting exceptionally low levels of harmonic and cross-over distortion. In addition to that, it comes equipped with a traditional thermal shut-down system. On two different PCBs designed for general use, assemble the two distinct circuits. The PCBs can be enclosed in a suitable enclosure.

152. Bicycle Guard

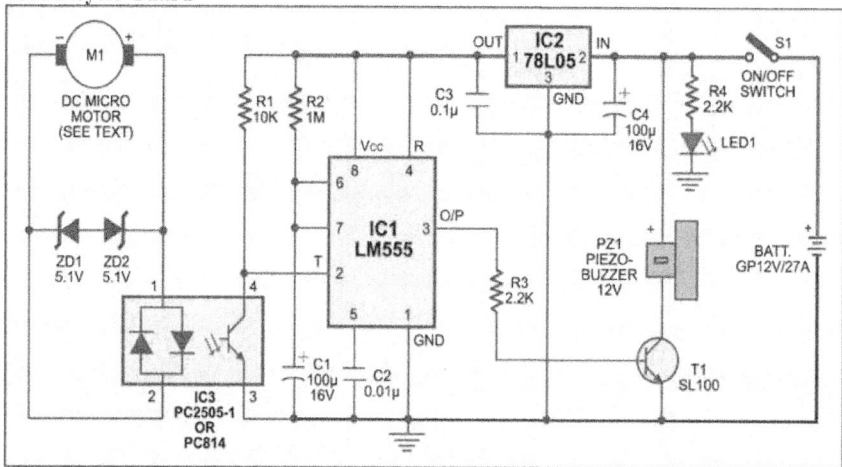

This anti-theft gadget for bicycles is simple, straightforward, and only requires a few components to put together. It also comes at a low cost. The circuit's most important component is a wheel rotation detector, which is implemented through the use of a DC micro motor. To accomplish this, you can make use of the micro-motor, sometimes known as the spindle motor, from an old local CD deck mechanism. You just need a little bit of know-how and some patience to successfully attach a little metal pulley that has a rubber washer covering it to the spindle of the motor. After that, secure the unit in the rear wheel of the bicycle in the same manner as the dynamo assembly that is now there.

Circuit for an anti-theft device

When you are utilizing this bicycle guard, the power supply switch labeled S1 should be kept in the "on" position. The power for the circuit comes from the 12V battery while it is in the "on" position, which is achieved by toggling the switch. Now LED1 is on, and the LED current is restricted thanks to resistor R4. The next step is to power the monostable that is built around IC1, which is the CMOS version of the timer LM555, by means of a low-current, fixed-voltage regulator called IC2 (78L05).

Functioning of the circuit

At the beginning of the experiment, while the bicycle is not moving, the monostable output at pin 3 of IC1 is low, and the circuit is in the idle state. In the case that an effort is made to steal something, either the forward or reverse spinning of the DC motor will create a minor voltage at its DC input terminals, which will cause the internal LED of the 4-pin DIP AC input isolator optocoupler IC3 (PS2505-1 or PC814) to light up. As a consequence of this, the intrinsic transistor of IC3 conducts, pin 2 of IC1 is pulled low by the optocoupler, and the monostable that was constructed around IC1 is activated. The piezobuzzer-driver transistor T1 is now being driven by the output at pin 3 of IC1. This causes the buzzer to begin ringing so that you are made aware of the situation. Within the confines of this circuit, the buzzer maintains a "on" state for close to two minutes. You are able to modify this time by adjusting the values of the component's resistor R2 and capacitor C1. Optocoupler IC3 is shielded from damage by Zener diodes ZD1 and ZD2, which each have a voltage of 5.1 volts. Because of its small size and high level of dependability, the pricey GP12V/27A battery has been selected for this application. This circuit is compatible with 12V active buzzers that produce a high-pitched tone when activated. These can be easily purchased from vendors in the market.

Note: To get a higher overall detection sensitivity, this particular optocoupler, as opposed to a bridge rectifier, is what's been incorporated into the circuit. Under no circumstances should a DC optocoupler ever be used in its place.

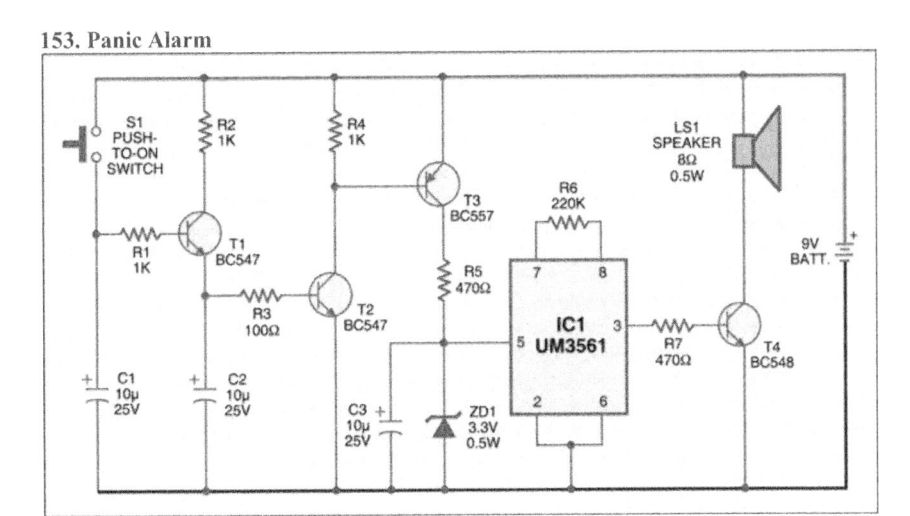

153. Panic Alarm

If you feel unsafe or need help right away, just set off this panic alarm. It will get the attention of people who can help right away. After three minutes, the alarm will stop going off. It is small enough to fit in a pocket or a handbag and is especially helpful for women who are traveling alone. The panic alarm circuit is simple and can be put together with easy-to-find parts on any general-purpose PCB. Basically, it is a transistorized timer that uses the ability of a capacitor to charge and drain to set the time delay.

Panic alarm circuit & operation

When switch S1 is briefly pressed, capacitor C1 fills up to full battery voltage and gives transistor T1 base current. When the voltage at T1's base goes up, the transistor begins to work. C2 gets power from T1, which is an emitter follower. T2 conducts when the voltage across C2 goes up. This lowers the voltage at the collector of T2 to zero, which makes the pnp transistor T3 work. The Zener diode ZD1 makes sure that the voltage going to IC1 is always 3.3 volts. The breakdown voltage of ZD1 is kept by capacitor C3. Limiters of current are R1, R3, and R4. The siren generator is IC1 (UM3561), which has an oscillator built in. When transistor T3 gives IC1 power, it oscillates with the help of resistor R6 (220 k). Transistor T4 then amplifies siren tone pulses at pin 3 to make an alarm sound. As capacitor C1 slowly empties, the base current of transistor T1 drops until it stops conducting. But T2 still uses the charge from C2 to do its work. When the charge on C2 goes down, T2 stops, making the base of T3 positive. T3 immediately turns off, taking the power away from IC1. This takes about three minutes, and the alarm will sound during that time. Make the circuit as small as possible on a matrix board or general-purpose PCB and put it in a cabinet. The circuit can run for a long time on a 9V battery. To make the unit small, use a small Mylar speaker.

154. Visual AC Mains Voltage Indicator

If someone tells you that the mains voltage can change from 160 volts to 270 volts, you shouldn't be surprised. Even though most of our electrical and electronic devices have some kind of voltage stabilization built in, power fluctuations cause more than 90% of the problems with these devices. This simple AC mains voltage indicator circuit shows the AC mains voltage from 160 to 270 volts in 10-volt steps. There are 12 LEDs with the numbers LED1 through LED12 that show the voltage level. All the LEDs stay off when the AC mains voltage is less than 160 volts. When the voltage reaches 160 volts, LED1 lights up, when it reaches 170 volts, LED2 lights up, and so on. With every 10 volts more, the number of LEDs that light up keeps going up. When the voltage coming in reaches 270 volts, all of the LEDs light up.

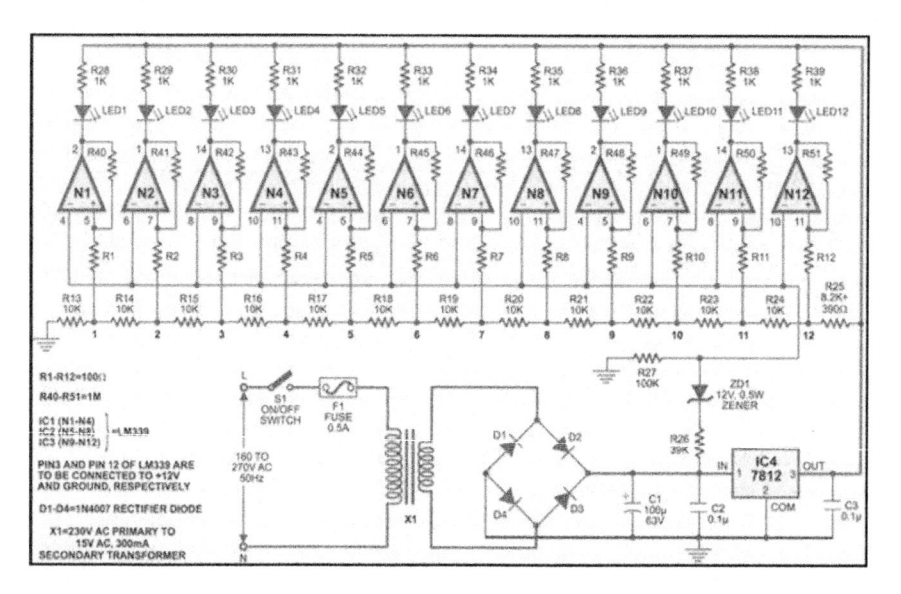

AC mains voltage indicator circuit

A 12V regulator and three LM339 comparators (IC1, IC2, and IC3) make up most of the circuit (IC4). It runs on 12V DC that has been controlled. Step-down transformer X1 converts 230V AC from the mains to 15V AC, which is then rectified by a bridge rectifier made up of diodes D1 through D4, filtered by capacitor C4, and controlled by IC4. For controlling the level of the AC, the input voltage of the regulator is also sent to the inverting inputs of gates N1 through N12. The LED display circuit is made up of four op-amp comparators (IC1–IC3). All of the comparators get their inverting inputs from the unregulated DC voltage, which is proportional to the mains input. The non-inverting inputs get their reference DC voltages from the regulated output of IC4 through a series network of precision resistors.

The resistors R13 through R25 are chosen so that the reference voltage at points 1 through 12 is 0.93V, 1.87V, 2.80V, 3.73V, 4.67V, 5.60V, 6.53V, 7.46V, 8.40V, 9.33V, 10.27V, and 11.20V, respectively. When the AC voltage at the input changes from 160V to 270V, the DC voltage at the anode of ZD1 also changes. When the voltage coming in changes from 160V to 270V, the voltage between the filter capacitors C1 and C2 changes from about 14.3V to 24.1V. Zener ZD1 is used to drop a fixed 12V and give all comparator stages proportional voltages (inverting pins). When the voltage at the comparators' non-inverting input goes up, the LED at the output lights up.

155. Simple Low-Cost White Noise Generator

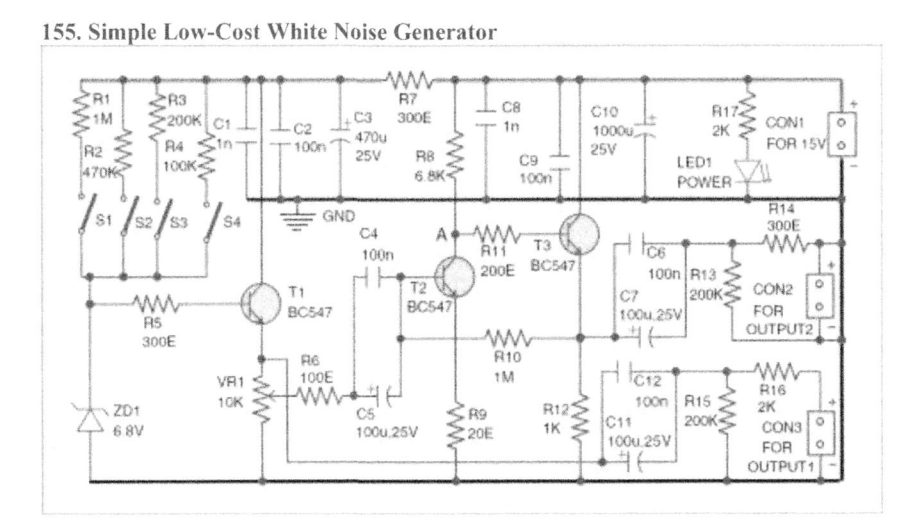

Figure: Circuit diagram of the white noise generator

When testing preamplifiers, filters, power amplifiers, etc., white noise is frequently employed. So, at least for the audio signal range, white noise generators are needed. There are generators like this on the market, but they usually cost a lot. Here is an easy-to-make, cheap white noise generator that uses a Zener diode and three common high-gain transistors. This noise generator can be used by itself for many different kinds of testing. It can also be combined with other filters to make "colored" noises like pink noise, grey noise, blue noise, etc. Up to about 100 kHz is where this noise generator is meant to be used.

Circuit and working

Fig. 1 shows a diagram of how the white noise generator works. It is made up of three BC547 transistors (T1–T3) and a 6.8V Zener diode (ZD1). It needs a power supply with 15V DC. The voltage at which Zener diode ZD1 breaks down should be less than the voltage of the power supply, which in this case is 15V. Most of the time, you don't have to choose the Zener diode based on how much noise you want to make. But if you need to, you can change the LED, Zener diode, and transistors to make the noise you want. The Zener diode makes noise based on how much current is going through it. Here, we have four options for that current. Switches S1, S2, S3, and S4 or jumpers can be used instead of them to save money. At low currents, the Zener diode has a high internal resistance, which is why the transistor T1 works as an emitter-follower. The circuit has two outputs: Output 1 is where the white noise from T1 is buffered, and Output 2 is where the white noise from T2 and T3 is amplified. Pot VR1 is used to change how loud the noise is that T2 and T3 are amplifying.

Transistor T2 acts as a voltage amplifier, which makes the noise bigger. Most of the time, the resistors connected to T2's emitter and collector determine how much voltage is gained. The voltage at point A, which is the collector of T2, should be about half of the voltage from the power supply. You might have to change the value of feedback resistor R10 to do that. Transistor T3 acts as an emitter-follower and gives Output 2 a low output resistance.

The Zener diode and the power supply determine how much the amplitude can go up to. The voltage of the power supply (Vcc) should be between 12V and 25V. Higher Vcc is better because it can send out a stronger signal if it needs to. Even if the power supply is 9V, the circuit will still work, but the Zener diode's breakdown voltage rating should be less than 6V in that case. (A 15V power supply and a 6.8V Zener diode were used in the tests at EFY Lab.)

156. Traffic Light Controller

This simple traffic controller can be used to teach youngsters the fundamentals of traffic laws. The circuit (illustrated in Fig. 1) makes use of components that are easily obtainable. It is primarily composed of rectifier diodes (1N4001), a 5V regulator 7805, two timers IC 555, two relays (5V, single-changeover), three 15W, 230V bulbs, and various discrete components. Transformer X1 takes the mains electricity and steps it down to produce a secondary output of 9 volts and 300 milliamps. The output of the transformer is rectified by a full-wave bridge rectifier, which is comprised of the diodes D1 through D4, filtered by capacitor C1, and regulated by integrated circuit 7805. (IC1). IC2 is hooked up to function as a multivibrator, and its "on" and "off" periods, which each last about 30 seconds, are determined by the component values that are selected.

The output of pin 3 of IC2 becomes high for a period of half a minute as soon as the mains power is turned on. This, in turn, energizes relay RL1 through transistor T1, and as a result, the normally-open (N/O) contact on the relay causes the red light (B1) to illuminate. At the same time, the power from the mains is cut off to the pole where relay RL2 is located.

When the "on" time of IC2 comes to a conclusion, a high-to-low pulse is generated at its pin 3, which then triggers IC3 through C5. Because IC3 is programmed to operate as a monostable and has a "on" time of approximately 4 seconds, pin 3 of IC3 will maintain its high state for the duration of this time and will energize relay RL2 by means of driver transistor T2. Therefore, the amber lamp (B2) will light up for a total of four seconds.

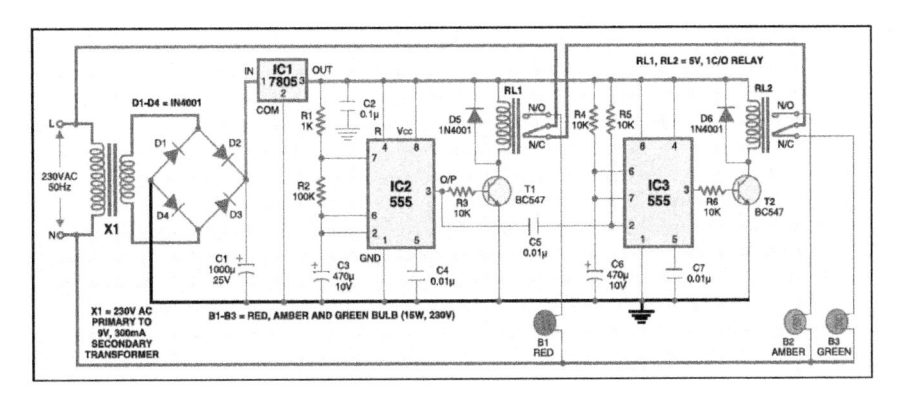

Fig. 1: Traffic Controller Circuit System

Following the expiration of the 4-second time period of timer IC3 at pin 3, the relay RL2 becomes de-energized, and the green lamp (B3) begins to glow for the duration of the remaining approximately 26-second 'off' period of IC2. The normally closed (N/C) contacts of relay RL2 are what are responsible for turning on the green bulb. Therefore, when the mains power is turned on, the red light will glow for 30 seconds, the amber light will glow for 4 seconds, and the green light will glow for 26 seconds. You can construct this circuit on a printed circuit board (PCB) designed for general use and then enclose it in an insulated box. In the box, there ought to be sufficient room for attaching both the X1 transformer and the two relays. Either it can be put on the PVC tube that was used in the building of the traffic signal container or it can be fixed close to a power supply that operates at 230V AC and 50Hz. Figure 2 provides an illustration of the building of the traffic light container box. To house the lamps, you will need a sturdy cardboard box of 30 by 15 by 10 centimeters in volume. Use a piece of plywood with a surface area of 10 by 45 centimeters and a thickness of 1.5 centimeters. Place three light sockets and the box onto the plate, and then fasten them with nuts, bolts, or screws.

Create three tubes out of the thin aluminum sheeting that can be found in most home improvement stores. It is recommended that the inner diameter of aluminum tubes be made to a size that allows them to fit securely on light sockets. Make holes carefully on the other side of the sockets using a knife that is very sharp. Connect the wires to the sockets on the back, then thread them through the PVC tube to exit.

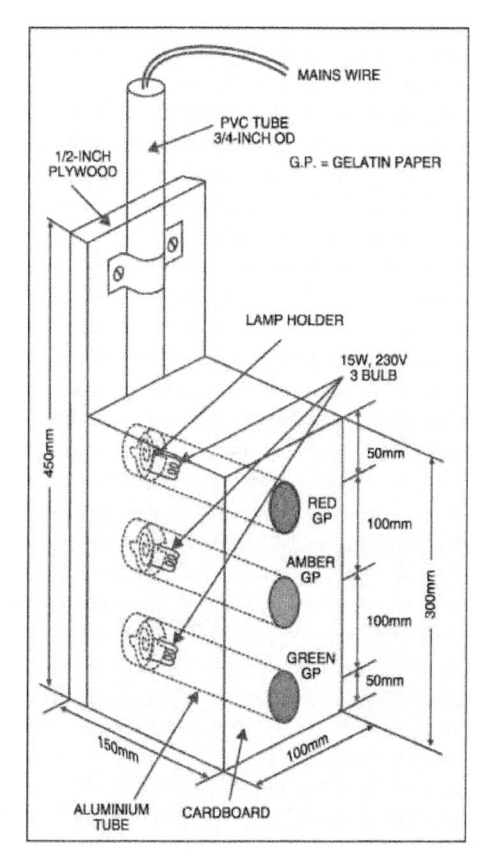

Fig. 2: Construction of the traffic light container box

Once that, you should press on the tubes after you have installed three 15W bulbs (B1 through B3). Place the remaining ends of the tubes in the holes that were cut out of the front panel of the cardboard box to provide support. Placing three sheets of colored gelatine paper between two pieces of cardboard and adhering them to the tubes will give you a rainbow effect. The mounting of the red, amber, and green lights on the tubular shape results in an improvement in the lights' visibility.

157. Weather Station Using STM32

This is an example of a basic "Weather Station using STM32 with ThingSpeak IoT." Powered by STM32 ARM Cortex (Blue Pill), this system detects environmental parameters like as temperature, humidity, and soil moisture. This example is a "Weather Station using STM32 with ThingSpeak IoT." The DHT11 is a temperature and humidity sensor that, as the name suggests, is used to measure the atmospheric temperature and humidity in a specific environment or in a confined closed space. A soil moisture sensor, on the other hand, is used to measure the level of soil moisture on the earth. Both of these sensors are used in conjunction with one another.

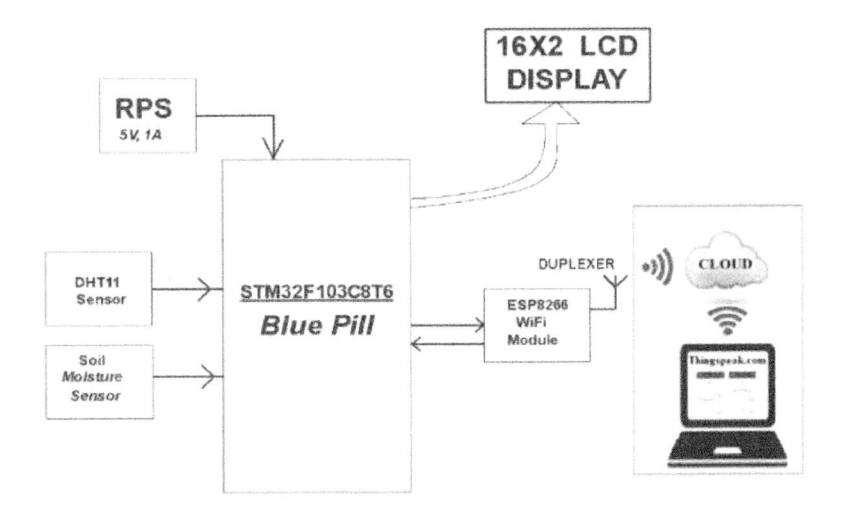

Now let's have a look at how to interface the widely used DHT11 temperature, humidity, and soil moisture sensors with the STM32 microcontroller by utilizing the Arduino IDE. The block diagram for this process is shown below. Microcontroller manufacturer ST Microelectronics' STM32F103C8T6 is included on the Blue Pill development board, which is also known as the STM32. This information is provided for individuals who are unfamiliar with the STM32. It is a 32-bit ARM Cortex M3 controller that features a high clock frequency and is designed to be used in high-speed applications that have power constraints.

Construction & Working principle

These are the primary materials for this project:

1. STM32 (Blue Pill) (Blue Pill)
2. DHT11 Sensor
3. Soil Moisture Sensor
4. ESP8266 WiFi Module
5. 16X2 LCD Display

The next section provides a full description of each element.

Working

The operation of the project is based on the continuous measurement of the weather monitoring parameters, the display of those parameters on a 16x2 LCD display, and the uploading of those parameters into a cloud platform. Temperature, humidity, and soil moisture values are some of the weather monitoring parameters that will be displayed. In this instance, I am utilizing the ThingSpeak IoT Platform in order to save the data in a setting that is real-time.

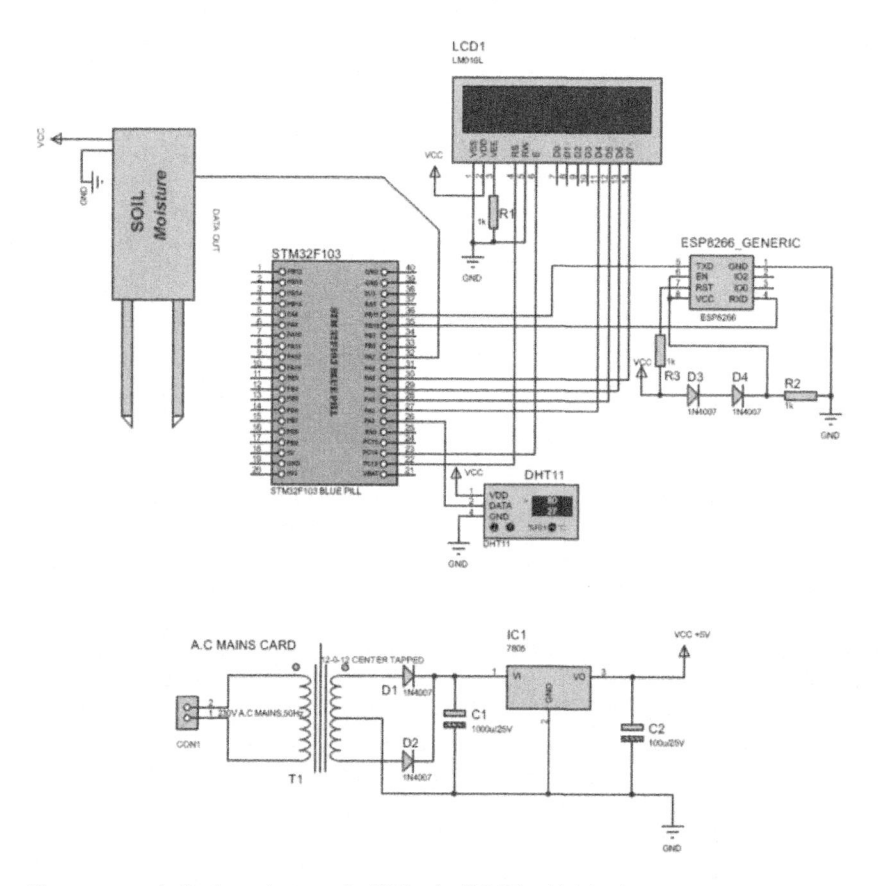

The power supply for the project must be 5V for the STM32, which is also referred to as the Blue Pill Development Board. 3.3V~3.6V for ESP8266 Wi-Fi Module. Here, I simply used a potential divider circuit to convert 5V to 3.6V. The diodes D3, D4, and a 1K ohm resistor are linked in series, and the cutin voltage for each silicon diode is 0.7V. Because I am connecting these two diodes in series, the overall voltage will be 1.4V less than 5V, which will be 3.6V.

158. Simple Pulse Generator
This pulse generator circuit is very useful for testing and running things like counters, stepping relays, and so on. It gets rid of the need to set a switch to make the right number of pulses. By pressing the right switches S1 through S9, one can get from 1 to 9 clock pulses that go in the wrong direction.

Pulse Generator Circuit

Schmitt trigger NAND gate N1 of IC2, resistor R1, and capacitor C1 are wired together to make clock pulses. These pulses are sent out through NAND gate N3, which is controlled by decade counter CD4017 (IC1). At first, none of the switches from S1 to S9 are pressed, so the LED is on. As resistor R3 pulls up pins 5 and 6 of NAND gate N2, its output pin 4 goes low. This stops the NAND gate N3 from setting its output pin 10 to high, so there is no pulse.

The Q outputs of IC1 are usually low because it is a decade counter. When clock pulses are sent to it, each of its Q outputs goes high in turn, from Q0 to Q1 to Q2 to Q3 to Q4 and so on. If any of the switches S1 through S9, say S5 (for five pulses), is briefly pressed, pins 5 and 6 of the NAND gate N2 go low. This makes its output pin 4 high, which fully charges capacitor C2 through diode D. At the same time, the high output of N2 turns on NAND gate N3, and clock pulses come out of pin 10. This is the number of pulses our device needs to check itself.

Circuit operation

When the clock pulses are sent to IC1's clock-enable pin 13, counting begins. As soon as the output pin 1 (Q5) of IC1 goes high, the input pins 5 and 6 of NAND gate N2 will also go high through switch S5. This is because the high-frequency clock allowed five pulses per momentary press. This high input to N2 gives a low output at pin 4, which turns off NAND gate N3 and leaves no pulse for counter IC1 to move forward. Before the next use, the counter IC1 must be in the "standby" state, which means that the output of Q0 must be "high." To do this, a time-delay pulse generator with a NAND gate N4, a resistor R4, a diode D, a capacitor C2, and a differentiator circuit with C3 and R5 is used. When NAND gate N2's output pin 4 is low, it slowly empties capacitor C2 through resistor R4. When the voltage across capacitor C2 drops below the lower trip point, the output pin 11 of NAND gate N4 goes high, and a high-going, sharp pulse is made at the junction of capacitor C3 and resistor R5. This sharp pulse resets the counter IC1, and its Q0 output (pin 3) goes high. This is shown by the way LED lights up.

Make sure the red LED is on before going on to the next step to get the next pulse. If you briefly press any of the switches, the LED will light up. If the switch is held down, the counter keeps counting and you can't tell how many pulses it has counted.

159. Auto Reset Over/Under Voltage Cut-Out

This over/under voltage cut-out will protect your expensive electrical and electronic appliances from the harmful consequences that can be caused by mains voltages that are extremely high or extremely low. The circuit has an automatic reset and uses parts that are easy to find. It uses the comparators that are built into 555 timer ICs. To make relays and control circuits work reliably, power is taken from different parts of the power supply circuit.

The circuit is controlled by comparator 2, while the output of comparator 1 (which is connected to reset pin R) is kept low by connecting pins 5 and 6 of the 555 IC together. The voltage on the positive input pin of comparator 2 is 1/3 of Vcc. So, as long as pin 2's negative input is less than 1/3 Vcc, the output of comparator 2 is high and the internal flip-flop is set, meaning that the Q output (pin 3) is high. At the same time, pin 7 is in a high impedance state, so the LED that is connected to it is off. When pin 2 is taken to be more positive than 1/3 Vcc, the output (at pin 3) changes direction and goes low. At the same time, pin 7 goes low (because the flip-Q flop's output is high) and the ED connected to pin 7 lights up. Both timers, IC1 and IC2, have been set up to work the same way.

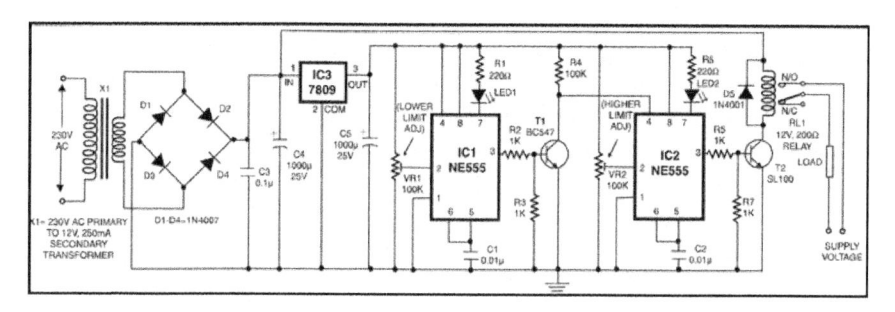

The cut-out for low voltage (let's say 160 volts) is set by noticing that LED1 just turns on when the mains voltage is a little higher than 160V AC. At this setting, the output of IC1 at pin 3 is low, and transistor T1 is in a state called "cut-off." Since RESET pin 4 of IC2 is connected to Vcc by 100 kilo-ohm resistor R4, it stays high. Setting VR2 for an overvoltage cut-out (let's say 270V AC) is done by noticing that LED2 goes out. When the voltage from the mains is just below 270V AC. Pin 3 is also high when RESET pin 4 of IC2 is high. So, transistor T2 conducts and turns on relay RL1, whose N/O contacts connect the load to the power supply. As long as the mains voltage is more than 160V AC but less than 270V AC, this is the case.

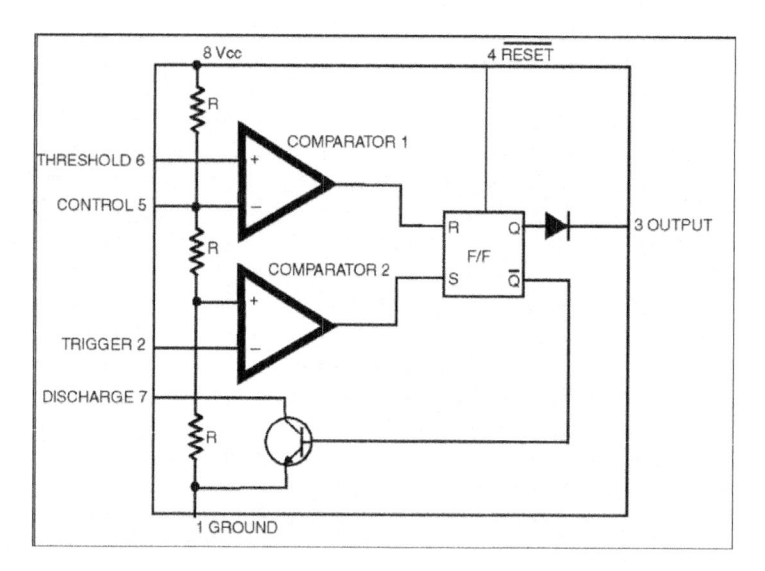

When the mains voltage goes above 270V AC, pin 3 of IC2 goes low. This turns off transistor T2 and turns off relay RL1, even though pin 4 of RESET has always been high. When the mains voltage drops below 160V AC, pin 3 of IC1 goes high, turning off LED1. When pin 3 has a high output, transistor T1 is turned on. As a result, both the collector of transistor T1 and RESET pin 4 of IC2 are pulled low. So, the output of IC2 goes low, and T2 doesn't conduct. Because of this, relay RL1 is turned off, which disconnects the load from the power supply. When the mains voltage goes back above 160V AC but is still below 270V AC, the relay turns on to connect the load to the power supply.

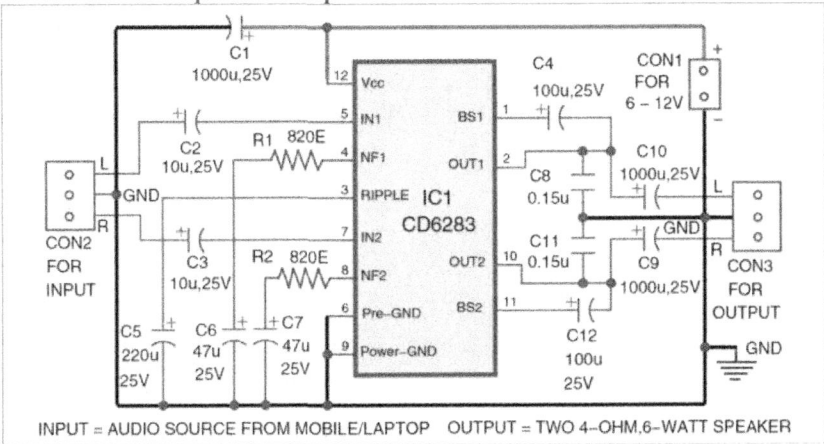

160. This Stereo Amplifier Is Simple to Make

INPUT = AUDIO SOURCE FROM MOBILE/LAPTOP   OUTPUT = TWO 4-OHM,6-WATT SPEAKER

You may construct this straightforward stereo amplifier by using a CD6283 audio amplifier and a few components that are considered to be passive. Using a power supply of 12 volts, it generates an output of 4.6 watts on each channel. Above figure presents the schematic representation of the stereo amplifier's internal workings. CON1 is the connector for the 6-12V input power supply, CON2 is for the input signal, and CON3 gives the output to two speakers that may be connected to it. Those speakers may be linked to it.

Capacitors C2 and C3 are used to link pins 5 and 7 of IC1 to CON2 so that audio inputs can be taken from those locations. Power supply is accessed using the CON1 connector, which is connected to Pin 12. The output of the left (L) channel is linked to pin 2 of IC1 through an electrolytic capacitor called C10, and the output of the right (R) channel is connected to pin 10 of IC1 through an electrolytic capacitor called C9. Left and right channels each have a speaker with an impedance of 4 ohms and a power output of 4 watts connected to CON3.

161. Simple Touch Sensitive Switch

A touch switch is a type of switch that can be activated simply by having an object come into contact with it. In addition to being utilized on public computer terminals, it is also utilized in numerous lamps and wall switches that have a metal appearance. This touch-sensitive switch is constructed using a NAND gate IC called CD4011 and a transistor called BC547. The RS flip flop, which consists of gates N1 and N2, is set whenever someone touches plate 1. Since plate 1 is connected between pin 1 of gate N1 and ground, this triggers the setting of the RS flip flop. The relay is activated by the high output that is produced at pin 3 of gate N1, which is then driven by the relay driver transistor. In turn, the relay will switch on the load that is functioning on mains power.

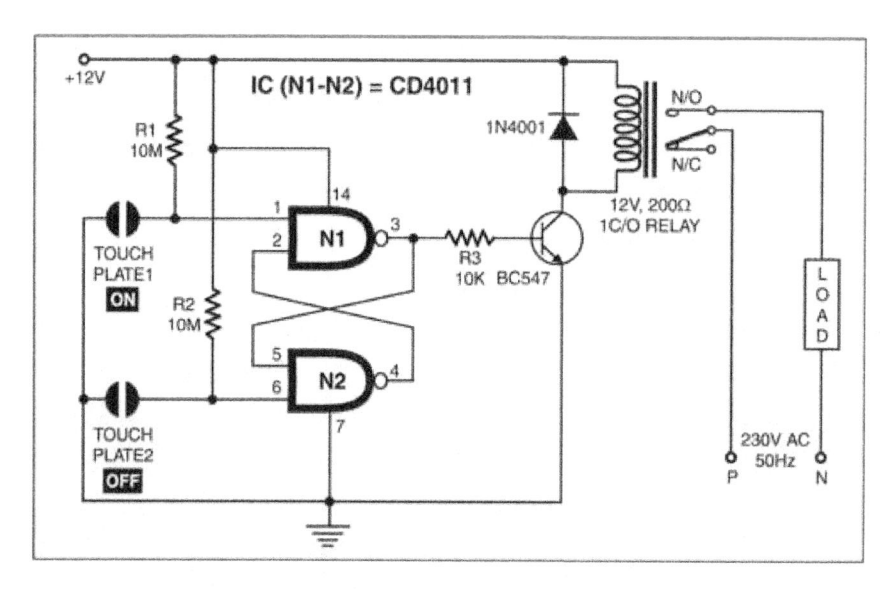

When someone touches plate 2, however (which is connected between pin 6 of gate N2 and ground), the RS flip-flop is reset. This is because plate 2 is connected between pin 6 of gate N2 and ground. The relay is de-energized by the relay driver transistor as a consequence of the low output that is produced at pin 3 of gate N1. In turn, the relay will cut off the load that is being powered by the mains. When the relay is de-energized, the back e.m.f. that is induced in the relay is mitigated by the diode that is connected across the relay coil.

A power supply of 12 V is required for the circuit to function. If you wish to have control over greater loads, you should adjust the current rating of the relay so that it can handle the additional demand.

162. Multi-way Switch Circuit

In the realm of the electrical system, the method that is most frequently encountered is known as multi-way switching. The user has the convenience of turning on and off any electrical item from more than one location thanks to this feature. The most typical application of multi-way switching is for lighting systems, and these systems are typically installed in the most frequently used parts of a building, such as corridors, reception areas, kitchens, living rooms, and a variety of other locations.

A multiway switch circuit can be created quickly and easily by using only a few wires, however in this example, we will use an integrated circuit (IC) CD4042. Because this IC may be latched at four outputs by supplying the clock pulse, we will control an AC light by pressing one of four distinct buttons located in one of four distinct locations.

Material Required

- CD4042 IC
- IC for the NE555 timer
- Bulb or CFL
- Resistor (1k-2; 10k-2) (1k-2; 10k-2)
- Capacitor (0.1uf-2) (0.1uf-2)
- 12v Relay driver module
- Toggle switches - 4
- 12v supply

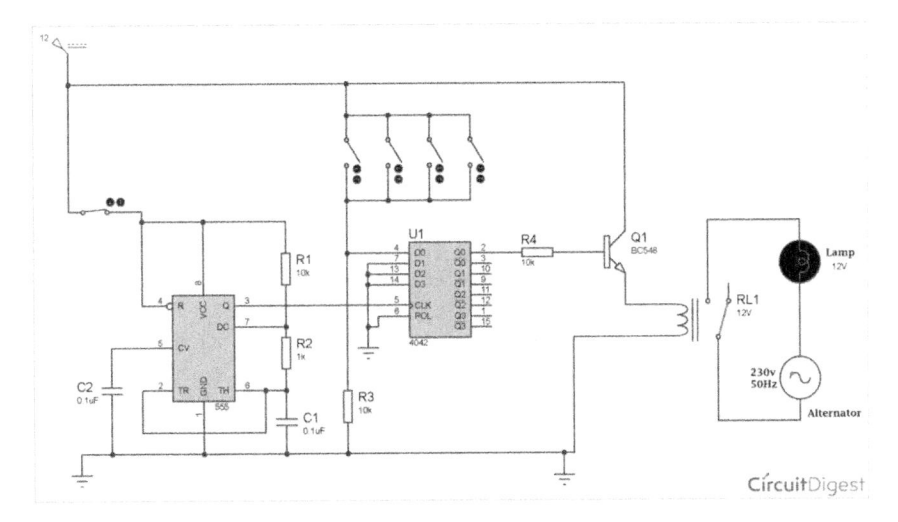

The CD4042 contains four flip-flops, all of which are strobed by the same internal clock. It accepts data from a total of four sources, and each data input has two outputs. The maximum supply voltage of the IC is 18 volts, and the DC input current can be as high as 10 milliamperes. The IC can function at temperatures ranging from -55 degrees Celsius to 125 degrees Celsius. The information that is accessible at the data input is moved to outputs Q and Q' based on the clock and polarity inputs that are used.

163. Car Reverse Horn

This automobile reverse horn is compatible with any vehicle that has a battery that can handle 12 volts. Beeping tone signals are produced by an astable multivibrator, also known as an AMV, which is formed by the transistors T2 and T3. A straightforward speaker amplifier is provided by the medium-power pnp transistor T1, which acts in this capacity. A npn medium power transistor with the model number BD139 (T4) Its base has a connection for a zener diode (ZD1) that provides 5.6 volts. The horn circuit, which is based around transistors T1 through T3, receives its power from a regulated 5.6V DC supply that is provided by ZD1. Protecting against the effects of reverse polarity is the role that diode D1 plays. Through the brake switch of the automobile or three-wheeler, 12V DC is introduced into the circuit.

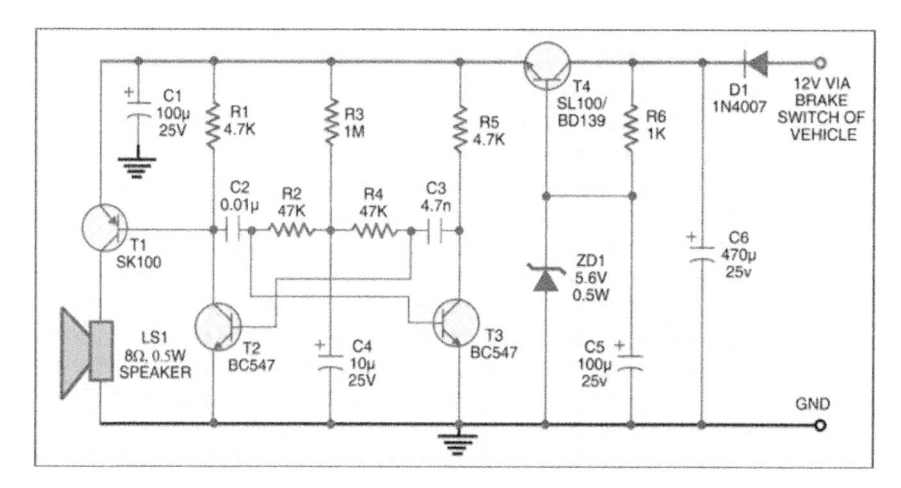

When you put your vehicle into reverse, the brake switch makes a connection to the battery of the vehicle so that the circuit can be completed. The AMV goes through a process of oscillation and generates an audio frequency. This frequency is then passed on to the speaker driver transistor T1, which causes a tone to be produced. Altering the values of the capacitors C2 through C4 in the multivibrator circuit will result in a different tone being produced.

Construct the circuit using a PCB designed for general use and encapsulate it in an appropriate enclosure. Establish a connection between the circuit's 12V DC power supply and the brake switch. When you put your automobile into reverse, the circuit receives the 12V it needs to function properly. Install the cabinet in the vehicle in such a way that it emits sound toward the back while the vehicle is in reverse gear.

164. Bicycle Indicator

The electronic turn signal unit for a bicycle described here is made of cheap parts and is a good alternative to many versions you can buy in stores. It has a completely different way of working and is easy to use. The circuit runs on a 9V PP3 (alkaline-type) battery and is made up of four low-power transistors and a few passive components that make up two independent free-running oscillators. Both square-wave oscillators (one built around T1 and T2 and the other around T3 and T4) drive four red LEDs (LED1 and LED2, and LED5 and LED6) that blink to show which way to turn. More steady-glow LEDs (LED3 and LED4) are added to show if the device is working.

Circuit operation

How the circuit works is easy to understand. When switch S1 is turned on, the battery's DC power is sent to the oscillator circuit, which is made up of transistors T1 and T2. Now, the oscillator on the left side starts to oscillate, and the front left (FL) and rear left (RL) lights start to blink at a rate set by timing capacitors C1 and C2. The amount of current that LEDs can use is limited by resistors R2 and R3 (LED1 and LED2). At the same time, the green LED (LED3) begins to light up to show the status of the current direction.

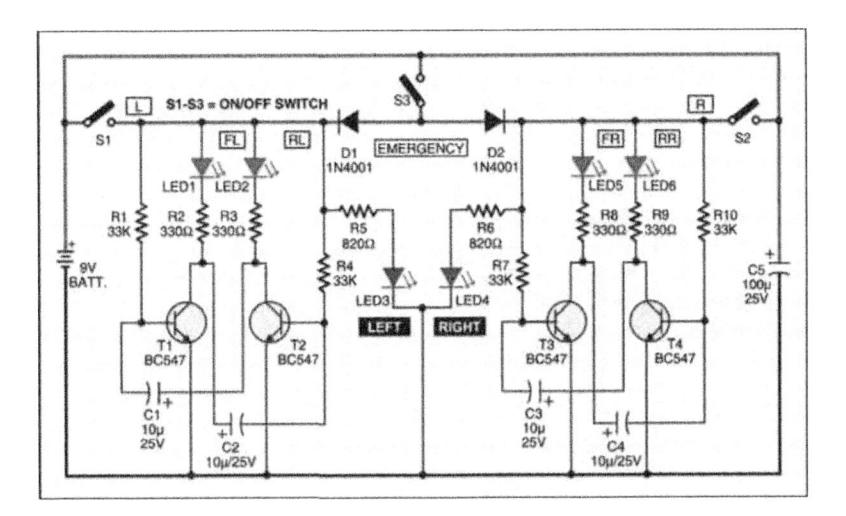

When switch S2 is turned on, the same thing happens in the next oscillator circuit, which is made up of transistors T3 and T4. The front right (FR) and rear right (RR) indicators start blinking, and the green LED (LED4) lights up to show the direction. Switch S3 is used to show that there is an emergency. When it is in the "on" position, diodes D1 and D2 give power to both oscillators. So, LED1 through LED6 all start working at the same time. In this state, all of the LED's flash, except for LED3 and LED4, which stay lit.

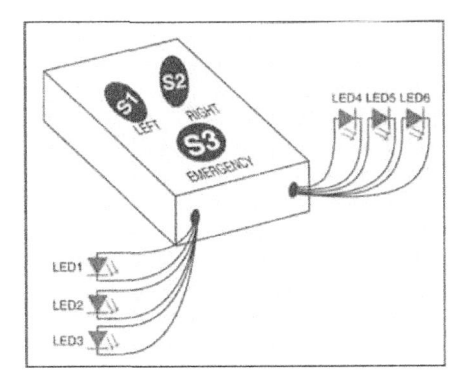

Fig. 2: Suggested enclosure of the master unit

Connect the front indicators (LED1 and LED5) to the left and right sides of the handle. The rear indicators (LED2 and LED6) can be mounted in the bicycle's carrier frame. You can use the symbol to show which way to go.

165. Fridge Door Alarm Circuit using 555 and LDR

The purpose of the circuit can be gleaned from its very designation. If the door of the refrigerator is left open for an extended period of time, this circuit will sound the alarm. The temperature inside the cabin will rise if the door to the refrigerator is left open for an extended period of time. The thermostat will detect the increase in temperature and work to bring down the interior temperature of the cabin. The temperature of the system will be kept as stable as possible by it at all times. As a result of the compressor having to operate continually in order to remove the heat from the cabin, the power consumption of the receptacle will grow. Additionally, continued use under these conditions would shorten the life of the compressor, and it would most likely fail to perform properly.

As a result, this Fridge Door Alarm Circuit is an excellent option that is designed to notify to the user that the door has been left open for an extended period of time. Additionally, we have the ability to specify a variety of pre-set times after which the auditory notification must be provided. This is accomplished through the utilization of the adaptable 555 timer IC operating in the astable multivibrator mode in conjunction with LDR. After a predetermined amount of time, the buzzers will begin to sound an alarm. This is because the LDR detects when the door of the refrigerator is opened as soon as we do, which causes the 555 timers to begin the countdown.

Required Components:

- 555 timer IC - two of them
- 5mm LDR – 1No.
- Buzzer – 1No.
- Diode (1N4007 or 1N4001) – 1No.
- Capacitor, 47uF(Electrolytic) – 1No.
- Capacitor, 0.1uF(Ceramic) – 1No.
- Resistors (10kΩ - 1; 470kΩ -1; 150kΩ -2; 100Ω -1)
- Breadboard
- Putting the wires together

A single 9V battery serves as the source of power for the entire circuit. It is dark inside the refrigerator when the door is closed, and the resistance of the LDR is very close to 1M, according to the datasheet. The output voltage of the potential divider is observed across the capacitor; however, the capacitor continues to have a charged state (Voltage greater than 2/3Vcc), which results in a LOW output voltage.

When we open the refrigerator, light shines on the LDR; this causes the resistance of the LDR to decrease, which in turn triggers the capacitor to discharge; the amount of time it takes for this particular RC combination is thirty seconds. After this (when the voltage is less than 2/3Vcc), the output begins to oscillate at a particular frequency, and the output level rises to a HIGH value. Once more, the capacitor is allowed to charge until it reaches a certain point, after which the process of discharging the capacitor continues. This process will continue until the LDR resistance becomes high, which will occur when there is insufficient light (door is closed).

Diagram of the Refrigerator Door Alarm Circuit:

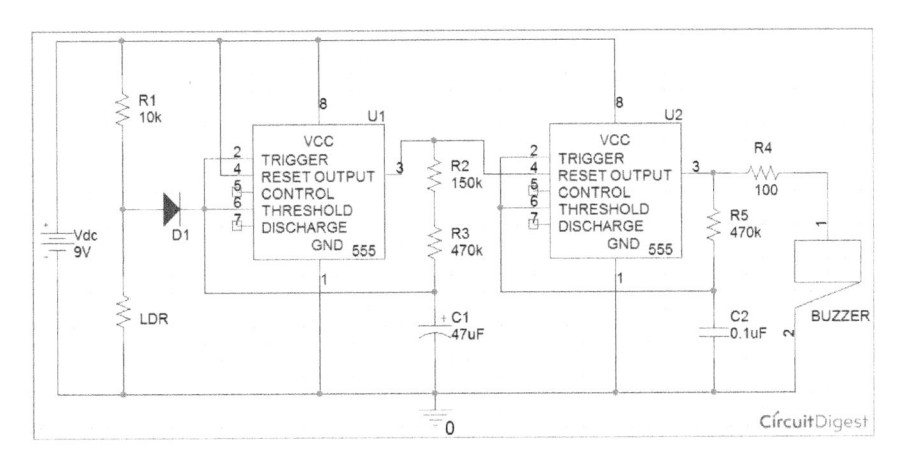

Because of this, the second 555 timer begins to oscillate, and the output alternates between HIGH and LOW. This causes the buzzer that is connected to the output to beep in a pattern, which is the combinational cause of the oscillations that are produced by the first timer and the oscillations produced by the second timer. In the event that the first timer output is in the HIGH state, the second timer master reset will take place. As a result, the voltage on the capacitor C2 rises (to a level higher than 2/3Vcc), and the output level drops. The capacitor begins to discharge in a short period of time, which causes the output to go HIGH (the voltage drops below 2/3 of Vcc). As a result, the sound produced by the buzzer that is attached to the output is a pulsed beep.

166. 3 Phase Motor Programmable on and Off Controller

With a programmable time, switch, you can make a 3-phase motor controller that automatically turns on and off. In this case, you can only set up to eight-time durations. The system has two programmable time switches that can be used to set when the motor will start and stop, as well as two control circuits that are connected to the start and stop switches of the 3-phase motor's starter. Fig. 1 shows a block diagram of the system.

Let's say that both time switches are set to the same time. So, if 8 a.m. is set as the start time for timer1 ON mode, 8.01 a.m. will be set as the start time for timer1 OFF mode in the start time switch. And if the stop time is set to, say, 9 AM for the timer2 ON mode, then the stop time switch will be set to 9.01 AM for the timer2 OFF mode.

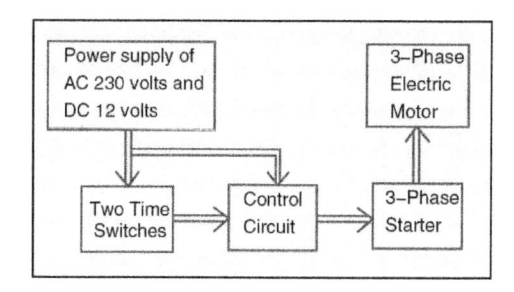

202 | 500 Electronic Project Ideas for Inventors

Fig. 1: Block diagram of the 3-phase motor programmable controller

When it's 8 a.m., the start time switch connects the primary of transformer X1 to 230V AC. The power supply's output is connected to IC1's reset pin 4. R4 and C3 are parts that turn on by themselves. The monostable's output at pin 3 is high for a time period of 1.1R5C4, which is almost five seconds.

3 phase motor programmable controller circuit

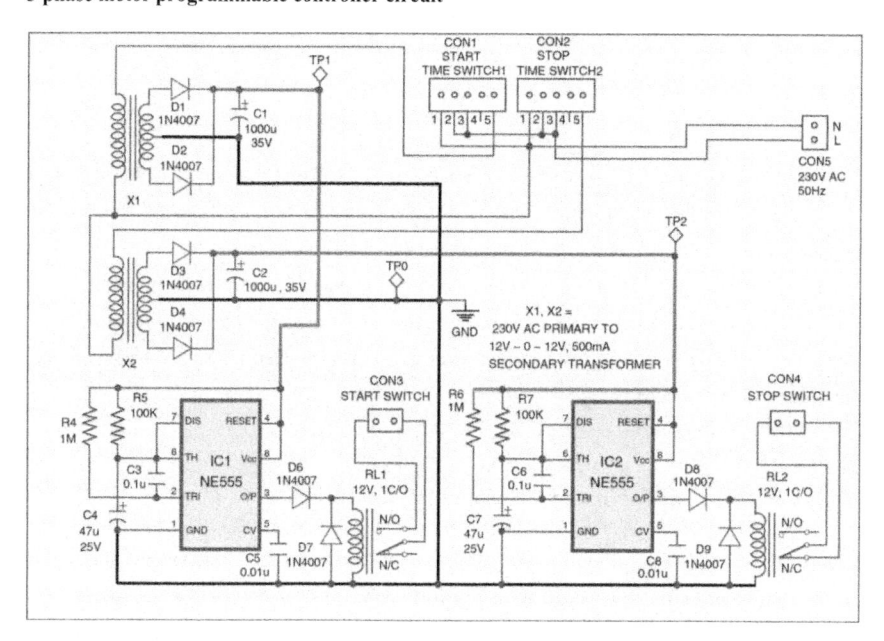

Fig. 2: Circuit diagram of the 3-phase motor programmable controller

As long as pin 3 of IC1 is high, relay RL1 is on for five seconds. This shorts the start switch, which lets the 3-phase supply to the motor go for longer. This is almost the same as pressing the 3-phase motor starter's start switch for five seconds. When it's 9 AM, the second time switch (stop switch) sends 230V AC to the primary of transformer X2 from the secondary. Again, 12V DC is sent to the second monostable circuit with relay RL2 by using a full-wave rectifier and filter circuit. The 3-phase motor's stop switch is connected in series with the relay's normally-closed (N/C) terminal. So, the relay stops the motor by breaking the circuit. This is an example of a time span from 8 a.m. to 9 a.m. So, the 3-phase electric motor can be set to turn on and off at one of up to eight different times. There is a way to choose which days of the week the controller will work. It can be set to work Monday through Friday, Monday through Saturday, all seven days of the week, or just one day of the week. This system can be used for many things, like turning on a water pump in a multi-story commercial building only five or six days a week to fill tanks on the roof. It can also be helpful for farmers, factories, or train stations that use 3-phase motors.

Circuit operation

As shown in Fig. 2, two power supply circuits are built around transformers X1 and X2 and the other parts that go with them. The setup gives 12V DC to two control circuits built around two monostable 555 timers, IC1 and IC2. Two TM-619-2 time switches made by Frontier are used in this system. These use 230V AC at 50Hz to work. Each switch has a 16A contact single changeover relay built in. It has an LCD screen with buttons like CLOCK, TIMER, DAY, HOUR, MIN, and MANUAL. With these buttons, you can set a real-time clock and program different time lengths.

The time switch is a digital device that can be programmed. It has a digital real-time clock and can be set for up to eight different time periods. The times can be for one day, every other day, Monday through Friday, Monday through Saturday, or Monday through Sunday.

167. Digital Frequency Meter Using Arduino

This project discusses the creation of a digital frequency meter using Arduino Uno to determine the sinusoidal frequency signal in the range of 20 Hz to 5 kHz. The frequency range covered by the meter is 20 Hz to 5 kHz. In order to follow the reasoning, you will need to count the total number of pulses that occur each second at a digital pin on Arduino.

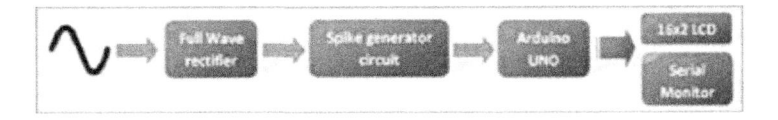

Circuit and the working

Fig.1 presents the schematic representation of the digital frequency meter's internal workings. In addition to a few other components, it has an Arduino Uno board, an optocoupler MCT2E (IC1), a 16x2 LCD, and a few more. In accordance with what is depicted in Figure 3, the sinusoidal signal whose frequency is going to be measured is fed into the signal input terminals (CON1). In order for this frequency meter to function properly, the signal being tested must be one that alternates, and its amplitude must not be more than the maximum forward diode current rating of the optocoupler for a given value of series resistance. (The series resistance that we chose was 1k in this instance.)

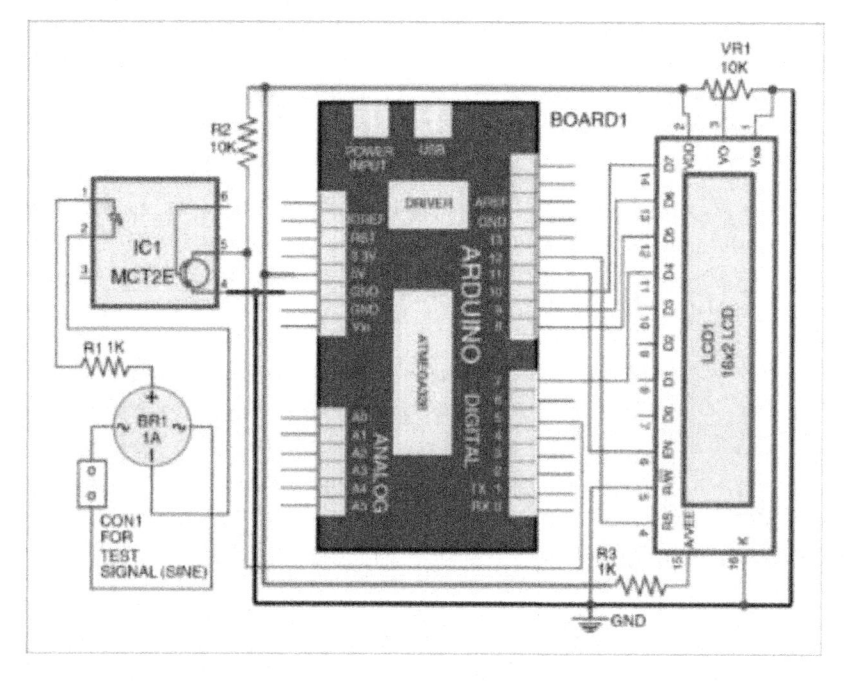

Fig. 2: Circuit diagram of digital frequency meter

The alternating signal is then passed via a bridge rectifier, which transforms it into a completely rectified pulsing direct current signal. On the side of the optocoupler that contains the diode, this pulsating DC signal is applied. This signal has been entirely rectified. The optocoupler is responsible for the production of spikes in the signal. The spikes have a frequency that is twice as high as the test signal input. These spikes are then applied to digital I/O pin 5 of the Arduino, and the program that is running in the CPU of the microcontroller (MCU) then calculates the frequency of the test signal and displays it on the 16x2 LCD as well as on the serial monitor of the personal computer.

Major components used in this project are:

- Arduino UNO
- Bridge rectifier
- Optocoupler
- 16×2 LCD
- Arduino IDE

168. Musical AF/IF checker

There are numerous engineers that have developed many circuits for signal generators, but only a handful of them are reliable over a range. The majority of the circuits are built for a frequency range that is either fixed or constant. Instead of producing an oscillation at 10 KHz, this circuit creates music. The frequency of 455 kilohertz is used to modify musical notes. When audio equipment is being serviced, the modulated signal is utilized in the process of testing and aligning the IFTs in the device.

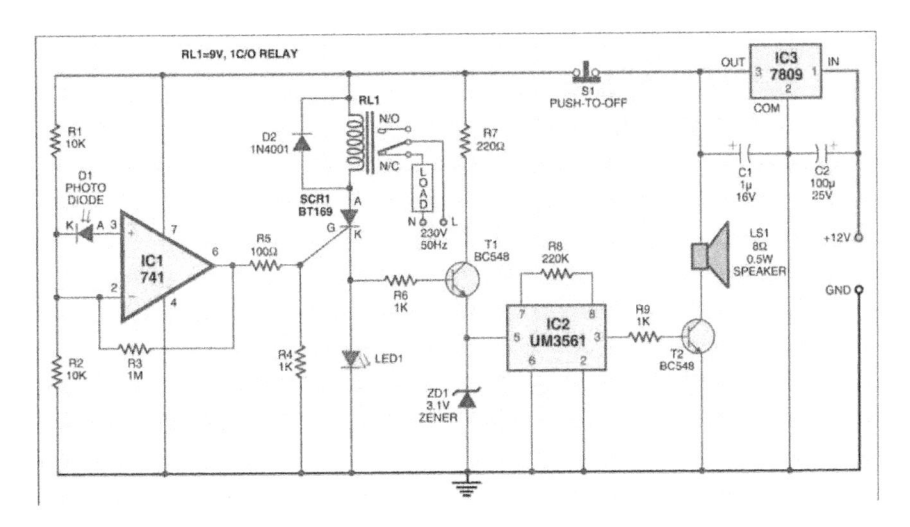

The audio tone generator, the RF oscillator, and the modulator make up the primary components of the circuit. The musical integrated circuit UM66 (IC1) was utilized for the audio tone generator. This integrated circuit includes ROM memory for 64 notes. The oscillator part is made out of a crystal oscillator with low background noise. Crystal operating at 455 kilohertz, used for controlling frequency. Because there is not a tuned circuit being used in the circuit, there will be no frequency drift. Because of this, the IFTs are able to be properly aligned. The AF and RF signals are both modulated by the circuit's modulator, which is located in the output section of the circuit. The signal that has been modified is recorded from the output jack. A modest box made of aluminum houses the entirety of the apparatus. On the front panel of the box may be found both the output connector and the switch labeled SW1.

We are able to obtain the modulated IF signal from the jack when the switch SW1 is set to the position A. We are able to get an AF signal from the jack while the switch is in the B position. For the purpose of adjusting the alignment of a two-band radio, we can switch out the crystal for one that generates frequencies of 550 kHz, 1600 kHz, 600 kHz, 5 MHz, or 16 MHz, and we can swap out the IFT for a tiny ferrite core transformer in its place (or an IFT without the tuning capacitor can also be used).

169. Flashing Light with twilight switch

A flashing light is particularly helpful for indicating that there is either an impediment or that work is currently being done. The project autonomous flashing light with a twilight switch illuminates when it is dark outside but goes off by itself during the day.

Circuit Description

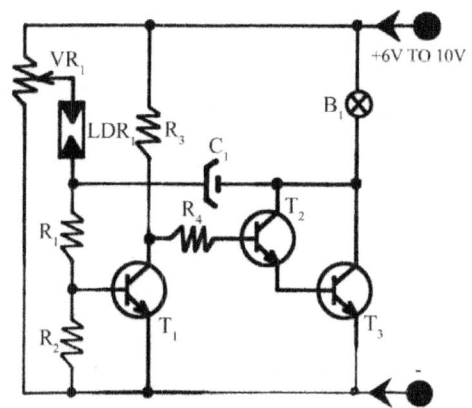

Figure 1 : Circuit Diagram of Flashing Light With Twilight Switch

The LDR is utilized as the sensor in the following example of a circuit schematic for an automatic flashing light that also includes a twilight switch. When there is light present, an LDR has a low resistance, but when there is no light, it has a high resistance.

When there is insufficient light, the LDR generates high resistances, which causes the transistor T1 to become inoperable. Because of this darlington pair built from a transistor, T2 and T3 are turned on, which causes the glow bulb to continue lighting up. As indicated in the circuit design, the feedback from its output is delivered to the junction where the resistor R2 and the LDR are connected. This circuit functions as an oscillator because of the feedback, which also allows it to function as a flasher. Adjusting the sensitivity of the LDR requires the usage of variable resistor VR1.

PARTS LIST

Resistors (all 14 watt, less than 5% carbon unless specifically indicated otherwise) R1 = 2.2 KΩ; R2, R3 = 1 KΩ; R4 = 3.3 KΩ; VR1 = 25 KΩ

Capacitors

C1 = 1 µF – 10 µF

Semiconductors

T1, T2, and T3 equal BC547B, and BEL187-P.

LDR for a variety of applications, B1 = 3V to 10V bulb

170. Automatic Temperature Controlled Fan

The temperature transducer AD590 serves as the centerpiece of the autonomous temperature-controlled fan's circuit, which also includes the operational amplifier LM324. The AD590 is a temperature transducer, meaning that it converts temperature to the appropriate voltage. The output of the transducer is connected to pin 2 of the LM324 integrated circuit. Pin 6 and pin 10 get the two reference voltages through the variable resistors VR1 and VR2 in the appropriate proportions. The value of these variable resistors is predetermined in accordance with the temperature at which they are working (i.e. RL1 energized when temperature is above 300C and RL2 energized when temperature is below 230C). Pin 7 and pin 8 are used to collect the output, which is then passed on to the base of transistors T1 and T2 through resistors R5 and R6 in the appropriate order. Relay driver transistors T1 and T2 are comprised of the transistor pair. Both of these inputs come from various stages of regulation at separate stages of the regulator.

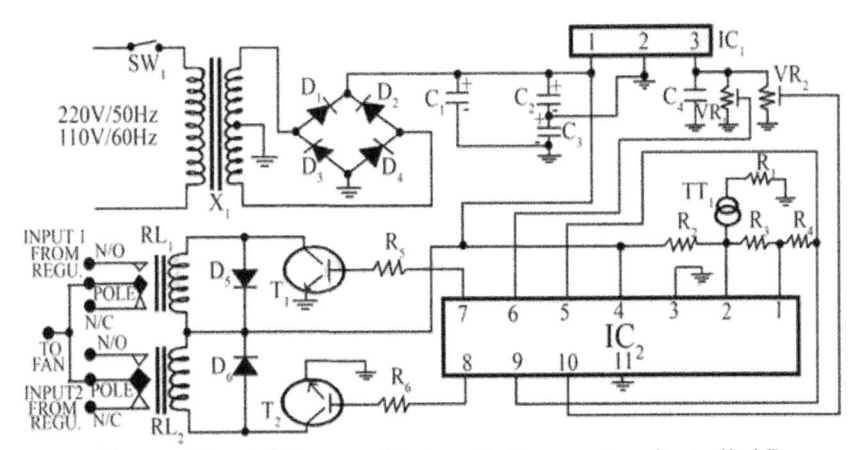

Figure : Circuit Diagram of Automatic Temperature Controlled Fan

Circuit for the power supply: By making use of transformer X1, the primary AC voltage is brought down to 12 volts, 0 volts, 12 volts, and then rectified using bridge rectifier (D1 through D4). The capacitor is used to filter the rectified output before the signal is sent to the input pin 1 of the voltage regulator IC 7812. (IC1). The grounded pin 2 of IC1 serves as the source for the regulated output, which is taken from pin 3.

PARTS LIST

Resistors (all of them are 14 watt, less than 5% carbon).

R1 = 100 KΩ; R2 = 56 KΩ; R3 = 39 KΩ; R4, R5, R6 = 1 KΩ; VR1, VR2 = 10 KΩ (Preset)

Capacitors

C1 = 1000 μF/40V; C2, C3 = 1000 μF/25V; C4 = 0.1 μF

Semiconductors

IC1 = LM7812 (12V regulator IC) (12V regulator IC) IC2 = LM324 (operational amplifier) (operational amplifier)
T1, T2 = SL100

D1 – D6 = 1N4001 (rectifier diode) (rectifier diode) Miscellaneous

TT1 = AD590 (temperature transducer) RL1, RL2 = 12V 200Ω

171. Sound Operated Light

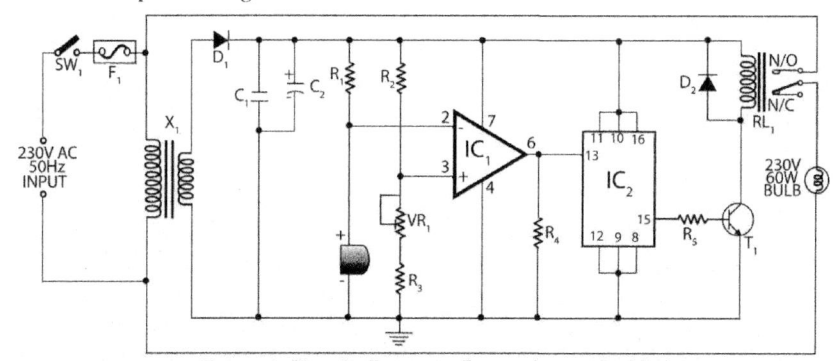

Figure 1: Circuit diagram of sound operated light

Theft is a common problem for people who live in rented housing or hostels since they have less security. The solution to this type of issue is a sound-operated light that is both easy and inexpensive. This type of light can be made to turn on when someone claps, tries to open your door, or even puts a key in the door lock. Here is the circuit for such a light. The light turns on and off alternately, which means that each sound pulse causes the light to either turn on or off. Clap operated light or clap switch are two more names for this device. The complete sound-activated light circuit is conceived and built around an integrated circuit that functions as an operational amplifier (IC1) and a JK flip-flop IC (IC2). The sound that is coming from the outside is picked up by the microphone, which then converts it into a matching electrical signal.

The potentiometer VR1 is used to determine how sensitive the microphone is to sound. In order to alter the output voltage coming from pin 6, the high value of the reference voltage is sent to IC1's pin 3. When sound is picked up by the microphone, the relay RL1 is activated since the IC2 (JK flip-flop) has been wired in this instance to function as a toggle flip-flop, and its output is coupled to the relay driver transistor T1 through the resistor R1. As can be seen in the circuit diagram, the relay contact is used to link the bulb from the sound-operated light to an AC supply.

PARTS LIST

Resistors (all of them are 14 watts, less than 5% carbon).

R1 = 22 KΩ; R2, R5 = 1 KΩ; R3 = 470 Ω; R4 = 10 KΩ; VR1 = 10 KΩ

Capacitors

C1 = 0.1 µF; C2 = 470 µF/35V

Semiconductors

IC1 equals A741 (an operational amplifier), IC2 equals CD4027, T1 equals 2N2222, and D1, D2 equals 1N4001 respectively. Miscellaneous

OR X1 is a 230-volt-alternating-current primary to 0-9-volt, 250-milliampere secondary transformer

(Main transformer with 110V AC output; secondary has 0-9V and 250 mA) RL1 is a 12-volt, 200-ohm, one-contact relay.

SW1 stands for the on/Off Switch.

F1 = Fuse, MIC = Condenser Microphone, Bulb 230V, 60W (110V, 60W)

172. Electronics Thermometer

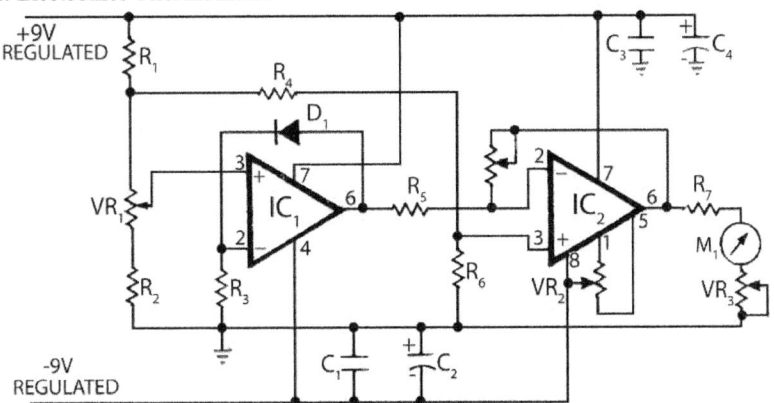

Figure 1: Circuit Diagram of Electronics Thermometer

Because it is difficult to read, clinical thermometers are only used by doctors. Presented here is a schematic for an electronic thermometer that can measure temperatures ranging from -1200 degrees Celsius to 12500 degrees Celsius. This electronic thermometer with a single circuit can be utilized in a variety of settings to measure temperatures. This circuit's ability to measure a broad temperature range contributed to its versatility. The "**Electronics thermometer**" circuit as a whole is constructed and manufactured around the silicon diode D1 (1N4148), which serves as the circuit's operational amplifier IC. Diode D1 is employed as a temperature sensor; the temperature determines the value of the voltmeter drop across the diode; for example, the voltage drop across the diode is 0.7V at room temperature and decreases by approximately 2mV for every degree Celsius.

In an electronic thermometer, a component known as an operational amplifier is utilized for the temperature-to-voltage conversion. VR1, R1, and R2 are responsible for maintaining a constant input voltage at the non-inverting pin 3 of IC1, which is also the location of the feedback path formed by sensor diode D1. The voltage that is measured across the diode has a direct influence on the output of IC1. Amplifier for operational purposes the output from IC1 is amplified thanks to the voltage amplifier that is IC1, which is employed. Ammeters, in the end, are what are utilized to indicate the temperature.

PARTS LIST

Resistors (all of them are 14 watts, less than 5% carbon).

R1 = 680 Ω; R2 = 1 KΩ; R3, R4, R5 = 1 KΩ; R6 = 6.8 KΩ; R7 = 10 KΩ; VR1 = 2.2 KΩ VR2, VR3, VR5 = 10 KΩ

Capacitors

C1, C3 = 0.1 µF; C2 = 10 µF/16V; C4 = 10 µF/16V

Semiconductors

IC1 and IC2 both equal A741; D1 is equal to 1N4148 (Sensor)

Miscellaneous

M1 = 1mA-0-1mA or 0-1mA Ammeter

173. Milli-Ohm Meter with 0.1 To 1-Ohm Range

The vast majority of shunt resistors have values that are lower than one ohm. Because of their limited range and resolution, measurement instruments are unable to properly measure resistances with such low values. This is because of the limitations of the devices. These are typically tested with resistance bridges, which require a specific configuration in order to function properly. A circuit that can test resistances ranging from 0.1 to 1 ohm is shown here for your convenience.

Circuit and the working

Figure show the schematic of the circuit that makes up the milli-ohm meter. The Arduino Uno board (Board1), a low-dropout regulator MIC5219 (IC1), and a 16x2 LCD display are the main components of this circuit (LCD1). A voltage drop takes place across a resistor whenever current flow through it. The magnitude of this drop is proportional to the amount of current that is flowing through the resistor. Using this circuit and using this principle, which is known as Ohm's Law, allows for the measurement of resistance. The test points A and B in the circuit are connected to the resistance that is to be measured, which is connected across those two test points. The voltage drop that occurs across the resistor may be seen on the Arduino board at the analogue pin labeled A0. In order to achieve a higher precision, the analogue reference voltage was changed to an internal reference of 1.1V. The Arduino

Uno serves as the circuit's central processing unit (CPU), calculating the circuit's resistance based on the analogue-to-digital (ADC) value and displaying the result on the liquid crystal display (LCD). A constant-current regulator is implemented in IC1's hardware configuration.

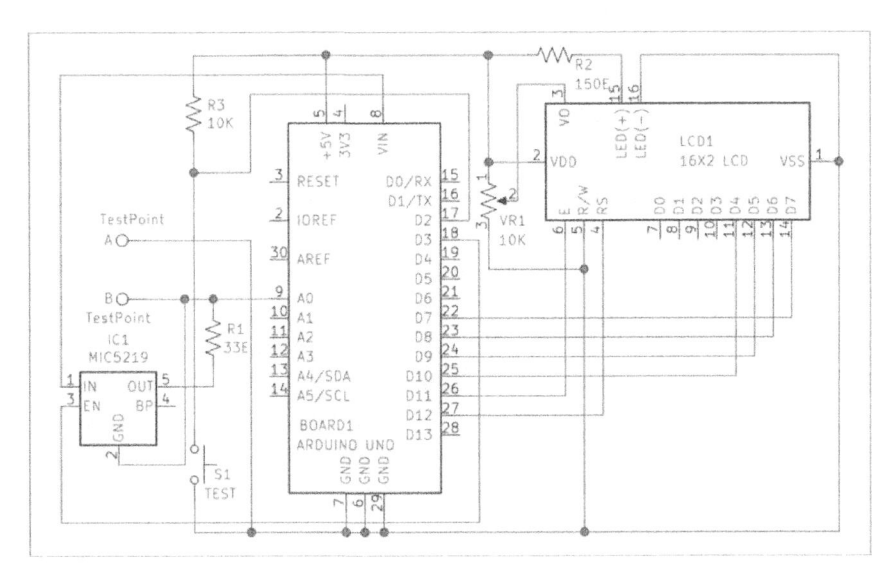

Fig. 2: Circuit diagram of milli-ohm meter

When the push-to-on switch S1 is activated, the Arduino enables IC1, and the integrated circuit then produces a constant current of 100mA across the test resistor. A voltage drop takes place across the test resistor, the value of which is then displayed on LCD1 after being calculated by the Arduino in accordance with Ohm's Law. The measured resistance shouldn't be higher than one ohm at any point in time. If it is greater than one ohm, a notice that reads "!OL OR NC!" will show on the LCD1 screen.

In addition, the component is removed from the circuit anytime the resistance value reaches a threshold that is greater than one ohm. After the measurement, IC1 will be rendered inoperable if the S1 button is pressed again. Atmega328P is the core component of the widely used and widely available open-source microcontroller development board known as Arduino Uno. It includes fourteen digital input/output (I/O) pins, six of which can be used for PWM outputs while the remaining six can be used for analog inputs. In addition, it has the capability of being outfitted with a power jack, USB connector, 16MHz crystal, and ICSP header. Since it already has the Arduino bootloader loaded onto it, there is no requirement for any additional hardware to be able to burn the Atmega chip.

The operation of the circuit is determined by the software program that is stored into the internal memory of the Arduino Uno. The Arduino programming language known as Sketch was utilized in the creation of the mR Meter.ino software. The application is compiled and uploaded using an Arduino integrated development environment (IDE).

174. 70/40 Watts Hi-Fi amplifier

This project will provide you with the schematic for the world's best high-fidelity (HI-FI) amplifier, which you may never need to upgrade from. This 70/40 watts hi-fi amplifier has a strong output, superb specifications, and a small size, all of which contribute to its versatility.

An explanation of the hi-fi amplifier's circuitry, rated at 70/40 watts

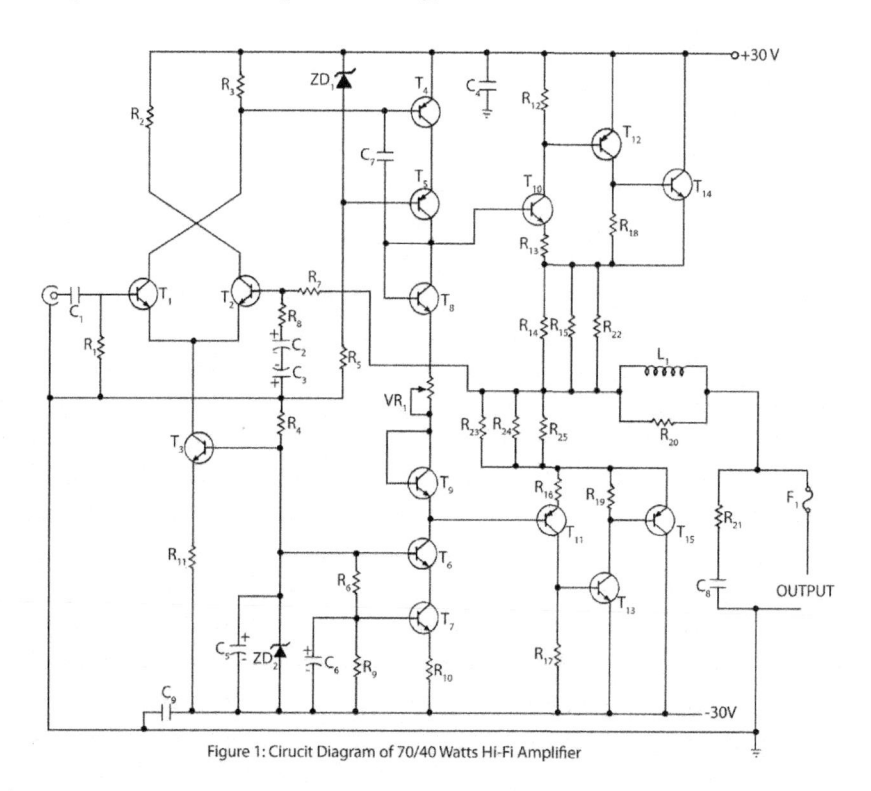

Figure 1: Cirucit Diagram of 70/40 Watts Hi-Fi Amplifier

The circuit for the 70/40 watts hi-fi amplifier is built around transistors, which may be set in a variety of different modes. The signal that is to be amplified is fed into the base of transistor T1, which is set to act as a differential amplifier with transistor T2. In order to keep the current of the differential amplifier constant, transistor T3 is utilized.

A cascaded pair is formed by connecting transistor T4 and T5, which is directly generated from transistor T1. For this amplifier circuit, the transistors T6 and T7 should be re-configured as a cascaded pair so that they can supply a consistent current source to the cascaded pair T4 and T5 and produce better results. Transistors T8 and T9, which behave similarly to diodes, are responsible for compensating the temperature coefficient of transistors T10 and T11. The transistors T10, T12, and T14 together with the transistors T11, T13, and T15 together make up a triple Darlington pair. Because the output of this circuit is made up of triple Darlington pairs, the current that flows through T4 and

T5 is limited to a value that is around 6 milliamps low. The output current of the hi-fi amplifier is quite sensitive to seemingly insignificant shifts in the VBE of both T10 and T11.

Coil L1 is utilized in the hi-fi amplifier rated at 70/40 watts so that distortion can be avoided in the event that capacitive loads are connected to the amplifier's output. A fuse, which is linked in series to the speaker, is utilized to prevent DC voltage from being applied across the speaker. This amplifier has a gain that is approximately equal to 32, and its value may be determined by dividing (R7 + R8) by R8.

PARTS LIST

Resistors (all of them are 14 watt, less than 5% carbon).

R1, R7 = 100 KΩ; R2, R3, R9 = 1.2 KΩ; R4, R5, R6 = 2.7 KΩ/2W; R8 = 4.7 KΩ R10, R13, R16, R18, R19 = 100 Ω; R11 = 3.3 KΩ; R12, R17 = 680 Ω

R14, R15, R22, R23, R24, R25 = 1 Ω/2W; R20 = 10 Ω/2W; R21 = 1o Ω/1W; VR1 = 100 Ω

Capacitors

C1 equals 1 microfarad per polyester; C2 and C3 equal 25 microfarads per 25 volts; C4, C8, and C9 equal 0.1 microfarads per polyester. C5 = 10 µF/60V electrolytic; C6 = 4.7 µF/10V electrolytic; C7 = 56 pF ceramic disc

Semiconductors

T1, T2, T3 = BC546B; T4 = BC558B; T5, T11 = 2N4033; T6, T10 = 2N3019; T7 = BC548B T8, T9 = BC147B; T12 = BD140; T13 = BD139; T14 = 2N3055; T15 = MJ2955

ZD1 = 3.3V 400mW zener diode; ZD2 = 3.9V, 400mW zener diode

Miscellaneous

F1 is a 3.5 Ampere fuse, while L1 is a coil that has 20 turns of 20 SWG wound over a thin pencil.

SPECIFICATIONS

73 watts into a 4-ohm load, and 44 watts into an 8-ohm load at the output (1 KHz, 0.7% THD).

175. Thermistor-Based Fire Alarm System

There are many fire alarm circuits provided across a variety of websites. However, you may find a straightforward and low-cost DIY for a fire alarm that uses a thermistor on this website. Where a thermistor is utilized within the fire alarm system to act as the temperature sensor. The thermistor operates using the same mechanism as the LDR (change their resistance with change in heat where LDR change their resistance with change in light fall on it).

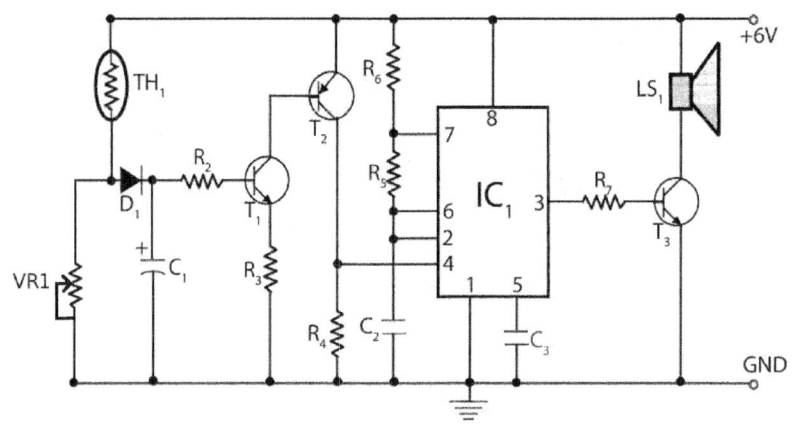

Figure 1: Circuit Diagram Of Fire Alarm Using Thermistor

The entire thermistor-based fire alarm circuit is constructed and assembled around the thermistor (TH1), the timer IC (IC1), and the driver transistor for both of these components. The astable multivibrator oscillator that is used in this circuit is the timer IC, which is denoted by IC1, and it oscillates in the audio frequency region. The two transistors T1 and T2, which are responsible for driving the IC timer (IC1). In order to produce sound, the output from pin 3 of IC1 is routed through transistor T3 and then into a loudspeaker. The frequency of IC2 can be set by adjusting the values of the resistors (R5 and R6) and the capacitor (C2).

When the thermistor TH1 reaches its operating temperature, a channel with low resistance that extends positive voltage to the base of the transistor is made available. In addition, the collector of transistor T1 is linked to the base of transistor T2, which thus supplies a positive voltage to the reset pin 4 of integrated circuit IC1. The thermistor circuit used in fire alarms is able to function across a broad input power supply voltage spectrum, from 6 to 12 volts.

PARTS LIST

Resistors (all of them are 14 watts, less than 5% carbon).

A variable resistor with a value of 10 kilohms that can be used to adjust the circuit's sensitivity. R3, R7, R8 = 470 Ω; R2 = 33 KΩ; R4 = 560 Ω; R5 = 47 KΩ; R6 = 2.2 KΩ

Capacitors

C1 = 10 µF/16V; C2 = 0.04 µF; C3 = 0.01 µF

Semiconductors

IC1 equals NE555, which is a timer IC; T1 equals BC548; and T2 equals BC558

T3 can be any medium-power, general-purpose NPN transistor, such as 2N4922, 2N4921, 2N4238, or FCX1053A. SL100B is one example of such a transistor.

D1 = 1N4001

Miscellaneous

TH1 is a 10-kiloohm resistor, while LS1 is an 8-ohm, 1-watt speaker.

176. Light sensitive switch

A light sensitive switch is a specialized kind of switch whose characteristics change depending on the amount of light that penetrates it. Here is a light sensitive switch that is uncomplicated, easy to operate, and does not cost very much. The detection of light is the driving force behind the operation of this circuit; more specifically, it will switch on or off automatically depending on how much light is falling on it.

Circuit Description

Because this is a switching circuit, we can break it up into two sections - the power supply and the switching circuit—in order to get a better understanding of how it works. In this part of the power supply, the function of the step-down transformer is performed by register R1, and the zener diode ZD1 is responsible for the additional rectification necessary to change the voltage into 10 V dc. The output voltage that is measured across the zener diode is subjected to an additional filtering process by capacitor C1.

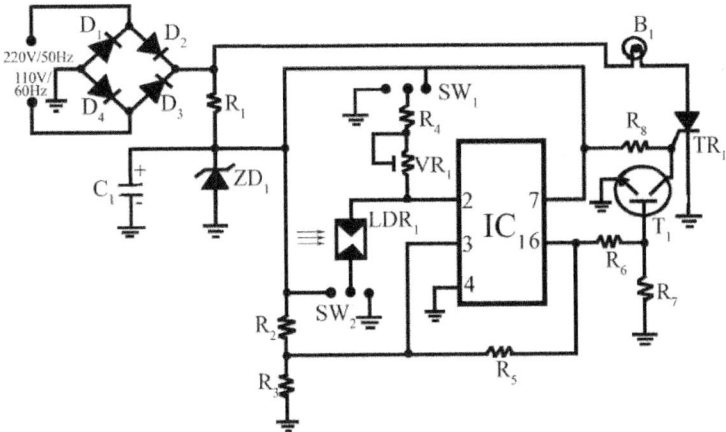

Figure: Circuit Diagram of Light Sensitive Switch

Another part of the circuit is the switching section, which is constructed around the light-dependent register LDR1 with the assistance of the operational amplifier IC 741. The LDR serves as the sensor for the switching circuit. Simply by moving the switch to a different position, we are able to make this circuit sensitive to both light and dark (i.e. turn on in light and turn in dark respectively). LDR detect the light and alter their resistance corresponding to the light, and the result is delivered to pin 2, where it is regulated further by preset VR1. The output is received from pin 6 of IC1, and it is then delivered to the base of transistor T1 by way of resistor R6. Resistor R7 serves the purpose of acting as a current limiter. The output from the collector of the transistor T1 is transferred to the gate of the TRIAC1, which then causes the light to turn on. It is possible to make the circuit function as a light sensor by connecting points 1 and 2 of switch SW2 to points 2 and 3 of switch SW1 (i.e. turn on the bulb when light fall on it).

As dark Sensor

The function of the circuit as a dark sensor can be activated by connecting points 2 and 3 of switch SW2 to points 1 and 2 of switch SW1 (i.e. turn on the bulb in absence of light).

PARTS LIST

Resistors (all of them are 14 watt, less than 5% carbon).

R1 = 100 KΩ/1W; R2, R3 = 100 KΩ; R4 = 4.7 KΩ; R5, R6 = 220 KΩ; R7 = 68 KΩ; R8 = 33 KΩ; VR1 = 100 KΩ (preset)

Capacitors

C1 = 100 μF/16

Semiconductors

IC1 equals LM741 (an operational amplifier), T1 equals BC547, TR1 equals 10GD (a triac), D1–D4 equals 1N4004, and ZD1 equals 10V. /100 mW

Miscellaneous

LDR1 = Light – Dependent – Resistor; B1 = 200W bulb

177. No-Load and Overload Protector for AC Motors

Sometimes the water pump runs when there is no water in the sump or underground tank or when the treadmill is accidentally left on. When this happens, the motors of these electrical machines run when they don't need to. If you use these for a long time, you might get tired of them. This no-load and overload protector circuit can be used to fix these problems. It can turn off the appliance if it gets too hot. Fig. 1 shows the prototype that the author made.

Circuit and how it works

Figure 2 is a diagram of the AC motorized circuits no-load and overload protector. It is made up of a

NE555 timer (IC1), a decade counter (IC2), a quad comparator (IC3), a quad NOR gate (IC4), a current transformer (CT1), BC547 transistors (T1, T3, T4), SL100 transistors (T2), and a few other parts. The circuit runs on a 12V DC power source. At connector CON1, you can use a 12V, 1A SMPS unit. In most cases, Q4 of IC2 will be in the low state, except when there is no load. If reset pin 15 of IC2 is high, the output of Q0 will be high. When reset pin 15 is in the low state, the counter can count. This means that each output (Q0 through Q3) will change from low to high in order. All four outputs (Q0–Q3) are logically ORed and used to drive T2 and turn on relay RL1.

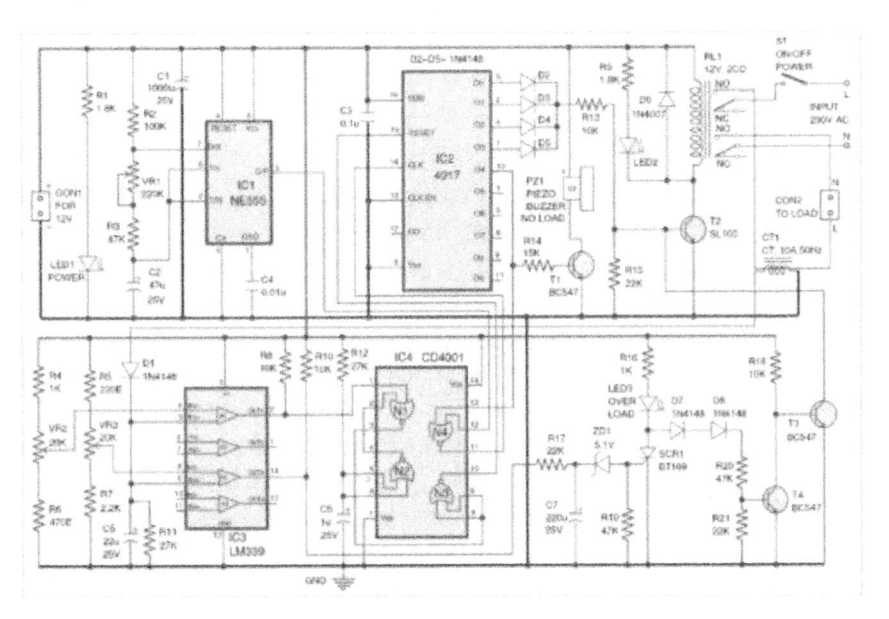

NOR gate N4 will not let the clock pulse into the counter if Q4 of IC2 goes high. So, T2 will be turned off and RL1 will lose power. This will keep happening until the next time the power supply is turned back on. AC voltage is sent through the coil of the 50Hz, 10A current transformer (CT1) to measure how much load current is going through the motorized electrical device. The load current of an electrical device is proportional to the sampled AC voltage in the CT coil. This 50Hz AC voltage across the coil is turned into a DC voltage by reversing it and filtering it. In two comparators, A1 and A3, of the LM339, the DC sample voltage is compared with the reference voltages that come from two resistive networks (IC3).

The voltage from potmeter VR2, which is used to check the no-load condition, should be set to be just a little bit less than the sample DC voltage when the load is normal. This gives pin 2 a high output. If the load current goes down, the reference voltage from VR2 goes above the sample voltage. This makes the output of A1 at pin 2 go low. In normal load conditions, the reference voltage from VR3 is set to be just a bit higher than the sample DC voltage. This makes pin 14 of A3 have a low output. If there is an overload, the sample voltage goes above the reference DC voltage, and the output at pin 14 of A3 goes high. Gate N2 of IC4 is wired as a NOT (inverter) gate so that an RC network made of R12 and C6 can be used to make an initial reset pulse circuit. This reset circuit is needed to reset counter IC2 when the power is turned on for the first time, so that IC2's output Q0 goes high and turns on T2. This, in turn, turns on relay RL1 so that the load can get AC power.

At the cathode of D1, you can get a DC sample voltage from the rectifier. It is used to compare and make gates work logically. Most of the time, the output of A3 at pin 14 is low and the capacitor C7 is empty. Zener diode ZD1 won't work, and SCR1 won't be able to do its job. As the base bias goes through R16, LED3, D7, D8, and R20, T4 turns on. T4 grounds the base of T3, which sets off T3. When T3 is off, it won't pull down T2's base bias voltage. The levels of output from Q0 to Q3 of IC2 control T2. The truth about the NOR gate is shown in the table.

178. Microphone Amplifier

If you have been hunting for the sensitive sound pick-up circuit, then you have found it right. The schematic presented on this website can be adapted to function as a variety of different devices. It is possible to combine it with a more complex device as a sound-activated alarm, in addition to using it as a straightforward microphone. This circuit works just as well for a bugging device as it does for other purposes.

Circuit Diagram

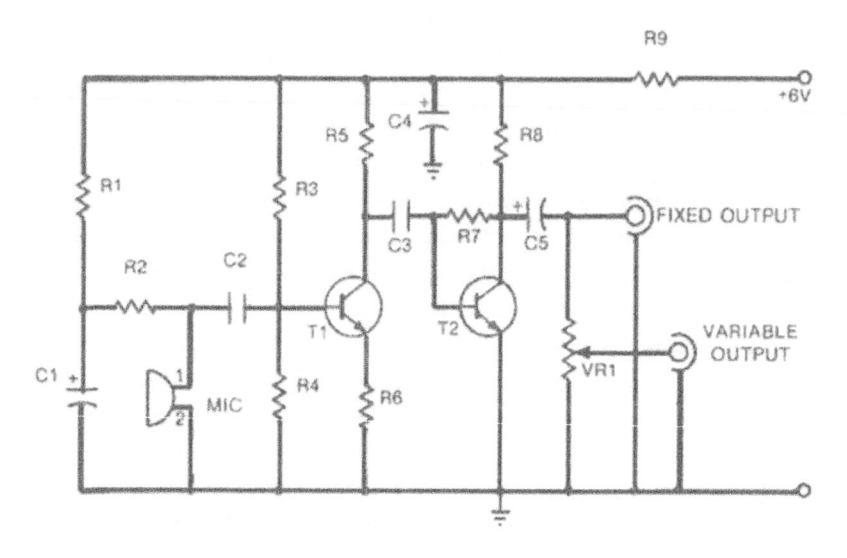

You can see that the transducer in this particular circuit is a microphone by looking at the circuit diagram. Due to the relatively low output of the condenser microphone, a FET amplifier is required in order to boost the signal. The R1,R2 resistor network supplies the necessary power for this amplifier circuit. A two-stage amplifier receives the signal from the output of a condenser microphone. The current series feedback is utilized in the first stage by the transistor T1 (BC149C). In the voltage shunt feedback design, the second stage, which is comprised of the transistor T2, is linked. The amplification provided by these two stages is adequate to enable the user to detect even the faintest whisper.

The amplifier design on the site calls for a supply of 4.2 volts, which may be achieved by connecting a resistor R9 [1k] in series with the circuit. It is possible to change the value of this resistor so that it is appropriate for a supply voltage that is not 6 volts. When a 10k potentiometer is connected to the circuit in the manner depicted in the diagram, the output of the microphone amplifier can be made variable. Depending on the input sensitivity of the primary amplifier system, increasing the gain of the circuit can be accomplished by lowering the value of resistor R6 to either 47 or 22 ohms. Additionally, an increase in gain was seen when the supply voltage was changed to 3V and R9 was removed entirely. A little enclosure in the shape of a circle should be employed to contain the microphone.

PART LIST

Resisters

R1=1.2 KΩ; R2=2.7 KΩ; R3=33 KΩ; R4=6.8 KΩ; R5=3.3 KΩ; R6=100; R7=560 KΩ; R8=4.7 KΩ; R9=10 KΩ; VR=11 KΩ

T1 is equal to BC149C, and T2 is equal to BC147B

Capacitors

C1=47μ, 10V; C2,C3=0.1μ; C4=220μ, 10V; C5=10μ

Miscellaneous

Battery of the 6 Volt Type BATT

Microphone with a Condenser (MIC)

OTHER

PCBs, connecting wires, hardware, and so on are all included.

179. Multi Switch Controlled Relay

With the help of electronic circuits, it's easy to control appliances these days, and everyone wants to control appliances in more than one step. The following diagram shows a control circuit for a relay, which can be particularly helpful for controlling appliances using more than one switch.

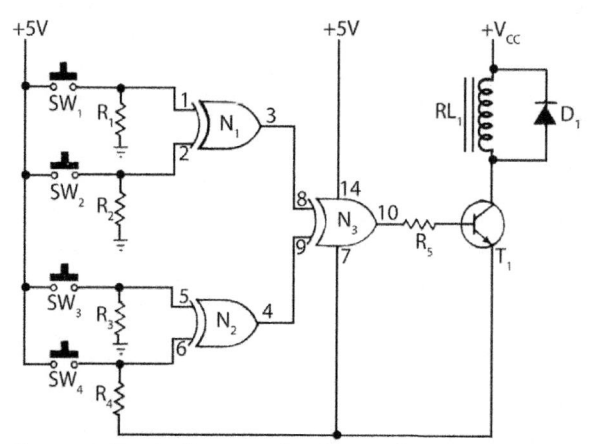

Figure 1: Circuit diagram of multi switch controlled relay

This entire multi switch controlled relay circuit is built around four 2-input EX-OR gates IC CD4077. This IC is the center of the circuit. Everyone should be able to grasp the theory behind a multi switch controlled relay because it is pretty straightforward. The truth table for EX-OR gates with various inputs is presented in the following table. There is a range of 9V to 15V available for the value of VCC.

Truth Table

A	B	OUT
0	0	0
0	1	1
1	0	1
1	1	0

PARTS LIST

Resistors (all ¼-watt, ± 5% Carbon) R1 – R4 = 4.7 KΩ; R5 = 15 KΩ

Semiconductors

IC1 = CD4077; T1 = BC547B; D1 = 1N4148

Miscellaneous

SW1 – SW4 = Push to on/off switch; RL1 = 6V 100 Ω relay

180. Walky-talky without using inductor or coil

This website has the world's first verified walkie-talkie project that doesn't use a coil. A walkie-talkie is a project that any electronics hobbyist would find to be incredibly interesting and attention-grabbing. Communication can take place even in the absence of a physical connection and within a range of up to 500 meters for mobile networks. The use of coil, which is included in nearly all communication equipment, might be challenging for electronics hobbyists. Therefore, not a single coil is included in the design of this circuit. The entirety of the walkie-circuit talkies may be broken down into its two primary sections: the transmitter part and the receiver section.

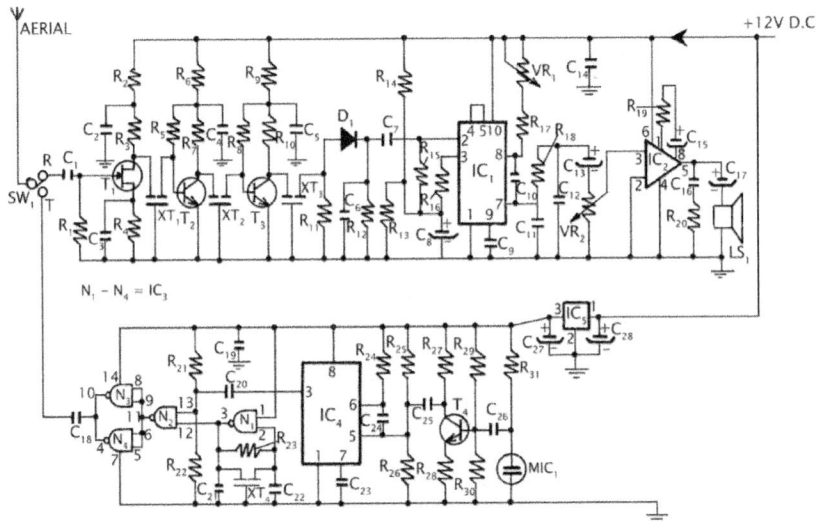

Figure 1: Circuit Diagram of Walky-Talky

The frequency of roughly 30 KHz is generated by the transmitter portion using IC NE566 (IC4) as a VCO, which stands for voltage control oscillator. For the purpose of determining frequencies, the resistor R24 and the capacitor C24 are both employed as frequency components. The mike (MIC1) picks up the speaker's voice and converts it into an electrical signal of the same value. The signal from the microphone is amplified by transistor T4, and the amplified signal is then sent to pin no. 5 of integrated circuit IC4. The output from pin 3 of IC3 is completed by the NAND gate N1, which is controlled by the crystal oscillator XT4. Finally, the signal from NAND N2 through N3 and N4 is passed on to antenna in order to be transmitted.

The signal that is transmitted from one walkie-talkie to another is picked up by the same antenna that is used for transmission. This is the receiver part. The received signal is amplified and strengthened by the field effect transistor T1, which is followed by transmission to the amplifier section, which is comprised of the transistors T2 and T3 and crystal oscillators XT1 through XT3. The detector part consists of a diode called D1, a capacitor called C6, and a resistor called R12. The detector component provides the frequency of thirty thousand hertz. Capacitor C9, resistor R17, and variable resistor VR1 are the components that work together to alter the frequency of the phase-locked loop IC NE565 (IC1). The signal is amplified by the amplifier IC LM386 (IC2) before being distributed to the speaker.

PARTS LIST

Resistors (all of them are 14 watts, less than 5% carbon).

R1 = 47 KΩ; R2 = 100 Ω; R3, R4, R11, R27 = 2.2 KΩ; R5 = 330 KΩ; R6, R10 = 560 Ω

R7 = 1 KΩ; R8 = 220 KΩ; R9 = 100 Ω; R12, R15, R16 = 4.7 KΩ; R13, R31 = 10 KΩ

R14 = 15 KΩ; R17 = 1.8 KΩ; R18 = 1.2 KΩ; R19 = 1 KΩ; R20 = 4.7 Ω; R21, R22 = 100 KΩ R23 = 120 KΩ; R24 = 5.6 KΩ; R25 = 22 KΩ; R26 = 150 KΩ; R28 = 330 Ω; R29 = 220 KΩ R30 = 47 KΩ; VR1 = 4.7 KΩ; VR2 = 22 KΩ

Capacitors

C1, C6, C10, C24 = 1 KpF; C2, C4, C5 = 47 KpF; C3 = 20 KpF; C7, C9, C23= 2.2 KpF

C8 = 4.7 µF/16V; C11 = 22 KpF; C12, C16 = 0.1 µF; C13 = 2.2 µF/16 V; C14, C19, C25, C26 = 0.22 µF; C15 = 10 µF/16V; C17 = 220 µF/16V; C18, C20 = 10 KpF; C21, C22 = 68 pF C27 = 1000 µF/16V; C28 = 10 µF/16V

Semiconductors

IC1 is a Phase Lock IC with the value NE565, IC2 is an Amplifier IC with the value LM386, IC3 is a Quad 2-input NAND Gate IC with the value CD4011, IC4 is a Voltage Controlled Oscillator with the value LM566, IC5 is a Voltage Regulator with the value LM7812, T1 is a BFW10, T2, T3, and T

Miscellaneous

XT1–XT4 equals a crystal with a frequency of 10.7 MHz; SW1 is a switch with a single pole and double throws; LS1 is an 8 speaker. MIC1 refers to a condenser microphone, and Areal refers to an

181. Ohm Meter

This site's ohm meter circuit schematic is really helpful for measuring the low resistance range from 0 to 1 and 0 to 10, as it covers both of those ranges. You have the ability to change the range to reflect your preferences.

The circuit for a low Ohm meter that is described here is straightforward, and it has several advantages over other meters, including the following:

1. You only need to set it up once and then you can forget about it for good. You won't ever have to look at it again.
2. This circuit has a capacity for reading scales that ranges from zero to a predetermined value

rather than infinity.

3. This meter has a low power consumption due to its utilization of a 1.5-volt penlight cell, two scales (0-1 ohms and 0-10 ohms) over a dial, and a push-to-on button that controls the circuit's huge power consumption.

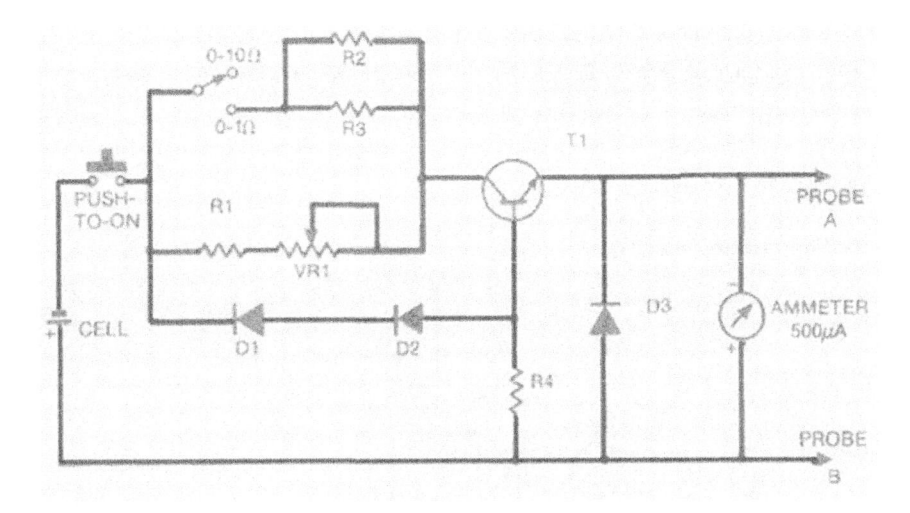

PART LIST

Resisters

R1=27K; R2=3.3K; R3=3.3K; R4=330K; VR1=100 OHM DIODES

D1=1N4001; D2=1N4001

Description Of the Circuit

The ohm meter, the circuit design for which is shown below, is capable of measuring resistance in the range from 0 to 10 ohm. Over there on the circuit diagram, you can see the selector switch, which can select several options. both 0 to 1 ohm and 0 to 10 ohms were included in the testing range. The function of the Transistor T1 is that of a constant current. Generator that sends a known current through the resistors whose resistance is being measured in order to get an accurate reading. If the highest voltage drop across the emitter of the transistor T1 will be greater than 100 mV, and the ground is displayed on the meter, and the meter's internal resistance is significantly higher than the testing resistance of 10 ohms, then the transistor T1 is not functioning properly. This ohm meter does not have the capability to load the circuit as a result. In order to prevent the ohm meter from being damaged by an overload when the testing resistor that is being measured is not present, a diode labeled D3 has been placed across the micro ammeter. This diode serves as a protective precaution. The transistor T1 is responsible for applying a bias to the resistors R1, VR1, R2, R3, D1, D2, and R4. Despite the fact that the battery's capacity is decreasing, the bias level can be maintained thanks to diodes D1 and D2. For the sake of this project, the scale of the meter should read 0-500 A. Any kind of standard resistance meter can serve as the shunt resistance for this project. The silicon npn that has a high gain factor is the transistor designated as T1.

Now you need to modify the meter by connecting the A and B probes together to create a short. If the meter is adjusted in advance, it will show that there is no resistance at all. The 0-to-10-ohm scale is the only one that needs to be adjusted, and all of the other scales will adjust themselves automatically after that. Constructing this will take no more than a few minutes at most. This is a great project for people who are just starting out with electronics.

182. Electronics Counter

Figure 1: Circuit Diagram of Electronics Counter LDR operator

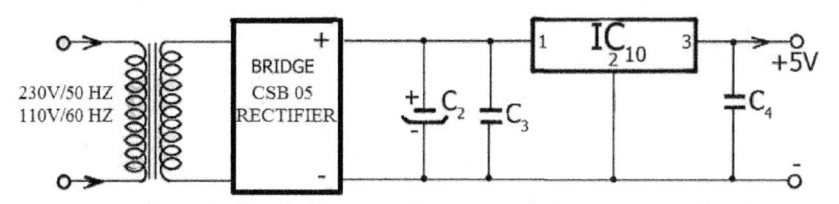

Figure 2: Circuit Diagram of Power Supply for Electronics Counter

Simple counting is something that anybody is capable of doing, but counting in intervals up to a large number is laborious and increases the risk of forgetting numbers. Owing to the fact that we have already published Counter Circuit | Digital Counter. The electronic counter is the second project in a series of counting-based projects that have been undertaken by dreamlover technologies. Bothe the counting circuit described on this website is capable of counting up to 10,000 with the assistance of four screens that only have seven segments each. The prior circuit used CMOS ICs, but the electronic counter uses TTL ICs. The discrepancy can be explained by this.

The entirety of the electronic counter's circuit may be broken down into three primary sections: the input portion, the display section, and the driver or decoder part. The LDR is the first component in the input circuit, which is then followed by a negative square wave generating circuit built around the IC (NE555). In this situation, the light source that is focused on the LDR is a bulb. When the light that is focused on the base of the LDR is blocked, the property of the LDR is that it gives a trigger, which causes a square wave to be formed, which is then delivered as an input signal to the counter circuit. Therefore, the items that need to be counted are organized in a row so that they can be moved between the LDR and the light source one at a time.

Pin 14 receives an input square wave, and IC2 displays any number between 0 and 9 in response to this wave. Following each successive negative pulse, the decoder IC generates a carrying pulse, which is then passed on to the subsequent one (i.e. from IC2 to IC3, IC3 to IC4, IC4 to IC5). BCD to 7-segment latch decoder drivers are IC5 and IC6, respectively. To return the electronic counter to its initial state of 0000, the reset switch labeled SW1 must be utilized.

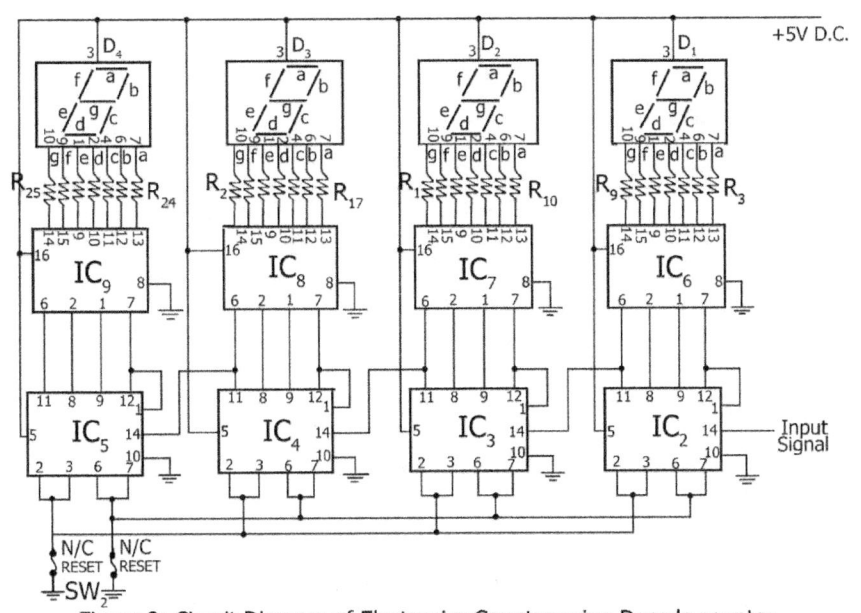

Figure 2: Circuit Diagram of Electronics Counter using Decade counter

PARTS LIST

Resistors (all of them are 14 watt, less than 5% carbon).

R1 = 1 KΩ; R2 = 100 KΩ; R3 – R30 = 180 Ω; VR1 = 100 KΩ preset

Capacitors C1 = 4.7 µF; C2 = 1000 µF/10V; C3, C4 = 0.1 µF

Semiconductors

IC1 = NE555 (Timer IC); IC¬2 – IC5 = 7490 (Decade and Binary counter)

IC6 – IC9 = 7447 (BCD to 7-segment decoder); IC10 = µA 7805 (Voltage Regulator) Display FND 507 if you subtract D1 from D4

Miscellaneous

Mic1 stands for microphone, B1 for bulb, and LDR for light-dependent resistor.

183. Clap operated Remote Control for Fans

This clap-operated remote control fan circuit is utilized to regulate not only the switching features of the fan, but also the speed of the fan. The fact that a clap-operated remote control for a fan may control up to ten-step speeds of the fan is the primary benefit of using such a control. Typically, a fan has between three and five step speeds.

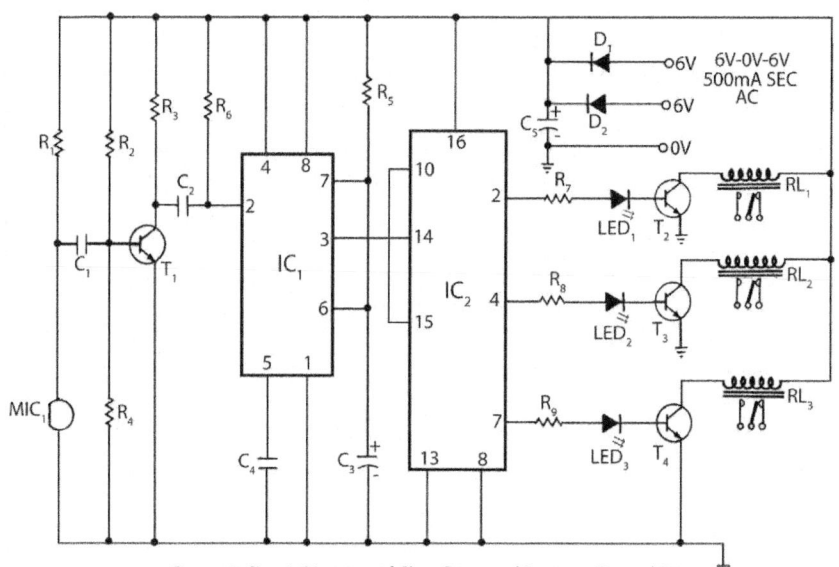

Figure 1: Circuit Diagram of Clap-Operated Remote Control Fan

This entire clap-operated remote control for fan circuit is broken up into four key sections, which are designated as the sound-operated trigger pulse generator, the clock pulse generator, the clock pulse counter, and the load operator respectively. Sound-operated trigger pulse: – The class-C amplifier mode that the transistor T1 BC148 is set up to operate in is the primary focus of this section. The voice signal is converted into the electrical signal that corresponds to it by the MIC1, and this electrical signal is then supplied to the base of the transistor T1 in order to amplify and boost the signal's intensity. A pulse generator for the clock: This portion is constructed around the timer IC NE555 and is set up to function as a monostable multivibrator. The trigger pulse that is produced by transistor T1 is supplied to pin 2 of integrated circuit IC1, and the time period (T) for when the output is high is determined using a formula. T = 1.1RC Pulse counter for the clock: This section revolves around the decade counter CD4017BC, which counts the clock pulse that is produced by the timer IC (IC1). The output of IC1 is connected to pin 14 of IC2, which in turn receives it. IC2 has ten outputs, which are labeled o, 1, 2, 3, 4, 9, respectively. Only three of the pin's outputs—numbered 1 through 3—are utilized in this application. 2, 4, and 7 in descending order. Pin 10's output, number 4, is connected in a straight line to pin 15's reset. This section is built around three transistors, which serve as relay drivers, so that it can operate three separate relays. As shown in the circuit diagram, the output from each pin of IC2 is given to the base of each transistor through a resistance of 100 and an LED. The transistor's collector serves as the source of the output, which is then connected to the relay. The three LEDs that are used to indicate gear or speed are labeled as follows: LED1, LED2, and LED3 respectively indicate gear 1, gear 2, and gear 3.

NOTE: -This circuit used to operate in 1st speed similarly, 2nd clap for 2nd speed, 3rd clap for 3rd speed, and 4th clap to switch off the fan. NOTE:-This circuit used to operate in 2nd speed similarly, 2nd clap for 2nd speed, 3rd clap for 3rd speed, and

PARTS LIST

Resistors (all of which are 14 watt, less than 5% carbon).

R1 = 10 KΩ; R2 = 1.2 MΩ; R3 = 2.2 KΩ; R4 = 150 KΩ; R5 = 220 KΩ; R6 = 10 KΩ; R7, R8, R9 = 100 Ω

Capacitors

C1, C2 = 0.1 µF/16V; C3 = 4.7 µF/16V; C4 = 0.01 µF (ceramic disc); C5 = 1000 µF/12V

Semiconductors

IC1 is a NE555, which is a decade counter. IC2 is a CD4017BE, which is a timer IC. T1 is BC148, T2, T3, and T4 are all BC148.

BEL187; D1, D2 = 1N4001 silicon diode

Miscellaneous

MIC1 is a condenser microphone with 34LOD, LED1 is green, LED2 is yellow, and LED3 is red, and the secondary transformer for 6V-0V-6V has 500mA.

184. Mobile cellphone charger

When traveling, charging a mobile device's battery can be quite difficult because power supply sources are not always easily accessible. Using AA cells to charge a mobile phone battery is covered in this straightforward project, which makes use of extremely common electronic components.

Circuit descriptions of mobile cellphone charger

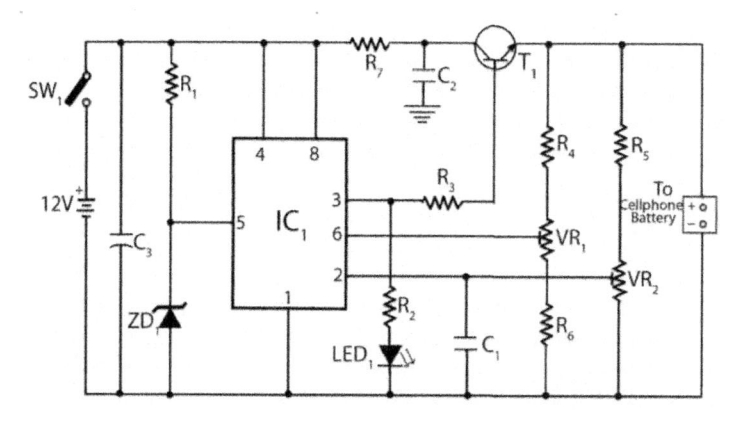

The timer IC NE555 is the most important component of the mobile phone charger circuit. It is responsible for charging the battery and monitoring the voltage level. IC1 get control voltage to pin 5 by Zener diode ZD1¬. A voltage that is set by VR1 and VR2 is provided to pin 6, which is the threshold, and pin 2, which is the trigger. When a discharge battery is attached to the circuit, the trigger pin 2 of IC1 will drop below 1/3VCC. This will cause the flip-flop of IC1 to activate, which will cause output pin 3 to be set to a high value. When the battery has reached its full capacity and another fully charged battery is connected, the procedure will then proceed in the other direction. Transistor T1 is employed in this situation to boost the charging current coming from output pin 3 of IC1. Potentiometers VR1 and VR2 should be adjusted appropriately.

LED status indicators for the various charging states

Load across the output	Output frequency (at pin 3)	LED1
No battery connected	765 kHz	On
Charging battery	4.5 Hz	Blink
Fully charged battery	0	Off

PARTS LIST

Resistors (each one a quarter watt, less than five percent carbon).

R1 = 390 Ω; R2 = 680 Ω; R3 = 39 Ω/1W; R4 = 27 KΩ; R5 = 47 KΩ; R6 = 3.3 KΩ; R7 = 100Ω/1W; VR1, VR2 = 20 KΩ

Capacitors

C1 = 0.001 µF (ceramic disc) (ceramic disc) C2 = 0.01 µF (ceramic disc) (ceramic disc) C3 = 4.7 µF/25V (Electrolytic) (Electrolytic)

Semiconductors

IC1 is the NE555 timer IC.

T1 can be SL100 or any other medium power general purpose NPN transistor such as 2N4922, 2N4921, 2N4238, or FCX1053A.

ZD1 = 5.6 V/1W LED1

Miscellaneous

SW1 is a switch that turns on and off the 1.5V*8 AA batteries mobile connection.

185. Test a Diode | Zener Diode

The diode tester circuit, which can be used for a variety of purposes, is available for download here. Utilizing this circuit also enables one to verify the functionality of a zener diode. The transistors and resistors that make up this diode tester's circuit layout are rather straightforward components. Because there is no IC of any sort utilized in this circuit, it will be simple for those who are just starting out in the field of electronics to comprehend how the diode tester in this circuit actually functions.

PART LISTS

Resistors

R1=2.2K; R2=10K; R3=680 Ohm; R4=1.2 K; R5=10K , 0.5W TRANSISTORS

T1 equals BC147B, while T2 is SL100.

Capacitors

C1=470μ 35V; C2=1μ 40V DIODES

D1=LED; D2 -D5=1N4001 OTHERS

S1=ON/OFF Switch; X1=12V-0-12V Transformer Volt Meter 30V

Probe 2 pieces

Description of the circuit

The 12-0-12 step-down 500mA power transformer was utilized for this purpose. The output of the transformer is sent into the bridge rectifier, which is comprised of the diodes D2, D3, D4, and D5, and is used to convert the alternating current supply into the direct current supply. The DC output is filtered by capacitor C1, which serves in this capacity. The capacitor that we used was 470 F, but you can use any value. The higher the capacitance value, the purer DC that can be obtained. As a bleeder, the resistor R2 has a value of 2.2K. You can see that the transistors T1 [BC147B] and T2 [SL100] are being used for the regulator compressor here. These transistors get the DC output through their connection. T1 acts as a series pass driver or a current regulator. Resistor R2 with a value of 10k serves as a base bleeder, while capacitor C2 with a value of 1 F filters base potential. The base bias for transistor T1 is obtained from the supply by way of resistor R3, which has a value of 680 ohms. When the test probe is completely open and there is no connection to the zener, the base potential of transistor

T1 is approximately 32V, which is measured across resistor R4 or capacitor C2. The net DC voltage equivalent to the potential of the base of transistor T1 (which is fed to the voltmeter) is provided by the transistor T2 (which serves as a series pass regulator) and is supplied by the transistor T1 (BC147B), which supplies the base potential.

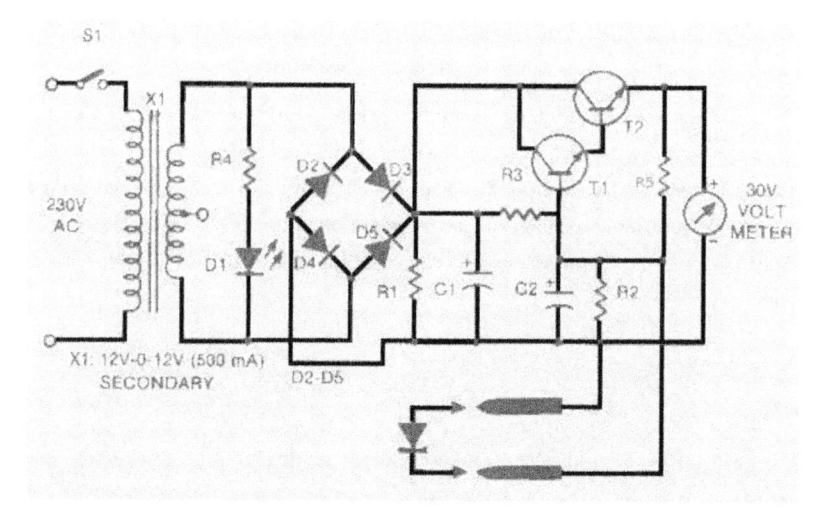

With no Zener diode attached across the probe, the voltmeter is currently reading somewhere about 30V. When a Zener diode is put across the test probe, the base potential of transistor T1 drops until it reaches the level at which the Zener diode breaks down. As a result of this, the base potentials of the transistors T2 and T1 are brought into parity with one another. The meter is now displaying the real voltage produced by the Zener. By moving the needle using the zero-adjustment screw on the meter, it is possible to make a modification to the scale of the meter that is 0.6 V in magnitude.

PARTS LIST

Resistors (all of them are 14 watt, less than 5% carbon).

R1 = 100 KΩ; R2 = 39 KΩ; R3 = 2.2 KΩ; R4 = 680 Ω; R5 = 100 Ω; VR1 = 4.7 KΩ; VR2 = 10 KΩ

Capacitors

C1 = 27 KPF (273); C2, C4 = 2.2 µF/16 V; C3 = 22 µF/16 V; C5, C10 = 100 µF/16 V; C6 = 10 µF/16 V; C7 = 100 KPF (104); C8 = 47 KPF (473); C9 = 220 µF/16 V

Semiconductors

T1 is a BC147B, and IC1 is an LM386. Both are power amplifiers.

Miscellaneous

L1 equals the pick-up coil on the speaker 8.

SW1 is the switch for on and off.

186. Sound Pressure Meter

The diode tester circuit, which can be used for a variety of purposes, is available for download here. Utilizing this circuit also enables one to verify the functionality of a zener diode. The transistors and resistors that make up this diode tester's circuit layout are rather straightforward components. Because there is no IC of any sort utilized in this circuit, it will be simple for those who are just starting out in the field of electronics to comprehend how the diode tester in this circuit actually functions.

PART LISTS

Resistors

R1=2.2K; R2=10K; R3=680 Ohm; R4=1.2 K; R5=10K , 0.5W TRANSISTORS

T1 equals BC147B, while T2 is SL100.

Capacitors

C1=470μ 35V; C2=1μ 40V DIODES, D1=LED; D2 -D5=1N4001 OTHERS

S1=ON/OFF Switch; X1=12V-0-12V Transformer Volt Meter 30V

Probe 2 pieces

Description Of the Circuit

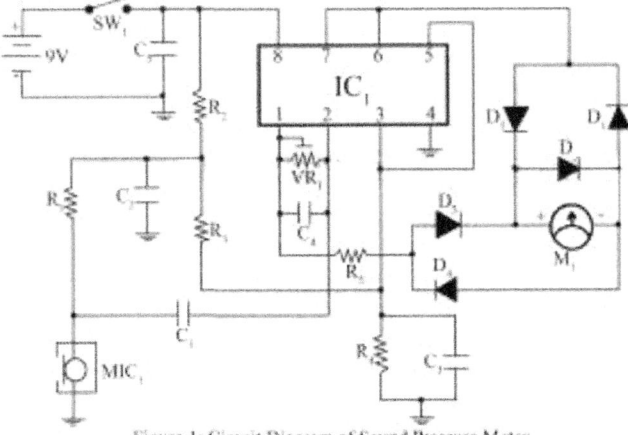

Figure 1: Circuit Diagram of Sound Pressure Meter

The 12-0-12 step-down 500mA power transformer was utilized for this purpose. The output of the transformer is sent into the bridge rectifier, which is comprised of the diodes D2, D3, D4, and D5, and is used to convert the alternating current supply into the direct current supply. The DC output is filtered by capacitor C1, which serves in this capacity. The capacitor that we used was 470 F, but you can use any value. The higher the capacitance value, the purer DC that can be obtained. As a bleeder, the resistor R2 has a value of 2.2K. You can see that the transistors T1 [BC147B] and T2 [SL100] are being used for the regulator compressor here. These transistors get the DC output through their connection. T1 acts as a series pass driver or a current regulator. Resistor R2 with a value of 10k serves as a base bleeder, while capacitor C2 with a value of 1 F filters base potential. The base bias for transistor T1 is obtained from the supply by way of resistor R3, which has a value of 680 ohms. When the test probe is completely open and there is no connection to the zener, the base potential of transistor T1 is approximately 32V, which is measured across resistor R4 or capacitor C2.

The net DC voltage equivalent to the potential of the base of transistor T1 (which is fed to the voltmeter) is provided by the transistor T2 (which serves as a series pass regulator) and is supplied by the transistor T1 (BC147B), which supplies the base potential.

With no zener diode attached across the probe, the voltmeter is currently reading somewhere about 30V. When a zener diode is put across the test probe, the base potential of transistor T1 drops until it reaches the level at which the zener diode breaks down. As a result of this, the base potentials of the transistors T2 and T1 are brought into parity with one another. The meter is now displaying the real voltage produced by the zener. By moving the needle using the zero-adjustment screw on the meter, it is possible to make a modification to the scale of the meter that is 0.6 V in magnitude.

PARTS LIST

Resistors (all of them are 14 watts, less than 5% carbon).

R1 = 100 KΩ; R2 = 39 KΩ; R3 = 2.2 KΩ; R4 = 680 Ω; R5 = 100 Ω; VR1 = 4.7 KΩ; VR2 = 10 KΩ

Capacitors

C1 = 27 KPF (273); C2, C4 = 2.2 µF/16 V; C3 = 22 µF/16 V; C5, C10 = 100 µF/16 V; C6 = 10 µF/16 V; C7 = 100 KPF (104); C8 = 47 KPF (473); C9 = 220 µF/16 V

Semiconductors

T1 is a BC147B, and IC1 is an LM386. Both are power amplifiers.

Miscellaneous

L1 equals the pick-up coil on the speaker 8.

SW1 is the switch for on and off.

187. Watch Man Watcher

This is a circuit that can be used at night in places like offices, stores, and warehouses, among other places, to determine whether or not the watchman who is responsible for your establishment is working. In order to function, it makes use of an existing telephone that is located as close as possible to the watchman's station (for example, in an office or store). You can send an audible alarm to the watchman by simply dialing the office or store's telephone number from your home or any other location, preferably by using your mobile phone. This should be done as quickly as possible. The provided circuit is able to detect the ring, and the watchman is also provided with a visible warning signal in the form of a bright bulb. Soon after the ring tone, the lamp continues to be "on" for a period of time that is close to sixty seconds. The watchman is given instructions to signify his presence by merely directing the light beam from his torch toward an LDR sensor unit that has been installed on the wall (without lifting the handset off-cradle of the ringing telephone). During the window of time in which the warning lamp is on, this task needs to be completed. In the event that he is unable to complete it within the allotted amount of time, the circuit will record his absence by increasing a count. If he does it, the count will not be affected in any way.

There is room for up to nine distinct alert bells in this scenario. The number that is displayed on the screen is the total amount of times that the watchman overlooked his presence. The timing details can be obtained by using the called number in conjunction with the displayed count on the mobile phone, which records both the called number and the call time.

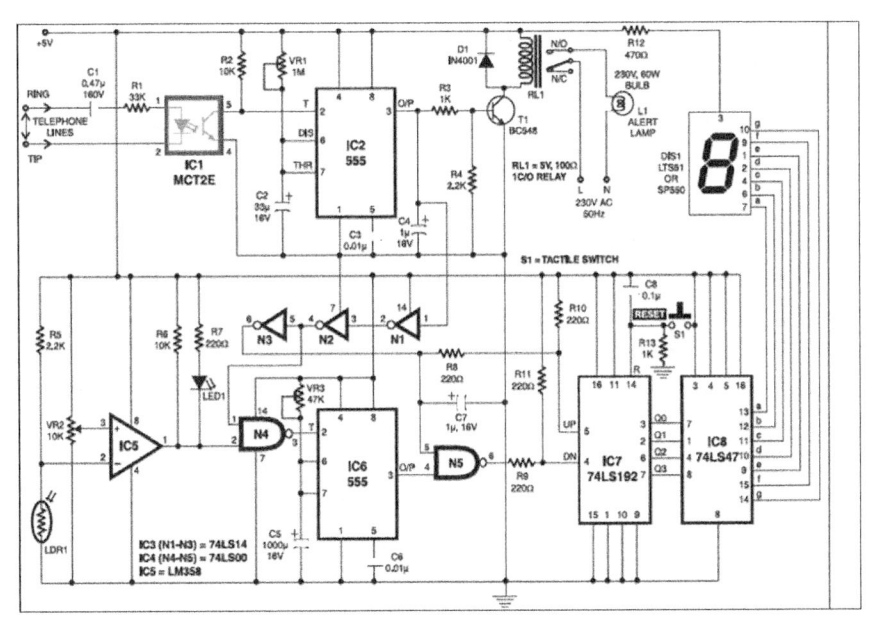

By setting its reset pin 14 to high using the reset switch S1, the telephone lines Counter 74LS192 (IC7) can be reset to the zero state. This applies to both the TIP and RING lines. The IC 74LS47 provides the power for the 7-segment, common-anode display known as DIS1 (IC8). After what feels like an eternity, the count of one is finally displayed as the phone rings. This occurs if the watchman is unable to concentrate the beam of the torchlight on the LDR1. If the LDR1 photoresistor detects light from the watchman's torch within the allotted amount of time, the down clock will remain high until the up clock also reaches its maximum. Since the counter first counts upwards and then counts downwards, the count, in practice, does not change.

Every component, with the exception of LDR1, is stored in a cabinet that is hermetically sealed and has a locking mechanism. Only LDR1 is mounted on the wall and may be seen from the outside. This is done to prevent the counter from having to be reset manually. In the event that the power goes out, the count will not have to be reset because the circuit will be powered by a battery.

188. Under-/Over-Voltage Beep for Manual Stabilizer

Manual voltage stabilizers are still popular because they are easy to make, cheap, and reliable because they don't have any relays. They can also handle a wider range of mains AC voltages than automatic voltage stabilizers. Most of the time, these are used in homes and businesses to power things like lights, TVs, and refrigerators. They are also used in places where the mains AC voltage goes from very low (during peak hours) to very high (during non-peak hours).

Some of the manual stabilizers on the market have a high-voltage auto-cut-off feature that turns off the load when the output voltage of the manual stabilizer goes over a certain high voltage limit that has been set. The output voltage may go up if the AC mains voltage goes up or if the rotary switch on the manual stabilizer is not used correctly.

One of the biggest problems with using a manual stabilizer in places where the voltage changes a lot is that you have to keep an eye on its output voltage, which is shown on a voltmeter, and keep changing it with its rotary switch. Or, the output voltage could reach the limit set by the auto-cut-off feature, which would turn off the load without the user knowing. To turn on the load again, the rotary switch on the stabilizer has to be used to change the voltage. This is a very annoying and inconvenient way to work for the user.

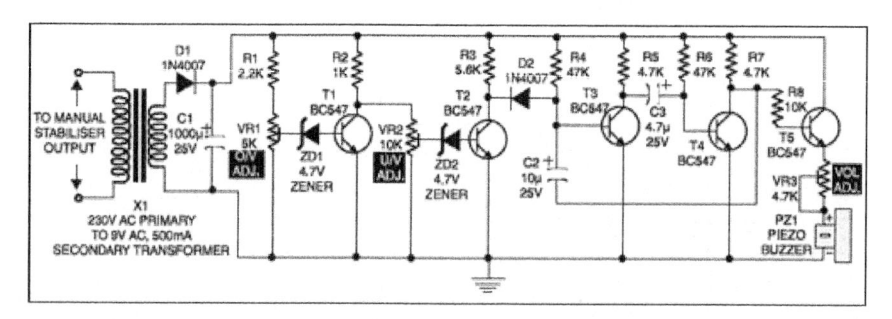

This under-/over-voltage audio alarm circuit, which is an add-on circuit for the existing manual stabilizers, solves the above problem. When the stabilizer's output voltage drops below a preset low-level voltage or rises above a preset high-level voltage, it makes different beep sounds for "high" and "low" voltage levels. For "high" voltage level, it makes short-duration beeps with short gaps between each beep. For "low" voltage level, it makes slightly longer beeps with longer gaps between each beep. With the help of the rotary switch and these two different types of beep sounds, it is easy to read just the stabilizer's AC voltage output. There's no need to check the voltmeter reading often.

It is best to set the high-level voltage 10V to 20V lower than the required high-voltage limit for auto-cutoff operation. In the same way, low-level AC voltage can be set 20V to 30V above the minimum operating voltage for a certain load. The output terminals of the manual stabilizer are hooked up to the primary winding terminals of step-down transformer X1. So, the 9V DC across capacitor C1 will change depending on the voltage at the output terminals of the manual stabilizer, which is used in this circuit to sense whether the voltage is high or low.

Together, the Zener diode ZD1 and the preset VR1 are used to sense and adjust the high-voltage level that makes the beep sound. In the same way, transistor T2 is used with Zener ZD2 and preset VR2 to detect and adjust the low voltage level for the beep signal. When the DC voltage across capacitor C1 goes above the preset high-level voltage or below the preset low-level voltage, the collector of transistor T2 goes high because transistor T2 is not conducting. But if the DC voltage measured across C1 is between the high-level and low-level voltages that have already been set, transistor T2 conducts and its collector voltage is pulled to the ground level. The astable multivibrator circuit, which is made up of transistors T3 and T4, uses these changes in the collector voltage of transistor T2 to start or stop oscillations. Through resistor R8, the base of transistor T4 is linked to the collector of transistor T5, which drives the buzzer. So, the buzzer goes off when the collector voltage of transistor T4 goes up. Set VR3 is used to change how loud the buzzer is.

When everything is normal, the DC voltage measured across capacitor C1 is within the window voltage zone. Because diode D2 and transistor T2 are conducting, the base of transistor T3 is pulled low. Because of this, capacitor C2 is drained. Because transistor T3 is in cut-off state, the astable multivibrator stops oscillating, and transistor T4 starts conducting. The buzzer doesn't make a beep sound because transistor T4 is on and transistor T5 is off.

When the DC voltage across capacitor C1 goes above or below the window voltage level, transistor T2 turns off. Its collector voltage goes up, and so does the voltage across diode D2. So, there is no way for capacitor C2 to be drained through diode D2. The astable multivibrator starts to beep, and the time between beeps is controlled by the DC supply voltage, which is low when low-level voltage sampling is taking place and high when high-level voltage sampling is taking place. For an astable multivibrator to work, it takes less time to charge capacitors C2 and C3 when the DC voltage is high and a little more time when the DC voltage is low. So, when the voltage is low, the buzzer goes off several times in a row, while it only goes off once when the voltage is high.

189. Solar Tracking System

In most cases, solar panels are fixed in place and do not move in response to the movement of the sun. A solar tracker system is presented here. This system follows the path of the sun as it travels across the sky and works to keep the solar panel perpendicular to the path of the sun's rays. This ensures that the solar panel receives the maximum amount of sunlight possible throughout the course of the day.

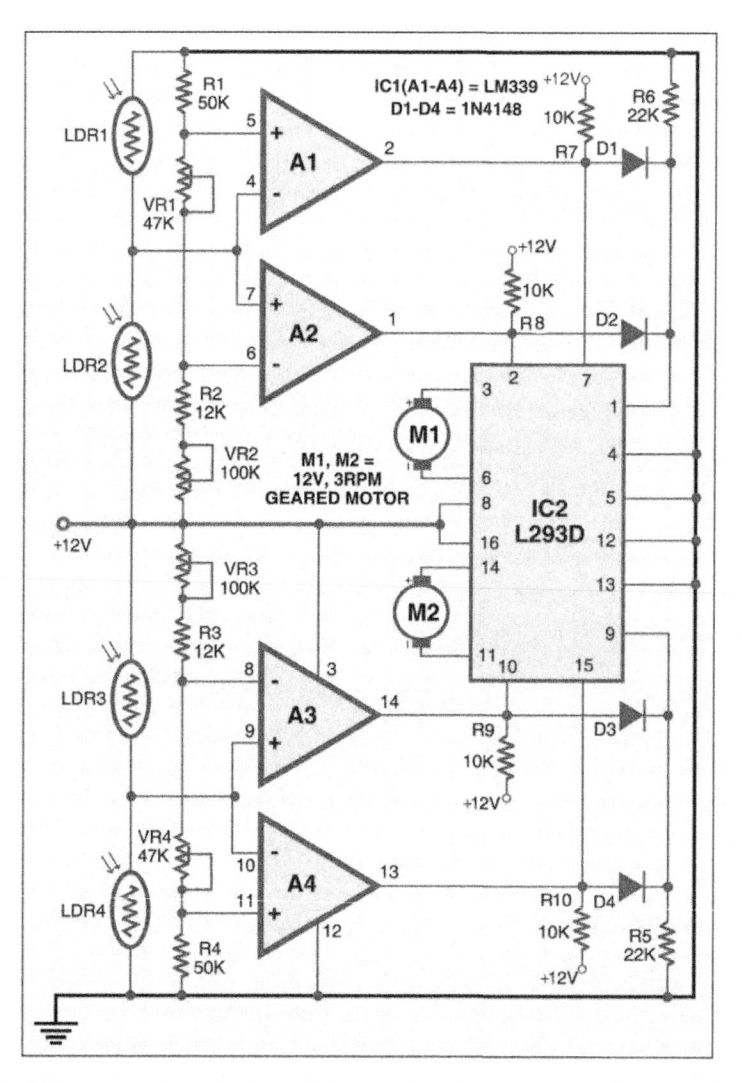

The solar tracker begins its pursuit of the sun at the crack of dawn, continues throughout the day until dusk, and then begins once more at the crack of dawn the next day. The schematic for the solar tracking system can be found in Figure 1. Comparator integrated circuit LM339, H-bridge motor driver integrated circuit L293D (IC2), and a few discrete components make up the sun tracker. Light-dependent resistors LDR1 through LDR4 are utilized here as sensors for the purpose of determining where the panel is located in relation to the sun. These send a signal to the motor driver IC2 that tells it to move the solar panel so that it is towards the sun. Comparators A1 and A2 are connected to the light intensity detecting resistors LDR1 and LDR2, which are attached to the edges of the solar panel along the X axis. To ensure that the motor M1 is stopped when the sun's rays are parallel to the solar panel, the presets VR1 and VR2 have been programmed to produce a low comparator output at pins 2 and 1, respectively, of comparators A1 and A2, respectively.

Comparators A1 and A2 receive a high input when pins 4 and 7 on their respective circuits receive a high value because LDR2 has a lower resistance after it has been exposed to lighter than its counterpart, LDR1. As a consequence of this, the output pin 1 of comparator A2 turns high, which causes motor M1 to revolve in a specific direction (let's say counter-clockwise), which in turn rotates the solar panel.

Comparators A1 and A2 will get a low input when LDR1 has a lower resistance than LDR2 because LDR1 has been exposed to a greater amount of light. This will occur at pins 4 and 7, respectively. The output pin 2 of the comparator A1 is now high because the voltage that is present at pin 5 is greater than the voltage that is present at pin 4. As a consequence of this, motor M1 begins to revolve in the other direction (let's assume clockwise), which in turn causes the solar panel to turn.

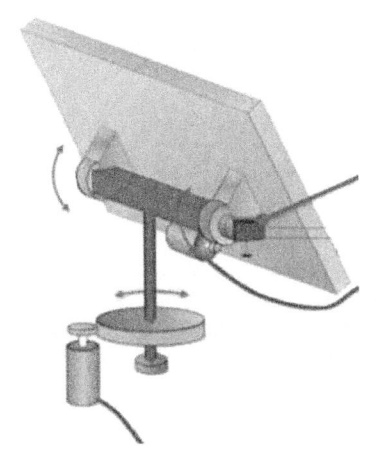

In a similar manner, LDR3 and LDR4 follow the path of the sun along the Y axis. The setup that has been suggested for the solar tracking system may be seen in Fig. 2.

190. Automatic Heat Detector

This circuit utilizes a complementary pair consisting of a npn metallic transistor T1 (BC109) and a pnp germanium transistor T2 (AC188) to detect heat (due to an outbreak of fire, etc.) in the surrounding area and energize a siren. However, the collector of transistor T2 is connected to relay RL1, while the base of transistor T1 is connected to the collector of transistor T2.

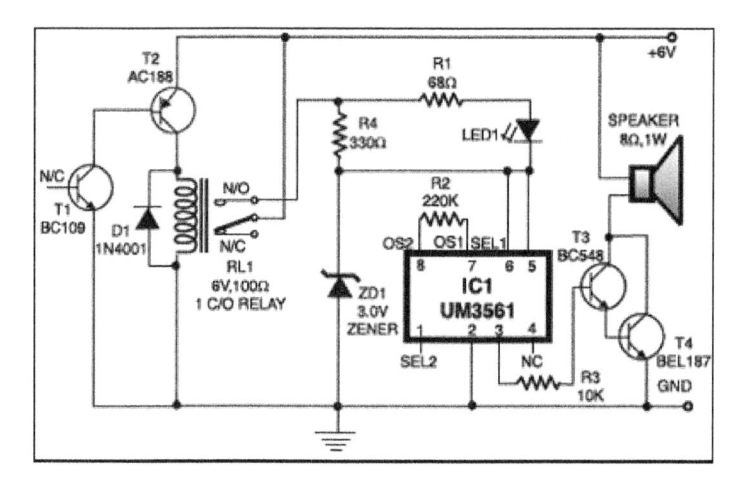

The second component of the circuit is a widely used integrated circuit known as UM3561, which is a siren and machine-gun sound generator IC. This component has the ability to generate the sound of a fire-brigade siren. When the relay is in the energized condition, pins 5 and 6 of the IC are linked to the supply of +3V, while pin 2 is connected to ground. The frequency of the oscillator that is integrated within the device can be adjusted by connecting a resistor (R2) between pins 7 and 8. Pin 3 provides access to the available output.

A Darlington design is used to connect two transistors, BC147 (T3) and BEL187 (T4), in order to magnify the sound that is produced by UM3561. When the relay is in its energized state, the 3V supply for UM3561 is provided by a resistor R4 that is connected in series with a zener rated at 3V. When the siren is activated, the LED1 that is linked in series with the 68-ohm resistor R1 and across the R4 resistor lights up. Bring a burning match stick close to transistor T1 (BC109), which will cause the resistance of its emitter-collector junction to drop owing to a rise in temperature. This will allow the transistor to begin conducting, which will test whether or not the circuit is functioning properly.

Pin Designation		Sound Effect
SEL1	SEL2	
No Connection	No Connection	Police Siren
+3V	No Connection	Fire Engine Siren
Ground	No Connection	Ambulance Siren
Do not care	+3V	Machine Gun

Additionally, conductivity may be observed from transistor T2 due to the fact that its base is connected to the collector of transistor T1. Because of this, relay RL1 becomes energized and turns on the siren circuit, which causes the loud sound of a fire department siren to be produced. Lab notes. We have included a table to make it possible for readers to get all of the attainable sound effects by returning pins 1 and 2 in the manner outlined in the table.

191. Automatic Water Pump Controller

The following diagram depicts a circuit that can be used to automatically control the motor of the water pump. When the water level in the overhead tank (OHT) drops below the lower limit, the engine starts up without any intervention from the user. In a similar fashion, it is turned off when the tank reaches its full capacity. The circuit is straightforward, condensed, and cost-effective due to the fact that it is built around a single NAND gate integrated circuit (CD4011). It operates on a power supply that is 12V DC and has a very low power need.

192. Little Power-Hila Vinegar Battery to power a calculator.

LCD calculators have an extremely low power consumption. These devices can be simply powered by this vinegar battery. Take the back off of a cheap calculator, remove the battery, and then stretch the two battery cables out the sides. After that, put the calculator back together. The creation of this basic battery is depicted in the graphic that can be found above. A zinc coated nail (2 "A galvanized nail and a length of copper wire, measuring six centimeters and having a gauge of fourteen, are submerged in vinegar, which contains 4% acetic acid. A film container that has been used and discarded will work wonderfully for this project. The "+" terminal of the battery is represented by the copper lead, whereas the "-" terminal is represented by the galvanized nail."

Our meter shows that the vinegar battery is producing .834 volts of electricity at the moment. Even after being charged, this battery can only produce a very small current. Increase the voltage and current flow via your circuit by connecting more batteries in series. When you put three batteries into an LED, it will glow very weakly.

193. Night Vision Enhancer

This is a straightforward green LED flashlight that is powered by a 3V battery pack and is based around a 555-timer integrated circuit (IC1). Students are disproportionately impacted by load shedding because it is a widespread issue in poor countries. The group of dreamlover technology posted a very simple, useful, and inexpensive project using ultra-bright white LEDs. These LEDs provide sufficient light for reading purposes while consuming very low power—specifically, 3 watts of power. Keeping this problem in mind, the group of dreamlover technology posted the project. It functions similarly to an emergency light in that if the AC mains fail, the battery backup circuit will instantly light up the LEDs. However, as soon as the power is restored, the battery supply will be cut automatically, and the circuit will once again function using the AC mains.

Circuit Description of LED-based reading lamp

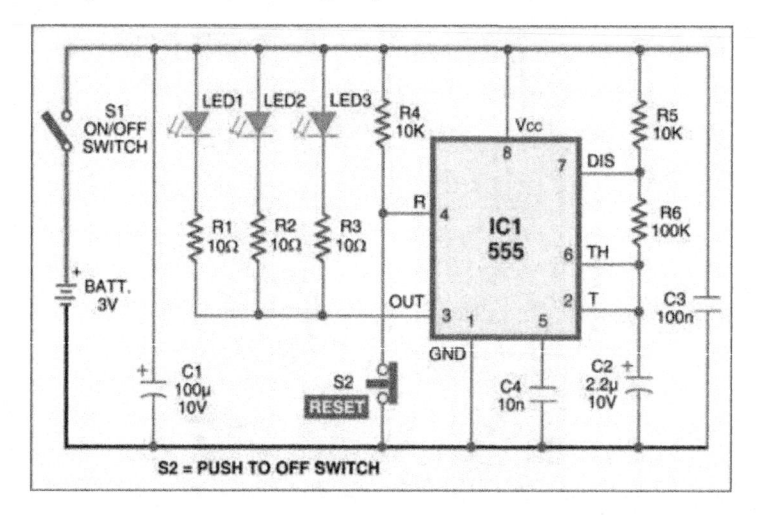

The power part of the circuit of an LED-based reading lamp consists of a bridge rectifier linked to the secondary coil of a step-down transformer X1 with a current rating of 500mA and 0-7.5V. In order to obtain pure DC output, the pulsating DC that is produced by the rectifier's output is fed into the voltage regulator IC1 input. Each individual LED, from LED1 all the way up to LED10, is connected in parallel across the voltage regulator's output. In this case, the current is restricted by connecting resistors R1 through R10 in series with the LEDs in the appropriate order.

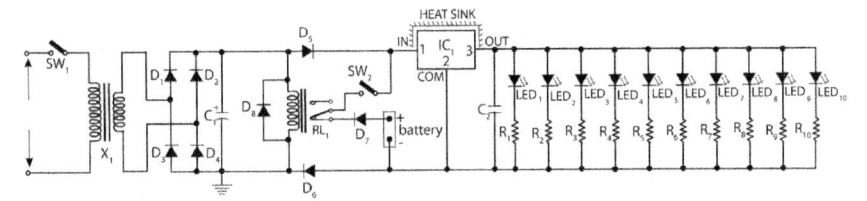

Figure 1: LED-based reading lamp

In this circuit, the lamp's intensity can be increased by the same amount by using five additional LEDs in the same way as was done before. When there was power from the AC mains available, relay RL1 was activated to cut power to the battery, and when there was no power from the AC mains, it did the opposite. A lead from the rectifier is connected in a straight line to the positive and negative terminals of the battery so that the battery can be charged. Diodes D5 and D6 are employed in this situation as reverse-current protection diodes; they prevent the current from the battery from flowing towards the supply section. Diode D7 is utilized to protect against reverse polarity.

COMPONENTS

Required Resistors (all ¼-watt, ± 5% Carbon)

R1- R10 = 56 Ω

Required Capacitors

C1 = 1000 µF/16V; C2 = 0.1 µF

Semiconductors that are required

IC1 = 7805 Voltage regulator

194. Emergency Photo Lamp

This emergency light is capable of being powered by either a non-rechargeable battery or a rechargeable battery (such as a 3.6V Ni-Cd battery) (3.0V CR2032). In the event that the power goes out and you are left in the dark, the white LED (LED1) will begin to illuminate automatically. The circuit has a very low quiescent current; thus, the battery is almost never drawn upon unless the LED is lit. This is the only time that the battery is used.

Emergency lights are devices that are not only incredibly helpful but also highly popular. These portable lights are equipped with battery backups that allow them to illuminate immediately in the event that the power goes out, ensuring that we are never left to fumble around in the dark. Building your very own emergency light at home might be a very different experience than purchasing one from the store, despite the fact that they are available for very little money on the market. Not only will it assist you in developing a design of high quality, but it will also get you familiar with the technical features of the device. In addition to that, you are provided the chance to personalize the circuit in accordance with the needs that you have. The current emergency lighting circuit was designed with the goal of being as space-efficient as possible while yet having the capability of being left plugged into a mains socket for continuous, hands-free operation. When the power from the mains is not available for an extended period of time, the light can also be charged using a solar panel.

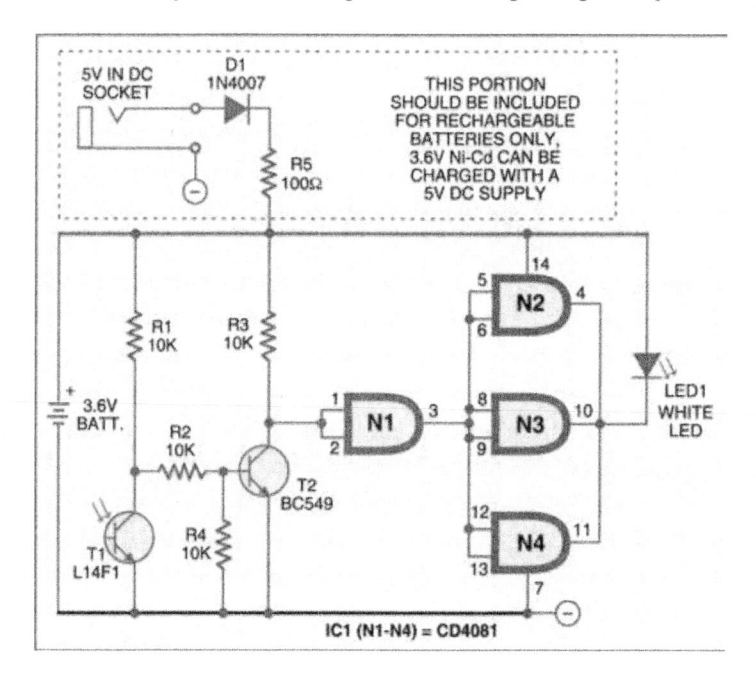

The backup battery is a 6V, 4.5AH Ni-cd battery, which is possibly the most widely used form of rechargeable battery. Aside from this, lead-acid batteries are highly affordable, have the ability to provide very high currents despite having an extremely low internal resistance, and can be purchased in large quantities.

195. 1W LED for Automotive Applications

This simple circuit lets you run a 1W LED from the battery of your car. IC MC34063 is used here as a buck converter. It is a monolithic switching regulator sub-system intended for use as a DC-DC converter.

Why do we require an LED Driver circuit?

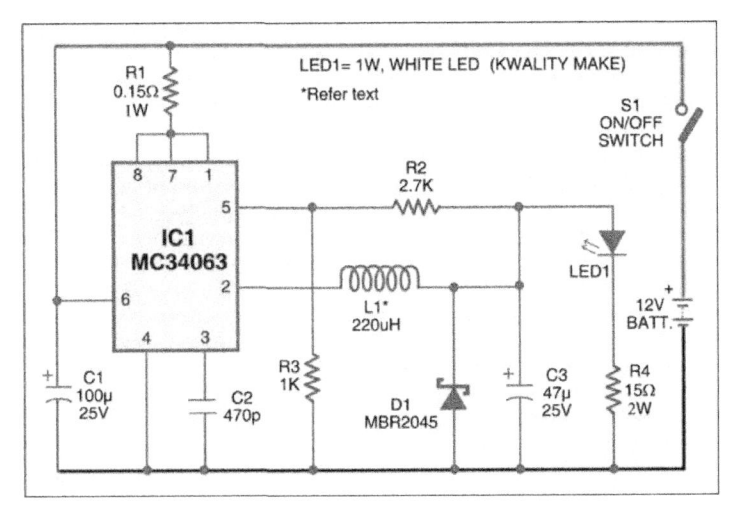

It's crucial to understand why an LED Driver circuit powers an LED before we get too far into this project. It is also possible to provide power to an LED by connecting it directly to a sufficient power source; but, in the majority of circumstances, the LED will pump in more current than it could manage, and it will eventually be destroyed. As a result, we make use of a driver circuit, which acts to limit the amount of current that is allowed to flow through an LED. The datasheet that comes with an LED should provide information on the maximum amount of current that it can draw. Using the LM317 integrated circuit (IC) as a current limiter, which I will describe how to do in this post, you will be able to construct the circuit depending on the current constraints that you have.

The device consists of an internal temperature-compensated reference, a comparator, a controlled duty-cycle oscillator with an active current-limit circuit, a driver and a high-current output switch. These functions are contained in an 8-pin dual in-line package. Another major advantage of the switching regulator is that it allows increased application flexibility of the output voltage.

196. Play with Robotic Eye (IR Sensor)

IR sensors are used in a wide variety of devices, including TV remote controls, burglar alarms, and object counters, among other things. Both an object-detection circuit and a proximity sensor for use with path-tracking robots have been fabricated here using infrared LEDs, which are referred to as IR sensors. The fundamental concept is to send an infrared light signal through an IR LED, which is then picked up by the receiving LED after being reflected by any obstacles in the path of the vehicle.

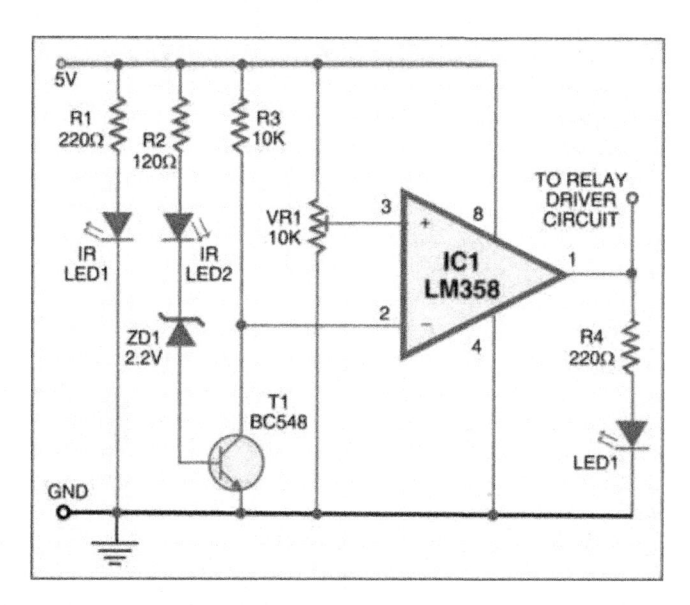

This module, which consists of two LEDs, an infrared transmitter, a receiver, and an IC that produces a frequency-modulated signal, is typically used in robots to identify impediments. When compared to a simple LED emitting LEDs with a photodetector, this module offers several benefits over its counterpart. The principle that applies in this scenario is that the emission frequency is modulated, thereby preventing the receiver from erroneously detecting the wrong signal due to the presence of ambient light. In addition, the building module that makes working with Arduino easier for both connecting and educational purposes.

The way the eye works is as follows: when there is an object in front of each sensor (up to 40 centimeters, adjustable), the sensor's state changes at the exit from Hi to Lo (of 5volts to zero). This causes the Arduino to realize that there was a change in state in one of its ports, which causes the servomotor to rotate the eye to the position that was programmed into it by the software. This was a practical example of using infrared sensors in a classroom setting for educational purposes. There are a multitude of other possible applications. Your imagination is the only limit here. Hope you enjoy.

197. Faulty Car Indicator Alarm

Car drivers are required to turn on their turn signal lamps prior to making any kind of turn, whether it be to the left or right, so that drivers of vehicles that are approaching them can take the appropriate safety measures. Should the turn indication lamps on your vehicle fail to shine for any reason, you significantly increase the risk of being involved in an accident. Presented for your consideration is a circuit that, should your turn indicator lamps fail to illuminate, will trigger an alert, so assisting you in preventing any potential mishaps.

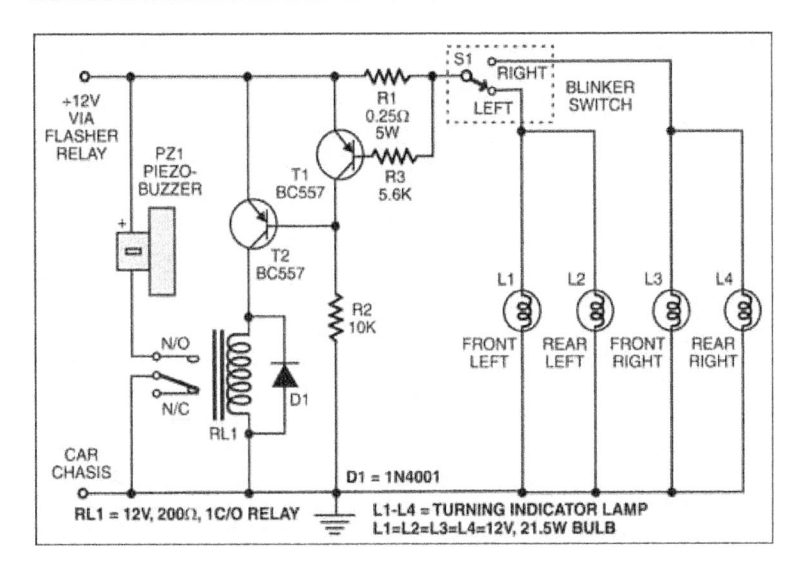

198. Long-range Burglar Alarm Using Laser Torch

Laser torch-based burglar alarms often only function when it is dark outside. But this long-range photoelectric alarm may work successfully throughout the daytime to warn you about intruders in your large compounds, etc., in addition to protecting you during the night. The laser transmitter and receiver components of the alarm are to be installed on the opposite pillars of the entry gate in order to complete the security system. The buzzer in the receiver circuit will raise an alarm any time someone walks in to disrupt the falling laser beam that is being broadcast.

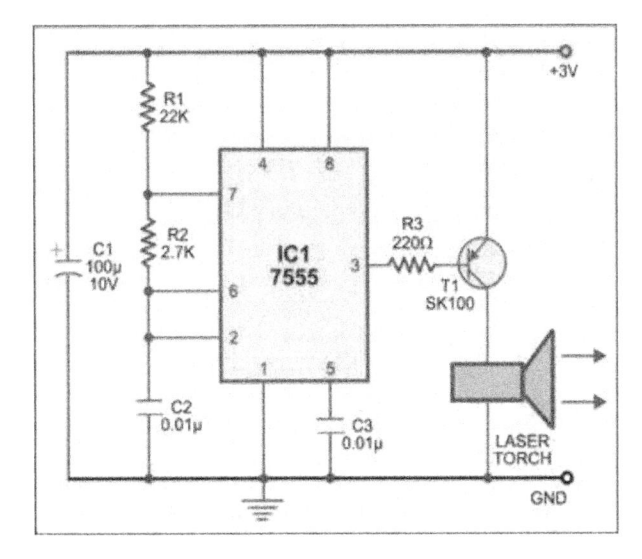

The circuit that makes up the transmitter is powered by 3V DC. The astable multivibrator that was constructed around timer 7555 (IC1) generates a frequency of 5.25kHz. When working with low voltage, the CMOS version of timer 7555 is what's needed. The base of the laser flashlight is wired to the emitter of the npn transistor T1, and the spring-loaded lead that protrudes from the interior of the flashlight is wired to the ground.

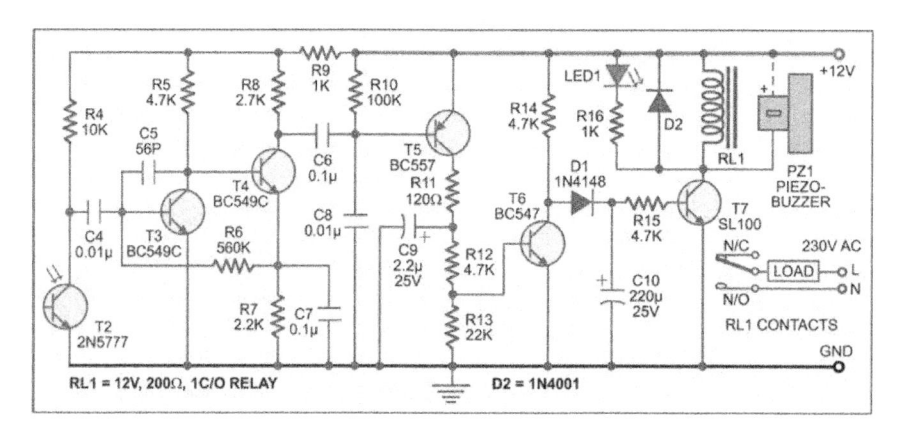

Receiver circuit

The circuit that makes up the receiver is powered by 12V DC. In order to detect the laser beam that is emitted from the laser torch, it makes use of a photodarlington 2N5777 (T2). The beam signals that are output by the photodarlington are sent to the two-stage amplifier, which is then followed by the switching circuit, etc. As long as the laser beam continues to illuminate photoDarlington T2, relay RL1 will not become energized, and the buzzer will not activate. Additionally, LED1 does not emit any light. When the laser beam that is falling on photoDarlington T2 is disrupted in any way, npn transistor T6 is driven out of conduction, and npn transistor T7 is driven into conduction in its place. As a direct consequence of this, LED1 begins to shine, and relay RL1 begins to energize, which causes the buzzer to ring for a few seconds (determined by the values of resistor R15 and capacitor C10). The normally opened (N/O) contact of relay RL1 allows a big indication load, such as a 230V AC alarm for louder sounds or any other device for brief indication, to be connected to 230V AC mains. This connection causes the large indication load to become active at the same time.

199. Soldering Iron Temperature Controller

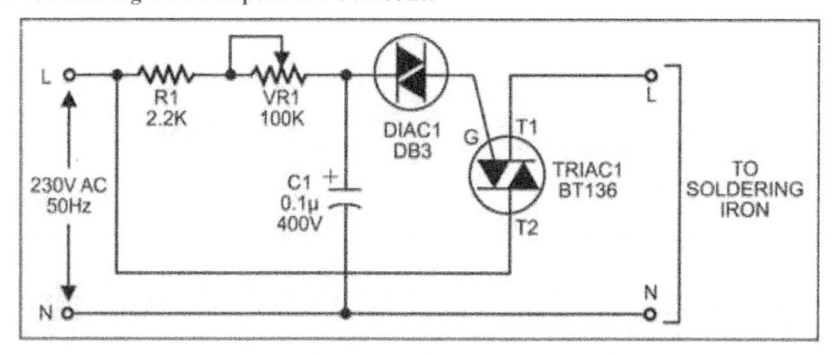

The following diagram depicts a straightforward circuit that can be used to regulate the temperature of a soldering iron. Because the heat dissipation from the iron can be controlled, it is especially helpful in situations in which the soldering iron will be left on for an extended period of time. When a soldering iron is turned on, it takes some time for the iron to reach the temperature at which the solder will melt. Connecting this circuit to the soldering iron in the manner depicted in the figure causes the iron to quickly reach the temperature at which the solder will melt.

200. Make your own Electric Bug Zapper

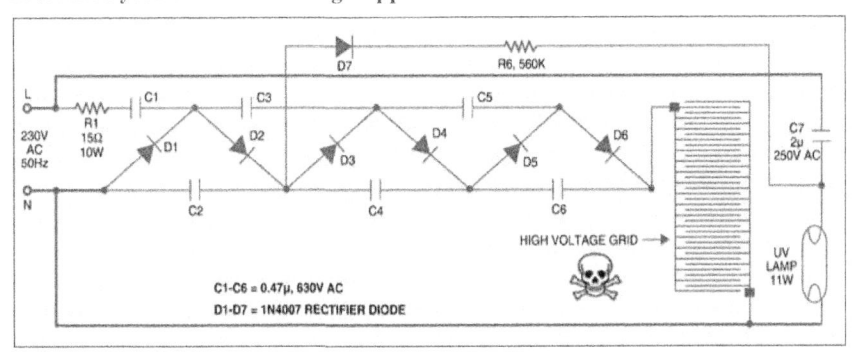

None of us likes bugs at home. To kill these flying insects, they should first be attracted and then electrocuted. Bug zapper is one such device with a high-voltage electrocuting circuit and an insect-attracting UV lamp of 365 nm wave-length. This ultraviolet fluorescent lamp is mounted in the middle of the cabinet and a pair of carefully spaced, electrically insulated, charged wire grids surround the light. When an insect comes close enough to the mesh pair, an electrical arc is formed, the dielectric breaks down and current flows through the insect's body. Electrocuting the insect doesn't require it to touch both the wires as an arc form in the air gap over 1800V

201. Timer for Mosquito Destroyer

In mosquito repellents that are heated electrically, a mat or liquid is heated by an electric vaporizer, which then causes the mat or liquid to release non-degrading compounds into the air. These chemicals keep insects away from enclosed areas. The following is a circuit for a Mosquito Destroyer that suspends normal machine function for a period of fifteen minutes. This does not compromise the device's ability to ward against mosquito bites in any way. Because the circuit in question (shown by Figure 1) does not employ any transformer-based power supply, it is sufficiently small to be housed within the switchboard. It is powered by 230V AC mains, which are passed through a voltage-limiting resistor known as R6 (22-kilo-ohm). In order to produce 12V DC, the low-voltage alternating current (AC) is first rectified by diodes D1 and D2, then filtered by capacitor C3, and finally regulated by Zener diode ZD1. Using a 14-state binary counter, IC HEF 4060 is what makes the timer action possible (IC1). The oscillations of the IC are controlled by capacitors C2, R2, and R3, and the blinking of LED1 coupled to the Q3 output (pin 7) of IC1 serves as an indicator of these oscillations. When the switch S1 is used to power on the circuit, the integrated circuit IC1 is reset by the capacitor C1 and the resistor R1, and it then begins oscillating. After 15 minutes, the output of its Q11 swings high (pin 1), which causes the triac BT136 (triac1) to be triggered through the resistor R5. When the triac starts conducting electricity, the neutral line travels through the M2 terminal of the device to the plug socket. After being powered on for 15 minutes, the vaporizer that is attached to the power outlet will then turn itself off. Until the power switch S1 is switched off, this cycle will continue to repeat itself. As a result, the circuit contributes to a reduction in the number of chemical vapors in the air that is breathed in. Because the vaporizer is only operational for fifty percent of the time, its power consumption is cut in half, effectively doubling the number of days it may be used.

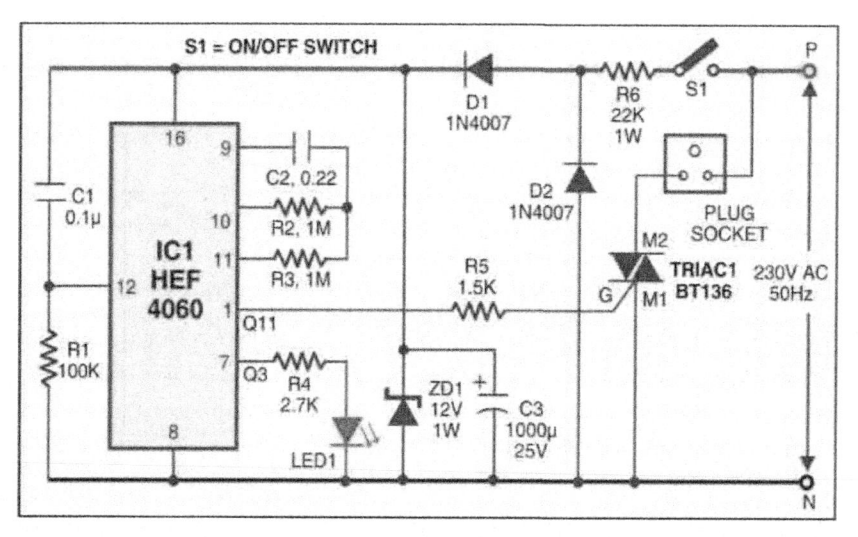

Fig. 1: Timer circuit for mosquito destroyer

Construct the circuit using a PCB designed for general use, taking care to leave sufficient space between each component. For the exposed leads of components, particularly triacs and diodes, sleeving is an essential precaution to take. In order to complete the circuit, enclose it in the switch box, and connect the plug socket to it in the manner indicated on the circuit diagram. Caution, the author's prototype, can be seen in Figure 2. Because the circuit contains 230V AC, checking and repairing it requires the utmost caution in order to prevent potentially fatal shocks. You should only assemble it if you have the necessary training and experience to work with high voltages.

202. Radiation Sensor

When you are in close proximity to an electronic equipment, such as a computer or television, your body is subjected to what is known as "electronic smog." For instance, the spot of electrons that sweeps over the screen in CRT-based monitors generates pulsed electromagnetic radiation (PEMR). Some of this energy is lost to the environment in the form of radiations that have very low frequencies and extremely low frequencies.

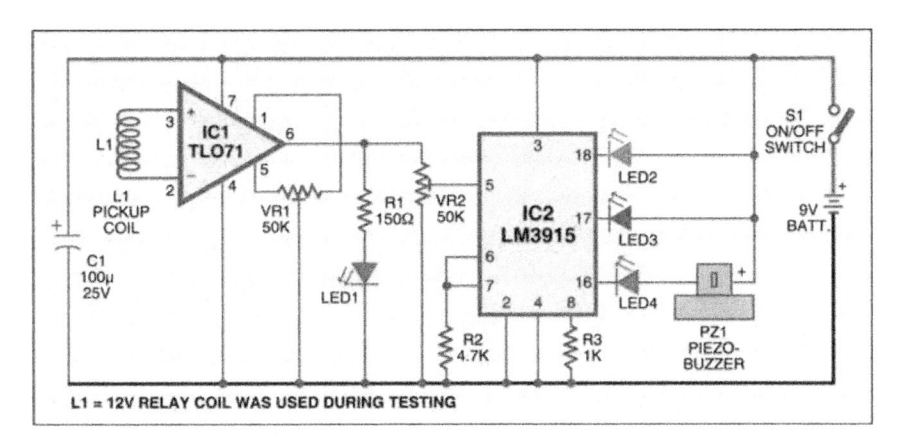

203. Handy Tester

Here is a low-cost multitester that can be used to test the state of practically all the electronic components, from resistors to integrated circuits (ICs). It is intended for use by beginners. It just requires a few components yet is able to detect polarity, continuity, logic states, and the activity of multivibrators despite its small size. The circuit takes advantage of the biasing property that bipolar transistors have and is extremely straightforward. In order to provide the results of the test, transistors T1 and T2 perform the function of transistor switches by driving the red and green halves of the bicolor LED1 individually.

Working of a Multitester

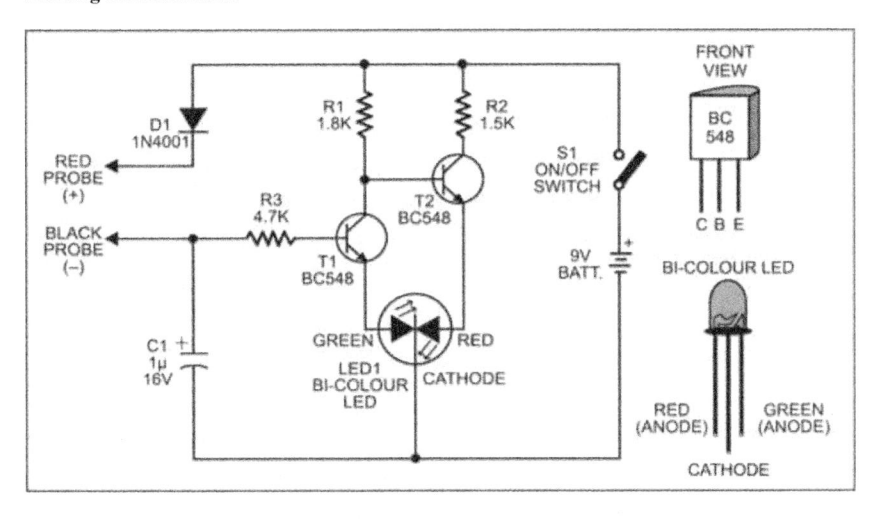

Because there is no longer any forward bias, the conductance of transistor T1 is terminated when power is supplied by depressing switch S1. Simultaneously, transistor T2 begins conducting once it has drawn base bias voltage from the battery and passed it through resistor R1. This makes it possible for the red half of the bicolor LED1 to light up.

When a positive voltage is applied to the base of transistor T1 via resistor R3, the transistor conducts, which lights up the green half of the bicolor LED1. When transistor T1 is conducting, its base is grounded, which causes transistor T2 to switch off, so turning off the red half of the bicolor LED1. The signal that is acquired at the base of transistor T1 is therefore essential to the operation of the circuit. The following table details the testing techniques for a variety of components, along with the indications and findings that should be obtained.

204. Strain Meter

This strain meter shows whether a change in the shape of an object, like a crane strut, is compressive (shortening) or tensile (lengthening). A strain gauge is stuck to the thing being tested and measures the strain. The reading on the meter changes when the resistance of the strain gauge changes. For this, a voltmeter or other analog or digital meter with a full-scale deflection of 1V DC can be used. But it would be better to use a digital multimeter.

The circuit shows how a single op-amp IC 741 (IC3) boosts the signal from a network of three resistors (R1–R3) and a strain gauge (SG1). Since the recommended strain gauge for this circuit has a nominal resistance of 120 ohms, each of resistors R1 through R3 should also be 120 ohms. The resistors should also have a tolerance of more than 0.1% so that changes in temperature don't cause the meter's reading to change in ways that aren't wanted.

Circuit operation

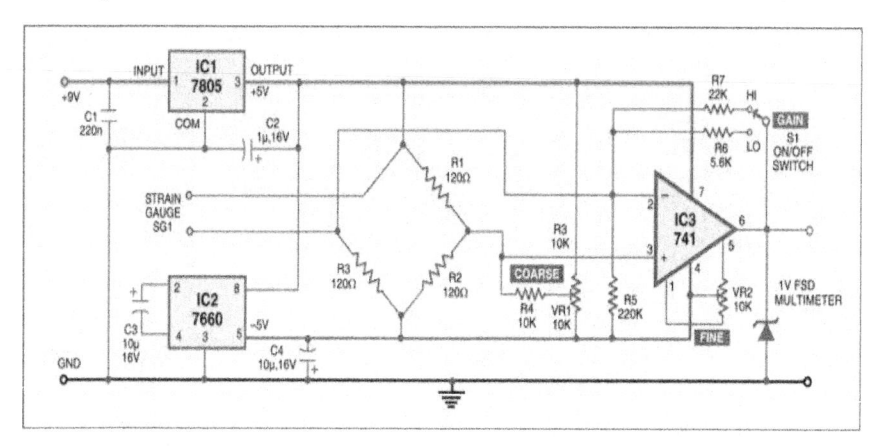

How the circuit works is easy to understand. When the strain gauge is not being pulled on, the variable resistors VR1 (coarse control) and VR2 (fine control) are set so that the multimeter reads zero. (These zero voltages could be any voltage that works for the circuit, like 1.5V. It doesn't have to be a zero-voltage supply.) If the meter reading goes up, it means that the length of the strain gauge is getting longer (tensile strain). If the number on the meter goes down, the strain gauge is getting shorter (compressive strain).

By changing the value of VR2, you can change how sensitive the op-amp is (IC3). It lets the circuit respond in different ways to different amounts of strain. The meter gives stable readings because voltage regulators IC1 and IC2 provide 5V and -5V power supplies. The output from the Wheatstone bridge is made stronger by the op-amp.

205. Water Pump Controller

Here is a simple circuit that can be used to control the level of water in an overhead tank. A step-down transformer, a 24V AC double changeover relay, two buoys, and two micro switches are the main parts of this pump controller. Any available relay can be used, no matter what voltage is in its coil. Obviously, the engine control should take into account the current rating of the contacts. It should have at least two contacts. A step-down transformer with a voltage that can be changed to match the relay's coil voltage is used.

AC is what makes the circuit work, so no change is essential. Microswitches S1 and S2 are placed on top of the water tank. They are controlled by two separate buoys, one that senses the bottom level and the other that senses the level at the top. To connect these switches to the relay, a three-center wire is used. Water is a valuable natural resource that has to be used more effectively. When using a manual method, users are responsible for physically checking the level of water in their own water tanks. The careless use of water ultimately results in the loss of usable water and contributes to the problem of water scarcity.

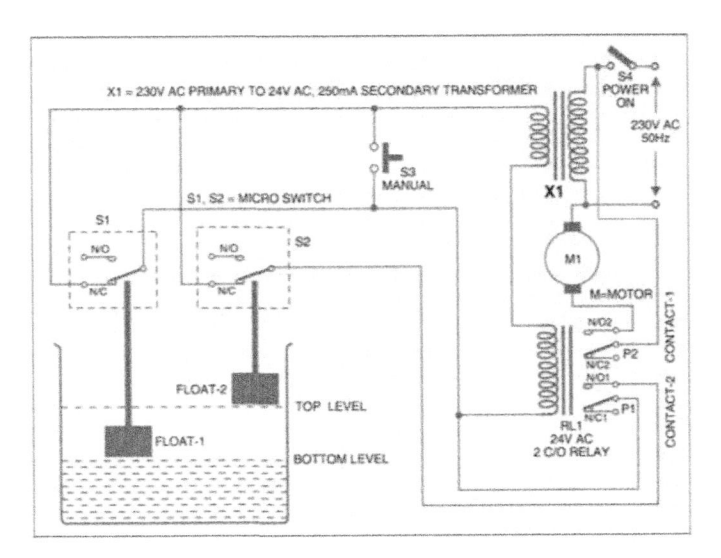

The process of automating the usage of water involves the utilization of an automatic system. The primary objective of water automation is to minimize the amount of manual labor required while maximizing efficiency in water use. It is used for a variety of functions, including irrigation in agricultural land, operating water pumps, monitoring water usage, and invoicing for water usage, among other things, in a variety of settings, including households, agricultural land, industries, hotels, and so on. Researchers have successfully executed a number of projects involving water automation by utilizing an android application. These projects include a water pump controller, water level sensor, water billing, and the detection and management of water leakage.

206. Timer with Musical Alarm

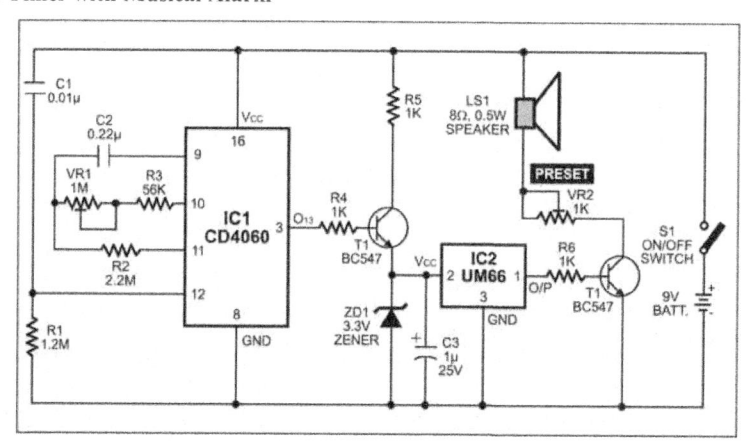

A timer is a special kind of clock that can measure how long something takes. Here, we show you a simple alarm clock with music that can be used for that. This inexpensive musical alarm timer can add a delay of between one minute and two hours. When the timer goes off, a musical song play. The popular CMOS oscillator/divider CD4060 is the center of the circuit (IC1). It runs on a 9V PP3 battery and draws very little power when it is not in use. By changing how preset VR1 is set, the delay time can be changed. After the time delay is over, pin 3 of IC1's output goes high, and npn transistor T1 conducts to give melody generator IC UM66 (IC2) pin 2 a positive power supply. Zener diode ZD1 cuts this power supply down to 3.3V, which is what IC2 needs to work. Through driver transistor T1, the output of IC2 is sent to the loudspeaker (LS1). With preset VR2, you can change how loud the loudspeaker is. When power is turned on by pressing switch S1, the timer starts. To stop the alarm from going off, you have to turn off the power supply.

207. Simple Key-Hole Lighting Device

In the dark, it may take you a few minutes to find the key-hole in your door so you can open it. Here is the circuit for a light that makes it easy to open the door when it's dark outside. In this circuit, a very bright red LED flashes instead of staying on all the time. The circuit board for this simple circuit was taken from a broken quartz watch. Most people who do hobbies might have one of these in their junk box. First, take the coil and its connections off of the pads it is attached to. Then, the other parts can be connected across the pads, just like the diagrams show. There are two types of circuits: manual (shown in Fig. 1) and automatic (shown in Fig. 2).

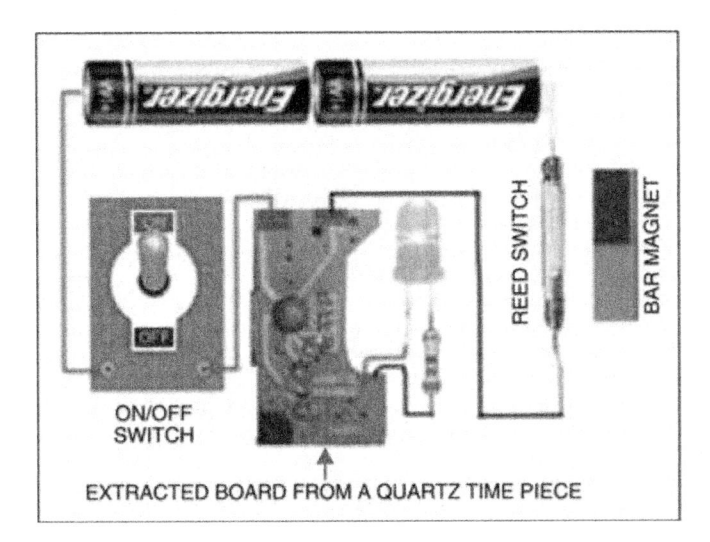

A couple of AA cells power the circuit (shown in Figure 1). When the door is shut, a magnet that is attached to the door frame moves close to the reed switch, which turns on the circuit. LED1 blinks every two seconds when the circuit is turned on. The unit has an on/off switch to turn it off when it's not being used. As shown in Fig. 2, a phototransistor L14F1 (T1) is added to stop LED1 from flashing during the day (when there is sufficient light).

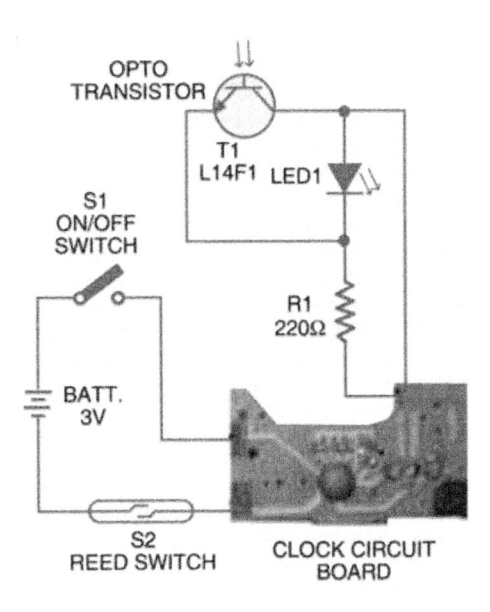

Fig. 2: Circuit for key-hole lighting device (automatic)

The unit has an on/off switch to turn it off when it's not being used. If you put this circuit together right, it will decorate your door for a surprisingly low cost.

208. Ball Speed Checker

This ball speed checker circuit uses the time it takes the ball to move from the bowling crease to the batting crease to figure out how fast the ball is going.

Ball speed checker circuit

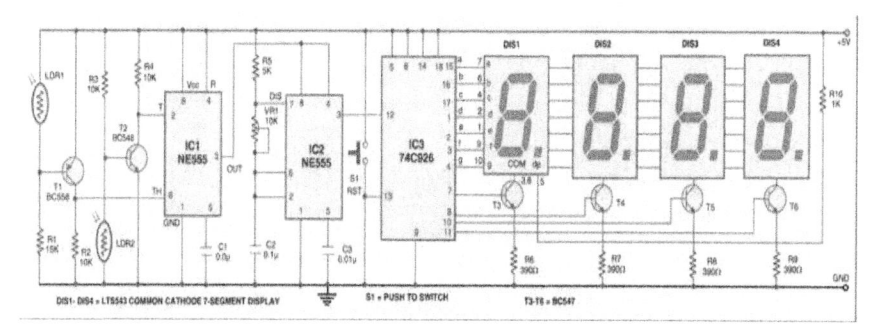

Figure 1 shows that the circuit is made up of two NE555 timer ICs (IC1 and IC2), a 74C926 (IC3), an LTS543 (display), and a few other parts.

In bistable mode, the first NE555 timer (IC1) is wired. The output of IC1 is connected to pins 4 and 8 of IC2, which is wired as an astable multivibrator and makes a 1 kHz frequency. IC3 counts and shows what comes out of IC2. Around LDR1 and LDR2, there are two light barriers that send out signals when the ball blocks the laser beam as it moves from the bowling crease to the batting crease.

Circuit operation

When the ball goes through the bowling crease, it briefly blocks the IR beam from reaching LDR2. This sends a pulse to IC1's pin 2, which then sets IC1. When IC1's output goes high, it gives +5V to IC2 and turns it on. IC2 then starts oscillating. IC3 keeps track of these pulses and shows them on 7-segment displays (DIS1 through DIS4).

When the ball goes past the batting crease, it stops the laser beam from reaching LDR1. This sends a pulse to IC1's pin 6 and turns it back on. IC1's output goes low, which stops +5V from going to IC2 and turns it off. IC2 stops moving back and forth, and IC3 stops counting. On 7-segment displays, the time is shown. From 0 to 9999, which is 10 seconds, the counter counts up. About 18 meters separate the bowling crease from the batting crease. Now you can figure out the speed of the bowling ball by plugging in the distance and time it takes the ball to travel from the bowling crease to the batting crease in the following formula:

$$\text{Speed} = \frac{\text{Distance}}{\text{Time}}$$

$$\text{Speed} = \frac{18}{\text{Time}} \times 3.6 \text{ km/hour}$$

Two mirrors that reflect light can be stuck to wooden bases to make a sensor. Figure 2 shows how the mirrors should be set up on the sides of the bowling crease. Attach a laser flashlight to one of the mirrors. After hitting the mirror in front of it, the laser beam must make a small angle so that it bounces back and forth between the mirrors until it reaches the LDR at the bottom of the mirror. In this way, the laser beam makes a barrier of light. When the ball gets in the way of the beam, a pulse is made. Put another of these sensors on the side where the batter is.

209. Halogen lamp Saver for Bikes

Because of their low cold current, halogen lamps tend to burn out. The thin filament inside the lamp melts when the lamp heats up quickly, which shortens the lamp's life. The circuit shown here makes the halogen lamp last longer by letting it turn on slowly. When you press switch S1, capacitor C1 charges through resistors R1 and R2. This makes MOSFET T1 work. As long as the load (bulb B1) doesn't draw too much current, it works like a normal switch. Resistor R4 controls how much current flows through the load and also checks how much current flows through MOSFET T1. As more current flows through MOSFET T1, the voltage drop across R4 goes up.

If the voltage at R4 gets to 0.65V, transistor T2 starts conducting and MOSFET T1 stops. Because of this, the lamp stops shining. When S1 is turned off, C1 will keep its charge. This leftover charge will get rid of itself through diode D1 and resistor R5. So the soft turn-on works right every time the light bulb is turned on.

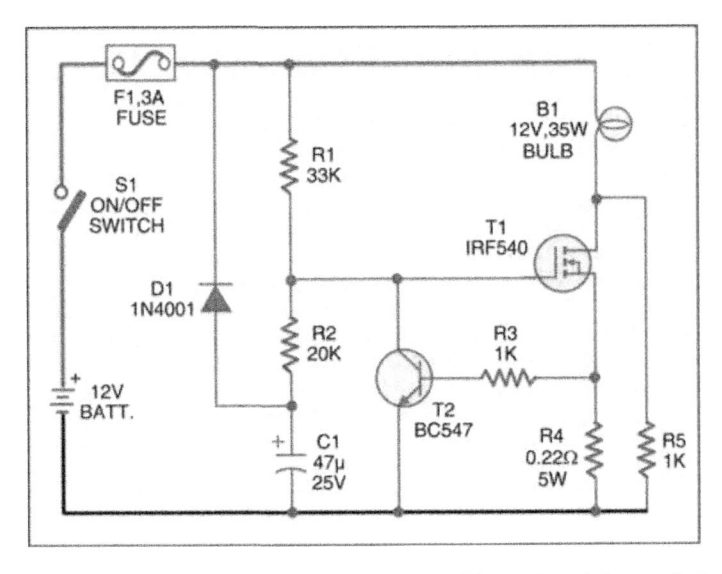

Put the circuit together on a PCB that can be used for many things and put it in a small plastic box. Use a heat sink to mount the MOSFET so that the heat can escape. The current-limiting resistor R4 protects the battery in case the output is short-circuited or the lamp draws more current than specified. Connect the bulb with two high-current gauge wires on the outside.

210. HDD Selector Switch

This add-on circuit utilizes the computer's switch-mode power supply (SMPS) to allow for the selection of three separate hard disk drives (HDDs) and the guarantee that you're secured HDD will remain secure at all times. It is a good way to keep hackers and spies from getting in. The block diagram of the HDD selector switch is shown in Figure 1. Within a certain amount of time (say, 20 seconds) after turning on the PC, you can choose between three different hard drives, such as "main," "net," and "backup." After the time has passed, you can't open your HDD without restarting your PC. Figure 2 shows that the HDD selector switch is made up of IC CD4093 (IC1) and a few other separate parts. The IC CD4093 has four Schmitt NAND gates with two inputs each. Pin 1 of gate N1 is connected to the positive supply through capacitor C1 and diode D1. It is also connected to the ground (GND) through resistor R1. Resistor R2 pulls pin 2 of gate N1 low. The master lock switch is also linked to pin 2 and +12V. Pin 3 of gate N1 is connected to the inputs of gates N3 and N4, which are pins 8–9 and 12–13, respectively. The outputs of gates N3 and N4 are sent to gate N2. A two-color LED (LED1) is connected to the output pins of N3 and N4 and the input pins of N2. Pins 1 and 2 make the LED light up green, while pin 4 makes it light up red. The point where pins 5, 6, 10, and 11 meet is linked to the point where switches S2 through S4 all meet.

Through diodes and resistors, switches S2 through S4 are used to turn on the gates of SCR1 through SCR3 (for 12V supply) and SCR4 through SCR6 (for 5V supply). Three green LEDs (LED2 through LED4) are attached between the cathodes of SCR1 through SCR3 and grounds to indicate the selected HDD. When the appropriate switches (S2 through S4) are pressed, the cathodes of SCR1 through SCR3 will get 12V and the cathodes of SCR4 through SCR6 will get 5V.

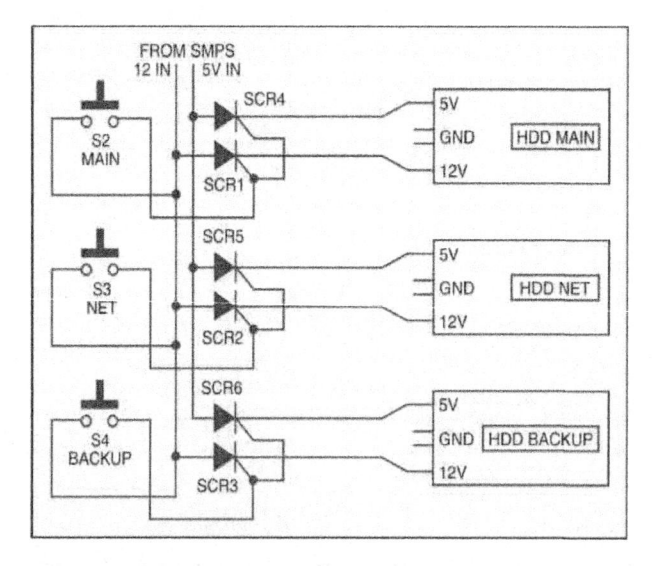

Fig. 1: Block diagram of hard disk drive selector (HDD)

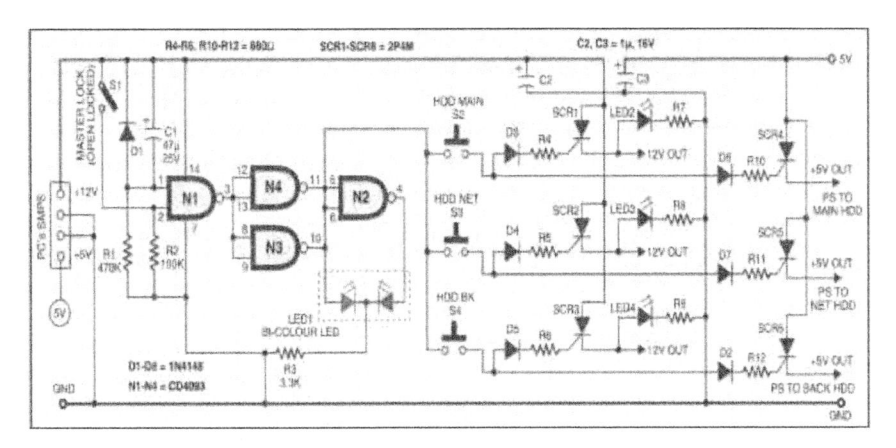

Fig. 2: Circuit of HDD selector switch

How the circuit works is easy to understand. When the SMPS is first turned on and the lock switch S1 is closed, the bicolor LED1 lights up green for about 20 seconds. In this state, switches S2 through S4 can be used to connect any HDD to the power supply. For example, if you want to turn on the net HDD, you need to briefly press switch S3 within 20 seconds. This will give this HDD 5V and 12V power supplies. When LED1 turns red, switches S2 through S4 stop working, so you can't turn on any HDD. On the other hand, LED1 glows red when the SMPS is "on" and switch S1 is open. No hard drive (HDD) may be powered by connecting to the supply through switches S2–S4 in this configuration. Put the circuit together on a PCB that can be used for many things and put it in a box. Since SCRs produce a lot of current, PCB tracks should be thick enough. The PCB and the wires that connect it should be made well. Install the components in the same manner as shown in the author's prototype, and then mount the PCB on the spare drive slot cover of the CPU cabinet measuring 13.4 centimeters or 5.25 inches.

211. Simple Key-Operated Gate Locking System

Only people who know the set code can open the gate with this simple key-operated locking system. To make the motor in the gate work, you have to type the code into the keypad within the time limit. If someone tries to open the gate and presses the wrong key on the keypad, the system is turned off and an alarm goes off to let you know someone has broken in.

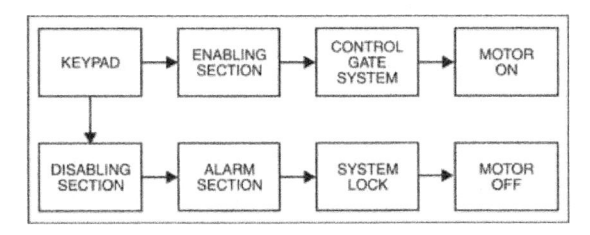

Fig. 1: Block diagram of simple key-operated gate locking system

The block diagram of the key-operated code locking system can be seen in Figure 1, and the circuit diagram can be seen in Figure 2. Establish connections between the keypad's various points and the circuit's points A, B, C, D, E, and F, as well as ground. Here, the S7, S16, S14, and S3 keys are used to enter the code, and the other keys are used to turn off the system. To make the code, it is very important to press the keys in that order. Press the switches S7, S16, S14, and S3 in order to start the motor of the gate. If the keys are pressed in a different order than what was set, the system will automatically lock and the motor will not start.

At first, pin 14 of AND gate IC6 doesn't have 6V, so no pulse gets to the base of npn transistor T1 to set off timer IC5, so the gate doesn't open. To turn on the system, you must first turn on IC4. When the switch S7 is pressed, the timer IC4 sends 6V to IC6 for about 17 seconds. You have to press the switches S16, S14, and S3 in order during this time. Because of this, the outputs of timers IC1, IC2, and IC3 go high in order. These high outputs are then sent to gates N1 and N2 of IC6 to trigger IC7 through npn transistor T1. IC1, IC2, and IC3 are all set to have high outputs for 13.5, 9.43, and 2.42 seconds, respectively.

When all four switches (S7, S16, S14, and S3) are pressed in order, timer IC7 starts the motor for the amount of time that was set, which opens the gate. When the time is up, the motor stops by itself. Adjusting preset VR5 lets you choose how long the motor is "on." Here, the shortest time it can be "on" is 5.17 seconds, and the longest time it can be "on" is 517 seconds. If a switch other than S7, S16, S14, or S3 is pressed, IC5 is triggered to turn on relay RL1, which cuts power to the second relay. This locks the system and makes the piezo buzzer PZ1 sound, so you know someone is trying to open the gate lock.

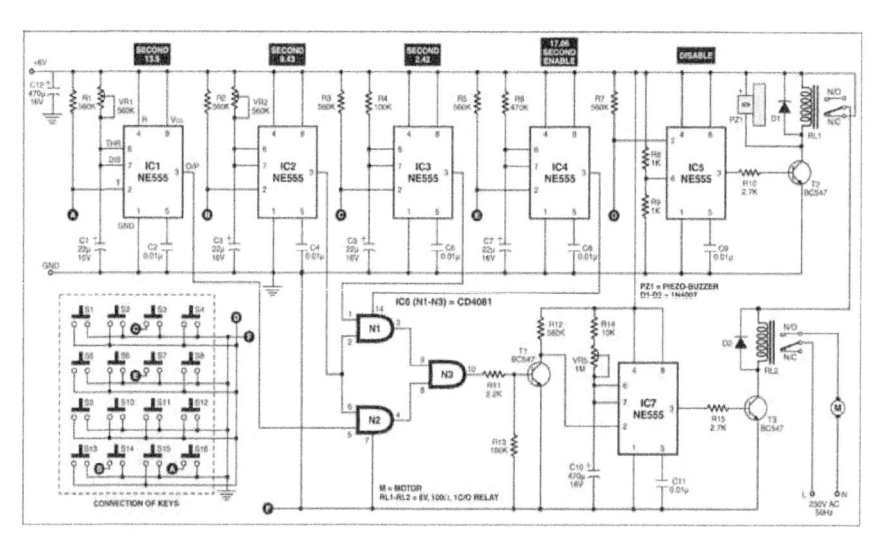

Fig. 2: Circuit of simple key-operated gate locking system

Now, press any key on the keypad (except S7, S16, S14, and S3) to turn off the sound and reset the system. The circuit runs on a regulated 6V DC power supply and is easy to put together on a PCB for general use.

212. Mains Box Heat Monitor

This basic main box heat monitor circuit continuously monitors the mains distribution box and sounds an alert when it detects a high temperature owing to overheating, assisting in the prevention of disasters caused by any sparking in the mains box due to short circuits. It also turns on a bright white LED by itself when the power goes out. In the dark, the LED gives off enough light to check the wiring or fuses in the mains box. When power goes out, the circuit beeps once. When power comes back on, it beeps again. Transformer X1 takes the mains AC power and steps it down so that the secondary output is 9V AC at 500 mA. The output of the transformer is fixed by diodes D1 through D4. The ripple is avoided by capacitor C1. LED1 shows that the device is turned on. LED1's current is limited by resistor R1. The temperature sensor is the Germanium diode D5 (1N34), which is hooked up in the reverse bias mode. At normal temperature, the diode's resistance is high, so transistor T1 conducts and keeps reset pin 4 of IC1 in a low state. The NE555 (IC1) has been configured as an astable multivibrator. When the fuse overheats, diode D5's resistance drops, and transistor T1 stops conducting. This turns on IC1 and starts the oscillator's alarm. You can set the temperature at which the alarm circuit goes off by adjusting preset VR1.

The circuit for mains box heat monitor

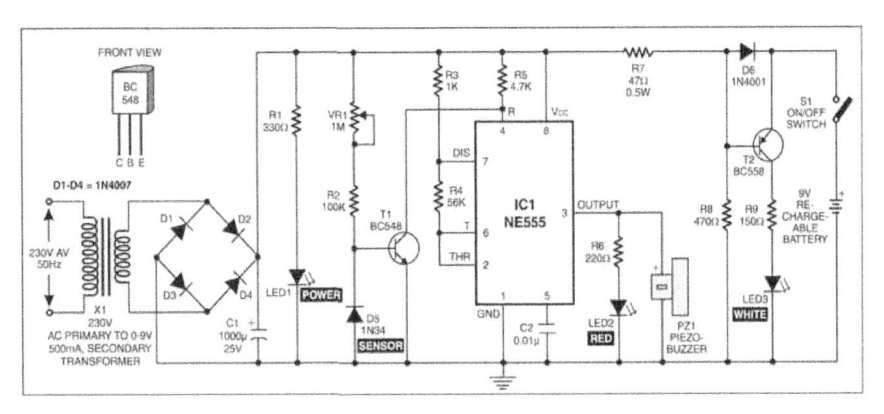

The emergency light circuit is made up of a pnp transistor BC558 (T2) and a few passive components. It is powered by a 9V rechargeable battery that is constantly charged by a forward-biased diode D6 when mains power is present. The charging current is brought down to a safer level by resistor R7. When diode D6 has a forward bias, it makes transistor T2 have a reverse bias. This turns off the white LED (LED3). When the power goes out, the transistor T2 is turned on, and the LED lights up. When the power comes back on, transistor T2 stops doing its job, so the LED stops lighting up.

213. Digital Soil Moisture Test

You can determine whether the soil is dry or wet by using this digital soil moisture tester, which is both simple and convenient to use. It measures the amount of moisture in the soil. Additionally, it can be utilized to determine if cotton, wool, or woven fibers are dry or damp.

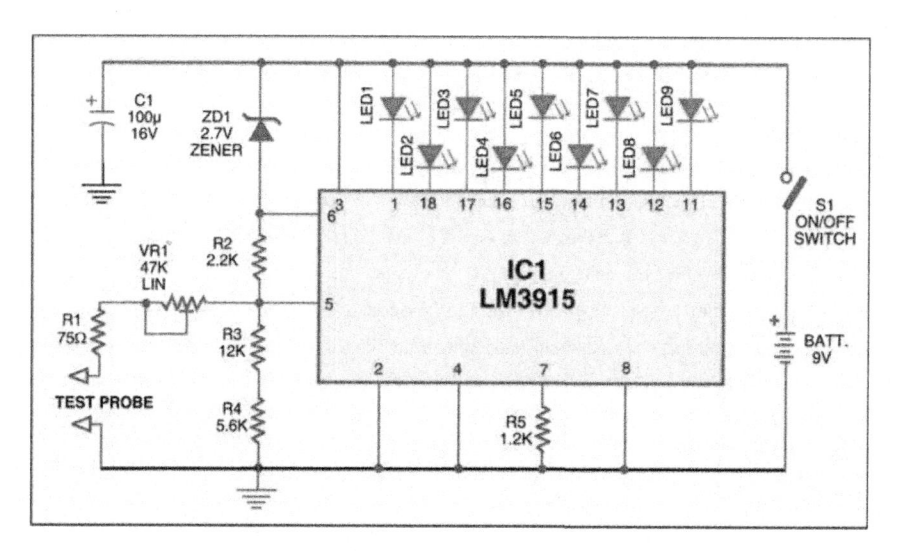

The common display driver integrated circuit (IC) LM3915 is responsible for lighting the tester's array of LEDs (IC1). After inserting both of the test probes into the soil, the display panel will indicate how much conductance, which is the reverse of resistance, there is between the two of them. The readings of soil resistance, which range from 0 to approximately 5 kilo-ohms, tell the tester whether the soil is dry or moist, and LED1 through LED9 light up to display this information in dot mode. When there is a significant increase in conductance, the first LED (LED1) will illuminate (resistance is almost nil).

The resistivity of soil is typically somewhere in the range of 0 to 5 kilo-ohms most of the time. In order to calibrate, therefore, connect a 5-kilohm potmeter in series with the two probes. Adjust the trimpot VR1 so that the resistance is zero (minimum). It is expected that LED1 will shine. In the same manner, adjust the resistance such that it is 5 kiloburs (maximum). At this point, LED9 ought to light up.

Building and testing

Assemble the components of the circuit on a multi-use printed circuit board. After the tester has been constructed, it and the battery should be stored in a compact plastic cabinet similar to the one seen here.

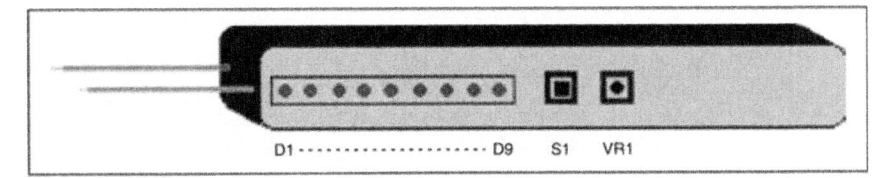

Proposed cabinet with probes

For the purpose of fabricating the probes, a fresh set of injection needles may be employed. It is recommended that the needles be firmly affixed to a piece of laminated plastic sheeting and spaced approximately 2.5 cm apart. When you need to connect objects, make use of the short length of a flat cable that has two wire leads. After the circuit has been assembled, turn it on using switch S1, and then gradually adjust the zero-set trimpot (VR1) until LED1 illuminates when the probes are brought into contact with one another. The most reliable source of power for the tester is a compact alkaline battery operating at 9 volts.

214. Over-Heating Indicator for Water Pipe

If you don't pay attention to a hot water pipe in your bathroom, it could burst if it gets too hot. This circuit checks the temperature of the water pipe to see if it is getting too hot. If the pipe's temperature goes above a certain point, an LED flashes. A small sensor with a thermostat built in is used in the circuit. It is actually a thermal switch that can be reset automatically when its temperature goes above the set limit. It comes in two different types: "normally open" (N/O) and "normally closed" (N/C).

This page talks about three different kinds of circuits. The LED1 on the first circuit shows that the pipe is too hot. When the temperature of the water pipe goes above the limit set by the thermostat, the N/O contact closes and current flows through LED1. When the temperature of the water pipe is too high, the LED1 lights up.

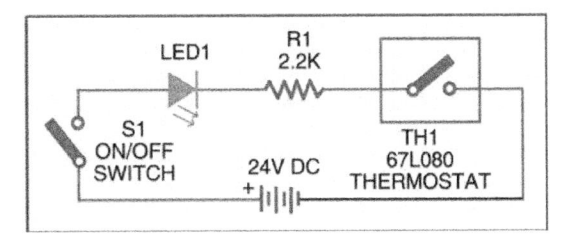

Overheating indicator for the pipe through LED

The piezo buzzer PZ1 in the circuit below lets you know if a pipe has gotten too hot. When the pipe temperature goes above the set limit, the N/O contact of the thermostat closes, making the piezo buzzer sound. When the limit is reached, the N/O version closes the switch, while the N/C version opens the switch. As soon as the temperature goes back to where it started, the thermostat resets itself. It can be put right on the part that needs to be watched.

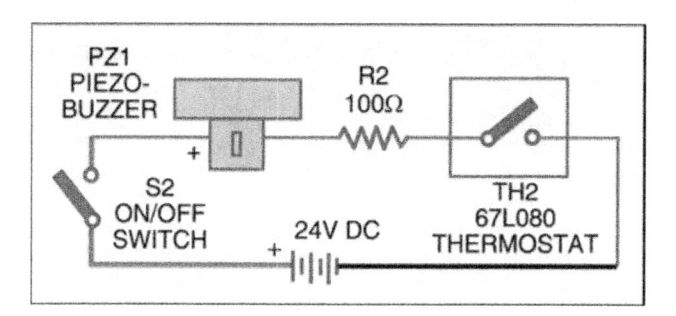

Circuit for aural indication of overheated pipe

The piezo buzzer PZ2 and LED2 in the circuit below make a sound and light signal. When the pipe temperature goes over the set limit, the N/O contact of the thermostat closes, causing LED2 to light up and piezo buzzer PZ2 to sound.

There are different temperature settings for the sensor. For each model number, the maximum temperature can be anywhere from 70°C to 110°C. For example, the type 67L080 will change state at about 80°C+5°C. You can choose the type that meets your needs. If you want to connect switches to the relay, use a three-core wire.

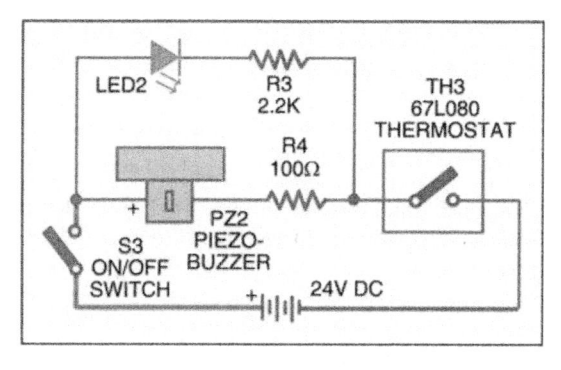

Circuit for audio-visual indication

The circuit only has a few parts. Also, it doesn't need its own power source and can be connected to the water heater's power source. Most of the time, the circuit is not being used and no current is flowing through it.

215. Linear Timer for General Use

This simple linear timer can be used to control any electrical appliance that needs to be turned off after a certain amount of time, like a small heater or a boiler, as long as the relay-switch parameters match the needs of that appliance. It has low-cost parts and a mix of digital accuracy and simple analog control. This lets you set long timing periods without having to use resistors or capacitors with high values.

The circuit is made up of NE555, CD4040, LM358, and CD4069. The timer is pretty easy to understand. It works by converting digital signals to analog ones (DAC). The counter (CD4040) controls the DAC, which is made up of the resistors R2-R23 and the operational amplifiers N1 and N2 that come after them. The second operational amplifier (N2) compares the output of the DAC to a variable voltage set by VR1 based on how much time is needed. The longer the time is, the higher the count, and the higher the count, the higher the DAC output voltage.

Circuit operation

As soon as the counter's count gets high enough to make a voltage slightly higher than the voltage set by potentiometer VR1, the second op-amp (N2) goes low to turn off transistor T1 and reset the clock generator built around NE555 (IC2). So, the clock generator doesn't make any more pulses, and the counter stays stuck at this count forever. When you press the reset button (S1), the counter starts over (IC1). So, the DAC's voltage drops to zero. This makes the second op-amp (N2) go high, which turns on clock generator IC2 and makes transistor T1 conduct. The timer goes off again, and at the same time, the relay turns on. A diode is put in parallel with the relay coil to stop the voltage from being made in the opposite direction. The frequency of the clock generator and the supply voltage determine how much time you can get:

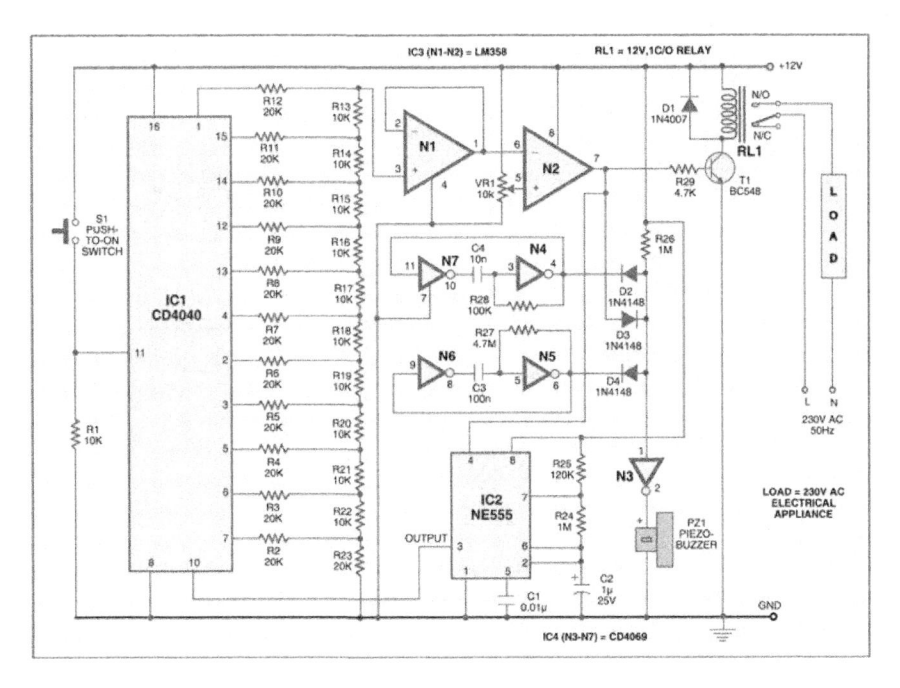

This is because the ratio Vopmax/V changes slowly as the supply voltage changes. Vopmax goes up as V goes up, so this ratio changes slowly with the supply voltage and is thought to be constant. So, if f is 1 Hz and V is 15 volts, the most time that can be used is 3686 seconds, which is just over an hour.

With ease, a 555 timer can make a frequency as low as 0.1 Hz. The timer can also handle changes in the voltage supply. At 12V, resistors R24, R25, and capacitor C2 in this circuit make a frequency of about 1.47Hz. After about 1.5 hours, the relay de-energizes to turn off the load. At the same time, the alarm built around IC4 goes off, showing that the appliance is turned off. Put the circuit together on an all-purpose PCB.

216. Noise Meter

Most of the time, sounds up to 30 dB are pleasant. Above 80 dB, it starts to bother me. And if it gets louder than 100 dB, it could affect your psychomotor performance, making it harder to pay attention and making you feel stressed. Noise pollution can also make it hard for you to hear. Around 47 dB is the level of noise in a home. But hi-fi music systems and TVs turned up loudly add to this noise, which is bad for your health. Here's a simple circuit for a noise meter that can measure and show how loud the noise is in your room. It also sounds an alarm when the noise level goes above 30dB.

A sound intensity sensor and a display unit are part of the circuit. The regulator circuit, which is built around regulator IC 7809 (IC1), gives the circuit a stable 9V power supply. The sound intensity sensor is made up of an op-amp IC CA3130 (IC2), a condenser microphone, and other parts. Op-amp IC2 is set up as an inverting amplifier with a high gain. Resistors R3 and R4 split in half the voltage that goes to IC2's non-inverting pin 3, which is also used as the reference voltage. The sensitivity of the condenser microphone is set by resistor R1.

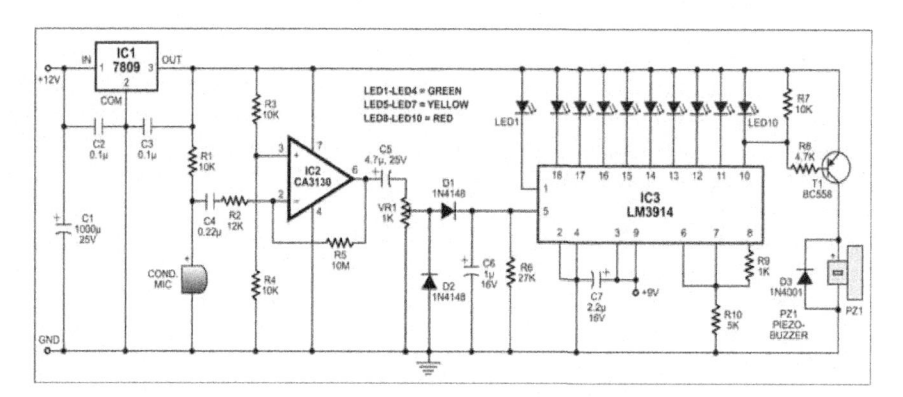

Circuit operation

The microphone picks up sound vibrations and turns them into electric pulses, which are then sent to the inverting input of IC2 (pin 2) through capacitor C4 and resistor R2. Capacitor C4 stops any DC from getting into the op-amp, since DC could mess with how the op-amp works. For negative feedback, the output of IC2 is connected to the input of the inverter through resistor R5, which has a value of 10 megaohms. Since IC2 has a very high input impedance, even a small current can turn on the op-amp. Through capacitor C5, the output of IC2 is sent to preset VR1, which is used to control the volume. Capacitor C5 stops DC from going through preset VR1, so only AC can go through. AC signals from VR1's wiper are sent to a diode pump, which is made up of diodes D1 and D2. The diode pump turns the AC around and keeps it at the level of IC2's output. Capacitor C6 acts as a storage area for DC, and resistor R6 is the way for it to flow out.

The display circuit is built around the monolithic IC LM3914 (IC3), which reads the analog voltage and drives ten LEDs to make a logarithmic analog display. The current going through the LEDs is controlled by the resistors inside IC3, so there is no need for resistors on the outside. IC3 has a built-in low-bias input buffer that accepts signals all the way down to ground potential and drives ten separate comparators. As the voltage going into IC3 goes up, the outputs go low in a decreasing order from 18 to 10.

Each LED connected to the output of IC3 represents 3 dB of sound level. When all ten LEDs light up, this means that the sound is 30 dB loud. To get the dot-mode display, pin 9 of IC3 is hooked up to 9V. In the dot-mode display, some of the segments overlap a little bit. This makes sure that all LEDs are always "on." When output pin 10 of IC3 goes low, pnp transistor T1 gets base bias (normally cut off by resistor R7), which makes the piezobuzzer (PZ1) connected to its collector sound.

217. Mains Failure and Resumption Alarm

When the AC mains go out or come back on, this mains indicator sounds an alarm. It is very useful in places like factories, movie theaters, hospitals, and so on. The mains detector circuit is made up of diodes D1 and D2 and capacitors C1 and C2. It gives enough voltage for the LED inside the optocoupler MCT2E to light up (IC1).

Mains Indicator Alarm Circuit

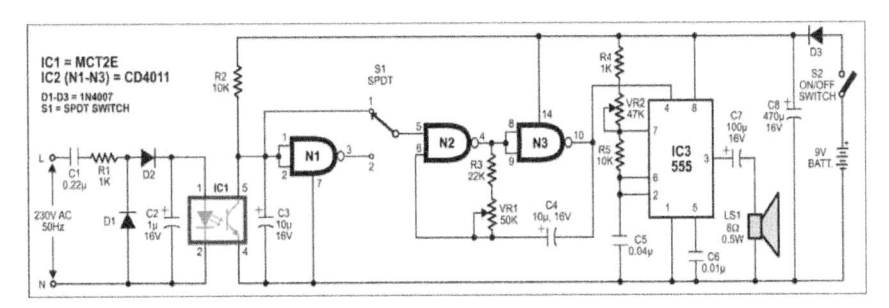

The SPDT switch S1 starts out in position 1. When the power goes out, pin 5 of gate N2 goes high, and the oscillator built around gates N2 and N3 of IC2 sends low-frequency oscillations to pin 10, which are then sent to pin 4 of IC 555. (IC3). With preset VR1, the oscillation frequency can be changed from 0.662 Hz to 1.855 kHz. IC 555 (IC3) is set up to make a sound tone. With preset VR2, you can change this oscillator's tone from 472 Hz to 1.555 kHz. The low-frequency input tells IC3 to make sounds, and the loudspeaker LS1 that is connected to its output pin 3 sounds an alarm when the power goes out.

Circuit operation

Slide the pole of switch S1 to position 2 to turn off the alarm. Now the circuit is ready to sense the return of power to the mains. When the mains power comes back on, pin 5 of gate N2 goes high, and the oscillator built around gates N2 and N3 of IC2 produces low-frequency oscillations at pin 10, which are sent to reset pin 4 of IC3. So, the loudspeaker LS1 sounds again to show that power has been restored. Slide the pole of switch S1 back to position 1 to turn off the alarm. Now the circuit is ready to sense when the power goes out again. A 9V battery powers the circuit. It can be put in a box and put wherever you want to keep an eye on the mains.

218. Multipurpose White-LED Light

Standard fluorescent lamps and their smaller versions, called compact fluorescent lamps (CFLs), give off light in all directions (360°) and tend to make the room warmer. The battery in emergency lights that use these lamps only lasts a few hours because power is lost when DC is turned into AC. These problems can be fixed by using white LEDs that are very bright. Here is a lamp with white LEDs that can also be changed to work as an emergency light in the bedroom. Its main features are that it works for a long time without stopping, it uses very little power, the light angle can be changed, it has a very long life, and it doesn't give off much heat.

Multipurpose lamp circuit

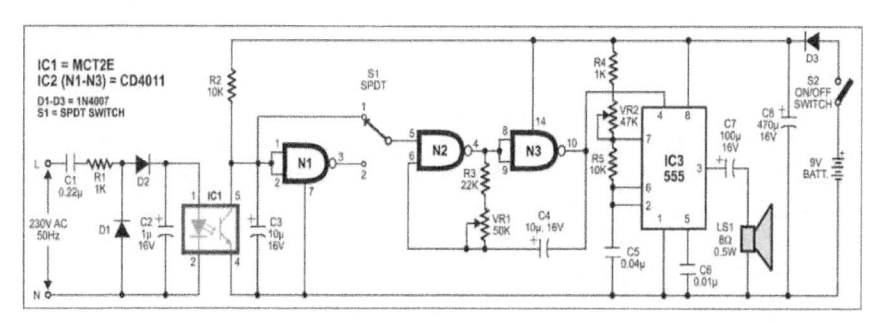

Fig. 1: Cluster LED multipurpose lamp

Fig. 1 shows the circuit of a white LED light that can be used in many different ways. The circuit is very simple. It uses a battery charger built around an integrated circuit (IC) called LM317 (IC1) and a few white LEDs. The current going through the battery is limited by the 4.7-ohm, 2W resistor R3. The angles of 60° and 20° are chosen for white LEDs. Three rows of LEDs (A, B, and C) are made on separate sheets of clear acrylic. Twelve LEDs are attached to each sheet, for a total of sixty-four LEDs. The LEDs in the left (A) and right (C) columns are 20°, but the LEDs in the middle (B) column are 60°. All twelve LEDs in each column are connected in series to separate 15-ohm current-equalizing resistors (R8 through R19) as shown in Fig. 2 and to current-limiting resistors R7 (10-ohm, 1W) and R6 (5-ohm, 1W) as shown in Fig. 1. The whole thing is powered by a 6V, 4Ah rechargeable battery that doesn't need to be maintained. In torchlight mode, the light stays on for about 7 hours, and in table lamp mode, it stays on for about 14 hours. This depends on the size and quality of the battery. Only the left and right LED columns are used for the torch mode. These LEDs can shine light as far as 6 meters. In table lamp mode (spread light), only the LEDs in the middle column are turned on.

Operating modes

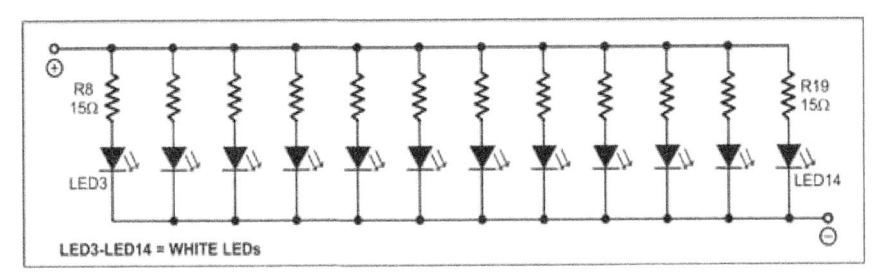

Fig. 2: Arrangement of LEDs for column A, B or C

The rotary switch S1 is a single-pole, three-way switch that lets you choose between the table lamp and torch modes. When the pole of switch S1 is in position 1, the C column of 60° LEDs light up and the system works as a table lamp. When the pole of switch S1 is set to position 3, columns A and C light up, and the system works as a flashlight. When the pole of switch S1 is in position 2, neither the table lamp nor the torch modes work.

When the power is turned on, LED2 lights up. Flip switch S2 to the "on" position to charge the battery. To see how the battery is doing, flip switch S3 to the "on" position. This will let you know how full the battery is. If LED1 goes out, it means the battery needs to be charged. Figure 3 shows the circuit for an emergency lamp with a brightness control. It is based on Figure 1, but the LEDs are put together in a slightly different way. Built around four multichip (MC) LEDs, it is small and easy to use, and it can work in two ways: as a bedroom lamp or as an emergency light.

In the mode for a bedroom lamp, only one blue LED lights up. This LED is mounted upside down at the top so that the blue light can't be seen straight on. The way the lights are set up gives off a nice, even light. In emergency lamp mode, the 8mm, 80° bright-white, multichip LEDs spread light over an area of 80°, which is enough for indoor use. PCBs for multichip LEDs that are round each have four internal connections. Solder LED17 through LED20 on the first PCB, LED21 through LED24 on the second PCB, LED25 through LED28 on the third PCB, and LED29 through LED32 on the fourth PCB, leaving 3 to 4 cm between each pair of LEDs. Lastly, put the four circular PCBs and the reflector in a small cabinet so that light can spread around the room. Each LED with more than one chip is as bright as 32 candles. Because of this, if you use four 8mm multichip LEDs, you will get a total of 128 candles. When the rotary switch S5 is turned to the emergency lamp mode, all four multichip LEDs (LED17 through LED32) light up. The DC power source is a 6V, 4Ah battery that can be charged. The charging circuit is based on the well-known IC LM317 (IC2). The battery's current limiter is R21, which is 2.2 ohms and 1 watt.

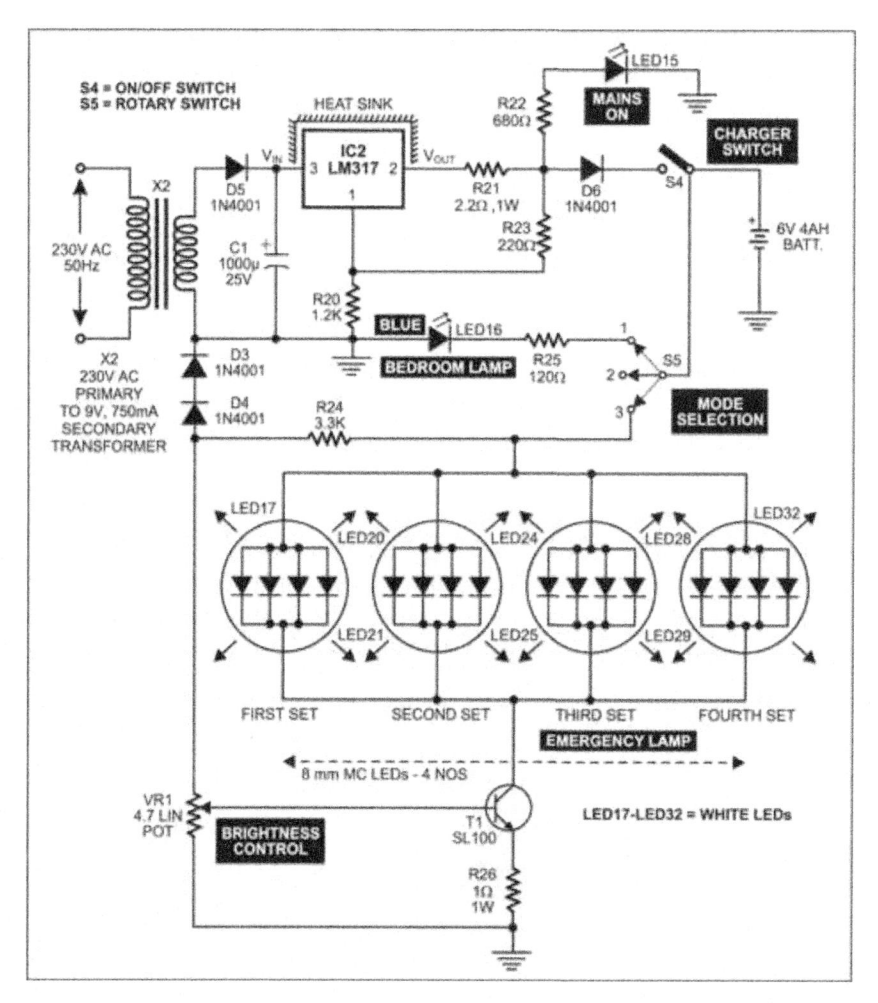

Fig. 3: Multipurpose lamp with brightness control

You can change how bright LEDs are (their candle power) to meet your needs. The candle controller is made up of the transistor SL100 (T1) and the parts that go with it (brightness controller). Resistor R24 and diodes N3 and N4 keep the voltage at the transistor's base steady (1N4001). The base of the transistor gets this constant voltage from a potentiometer called VR1 (4.7k lin.). You can change how bright the multichip LEDs are by adjusting the potentiometer. The transistor doesn't need a heat sink.

219. IR Based Light Control

When a car or motorbike comes through the gate and crosses the sensing area, this IR-based light control circuit turns on the lights in the portico, car parking, or other areas. It can also be used to keep an electronic eye on your house by setting off an alarm at the same time. The system is made up of a pair of transmitters and receivers and a switching circuit. The transmitter and the receiver are put on opposite pillars of the gate so that the infrared (IR) beam from the transmitter hits the IR sensor of the receiver directly. When someone walks through the gate and blocks the IR beam from hitting the sensor, the system is turned on.

Transmitter

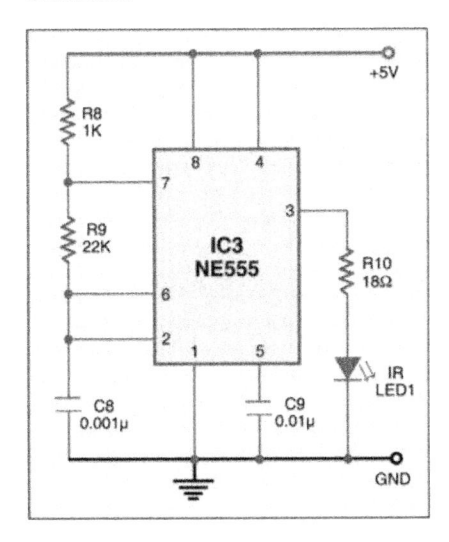

IR based light control: Transmitter circuit

Figure 1 shows the circuit for the transmitter. It is made up of an infrared (IR) LED and a NE555 timer (IC3), which is wired as an astable multivibrator with a frequency of about 38 kHz. Through IR LED1, the 38kHz infrared beam is sent out.

Receiver

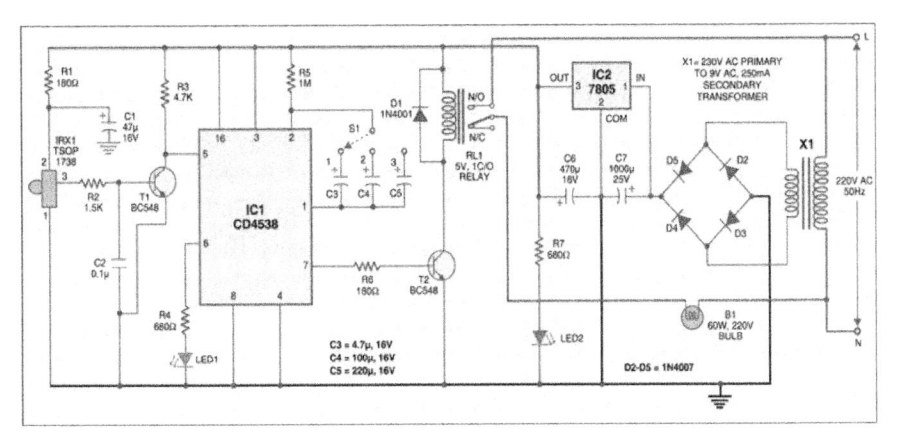

IR based light control: Receiver circuit

The receiver circuit is shown in Figure 2. It is made up of a TSOP1738 IR sensor, a CD4538 dual monostable multivibrator (IC1), and a few separate parts. When the IR beam hits the sensor, pin 3 of its output stays low, and pin 5 of IC1 goes high because T1 is conducting. When someone blocks the IR beam from hitting the sensor, pin 3 of the sensor's output goes high, which triggers IC1 (at pin 5 via transistor T1) As a result, pin 7 of IC1's output goes high for the set amount of time. When BC548 (T2) conducts, it turns on relay RL1 and turns off LED1. When the monostable time is up, LED1 starts to light up again.

By choosing the right value for the capacitor connected to IC1's pins 1 and 2, the time period of IC1 can be changed. IC1's time period is set by a small rotary switch with three positions and the capacitors C3, C4, and C5. Diode 1N4001 (D1) works as a diode that can move around freely. Between the pole of relay RL1 and the neutral terminal of AC mains is the switchable bulb (B1). When the relay is turned on, the normally-opened (N/O) contacts of the relay connect it to the live terminal of the AC mains supply. From the AC mains, the step-down transformer X1, the bridge rectifier (made up of diodes D2 through D5), and the regulator IC 7805 get the 5V power for the transmitter and receiver (IC2).

220. Sequential Device Control using TV Remote Control

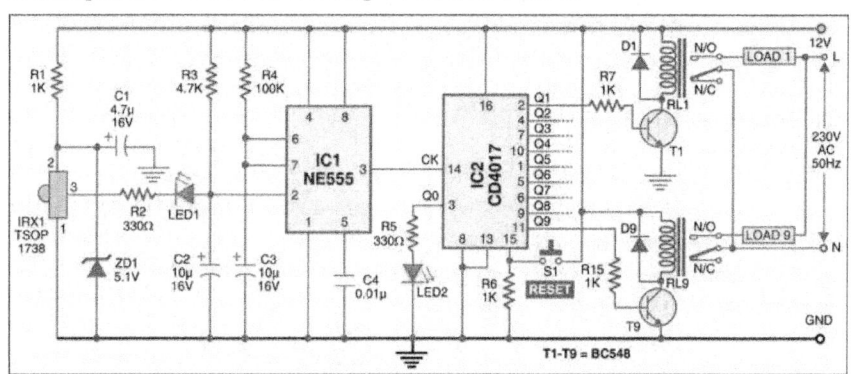

With this circuit, you can turn on and off up to nine devices at the same time with your TV remote. The circuit is made up of a TSOP1738 IR sensor module, which is the same as a TSOP1238. LED1 connects the output of TSOP1738 to the trigger input of IC NE555 (IC1) at pin 2. The output of IC1's monostable mode is sent to the clock input of IC CD4017 (IC2). Resistors R7 through R15 are used to connect each of the nine outputs of IC2 to the bases of relay-driver transistors T1 through T9, respectively. The positive end of 12V is connected to the collectors of RL1 through RL9, which are relays. A 12V battery or a 12V adaptor is used to power the circuit. Since TSOP1738 only needs up to 5 volts, a zener diode with 5.1V is used. How the circuit works is easy to understand. When the power is turned on, LED2 lights up to show that the circuit is ready to be used. Now, use the TV remote to turn on the first device by pressing any button. When you press any button on the remote a second time, the first device turns off and the second device turns on. As you press the button on the remote, the devices turn on and off one by one. When the ninth button is pressed, the eighth device will turn off and the ninth one will turn on. When you press any button on the remote again, IC2 is reset, which makes all of its outputs except for Q0 go low. On the other hand, switch S1 can be used at any time to reset IC2. Keep in mind that only one of the nine devices can be turned on at a time. Put the circuit together on a PCB that can be used for many things and put it in a suitable cabinet. Install the sensor in the middle of the cabinet's front panel. Finish making all the connections for the relays on the back of the cabinet.

221. Resistor Calculator

Here is some free software for Windows that is easy to use and saves a lot of time and effort when figuring out the color code of resistors and the resistance values needed for LED circuits. There are color bands on every resistor that show its resistance, tolerance, and sometimes its temperature coefficient as well. You may know the mnemonic for remembering the resistor color codes: B. B. ROY of Great Britain has a Very Good Wife, where the capital letters stand for black, brown, red, orange, yellow, green, blue, violet, grey, and white.

Resistor Calculator 1.0.6

Presented here is a simple, easy-to-use freeware for Windows that saves a lot of time and effort in determining the colour code of resistors and resistance values required for LED circuits

DILIN ANAND

Every resistor is marked with colour bands that indicate its resistance, tolerance and sometimes temperature coefficient as well. You might be familiar with the mnemonics for memorising the resistor colour coding: *B. B. ROY of Great Britain has a Very Good Wife*, where the capitalised letters stand for black, brown, red, orange, yellow, green, blue, violet, grey and white, respectively.

Quite often, the whole mental calculation process of resistor codes can be an exhausting task. It also consumes much time. Sometimes you may not remember the mnemon-

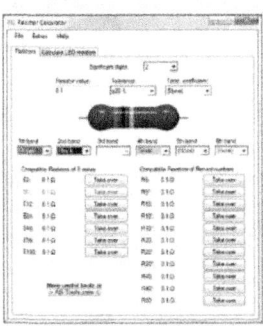

Fig 1: Home screen of resistor calculator

License type	Freeware
Developer	Andreas Breitschopp, AB-Tools.com
Operating system	Windows 8, Windows 7, Windows Vista and Windows XP
Latest version	1.0.6
File size	1.29 MB

interface is clear and plain, which

is a freeware. The users get the latest updates of this program automatically. All updates are absolutely free.

The integrated help system is simple and easy to understand. It also has an intuitive program interface. Further help and support is provided by e-mail and is also available at their website. The program is currently

The whole process of figuring out resistor codes in your head can be a lot of work. It also takes a lot of time. You might not always remember the mnemonics the right way. What if the value is not covered by the mnemonics, like 0.1 or 1? Sometimes it can be hard to remember whether the first, second, or third B stands for black, brown, or blue. Another common problem is that you might not know what standard values of resistors are on the market. In situations like these, this Resistor Calculator software can be a huge help. It tells you the resistor value of the color code you enter, figures out the resistor value of the color code you enter, and gives you a list of all the standard resistor values that match your required resistance. It can also figure out how much of a resistor needs to be added to a circuit with LEDs.

Resistor Calculator is one of the many free programs that AB-Tools.com has made. The Windows operating system is needed to run this software. The user interface is clear and simple, making it easy for even beginners to use.

Resistor Calculator Online https://www.digikey.com/en/resources/conversion-calculators

Download PC Software for free https://www.ab-tools.com/en/software/resistorcalculator/

Features of Resistance calculation. By putting in the resistance value, we can figure out the color code of the resistor to be used. You can also set the needed tolerance and/or temperature coefficient and get the color bands that go with them. One can find out the value of the resistor, the tolerance, and the temperature coefficient for the given color combinations.

222. Triple-Mode Tone Generator

Here is a simple circuit for a triple mode tone generator that makes three different tones. You can use it as a burglar alarm, a call bell, or any other kind of security alarm. Fig.1 shows the circuit for the three-mode tone generator, and Figure 2 shows how it can be used to make a car horn. A 12V battery is used to power the circuit. The LM556 dual timer IC is the heart of the circuit. It has two separate LM555 timers built into it. The first timer is set up to work as a wide-range oscillator. The first timer tells the second timer what to do by what it does. Depending on where the rotary switch is, the circuit makes one of three sounds (S1). S1 is a three-way switch with one pole.

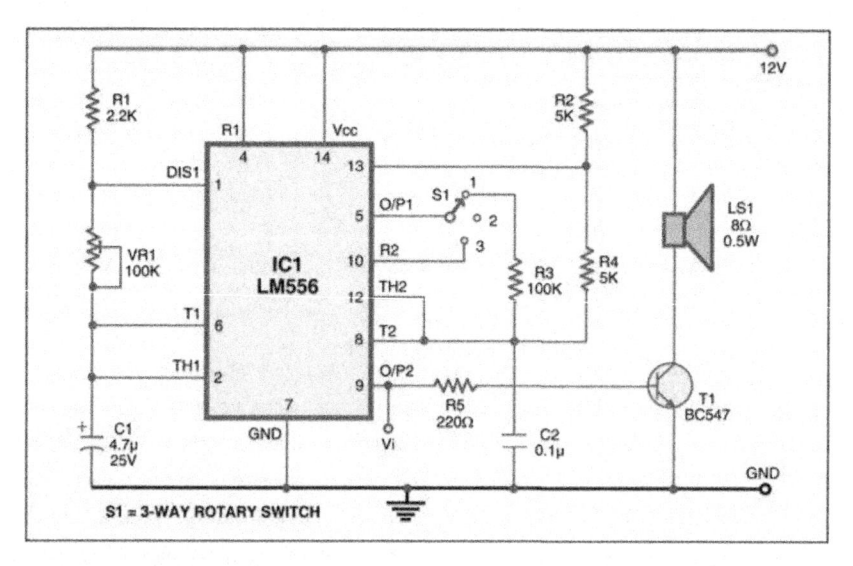

Fig.1 Triple Mode Tone Generator Circuit

Circuit operation

When switch S1 is in position 1, a two-tone sound comes out of pin 9 of the second timer. When the switch S1 is in position 2, a continuous tone comes out of pin 9 of the second timer. When the switch S1 is in position 3, a tone burst comes out of pin 9 of the second timer.

Since the LM556 can only sink 200mA of current, the second timer's output is amplified by the transistor T1. Depending on the tone, the speaker sounds. To get a louder sound or to make a car horn, connect the output of the second timer at pin 9 to the audio signal Vi. Do this after disconnecting resistor R5 (Fig. 2). The TDA 2030 amplifier has a high output current and very little distortion from harmonics and cross-over. It also has a standard system for shutting down when the temperature gets too high.

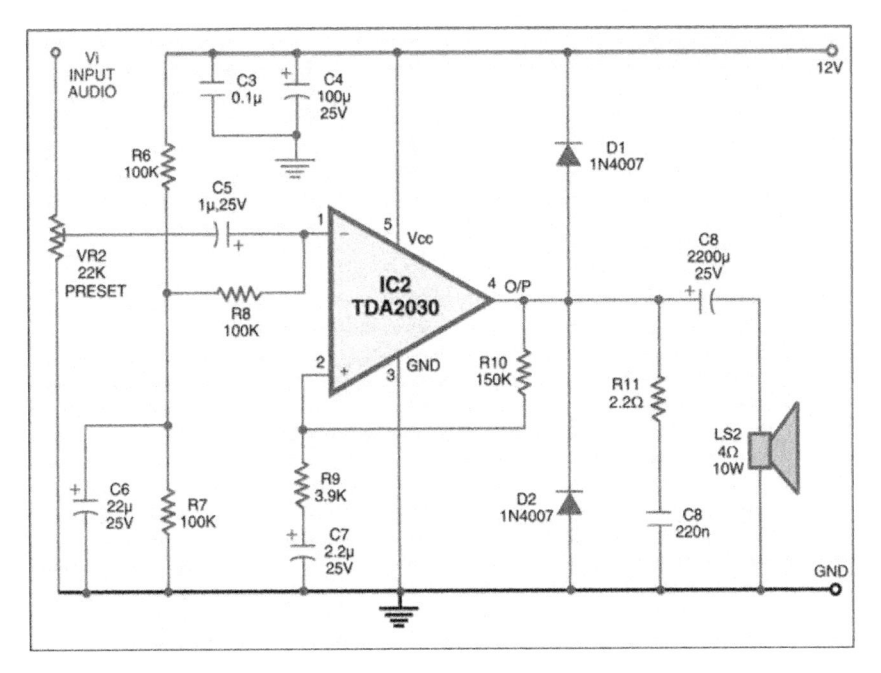

Fig. 2: TDA 2030A amplifier for automobile horn

223. IR-Controlled Water Supply

This circuit can be used for any type of water supply unit, like a toilet flush or washbasin tap, in homes or small restaurants. The part of the power supply that is made up of IC 7805 (IC1). The input power for IC1 can come from either a 12V solar panel (SP) or a 9V battery. Voltage regulator 7805 provides constant 5V DC to operate the circuit. The transmitter part is made up of the timer 555 (IC2), the npn transistor BC548, the infrared LED1, and a few other separate parts. The NE555 timer is set up as an astable multivibrator. By adjusting the preset VR1, it can make 38kHz pulses. The IR LED1 is powered by the transistor T1, and it sends modulated IR signals up to a few meters away.

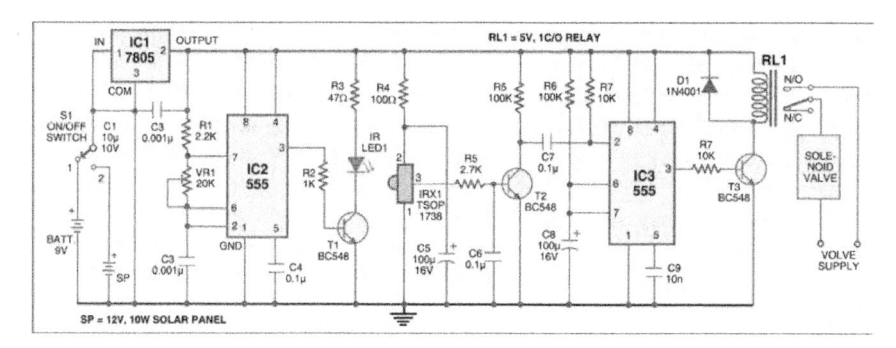

274 | 500 Electronic Project Ideas for Inventors

Figure. The circuit can be divided into three sections: power supply, transmitter and receiver.

The receiver section is made up of the TSOP1738 chip, the timer 555 (IC3), and a few other parts. IR LED1 sends out an IR signal, which is picked up by IR receiver TSOP1738. The IR LED and the TSOP1738 are facing each other, so the IR beam from the IR LED goes straight to the TSOP1356. The output of the IR receiver module is sent to trigger pin 2 of IC3. When IC3 is turned on, its output goes high for about ten seconds, depending on how much R6 and C6 weigh. As long as the monostable output is high, relay RL1 stays on. So, the normally-opened (N/O) contacts of relay RL1 supply the solenoid valve with the power it needs.

Depending on the kind of solenoid valve, the power supply needs can vary. If a 5V solenoid valve is used, you can get rid of the relay. In this case, you can connect the valve between the collector of transistor T3 and the positive side of 5V. So, when a person's hand gets in the way of the IR beam coming from the IR transmitter (LED1) on IRX1, the output of the TSOP1738 goes high, which turns on the monostable IC3. The relay turns on for about 10 seconds to open the water supply by moving the solenoid valve. After 10 seconds, the valve shuts again, stopping the flow of water. When the valve is opened, water flows for 10 seconds. If the hand is placed again during that time, the monostable circuit does nothing. Instead, it keeps doing what it's supposed to do. Put the circuit together on a PCB that can be used for many things and put it in a suitable cabinet. Install the IR LED and IRX1 on the side walls of the toilet flush or washbasin so that the IR beam that is sent to the receiver goes straight to it.

224. Twilight blinker lamp

When the sun goes down or comes up, there isn't enough light to see your way through an open door or around obstacles. Here is a twilight blinker lamp that you can put near obstacles to keep things from going wrong.

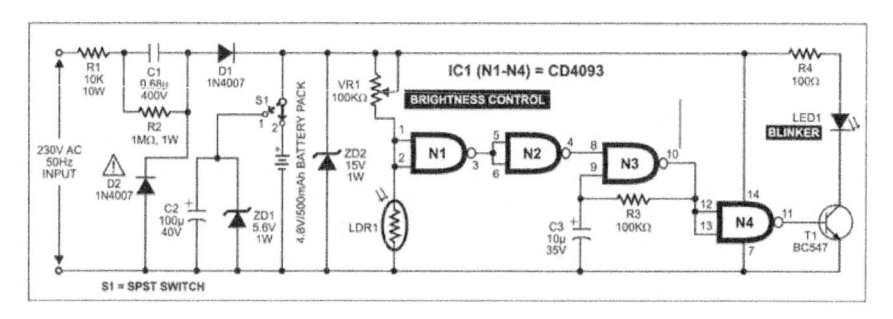

Fig. 1: Circuit diagram of twilight blinker lamp

In Figure 1, you can see how the twilight blinker lamp works. By using resistors R1 and R2, capacitor C1, and diodes

D1 and D2, the mains power (230V AC) is turned into a DC voltage that is low enough to safely charge the backup battery pack. A bleeder resistor is the resistor R2 that is placed across capacitor C1. Zener diode ZD2 protects against over-voltage. Mini Ni-Cd battery packs for cordless phones are easy to find and don't cost too much. Use a battery pack like this with a rating of 4.8V and 500mAh for a back-up that works well and lasts a long time. If you use a battery, the pole of switch S1 should be in position 2. If you don't want the back-up feature, move the switch S1 to position 1. The rest of the circuit is made up of a light detector resistor (LDR1), an IC CD4093 (IC1), and a brightness control preset (VR1). LDR1 is used as a sensor because its resistance is low during the day and high at night.

Circuit operation

When light hits the LDR, its low resistance gives NAND gate N1 a low level at its inputs. The high input from N1 makes the output from N2 low, so the relaxation oscillator (made up of NAND gates N3 and N4 of IC1, capacitor C3, and resistor R3) does not oscillate. So, transistor T1 doesn't work, and LED1 doesn't flash. On the other hand, when it's dark, the high resistance of LDR1 gives NAND gate N1 a high level at its input pins. Because N1's output is low, N2's output is high, and the relaxation oscillator keeps going back and forth. Because of this, transistor T1 works and LED1 blinks.

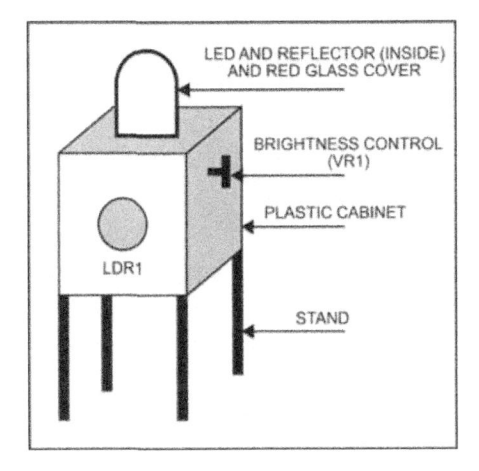

Fig. 2: Proposed enclosure

The LED driver is transistor T1. Resistor R4 controls how much current flows through LED1 and, by extension, how bright it is. To get lighter, you can connect one or two more LEDs in series with LED1. Because LED1 is not very bright, the battery back-up time will be longer.

225. Electronic Street Light Switch

Here is a simple and cheap switch for a street light. This switch makes the light come on at night and go off in the morning. The automatic function not only saves time, but it also saves electricity. The circuit can be broken down into two main parts: the power supply and the switching. When you press the S1 switch, the power comes from the mains. By using resistor R1, diode D1, and zener diode ZD1, the mains voltage is stepped down to 9.1V DC. Capacitors C1 and C2 filter the output across ZD1. By changing the value of zener diode ZD1, the output voltage can be raised to up to 18V or dropped to 5V.

276 | 500 Electronic Project Ideas for Inventors

The light-sensitive resistor LDR1, the transistors T1 through T3, and the timer IC1 are the parts that make up the switching circuit. LDR1's resistance stays low during the day and high at night. Timer IC1 is made to work as an inverter, so a low input at pin 2 gives a high output at pin 3, and vice versa. Triac 1 and street light B1 are turned on by using the inverter.

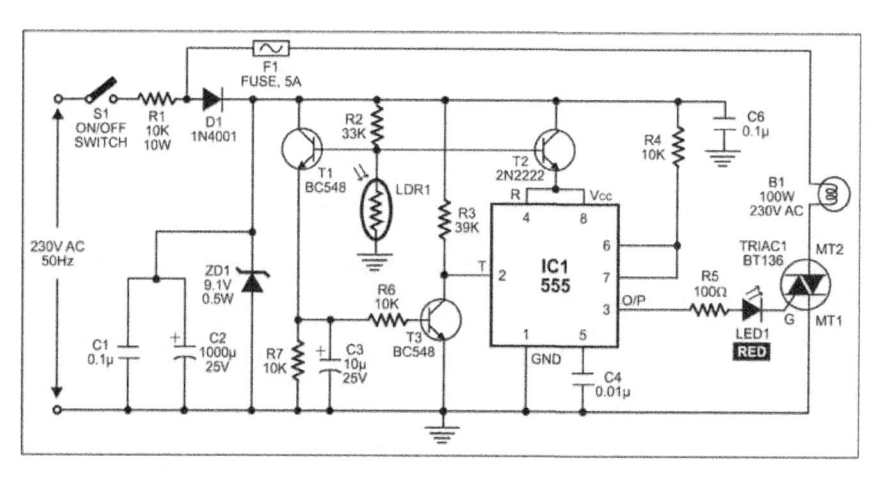

Street light switch circuit

Circuit operation

During the day, light hits LDR1 and keeps transistors T1 and T2 from turning on. This keeps pins 4 and 8 of IC1 from being high. Since T3 is also turned off, IC1 is not set off. Because of this, the output pin 3 of IC1 (which is connected to the gate of triac 1 through resistor R5 and red LED1) stays low, so the street light doesn't turn on. At night, LDR1 doesn't get any light, so transistors T1 and T2 work, making pins 4 and 8 of IC1 high. Since the transistor T3 is conducting, the trigger pin 2 of IC1 stays low. The street light turns on when the output of IC2 at pin 3 is high.

226. Standby Power-Loss Preventer

Electronics still use some power even when they are in the standby mode, which is when they have been turned off with a remote but not the mains power switch. For example, a CRT TV or PC monitor uses 80 to 100 watts of power when it is on. Even when it's not doing anything, it uses a few watts of power. So, if you leave these devices on standby for a long time, they may cause your electricity bill to go up. The circuit described here can help you save money on your electricity bill by automatically cutting off the power to a device (like a CRT TV) when you turn it off with the remote. (But you might need to get up and turn the TV back on!) The circuit is turned on by the idea of electromagnetic induction.

When the CRT monitor is "on," the circuit can pick up the high-energy electromagnetic radiation coming from it. A relay coil (L1) connected between the inverting pin 2 and the non-inverting pin 3 of IC CA3130 (IC1) picks up the monitor's electromagnetic radiation to make a small current. IC

CA3130 is a CMOS operational amplifier with p-channel MOSFETs that have their gates protected. It needs a very low input current and has a very high input impedance (typically, 5 pA). IC1 is a good choice for this application because it has a high input impedance and a low input current. The 5 pA internal bias of IC1 is enough for it to work, so it doesn't need an external bias. Offset null adjustments are made with VR1, so that when the TV is in standby mode, the output of IC1 is low.

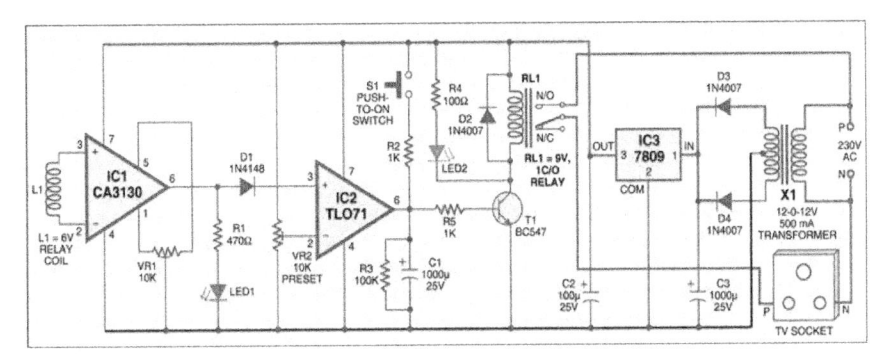

When the TV is first "off," pin 6 of IC1 is low, so LED1 doesn't light up. Press the S1 switch for a few seconds to turn on the TV. Through resistor R2, capacitor C1 gets charged up, and when transistor T1 conducts, it turns on relay RL1. The relay's normally-open (N/O) contacts send power to the TV. When the TV is turned on, electromagnetic radiation hits the sensor coil. This makes the output of IC1 go high, which makes LED1 light up. Diode D1 connects the high output of IC1 to the non-inverting input of IC TLO71 (IC2), so the TV stays "on." IC2 is a low-noise op-amp with a JFET input. When you use the remote to turn off the TV, the electromagnetic field around it goes away, and IC1's output goes down. So, the output of IC2 also goes low, and transistor T1 stops conducting. The time it takes for capacitor C1 to discharge through resistor R3 gives a delay of a few seconds to de-energize the relay, after which the mains power to the TV is cut off. The only way to turn the TV back on is to briefly press switch S1. The circuit gets its power from a step-down transformer that goes from 12V to 0V to 12V. (X1). The secondary output of the transformer is fixed by diodes D3 and D4, and ripples are smoothed out by capacitor C3. The IC 7809 (IC3) gives the circuit the 9V it needs to work. Put the circuit together on a general-purpose PCB and put it in a box with an AC plug for the TV. Place the unit close to the TV, preferably on top or on the side, and plug it in. Set VR1 so that LED1 goes out. At this point, the relay shouldn't be turned on. Press switch S1 for a few seconds, until the relay turns on and LED1 lights up. This turns on the TV. If you now let go of switch S1, the TV will stay "on." In standby mode, use the remote to turn it off.

227. Touch Sensitive Alarm

People have cared a lot about their safety for a long time. Any personal item that is out of reach is at risk of being stolen, and sometimes even the things we are using get stolen. Here is a touch-sensitive alarm that goes off whenever one of your things is touched. Radiation signals are the idea behind the touch-based alarm. Mains wiring can send out signals that can travel a few meters. These can also be caused by the body's own electromagnetic field. The AC hum signal is what makes this alarm work when you touch it. When someone touches the touch plate, a low-power AC hum is made. This is the same hum that is made by the AC wiring in the house.

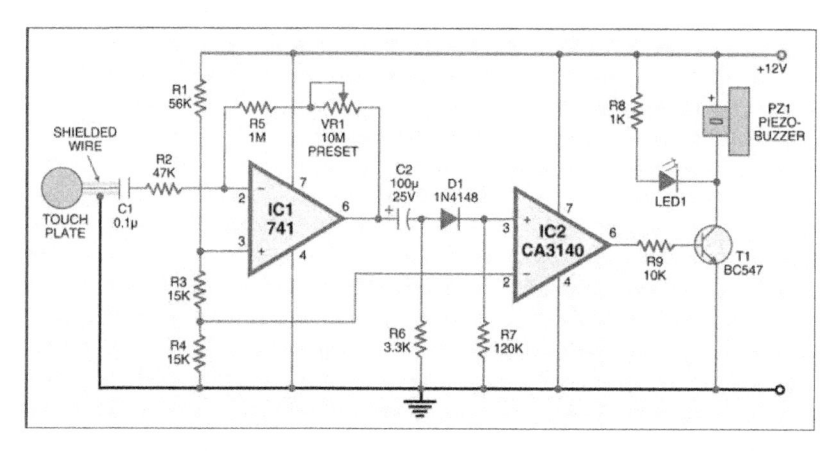

Touch sensitive alarm circuit

The low power AC hum signal is first sent to an IC 741-based high-gain preamplifier (IC1), which boosts the signal. The signal is then sent to an IC3140-based voltage comparator (IC2), which compares the signal to a known voltage. When the signal is amplified and sent to pin 3 of IC2, the output goes high. Because of this, transistor T1 conducts, which makes the buzzer go off. LED1 lights up at the same time. For this unit to work well, it needs to be powered by 12V from the mains.

228. Hum-Sensitive Touch Alarm

The mains hum signal from an intruder's body is picked up by the touch alarm circuit. This makes the alarm sound. This happens when an intruder touches the doorknob or any other object that needs to be protected and is set up as a sensor. If the circuit is connected to the door knob, it stays in the standby state no matter what else is going on around it. When someone touches the door now, the circuits turn on and the alarm goes off. This article talks about a few alarm systems that are based on old ideas, but they are not the only ones. Also, type numbers and values are shared between real-world circuits. Electronic hobbyists can build these circuits with just a little bit of work.

The main benefit

The main benefit of this touch alarm circuit is that it will only pick up the 50 Hz or 60 Hz hum that comes from a human touch. It will ignore any other electrical noise. This means that the circuit is foolproof and can't be set off by electrical interference in the air or even by lightning. There are a lot of simple circuits online that show how to make a touch alarm with a couple of transistors, but these ideas are not at all reliable. Electrical noise or interference can easily set off these transistorized systems, making the alarm go off when it shouldn't. Also, these regular transistor-based circuits have to be set up right where the detection is made. The proposed circuit, on the other hand, can be set up far away, with only the touch sensor connected to the detection area through a cable.

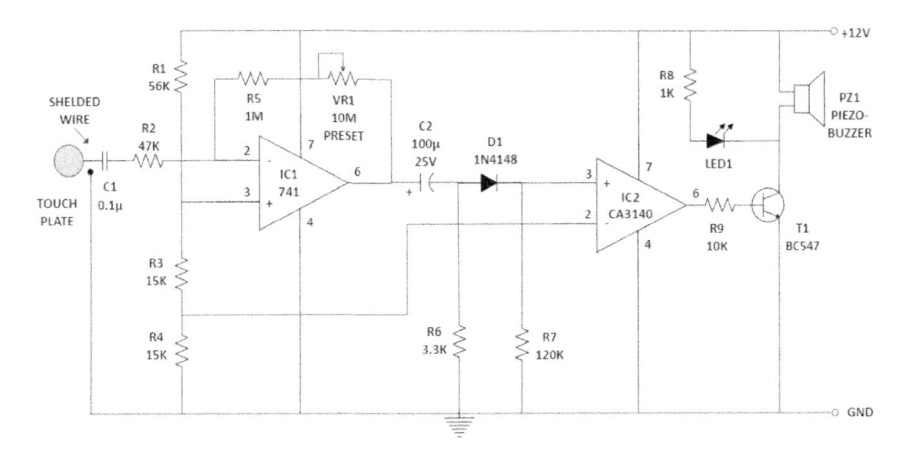

Mains Hum Sensor

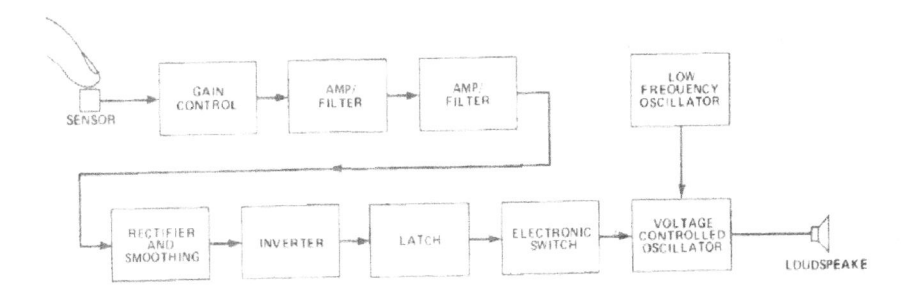

First, we'll look at the touch alarm circuit, which can tell when someone touches a metal object because of the "mains hum." The transducer can be anything from the handle on a door to a cabinet door with valuables inside. Even though it is called "independently operational," changing the circuit to fit a large alarm system is not too hard. A block diagram of how the unit works is shown in Figure 1.

In almost every building with mains wiring, the "mains hum" can be felt by any part made of a material that conducts electricity. The human body is included because it is big enough to be able to pick up a hum signal. In the touch detector circuit, the metal sensor at the input must be small and connected to the rest of the parts with a short wire that is 300 to 500 mm long. For longer connections, use a shielded wire. The signal from the sensor goes into a gain control, which is a standard volume regulator with a variable attenuator that can be adjusted so that the alarm doesn't go off when the sensor sends a signal that is affected by the weather. If someone touches the sensor, the fairly large signal picked up by their body is sent to the sensor. This gives the sensor a strong input signal that turns on the unit.

229. Room Sound Monitor

Using this simple sound monitoring circuit, it is possible to listen in on discussions in a room without being detected. A 3V battery runs the circuit, which is very sensitive.

Sound monitors circuit operation

The way the circuit works is easy to understand. When a conversation is going on near the condenser microphone, transistors T1 and T2 boost the weak sound signals. The signal is sent to the audio driver transformer X1, which boosts it even more. Through a capacitor, the output of the transformer is sent to a speaker.

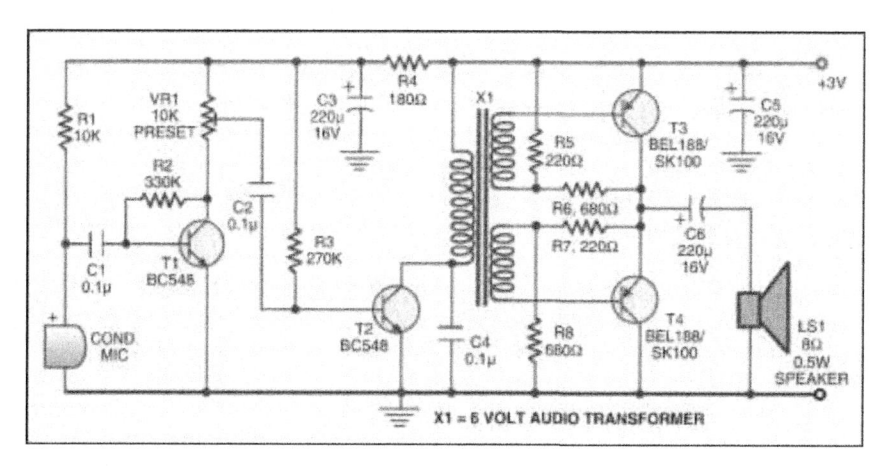

Sound monitor circuit

In short, when the microphone picks up sound in a room, it can be heard through the speaker outside the room. Keep the speaker away from the mic to stop feedback from making the speaker howl. Transformer X1 is a six-lead driver transformer that is easy to find. It has one primary winding and two secondary windings. Change the preset VR1 for the best gain and clarity

230. Security System Switcher

Any security system can be controlled by using an audio signal as an input. For instance, an automatic security camera can be set up to answer the door when someone knocks on it. With the switcher circuit for the security system that is described here, the security system can automatically be in the "on" state. It uses a transducer to find intruders and gets power from a 5V regulated DC power supply.

Security system switcher circuit

As shown in Fig. 1, a condenser microphone is connected to the input of a small signal pre-amplifier built around transistor T1. The biasing resistor R1 has a big effect on how sensitive the microphone is. Most microphones have a FET inside that needs a bias voltage to work. The sound picked up by the microphone is amplified and sent to the monostable configuration input pin 2 of IC1 (LMC555). IC2 (CD4538B) is a dual monostable precision multivibrator with separate controls for the trigger and reset. Switch S1 connects the output of IC1 to the first trigger input pin 4 of IC2(A). If an intruder opens or breaks the door, sound signals trigger IC1. The timer output pin 3 of IC1 goes high, which turns on the first monostable multivibrator IC2 (A). Preset VR1 lets you change the time range of IC2(A), which is between 5 and 125 seconds.

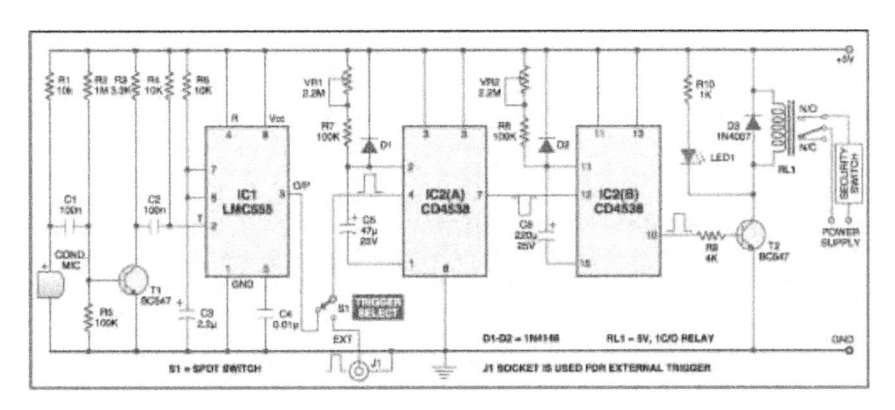

Another monostable multivibrator, IC2(B), also has a time range of about 25 to 600 seconds that can be changed with VR2. With the help of the output of IC2(B), relay RL1 is turned on. There is an indicator LED1 to show that the relay is working. A security switch is used to turn on or off any AC/DC-powered security device. So, the device's safety switch is hooked up to the n/o contacts of the relay. For any AC/DC-powered security device, you can also use high-powered beacons, sirens, or hooters instead of the security switch.

Construction & testing

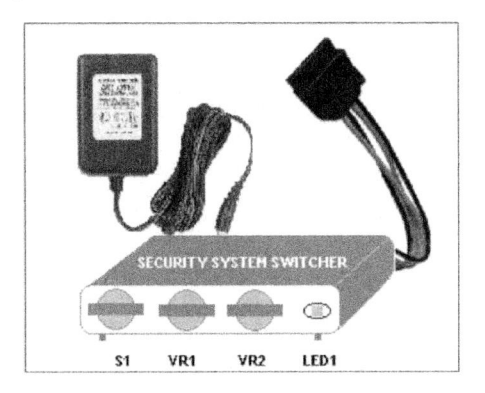

Assemble the circuit on a general-purpose PCB and put it in a cabinet, as shown in Fig. 2, along with a 5V adaptor to power the circuit. Connect the security switch according to the diagram, and use the right AC/DC power supply to make the security device work. Warning! When connecting the mains power supply to the relay contacts, it's important to follow all electrical safety rules. Using the single pole double throw (SPDT) switch S1, the active high signal (trigger input) can come from either the inside or the outside.

231. Doorbell-controlled Security Switch

Ringing the doorbell is one way to see if anyone is home. Even thieves use this same method. In this kind of situation, the circuit shown here comes in handy. It is a simple, multi-purpose security switch that can be controlled by a doorbell. It turns on a connected security device, like a night-vision door camera, instantly. The circuit works with a power supply of 9V DC. The 230V AC electric doorbell is connected to the circuit's input in the same way. At the end of the circuit, an electromagnetic relay is used to turn on the security device that is connected to it.

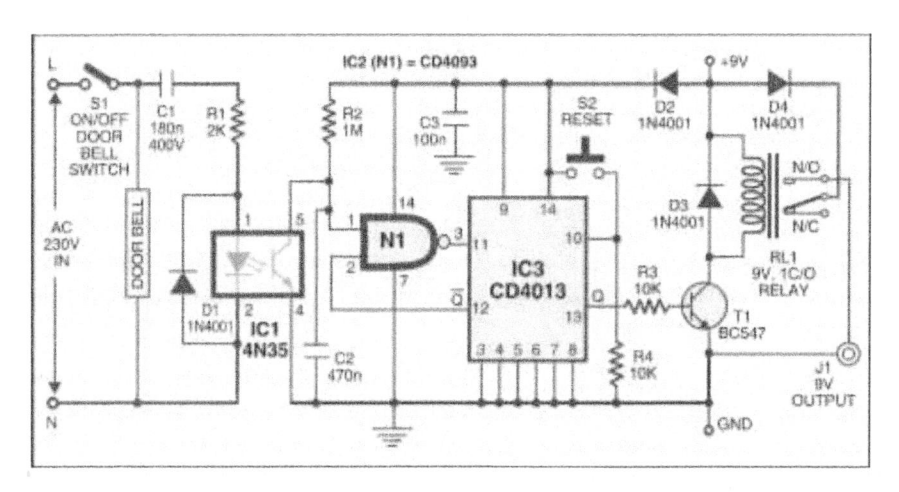

Figure 1 shows how the doorbell works with the security switch. The way the circuit works is simple: When someone presses the doorbell switch S1, mains voltage is sent to the doorbell and opto-isolator IC 4N35 (IC1). The doorbell rings, and as the transistor inside opto-isolator IC1 conducts, it sends a high-to-low pulse to pin 5. Since pin 5 of IC1 is connected to pin 1 of IC2, a high-to-low pulse is also made at pin 1 of gate N1, which turns on IC3 at pin 11. So, the Q output (pin 13) of IC3 goes high, and relay driver transistor BC547 turns on the relay (T1). Now, the 9V DC power supply is available at socket J1 to turn on the door camera with night vision.

Note that the output socket (J1) is designed to turn on the 9V security device and works as a security switch (night-vision door camera). In the same way, you can use this 9V supply to turn on another switching relay in a home security system that is already set up. If you press switch S2, you can reset the circuit. Put the circuit together on a PCB that can be used for many things and put it in a suitable cabinet. Before soldering parts onto a PCB, look at Fig. 2 to see how the pins of the opto-isolator 4N35 and the transistor BC547 are set up. Connect the doorbell to the circuit and the night-light to the socket J1.

232. Pencell Charge Indicator

AA cells and button cells that are used in small electronics and have an end voltage of 1.5V are usually rated at 500 mAh. As the cells lose power, their internal impedance goes up. This, along with the load, makes a potential divider, which lowers the voltage at the battery's terminals. This causes the gadget to work less well, so we have to replace the battery with a new one. But the same battery can be used for something else that needs less current. Here's a simple tester (Pencell Charge Indicator) that you can use to quickly check if pencells and button cells are dead before you throw them away. The tester checks the battery's holding charge and terminal voltage to tell if the battery is right for a certain device or not. With the right amount of voltage and current, a 9V battery can power the circuit. When you close switch S1, it gives the circuit a stable 6V DC.

The op-amp CA3140 (IC1) is used as a voltage comparator in the circuit. Even a small difference in voltage between its inverting and non-inverting inputs can be picked up by it. The voltage from the battery being tested is fed into IC1's non-inverting input (pin 3), while its inverting input (pin 2) is given a reference voltage of 1.4V from resistor R4 and the series combination of diodes D1 and D2. For testing the charge capacity, resistors R1 and R2 provide a loading of 10 mA and 100 mA, respectively.

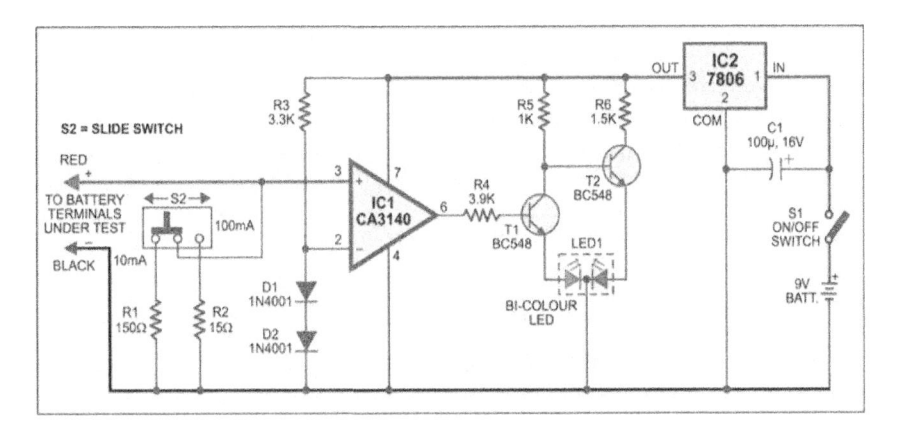

When a new battery is connected to the test terminals, the non-inverting input of IC1 gets 1.5V, which is more than the voltage of the inverting input, so the output of IC1 goes high. This high output gives forward bias to transistor T1 through resistor R4, and it conducts to light up the green half of the bicolor LED (LED1). At the same time, the base of transistor T2 is pulled down, turning it off, while the red half of bicolor LED1 stays off. When a partially discharged battery with a terminal voltage of less than 1.4 V is connected to the test terminals, the output of IC1 goes low to turn off transistor T1. This lets the bias voltage go through resistor R5 and forward bias transistor T2. This makes the red LED in bicolor LED1 light up. The S2 slide switch is used to see if the battery has enough current to power a 10 mA or 100 mA load. If the discharged battery can still hold more than 100mA of current, the green LED in bicolor LED1 will light up. This means that the battery can be used again in a low-drain circuit. The circuit can be easily put together on a perforated board with components that are easy to find. Put it in a small case with probes or a place to hold batteries so you can test it.

233. Power Resumption Alarm and Low Voltage Protector

The low voltage protector circuit described here keeps your AC motors and other electrical devices from getting damaged by low voltage when they turn on. When power comes back on, it stays in standby mode without giving power to the load. The load can only be turned on by hand. This keeps the device from getting broken if it is "ON" when power comes back on. The unregulated power supply comes from a step-down transformer that goes from 12V to 12V at 300 mA and two rectifying diodes, D2 and D3. Capacitor C3 gets rid of the ripples in the rectified DC. Along with the power supply for power resumption, there is an audio/video indicator (a piezobuzzer and an LED3).

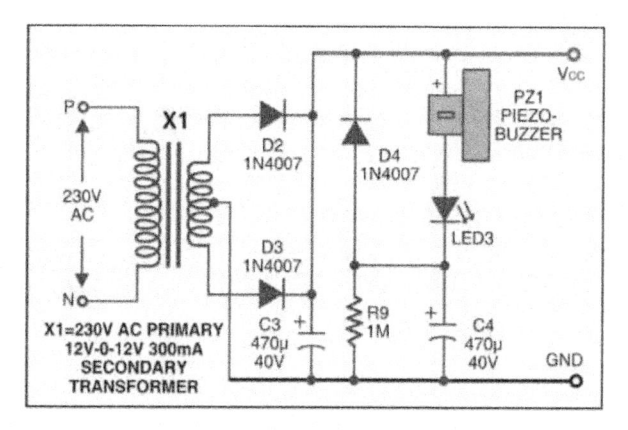

Fig. 1: Power supply circuit with resume indicator

When the power is turned on, capacitor C4 charges through the piezobuzzer and LED3, making both of them work. The piezobuzzer beeps, and for a few seconds, LED3 lights up. When capacitor C4 is fully charged, the LED's cathode goes high. This stops any more current from going through the buzzer. When the power is turned off, capacitor C4 drains through resistor R9.

The voltage comparator IC CA3140 (IC1) is used in the circuit to find changes in the voltage of the unregulated power supply caused by AC mains. The mains voltage changes in both the primary and secondary windings of the transformer. IC1 picks up on these changes to turn on or turn off the relay. To make transistor T1 work, Zener diode ZD1 sets a reference voltage of 3V. ZD1's breaking point is changed by the setting VR1.

When the voltage is normal, the zener diode ZD1 breaks down and the transistor T1 is biased in the forward direction. Capacitor C1 gives the device a few seconds of time to start up before any changes can affect it. When transistor T1 is on, the inverting input of IC1 (pin 2) is low. But IC1 doesn't have a high output because its power comes from the way SCR1 works (BT169). To turn on the relay, you have to do something by hand. When the push-to-on switch S1 is pressed, SCR1 fires and sends power to IC1's pin 7 through the switch. Since the voltage at the non-inverting input (pin 3) of IC1 is half of the supply voltage, its output goes high, and the relay (RL1) turns on. When IC1's output is high and the relay is turned on, LED2 lights up.

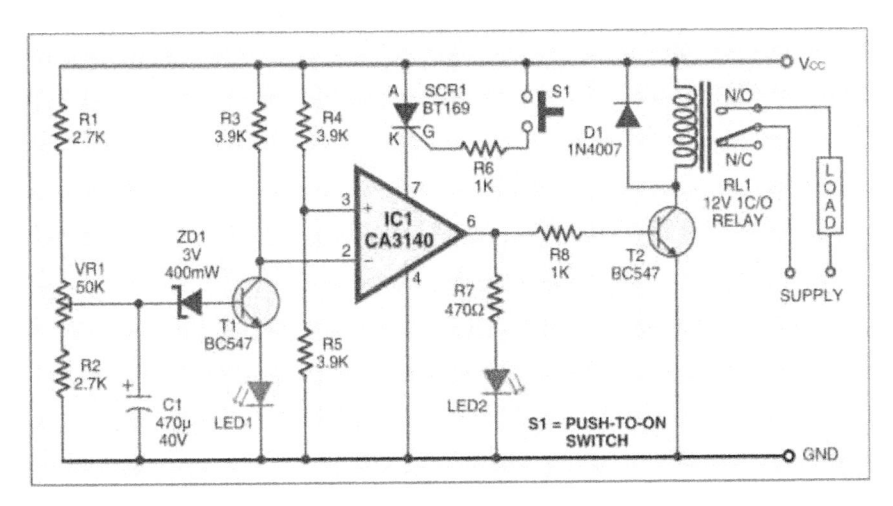

Fig. 2: Circuit of low voltage protector

When the line voltage drops below 180V, the secondary voltage of the transformer also drops, say to below 12V. When this happens, ZD1 stops conducting and the collector of T1 goes high. The output of IC1 is low because its inverting input (pin 2) has a high voltage. The power to the device is cut off when the relay loses its charge.

234. Flashing LED Light

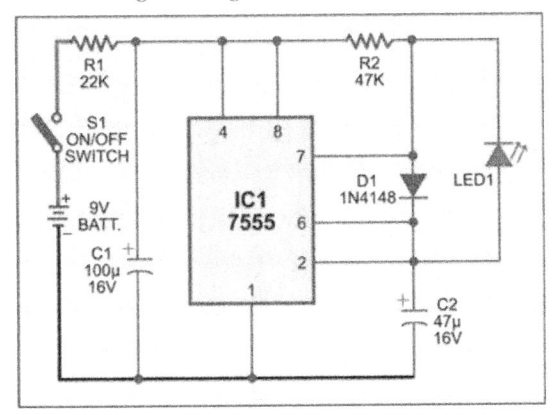

If you put a flashing LED at the front door of your garage or house, thieves will think that you have a high-tech security system. This circuit is just a low-current drain flasher that makes an LED light up and go out. It uses a single CMOS timer that is set up with a few extra parts to work as a free-running oscillator. Since the LED flashes very briefly, the average current through it is about 150 A, with a high peak value that is enough for normal viewing. It's a real miser because of this.

The 9V battery source is linked to the circuit by the "on"/"off" switch S1. When switch S1 is closed, the IC gets power from capacitor C1, which is constantly charged by resistor R1. By giving power to IC1 from capacitor C1, the battery doesn't have to be used up. Most LEDs use 20 mA of current, which is often more than the power used by the rest of the circuit. This is not good if the device runs on batteries. In this circuit, the LED only uses a small amount of the normal amount of power.

Through resistor R2 and diode D1, C2 gets charged. When the voltage across C2 reaches two-thirds of the supply voltage, threshold pin 7 of IC1 turns on as a current sink. Through LED1, the capacitor quickly drains into pin 7. Diode 1N4148 (D1) and resistor R2 make up the one-way charging path for capacitor C2. With the charges in C2, LED1 lights up briefly for a short time. The charging cycle starts over again. This is how LED keeps flashing. This job can be done just fine with a 9V PP3 battery.

235. Automatic Soldering Iron Switch

Automatic Soldering Iron Switch Circuit

Most of the time, soldering irons are used in electronics assembly for installation, repairs, and a small amount of production work. Other ways of soldering are used in high-volume production lines. One could be on your own test bench. When we use a soldering iron, we often leave it on for a long time. Not only does this make the iron smoke, but it also wastes electricity. To solve this problem, here is a circuit for an automatic soldering iron switch that turns off the iron after a set amount of time. Here, the time is set to 18 minutes, but it can be changed if necessary. When the circuit is not in use, it doesn't use any power. The circuit can also be used to control an electric iron, a kitchen timer, or other appliances.

A monostable multivibrator built around a timer IC 555 is the heart of the circuit. To turn on the soldering iron when the circuit is in sleep mode, briefly press switch S1. When the multivibrator is turned on, pin 3 of its output goes high for about 18 minutes, which keeps relay RL1 on through transistor T1. At the same time, capacitor C3 charges, and the normally open (N/O) contacts of relay RL1 send AC power to the soldering iron to turn it on. The soldering iron stays "on" for as long as resistor R1 and capacitor C2 say it should. Here, 18 minutes are set for this time. When LED1 flashes, it shows how hot the soldering iron is getting. When the time is up, relay RL1 turns off, which turns off the soldering iron, and the buzzer goes off until capacitor C3 is drained. Use a bell push switch or a similar switch that can handle the right amount of current to turn on the circuit.

236. White LED Light Probe for Inspection

This white LED light probe circuit can be used to look inside CPUs, monitors, PCB modules, and other electronic devices with tight spaces. The light source is a skinny tube with an extremely bright white LED at the end.

White LED Light Probe Circuit

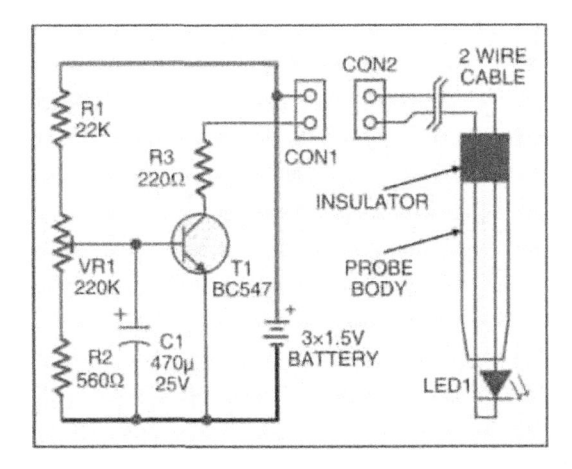

The brightness of LED1 is controlled the brightness of LED1 is controlled by T1, which is the center of the circuit. VR1 is used to set the forward-bias voltage of the transistor. The base of transistor T1 is linked to the middle pin of preset VR1. The brightness of LED1 is controlled by a voltage-divider circuit made up of resistors R1 and R2 and a preset VR1. By changing preset VR1, the forward bias voltage and, by extension, the collector current of transistor T1 are changed. This is how you can change how bright white LED1 is. In the collector circuit of T1, R3 limits the amount of current that flows through LED1.

For the LED to be as bright as it can be, it needs up to 20 mA. So there's no need for a separate switch to turn it on and off. When not in use, disconnecting CON1 from CON2 is enough.

237. Calling Bell Using an Intercom

Here is a simple calling bell circuit that can be used in small offices with an existing intercom system to call the office boy. There are up to nine places where extension lines can be used to call the office boy. The system is linked to the office boy's own phone line. When someone needs the office boy's help, they can use the intercom to call the office boy's extension number and then press a key to let the office boy know where they are (say, 5). This number will be shown on a seven-segment display, and an office boy will be notified by a bell. If you press a switch, the display will go blank.

The whole circuit can be broken up into two parts: the ring stop/bell timer and the decoder/display. The ring stops and bell timer section is made up of an optocoupler MCT2E (IC3), two NE555 timers (IC4 and IC5), and a few discrete components. The optocoupler is what makes the timers go off (IC4 and IC5). Both timers are set up to work in a single-stable mode. The time period of IC4 is always between two and three seconds, while the time period of IC5 can be set with preset VR1 to be between three seconds and one minute. Increasing the value of capacitor C6 or resistor R16 will make the time period last longer.

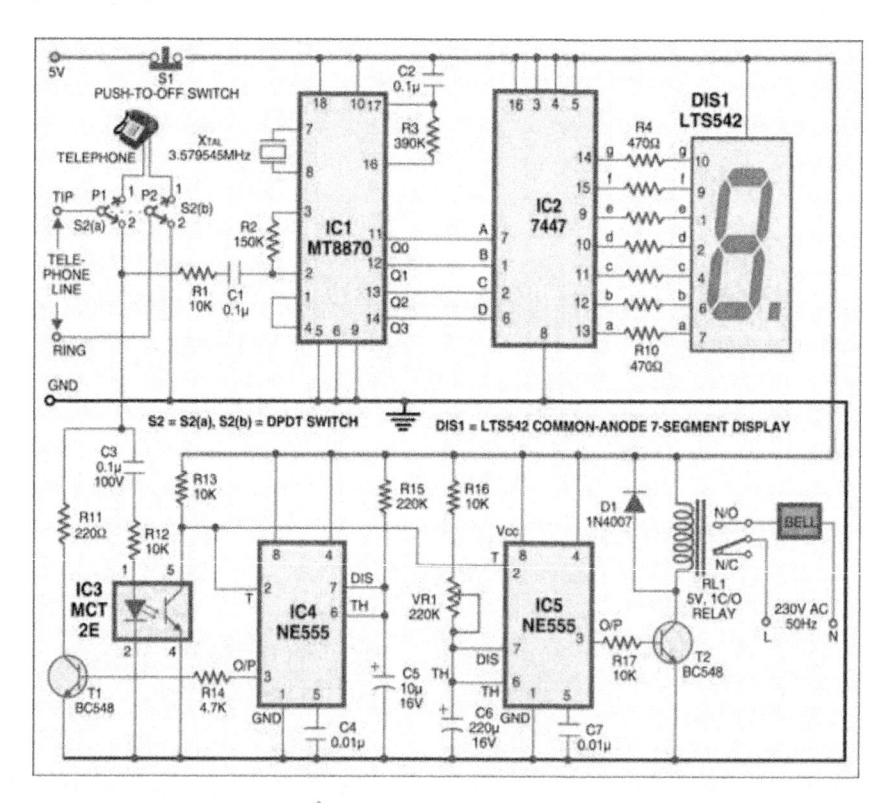

When the office boy's phone rings, the LED inside optocoupler IC3 lights up. This turns on timers IC4 and IC5. When IC4's pin 3 has a high output, transistor T1 conducts, which puts resistor R11 across the phone line to stop the ringing. At the same time, the high output of IC5 (pin 3) turns on the bell for a set amount of time through relay RL1.

The decoder cum display section is made up of a DTMF IC 8870 (IC1), an IC 7447 (IC2), a common-anode 7-segment display LTS542 (DIS1), and a few discrete components. The most important IC in this part is MT8870. It is used a lot in phones and a wide range of other things. MT8870 turns the DTMF (dual-tone, multi-frequency) tone pair that is made when a phone button is pressed into binary values. The binary values are sent to the driver IC 7447, which then tells the seven-segment display what number to show.

How the circuit works is easy to understand. Connect the circuit you just put together to the office boy's extension. Use a 5V power adaptor to connect 5V to the circuit. Now, when you call this dedicated extension number, the ringing will stop in a few seconds. After the phone stops ringing, press a number on the phone to tell them where you are (say, 5). This number will show up on the seven-segment display, and the office boy will hear the bell for a set amount of time. The office boy can wait for the next call after helping out in this room. If you want to use the phone as a regular intercom, just connect switch S2 to position 1. The circuit will no longer work. You can now use the phone to call any extension number. If you want to use the call bell again, just move switch S2 to position 2. Put the circuit together on a PCB that can be used for many things and put it in a suitable cabinet. At the back of the cabinet, connect the plugs for the phone line and the AC power. Install the 7-segment display, the buzzer, and the S2 switch on the front of the cabinet.

238. FM Bug

This FM transmitter circuit will allow you to eavesdrop on individuals. The transmitter can be put in any room, and a regular FM radio can be used to listen to the conversation from a long way away. The circuit is built around a single transistor 2N3904 (T1), a custom-made coil (L1), three capacitors (C1 through C3), a trimmer (VC1), two resistors (R1 and R2), and, of course, a condenser microphone (MIC1). The circuit sends signals between 88 and 105 MHz. The range of transmission is 100 meters.

The way the circuit works is simple. It uses analogue modulation, in which the message signal changes the carrier signal. The microphone picks up the sounds around it to make an electrical signal that matches the sound. This is the signal for the message that needs to be sent on the FM band. The signal for the message is sent to the base of transistor T1. The carrier frequency is made by the tank circuit, which is made up of trimmer VC1 and coil L1. Using the trimmer, you can change this frequency. The audio signal from the condenser microphone is added to the carrier signal made by the tank circuit by the transistor T1. This signal with changes is sent through the antenna (ANT.). Using VC1 to adjust the carrier frequency in the FM band and an oscilloscope to check it, when you tune the frequency of the FM radio set to match the frequency of the carrier, you will hear the conversation that MIC1 picked up. Make the coil L1 with a 25SWG wire length of about 25cm. Wrap the wire eight times around a cylinder with a diameter of 6 mm, then take it out. Put the circuit together on a general-purpose printed circuit board (PCB) and put it in a suitable cabinet. Fix the S1 switch on the front of the cabinet after the soldering is done right. Make sure the oscillator is tuned right. By adding a dipole antenna to the FM bug transmitter, the range will be increased. A 3V battery powers the circuit.

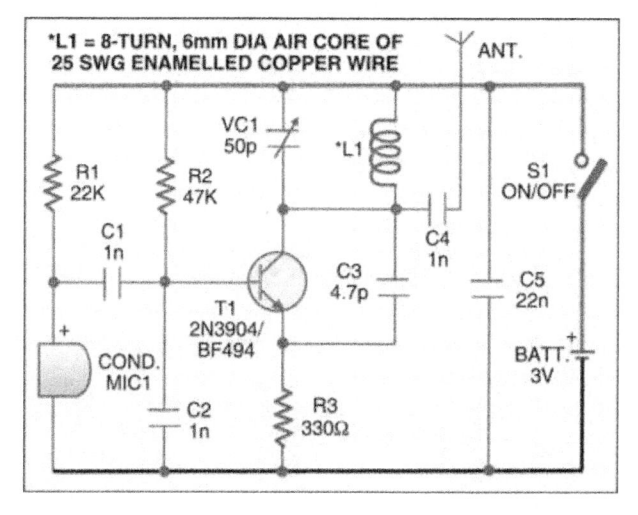

FM Bug circuit

239. Digital Frequency Comparator

Here is a digital frequency comparator for oscillators with a 7-segment display and a light-emitting diode that shows the result (LED). When an oscillator's frequency count is less than "8," the LED that goes with it stays off. When the number "8" is reached, the LED turns on and the 7-segment display shows "8." This demo circuit compares the frequencies of two NE555 timers set up as astable free-running oscillators. The digital frequency comparator part of the circuit has two 74LS90 decade counter ICs (IC2 and IC6), two 74LS47 7-segment display driver ICs (IC3 and IC7), a 74LS74 set/reset flip-flop (IC4), a 74LS00 NAND gate (IC8), and two 7-segment displays (DIS1 and DIS2). The counters get their frequencies from the oscillators that are built around the timers and are not stable but can run on their own.

How a circuit works

When the power is turned on to the circuit, resistor R1 and potmeter VR1 start to charge timing capacitor C1. As soon as the voltage of the capacitor reaches 2/3Vcc, the internal comparator of IC1 turns on the flip-flop, and the capacitor starts to discharge through VR1 towards ground. When the voltage of the capacitor reaches 1/3Vcc, the lower comparator of IC1 is triggered, and the capacitor starts charging again. Charge and discharge happen again and again. This means that the capacitor charges and drains between 2/3 and 1/3 of the power supply every so often (Vcc). The output of NE555 is high when capacitor C1 is being charged and low when it is being drained.

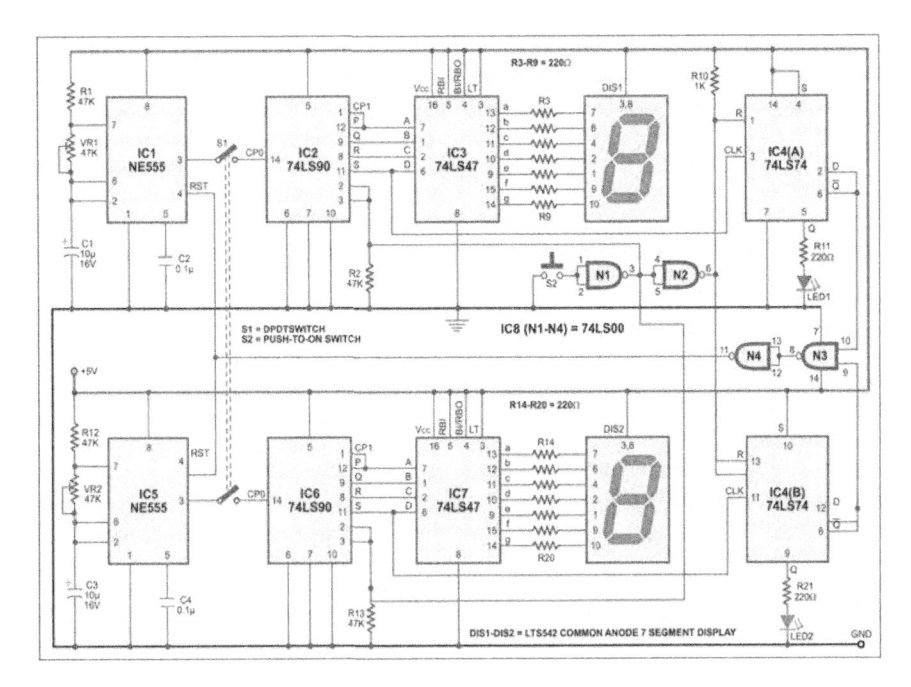

The second oscillator, IC5, works the same way. The potentiometer lets you change the frequency of the oscillator (VR1 or VR2). Through the DPDT switch, the output pins (pin 3) of the oscillators (IC1 and IC5) are linked to the decade counters (IC2 and IC6). IC2 and IC6 both keep track of the first eight cycles. IC 74LS90 is a 4-bit ripple decade counter. It has a counter to divide by two and a counter to divide by five. Each part has its own input for the clock. The P output (pin 12) of the divide-by-two section is connected to the CP1 input of the divide-by-five section (CP0). When the clock pulse hits the divide-by-two section, it changes into a divide-by-ten counter.

A high pulse at pins 2 and 3 resets the 74LS90 decade counter. At first, resistor R2 pulls down pins 2 and 3 so they don't move. The outputs of IC2, P through S, are connected to the inputs of IC3, A through D. The clock pulse is also sent from pin 11 (S) of IC2 to pin 3 of IC4(A). The number of items is shown on the seven-segment display.

Display

The 7-segment decoder/driver (74LS47) takes four binary-coded decimals (8421), makes their complements internally, and decodes the data using seven AND/OR gates with open-collector outputs to drive the display segments directly. In the "on" state, each segment-driver output can sink 40mA of current. The ripple-blanking input (RBI), blanking input (BI), ripple-blanking output (RBO), and lamp test are turned off by connecting pins 3, 4, and 5 of the display driver to Vcc (LT). With the help of current-limiting resistors R3 through R9, IC3 sends segment data to the 7-segment display (each 220 ohms). The reset pin (RST) of NE555 is controlled by IC4, which is called IC4. It is a dual D-type flip-flop that has direct clear and set inputs and outputs that are the opposite of each other. On the rising edge of the clock pulse, the data from the inputs is sent to the outputs. The flip-flops work in toggle mode because the Q output is connected to the D data input.

240. Bhajan and Mantra Chanting Amplifier

The practice of chanting a variety of mantras is popular in India due to the widespread belief that doing so will bring the practitioner good fortune, mental tranquility, and an enhanced capacity for concentration. We are going to show you the circuit for an electric chanting amplifier that has one mantra and nine other bhajans that you may choose from. The device is compact and convenient, making it possible to carry it anywhere and use it while traveling. It is suitable not only for the pooja room but also for the living room, making it an appropriate present for any event that one would attend.

Circuit and the working

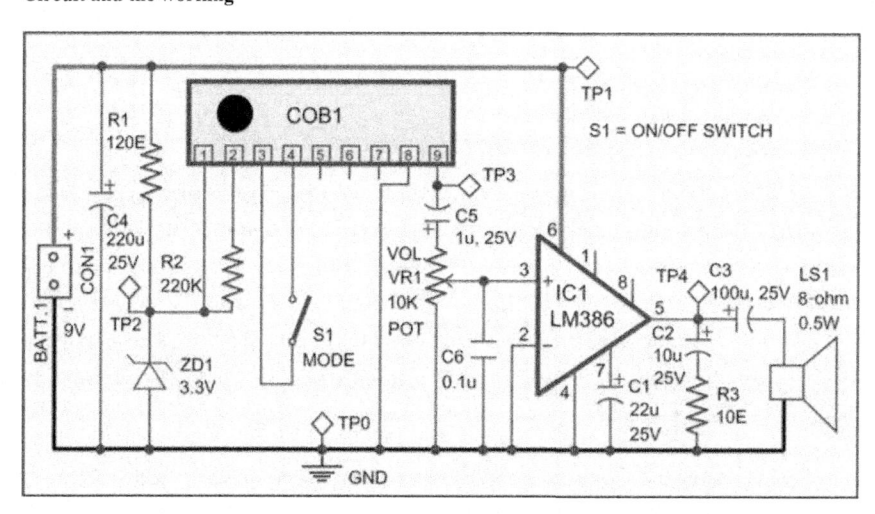

Figure 1 presents the bhajan and mantra chanting amplifier circuit for your perusal. A 9V PP3 battery provides the necessary power for it. The chip-on-board (COB) and LM386 audio amplifier serve as the primary components of the circuit (IC1). The chanting IC, also known as COB1, can be purchased in the Indian market under the name Ganpati 101. COB1 receives 3.3V from the Zener diode, which allows it to function properly. The ten previously recorded songs are already saved in COB1's database. The part in the COB that is responsible for generating the right sound frequency is the resistor R2 that is linked to pin 2 of COB1. Pin 9 of the COB can be used for audio output if it's needed.

S1 is a mode-selection switch that is used to select the mode that will be used to play a specific song. The bhajan starts playing once the mode switch S1 is turned on. You can select the next bhajan by toggling the switch in the appropriate direction. If the power is not turned off while the S1 switch is closed, a specific song will play over and over again. The LM386 is an audio amplifier that has a low power output. Capacitor C5 and volume control VR1 are the two components that deliver audio input to its pin 3. Pin 5 is where you can access the device's audio output.

241. Low-Cost Automation Using PIC16F676

Home and industrial automation systems abound. IR remote control, Bluetooth home automation, DTMF home automation, Wi-Fi home automation, RF home automation, and voice-controlled home automation. Here's a cheap automation system. It can turn devices on and off at home or in industry.

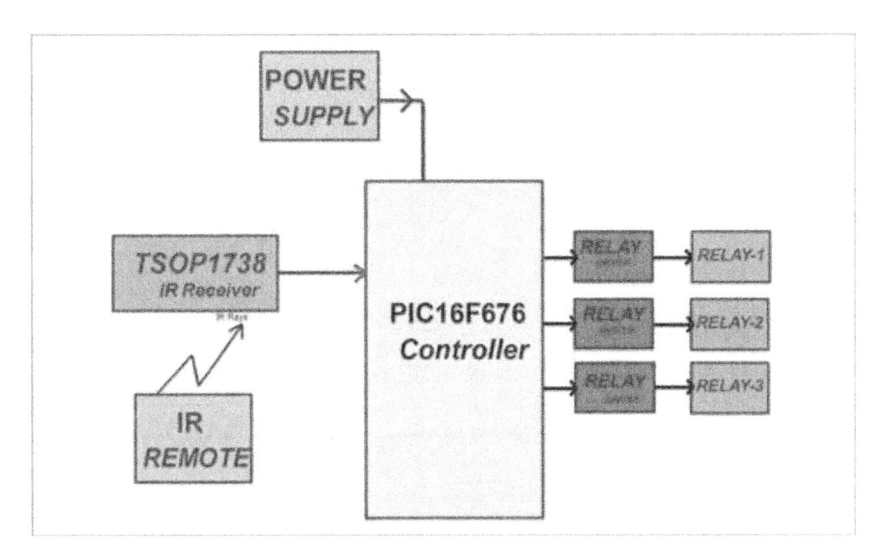

Fig. 1: Block diagram

IR remote controllers and receivers use conventional data standards. NEC, PHILIPS RC5, JVC, and SIRC are standard protocols (Sony infrared remote control). This project uses NEC protocol. After understanding IR remote frame format, we'll interface IR receiver TSOP1738 with PIC16F676 to decode NEC remote key presses. Fig. 2 shows how each bit is sent using pulse distance. Logical '0' is a 562.5s pulse burst followed by a 562.5s gap, taking 1.125ms to broadcast. Logical "1" is a 562.5s pulse burst followed by a 1.6875ms gap, for a total of 2.25ms.

When a remote button is pressed, the following message is sent:

1. 9ms pulse lead (16 times the pulse burst length used for a logical data bit)
2. 4.5ms space
3. Receiver's 8-bit address
4. 8-bit address logical inverse
5. Command 8-bit
6. Command's 8-bit logical inverse
7. A 562.5-s pulse burst to cease message transmission.

Each byte's least significant bit is supplied first. Fig. 3 shows a NEC IR transmission frame for address 00h (00000000b) and instruction ADh (10101101b). Frame transmission takes 67.5ms. It takes 27ms to send 16 bits of address and command.

Microchip's PIC16F676 is an 8-bit CMOS flash microcontroller. Its 14-pin size and high-performance RISC CPU make it suited for embedded or industrial automation applications. This little chip includes everything needed to develop projects. Features include:

- The PIC16F676's flash memory increases the microcontroller's processing speed.
- It's available in 14-pin PDIP, SOIC, and TSOP.
- PIC16F676 has 1.7kB of program memory, 64 bytes of RAM, and 128 bytes of EEPROM.
- The device's 10-bit ADC has eight analog channels. This module converts analogue sensor values to digital.
- Device features include power-on reset, comparator, in-circuit serial programming, and master clear reset. These allow it outperform other onboard chips and eliminate the need for external components. Fig. 2 shows PIC16F676 pinout.

Controller PIC16F676

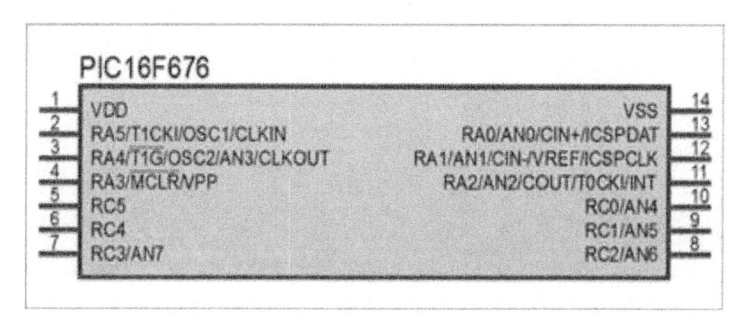

Fig. 2: Pin diagram of PIC16F676

Working circuit

The project uses a PIC16F676 microcontroller and a NEC IR remote to control any AC load, including lights and fans. The microcontroller receives the remote's IR signals for the lights and fans and controls the relays with a relay driver circuit. Relays control lights and fans. In this project, digital IO pins RC0, RC1, and RC2 of Port C of PIC16F676 microcontroller control relays to switch appliances on/off. Programmatically, these pins are outputs. The IR receiver is attached to PIC16F676 input pin RC4. PIC16F676 runs on +5V. The 230V AC mains supply is stepped down with a transformer and rectified with a full-wave rectifier. IC 7805 regulates rectified voltage to +5V. This project is straightforward. When an IR remote button is pressed, it sends 38kHz modulated coded pulses. Microcontroller reads pulses from TSOP 1738 sensor. The microcontroller decodes the pulse train into hex and compares it to the program's predetermined values. When a key is pressed on an IR remote, the program decodes the delivered IR signal. This project decodes a 32-bit string (4 NEC protocol bytes). The lights/fans software compares the third byte of this 32-bit string. If a match occurs, the controller triggers the respective relay using transistor BC547 and an LED indicates the outcome. The circuit's three LEDs display relay status. IR remote buttons 2, 4, and 6 control three relays. Key 2 toggles RL1, 4 toggles RL2, and 6 toggles RL3.

Software

The circuit uses PIC16F676's native software. The software is compiled with Mikro C PRO 7.2.0 for PIC and uploaded to PIC16F676. The embedded C language in main.c enables for short coding. PIC K150 programmer board burns hex code onto MCU chip. IR signal is detected without a header, interrupt, or capture and compare mode. RC4 reads data like a push button. When a signal goes high or low, debouncing and the timer start. Time values are recorded in an array whenever the pin changes state. IR remote sends logic 0 as 562.5s and logic 1 as 2250s. The software assumes 562.5's pulse is 0 and 2250's pulse is 1. It's then converted to hex. 32-bit remote signal (4 bytes). Program stores all array bytes and decodes the third to compare. The simple programming statement 'switch' controls home appliances. Before programming PIC16F676 with PIC K150, set fuse bits. Without fuses, the program won't work.

242. Mains Box Heat Monitor

This simple main box heat monitor circuit performs continuous monitoring of the mains distribution box and triggers an alarm whenever it detects a high temperature as a result of overheating. By doing so, it helps to prevent catastrophes that could be brought on by sparking in the mains box as a result of short circuits. It also turns on a bright white LED by itself when the power goes out. In the dark, the LED gives off enough light to check the wiring or fuses in the mains box. When power goes out, the circuit beeps once. When power comes back on, it beeps again.

Mains Box Heat Monitor Circuit

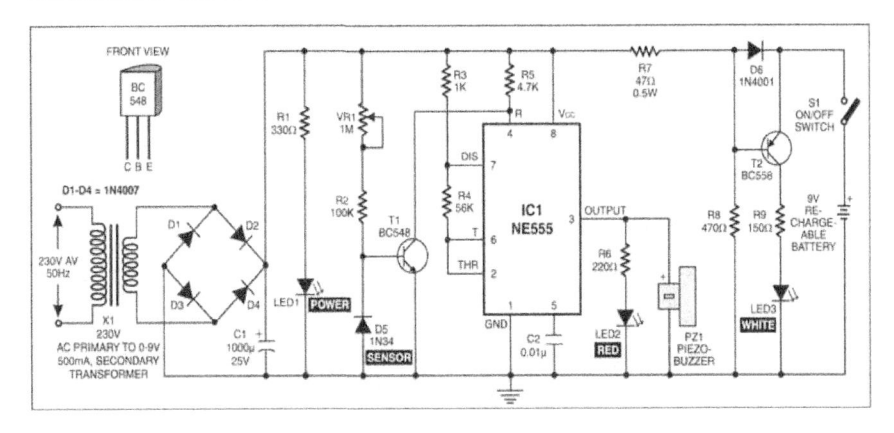

Transformer X1 takes the mains AC power and steps it down so that the secondary output is 9V AC at 500 mA. The output of the transformer is fixed by diodes D1 through D4. The ripple is avoided by capacitor C1. LED1 shows that the device is turned on. LED1's current is limited by resistor R1. The temperature sensor is the Germanium diode D5 (1N34), which is hooked up in the reverse bias mode. At normal temperature, the diode's resistance is high, so transistor T1 conducts and keeps reset pin 4 of IC1 in a low state.

As an astable multivibrator, NE555 (IC1) is wired. When the fuse gets too hot and the temperature around diode D5 goes up, the resistance of the diode goes down, and transistor T1 stops conducting. This turns on IC1 and starts the oscillator's alarm. You can set the temperature at which the alarm circuit goes off by adjusting preset VR1. The circuit for the emergency light utilizes a pnp transistor designated as BC558 (T2) in addition to a few passive components. It is powered by a 9V rechargeable battery, which, while mains power is available, is continuously charged through forward-biased diode D6. The charging current is brought down to a safer level by resistor R7. When diode D6 has a forward bias, it makes transistor T2 have a reverse bias. This turns off the white LED (LED3). When the power goes out, the transistor T2 is turned on, and the LED lights up. When the power is turned back on, the transistor T2 stops conducting, therefore the LED is no longer illuminated.

243. Tachometer

In a motor or other type of equipment, the rate of rotation of a shaft or disk can be determined using an instrument called a tachometer. Here is the basic tachometer with a digital display that shows the number of turns per second (RPS). Figure 1 shows the tachometer's block diagram. Figure 2 shows the tachometer's circuit, which is made up of a timer IC 555 (IC1), an IC 7811 (IC2), an IC CD4081 (IC3), an IC CD4069 (IC4), two ICs CD4033 (IC5 and IC6), two common-cathode displays (each LTS543), and some other discrete parts. When switch S2 turns on IC1 in monostable mode, it makes a one-second pulse.

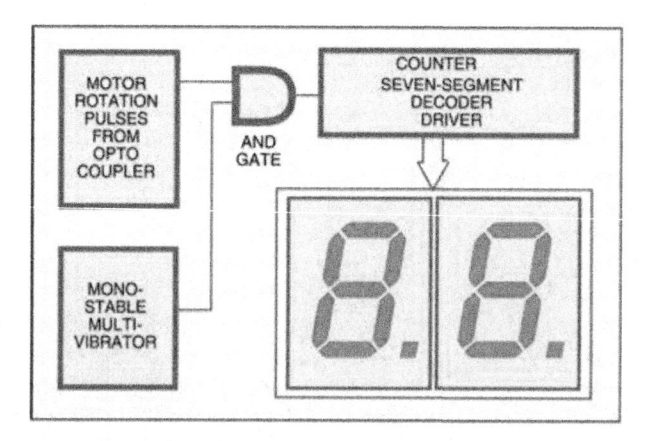

Fig. 1: Block diagram of the tachometer

IC1's time period is set by the variable resistor VR1. When the shaft is moving, IC2 sends out pulses.

IC5 (CD4033) is a decade counter that has a seven-segment decoded output. It is coupled to DIS1 so that the place of the unit can be displayed on the screen. Pins 2 and 14 of IC5 are connected to the ground. When the count goes too high, IC5's Pin 5 (CO) sends a pulse (count reaches from 0 to 9). Through NOT gate N3, this pulse is used as the clock input to IC6. That is, IC6 will go up by one pulse after IC5 has pulsed ten times. So, when the unit's count on DIS1 is done, the count continues on DIS2. On DIS1 and DIS2, you can count and show up to 99 digits.

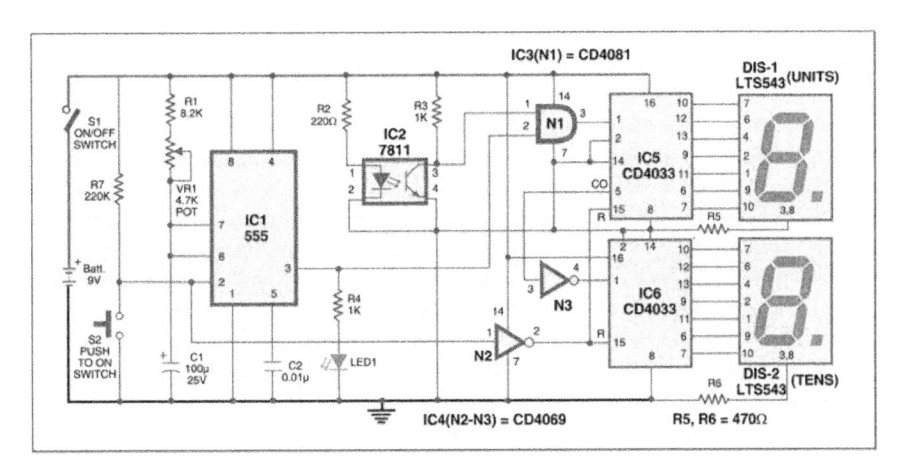

Fig. 2: Circuit of the tachometer

How the circuit works is easy to understand. Set up IC2 so that when the motor shaft turns, it cuts the IR signal and sends pulses to pin 3 of its output. In other words, when the motor turns, the shaft cuts the light beam inside the optocoupler (IC2). Pulses are made because of this. This series of pulses is sent to pin 1 of the AND gate N1. The output of IC1 is linked to pin 2 of the AND gate N1. To determine the RPS of the shaft, briefly engage the switch S2 to make gate N1 active through the monostable multivibrator for a period of duration equal to one second (IC1). The pulses are used to tell IC5 what time it is. The counter adds one to itself and makes the code that tells the seven-segment displays what to do. The screens show how many counts are made in one second. So, you can easily figure out the number of turns per minute (RPM).

244. Timer from Old Quartz Clock

From the circuit of an old quartz clock, you can make a low-cost timer that works well and doesn't cost much. This timer can be set for up to two hours, which is long enough for most everyday tasks. The outputs of the quartz clock circuit drive a coil. One of them is used to power the counter CD4040, which can count up to 4096 times. It does this by sending a pulse every two seconds. Using DIP switches, you can set the length of time to fit your needs. Each switch has a set amount of time that it works. With DIP switches 1 through 8, Q5 to Q12 outputs, you can set the time you want (refer the table). For example, to set the time to 68 minutes, you would move shift switch 8 to the "on" position. By doing this, Q12 will go up. If two switches are "on," the amount of time set is equal to the sum of the times each switch is "on." By turning all the switches on, you can set a maximum time of 136 minutes, which is the sum of all the switches' outputs. The counter's outputs go through diodes D5 to D12.

How the circuit works is easy to understand. When all of the outputs that have been chosen go high, transistor T3 conducts. So, the buzzer sounds until it is turned off and on again. At the same time, transistor T2 also conducts to stop the counter from counting by blocking the clock input. Put the circuit on a general-purpose PCB and put it and the quartz-clock PCB in a small plastic cabinet. Installing the DIP switch on the front of the panel makes it easy to set the time needed.

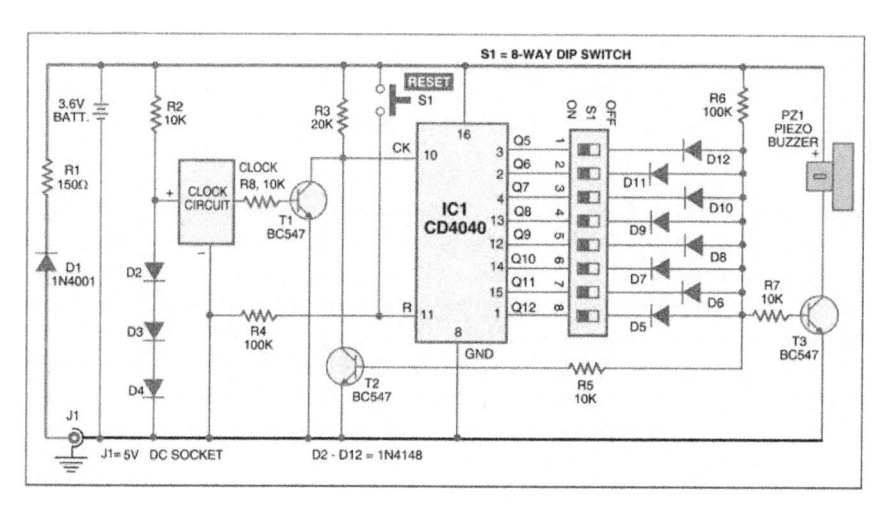

A 3.6V Ni-Cd rechargeable battery powers the circuit, which makes the unit small. As shown in the circuit, 5V is used to charge the battery. The timer has a very low current when it is not running. At this rate of power use, a 60-mA battery will last long enough. Some quartz clocks have an alarm that has a pin to turn it on. This gets rid of the need for a separate buzzer, which makes the circuit smaller.

245. Keep Away Ni-Cd from Memory effect

Due to partial discharge, Ni-Cd batteries have a memory effect that is not good. The solution is to fully drain the battery before charging it again. A simple resistor will do a good job of discharging the cell, but if the cell discharges past a certain voltage, the polarity of the cell could change. Here is a simple circuit that discharges the cell to a level of 600 mV. This makes sure that the cell is correctly discharged without the risk of polarity reversal. The current going through the battery is not always the same. This gives the battery a chance to recharge during the breaks, which makes it last longer.

As shown in Fig. 1, the circuit is a simple flyback oscillator with a frequency in the kHz range that depends on how X1 is wound. Wrap 30 turns of 36SWG wire around a bobbin core for the primary and 15 turns for the secondary. Figure 2 shows more about the size of the bobbin. It's easy to work. When the circuit is turned on, current flows through inductor X1 and through transistor T1. When the switch is off, the LED1 lets out the energy stored in the inductor. Once the end-of-cell-discharge voltage is reached, the LED stops glowing because the voltage at the base drops below the forward conduction voltage (0.6V) of the transistor.

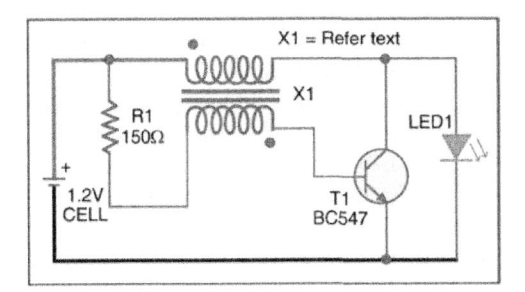

Fig. 1: Circuit to keep Ni-Cd away from memory effect

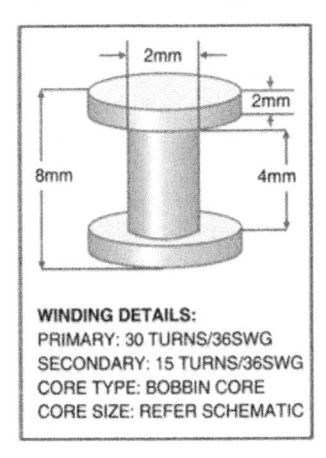

Fig. 2: Winding details

If you think a cell has the memory effect, you should empty it and then charge it a few times in a row. This will bring the cell back to its full size. Connect a few LEDs in parallel to up the load and speed up the discharge.

246. Periodically on off Mosquito ad hoc circuit

Some mosquito repellents on the market use a poisonous liquid to make poisonous vapors that make mosquitoes leave the room. Due to the constant release of poisonous fumes into the room, the natural balance of the air's components for good health reaches or passes the critical level after midnight. Most of the time, these vapors get into the brain through the lungs. They also have a small or large percentage of anesthetic effects on mosquitoes and other living things. If you are exposed to these toxic vapors for a long time, you may have problems with your brain or nerves. Here is a circuit that turns the mosquito repellent on and off after a certain amount of time. This lets you control how much toxic vapor gets into the room.

The circuit periodically turns the mosquito repellent "on" and "off" for about 20 minutes each time. So, if you leave the mosquito repellent on from 10 pm to 6 am (eight hours), it will be "on" for four hours and "off" for four hours of the total time. During "off" time, the air in the room tries to get back to its natural state. Another important thing about the circuit is that when it goes from "on" to "off," it doesn't make any noise or click like a relay does, so it doesn't wake you up when you're sleeping.

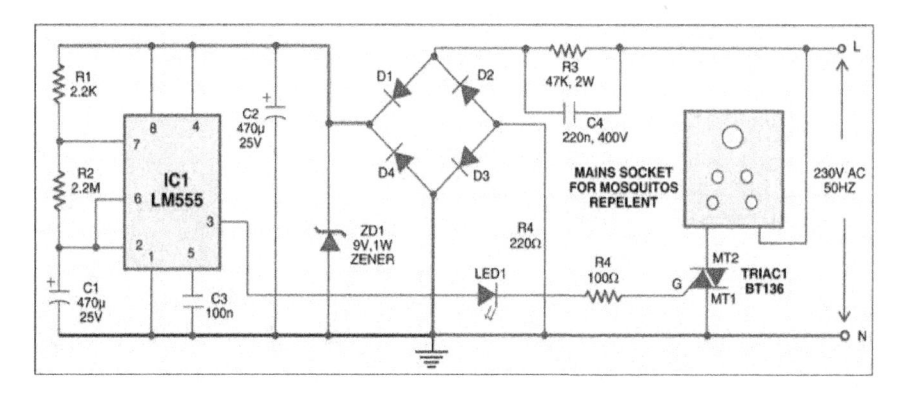

Circuit operation

The circuit is made up of IC 555 (IC1), which is a timer, and triac BT136, which is an automatic switch (TRIAC1). The power for the circuit comes from the AC mains. This is done by stepping it down and reversing it. By getting rid of the transformer, space and money are saved. Timer IC1 gets a steady 9V DC power supply from Zener diode ZD1 and capacitor C2.

The resistors R1 and R2 and the capacitor C1 make up the timer section. Through LED1 and R4, the output of timer IC 555 is sent to the gate terminal of BT136. When the timer output goes high, it turns on TRIAC1's gate, and LED1 shows the "on" time. During "off" time, IC1's output is low, so TRIAC1 isn't turned on and LED1 doesn't light up.

247. Crystal AM Transmitter

The following diagram illustrates the circuit for a medium power AM transmitter that can deliver between 100 and 150 mW of radio frequency (RF) power. A crystal oscillator is located in the most important part of the circuit. The extremely stable carrier frequency is generated with the help of a 10 MHz crystal. An amplification process is performed on the audio signal that is received from the condenser microphone by an amplifier that is constructed using transistors T1, T2, and T3. The audio signal that has been amplified serves as a modulator for the RF carrier that is produced by the crystal oscillator that is built around transistor T4. In this case, modulation is accomplished by means of the power supply line. At the collector of oscillator transistor T4, the amplitude-modulated signal, abbreviated AM, can be obtained. It is possible to extend the reach of a signal transmission by utilizing a matching dipole antenna in conjunction with coaxial cable. Utilize a radio with a sensitive receiver and an external wire antenna to achieve the best possible range. A 9V-12V battery is required for the circuit to function properly. In order to make oscillator coil L1, wind 14 rounds of wire with a gauge of 30 AWG around a radio oscillator coil former with an 8 mm diameter that contains a ferrite bead.

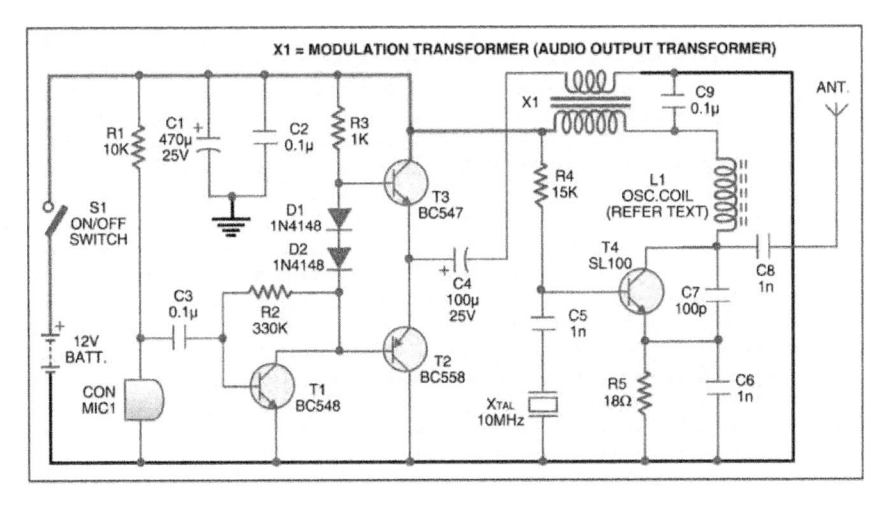

Crystal AM transmitter circuit

You can make do with the audio output transformer from your old transistor radio set in order to serve as the modulation transformer X1. Alternately, you may create it out of E/I section transformer lamination, with the inner winding having 40 turns of 26 SWG wire and the outer winding having 200 turns of 30 SWG wire, as demonstrated in this image.

248. Programmable Electronic Dice

This is an example of a straightforward programmable electronic dice with a numeric display. These dice have a 4-way DIP switch that may be used to display any random number between "1" and "2," "1" and "3,".... or "1" and "9."

Programmable electronic dice circuit

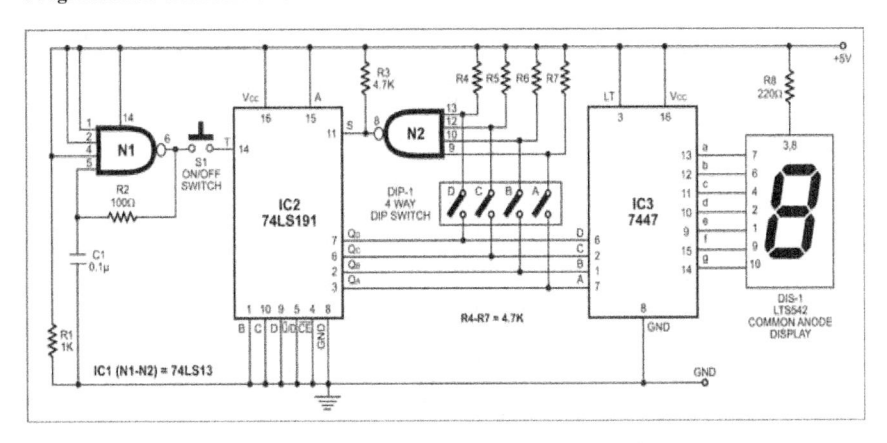

IC1 is a dual 4-input Schmitt trigger NAND gate with the model number 74LS13. In order to establish a clock frequency of roughly 70 kHz, which is then sent to IC2, Gate N1 is utilized as an oscillator. This oscillator is constructed using resistor R2 and capacitor C1. The inputs of IC2 are loaded with data by Gate N2. The 74LS191 in IC2 is a presettable binary counter, and it has a facility for parallel loading. Whenever its pin 11 goes low, the data that is now present at its inputs D through A (which is '0001') appears at its outputs QD through QA. This occurs when all of the inner switches of the DIP switch are open and DIS1 displays the minimum count as '1' rather than '0'.

Circuit operation

The DIP switch's inner switches A, B, C, and D must be set in accordance with the table in order to get the appropriate dice range. For instance, if you want the electronic dice to count from 1 to 8, you would need to close switches A and D while leaving switches B and C open. The display quickly changes between the numbers '1' and '8' when the switch S1 is pressed. As soon as you let go of S1, the display will stop bouncing around, and it will stay on the most recent number.

When switch S1 is pressed, the count output can go from '0001' all the way up to the maximum count stated in the table under the heading 'Dice Range.' This occurs with the inner switches of the DIP switch in the locations specified in the table. As soon as you let go of switch S1, the most recent count that falls within the dice range will be displayed.

The outputs of IC2 are displayed on a 7-segment display with a shared anode (LTS542) (DIS1). The BCD-to-7-segment decoder integrated circuit 7447 (IC3) is what drives the display. Resistance R8 controls the amount of current that flows through DIS1.

249. PC Based Candle Igniter

PC Based Candle Igniter Circuit

Here is a PC-based candle-igniter system that lets you light a candle with matchsticks by pressing the

"Enter" key on the PC's keyboard. It is especially helpful for special events like birthdays and anniversaries. As shown in the figure, the number of matchsticks needed to light the candle are placed along its wick. The heating coil that is used to light the matchsticks is kept close by. For the candle-matchsticks setup, the circuitry between the PC and the heating coil is made up of an inverter, a monostable, and a relay driver. The inverter is the transistor BC548 (T1), the monostable circuit is the IC 555 (IC1), and the relay driver is the transistor SL100 (T2).

When you press the "Enter" key on the keyboard, the inverted output at the collector of transistor T1 goes low, triggering IC1 through its pin 2. The monostable's output pin 3 goes high, and transistor T2 works for about 50 seconds.

When transistor T2 is turned on, relay RL1 is turned on. The normally open (N/O) contact on relay RL1 then connects the heating coil to 230V AC. You can also use an electric cigarette lighter instead of the heating coil. When the heating coil is connected to the 230V AC, it gets very hot and lights the matchsticks. The matchsticks' flames light the candle.

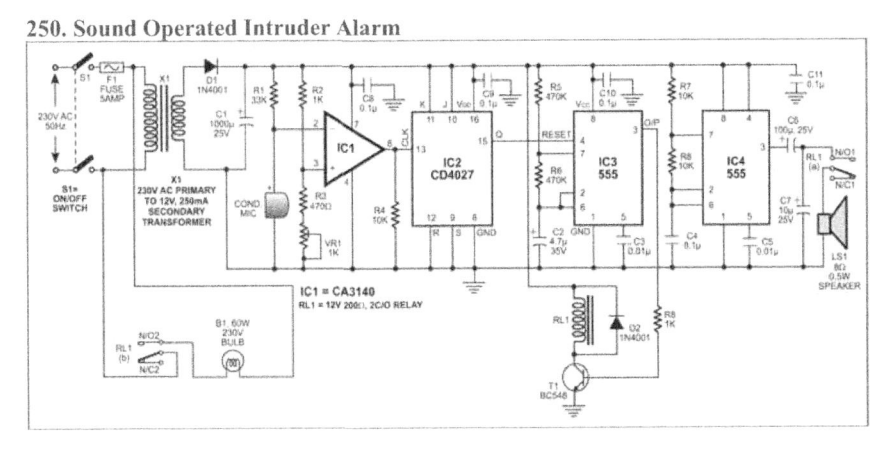

250. Sound Operated Intruder Alarm

When this sound-activated intruder alarm hears a sound, like when a door is opened or a key is put into a lock, it starts flashing a light and making a pulsing sound to warn you of an intruder. The next sound pulse automatically turns off both the light and the alarm. Stepping down 230V AC mains with transformer X1, reversing it with diode D1, and filtering it with capacitor C1 gives 12V DC. The voltage at the op-amp CA3140 (IC1non-inverting) input (pin 3) is used as the reference voltage, and it can be set with preset VR1. The voltage across the condenser microphone is the same as the voltage at the inverting input (pin 2).

Care should be taken to set up the condenser microphone so that it is sensitive to sound. A high reference value means that IC1's output at pin 6 can be changed by a quiet sound. Set the reference voltage so that the output doesn't change if the switch is accidentally turned on. When there is no sound, the voltage at pin 2 of IC1 is almost the same as the full DC voltage, so the output of IC1 stays low. Because the IC CD4027 is wired in toggle mode, pin 15 of its output is also low. This makes pin 4 of IC3's reset pin low so that the astable multivibrator built around timer 555 can be reset (IC3). So, transistor T1 is turned off and relay RL1 stays turned off. When RL1 is turned off, both of its N/O contacts, RL1(a) and RL1(b), remain open. RL1(a) keeps the light off, and RL(b) turns off the speaker by cutting the output of the astable multivibrator built around IC 555 (IC4).

Circuit operation

When there is noise, a current flow through the microphone, which lowers the voltage at pin 2 and makes the output of IC1 high. The pulse at pin 13 of IC2 turns it on, and when its output at pin 15 goes high, it turns on astable multivibrator IC3. The output of IC3 is high for three seconds and then low for 1.5 seconds. This happens again and again until pin 15 of IC2 stays high. When the output of IC3 is high, it turns on the relay through the driver transistor T1. When the output is low, it turns off the relay. When relay RL1 is turned on, relay contact RL1(a) sends AC power to bulb B1 so that it can light up. At the same time, relay contact RL1(b) lets the output of astable multivibrator IC4 go to the speaker, which makes an audio tone. About 480 Hz is the frequency of this sound tone. As long as the output of flip-flop IC2 stays high, both the light bulb flashes and the tone plays.

Now, if the circuit hears another sound, the flip-flop IC2's output goes low. This makes IC3's reset pin 4 go low, which stops IC3 from oscillating. When the output of IC3 is low, the relay is turned off, which turns off the bulb and the tone.

251. Versatile LED Display

Figure: Versatile LED Display Circuit

This flexible LED display circuit uses an EPROM (erasable programmable read-only memory) to show different patterns of light on LEDs. Since green and red LEDs have been used, it is now possible to display in three colors (green, red and amber). 5V DC is used to power the circuit. IC 555 (IC1) is set up as an astable multivibrator, and VR1 can be used to change the frequency at which it oscillates. The output of IC1 clocks the 12-stage binary counter IC CD4040 (IC2), which then sends address data to EPROM IC 2716. (IC3).

When the logic at any data pin is high, the LED that goes with that pin lights up. The red LED on LED1 lights up when the data at address location 00H is read. The green and red LEDs on LED2 light up when the data byte 44H at address location 09H is read (refer the table). This flexible LED display circuit uses an EPROM (erasable programmable read-only memory) to show different patterns of light on LEDs. Since green and red LEDs have been used, it is now possible to display in three colors (green, red and amber).

Versatile LED Display Circuit

5V DC is used to power the circuit. IC 555 (IC1) is set up as an astable multivibrator, and VR1 can be used to change the frequency at which it oscillates. The output of IC1 clocks the 12-stage binary counter IC CD4040 (IC2), which then sends address data to EPROM IC 2716. (IC3). The code for the screen is in IC3 (see Table I). When the logic at any data pin is high, the LED that goes with that pin lights up. The red LED on LED1 lights up when the data at address location 00H is read. The green and red LEDs on LED2 light up when the data byte 44H at address location 09H is read (refer the table).

252. Multiutility flash light

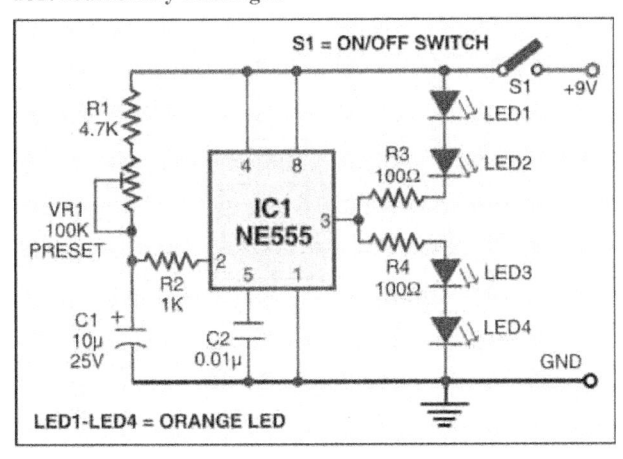

Flasher circuit

This multi-purpose flashlight has three different functions: a flasher, a sound-to-light display, and a white LED-based spotlight. Each of these functions performs a different function.

Circuitry for a flashlight with multiple uses

This multi-purpose flashlight has three different functions: a flasher, a sound-to-light display, and a white LED-based spotlight. Each of these functions performs a different function.

Multi-utility flashlight circuit

The sound-to-light display circuit that was shown above has a straightforward method of operation. A low voltage is produced by the condenser microphone (MIC), which is responsible for picking up sound in the surrounding area. The first part (A1) of the dual operational amplifier LM358 provides a boost to the low output of the microphone. The output of operational amplifier A1 is connected to the second portion (A2) of LM358, which in turn drives transistor T1 (2N2222). The three RGB LEDs flash in the appropriate manner. (An RGB LED is actually a two-pin white LED package that, when powered, emits red, green, and blue light in a sequential pattern.)

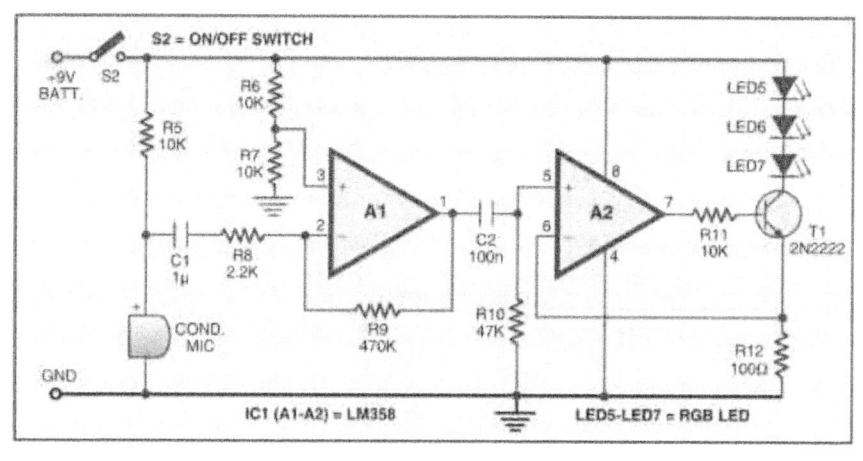

The sound-to-light display circuit

The circuit for the flashlight is made up of a quartet of white LEDs that are wired together in series. The amount of current that can pass through the white LEDs is controlled by resistor R13 (LED8 through LED11).

253. Twi-light using white LEDs

This lamp that is activated by natural light has a total of 25 white LEDs that are quite bright and employs a light-dependent resistor, also known as an LDR, as its sunshine sensor. Each row of LEDs has its own individual resistors that are connected in series with it. The operation of the circuit is a relatively straightforward process. The LDR1 is exposed to light during the day, which causes its resistance to drop significantly. As a direct consequence of this, neither of the transistors (T1 nor T2) transmit electricity, and none of the LEDs (LED1 through LED25) light up. On the other hand, the light does not hit LDR1 during the night, therefore it has a high resistance. This occurs because of the absence of light. Because of this, transistors T1 and T2 conduct, and all of the LEDs, from LED1 all the way up to LED25, light up.

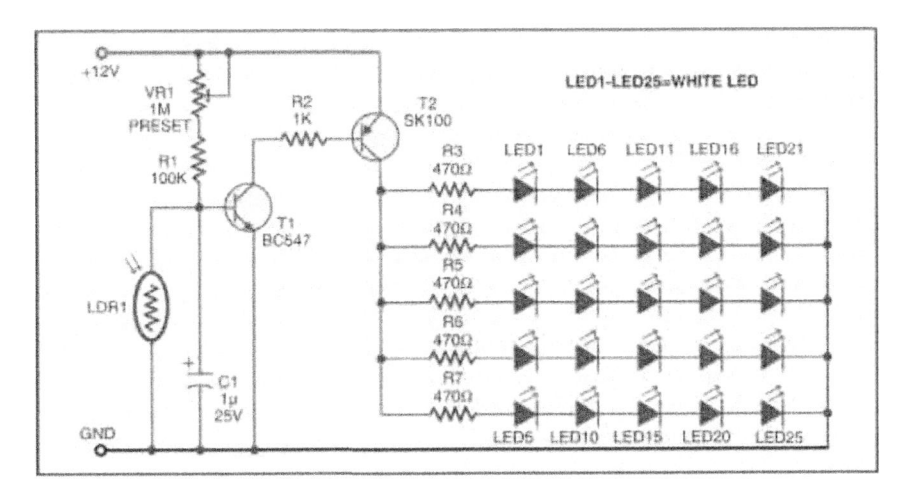

Put together the circuit using a PCB designed for general use, and then enclose it in a cabinet. On top of the box, connect the LEDs (numbered LED1 through LED25) as well as the LDR1. Position the machine so that, throughout the day, sunlight will fall directly on the LDR1 sensor. You can power the circuit with a 12V battery or any adaptor that can handle 12V.

254. PC Timer

This flexible PC timer with an adjustable time output is designed to be installed in a desktop PC. It controls how long PC peripherals like printers, scanners, and desk lamps stay on. Since it is made to work with 12 volts, it might also be useful in your lab. If you plan to use the circuit as a portable timer most of the time, it is best to put it in its own case. If the circuit is going to be used as a PC's internal timer, you could also build it into the PC's case (system box). It is easy to hold the circuit board in place with one of the metal panels that pull out from the back of the PC. What's left to do is drill holes in the metal bracket for the switch (S1), the indicator (LED1), and the connectors for the relay output.

Connect the circuit's input supply points to the PC's internal power supply cables to connect the input voltage. Any ATX SMPS 4-way drive live power connector that isn't being used (+5V/0V/0V/+12V) can be used to power the circuit. As shown in Fig, the PC's SMPS sends 9V to the circuit through the fixed three-terminal voltage regulator IC 7809 (IC1). Diode D1 protects against accidental polarity reversal. LED1 shows whether the power is on or off. A 14-stage ripple-carry binary counter CD4060 (IC2) and a quad 2-input NAND Schmitt CD4093 are the main parts of the circuit (IC3). After the time set by the R-C time constant components R2 and C5, the timer-controlled relay turns on. With the values of the parts shown (R2 = 470 kilo-ohms and C5 = 100 nF), the relay turns on and latches after 10 minutes. Change the value of R2 and/or C5 to change the time-out period. You can set the time by putting a preset in line with R2 and pin 10 of IC2.

PC timer circuit

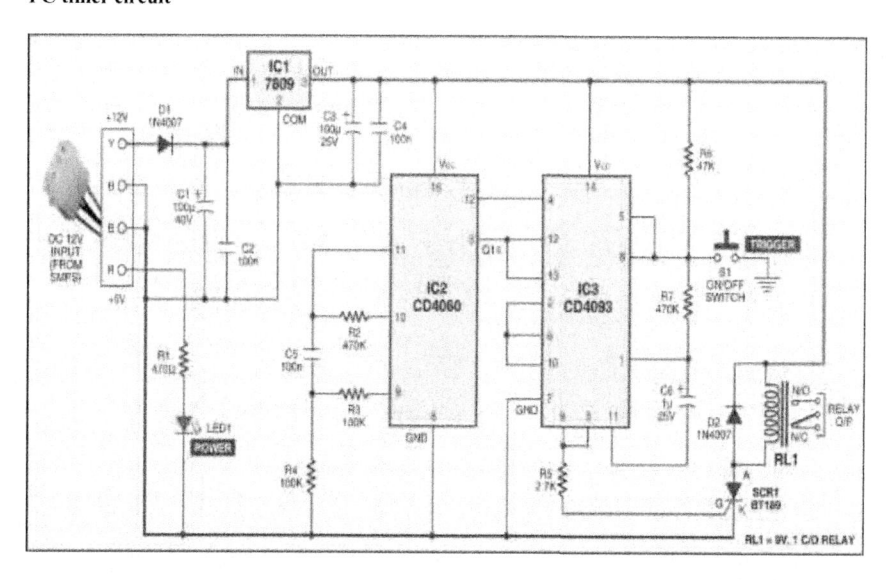

To start the timer, just briefly press the S1 switch. IC2 begins to oscillate, and after six seconds, the output of IC3 at pins 3 and 9 gives SCR1 the gate voltage it needs to turn on. So, the relay gets powered up and locks to turn on the connected load. The relay will stay latched until you press switch S1 again.

255. Infrared Object Counter

This infrared object counter can be put at the entrance gate to keep track of how many people are coming in. For instance, it can be used at bus stops or train stations to keep track of how many people arrive each day or week. A pair of infrared transmitters and receivers and a simple, low-cost calculator are used to make the counter. It works even when there is regular light around. The farthest it can find something is about 10 meters. That means that the transmitter and the receiver can't be more than 10 meters apart (at the opposite pillars of the gate). There is no need for a focusing lens. With an 8-digit calculator, the counter can easily reach 99,999,999, and with a 10-digit calculator, it can reach 9,999,999,999. The transmitter circuit (see Fig. 1) is powered by a 9V battery and has two infrared light-emitting diodes and an IC 555 wired as an astable multivibrator with a center frequency of about 38 kHz (LEDs). The 5V regulated power supply built around transformer X1, bridge rectifier with diodes D1 through D4, and regulator IC2 powers the receiver circuit (see Fig. 2). It has a simple calculator, an IR receiver module (RX1), and an optocoupler (IC3).

Circuit operation

When switch S1 is in the "on" position, the transmitter circuit turns on and sends out a square wave at pin 3. The two infrared LEDs (IR LED1 and IR LED2) connected to its output send out modulated IR beams at the same frequency (38 kHz). The frequency of the oscillator can be changed with preset VR1.

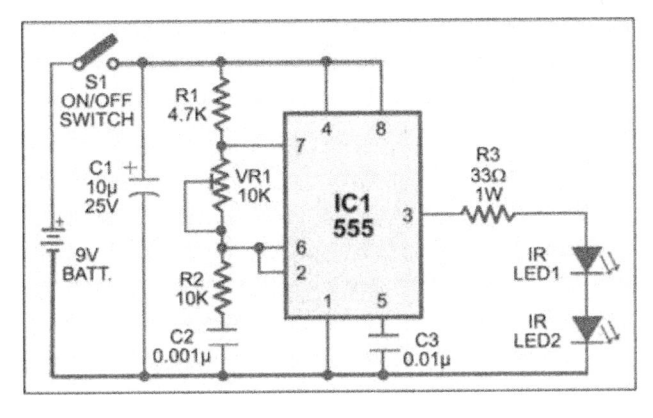

Fig. 1: Infrared Object Counter: Transmitter circuit

In the receiver circuit, the sensor is an IR receiver module TSOP1738, which is often used in color TVs to pick up the IR signals sent from the TV remote. When the IR beams from IR LED1 and IR LED2 hit the IR receiver module IR RX1 of the receiver circuit, pin 2 gets a low signal. This keeps transistor T1 in a state where it is not conducting.

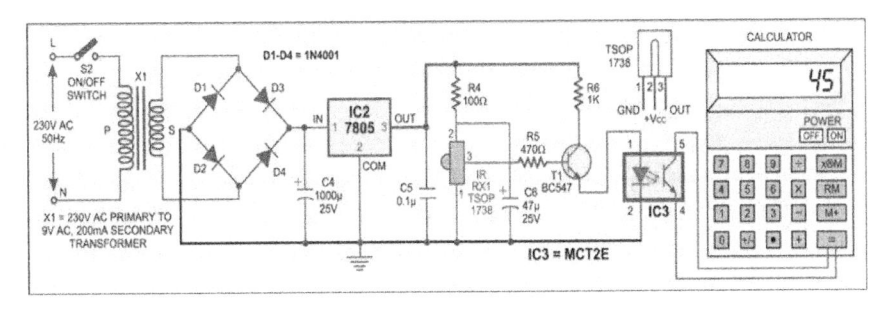

Fig. 2: Infrared Object Counter: Receiver-cum-counter circuit

Now, if someone walks through the gate and breaks the IR beam, the IR receiver module sends a high output pulse to pin 3 of the modules. So, T1 conducts to turn on IC3, and its internal transistor shorts out the '=' key on the calculator to move the count up by one.

256. Pushbutton Control for Single-Phase Appliances

With the use of two distinct push switches, this circuit gives you the ability to turn a single-phase appliance on and off. When the power and control circuits of industrial motors, which are almost always three-phase, need to be isolated from one another, a setup like this is frequently used. If the relay is located in a remote location, it will protect individuals from danger in the event of malfunctions. Additionally, the circuit protects expensive gadgets from being damaged by frequent power outages by having the device automatically turn off in the event of a power failure and remain off until it is manually turned back on again.

The relay's pickup voltage determines whether or not the device that is attached to the relay will activate when it is switched on. When the voltage from the mains supply is very low, the output across capacitor C1 drops to a level that is lower than that which is necessary to trigger the relay. When you press switch S2 in this scenario, the relay does not energize to supply the device with mains power since it is not in the correct state. As a result, the circuit provides some form of protection against low voltage. The voltage that is necessary for switching on relay RL1 is generated by a step-down transformer, a rectifier, and a filter capacitor working in conjunction with one another. The mains power supply can be disconnected from the load through the use of the push-to-off switch S1 (switch off the load). Switch S2 is a push-to-on switch, and it is used to connect the load to the mains power supply (switch-on the load). Back emf can be avoided by placing a freewheeling diode across the relay coil, while LED1 serves as an indicator of whether or not the relay has been energized.

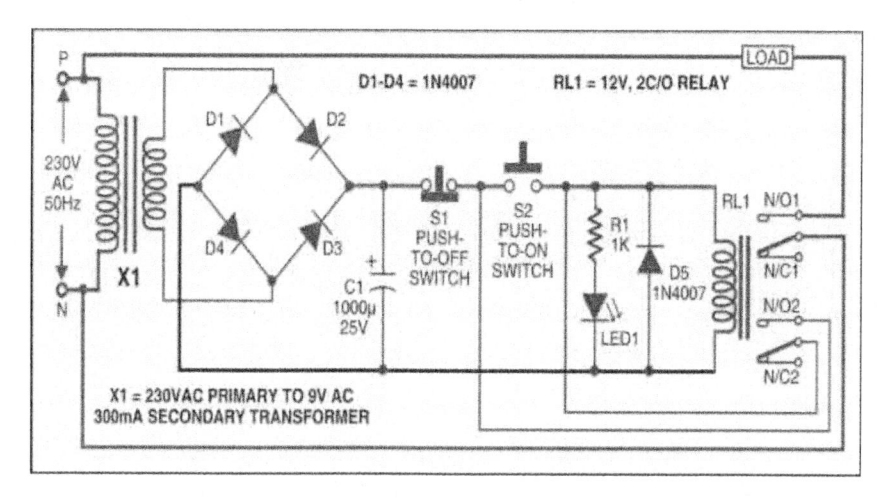

The operation of the circuit is straightforward. When you briefly press switch S2, current will flow through the relay coil, which will cause it to become energized. As a consequence of this, the relay's normally-open (N/O) contact will close, which will connect the load to the power. If the power goes out or you briefly press switch S1 to cut the current path of the relay, the relay will lose its ability to conduct current and will de-energize. Construct the circuit using a PCB designed for general use, and then enclose it in an appropriate cabinet. Install switches S1 and S2 on the front panel, and place the relay on the back side of the enclosure, so that the load can be connected.

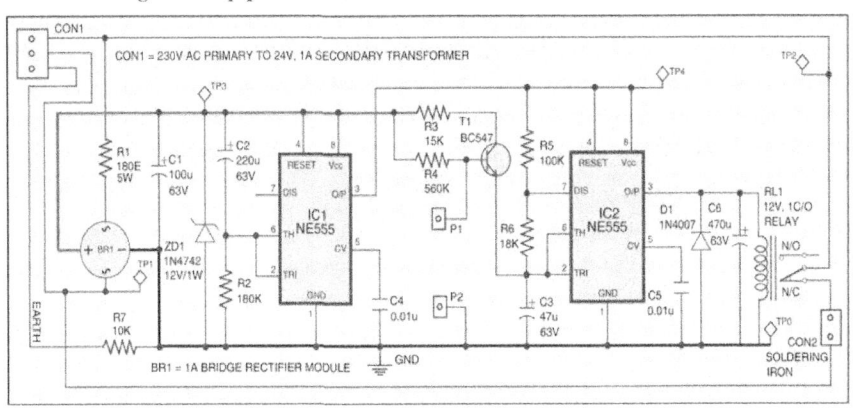

257. Soldering Iron Tip preserver

Although 60/40 solders melt at approximately 200 degrees Celsius, the temperature of the tip of a soldering iron should be approximately 370 degrees Celsius. It is vital to do this in order to swiftly establish a good connection without running the danger of scorching delicate components. When working with temperatures of this magnitude, it is important not to keep the point of the iron pressed on the joint for an excessive amount of time. Unfortunately, at this temperature, the tip oxidizes very quickly, and as a result, it needs to be cleaned often. This circuit can be of assistance in that regard since it maintains a temperature on the soldering tip that is slightly lower than 200 degrees Celsius even when the iron is not in use. When this occurs, oxidation is reduced to a negligible level, and the soldering iron can be brought back up to temperature in a matter of seconds whenever it is required. Due to the thermal inertia of iron, the temperature at the tip of the iron does not change during a regular soldering operation, even while the iron is allowed to return to a state of rest for only a little moment.

The circuit for the soldering iron tip preserver is depicted in Figure 1. The circuit is driven by two 555 timers that are referred to as IC1 and IC2. IC1 is wired in a monostable design, which allows for an initial warm-up time of around forty-five seconds. This allows the iron to reach the necessary soldering temperature. At the conclusion of this time period, its output pin (3) goes high, which makes it possible for IC2 (which is wired in an astable configuration) to turn the iron on by using relay RL1. At this location, the temperature rises rapidly for one second out of every six. When the iron is in the stand, the values of resistors R5, R6, and capacitor C3 determine the duty cycle. Electrical connections P1 and P2 are able to detect whether or not the iron is positioned properly within its stand. Because the base of T1 is drawn down when the iron is in its resting position, T1 does not conduct when the iron is in its resting position. On the other hand, if the iron is not on its stand, T1 will conduct, which will pull pins 2 and 6 of IC2 high and prevent it from functioning properly. During this time, pin 3 of IC2 is low, and as a result, the iron receives constant power through the normally-closed contact of relay RL1.

258. Overspeed Indicator

This speed indicator circuit was developed to indicate over-speed as well as the direction of rotation of the motor that is utilized in various micro hand tools, water pump motors, toy motors, and other types of appliances. A suitable mounting solution is used to attach a 12V DC motor, denoted by M1, to the part of the machine that does the rotation. A voltage is produced by the motor each time it completes a rotation. This speed indicator is constructed around an operational amplifier known as CA3140 (IC1). By setting preset VR1 at pin 2 of IC1, you can set the reference voltage (depending on the speed you want). When the voltage that is produced at pin 3 of IC1 is greater than the reference voltage that is produced at pin 2, the output pin 6 of the comparator IC1 becomes high, which causes the piezobuzzer PZ1 and the LED3 to light up.

Speed indicator circuit

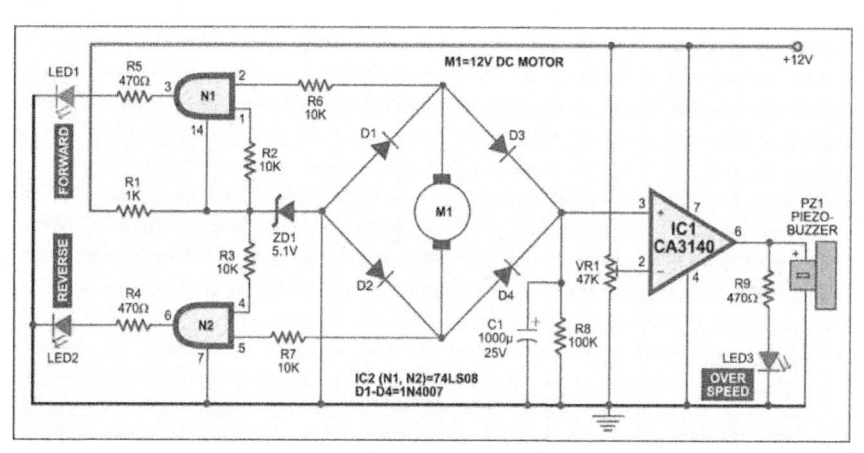

AND gate 74LS08 serves as the focal point of the rotation indication circuit (IC2). Pin 2 of gate N1 goes high when the motor is rotating in the forward direction, whereas pin 1 of gate N1 is pulled high by resistor R2 when the motor is rotating in the reverse direction. When both pins 1 and 2 are high, the output pin of gate N1 turns high, which causes LED1 to become lit. When the motor is turned counterclockwise, the same thing happens to pin 5 of gate N2: it goes high. When both pins 4 and 5 are high, the output pin of gate N2 turns high, which causes LED2 to become lit.

259. Automatic Washbasin Tap Controller

Set up your washbasin tap so that it turns on when your hands are just below the water outlet. This automatic washbasin tap controller is based on infrared technology. When your hands or a tool block the IR rays, water starts coming out of the tap automatically. The circuit is made up of 555 timers and has sections that send and receive signals. Both the sender and the receiver are powered by 5V DC. The transmitter always sends out IR rays, which fall on the receiver. When something gets in the way of the IR rays between the receiver and the transmitter, the output of the IR sensor briefly drops. This sets off a timer circuit in the receiver, which let's water out of the tap for eleven seconds.

The transmitter is built around the timer IC 555, which is used as an astable multivibrator to make a 38 kHz frequency (see Fig. 1).

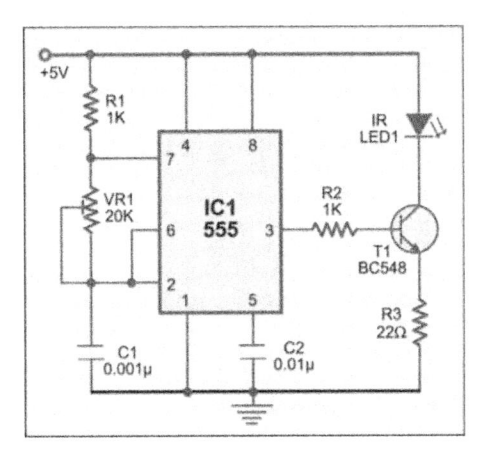

Fig. 1: Automatic Washbasin Tap Controller: Transmitter circuit

The output of the timer is sent to the transistor T1, which turns on the IR LED (LED1). Keep in mind that IR LED1 must be aimed at the IR sensor module of the receiver circuit. Its transmission wavelength is between 900 and 1100 nm, which is in the range where the TSOP1738 receiver module works best. The sensor module, monostable timer, and relay driver circuit are all parts of the receiver circuit (see Fig. 2). The TSOP1738 sensor module can pick up IR radiation that is changed at 38 kHz. When any IR radiation is detected or stopped, its normally high output drops for a short time.

Circuit operation

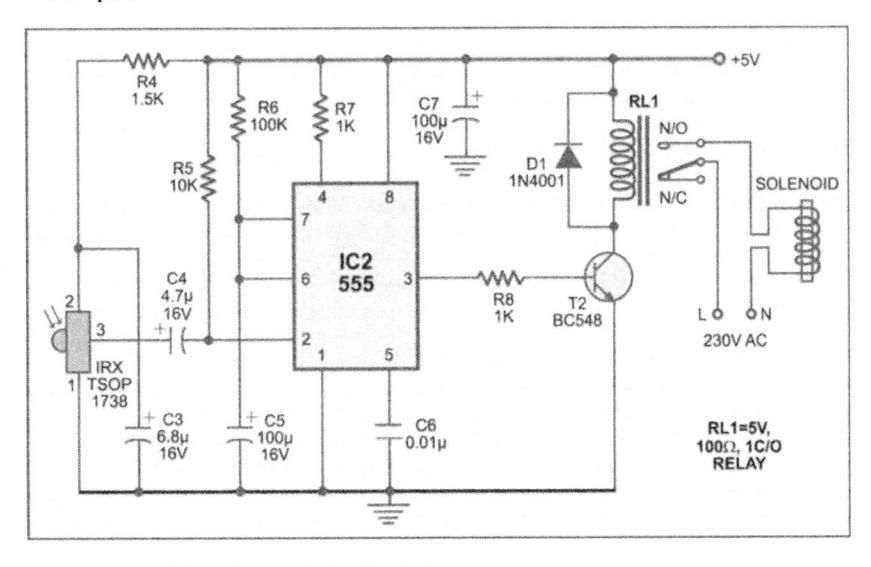

Fig. 2: Automatic Washbasin Tap Controller: Receiver circuit

When IR rays don't reach the receiver, the sensor output goes low for a short time, which turns on timer IC2. For eleven seconds, the timer's output is high, and the relay turns the solenoid. During this time, when the solenoid is turned on, it lifts the valve in the pipe, which let's water out of the tap. Solenoid valves are used for this specific job. The relay driver circuit is made up of diode D1, resistor R8, and transistor BC548 (T2). When the relay is turned off, the back emf creates high voltages that could damage the relay. This is where diode D1 comes in. Here's how you can figure out how long the timer will be on high: Ton=1.1 R6C5=1.1×100×103×100×10–6=11 seconds

260. 1.5W Power Amplifier

Here, we put all the theory to work and show you how to make a simple 1.5W power amplifier module with parts that are easy to find. Fig. 1 shows a block diagram of the amplifier. It's like most audio amplifiers, but the circuit is a little different. The audio input, the amplifier, the driver, the output, and the power supply are all parts of a power amplifier. Most of the voltage gain comes from the amplifier section. Between the amp section and the output stage is the driver stage, which acts as a buffer. Most of the time, the output stage has to drive a low-impedance load, like a speaker. Power comes from the power supply, and the signal that comes out of the load should, ideally, be the same as the signal that went in. In other words, the output stage gets power from a DC supply to boost the signal so it can drive a load.

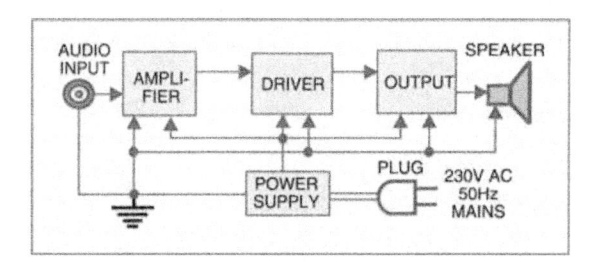

Fig. 1: Block diagram of 1.5W power amplifier

How a Power Amplifier Circuit Works

Figure 2 shows the parts of the circuit that are the amplifier, driver, and output. The JFET VHF/UHF amplifier 2N5484 (T1) and the NPN transistor BC548 make up the amplifier section (T2). The driver part is made up of the transistor BC639 (T3), and the output part is made up of the transistors BD139 and BD140 (T4 and T5). The signal from the input is sent to the volume control VR1 through the capacitor C1. It says that the value of VR1 is 1 megaohm. Since the gate terminal of the FET (T1) can be thought of as an open circuit, the value of VR1 is equal to the circuit's input impedance. Like all audio volume controls, VR1 needs to have a logarithmic taper, which is usually denoted as "type C," so that turning the control and the volume level seem to go in a straight line. Human hearing has a logarithmic response, which means that a change in output power by a factor of 10 sounds like a change by a factor of 2. In the amplifier section, the FET stage is used to make the input impedance high. The next step is a common emitter amplifier, which is made up of transistor T2. Set VR2 is used to change the amount of amplification and keep transistor stages T2 through T5 from getting too close to each other. This means that the collector voltage at T2 sets the DC voltage at T3, T4, and T5. The voltage at the emitters of T4 and T5 is the most important. This can be set to half the supply voltage with the help of VR2. Resistor R13 sends negative feedback from the output to the emitter of transistor T2 to stabilize this and other voltages in the circuit. If you don't include capacitor C8, the feedback will be for both DC and AC voltages. If C8 is added, it will only be for DC. When the voltage at the emitters of transistors T4 and T5 goes up, say because of a change in temperature, the voltage at the emitter of transistor T2 also goes up because of R13. This will make T2 pass less current, which will raise the DC voltage at its collector. So, transistor T3 will carry more current, and the voltage at its collector will drop. The voltages at the bases of T4 and T5 are then lower, and so are the voltages at their emitters.

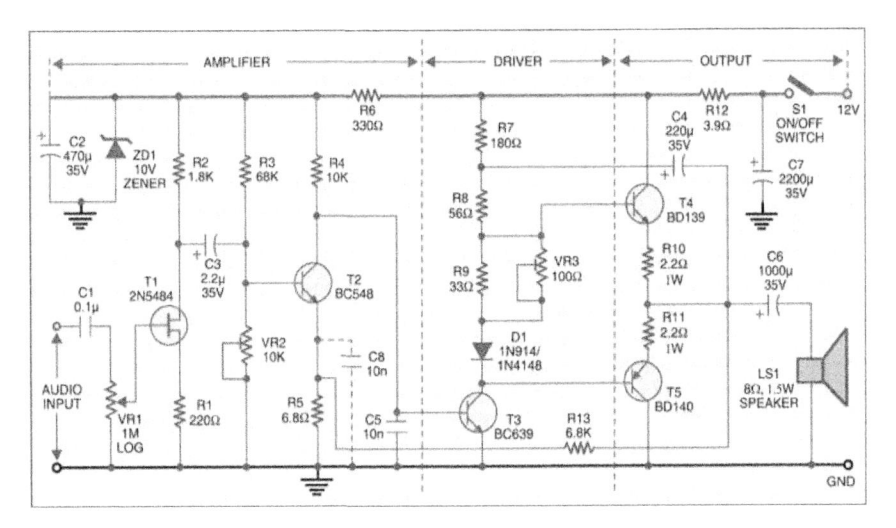

Fig. 2: 1.5W power amplifier circuit

The base circuitry for T4 and T5 is connected to T3's collector load, which is the driver transistor. In effect, T3 is set up as an amplifier with a common emitter. Through diode D1 and the parallel combination of resistor R9 and preset VR3, the output signal from T3 is sent to the base of T4. The collector of T5 is directly connected to the base of T3. So, since the driver stage drives a load with a low resistance, it needs a transistor that can handle a lot of power. The output transistors are the NPN transistor T4 and the PNP transistor T5, which are wired together as a class-AB output stage with complementary symmetry. In this set-up, you need a complementary NPN transistor and a symmetrical PNP transistor with the same amount of current gain. So, if you measure them, the DC current gains of T4 and T5 should be the same. One diode and two resistors connected in parallel make up the DC biasing circuit for T4 and T5. With VR3, you can change the quiescent collector current of T4 and T5 and, by extension, the operation class.

Capacitor C4 is also part of the output stage. It is called a bootstrapping capacitor. Bootstrapping is there so that the output voltage swing can be higher. If capacitor C4 is not used, biasing resistors R7 and R8 are combined into a single resistor. In an ideal situation, the output signal should be able to go from 0V to the supply voltage. But this can't happen because the base-emitter junctions of the output transistors need 0.6V of forward bias and there are losses. For the positive half cycle, the voltage at T4 must be at least 0.6V higher than the supply voltage if the output is to reach the supply voltage. In the same way, T5 can only give off 0V if its base voltage drops to –0.6V.

By adding the "bootstrap" capacitor C4, the swing in the output voltage is added to the DC bias voltages. So, during the positive half-cycle, the positive change adds to the bias voltage at T4, making it conduct more current and produce a higher output voltage. In the same way, during the negative half-cycle, the negative-going swing lowers the quiescent bias voltage. This makes T5 turn on harder and produce a lower output voltage. Power supply decoupling is an important part of designing an amplifier. When the output stage is putting out its maximum amount of power, the power supply sends out large peak currents. Under these circumstances, it's possible that some of the sound signals will show up on the power line.

So that this signal doesn't affect how the rest of the circuit works, it needs to be taken out of the part of the power supply that goes to the voltage amplifier. So, resistor R6 is added to capacitor C2 and diode ZD1 to keep the supply voltage of the amplifier at a steady 10V. Put the circuit together on a PCB that can be used for many things and put it in a suitable cabinet. Mount the diodes, electrolytic capacitors and transistors with the correct polarity. The metalized side of output transistors T4 and T5 should be facing the middle of the board.

For both of the transistors, you need a heat sink. Either a small piece of aluminum (20 mm2) or a heat sink that can be bought in stores can be used. The heat sinks on T4 and T5 should be separated from the transistors with a piece of Mylar or something similar, since this transistor (and, by extension, the heat sink) connects directly to the power supply.

261. Wireless Stepper Motor Controller

Infrared signals are used in this straightforward and easy-to-use wireless controller for stepper motors. With the help of this circuit, you will be able to exert control over the stepper motor from a distance of up to four meters. The circuit has both a transmitter and a receiver in its configuration. Infrared signals are used to carry out the communication that needs to take place between the transmitter section and the reception section.

Wireless Stepper Motor Controller Circuit

The timer NE555 integrated circuits (IC1 and IC2) in the transmitter part are set up to operate as astable multivibrators, and their frequencies are around 1 Hz and 38 kHz, respectively. Therefore, the 38 kHz carrier signal is modulated by a 1 Hz modulating signal because the output of IC1 is sent to reset pin 4 of IC2. The infrared LED is responsible for transmitting the modulated signal that comes from pin 3 of IC2. The amount of current that can go through the IR LED is controlled by resistor R5.

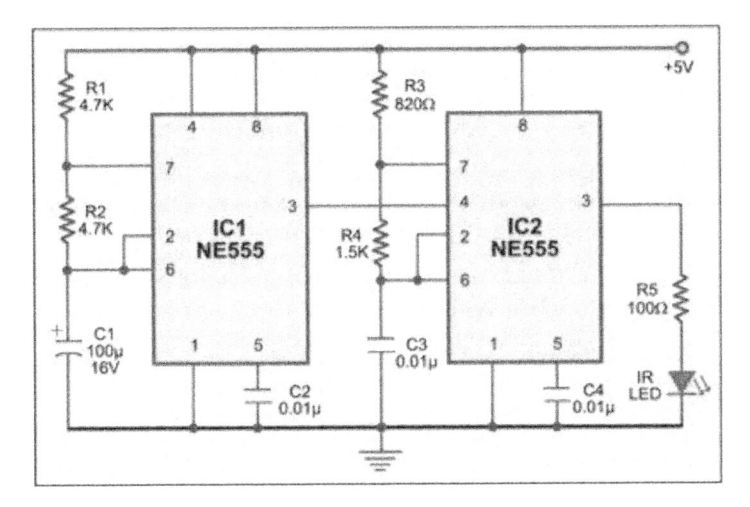

Fig1. Wireless Stepper Motor Controller: Infrared transmitter

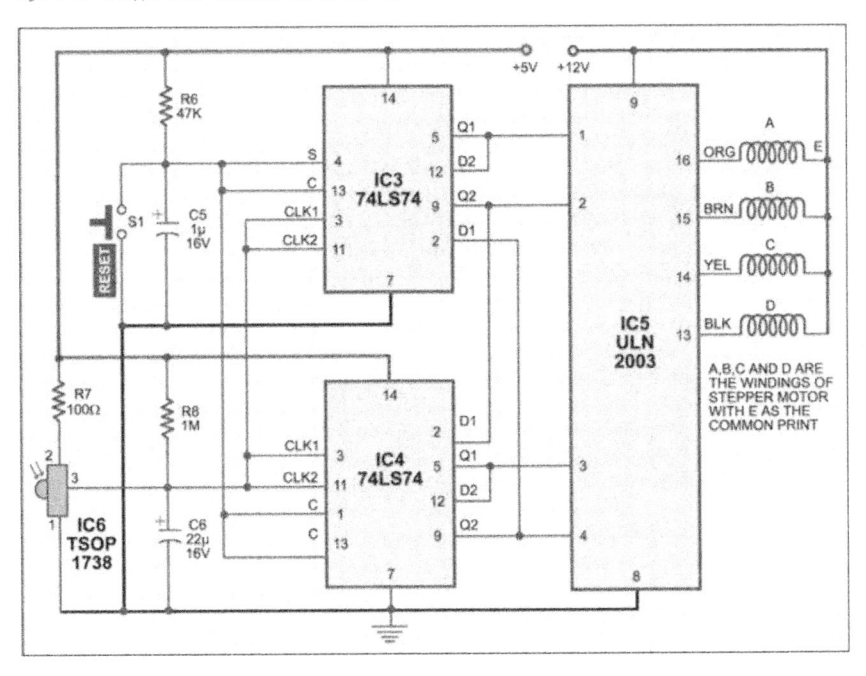

Fig2. Wireless Stepper Motor Controller: Infrared receiver and stepper motor driver circuit

The received signal is detected by the infrared receiver module TSOP1738 (IC6) of the receiver section, and the output of this module, which is located at pin 3, is used to provide clocks for the dual flip-flop 74LS74 Ics (IC3 and IC4), which are set up in the configuration of a ring counter.

When the power is turned on, the first flip-flop is initialized, and the Q1 output of that flip-flop goes high. In contrast, the other three flip-flops are reset, and the outputs of those flip-flops go low. When the first clock pulse is received, the high output of the first flip-flop is transferred to the second flip-flop. This occurs immediately. Therefore, upon receipt of each clock pulse, the high output will continually move in a ring-like pattern. The outputs of the flip-flops are connected to the stepper motor windings that are labeled 'A' through 'D' after being amplified by the Darlington transistor array included within ULN2003 (IC5). The point in the windings that is common to all of them is connected to a source of +12V DC. Flip-flops can be manually reset by hitting the S1 reset switch, which will result in the motor coming to a stop. The stepper motor immediately begins moving once more upon the release of the reset switch. The motor will stop working if there is any disruption in the connection between the transmitter and the receiver.

262. Battery Low Indicator

It is not safe to discharge rechargeable batteries below a specific voltage level. This lower limit on voltage is different for each type of battery. This simple battery low indicator circuit can be used with 12V batteries to show when the voltage drops below the value set by the user. A flickering LED is used to show what is going on.

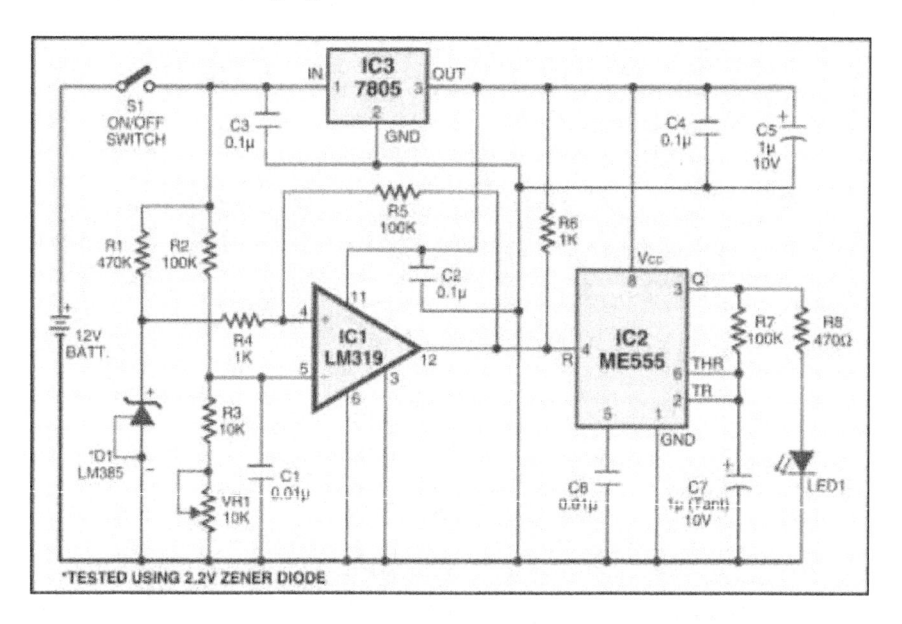

Voltage comparator IC LM319 is the heart of the circuit (IC1). It is a dual comparator that has an output that works with TTL. We have only used one comparison in this case. A reference voltage of 1.2 volts is sent to the comparator's non-inverting input (pin 4) from the band-gap reference diode D1 (LM385). A voltage is sent to the comparator's inverting input (pin 5) from the potential divider made up of resistors R2 and R3 and the preset VR1. So, if you are using a 12V battery and want to know when the voltage drops below 10.5V, adjust the voltage at the inverting input using reset VR1 to get a voltage of 1.2 volts (with battery voltage at 10.5V). When the battery is first fully charged, the voltage at the inverting input of IC1 is higher than the voltage at the non-inverting input, so IC1's output pin 12 stays low. The reset pin (pin 4) of IC2 is connected to pin 12 of IC1 and stays low. The astable multivibrator built around IC2 does not oscillate. Because of this, LED1 doesn't blink. When the battery voltage drops below 10.5V, the voltage at the inverting input of IC1 becomes lower than the voltage at the non-inverting input, and the output of IC1 goes high. When pin 12 of IC1 goes high, the reset pin of IC2 also goes high, and the astable multivibrator built around IC2 begins to oscillate. When LED1 flickers, it means that the voltage of the battery is low and that the battery needs to be charged before it can be used again. Both IC1 and IC2 run on +5V DC that is controlled by IC 7805's voltage regulator (IC3). Put the circuit together on a PCB that can be used for many things and put it in a suitable cabinet. On the front of the case, put LED1 and switch S1. Connect a 12V battery to a voltmeter to see how much power it has.

263. Speed Checker for Highways

When driving on highways, drivers shouldn't go faster than their vehicle's top speed limit. But drivers keep getting into accidents because they don't pay attention to their speedometers. This speed checker will be useful for highway traffic police because it will not only show the speed of a car on a digital screen, but it will also sound an alarm if the car goes faster than the speed limit for the highway. The system is mostly made up of two sets of a laser transmitter and an LDR sensor. Each set is put on the highway 100 meters apart, with the transmitter and the LDR sensor on opposite sides of the road for each pair. Fig. 1 shows how lasers and LDRs are put together. The system shows, with a resolution of 0.01 second, how long it takes the vehicle to go from one pair of sensors to the next. From this, you can figure out how fast the vehicle is going:

$$\text{Speed (kmph)} = \frac{\text{Distance}}{\text{Time}}$$

$$= \frac{0.1 \text{ km}}{(\text{Reading} \times 0.01)/3600}$$

or,

$$\text{Reading (on display)} = \frac{36000}{\text{Speed}}$$

Based on the above equation, the display will show 900 (or 9 seconds) for a speed of 40 km/h and 600 for a speed of 60 km/h (or 6 seconds). Note that the LSB of the display is 0.01 second and that each digit after the first one is ten times the one before it. You can also figure out the other readings in the same way (or time).

Speed checker circuit diagram

In the picture below, you can see how the speed checker works. It was made with the assumption that the speed limit on highways is either 40 km/h or 60 km/h, depending on the rule. Five NE555 timer ICs (IC1–IC5), four CD4026 counter ICs (IC6–IC9), and four 7-segment displays make up the circuit

(DIS1 through DIS4). IC1 through IC3 all work as monostables. The outputs of IC1 and IC2 control IC1 as a count-start mono, IC2 as a count-stop mono, and IC3 as a speed-limit detector mono. The outputs of IC1 and IC2 also control the set-reset bi-stable timer IC4, which in turn controls the on/off switching of the 100Hz (period = 0.01 second) astable timer IC5.

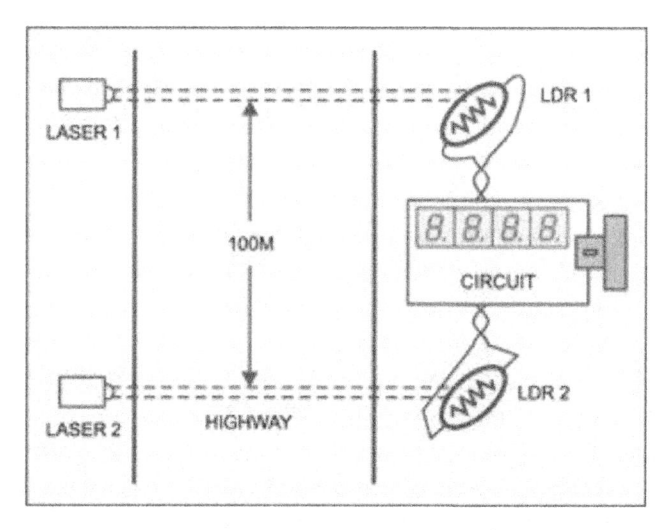

Installation of lasers and LDRs on highway

Circuit description

The preset VR1 or VR2 and capacitor C1 can be used to change the time period of the count-start monostable multivibrator timer NE555 (IC1). For a speed limit of 40kmph, preset VR1 sets the time limit to 9 seconds. For a speed limit of 60kmph, preset VR2 sets the time limit to 6 seconds. Slide switch S1 is used to set the time according to the speed limit (40 kmph and 60 kmph, respectively). Pin 2 of IC1 is connected to the point where LDR1 and R1 meet. Usually, the laser light keeps shining on the LDR sensor, so the LDR has a low resistance and pin 2 of IC1 is high. When a car blocks the light from getting to the LDR, the LDR resistance goes up, which makes pin 2 of IC1 go low and set off the monostable. Because of this, output pin 3 goes high for the amount of time that was set (9 or 6 seconds), and LED1 lights up to show this. When the power is turned on or when the reset switch S2 is pressed, pin 4 is controlled by the output of NAND gate N3. For IC2, the monostable is set off in the same way as for IC1 when the vehicle crosses the laser beam hitting LDR2. This makes a small pulse that stops counting and is used to figure out how fast the vehicle is going. LED2 lights up for as long as pin 3 of IC2 is high. The outputs of IC1 and IC2 are connected to NAND gate N1's input pins 2 and 1. When the outputs of IC1 and IC2 both go high at the same time, which means the car has gone over the set speed limit, pin 3 of gate N1 goes low, which turns on monostable timer IC3. The output of IC3 is used to power the piezobuzzer PZ1, which sounds an alarm if the driver goes over the speed limit. The piezobuzzer sounds for a certain amount of time, which is set by the resistor R9 and the capacitor C5.

How a circuit works

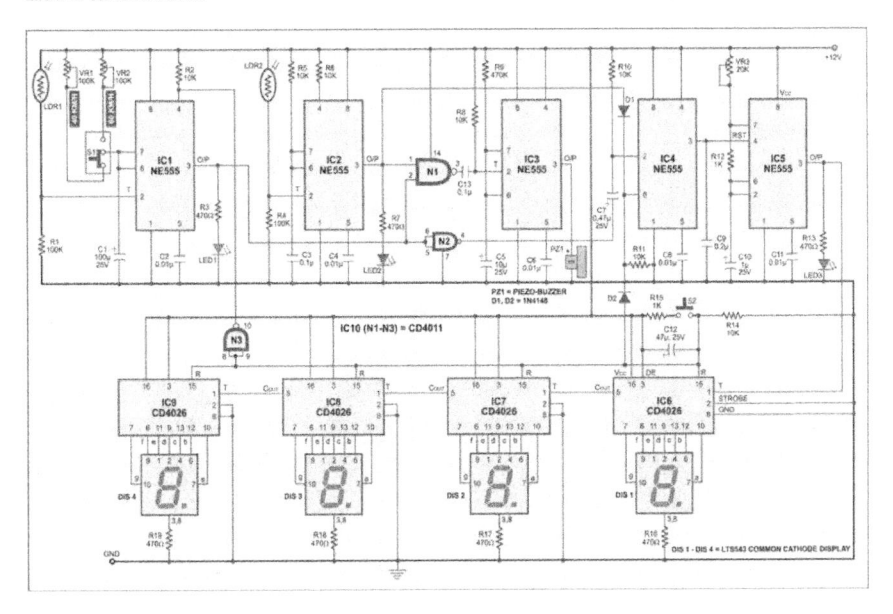

Circuit of speed checker for highway

At the beginning of the count-start pulse, the output of IC1 turns on the bistable (IC4) through gate N2. When IC4's pin 2 goes low, the high signal on its pin 3 turns on IC5's astable clock generator. Since IC2's count-stop pulse output is linked to IC4's pin 6 through diode D1, it resets IC5's clock generator. IC5 can also be reset by diode D2 when the power is turned on or by pressing reset switch S2. IC5 is set up as an astable multivibrator, which means that VR3, R12, and C10 are used to set the time period. Using preset VR1, the astable multivibrator is set to have a frequency of 100 Hz. The output of IC5 goes to the clock pin 1 of IC6 CD4026, which is a decade counter/7-segment decoder.

The IC CD4026 is a 5-stage Johnson decade counter and an output decoder that turns the Johnson code into a 7-segment decoded output for driving the DIS1 display. At each positive clock signal, the counter moves forward by one. In a chain of decade counters, the carry-out (Cout) signal from the CD4026 sends one clock to the next decade counter after every ten clock inputs. To do this, pin 5 of each CD4026 is connected to pin 1 of the next CD4026. When the reset signal is high, the decade counter is reset to 0. When the switch S2 is pressed, pin 15 of all CD4026 ICs, as well as IC1 and IC4, receive a reset signal. The power-on-reset signal is made by capacitor C12 and resistor R14.

264. Simple Stereo Level Indicator

Most low-cost home stereo power amplifiers don't have indicators for the level of the output. Each channel of these stereo power amplifiers can have an output power level indicator added to it. Low levels of output power don't bother or hurt people, so you don't need to add a preamplifier and low-level detector before IC LM3915. But you should know when the power output gets to be a lot. Here, we show you a very simple, low-cost stereo-level indicator circuit for home power amplifiers with a power rating of about 0.5W. Two LM3915 dot/bar display driver ICs are the main parts of the circuit (IC1 and IC2). The LM3915 measures analog voltage levels and drives ten LEDs to make a logarithmic 3dB/step analog display.

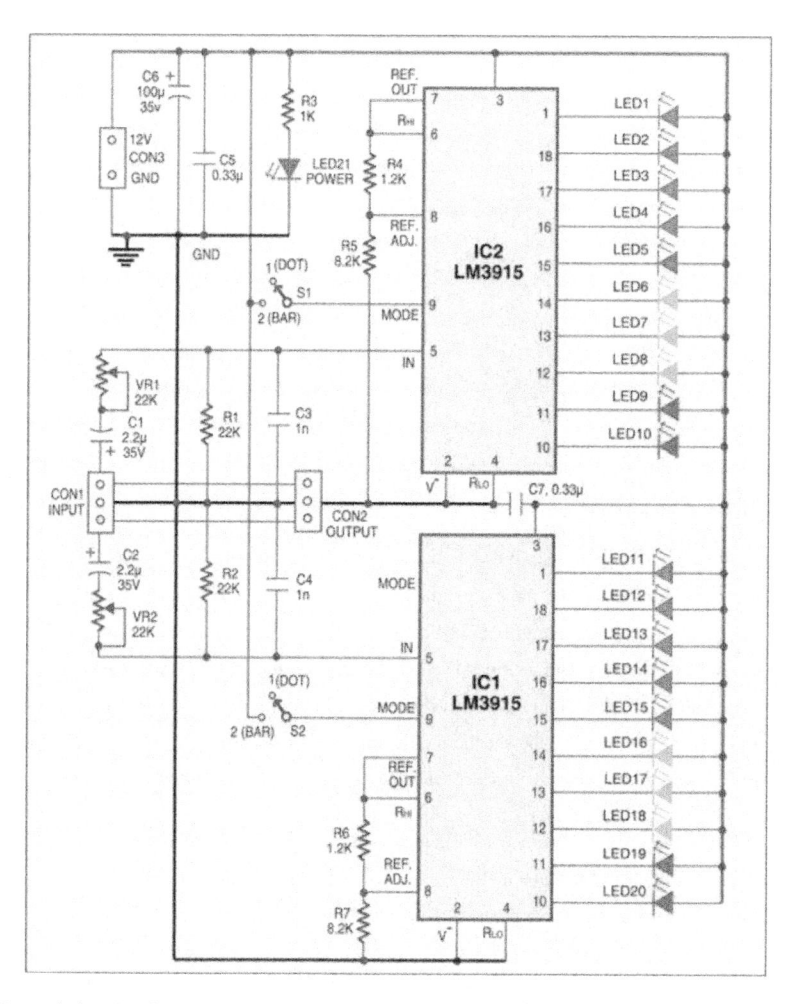

Voltages below 1V don't matter because they represent low levels of the audio signal. In the same way, input voltage levels above 30V result in output power levels that are too high for home power amplifiers. So, we're interested in voltage levels from 1V to 30V, which LM3915 can handle right away. The circuit is easier to build because the LM3915 does not need protection against inputs of up to 35V. Most audio power amplifiers can power loads from 2 ohms to about 32 ohms. A load of a few thousand ohms won't change the way the amplifier works. CON1 is the input connector for the speaker or headphones, and CON2 is the output connector. Each channel has its own LM3915 and ten LEDs that show how much power is being used. You can choose between three different colored LEDs to show the different sound levels. For instance, you could have 5 green LEDs, 3 yellow LEDs, and 2 red LEDs. If you don't have the right signal generators and measuring tools, you can calibrate the level indicator based on what you see. For example, the audio signal should be in the green LEDs zone when it is loud enough but not annoying, in the yellow zone when it is annoying or starting to get distorted, and in the red zone when it is heavily distorted or too loud to listen to in the room. Potentiometers VR1 and VR2, which are not required, can be used to calibrate. With the S1 and S2 switches, you can choose between the bar mode and the dot mode. If you don't need them, you can take them off or replace them with jumpers. Leave pins 9 of IC1 and IC2 open when the switches are taken out. The circuit works with 12V that has been fixed. You can also power it with a 12V battery that you can charge. Put the circuit together on a PCB that can be used for many things and put it in a suitable cabinet. Put all of the LEDs in two rows on the front of the cabinet. Attach the two ports for input and output to the back of the cabinet as well.

265. Manual EPROM Programmer

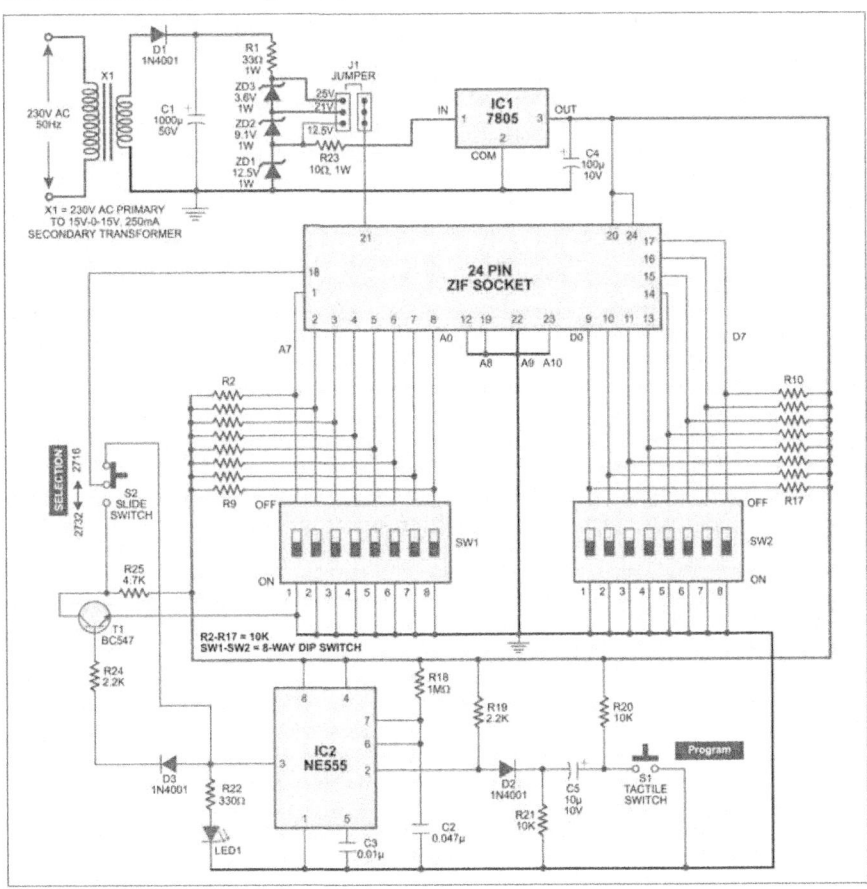

Most electrically programmable read-only memories (EPROMs) cost a lot to program because they need special devices called programmers. Here is a low-cost circuit for programming binary data into EPROMs 2716 and 2732.

EPROM Programmer Circuit

The timer NE555 (IC2) is wired as a monostable for the circuit. When the push-to-on switch S1 is pressed, IC2 makes a 50ms pulse that is sent through switch S2 to the program pin 18 of the ZIF SOCKET. For programming, the EPROM is put into the ZIF SOCKET. When the programming pulse is sent to the EPROM, LED1 lights up to let you know. Before sending the programming pulse to the EPROM, use jumper J1 to choose the programming voltage (25V, 21V, or 12.5V, depending on what the manufacturer says). This voltage will be sent to pin 21 of the ZIF SOCKET. The voltage that an EPROM needs to be programmed with is sometimes written on its body. The address and data for the EPROM (ZIF SOCKET) are set by the DIP switches SW1 and SW2, whose pins are initially pulled high by 10-kilo-ohm resistors.

The AC mains are stepped down by transformer X1 so that the secondary gives out 30V and 250 mA. The secondary output is blocked by capacitor C1 and turned back on by diode D1. With the help of Zener diodes ZD1, ZD2, and ZD3, programming voltages of 25V, 21V, and 12.5V are made. IC1 is used to give the circuit a steady +5V supply.

266. Noise Muting FM Receiver

When a frequency modulated (FM) receiver is tuned to an FM radio station's frequency, there is a lot of "hissing" noise in between the stations, which is very annoying for the operator and, therefore, not a good thing. Digital FM receivers don't have this problem because the output is automatically cut off when there are gaps between stations. Digital FM receivers, on the other hand, are much more expensive than their analogue counterparts. Analog FM receivers use an LC tuning system that makes "hiss" noise at the output. This noise doesn't get cut off when the receiver is tuned off-station. But if the received signal level is good (i.e., if the receiver frequency is close to that of an FM station), the limiter circuit that comes before the ratio detector circuit in an analog FM receiver will cut off any noise that is riding the FM carrier. The simple noise-muting FM receiver circuit described here stops this "hissing" noise from coming out of the output of an analog FM receiver circuit when it is not tuned to the frequency of an FM station. In other words, when the tuner circuit of an analog FM receiver is switching between FM stations, this circuit mutes the output of the receiver.

The circuit for the noise-muting FM receiver is built around a Sony CXA1619S AM/FM receiver chip, which comes in a 30-pin PDIP package. Sony FM receiver chips are known for having better features and being more affordable. Fig. 1 shows a block diagram of how the chip works. Fig. 3 shows the whole circuit of the FM receiver, including the power supply and the circuit that mutes the sound.

Most FM kits on the market also use the Sony CXA1619S, which is an AM/FM receiver IC that doesn't need many extra parts. It can work with as little as 3V of power. Here, we've used a 6V supply so that we can connect the receiver's output directly to a 32-ohm headphone or a low-power AF amplifier built around IC LM386. A 3-pin jumper is there to let you choose what you want.

Circuit Description

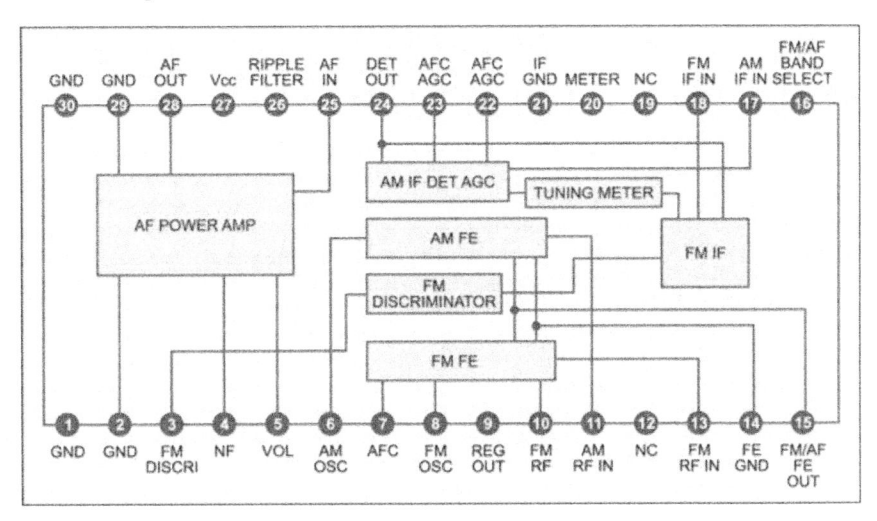

Fig. 1: Functional block diagram of Sony CXA1619S

The tuning meter output at pin 20 of the CXA1619S is used by the circuit that mutes the sound. Fig. 2 shows the chip's internal circuit around pin 20. Usually, a light-emitting diode is used to show that the tuning is correct (LED). The cathode of the tuning indicator LED is connected to pin 20, and the anode is connected to the positive supply through a current-limiting resistor.

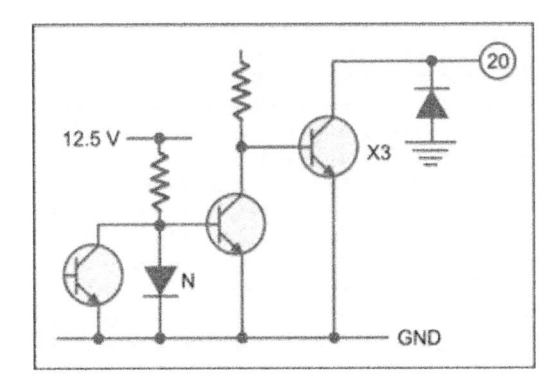

Fig. 2: Internal circuit schematic around pin 20 of CXA1619S

When the receiver is tuned to the center of the FM carrier, pin 20 of the CXA1619 goes very low and the LED lights up the brightest.

Circuit connections

Diode D1 was used to connect pin 20 of the CXA1619S to the input of a CMOS inverter gate instead of an LED. In the "off" tuning condition, when the receiver is not tuned to any FM station, input pin 1 of CMOS inverter gate N1 (1/6 CD40106) is biased to 2/Vcc using a potmeter arrangement made up of resistors R6 (33 kilo-ohms) and R7 (68 kilo-ohms). So, between FM stations, the output pin 2 of the inverter gate stays low, so transistor T1 (BC548) stays off and the 5V inverter relay connected to its collector stays off. But when an FM station is selected, the receiver's output pin 20 and the input pin 1 of the CMOS inverter gate both go low. So, the output of the inverter gate goes high, which makes transistor T1 turn on the relay.

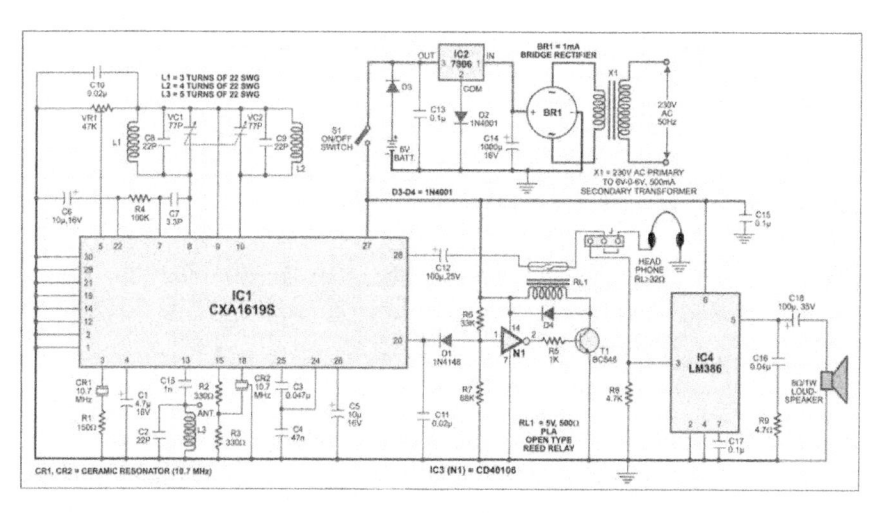

Fig. 3: Complete circuit of the noise muting FM receiver including the power supply

The audio output from the CXA1619S, which is available at pin 28, is sent to either the headphones or the AF amplifier circuit via the 100F capacitor C12 and the contacts of the reed relay. This happens when the relay is energized, which only happens when the receiver is tuned to the carrier frequency of an FM station. So, the annoying noise the receiver makes when it is tuned to a "off" station doesn't come out of the headphones or the loudspeaker. In other words, the noise is muted when the receiver is not tuned to an FM station because of the way the circuit is set up around the CMOS inverter gate.

Power supply: The output of IC 7806 is used to power the inverter gate N1 (IC3), transistor T1, and FM receiver chip/kit. This keeps the voltage in the whole circuit at a constant level. If the mains power goes out or if the device needs to be portable, four 1.5V cells can be used instead. Diode D3 in the power supply circuit makes sure that the battery output is not used when the mains power is on.

Antenna: You can use either a 75cm telescopic antenna or a wire of the same length as an antenna.

Circuit operation; When there is no FM station being picked up, diode D1 doesn't conduct, so pin 1 of inverter gate N1 stays high and pin 2 of the inverter stays low. So, transistor T1 doesn't work, and relay RL1 stays off. So, whether the load is a headphone or an audio power amplifier stage driving a loudspeaker, as shown in Fig. 3, no sound will come out of the headphones or loudspeaker. In other words, the CXA1619S doesn't play sound when the receiver isn't tuned to an FM station's frequency.

267. PC Based Stepper Motor Controller

This PC-based stepper motor controller is probably the smallest, least expensive, and easiest to use. A bipolar stepper motor with a step resolution of 18 degrees per pulse is controlled by a pair of H-bridges and a software program written in "C++." The PC-based stepper motor controller consists of both a driver circuit and a switching circuit. The stepper motor is driven by a circuit called the "driver," and the switching circuit decides how the motor is driven. So, the switching circuit is the main thing that controls the motor. The switches are the transistors (T1 through T8). Through the data pins D0 through D7, the software controls how these transistors turn on and off.

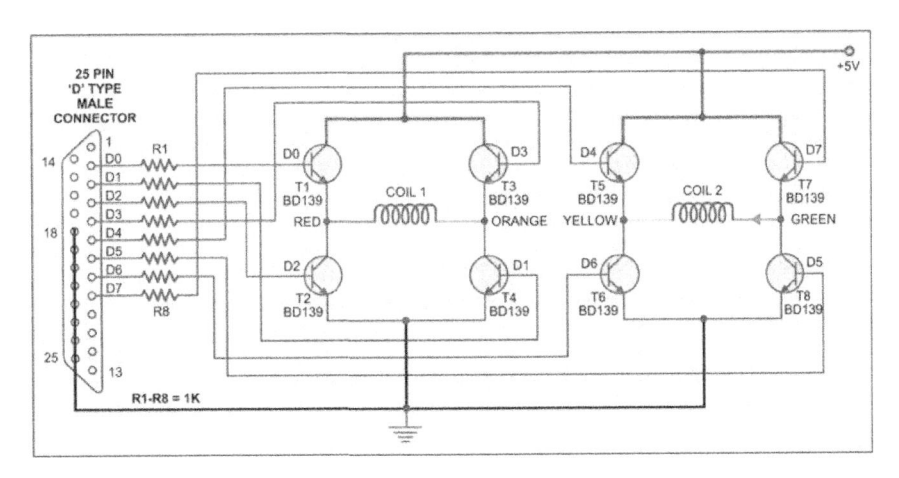

Fig. 1: Circuit of PC based stepper motor controller

You can change the speed, the direction, and the number of steps of a stepper motor. To change the speed of the motor, you have to change how often the pulses come out (PRF). To change the direction of the motor, you have to change the order of the pulses that are sent to its coils. By limiting the number of pulses that are sent to the motor, you can make it move only the number of steps you want.

Details about the stepper motor

Stepper motors with different ratings and specifications can be bought on the market and used for different tasks. Here, the 8.9cm (3.5-inch) floppy drive stepper motor is used. It is a bipolar stepper motor with a 5V DC rating and an 18o per pulse step resolution. The motor has two coils on the inside and four terminals on the outside, which are sometimes color-coded. Stepper motors with different step sizes (like 1.8o per pulse) can also be used with this circuit and control software. These motors should be rated for 5V and up to 1A of current.

Circuit description

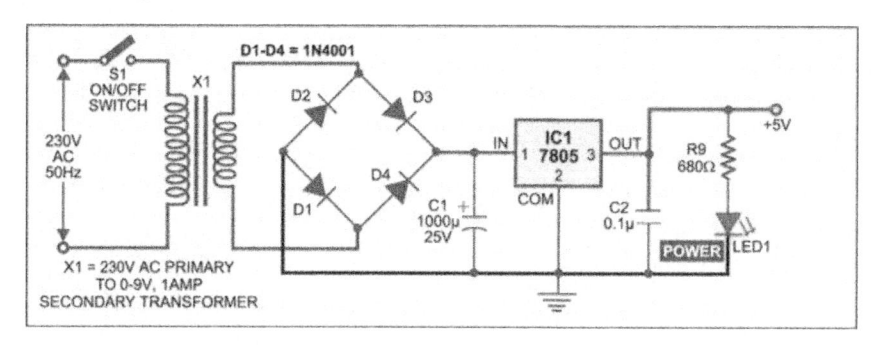

Fig. 3: Power supply for the circuit

H-Bridge driver: H-Bridge is a common, well-known circuit that is often used to drive a stepper motor. There are four transistors connected in a bridge (see Fig. 1). The bipolar stepper motor has two coils, so it uses two H-bridge circuits, one for each coil. The transistors T1 through T4 make up one H-bridge, and the transistors T5 through T8 make up the other bridge. The BD139 type transistors T1 through T8 should be used with heat sinks. Figure 2 shows the pins of BD139 and the regulator IC 7805. Through current-limiting resistors R1 through R8, the bases of all eight transistors are connected to the data pins (D0 through D7) of the 25-pin, D-type male connector.

Through resistors R1 and R2, the bases of transistors T1 and T4 are connected to parallel-port pins 2 (D0) and 3, and the bases of transistors T2 and T3 are connected to parallel-port pins 4 (D2) and 5, respectively. As shown in Fig. 1, the red and orange ends of the first coil (COIL1) are connected to the first H-bridge. Resistors R5 and R6 connect the bases of transistors T5 and T8 to pins 6 (D4) and 7 (D5), respectively. Resistors R7 and R8 connect the bases of transistors T6 and T7 to pins 8 (D6) and 9 (D7), respectively. As shown in Fig. 1, the yellow and green ends of the second coil (COIL2) are connected to the second H-bridge.

Getting power; Fig. 3 shows the part that has to do with the power supply. It has a 230V AC to 9V AC, 1A secondary transformer (X1), a filter, bridge rectifiers, and a 5V DC regulator (7805). (IC1). The H-bridge circuits are linked to the 5V DC that has been regulated. The ground of the circuit is shorted to pins 18 through 25 of the D-type parallel-port connector. When the switch S1 is shut, LED1 lights up to show that there is power in the circuit.

Operation: To make the motor turn either clockwise or counterclockwise, a specific sequence of pulses is sent to the red and orange terminals of COIL1 and the yellow and green terminals of COIL2. This is explained in the next paragraph.

Direction control: In Tables I and II, a "0" means that the logic is low and a "1" means that the logic is high. We know that the direction of the current is from high to low. To change the direction of rotation, all you have to do is change the way the current flows through the coils.

Handle speed: To change the speed, you have to change the number of times a pulse is sent out (PRF). With a PRF of 20 Hz, the stepper motor will get 20 pulses in one second. Since the motor's step resolution is 18o/pulse, it will turn 20 times 18o, which is 3600, or one full turn, in one second. So, the motor turns once every second, or 60 times per minute. Now, if you double the PRF from 20 Hz to 40 Hz, the RPS will also double to 2 RPS (120 RPM).

268. Digital Audio/Video Input Selector

Do you want to hook up more than one audio/video (AV) source to your color TV? Here's an AV input expander for your TV, so don't worry. It is cheap and easy to put together. The way the circuit works is simple and easy to understand. If 12V DC is put into the circuit, the power-on LED1 will light up. Now, press the switch S2 for a moment to make the Q0 output of IC1 high and reset the decade counter. When LED2 lights up, it means the circuit is ready to go. With switch S1, you can choose a certain audio-video (AV) signal. Press switch S1 once to choose the first AV signal. Press switch S1 twice to choose the second AV signal. You can choose the other two signals in the same way.

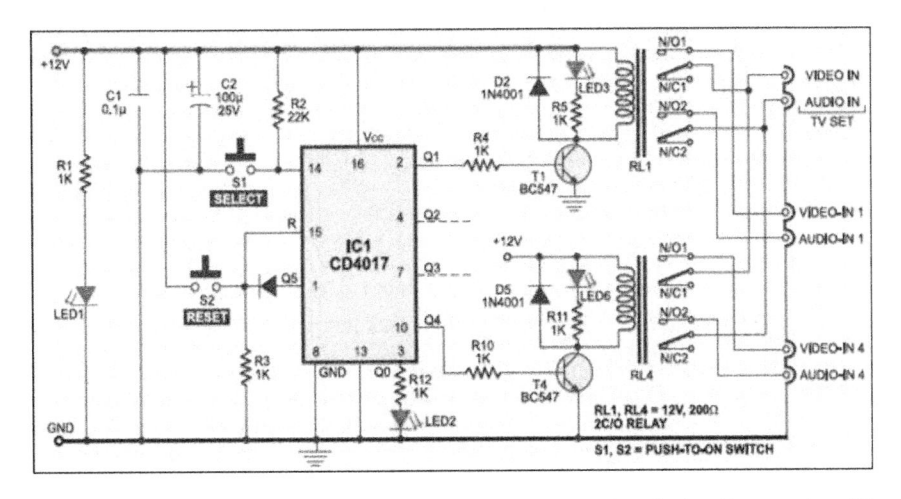

When switch S1 is pressed briefly once, the decade counter is started, and relay driver transistor T1 conducts to turn on relay RL1. Now, two-changeover relay RL1's normally open (N/O) contacts connect the TV's inputs to the first AV signal (marked as Video-In 1 and Audio-in 1). This is shown by the glow of LED3. When the switch S1 is pressed twice, IC1's Q2 output goes high. So, 2C/O relay RL2 (not shown in the circuit) is turned on, and the second AV signal is connected to the TV inputs (not shown in the figure). This is shown by the glow of LED4, which isn't shown in the figure. In the same way, when you press switch S1 three times, the Q3 output of IC1 goes high. So, 2C/O relay RL3 (which isn't shown in the figure) gets turned on, and the TV inputs are connected to the third AV signal source. This is shown by the glow of LED5, which is not shown in the picture. Again, when you press switch S1 four times, IC1's Q4 output goes high. So, 2C/O relay RL4 is turned on, and the fourth AV signal source is connected to the TV inputs (marked as Video-in 4 and Audio-in 4). This is shown by the glow of LED6. When you press switch S1 again, the decade counter starts over and LED2 lights up again. After that, the cycle starts over again. The circuit is set up to accept four inputs, so the Q5 output of IC1 is connected to IC1's reset pin 15.

269. Automatic Bathroom Light with Back-up Lamp

There are instances when we are unable to remember to turn off the light in the restroom, and as a result, it stays on for extended periods of time. This automatic bathroom light circuit stops electricity from being wasted by turning the lamp off after 30 minutes. When the main power goes out, the back-up LED lamp stays on for three minutes. This is especially helpful if you take a shower at night.

Automatic bathroom light circuit

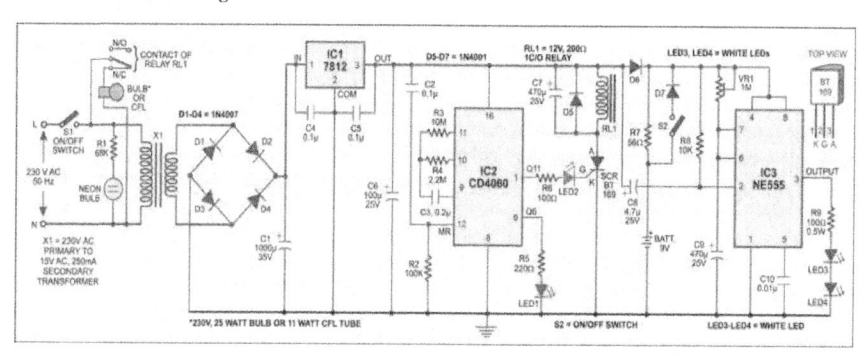

The circuit is constructed on a binary counter called CD4060 (IC2), which features an oscillator that is built right in as well as 14 cascaded bistable multivibrators. Based on the values of resistors R3 and R4 and capacitor C3, the oscillator sends out clock pulses. For the values given, IC2's Q11 output goes high 30 minutes after it is turned on. Resistor R2 resets the IC so that it can work right. The output of IC2 goes to the gate of the SCR through resistor R6 and LED2, which both lower the voltage and show the status of the output.

Circuit operation

When the gate drive is sent to the SCR, it fires to turn on relay RL1. The SCR's latching feature keeps the relay on until switch S1 is used to turn off the power to the circuit. When the relay is turned on, its normally closed (N/C) contacts open, turning off the light. The oscillator is working if LED1 is on. In case the mains power goes out, the back-up white LED lamp with LED3 and LED4 gives off plenty of light. It is powered by a 9V rechargeable battery. When the circuit is turned on, diode D6 and resistor R7 charge the battery at a rate of about 200mA.

The circuit for the back-up lamp is built around a timer called NE555 (IC3), which is set up as a monostable. Based on the settings for VR1 and C9, the output of IC3 stays high for three minutes. When the circuit is turned on, diode D6 sends power to IC3, and resistor R8 keeps the trigger pin 2 high. So, as long as the mains are on, its output will stay low. When the power goes out, pin 2 of IC3 is turned on by capacitor C8. This makes the monostable output go high, which turns on the white LEDs (LED3 and LED4). Resistor R9 makes sure that the amount of electricity going through the LEDs is safe. When the power goes out, the forward bias on diode D7 gives the monostable full voltage. A 15V AC, 250mA transformer provides the power for the circuit. A full-wave rectifier made up of diodes D1 through D4 fixes the second output. The DC is smoothed out by capacitor C1. The regulator IC 7812 (IC1) and the capacitors C4 and C5 give the circuit a stable 12V.

270. Simple Low-Power Inverter

Presented for your consideration is a straightforward inverter with a low power output that operates between 230- and 250-volts AC (DC to AC Converter). It can be used to power very light loads such as window chargers and night lighting, or it can simply be used to produce a shock to deter intruders from entering the building. The entire circuit is comprised of just two integrated circuits, specifically an IC CD4047 and an IC ULN2004.

Low Power Inverter Circuit

A monostable and astable multivibrator is represented by IC CD4047 (IC1). It is configured so that it operates in an astable mode and generates symmetrical pulses with frequencies ranging from 50 to 400 Hz. These pulses are sent on to IC2 using the resistors R1 and R2. The integrated circuit known as IC ULN2004 (IC2) is a common Darlington array with seven channels. The three Darlington stages are connected in parallel at this point in order to enhance the frequencies that have been received from IC1. The output of IC2 is connected, through resistors R3 and R4, to the input of transformer X1.

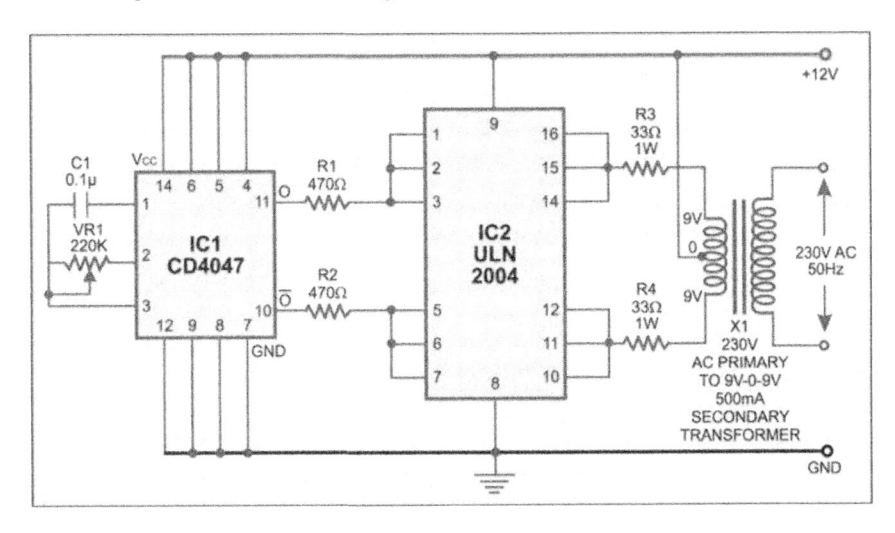

A typical step-down transformer with a secondary current rating of 500 milliamperes, Transformer X1 has a voltage range of 9 volts to 9 volts and a step-up function here. This indicates that it generates a high voltage. It is possible to keep the output current from the ULN within safe parameters by utilizing resistors R3 and R4. The output of 230-250V AC is accessible over the high-impedance winding that is part of the primary windings of the transformer.

271. Mains Interruption Counter with Indicator

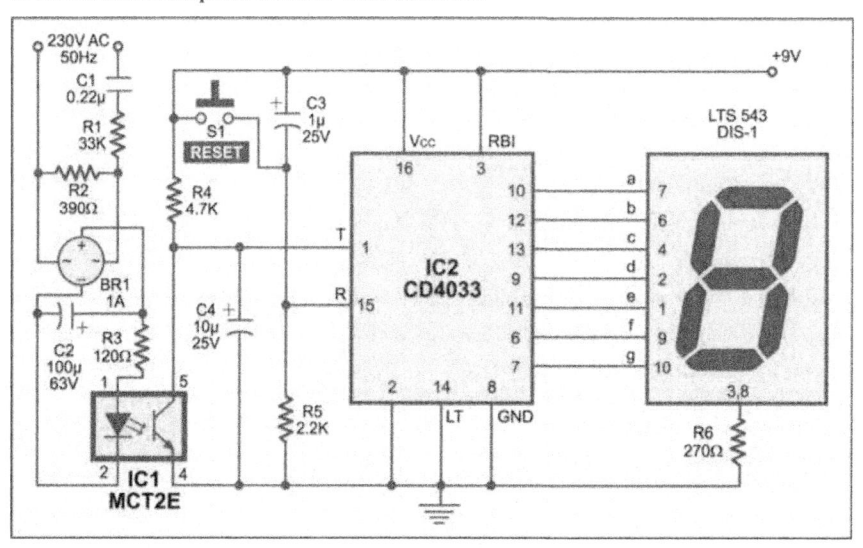

Fig. 1: Circuit of mains interruption counter with indicator

This mains interruption counter circuit keeps track of how many times the power from the mains goes out (up to 9) and shows that number on a 7-segment display. It is very useful for battery chargers for cars. Lead-acid batteries can take longer to charge depending on how many times the power goes out.

The circuit of the interruption counter with indicator is shown in Fig. 1. The whole circuit is powered by a 9V (PP3 or 6F22) battery. Figure 2 shows a block diagram of the mains interruption counter circuit, a battery charger, and a lead-acid battery, all of which are used in shops that charge car batteries. When 9V is put into the circuit, the power-on-reset signal from capacitor C3 and resistor R5 resets IC2, and the 7-segment display (DIS1) shows the number "0." The 230V AC mains are sent to the mains-voltage detection optocoupler IC MCT2E (IC1) through capacitor C1 and resistors R1 and R2.

This is followed by the bridge rectifier BR1, the smoothing capacitor C2, and the current-limiting resistor R2. When the LED inside optocoupler IC1 is turned on, it turns on the phototransistor inside, which pulls down the clock input pin 1 of IC2 to a low level. IC CD4033 (IC2) is a decade counter/7-segment decoder. Its pin 3 is held high, which makes the display show "0" at first. Clock pulses are applied to clock input pin 1 and clock-enable pin 2 is held low to enable the counter.

The number of power interruptions is shown on the seven-segment, common-cathode display DIS1 (LTS543). Capacitor C2 provides a small turn-on delay for the display.

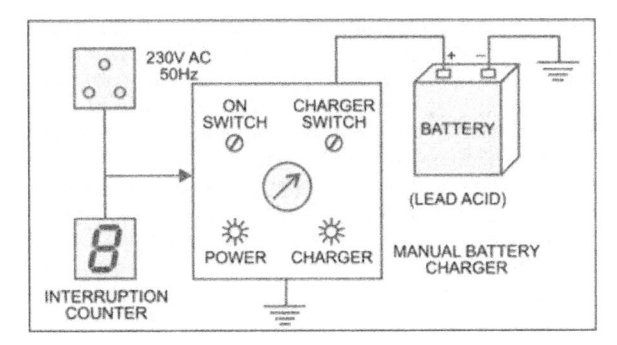

Fig. 2: Block diagram of the arrangement used in automobile battery charger shops

When mains fail for the first time, clock input pin 1 of IC2 again goes high and display DIS1 shows '1.' When the mains power comes back on, pin 1 of IC2 goes low, and DIS1 keeps showing "1." When the power goes out for the second time, pin 1 of IC2's clock input goes high, and the DIS1 display shows the number 2. When mains resume, pin 1 of IC2 again goes low and DIS1 continues to show '2.' This way, the counter keeps incrementing by '1' on every main interruption. Note that this circuit can count up to nine mains interruptions only.

272. FM Adaptor for Car Stereo

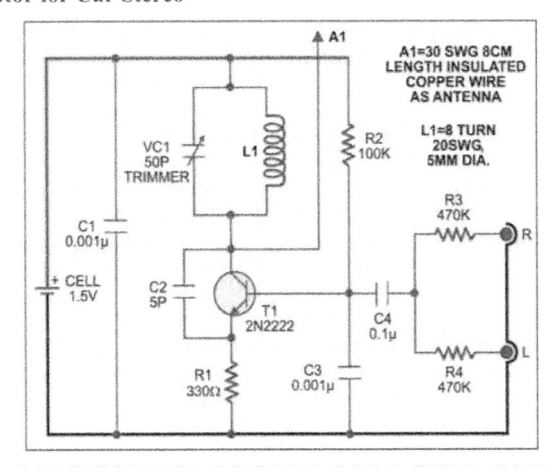

Fig. 1: FM rebroadcast circuit

This circuit can be used to play your favorite songs from your personal audio player through the FM-stereo vehicle radio if your automobile has an FM radio with stereo output but no built-in cassette player. Your individual music player needs to have a provision for a stereo output socket in order for the circuit to function properly. When this FM adaptor is connected to the output of a personal audio player, it can modulate audio over the 88-108 MHz FM band. The FM receiver in the car stereo can easily pick up the frequency that has been changed. The FM adapter is essentially a low-power transmitter consisting of a single transistor that has a range of between 15 and 20 meters. It is also possible to use it to rebroadcast TV audio or your preferred music from an audio system to a pocket or headphone FM receiver without bothering other people in the room or outside. Using any general-purpose npn transistor and configuring it as a voltage-controlled oscillator is all that is required to rig up this FM modulator (refer to Fig. 1). Frequency modulation is done by changing the base-collector voltage. This changes the depth of the depletion layer at the reverse-biased base-collector junction, which changes the capacitance at the collector and, in turn, the frequency at which the collector circuit resonates.

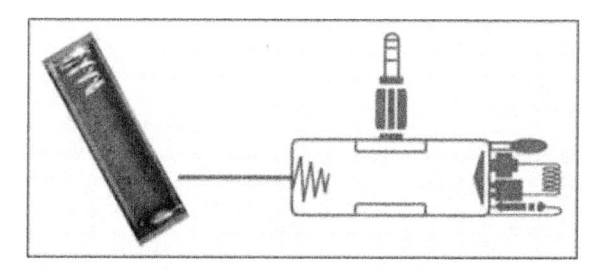

Fig. 2: AA-size single-battery holder and assembled FM modulator

To make inductor coil L1, wrap eight turns of 20SWG enameled copper wire around a 5mm dia. Solder the whole circuit inside a 1.5V, AA-size, single-cell battery holder that you can get from a toy. Make sure the trimmer capacitor (VC1) is securely soldered and easy to reach (with a non-metallic screwdriver) so that the capacitance can be changed.

Use a straight 8cm-long 28SWG copper wire or a telescopic antenna as an antenna (which can be retrieved from a pocket radio). The stereo pin should be soldered directly to the battery holder (see Fig. 2). The power switch is not necessary because the current drawn is only a few milliamperes and the power used is about 125 mW. When the device is not being used, it is easy to take out a typical AA-size alkaline battery.

To fit on the battery holder, make the circuit as small as possible. Check the connections in the circuit you just put together. If everything looks good, solder it to the battery holder and put the battery in. Use a battery-powered FM receiver for testing, not a system that runs on AC power, and do the following: Tune the FM radio to "dead space," which is a term for frequencies in the FM radio band that are either silent or only make a hissing sound. Most of the time, there are no signals near 108 MHz. Connect the FM modulator to the personal audio player's audio output. Extend its antenna and keep the transmitter about a meter away from the FM radio. Play some music on the MP3 player and slowly move the variable capacitor (trimmer) with a screwdriver that isn't metal until you hear music from the FM radio. This process needs careful tuning of the capacitors. Increase the distance between the transmitter and the radio and adjust the trimmer to get the best sound level. When the music on the

FM radio is clear, you can fix its position by putting some hot wax on the trimmer capacitor. The FM modulator is now ready to be used. Plug it into your personal audio player and you can send the sound from your car stereo or any other FM receiver up to 10 to 20 meters away.

273. Panic Plate

This touch-sensitive panic button circuit makes a panic alarm that can be used by the elderly or people who are sick to get help right away. The touch plate is mounted on the wall next to the bed, making it easy for the person who needs to stay in bed to call for help. The call is shown by the yellow LED3 on the panel, and the red LED means that you need to pay attention right away.

Panic button circuit

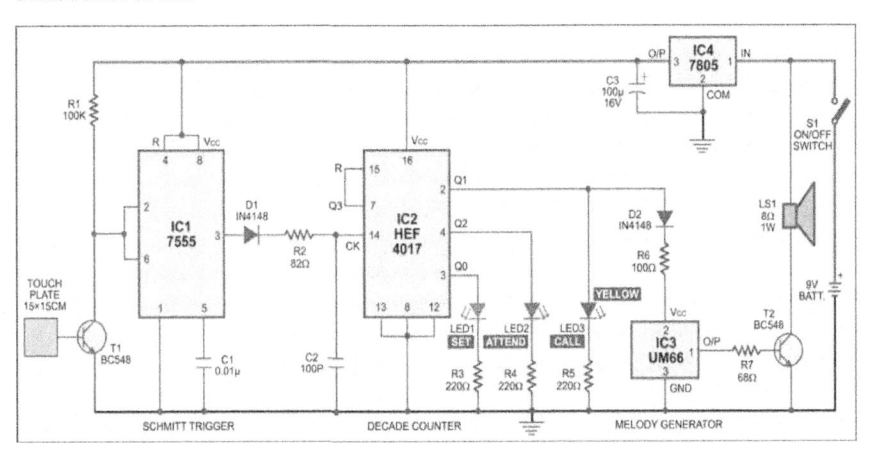

The circuit is based on the electric charge in the body, which can come from inside the body or from the electric field around it. When the sensor plate is touched, the body's electric charge turns on transistor T1 and makes it work. This turns on IC 7555 (IC1), and for a while its output is high. The common timer IC 555 has a CMOS version called IC 7555. The high-level output pulse from IC1 triggers IC HEF4017 (IC2), which is a Johnson decade counter IC with ten decoded outputs. If clock-inhibit pin 13 is low, the counters inside the IC move forward one at a time when the clock pulse on pin 14 goes high. When a signal is sent to pin 15, the IC is reset.

Circuit operation

When IC2 gets the clock pulse from IC1 through diode D1 and resistor R2, its Q1 output goes high to power LED3 and melody generator IC3. T2 boosts the tone from IC3 so that it can be heard from the speaker. IC3 gets power from the forward-biased diode D2 and the resistor R6. When IC2 gets another clock pulse, the Q2 output goes high, the call tone stops, and a red LED light up to let you know that you need to pay attention right away. When the power is turned on, the green LED lights up to show that the device is in standby mode. Capacitor C2 keeps the alarm from going off by accident. The output of Q3 (pin 7) of IC2 is connected to reset pin 15, so when the third clock pulse comes in, IC2 is reset. In short, the first touch will make a call, the second touch will end the call and turn on the red

LED, and the third touch will reset the IC and turn on the green LED after getting help. So, if the red LED stays on after a call, it means that no one has come to help the person. The 5V DC power for the circuit comes from IC4 and C3.

274. Twinkle Twinkle X'mas Star

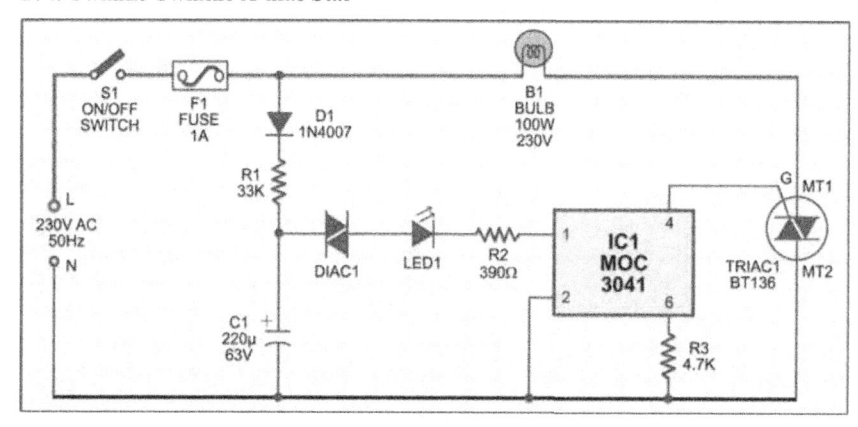

Fig. 1: Twinkle-twinkle X'mas star

If you don't decorate your Christmas tree with a blinking star, Christmas just isn't the same without it. This is how the circuit of a dazzling star looks like. The schematic for the blinking Christmas star may be found in Figure 1. A diac sits at the center of the circuit and is responsible for regulating the charging and discharging of an electrolytic capacitor, which in turn controls the blinking of the star. The remaining portion of the circuit is a solid-state AC relay, and its purpose is to light the bulb that is installed inside the Christmas star. When the "on" position of switch S1 is selected, the mains voltage is rectified by diode D1, and the capacitor C1 charges by the action of resistor R1. When the voltage across C1 is greater than the diac's breakdown potential, the diac will conduct, and the capacitor will discharge through LED1, R2, and the internal LED of optocoupler MOC3041 (IC1). A brief burst of light is produced as a result of this discharge of energy.

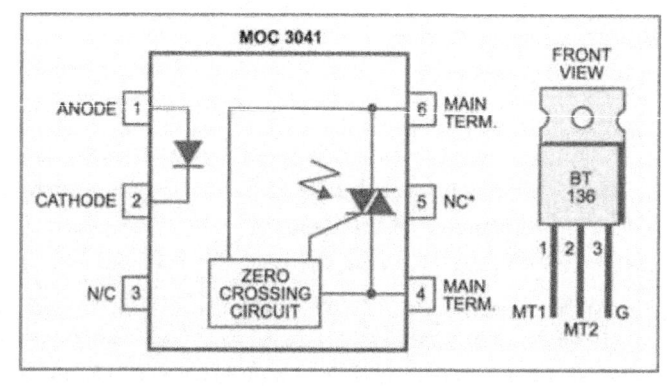

Fig. 2: Pin configurations of MOC3041 and TRIAC BT136

When driving triac BT136 with a zero-crossing optocoupler (IC1), practically all radio frequency interference is eliminated. The light bulb illuminates and the triac operates each time the optocoupler is provided with a pulse. Because this circuit is powered directly by the mains, you must exercise extreme caution to prevent a potentially fatal shock. It is recommended that you make use of a capacitor with a rating greater than 63 volts that has a low leakage rate. The rate of flashes can be altered by adjusting the value of the capacitor. Refer to Figure 2 for the pin configurations of the MOC3041 and the triac BT136 before putting the circuit on a general-purpose printed circuit board (PCB). MOC3041 should be used with an IC base.

275. Car Fan Speed Controller

You may regulate the speed of the 12V DC fans used in automobiles by using this fan speed controller circuit. The timer 555, which is wired as an astable multivibrator, is the center of the circuit.

Circuit operation

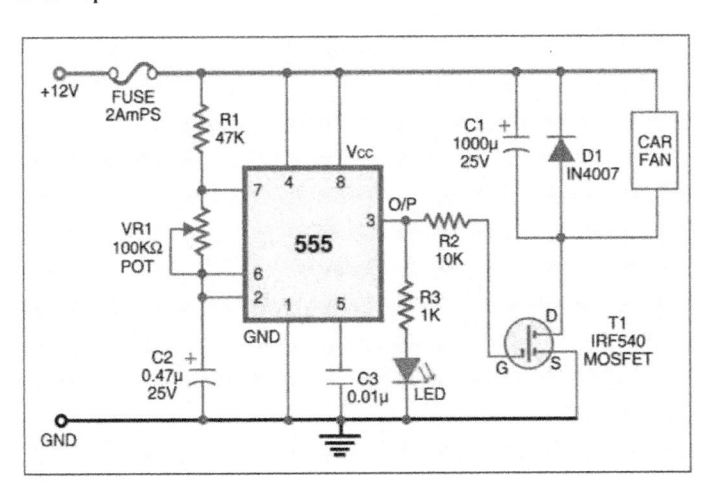

The multivibrator output is connected to an IRF 540 MOSFET. The positive end of the battery and the drain (D) of MOSFET T1 are linked to the fan. The speed of the fan is kept steady by connecting capacitor C1 in parallel with it. Back emf is stopped by the diode D1, which can spin freely. For safety, a fuse is included. The fan speed is changed by changing the duty cycle of the multivibrator, which is controlled by the potmeter VR1. If you think that the low or high level of fan speed is not enough, you can change the value of C2 (0.47 F) to slow or speed up the fan.

276. In Car Food and Beverage Warmer

People who are always on the go use this device. It is used to keep your tea, coffee, or food warm while using little power. A common timer, the 555, is used as a free-running astable multivibrator in the simple circuit. Diodes 1N4148 are connected backwards so that the duty cycles can change as much as possible. The goal of this work is reached by connecting up to five 10W heating elements in parallel, for a total of 50 watts. If fewer coil elements are connected in parallel through toggle switches S2 through S4, a lot less current will be used. Every one of these switches can handle 6A.

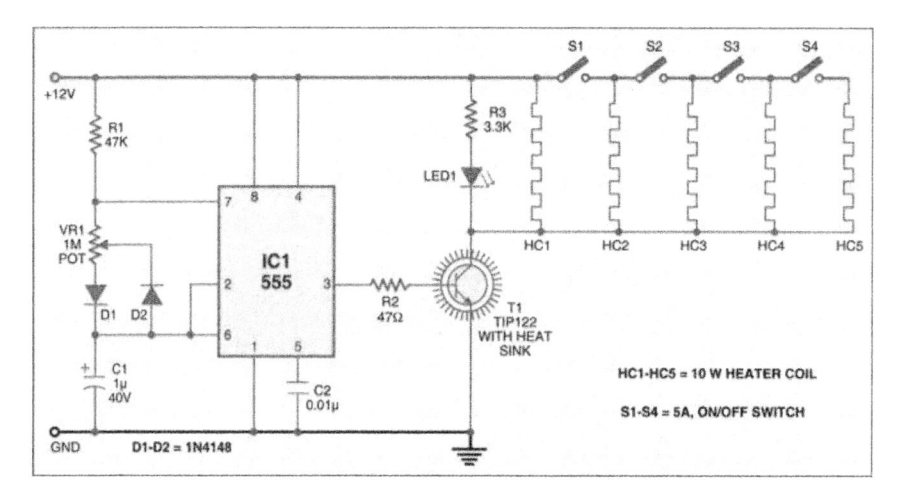

Food and drink warmer circuit diagram for a car: The T1 power transistor is a Darlington type that can handle up to 5A and puts out more than 60 watts. The discrete parts that were chosen ensure a fixed frequency of about 1 Hz at pin 3 of timer IC1 (555). The duty cycle can be changed with the resistor R1 and the potmeter VR1 (1 mega-ohm). The heater's output goes up as the duty cycle goes up. You can link up to five 10W heating elements in parallel, giving you a total of 50W. If fewer coil elements are connected in parallel through toggle switches S2 through S4, a lot less current will be used. Every one of these switches can handle 6A.

Put the circuit together on an all-purpose PCB. Power transistor TIP120 should be put on a thick heat sink. Use only two wire connections to separate the circuit from the heating elements. Use wires that can carry currents of more than 6A. Attach the coil elements below a rectangular plate made of aluminum or steel that is at least 1 mm thick. Don't forget to put something between the elements and the heating plate. Use the car battery as the power source with a wire that can carry enough current.

277. Flasher

Since there are only three parts to the flasher system, it is easy to build and put in place. It can be used to flash a signal, warn of danger, or flash in different patterns. Because it is simple and useful, the basic two-transistor flasher shown below has been used in a lot of different ways. Circuits like a micropower low battery indicator, a lightning detector, an off-line switching power supply, a micropower high voltage supply, a strange beeping capacitance probe, a windshield wiper controller, a lamp dimmer, a police siren, and a few others have been used in a variety of ways. With careful choice of transistors, the simple circuit can be used at very low frequencies, RF frequencies, low voltages, or even very high voltages. Both the amount of power it can handle and the amount of power it uses can be changed easily to fit the need.

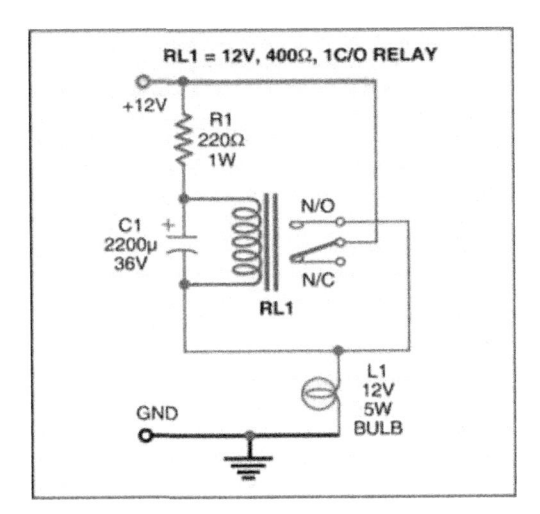

Fig. Three Component Flasher circuits

This circuit is a great place to start. It will flash if you build it. And it's easy to change the on time and how often the lights flash. Here is a picture of the basic flasher. Notice that it is a "two-wire" circuit and that the load and battery just connect in series. When power is turned on, the two resistors on the base of the PNP set a threshold voltage, and the capacitor starts to charge toward this voltage. When the voltage on the capacitor is high enough, the two transistors start to do their job. The flow of current makes the voltage across the circuit drop a little, and this drop makes the threshold voltage drop as well. Because the threshold voltage is lower, there is even more current, and this positive feedback makes the circuit turn on quickly. It stays on until the capacitor runs out. When that happens, a process in reverse turns the circuit off quickly.

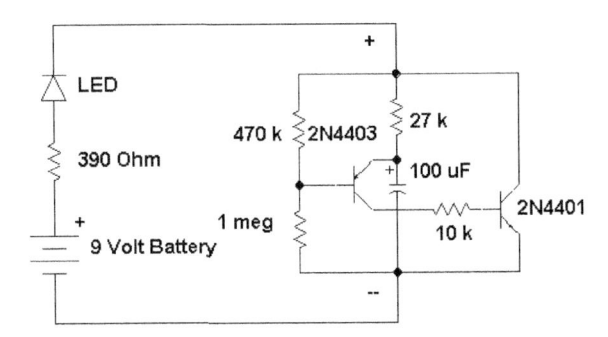

Fig.2: Two-transistor Lamp Flasher

Power transistors can be added so that the circuit can handle more current. The two connections shown below are typical. In the first circuit, a power transistor is turned on by a flasher circuit that is linked to a 220-ohm resistor. In the second circuit, the NPN is replaced by a power FET. When the circuit is turned off, the gate is pulled down by a pull-down resistor.

278. Home Automation Using Apple HomeKit and ESP8266

It's an Internet of Things (IoT) project that lets you control four home appliances with Apple HomeKit, the Hey Siri app, and an ESP8266 Wi-Fi module. We thought about building a home automation system a few months ago. We looked on the internet and found that most of these projects used Google Assistant or Alexa. Because of this, we had the inspiration to develop a home automation system that could be operated by the assistant offered by Apple, known as the Hey Siri function. We looked for project on this subject, but most of them required an Apple laptop and the Homebridge app. We didn't want to use these tools for the project. Then we found a great library that could make it possible, though it only showed how to control one relay as an example. With this home automation project, you can use four relays to control four different appliances.

Circuit

Circuit connections are very simple. The table shows how the relay module and NodeMCU are linked. Figure shows the full wiring diagram for controlling a light bulb with a relay module. You can connect the circuit in a similar way to control the other three devices.

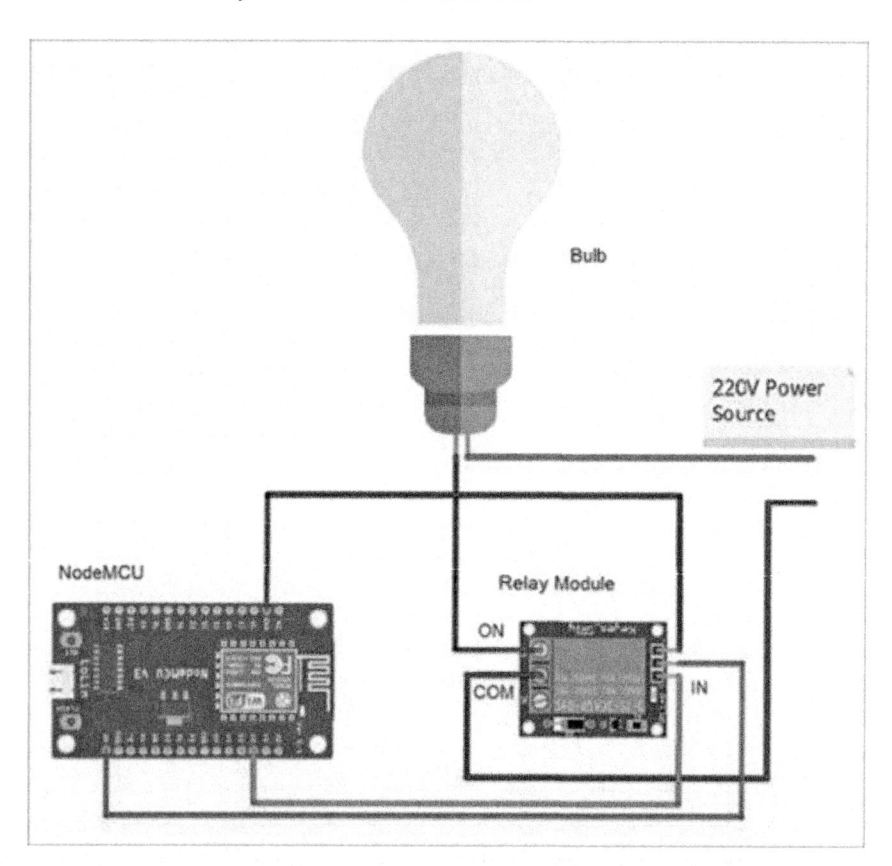

To make the project, you'll need the following tools:

- 5V relay module
- NodeMCU board
- Jumper wires
- Bulb holder (for 230V AC)
- 60W/100W bulb
- iPhone SE or any iPhone that supports Apple HomeKit application

279. Hot-Water-Ready Alarm

When the water in an electric kettle boil, it turns off by itself. What if the boiler beeps when the water is done boiling? You might not think of the sound the thermal switch makes when it trips as an alarm. Here's an example of such an add-on: a unit that sends out short beeps when the water has boiled. It's good because it has a very small number of parts, is cheap, small, and light. In this circuit, a reed switch is turned on and off by a magnetic field made by the current flowing through the coil. The current flows through the coil, which is made of twelve turns of 20SWG enameled copper wire. The number of turns will depend on how much power the kettle has and what kind of reed switch it has.

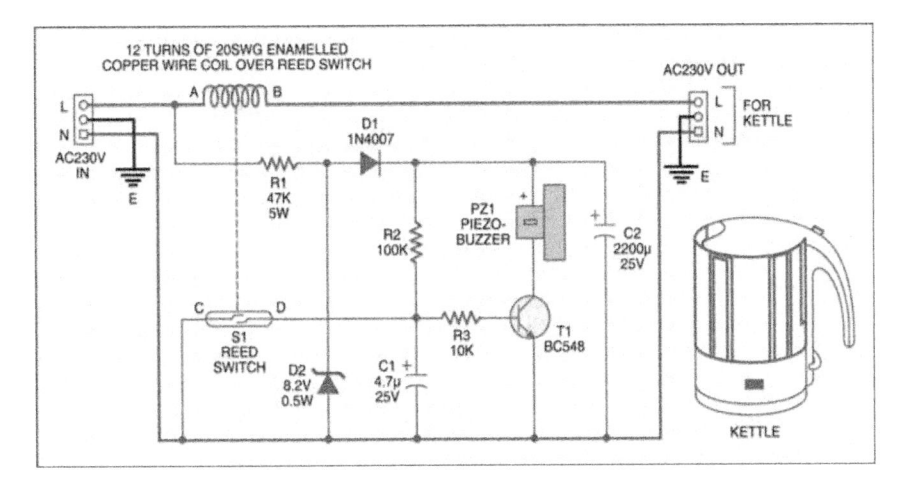

In this circuit, a reed switch is turned on and off by a magnetic field made by the current flowing through the coil. The coil is made of twelve turns of 20SWG enamelled copper wire to carry the current. The number of turns will depend on how much power the kettle has and what kind of reed switch it has. Construct the circuit on a standardized printed circuit board (PCB), and then encase it in a plastic housing. Connect the power from the mains to the circuit and the kettle to the 230V AC output. Put the thing where the kettle is. Connect the reed switch and its coil like the diagram shows.

280. Optical Smoke Detector

Fig.: Circuit of optical smoke detector

This optical smoke detector has a through-scan, slotted, infrared photo-switch that is cheap and easy to find. When smoke is found, the relay is turned on, which starts the sound and light alarm. The current going through the LED inside the sensor is limited by resistor R1. Most of the time, the phototransistor inside the sensor is on, so the base terminal of transistor T1 has a low voltage. Set VR1 determines the threshold level, and resistor R2 gives transistor T1 and the sensor unit extra protection. The resistor R3 pulls the trigger input (pin 2) of the monostable (IC2) that is wired around NE555 high. When there is no smoke near the sensor, pin 2 stays high because transistor T1 is conducting. As a result, relay RL1 turns off the audio/video alarm connected to its normally-opened (N/O) contact. This is shown by the red LED shining (LED1). LED1 can only get so much power through resistor R7.

When the sensor senses smoke, the internal phototransistor doesn't get enough light from the internal LED. As a result, the forward bias of transistor T1 makes trigger pin 2 of IC2 low, which turns the sensor off. Right away, LED1 goes out and relay RL1 turns on. The counter emf is stopped by the freewheeling diode D1. Here, the IC2-based monostable is wired as a missing-link detector. So, the relay stays on as long as the first input condition (no smoke) is still true. The time-out for the monostable is set by the components R6 and C2.

281. Capacitance-Multiplier Power Supply

High-quality audio amplifiers need a power supply that is very efficient and has almost no ripple, hum, or buzz. Normal bridge-type power supplies have a problem in that when the load is connected, the voltage drops to 3 to 4 volts. A power supply that is controlled by a transistor or IC is a much better choice, but it still has some ripple and hum that can be annoying to people with good hearing.

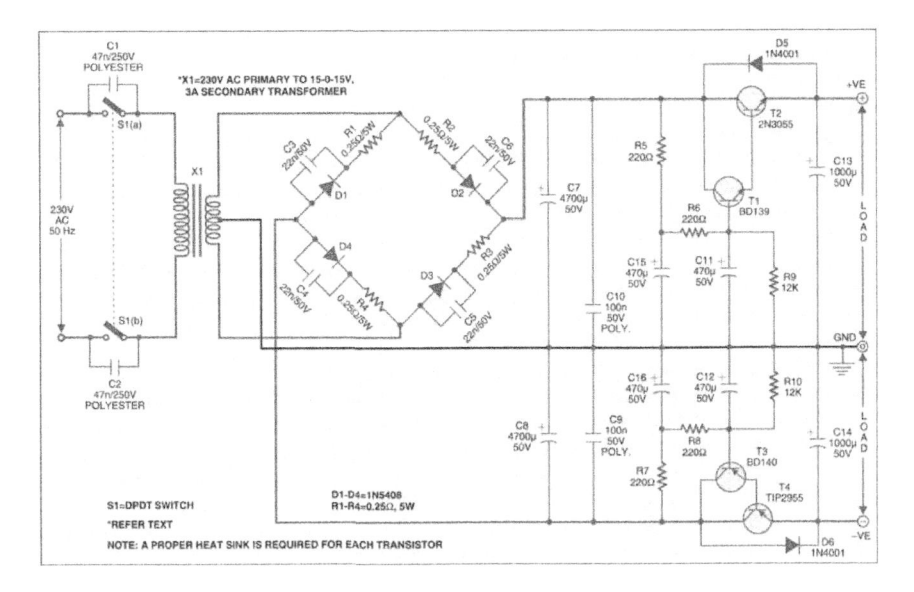

Here is the circuit for a power supply with regulation that uses a center-tapped transformer and is very efficient. After going through the rectifier bridge, the 15V AC output from the secondary drops by 0.6V x 4 = 2.4V. Also, the resistance of the transformer coil drops to 0.5V, and the ripple is 1V. So the drop is about 4V DC as a whole. Because of this, the secondary output of the transformer should be at least 4 V higher than the regulated output. As can be seen in the diagram, the regulators consist of two pairs of transistors connected in a Darlington configuration: T1-T2 and T3-T4. The current gain goes up because of the Darlington pairs of transistors. The 470F capacitors connected to the bases of transistors T1 and T3 reduce hum and also lower higher-order harmonics. The 12-kilohm resistors (R9 and R10) make up for any changes in the gain of the transistors. The peak current at power-on is limited by putting resistors R1 through R4 (0.25-ohm, 5W) in series with diodes.

The internal thermal noise of diodes C3 through C6 is quieted by the 22nF, 50V capacitors C3 through C6. To cut down on RF noise, the mains filter capacitors C7 and C8 (4700F) are switched by capacitors C9 and C10 (100nF, 50V). When capacitors C1 and C2 (47nF, 250V) are placed across the double-pole, double-throw switch S1, they stop the switch from bouncing. Resistors R5 through R8 (220 ohm) and R9 and R10 (12 kilo ohm) are either 0.25W metal-film or 0.5W carbon resistors. If 10,000F capacitors are used instead of 4700F capacitors C7 and C8, the DC output will not be affected by changes in the mains (200V-260V AC).

All of the electrolytic capacitors should be grounded at the same point. As the common ground, use a copper bar or a small copper plate. Heat sinks must be used with transistors T1 through T4. Use copper wire that is at least 16 SWG to connect the other parts of the circuit to the common ground. EFY note. In the EFY lab, a 230V AC primary to 15V-0-15V, 3A secondary transformer was utilized to get 15V DC, 2A. Choose the voltage and current ratings of the secondary coil between 18V and 22V without modifying any components.

282. Wireless PA for Classrooms

Most of the time, students in the back rows of a large classroom can't hear the teacher. So, the teacher has to literally yell for everyone to hear her. Here is a Wireless PA circuit that can be used by teachers as a wireless speech aid so that their voice can reach every student in a big classroom. The circuit is made up of a wireless microphone and an amplifier for the receiver. The teacher speaks into a wireless microphone, and what he says is sent over the FM band. You can cover the whole room by putting several receivers with speakers in different places.

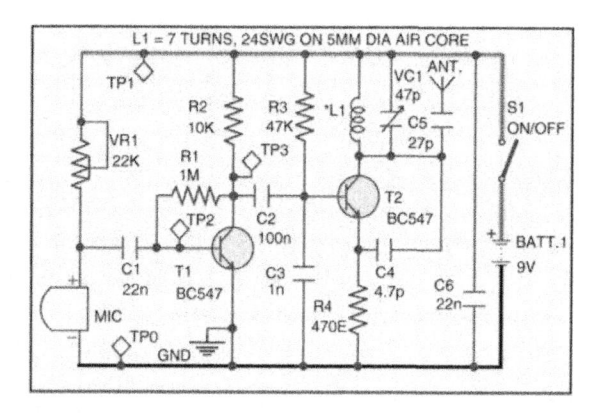

Fig. 1: Circuit of wireless microphone

Part of a wireless microphone

The transmitter part of a wireless PA for classrooms is made up of an electret microphone (MIC), two BC547 npn transistors (T1 and T2), and a few other parts. The electret microphone MIC picks up the sound signal, and the transistor T1 boosts it. The signal that has been amplified is sent to the base of transistor T2. Here, the signal at the base of transistor T2 changes its junction capacitance, which affects the L-C tank circuit made up of L1 and VC1. A Colpitts oscillator is similar to the tank circuit. The 27pF capacitor attached to the antenna keeps the L-C circuit from being affected by the capacitance of the antenna. The inductor L1 is made by wrapping seven turns of 24 SWG copper wire around an air core that is 5mm in diameter. Use any wire that can bend as an antenna.

Section of the receiver amplifier

The receiver part of a wireless PA system for classrooms is made up of an audio amplifier called TDA2030 (IC1), an FM receiver kit, a loudspeaker called LS1, and a few other parts. There are many

Sony ICs like CXA1019 and CAA1619 in FM receiver kits. Some of them also have Philips ICs. Both 6 V and 12 V can be used to power an FM kit. Choose 12 V for this project and power it with a 12V, 1A adapter.

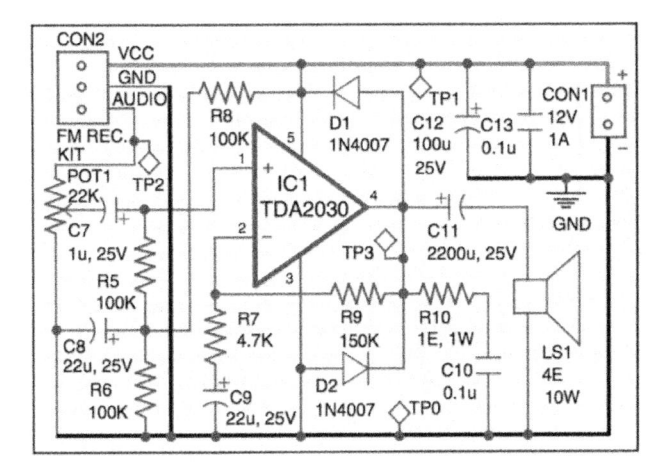

Fig.2.Circuit of receiver amplifier section

Choose a kit that doesn't have a gang capacitor but does have trimmers instead. Even a kit with ganged capacitors will work. But once the circuit is aligned with the FM transmitter/wireless microphone, it is hard to keep the gang in one place. The FM kit's output is hooked up to the sound amplifier made with IC1. A pair of diodes, D1 and D2, and a 100-kilo-ohm resistor, R8, are used to create a "virtual earth" that lets IC1 work with just one power source. Use a heat sink to keep IC1 from shutting down because it's too hot. How the circuit works is easy to understand. Once the transmitter and receiver are tuned (see the section on calibration), the microphone picks up the teacher's voice and sends it over the FM band. The sound is picked up by the FM kit, and the amplifier makes it louder. This louder sound can be heard at loudspeaker LS1.

Calibration

Keep the transmitter and receiver about three meters away from each other. First, turn on the receiver and tune it to a fixed frequency that is not on any of the regular channels. Put the volume in the middle. Now, use switch S1 to turn on the wireless mic. Using a small plastic screwdriver, turn trimmer VC1 a little bit at a time (no more than 5 degrees) until the radio stops "hissing" and the null point is found. This means that the frequency of the wireless microphone is now the same.

283. Low-Cost Battery Charger

A charger for lead-acid batteries with 12 volts and 7 ampere hours that is both easy to use and inexpensive is presented here. It can also be used to power car engines and lighting systems for emergencies. Diodes 1N4001 (D1) and 1N4001 (D2) make up a full-wave rectifier, as shown in the figure. The voltage that has been changed appears between SCR1 and the battery that needs to be charged. When the battery voltage is low, SCR2 shuts off. When the gate current flows through resistor R1 and diode D3, SCR1 is turned on. This starts charging the battery.

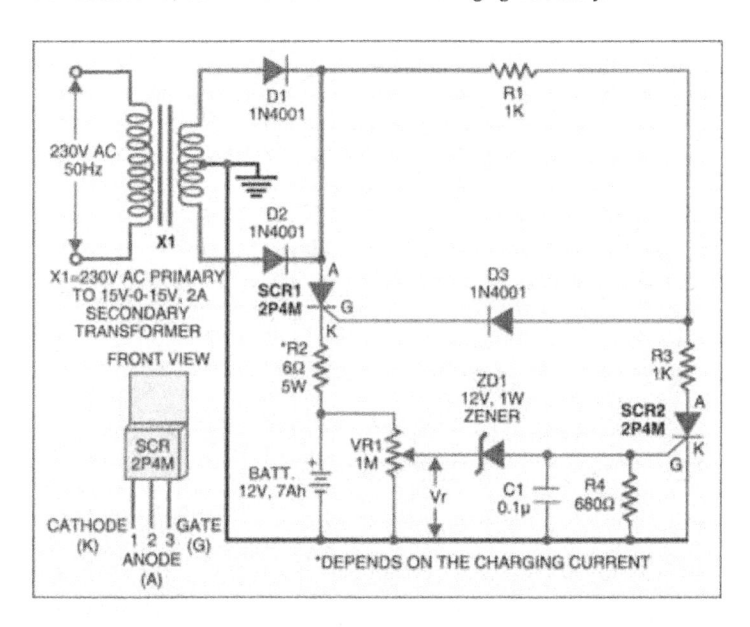

Here is a very simple and cheap way to charge lead-acid batteries that are 12V and 7Ah. It can also be used to power car engines and lighting systems for emergencies. As charging continues, the battery voltage rises until it is high enough to turn on the zener diode and trigger SCR2. SCR2 lowers the voltage at the junction of R1 and R2 until it is too low, which turns off SCR1. This happens when the battery is fully charged and the charging current stops because SCR1 is in an open-circuit state. When the reference voltage Vr drops below the Zener diode's breakdown voltage, the regulator starts to charge the battery. When voltage Vr is equal to the Zener diode's breakdown voltage, the regulator stops the battery from being overcharged.

284. Simple Automatic Water-Level Controller

These days, water level controllers are common. The automatic water level controller described here is made up of a timer NE555 and an inverter buffer CMOS IC CD4049. It uses easy-to-find, cheap parts and is easy to build and put on the over-head tank (OHT) to stop water from going to waste. The circuit can be powered by a 12V battery or by 230V AC mains with the help of a 12V adaptor. The three sensors, which are made of metal that doesn't rust, are attached to the OHT as shown in Figure 2 and connected to the circuit as shown in Figure 1. The power supply terminal Vcc is at the bottom of the tank, the sensor terminal L is just above the bottom of the tank, and the sensor terminal H is at the top of the tank. After you put the sensors in the OHT the right way and hooked up the power supply, the circuit is ready to be used.

Since the Vcc terminal is at the bottom of the tank, when the water level drops below sensor L, inverters N1 and N2 cause pin 2 of timer IC2 to go high. Because of this, the output of timer IC2 goes up. When relay RL1 is turned on, the motor starts to fill the tank with water. Even when the water level goes above sensor L, the motor stays "on."

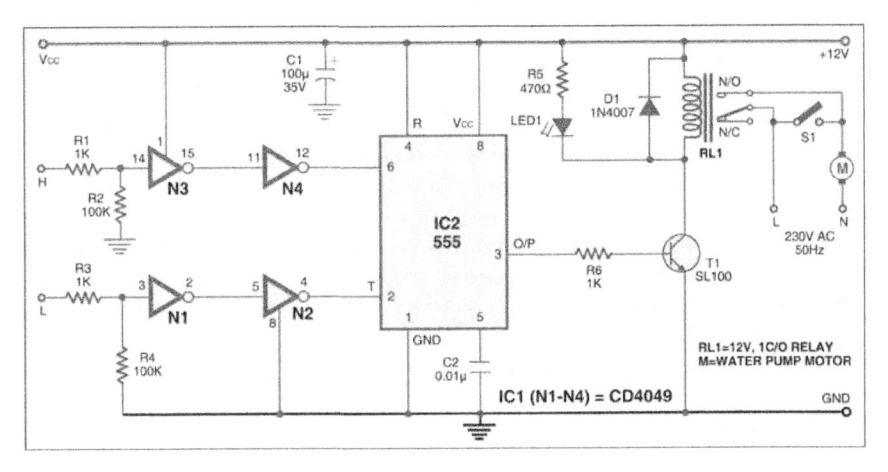

Fig. 1: Simple automatic water level controller

When the water in the tank gets high enough to touch sensor H, inverters N3 and N4 retrigger timer IC2 at pin 6, which makes its output go low. When the relay loses power, the motor that fills the tank with water stops. Even when the water level drops below sensor H, the motor stays "off." As the water is used up and the level goes below sensor L, the motor starts up again. After that, the cycle starts over again.

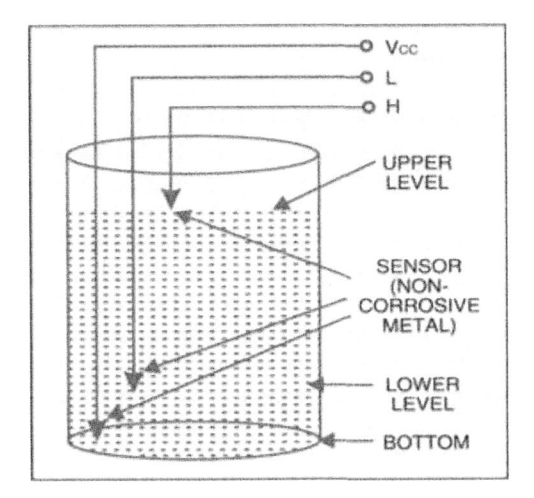

Fig. 2: Sensor installation in the overhead tank (OHT)

285. Touch Based Doorbell

Along with a detector and a Darlington driver stage, enhancement-mode MOSFETs are used in this touch-based doorbell. These MOSFETs are a component of a CMOS quad NAND gate called CD4007B. A dual complementary pair and an inverter make up the IC CD4007B. As shown in Figure, it has three p-channel and three n-channel enhancement-mode MOSFETs. The devices can also be used as linear amplifiers, current sinks/sources, and crystal oscillators. Each one can also be used as a very high-input op-amp.

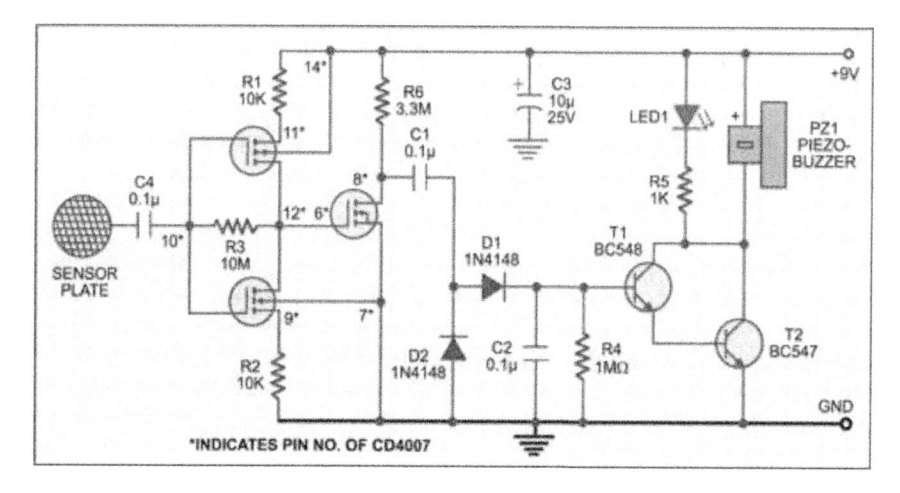

How the touch plate doorbell works Here, we used the complementary pair with a common gate (pin 10), separate drains (pins 9 and 11), and a common output (pin 12) as a very high input micropower linear amplifier that could increase the hand/finger-induced AC voltages by a factor of 40.

A single-stage n-channel MOSFET (pin 6 is the gate, pin 7 is the source, and pin 8 is the drain) amplifies the output of this complementary stage. This stage's output is picked up by the detector circuit, which is made up of diodes D1 and D2, and R4, which acts as a "bleeder resistance" across C2. When a person touches the metal sensor plate, the detected voltage is amplified by the Darlington stage, which is made up of transistors T1 and T2. This drives an LED-buzzer combination called LED1-PZ1.

286. Electronic Ludo

In the classic board game Ludo, players throw a dice by hand and move their pieces on the board by the number of squares shown on the dice. The number of dots on the dice's six flat sides determines how far you can move forward. In this electronic version of Ludo, instead of throwing the six-sided dice, players have to push a button. The 7-segment digital counter shows a number right away when the switch is pressed briefly. As with the manual dice, the numbers are randomly shown between "1" and "6," depending on how long the player presses the switch S1.

Electronic Ludo Circuit

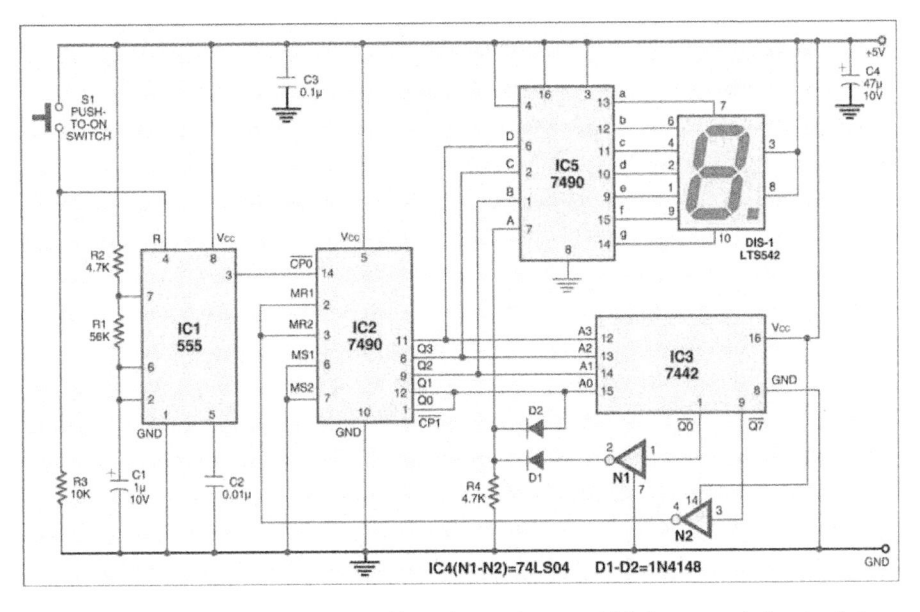

The Timer IC 555 (IC1) is wired in the astable mode to make a very high-frequency clock pulse that depends on the values of the capacitor C1 and the resistors R1 and R2. To turn off timer IC1, pin 4 is pulled to the negative power supply through resistor R3.

Circuit operation

When switch S1 is briefly pressed, IC1 begins to make a clock pulse. When S1 is set free, IC1 stops counting. The output pin 3 of IC1 is wired to the clock input pin 14 of IC 7490, which is a binary decade counter (IC2). Pins 11, 8, 9, and 12 of IC2's output are connected to pins 12, 13, 14, and 15 of IC 7442's BCD input (IC3). Inverter N2 flips the signal from pin 9 of IC3 and sends it back to pins 2 and 3 of IC2. This resets IC2 after every six counts. To operate the 7-segment common-anode display LTS542, the outputs from IC2's pins 11, 8, and 9 are additionally linked to the respective BCD input pins 6, 2, and 1 of the BCD-to-seven segment decoder/driver IC 7447 (IC5) (DIS1).

The clock pulses from IC1 would normally cause IC2 to produce binary code from 0 to 9, but the output from pin 9 of IC3 resets it after every sixth clock pulse. The count output will always be between 0 and 6. Diodes D1 and D2 make an "OR" gate and change the first "0" to a "1." This keeps the count on the screen between "1" and "6."

287. Motorbike Alarm

This alarm is easy to make and can be put on bikes to keep them from being stolen. The tiny circuit doesn't have any complicated wiring, so it can be hidden anywhere. Almost any bike can use it as long as it has a battery. The standby current is zero, so it doesn't drain the battery.

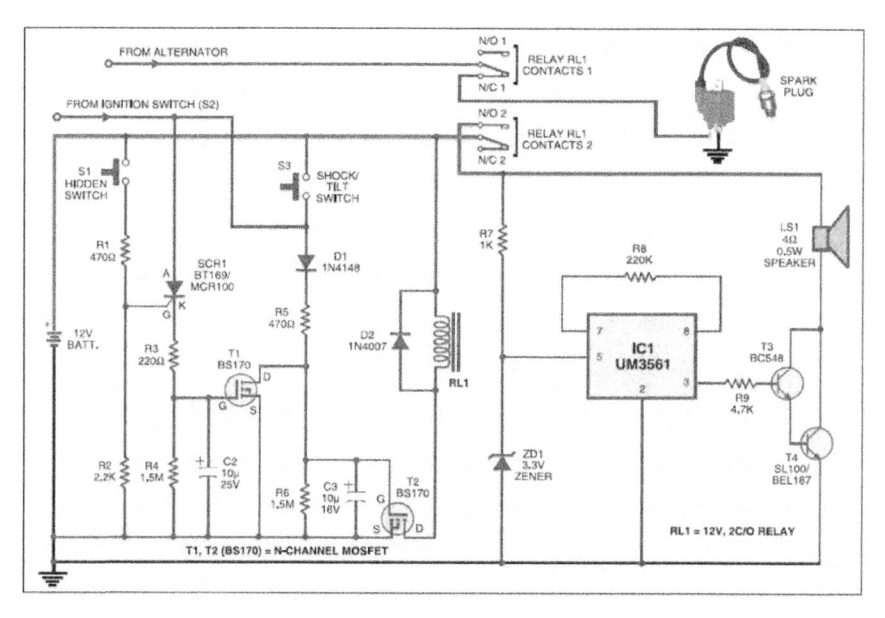

The hidden switch S1 can be a small push-to-on switch, a reed switch with a magnet, or any other simple device that works the same way. The circuit is built around two low-voltage MOSFETs that have been set up as monostable timers. Key S2 is an ignition switch on a motorcycle, and switch S3 is a tilt switch. When the motorbike key S2 is turned on, it sends power to the gate of MOSFET T2. When you use key S2 to turn off the bike's ignition, you have about 15 seconds to get off. This is done by resistor R6 to discharge capacitor C3. After that, if someone tries to get on the bike or move it, the alarm goes off for about 15 seconds and the ignition circuit is cut.

During parking, the hidden switch S1 is usually open, so it doesn't let the mosfet T1 turn on. But when someone starts the bike with the ignition switch S2, diode D1 and resistor R5 turn on MOSFET T2. When relay RL1 (12V, 2C/O) is turned on, the alarm (built around IC1) goes off and the ignition coil is taken out of the circuit. When the ignition coil is disconnected, the spark plug can't make a spark. Usually, there is a wire going from the alternator to the ignition coil. This wire has to go through one of the N/C1 contacts of relay RL1, as shown in Figure.

Also, when the coil is disconnected, power is sent to the sound generator IC UM3561 (IC1) through the N/O2 contact of relay RL1. This turns on the T3 and T4 Darlington pair, which makes the siren sound through loudspeaker LS1. To start the car, you need to turn on both the hidden switch S1 and the ignition key S2. If you don't, the alarm will go off. When S1 is turned on, SCR1 is set off, which sets off MOSFET T1. MOSFET T1 is set up so that it stops MOSFET T2 from working. So, MOSFET T2 doesn't turn on, relay RL1 stays off, the alarm stays off, and the ignition coil stays connected to the circuit. When the spark plug is connected to the ignition coil, the spark plug is better able to make a spark. The bike can't be stolen if the owner is the only one who can use the hidden switch S1. With the S3 tilt switch, you can't move the vehicle without turning it on. Versions of the switch made of glass and metal don't bounce and break quickly, even when tilted slowly. Unless otherwise stated, the angle at which the switch must be tilted for the contacts to work (the "operating angle") must be about 1.5 to 2 times the angle given for the "difference." The differential angle is the difference between where the door is "just closed" and where it is "just open."

The tilt switch has features like contacts that make and break when the switch is shaken, contacts that go back to the open position when the switch is still, non-position sensitivity, inert gas and hermetic sealing to protect the contacts, and a steel housing that is tin-plated. If you can't get the tilt switch, you can use a reed switch (N/O) and a piece of magnet to replace it. The magnet and reed switch should be set up so that the switch contacts close when the bike stand is moved from its resting position.

288. Dual Motor Control for Robots

Here, we present a basic circuit that can power two motors for a microbot, allowing it to move around obstacles. Two light-dependent resistors (LDRs) are used to see what's in the way, and the motors are then driven in the right way to automatically avoid the obstacles. With two H-bridge motor circuits, each motor can be driven forward, backward, or stopped on its own.

Circuit and how it works

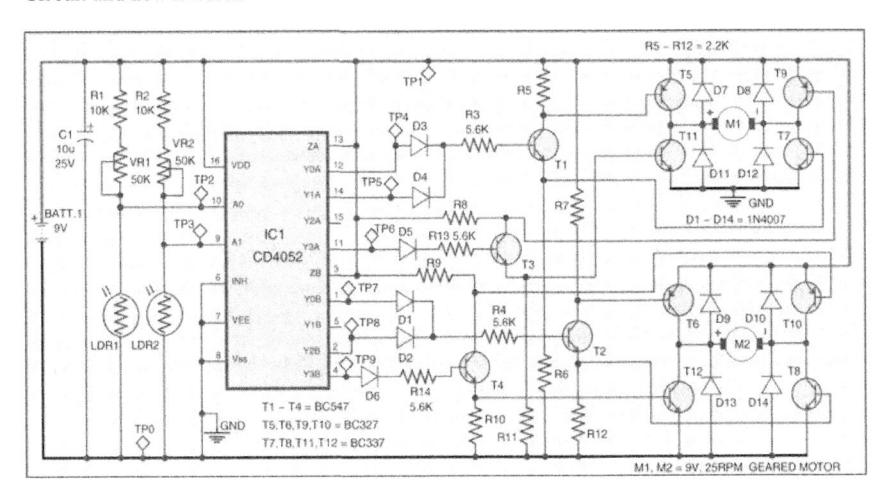

Figure shows the circuit for controlling two motors at once. The circuit is made up of a four-channel multiplexer (IC1), two light-sensitive resistors (LDR1 and LDR2), four BC547 npn transistors (T1 through T4), four BC338 transistors (T7, T8, T11, and T12), four BC327 pnp transistors (T5, T6, T9, and T10), and a few other parts. As was already said, the two motors are driven by two H-bridge circuits. The left side is driven by motor M1, and the right side is driven by motor M2. As shown in Figure, each H-bridge circuit is made up of a pair of npn and pnp transistors. Between the collector and the emitter of each driving transistor is a diode that protects against the motor's back EMF spikes when the transistor is "off." The output of IC1 drives the motor-control transistors (T1 through T4) through the diodes (D1 through D6). The light-sensitive resistors LDR1 and LDR2, which are attached to the front of the robot, control IC1. On the left side is LDR1, and on the right side is LDR2. When light hits both sensors, their resistances drop, which pulls the A0 and A1 inputs of IC1 towards 0V. In this case, IC1 connects output ZA to Y0A and output ZB to Y0B, letting current flow through diodes D1 and D3. This turns on both transistors T1 and T2, which makes both motors move forward. This is when neither of the sensors picks up anything and the robot can move forward without stopping.

When both sensors see an obstacle, the resistances of both LDRs go up, which pulls the A0 and A1 inputs of IC1 up. In this case, output ZA is connected to Y3A and output ZB is connected to Y3B. This lets current flow through diodes D5 and D6. This turns on transistors T3 and T4, which turn both motors in the opposite direction and move the robot away from the obstacle in front of it. If something blocks a lot of light from getting to LDR1 on the left side of the robot, the A0 input of IC1 goes high. ZA is linked to Y1A, and ZB is linked to Y1B. Since T1 is still on, the motor M1 keeps moving forward. But when the flow of electricity stops through D1, T2 turns off and motor M2 stops. So, the robot turns to the right and moves away from the obstacle. In the same way, if the robot sees an obstacle on the right, it will turn to the left. How the circuit works is easy to understand. The robot can move forward without stopping as long as it doesn't sense anything in its way. It moves away from any obstacle and uses the information from the LDR sensors to find its way.

289. Environment and Weather Monitoring System

In many regions, it is necessary to perform constant monitoring of the air pressure, temperature, and humidity in order to assist in the prediction of any changes in the weather and the circumstances of the surrounding environment. This allows us to organize our work in accordance with any resulting shifts. Continuous monitoring of air pressure, humidity, and temperature levels is required in a variety of settings, including but not limited to small production units, factories, and sporting events.

Because of this, today we are going to construct a sophisticated device that is capable of providing us with this information as well as data that is related to it in real time. We are going to make use of the LPS22HH sensor in order to get them. After the connection has been made, a battery with a voltage of 5V is needed to power the Arduino board. Please be patient as it will take a few minutes before it begins to display the temperature, humidity, and pressure of the room.

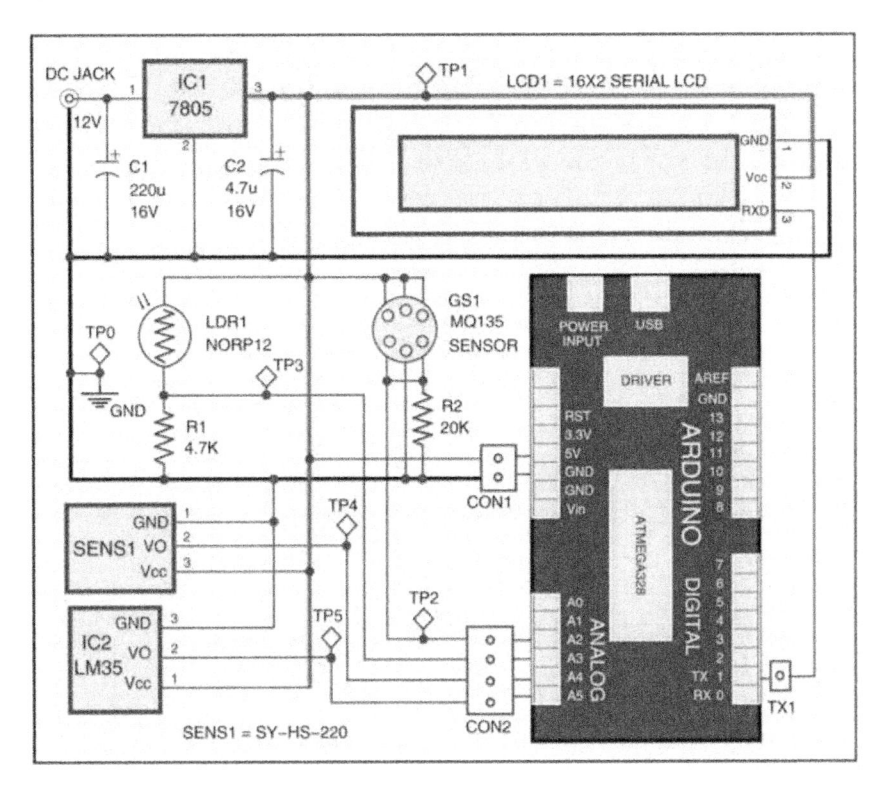

Fig Environment and Weather Monitoring System

290. Long-Range IR Transmitter

Within a range of five meters, the vast majority of infrared remote controls function dependably. If you want to create a long-range IR transmitter that is capable of dependable functioning over a longer range, say 10 meters, the complexity of the circuit will rise. It is necessary to raise the transmitted power by a factor of four in order to expand the range from 5 meters to 10 meters. Using an infrared (IR) laser pointer as the IR signal source is a good idea if you want to be able to produce a highly directional infrared beam (beam that is very narrow). You won't have any trouble finding the laser pointer in the marketplace. However, because the laser pointer emits such a narrow beam, you need to exercise extreme caution. Even the slightest jolt to the device could cause the beam's alignment to shift, which would result in the loss of contact.

Long range IR transmitter

Here is a straightforward circuit that will provide you with a rather extensive operating range. The radiated power is amplified by connecting three LEDs that are capable of transmitting infrared light (IR1 through IR3) in series. In addition, you can assemble the IR LEDs inside the reflector of a torch in order to raise the level of directivity and, consequently, the power density.

In order to improve the performance of the circuit, a metal oxide semiconductor field effect transistor, or MOSFET, model BS170, has been used. This MOSFET performs the function of a switch and, as a result, minimizes the amount of power that would be lost if a transistor were used instead. A 100 F reservoir capacitor named C2 is installed across the battery supply in order to prevent the voltage from dropping during the 'on' and 'off' operations of the device. When the infrared transmitter is powered by regular batteries, the benefit of this feature will become more apparent. During the "switching on" phase of operations, capacitor C2 provides an additional charge.

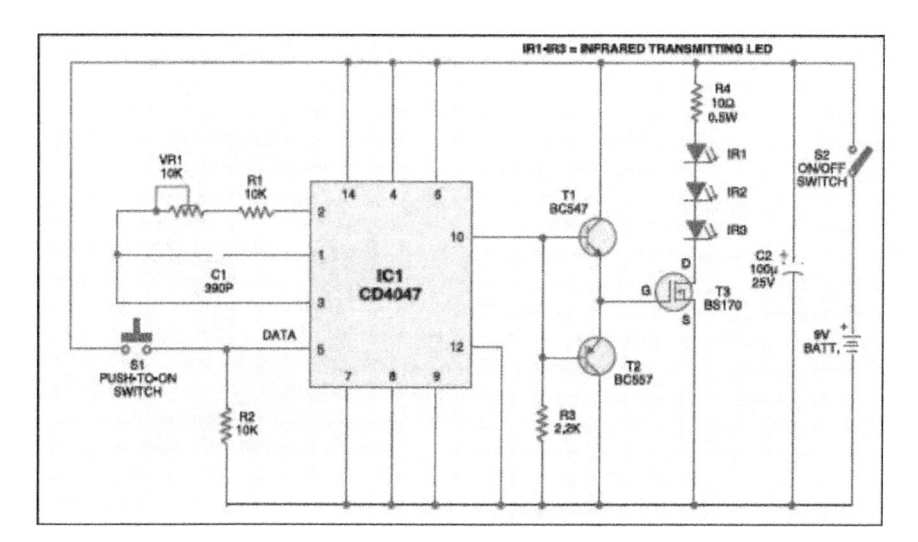

Because the MOSFET has a large capacitance across the gate-source terminals, a specialized drive arrangement has been made using a npn-pnp Darlington pair of BC547 and BC557 (as emitter followers) in order to prevent distortion of the gate drive input. This was done in order to keep the gate drive input from being altered in any way. The data that is going to be transferred is CMOS-compatible, and it is used for modulating the frequency of 38 kHz that is created by CD4047 (IC1). On the other hand, in the circuit that is being displayed here, the IR signal has been modulated and transmitted using the tactile switch S1.

291. Bench Power Supply Using a Computer's Power Supply

For electronics engineers and hobbyists, the most important tool is a good bench power supply. But bench power supplies, which are used in R&D labs, are expensive. The rating of output is also a big deal. The most important thing that labs need is a variable DC supply that can go up to 30V and has a few amperes of current. Because the circuit is so complicated, it's not easy to make a power supply for a lab. To give out a lot of current, it needs big, heavy transformers and semiconductors. Use a computer power supply unit as an easy alternative (PSU). The PSU can be turned into a good bench power supply with a few changes.

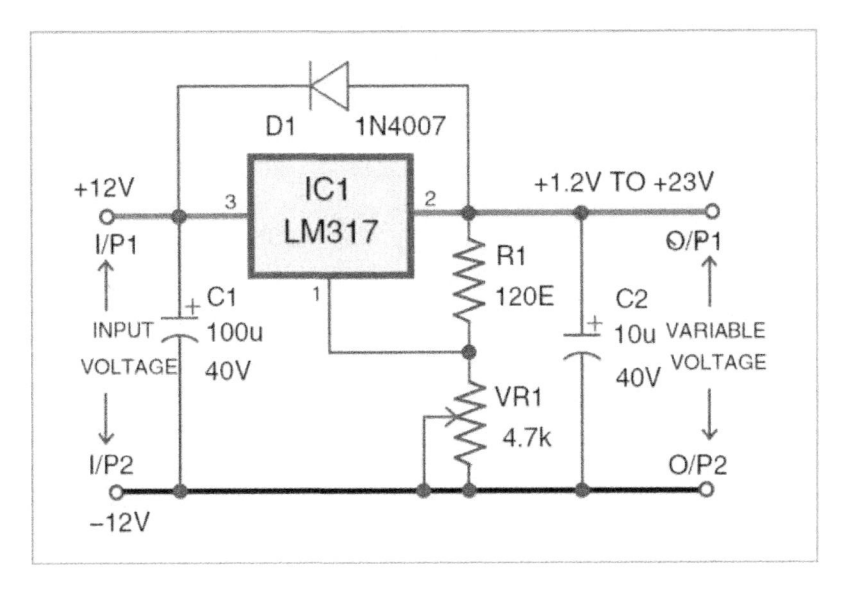

Fig: Adjustable regulated power supply circuit

Power up and testing: To test a computer PSU without connecting it to the motherboard, connect the green wire to any of the black wires (ground wires). When the power supply is "on," the cooling fan should turn on. You can now use a multimeter to check the voltages of all the different wires in relation to the black wire. If all the voltages match the color code table, the power supply is working well. Checking the voltage of the grey wire, which is a PGS or power-good signal, is another way. It should be about +5V.

When the power supply unit is opened: You already know that computers use 230V AC power. Before you can open the PSU enclosure, you have to unplug its 230V power cord. After you take the cover off the PSU and unscrew the case, watch out for the big capacitors on the PCB. Wait a little while or use the right tool to discharge them. If you don't do this, you could get a shock from capacitors.

Power supply with variable output: Use an LM338K or LM317 adjustable voltage regulator IC, as shown in Fig. 1, to turn the PSU into a variable power supply. 6. Connect +12V and -12V from the PSU to this circuit so that the maximum output ranges from 1.2V to +23V. But because the output current of -12V is only 500mA, the current output will drop to 500mA.

292. Leakage and Continuity Tester

Here is a simple leakage and continuity tester that costs money and can be used in two ways: normal mode and gain mode. Normal continuity testers can't find small leaks in electrolytic capacitors or large resistors, but the gain mode can. This is because the continuity current is amplified in the gain mode. The normal mode is used to test if the device is still working.

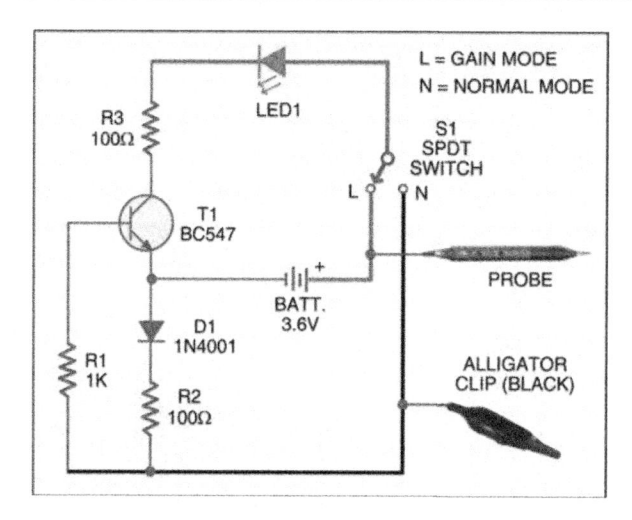

Fig. 1: Circuit of leakage and continuity tester

Circuit to test for continuity

In Figure 1, you can see how the leakage and continuity tester works. Because it uses a rechargeable battery (3.6V, 60mAh Ni-Cd), the instrument is small and light enough to fit inside a glue-stick tube.

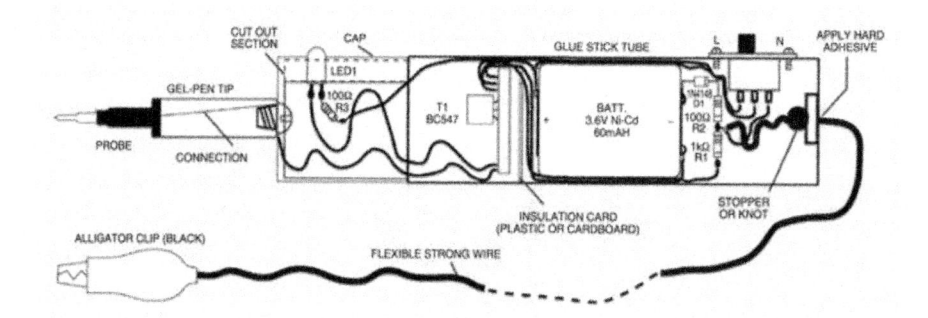

Fig. 2: Proposed arrangement for leakage and continuity tester

Using a transistor and a small base resistor R1 (1 kilo-ohm) that doesn't make much of a difference in a closed circuit makes the current flow stronger. Together, the 100-ohm resistor R2 and the signal diode D1 limit the direction and amount of charging. If you connect the probes to a 5V source, you can charge the battery through them. Use an LED1 that is red and see-through to find continuity quickly.

Circuit operation

To check for leakage, switch S1 to gain mode, attach the alligator clip to one of the component's terminals, and touch the probe to the other terminal. If LED1 lights up, it means there is a leak. In the same way, flip S1 back to normal mode to check for continuity. Clip the alligator clip to the circuit's ground terminal and touch the probe to the terminal you want to test for continuity or a short. If LED1 shines brightly, it means that the power is still on.

293. 5-Watt Audio Amplifier Using TA7222

Here is a simple 5-watt mono audio amplifier with IC TA7222 that uses a 9V battery or a 9V DC adapter to power a 4-ohm speaker. If you use two identical circuits, you can also make a stereo amplifier.

Circuit and working of Mono Audio Amplifier

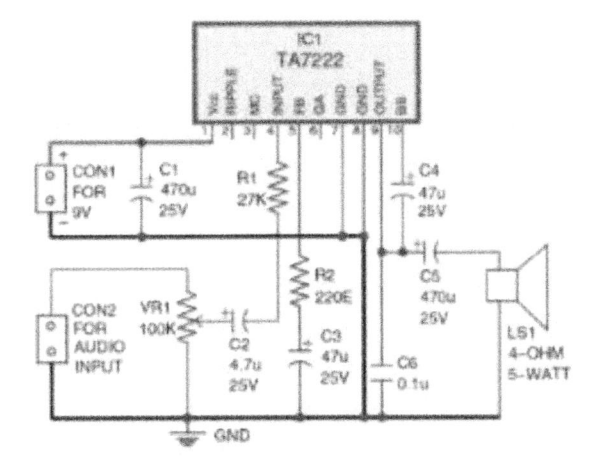

Figure 1 shows a diagram of how the 5-watt mono audio amplifier works. Aside from the audio amplifier TA7222 (IC1) and the 4-ohm speaker (LS1), this 5-watt AF amplifier uses a few resistors, capacitors, a potmeter, and two connectors.

The most important part of the circuit is IC TA7222 (IC1). Ten pins are on this IC. Pin 1 is hooked up to a power supply with +9V. A filter is made by putting a 470F, 25V capacitor (C1) between +9V and ground. The IC's pins 2 and 3 are not used. Pin 4 is connected to the audio input terminal by potentiometer VR1, capacitor C2, and resistor R1. The audio input signal is controlled by turning the pot on VR1.

Through resistor R2 and capacitor C3, pin 5 of the IC is connected to ground. No one uses pin 6. 7 and 8 are plugged in. Pins 9 and 10 are linked to capacitor C4. Through capacitor C6, pin 9 is linked to the ground. It is also connected to the 4-ohm speaker (LS1) through the capacitor C5.

294. Battery-Powered Night Lamp Using an Old LED Bulb

This circuit makes a battery-powered night light out of a cheap 3W LED board. This board has eight SMD2835 LEDs that are all wired together in a series. For proper lighting, brand-name LED bulbs have SMPS drivers and protection circuits. But some cheap LED bulbs get their power from capacitors and don't have any safety circuits. In this project, a switching regulator system called a 78S40 IC is used.

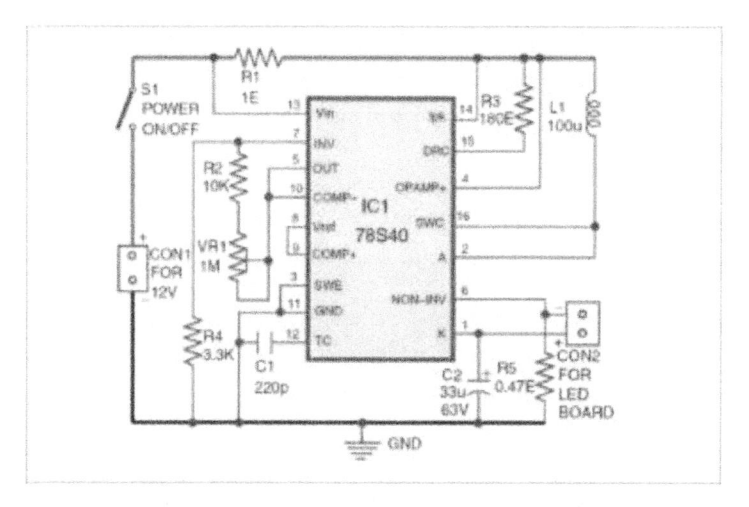

Fig. 1: Circuit diagram of battery-powered night lamp using an old LED bulb

SMD2835 LED module; SMD2835 is a new packaging structure with a self-heat-sink design and a high-efficiency LED module. It is used in many different kinds of lighting, including general, commercial, industrial, city, and car lighting. Here are some of the things it has.

Vertical structure: Chips with direct conduction fins are better at getting rid of heat, so they can handle more current (40-60mA).

A big heat sinks: The heat sink is 2-3 times bigger than SMD3528 and has a great design for getting rid of heat. It has large light-emitting surface. This improves the efficiency of light up to 90 per cent.

Cost: Even though SMD2835 is 25% less expensive than SMD3528, it can produce the same amount of light. The circuit in Fig. 1 is made up of a 78S40 DC-DC converter IC (IC1), eight high-bright SMD2835 LEDs (from an old LED bulb), and a few other parts. In this case, IC1 is set up as a changed boost converter. It does both boosting and driving a constant current. IC1 is in a 16-pin dual inline package that lets a reference and a non-inverting input of the comparator be pinned out. These extra features give this part a lot more flexibility and make it possible to use it in more complex applications.

Voltage boosting is done by IC1, L1, and the circuitry on the LED board itself. Sense resistor R5, which is connected in series on the LED board, is used to measure the voltage drop across it. This is done to control the current. Since the maximum voltage across the sense resistor is less than one volt, the on-chip operational amplifier is used to boost the sense voltage for feedback. The frequency at which the circuit works is set by capacitor C1. With preset VR1, you can change how bright the LED is. With switch S1, the LED board can be turned on or off.

295. Touchscreen and GLCD-Based Home Automation

This project uses a touchscreen to control electrical appliances in the home. The system has two relays that can control things like a light bulb and a fan. The control panel consists of a touchscreen that has been mounted on top of a graphical LCD. This touchscreen is used to send control commands via a pair of wireless radio frequency (RF) communication modules. The touchscreen panel on the transmitter side is connected to a microcontroller, which sends on/off commands to the receiver, where the loads or appliances are connected. By touching the right spot on the touchscreen panel, you can use the wireless RF modules to turn the loads on or off from a distance.

Circuit and how it works

The system is made up of two parts: the transmitter and the receiver.

The transmitter parts

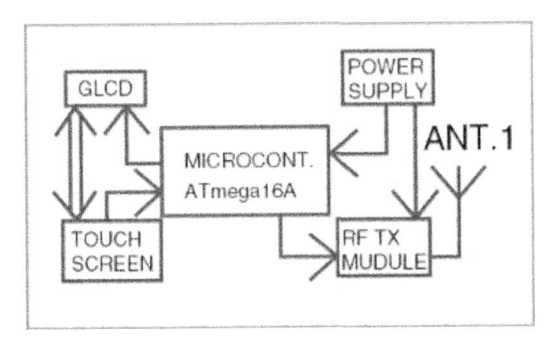

Fig. 1: Block diagram of the transmitter

In Fig. 1, you can see a block diagram of the transmitter part. The microcontroller, RF Tx module, and graphics LCD are all powered by the power supply unit (GLCD) The microcontroller is linked to the GLCD. Since the touch screen is clear, it is put on top of the GLCD. The microcontroller takes the information from the touchscreen and turns it into a four-digit binary code. The encoder takes this code and turns it into a code that can be sent through the RF transmitter and antenna.

Figure 2 shows a circuit diagram of the transmitter part. It is composed of several parts, including a transformer (X1), a bridge rectifier (BR1), a 128x64 GLCD (LCD1), a 4-wire touchscreen (TS1), a 5V voltage regulator (IC1), an ATmega16A microcontroller (IC2), an encoder (IC3), a 433MHz RF transmitter module (TX1), an antenna (ANT.1), and a few other things.

The microcontroller is directly linked to TS1. There are four wires on the touchscreen: X+, X-, Y+, and Y-. All of these wires are connected to the four analog-to-digital converter (ADC) pins on Port A of IC2. Wire X+ (pin 1) goes to PA0, wire X- (pin 2) goes to PA1, wire Y+ (pin 3) goes to PA2, and wire Y- (pin 4) goes to PA3.

ATmega16A's Port B pins PB0 through PB7 are connected to GLCD's data pins D0 through D7. Since the ATmega16A has a 1MHz clock built in, there is no need for an external crystal. Switch S1 connects the RST pin 9 of IC2 to ground.

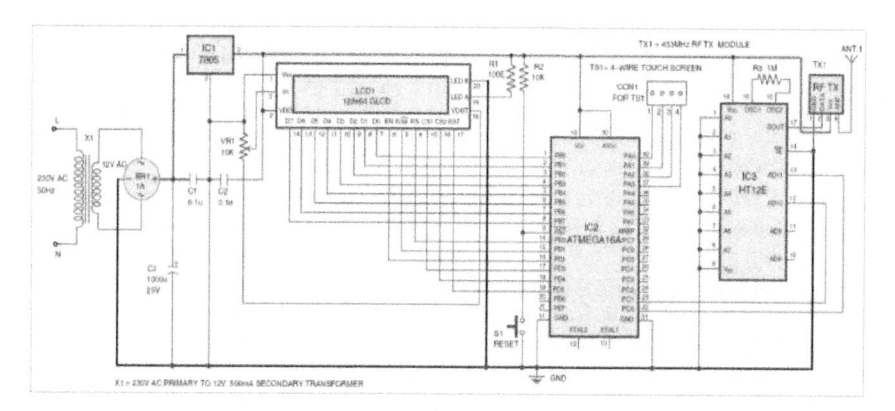

Fig. 2: Circuit diagram of the transmitter

When you briefly press S1, IC2 will be reset. The GLCD's control pins (RS, R/W, EN, etc.) are hooked up to Port D of the ATmega16A. Two chips in GLCD are controlled by control pins CS1 (pin 15) and CS2 (pin 16), which are each connected to pins 17 and 18 of IC2. The GLCD's reset pin (RST, pin 17) is connected to IC2's pin 19.

Since the touchscreen's output is analog, the user has to turn it into digital. This is done by the ATmega16A's built-in ADC. Port A has an ADC built in, so you don't need an external ADC chip (like ADC 0804). Encoder IC3 is linked to IC2's PC0 and PC1 ports. In both the sender and receiver circuits, all of the addresses from A0 to A7 of encoder IC3 and decoder IC5 are connected to ground. Pin 14 of IC3 is connected to ground for transmission. The bits are sent over RF signals from the output (DOUT) of IC3 to the data pin 2 of TX1.

The transmitter part is easy to understand. When the user touches the "light on" symbol on the touchscreen, the resistance of the touchscreen changes, which changes the voltage across the touchscreen output. The ADC of the ATmega16 microcontroller takes this difference in voltage and turns it into a digital signal. ATmega16A looks at the range of X and Y coordinates. If it is in the range that has been set, it sets the output at Port C to what the user wants. This signal is sent through an antenna with a 433MHz RF Tx module (ANT.1).

Parts of the receiver

Figure 3 shows a block diagram of the receiver part. The decoder and RF receiver module are powered by the power supply unit. The ANT.2 part of the receiver circuit gets the four-digit binary codes sent by the transmitter section.

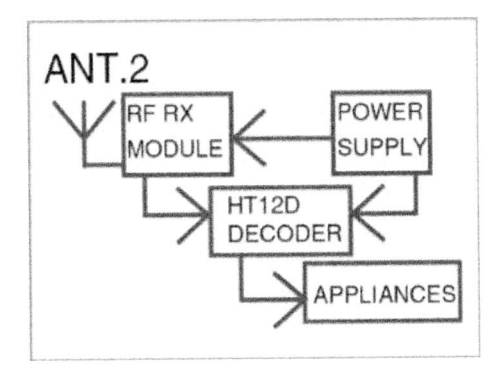

Fig. 3: Block diagram of a receiver

Here, the decoder is used to make sense of the signal that was received. The devices or appliances connected to the receiver section are switched on or off by the microcontroller in the transmitter section. Figure 4 shows a circuit diagram of the receiver part. It is made up of a transformer X2, a bridge rectifier BR2, a 5V voltage regulator 7805 (IC4), a decoder HT12D (IC5), a 433MHz RF receiver module (RX1), an antenna (ANT.2), a relay driver ULN2003 (IC6), a 12V voltage regulator 7812 (IC7), two 12V single-changeover relays (RL1 and RL2), and a few other

In the receiver part, the data pins of RX1 are connected to pin 14 of the HT12D decoder (IC5). All of IC5's address pins are connected to ground, just like IC3's address pins are connected to ground. Two of IC5's output pins, D10 and D11, are connected to two of ULN 2003's input pins, IN1 (pin 1) and IN2 (pin 2). (IC6). The relays RL1 and RL2 are connected to the output pins 16 and 15 of IC6. There are two voltage regulators used here, one for the circuit (5V) and one for the relays (12V).

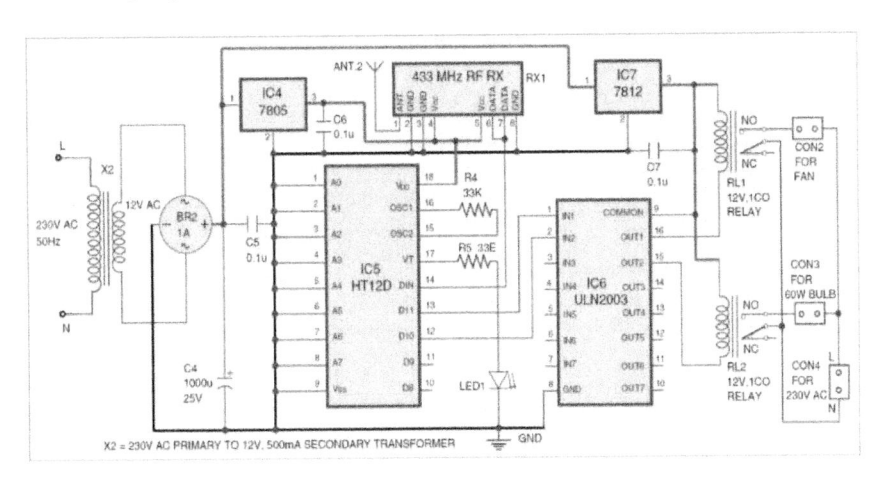

Fig. 4: Circuit diagram of the receiver

The receiver section is easy to use. When the user presses the "light on" symbol on the touchscreen on the transmitter side, 1s and 0s are sent through ANT.1 and ANT.2 to the receiver. The 433MHz RF

Rx module sends this information to the decoder IC5. The data is read by IC5 and sent to IC6. Through IC6, the respective relays are turned on. The relays' contacts turn on the appliances. When the user presses the "light off" icon on the touchscreen, the D10 output of IC5 goes down. Pin 2 (IN2) of IC6 gets this low output, which makes pin 15 (OUT2) high. This turns off relay RL2, which turns off the bulb. The same is true for the fan.

Software: The code is written in C for the AVR, and Atmel Studio 6.2 is used to compile it. STK500 is used to send the program to the microcontroller. At EFY Lab, the hex code was burned using a ProgISP programmer. Set the following fuse bits before you send the hex code (touchscreen.hex): set the following fuse bits: LOW = C1, HIGH = 99

296. GPS Clock using Arduino

Clocks that are synchronized with the Global Positioning System (GPS) show the correct time. These clocks are used everywhere, like at airports, bus stops, and train stations. These are also used a lot in the military. Here, we talk about a GPS clock made with an Arduino Uno R3, which is a microcontroller board based on the AVR ATmega328 chip and has six analog input pins and 14 digital input/output (I/O) pins. It has 32kB of ISP flash memory, 2kB of RAM, and 1kB of EEPROM. The board lets you talk in a serial way using UART, SPI, and I2C.

Circuit and how it works

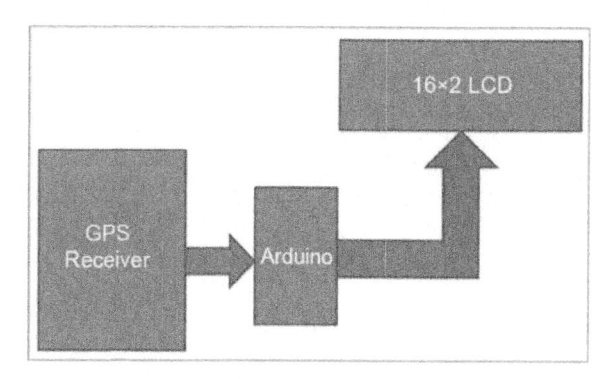

Fig. 1: Block diagram of GPS clock using Arduino

Figures 1 and 2 show, respectively, the block diagram and circuit of the GPS clock made with Arduino. In addition to the Arduino Uno board (BOARD1), the circuit uses a SIM28M GPS receiver module (GPS1), a GPS antenna (ANT.1), a 9V DC power supply adaptor, and a few jumpers for header connections. Here, we tried to get GPS time and date from a string ($GPRMC) that comes from GPS. About 70 characters make up this string. Figure 2 shows that Arduino is in charge of everything and receives GPS output signals. When Arduino gets the GPS output, it reads all of the strings and puts the ones it needs in a string or an array in the Arduino program. After storing the required string, Arduino pulls the time and date from the stored string and sends it to LCD1 so that the time and date can be shown. The data pins D4 through D7 of LCD1 are connected directly to Arduino pins 5, 4, 3, and 2. The EN and RS control pins of LCD1 are connected to Arduino pins 11 and 12, respectively. The Tx pin of the GPS receiver is directly hooked up to the Rx pin of the Arduino board. Keep in mind that the ground pins on both Arduino and GPS should be connected. Here, a 9600bps baud rate is used to run the GPS module. In the Arduino sketch, the class Serial.begin(9600) function is used to set the baud rate to 9600bps.

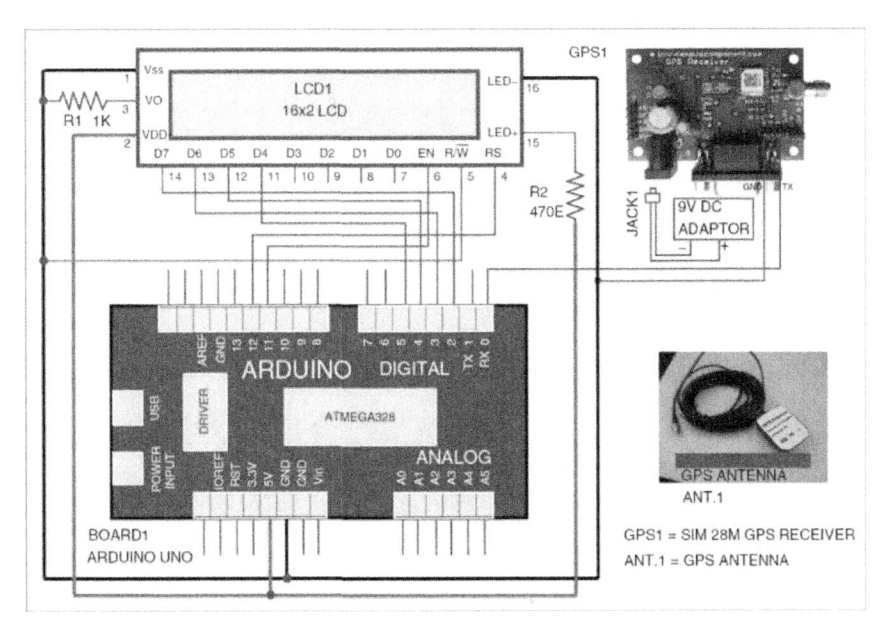

Fig. 2: Circuit diagram of GPS clock using Arduino

How GPS works

As shown in Fig. 3, the GPS receiver's output can be seen on the Arduino's serial monitor. Use the same connections as above to get this GPS data, but take the microcontroller ATmega328 off the Arduino board. Then, open the Arduino IDE and choose the Serial Monitor option to see the format of the GPS output. This window has a lot of strings, but you only need to use $GPRMC. Here, you can see the date after the ninth comma and the time in 24-hour format after the first comma. Date and time are taken from this $GPRMC string by an Arduino program, which then processes them and shows them on LCD1.

297. Optical Slave Flash Trigger

A "slave flash," which is also called a "speed light" or "strobe," is a flashgun that sits still until it sees a bright light. The slave flashes when it sees a flash. When you use flashguns that aren't attached to the camera, you can get great, dynamic lighting that would be hard to get any other way. The "Optical Slave Flash Trigger" is an off-camera flash trigger accessory that is shown here. It is very useful if you use a speed light. It's just a very good light sensor that picks up light pulses and wakes up the electronics, which then turns on the speed light that's attached.

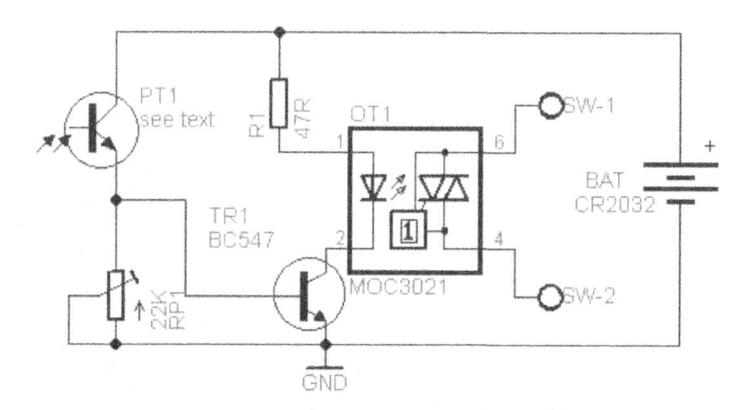

Project design

The main idea is to make things small, cheap, separate, and able to work on their own. The best way to do this is to make the heart of the circuitry out of cheap and easy-to-find parts. For such a simple design, I'd rather not use a microcontroller, even a cheap one (it would be overkill). The design is based on a very simple idea:

- Use a standard light sensor to pick up the light pulse from the master flash.
- Make sure the light intensity is higher than a predetermined baseline (which can be altered using a potentiometer to tweak the detection sensitivity)
- Use an electronic switch that doesn't make contact to turn on the save flash.

The Wheatstone bridge is the idea behind how this little circuit works. When light hits the phototransistor (PT1), its conductivity (and current) go up. This causes the driver transistor (TR1) to be biased in the forward direction. To initiate the slave flash, the resulting electrical current activates the opto-isolated triac (OT1) (thanks to an optical coupling, it enables safely to use even the oldest flashgun that could otherwise damage your circuitry by high voltage ignition). But you should know that the connected slave flash doesn't work when there's too much light around, like when it's bright outside. In the same way, it may misfire if you put a very bright artificial light source in its path. Here, the potentiometer (RP1) helps you set the right light detection threshold.

Parts Selection: A 5mm round phototransistor is used as the light sensor in this case. The type is a silicon NPN phototransistor that works quickly and well. Because the device is made of clear epoxy, it works with both visible light and infrared radiation. The opto-isolated triac in the circuit is a type MOC3021, which is made up of a gallium arsenide infrared emitting diode and a silicon bilateral switch that are connected optically. The whole circuit can be powered by the CR2032 Lithium Manganese Dioxide battery, which has a nominal voltage of 3V and a nominal capacity of 220mAh.

Interconnections: A female hotshoe is needed to connect the slave flash trigger circuit that has been put together to the slave flashgun. When the flash is in the hotshoe, all you have to do is connect the two trigger contacts. So, just use a (built-in or freestanding) sync cable to connect the switch outputs (SW1 and SW2) of the circuit to the electrical contacts (triggers) of the hotshoe.

Flash and the Camera; In the range of artificial light sources, 380-750nm, Xenon is known to come close to the "natural" colour spectrum of the sun. Xenon is used a lot in camera flash systems because of this. In an electronic flashgun, an electric charge is stored in a high-voltage capacitor until it is needed. When the flash is fired, the capacitor's stored energy is released through a flashtube. When Xenon gas is put into the flashtube, it gives off a very bright light for a very short time.

298. Automatic Water Refilled for Air-Coolers

Air coolers make a room feel cooler by adding water to the air. An exhaust fan or blower blows this moist air into the room, which makes the room temperature drop. These coolers need water added to them often. Here is a circuit that automatically adds water to an air-cooler tank when the water level falls below a certain level. Fig. 1 shows the circuit diagram for the automatic water refilled for air-coolers. It is made up of a NE555 timer (IC1), a BC547 transistor (T1), two reed switches (S1 and S2), and a few other parts. IC1 is set up like it would be in bi-stable mode, but switch S1 is connected to pin 6 instead of pin 4. As shown in Fig. 2, the two reed switches (S1 and S2) are placed at the top and bottom of the tank. How the circuit works is pretty easy to understand.

If the water level drops below what was set, the magnetic float closes the reed switch S2, which pulls pin 2 of IC1 to ground. This causes the voltage on pin 2 to drop below 1/3Vcc, which makes the output of IC1 go high. This gives power to the relay, which turns on the solenoid valve. So, the water starts to flow into the tank. When the tank is full, the magnetic float closes the reed switch S1. The output of IC1 goes low when pin 6 goes above 2/3Vcc. This turns off the power to the relay, which turns off the solenoid valve and stops water from flowing into the cooler tank. An air cooler's water tank has two holes: one for draining and one for overflowing. The level-sensing system for the water tank works like this: Attach a small PVC pipe to the tank between the two outlets (see Fig. 2). In the PVC pipe, put a magnetic float. Attach one reed switch to the pipe close to where the drain outlet is, and attach the other reed switch close to where the overflow outlet is.

An inexpensive speaker magnet and a hollow float ball can be used to create the magnetic float. The magnetic float should have a slightly smaller diameter than the PVC pipe so it can float freely from the bottom to the top or the other way around, depending on the water level. If you can, put the pipe in the inside wall of the tank and put the reed switches on the outside.

Circuit and working

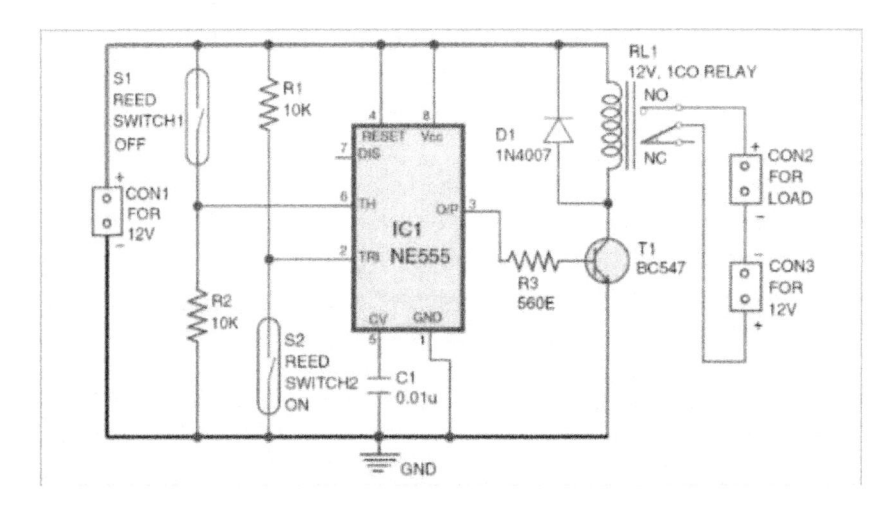

Fig. 1: Circuit diagram of automatic water refilled for air-coolers

Note: If the load is a 12V solenoid valve, as shown in Fig. 5, power the circuit and the load with a 12V DC supply. If the load is a 230V centrifugal pump, you can also use 230V AC power at CON3 and an external 12V adapter at CON1.

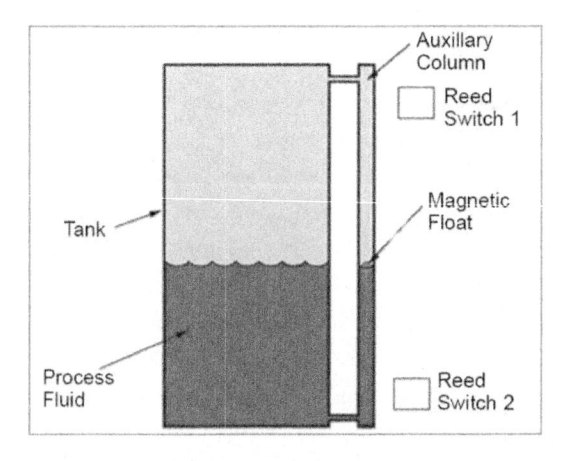

Fig. 2: Reed switch arrangement on the cooler tank

299. Electronic Horn

Here is a simple electronic horn circuit that uses the quadruple op-amp IC LM3900 (IC1). The four separate op-amps (A1 through A4) in IC LM3900 each have a wide output voltage swing. It can work at DC voltages up to 32V.

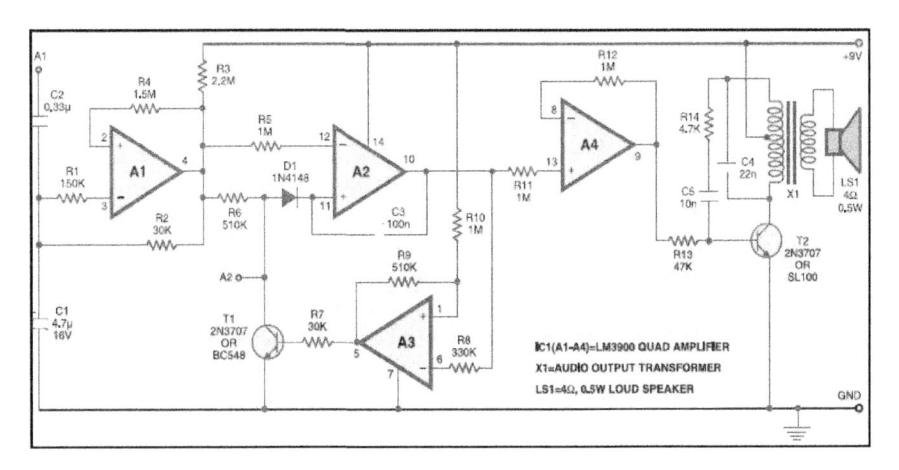

The first op-amp (A1) is set up to make a square wave with a low frequency. Op-amp A3 is a comparator, and op-amp A2 is an integrator. Together, A2 and A3 make an op-amp called a "wandering voltage generator." The op-amp A4 is set up as a buffer, and the base current for the npn transistor T2 comes from its output. A voltage-controlled oscillator is made up of the npn transistor T2 and the audio output transformer X1.

When the power is turned on, transistor T2 and transformer X1 make a basic tone. The frequency of this tone is changed by the wandering voltage generator, which is affected by the low-frequency squarewave generator. The circuit works with 9V that has been fixed. Connect point A1 to pins 1, 3, 4, 5, 8, 9, 10, 11, 12, and 13 of IC1 and point A2 to pins 1, 2, 3, 6, 8, 11, and 13 to make a variety of tones. When combined with an audio amplifier of around 10W, the circuit can function as a horn for a vehicle.

300. LME49710 Based Audio Amplifier

It's great to listen to music on a system you made yourself. Good-quality audio amplifier circuits can be made with a small number of parts. An LME49710-based audio amplifier is being talked about here. Fig. shows the prototype that the author made. 1.

LME49710 based audio amplifier circuit

Figure 1 shows that, the power amplifier based on the LME49710 chip has a power supply unit and an audio amplifier unit.

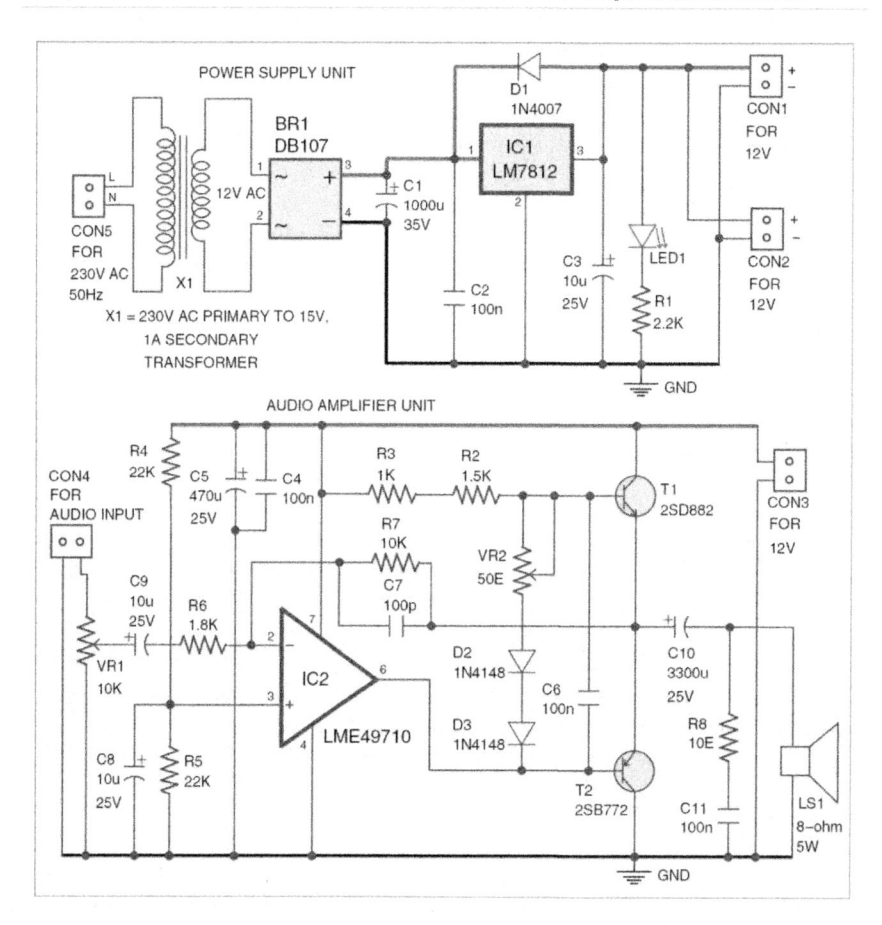

The linear power supply is made up of the transformer X1, the bridge rectifier DB107 (BR1), the 12V voltage regulator LM7812 (IC1), and a few other parts. The primary of X1 goes from 230V AC to 15V, 1A. AC is turned into pulsating DC by DB107, which is then filtered by capacitors C1 and C2 and sent to a 12V regulator (IC1). Diode 1N4007 (D1) is used to protect IC1 from voltage going in the wrong direction. LED1 shows if the power is on or off. LED1 is limited by the amount of current that flows through resistor R1. The LME49710 (IC2) is a high-performance, high-fidelity, audio-operational amplifier that is the center of the audio amplifier circuit. The 12V power supply is hooked up to the power pins on the op-amp. For the single-supply operation to work, the non-inverting terminal of op-amp IC2 must have a virtual ground with a voltage of $Vcc/2 = 12V/2 = 6V$. The circuit can be made with a simple voltage divider made of resistors R4 and R5 and a capacitor C8 that lets the voltage flow around the divider. The noise is decoupled at the IC level by capacitors C4 and C5. C5 is an electrolytic decoupling capacitor that cuts down on low- and mid-frequency noise, and C4 cuts down on high-frequency noise in the power supply lines. Connector CON4 is where audio comes in. At the input stage, capacitor C9 blocks DC voltage, and potmeter VR1 controls how loud the sound is.

The gain of amplifier IC2 is set by R7 and R6. R7 and C7 limit the bandwidth of the amplifier and keep it linear. Resistors R2 and R3 set the bias current of the diodes D2 and D3 that make up the Vbe multiplier, and the potentiometer VR2 helps adjust the bias current. Vbe multiplier is the DC bias voltage that makes up for the drop in voltage between the base and emitter when output transistors T1 and T2 turn on.

DC bias also keeps T1 and T2 on even when the speaker or load isn't getting any current. T1 and T2 power the speaker through C10, which also stops DC from getting in. High-frequency noise that comes from speaker LS1 is sent on a low-impedance path by resistor R8 and capacitor C11. Make it a stereo amplifier by making a second copy of the same circuit. The second amplifier unit can use the same power source at CON2.

Summary:

This chapter will walk you through the fundamentals of electronic components and show you 300 different ideas for electronic projects. In the next chapter, we'll look at what Arduino is and how it works. We'll also look at the different kinds of Arduino and how they work. There will be 100 ideas for Arduino projects. Using them will help you think of ideas that will help you see your dream project come to life.

CHAPTER 2:

TOP 100 ARDUINO PROJECTS

Introduction

The first ever Arduino controller board was born in 2005 at the teaching space of the Interactive Design Institute in Ivrea, Italy. The Interactive Design Institute has an article about a wiring design by a Colombian scholar named Hernando Barragan. The name of the proposal thesis was **"Arduino: The Revolution of Open Hardware."** Of course, it sounded slightly different from the typical proposal, but nobody would have believed that it would carve a niche in the domain of electronics.

The Arduino software IDE was developed by David Mellis and was based on Wiring. Previously, Gianluca Martino and Tom Igoe joined the development of Arduino Mission, and all five are well-known as the actual creators of the Arduino board. They needed a controller that was straightforward, easy to associate with different kinds of modules and components (such as LEDs, motors, relays, and sensors), considerably weightless, also easy to access in the open-source community, and simple to program. It also wanted to be affordable and easily accessible because students and artists aren't known for having a lot of money. They chose the Atmel AVR type of 8-bit microcontroller (MCU or C) devices and aimed a self-sufficient circuit board with easy-to-use connections, wrote bootloader firmware for the microcontroller on paper, and finished it all in a basic integrated development environment (IDE) that used programs called **"sketches."** The result was the Arduino hardware.

Microcontroller

A minicomputer on a single chip, having a processor, input/output memory, an analog to digital converter (ADC), and a digital to analog converter (DAC), is generally "embedded" inside some micro device that they control. A microcontroller is often small and cost-efficient. Microcontrollers are designed to perform a specific low-power application, making them ideal for embedded systems, whereas microprocessors are better suited for general computing applications that require more complex and versatile computing operations.

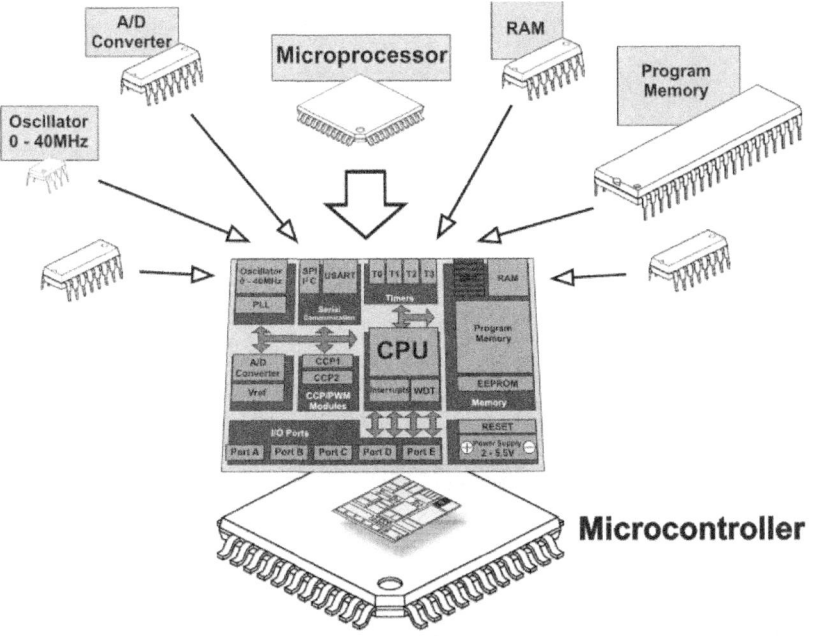

Fig. 0-1 Microcontroller versus Microprocessor

Development Board

A printed circuit board that contains a microprocessor and little or no hardware dedicated to a user interface and is designed to facilitate work with a specific microcontroller is a development board. On the development board, typical components include:

- Power circuit
- Simple input; usually buttons and LEDs
- Programming interface
- I/O pins

Here are some popular development boards listed:

1. Arduino
2. BeagleBone Black
3. Raspberry Pi
4. Intel Galileo
5. Goldilocks
6. pcDuino
7. Uruk
8. ExtraCore

What is the Arduino?

The word "Arduino" can mean 3 things

A physical piece of hardware

A programming environment

A community & philosophy

Why Arduino Developed?

- Physical Computing - by means of components which able to interact with people, besides by the world around us
- The Arduino was initially developed for artists, inventors and designers, to make prototype interactive displays
- Intended for non-scientists, less knowledge is required to learn.
- Minimalist programming a "**Forgiving**" microcontroller board capable of dealing with a wide range of wiring-connection errors

What can Arduino be used to teach?

- To understand preliminary electronics (voltage, current, resistance)
- What electronic components, sensors, and actuators work
- Elementary programming and troubleshooting
- Design of simple scientific devices and equipment
- Overcome the challenges of interacting with users via a DIY project (e.g., messages, formatting numbers, ease of use, etc.)
- Statistics and difference in data collecting and visualization

The Arduino will open up new opportunities for countries in development. In areas of medicine and manufacturing where regulations aren't as strict, the Arduino is going to open up a whole host of options and capabilities, ranging from low-cost PLC controllers to medical devices. These will be particularly useful in areas.

DIFFERENT TYPES OF ARDUINOS

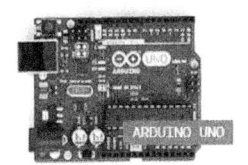

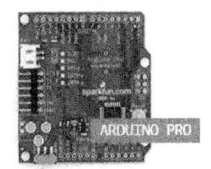

The microcontrollers used in the various models of Arduino

Name	Processor	Operating/Input Voltage	CPU Speed	Analog In/Out	Digital IO/PWM	EEPROM [kB]	SRAM [kB]	Flash [kB]	USB	UART
101	Intel® Curie	3.3 V / 7-12 V	32MHz	6/0	14/4	-	24	196	Regular	-
Gemma	ATtiny85	3.3 V / 4-16 V	8 MHz	1/0	3/2	0.5	0.5	8	Micro	0
LilyPad	ATmega168V ATmega328P	2.7-5.5 V / 2.7-5.5 V	8MHz	6/0	14/6	0.512	1	16	-	-
LilyPad SimpleSnap	ATmega328P	2.7-5.5 V / 2.7-5.5 V	8 MHz	4/0	9/4	1	2	32	-	-
LilyPad USB	ATmega32U4	3.3 V / 3.8-5 V	8 MHz	4/0	9/4	1	2.5	32	Micro	-
Mega 2560	ATmega2560	5 V / 7-12 V	16 MHz	16/0	54/15	4	8	256	Regular	4
Micro	ATmega32U4	5 V / 7-12 V	16 MHz	12/0	20/7	1	2.5	32	Micro	1
MKR1000	SAMD21 Cortex-M0+	3.3 V / 5V	48MHz	7/1	8/4	-	32	256	Micro	1
Pro	ATmega168 ATmega328P	3.3 V / 3.35-12 V 5 V / 5-12 V	8 MHz 16 MHz	6/0	14/6	0.512 1	1 2	16 32	-	1
Pro Mini	ATmega328P	3.3 V / 3.35-12 V 5 V / 5-12 V	8 MHz 16 MHz	6/0	14/6	1	2	32	-	1
Uno	ATmega328P	5 V / 7-12 V	16 MHz	6/0	14/6	1	2	32	Regular	1
Zero	ATSAMD21G18	3.3 V / 7-12 V	48 MHz	6/1	14/10	-	32	256	2 Micro	2
Due	ATSAM3X8E	3.3 V / 7-12 V	84 MHz	12/2	54/12	-	96	512	2 Micro	4

Arduino LilyPad

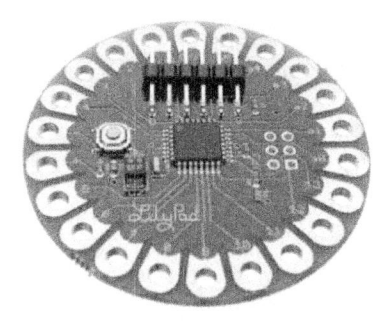

The LilyPad Arduino is a microcontroller board intended for wearables and e-textiles. LilyPad is well known for its clothing-based projects. It can be stitched to textile material, fabrics, and cloth, and conductive textile fiber can also be used to mount power sources, sensors, and actuators.

Arduino BT

The Arduino BT is an Arduino board that has a Bluetooth module built in. This module can be used for different types of wireless communication and remote control.

Arduino Esplora

Esplora is an Arduino board that can be used as a game controller and is made for easy gaming. The Esplora Arduino board consists of a joystick, linear potentiometer (slider), microphone, buttons for control, a temperature sensor, a light sensor, and a three-axis accelerometer, not the typical set of input/output pins.

Leonardo

- Compared to the Arduino Uno, Leonardo is a minor upgrade.
- It has built in Micro USB compatibility
- Utilized to PC as a mouse or keyboard

Arduino Due

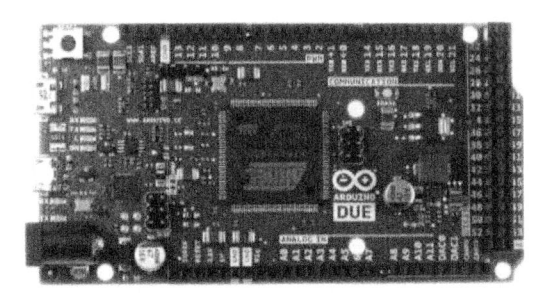

- Compared to the Arduino Uno, the Arduino Due has a much faster processor and a lot more analog and digital pins.
- Due is similar to the Arduino Mega, Furthermore, it operates on both 3.3 and 12 volts.

Arduino Nano

The Arduino Nano is a small microcontroller breadboard embedded model with a Micro USB port built in. It is a breadboard-friendly, full-featured microcontroller with all of the electrical features of the Die/Due/Uno boards plus extra digital and analog I/O pins and an onboard +5V AREF jumper.

Arduino Micro

When size matters: the three smallest boards ever created by Arduino developers, the Micro, Nano, and Mini.

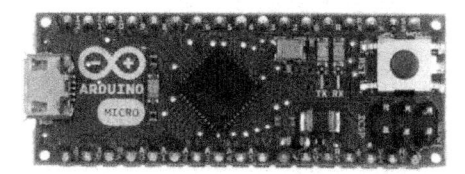

- ❖ Arduino Micro comprises all functionality of Uno and Leonardo
- ❖ Arduino micro is simply working on a breadboard

What are the benefits of using Arduino UNO?

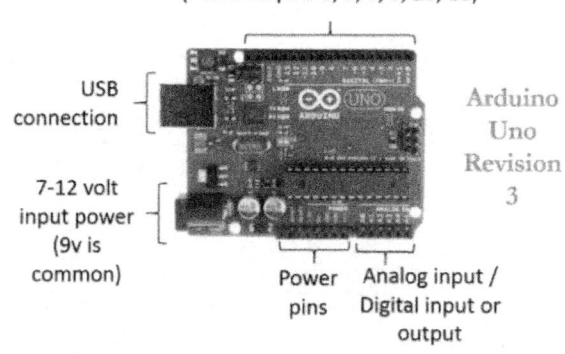

Figure: Overview of the Arduino UNO.

- ❖ Arduino is an open-source electronic prototyping board based on flexible, easy-to-practice hardware and software.
- ❖ We can get Arduino for less than $10.00 if you assemble your own board or buy a clone.
- ❖ Cross Platform IDE (found in operating systems like Windows, MAC, and Linux), Open-source IDE, and extensions.

What approach have I followed to learn Arduino?

1. **Start simple** - build confidence and **learn by doing** small projects before going to difficult ones.
2. Practice components that will capture the thoughtfulness and imagination of the students
3. Build a new project by modifying the previous one.
4. Make a "problem" for pupils to resolve that THEY will understand through practical

however not too complicated
5. Find problems in your society, gather knowledge from it, make an idea, then instantly do a development with it, and give a solution with your project.
6. Teach pupils just how to find required info from datasheets (e.g., tolerances, current limits, etc.) and, likewise learn from the internet

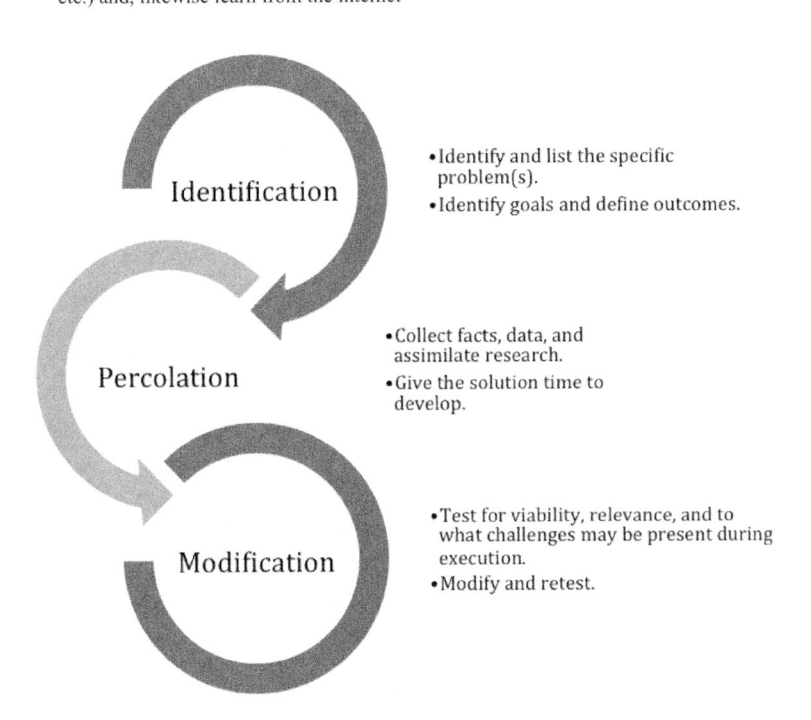

Identification
- Identify and list the specific problem(s).
- Identify goals and define outcomes.

Percolation
- Collect facts, data, and assimilate research.
- Give the solution time to develop.

Modification
- Test for viability, relevance, and to what challenges may be present during execution.
- Modify and retest.

Identification The first step in beginning a project is identifying, defining, and articulating the problem. The use of brainstorming and mind maps is valuable for some of the more difficult problems. **Percolation** Data collection, research, and idea exploration are the next steps. The main reason that **innovation** is so difficult is because many ideas in a vacuum may not seem realistic or appropriate.

LIST OF 100 ARDUINO PROJECTS IDEAS

301. Arduino Based Autonomous Fire Fighting Robot

Firefighters can independently detect fires with this sophisticated firefighting robotic system. Fire fighters face an increasing risk of death as technology leads to an automated system and self-traveling vehicles. If a fire is not controlled, it will spread rapidly. Even an explosion is possible in the event of a gas leak. The system delivers by overcoming this issue, safeguarding the lives of our heroes. An Arduino Uno is used to power this firefighting robotic system which is comprised of an ultrasonic sensor positioned on a servo motor for obstacle detection and free route navigation. Despite its small size, it has the capability to detect and extinguish fires. It also has water tank and spray mechanism for extinguishing flames. Servo motor is used to cover maximum area with water spraying nozzle. The 12V pump uses an electric motor to pump water from the main tank to the water nozzle. Because of its constant current consumption, this pump needs driver circuit.

Block Diagram:

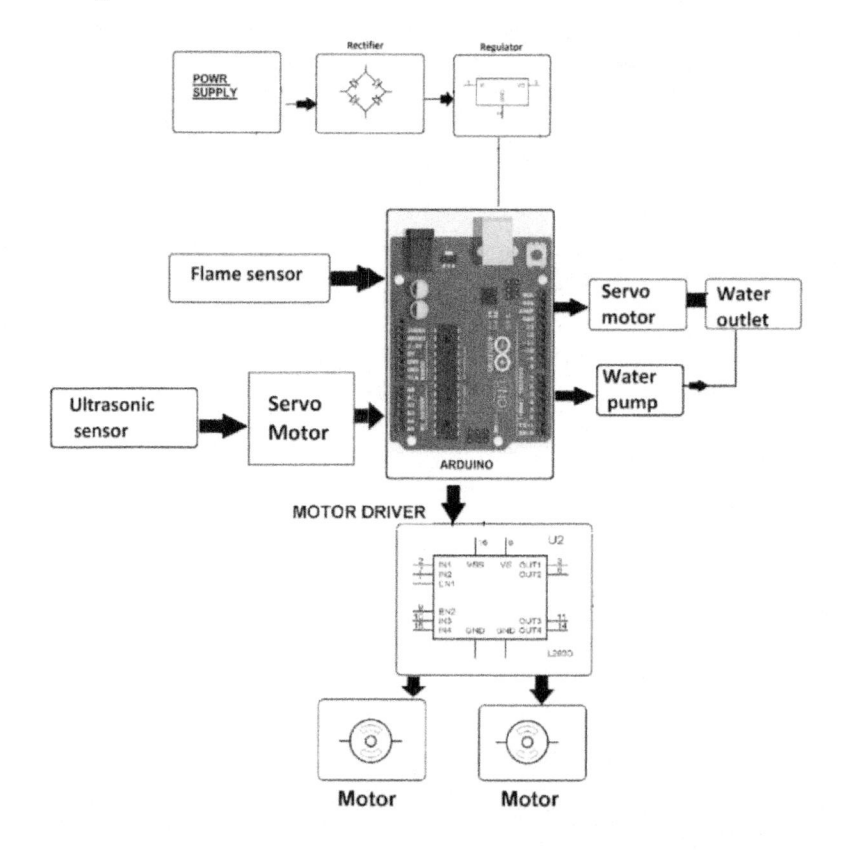

Hardware Specifications

- Arduino Uno
- Ultrasonic Sensor
- Fire Fighter Robot Body
- Fire Sensor
- Buzzer
- LCD Display
- Resistors
- Capacitors
- Transistors
- Cables and Connectors
- Diodes
- PCB and Breadboards
- LED
- Transformer/Adapter
- Push Buttons
- Switch
- IC
- IC Sockets

Firefighting robotic systems are designed with specific tasks in mind. The primary aspects of fire control and suppression are analyzing and locating fires as well as conducting search and rescue operations. Robotic systems for controlling fire, such as automatically activating fire alarms and sprinklers, can quickly extinguish anything in a heavily populated or hazardous area. These systems are usually simpler and primarily rely on UV or infrared sensors. Since they are fixed, they don't typically change over time.

Software Specifications

1. Arduino Compiler
2. Programming Language: C

302. Robot Snake based on Arduino controlled by Android

Twelve servo motors drive the snake's segments, which are joined with metal brackets. A 7.4-volt battery pack powers the servos and controls them with an Arduino Mega. An android app can be used to control the snake using Bluetooth.

Block Diagram:

In addition to autonomous movement, the snake is additionally capable of passive activity. Various different types of servos and brackets can be used to construct such a robot. There are 12 segments of the robot, each containing a servo motor, a C-bracket, a side bracket, a set of Lego wheels, and a wire clip. The Lego wheel axle must have two screw holes for the C-bracket to be attached to it. In spite of the fact that nine segments are already connected, they need to be expanded with two tail segments in order to accommodate the Arduino and batteries. The side brackets and C-brackets are connected to the side brackets. A 5AA battery holder is used to power the Arduino and therefore the tail of the snake. A separate energy supply powers the servos. That is the 7.4-volt battery pack. The voltage pin is attached to a five-volt Arduino pin. On the receiver, the lower pin is attached to ground.

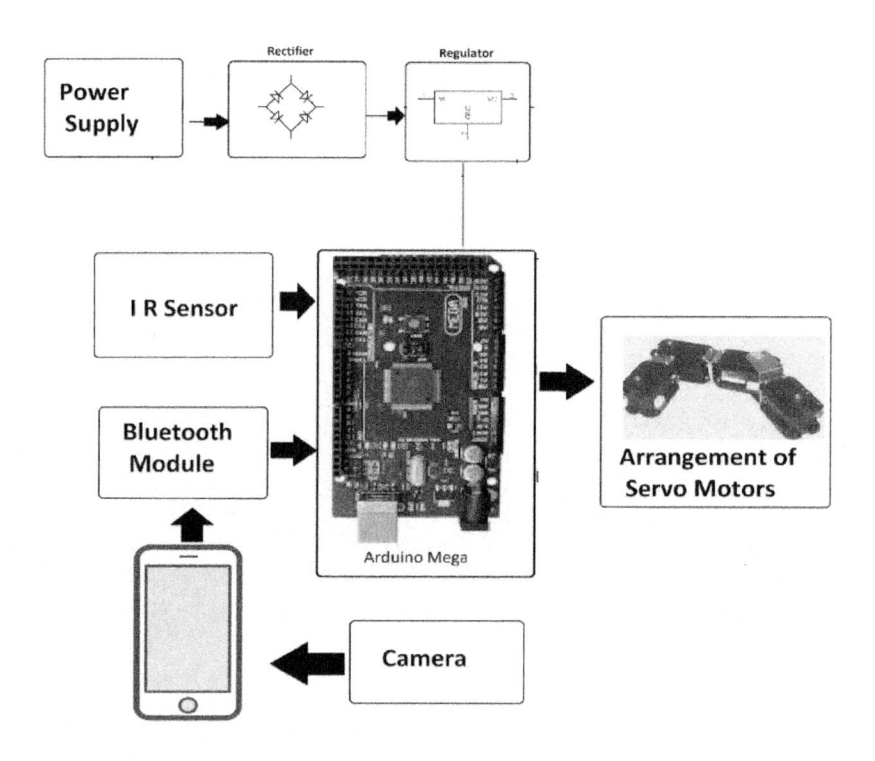

Hardware Specifications

- Arduino mega
- IR Sensor
- Servo Motors
- Bluetooth module
- Camera
- Cables and Connectors
- PCB and Breadboards

- Push Buttons
- Switch
- IC

Software Specifications

- Arduino Compiler
- Programming Language: C

303. Intelligent Gas Leakage Detector based on IoT

LPG leaks, which result in explosions, are common occurrences in day-to-day life. If leakage is not detected early, it can cause major damage. The MQ5 gas sensor can detect gas leakage but we were not able to use it before. In this IOT gas leak sensor, the machine will get attached to Wi-Fi, and the device will enable you to adjust parameters accordingly. Installed in gas storage areas, hotels, homes, and hotels, such IoT and Arduino systems detect leaks of LPG.

It uses a gas sensor called the MQ5 to detect the presence of LPG gas. LPG gas present in the air will be monitored continuously by this device. A green LED on the control circuit will light up if the value of LPG gas in the air is within the set limit, thereby giving a safe signal. In addition, if excessive gas levels are detected in excess of the predefined boundary, the RGB LED will turn red, and the solenoid will shut off and update the IoT value. Detecting gas leakage in the surrounding will be easy and effective with this Arduino and Internet of Things project.

Block Diagram:

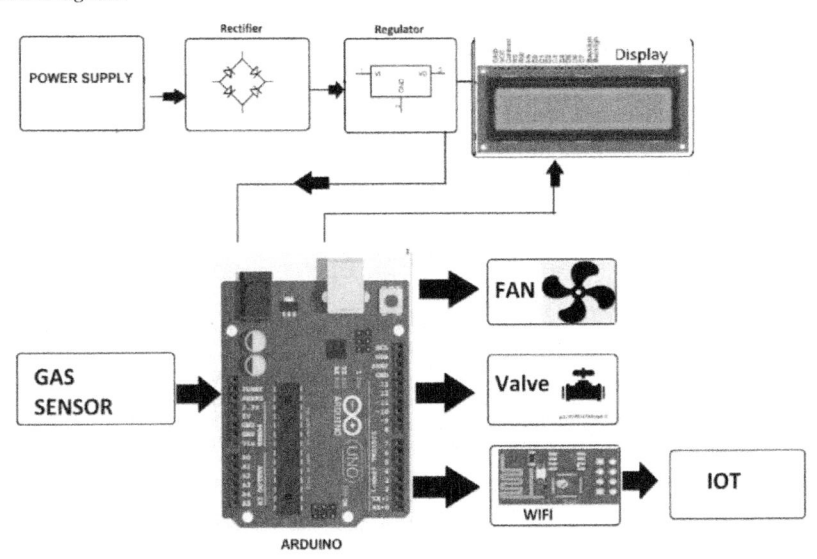

Hardware Specifications

- Arduino
- LCD
- Wi-Fi module
- Dc fan
- Gas sensor
- Buzzer
- Regulator
- Crystal Oscillator
- Resistors
- Capacitors
- Transistors
- Cables and Connectors
- Diodes
- PCB and Breadboards
- LED
- Transformer/Adapter
- Push Buttons
- Switch
- IC
- IC Sockets

Software Specifications

- Arduino Compiler
- Programming Language: C

304. Wireless Black Box for Cars

The project is about "Wireless recording systems for cars". This project has a primary objective of developing a vehicle black box system that would allow the installation of it into any vehicle all over the world. These paradigms are usually designed with a minimal range of circuits. Wi-Fi black boxes tell us about crashes and store information like the time, date, temperature, vibration, alcohol level, etc. in real-time every three seconds.

Block Diagram:

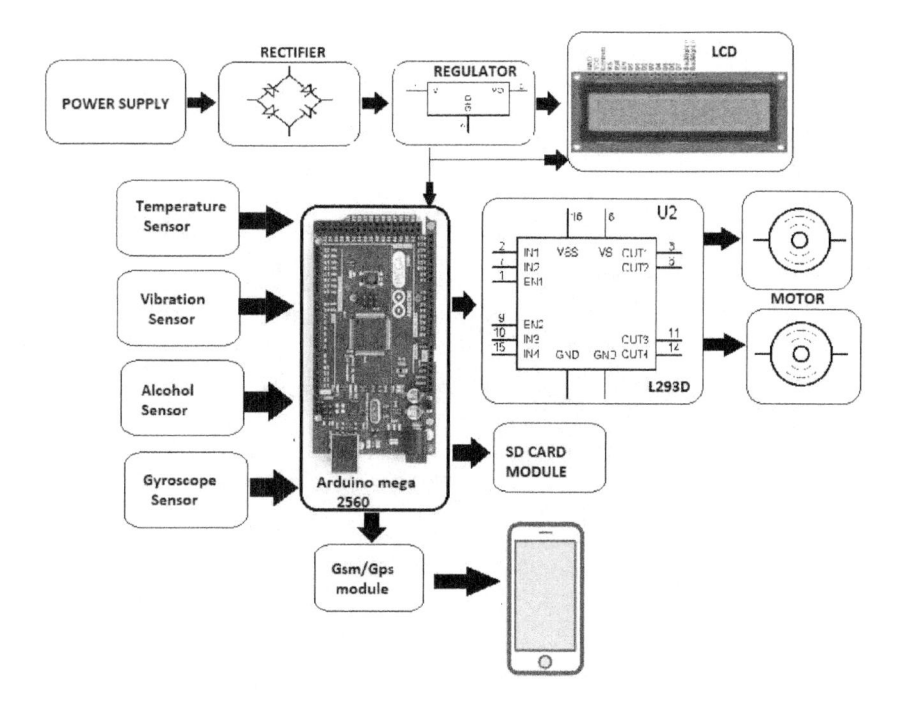

The system built into the car will send a message to the registered telephone numbers, such as emergency numbers of the police station, hospitals, relatives, and vehicle owners, depending on their location. Sensors including temperature and humidity sensors (DTH11) have been used. Vibration sensors from cars monitor vibrations during accidents. A steering wheel mounted alcohol sensor would tell if the driver was drunk. The tilt is detected by the gyroscopic sensor during an accident. Arduino mega2560 will receive a signal from all the sensors, which will then be sent to the microcontroller.

These projects used GSM, SD card, GPS, and the like. Their contributions were vital to the completion of the project. This project may be enhanced by adding video cameras, voice recorders, voice-controlled systems, and automatic warning systems in the future.

Hardware Specifications

- Arduino 2560
- GPS/Gsm module
- Temperature Scanner
- Alcohol Scanner
- Vibration sensor
- Gyroscope
- SD card module
- LCD
- L293D
- DC motor
- Power supply
- Buzzer
- LED
- Switch
- LCD
- Crystal
- Push Buttons
- Capacitors
- Resistors

Software Specifications:
1. Arduino ide
2. MC Programming Language: C

Project Implementation:

As a part of this project, there are specific sensors, like the temperature sensor (DTH11), which calculates temperature and humidity. Vibration sensors pick up sensations felt by drivers throughout an accident. If a driver is drunk, indicators on the steering wheel will indicate this. A gyroscopic detector is used during a crash to show tilt. All parameters are measured and sent directly to the Arduino. By uploading all data to the fire department's server, the ESP8266 module collects data. This model utilizes the GSM module, the SD card module, and the GPS module to achieve the desired outcome.

305. Smart Charger Monitoring System using Arduino

Batteries are charged or recharged by transferring energy into them via the use of a device called a battery charger or recharger. There are a variety of charging protocols available for batteries of different sizes and types. Smart battery chargers are primarily switch-mode power supplies that function in concert with battery handling and storage devices to control and monitor charging processes.

An Arduino is used to power this smart charger this intelligent charging system charges three batteries with 12V power simultaneously. During full charge, a battery is automatically disconnected from the mains. It has an automatic power cut-off system. A smart charger is mainly an inverter for switching on and off power supplies. It also communicates with the smart battery packs. Moreover, the LCD display module periodically displays the charge level for the battery.

Hardware Specifications

- Arduino Uno
- Relay
- Relay Drivers
- LCD Display
- Crystal Oscillator
- Resistors
- Capacitors
- Transistors
- Cables and Connectors
- Diodes
- PCB and Breadboards
- LED
- Transformer/Adapter
- Push Buttons
- Switch
- IC
- IC Socket

Block Diagram:

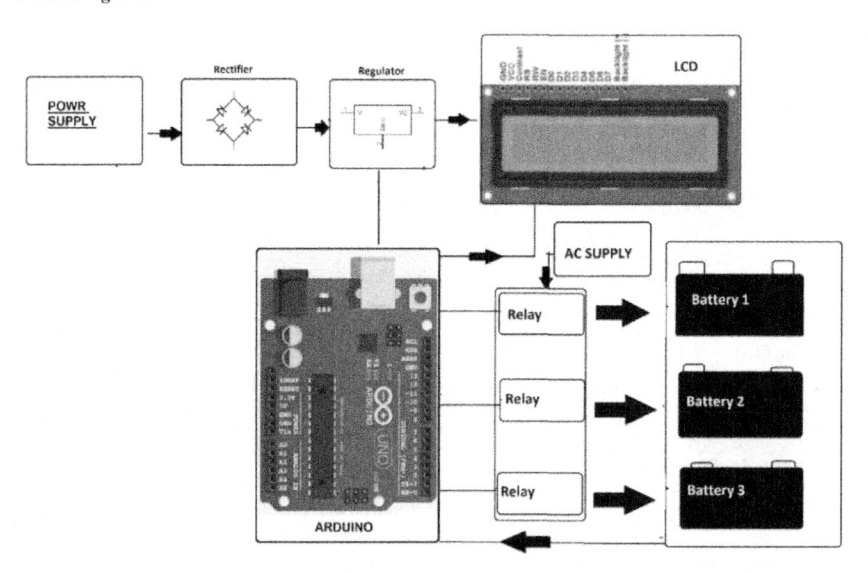

306. Arduino Based Autonomous Fire Fighting Robot

This sophisticated firefighting robotic system independently detects and extinguishes fire. Today, as the world slowly grows toward the automation of systems and self-driving cars, firefighters continue to run the risk of dying in the line of duty. When fire is not controlled it spreads rapidly. An explosion may occur in the case of a gas leak. Therefore, to meet the challenge of overcoming this issue and protecting our hero, the systems we have in place come to the rescue.

An Arduino Uno development board is used to power the firefighting robotic system, which utilizes an ultrasonic sensor and servo motor for obstacles sensing and free route guidance. Fire flame sensors ensure that the sensor can detect and move in close proximity to the fire. Firefighters extinguish the fire with water tanks and spray mechanisms. Servo motors drive water spraying nozzles to cover maximum area. The primary water tank has a 12V pump which pumps water up to the water nozzle.

As such a pump consumes much more current than a controller can handle it needs its own driver circuit.

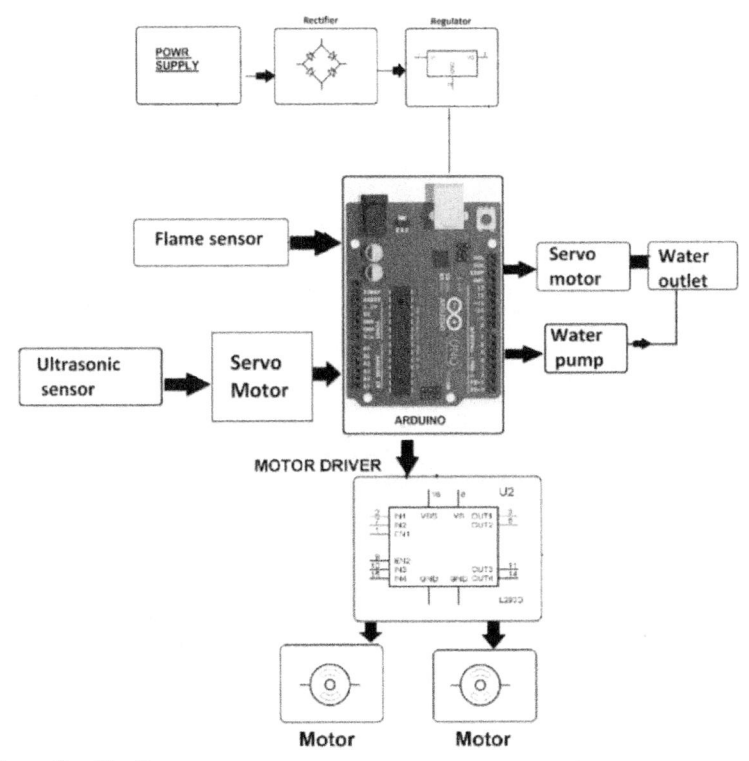

Hardware Specifications

- Arduino Uno
- Ultrasonic Sensor
- Fire Fighter Robot Body
- Fire Sensor
- Buzzer
- LCD Display
- Resistors
- Capacitors
- Transistors
- Cables and Connectors
- Diodes
- PCB and Breadboards
- LED
- Transformer/Adapter
- Push Buttons
- Switch
- IC
- IC Sockets

Software Specifications

1. Arduino Compiler
2. Programming Language: C

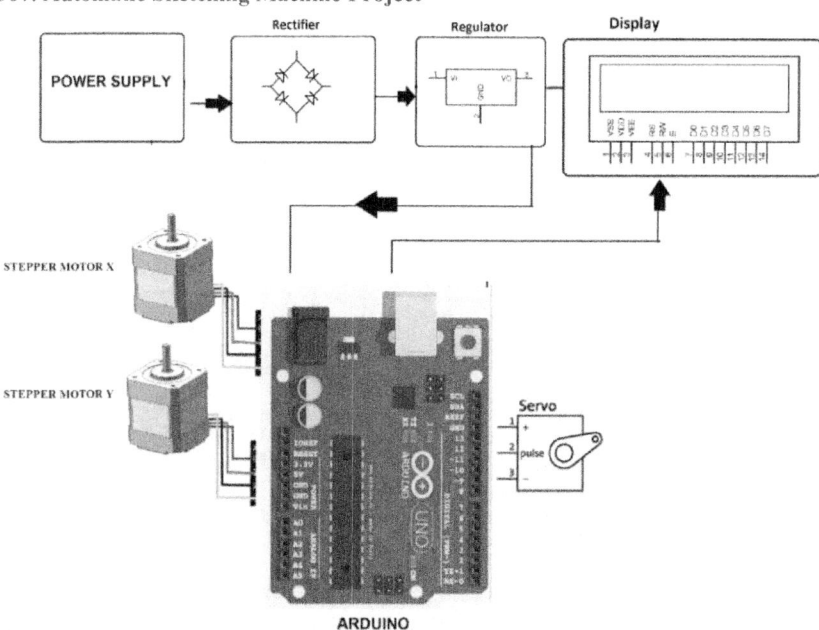

307. Automatic Sketching Machine Project

With the advent of machines, machines can now also draw perfect sketches. Using the process proposed here, a machine can sketch pictures accurately and fast just like a human. With an Arduino based circuit, which is attached to motors and belt-based machinery, the designed system will be able to draw a sketch with a pen.

Block Diagram:

In order to transmit movement commands according to the image fed into the Arduino based circuit, two stepper motors are interfaced with it. The sketching process is controlled then with a well-planned mechanism to achieve this task. The motor raises the pen above the page where it is not needed, so the pen touches the paper only where a dot has to be placed. A 2D sketching mechanism has been created using this mechanism in conjunction with the motion of the x and y axes.

Hardware Specifications:

- Arduino UNO
- Stepper Motors
- Servo Motor
- LED's
- Resistors
- Capacitors
- Diodes
- Connectors & Cables
- Connecting Rods
- Pulley
- Rubber Belts
- Bed Frame

- Bearings
- Screws & Joints

Software Specifications:
- Arduino IDE
- MC Programming Language: C

308. Arduino based Sun Tracking Solar Panel

The future of mankind depends upon harnessing solar energy properly, in place of the traditional energy sources it has used for a long time. We branched out from the existing project to design this so solar energy can be harnessed even more efficiently.

This project is designed to be controlled by a solar panel using a controller board based on the Arduino controller board. Solar panels harness the power of the sun. The solar panel is attached to a motor so it can gather more solar energy since it is incident on the sun. Electrical connections are made between this motor and the controller board. Checking on the availability of solar energy constantly from one horizon to another, the system makes sure that this is happening. In the scan, the scanner determines which direction receives the greatest amount of solar energy and therefore captures the brightest incident sunlight. As a result, the system utilizes the maximum amount of power it can generate with the Solar Panel.

Block Diagram:

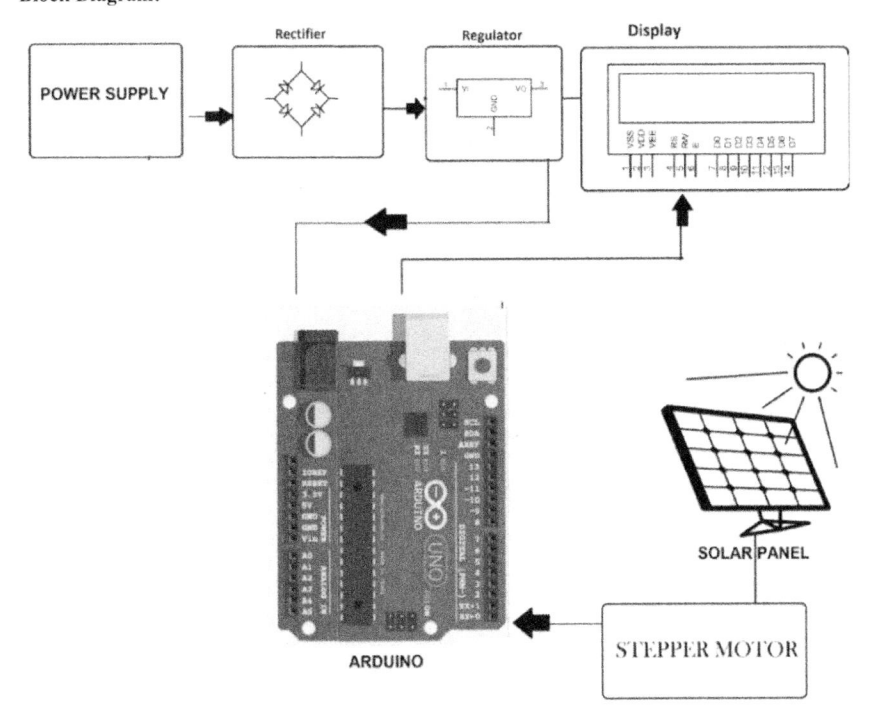

Hardware Specifications

- Arduino Uno R3

- Solar Panel
- Stepper Motor
- Crystal Oscillator
- Resistors
- Capacitors
- Transistors
- Cables & Connectors
- Diodes
- PCB

- LED's
- Transformer/Adapter
- Push Button

Software Specifications

- Arduino Compiler
- Programming Language: C

309. Fire Department Alerting System using Internet of Things and Arduino

Accidental deaths occurring due to fires are among the most common. Fire departments need to be alerted instantly in order to ensure immediate response. Every second can make all the difference in these situations. The system lets the fire department know about the situation at any time instantaneously and automatically, so instant activity may be taken.

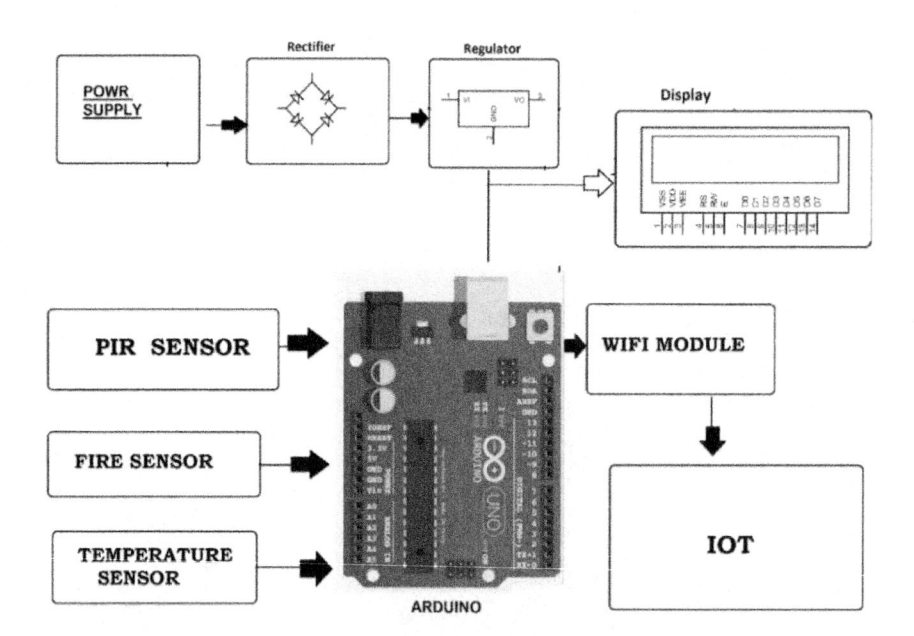

Fire sensors are used in conjunction with a PIR system to sense flames and alert fire departments through the Internet of Things. Arduinos are used to check if a sensor has been triggered. Then it uses temperature sensors to confirm that there is truly an outbreak of a fire. The system connects to an internet-connected server via Wi-Fi and transmits data about this incident over the Internet. IoT Gecko is the platform we here use to develop the IoT interface. It displays device id (named after area/flat id)

data immediately upon receiving sensor data from IOT Gecko. In the 21st century, the fire department begins to receive alerts via the internet about fire incidents so it can act quickly.

Hardware Specifications:
- Rectifier
- Regulator
- Power Supply
- LCD Display
- PIR Sensor
- Arduino Uno
- Wi-Fi Module

Software Specifications:
- Arduino Compiler
- MC Programming Language: C

310. Internet of Things based Irrigation Monitoring & Controller System using Arduino

A farmer is typically a person who works on a huge plot of land in order to grow several kinds of crops. Not all farmlands can be monitored by one person at any one time. There are times when a particular patch of land can get so much water that it becomes water-logged, or it might get so little water that it becomes dry.

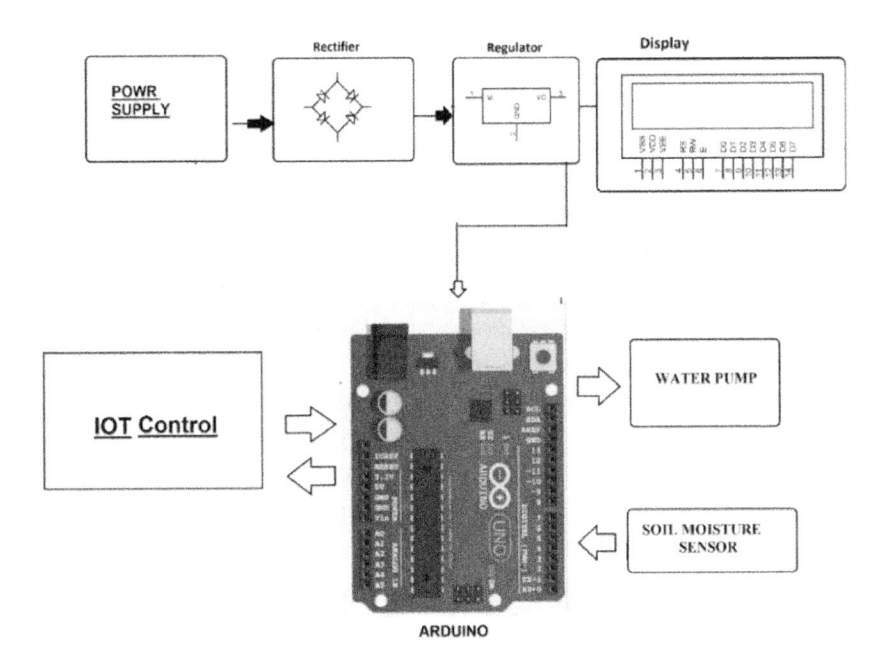

A farmer could suffer losses in either case if his crops are damaged, or if the crops are damaged by a storm. We propose an "Internet of Things Irrigation Monitoring and Control" project to solve this problem. One of the features of this project is that the utility company may monitor and regulate the supply of water from a faraway place. The Internet of Things concept is used in this system. As such, our system uses a wireless module to connect to the internet. A web server is connected to our desired website using an Arduino Uno board.

In these project, two concepts are shown; a) Motor status b) Moisture level a moisture sensor is equipped in the circuit, which keeps an eye on the soil moisture content. Users can then control the water supply remotely by checking the current moisture level on the website. Using the motor control switch, the water pump can be switched from 'ON-OFF' to 'OFF-ON'. Therefore, the issue of 'soil hydration' can be monitored and the 'supply of water' manipulated just by turning on or off the 'motor'. Thus, there is no need for the user to worry about his crops getting damaged because of 'waterlogging' or 'drought'. A person may not be able to constantly be present at their garden for people having small gardens. This project could be used to keep track of "soil-moisture" and supply water even from a distance.

Hardware Specifications

- Rectifier
- Regulator
- LCD Display
- Power Supply
- Wi-Fi Modem
- Water Pump
- Soil Moisture Sensor
- Arduino Uno

Software Specifications

- Arduino Compiler
- MC Programming Language: C
- IOT Geck

311. Internet of Things based Smart Agriculture Monitoring System Project

Since ancient times, agriculture has been practiced in every country. Science and art are both involved in cultivating plants. In human civilization's rise to sedentary civilization, agriculture was paramount. Farmers have been cultivating crops by hand for ages. Agriculture needs to adopt new technologies and implement implements to keep up with the trending of the world. Agriculture is becoming smarter with the use of IoT. Technology such as IoT sensors has the potential to provide valuable information about agricultural fields. By automating IoT-connected smart agriculture systems, we propose a new model. In order to monitor agriculture using IoT, wireless sensor networks were deployed throughout the system and sent the collected data to remote nodes using a wireless protocol.

IoT technology used in this smart agriculture consists of an Arduino, a Temperature, and Water level, GPRS, and Moisture sensor. Monitoring the water level, moisture, and moisture content of the soil is part of the IoT-based agriculture monitoring system. Whenever there's a problem, it sends an alert to the user's phone. Water level sensors sense a fall and start the pump automatically if necessary. The fan starts when the temperature reaches that level. The LCD display module displays all of this. This is also seen in the Internet of Things, which provides information on Humidity, Moisture and water level by the minute based on the date and time.

Block Diagram:

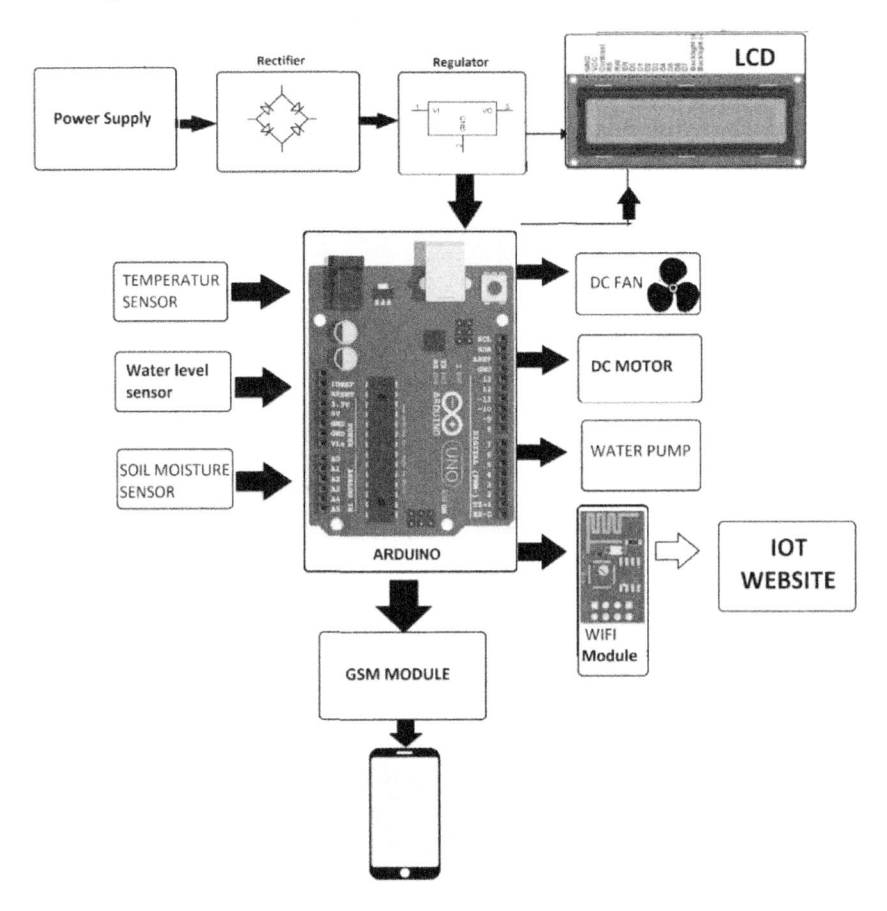

Different crops demand different temperatures; they are cultivated at different altitudes and temperatures. If it is desired to forcefully stop the water flow using the IOT, a button is available from which it can be forcibly stopped.

Hardware Specifications

- Arduino
- GSM Modem
- Wi-Fi Modem
- Temperature Sensor
- Humidity Sensor
- Water Sensor
- Mini Exhaust Fan
- Water Pump
- Crystal Oscillator
- Resistors
- Capacitors
- Transistors
- Cables and Connectors
- Diodes
- PCB and Breadboards
- LED
- Transformer/Adapter
- Push Buttons

- Switch
- IC
- IC Sockets

312. Arduino Ultrasonic Sonar/Radar Monitor Project

These advanced Arduino sonar technologies can be utilized to screen the patch area as well as detect suspicious objects. A car that has explosive material in it can be controlled remotely. We are able to prevent enemies from entering the public with this Arduino sonar radar, which in turn will save many lives.

Block Diagram:

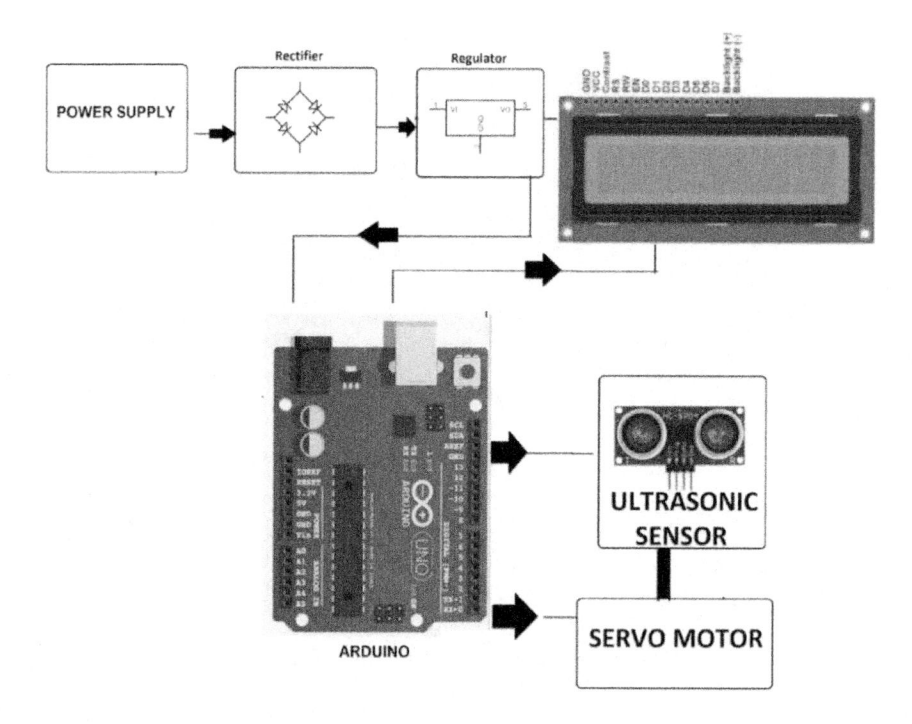

The Sonar Arduino system continuously scans the surrounding area and produces a beep upon detecting a moving target that is within our range. Moreover, the radar measures the angle and distance of the target from our source. Our system enables us to track the exact position of the object in real time and traces its path.

Radar: How does it work?

Radio detection and range technology is used in RADAR systems. Microwaves are used by radar to determine the range, altitude, direction, and speed of objects within a radius of about 100 miles of their location. By using a radar antenna, radio wave/microwave signals are transmitted and bounced off various objects on their path. As a result, we can estimate the proximity of a certain object.

The operating principle is:

With the use of **electromagnetic sensors**, a radar can detect and locate objects. Radiation from a radar is in the form of microwave waves or radio waves. Reflections from objects around them can intercept the waves.

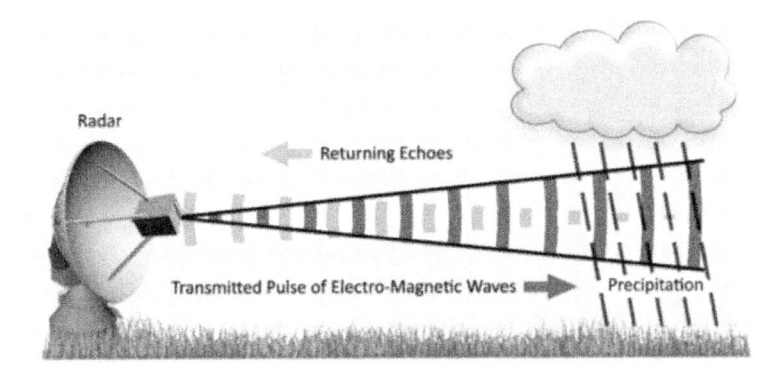

Radio waves intercepted by radar are reflected in many directions after they reach their target. The radar can direct these waves back to the receiver after receiving and amplifying them. Once again receiving these waves at their origin indicates that an object is in the propagation direction. In addition to air traffic control and air defense, radar astronomy, antimissile systems, and outer space surveillance are some applications of modern radar systems.

Hardware Specifications

- Arduino Uno
- Ultrasonic Sensor
- Servo Motor
- LCD Display
- Resistors
- Capacitors
- Transistors
- Cables and Connectors
- Diodes
- PCB and Breadboards
- LED
- Transformer/Adapter
- Push Buttons
- Switch
- IC
- IC Sockets

Software Specifications

1. Arduino Compiler
2. Programming Language: C

313. Smart Dustbin with IOT Notifications

Increasing populations result in an increase in trash in urban areas. With IoT and sensor-based circuitry, we present here an intelligent dustbin that can assist in solving this problem In the normal dustbin, you must open it with your foot and throw garbage. Also, a person must remember to empty their trash cans when they are at capacity, so they do not overflow. In this paper we develop a smart dustbin which can do all of these tasks without any human involvement. Essentially, our system pairs a clap sensor with a foot switch. In response to the clap or foot tap, the door opens and closes automatically by itself.

Block Diagram

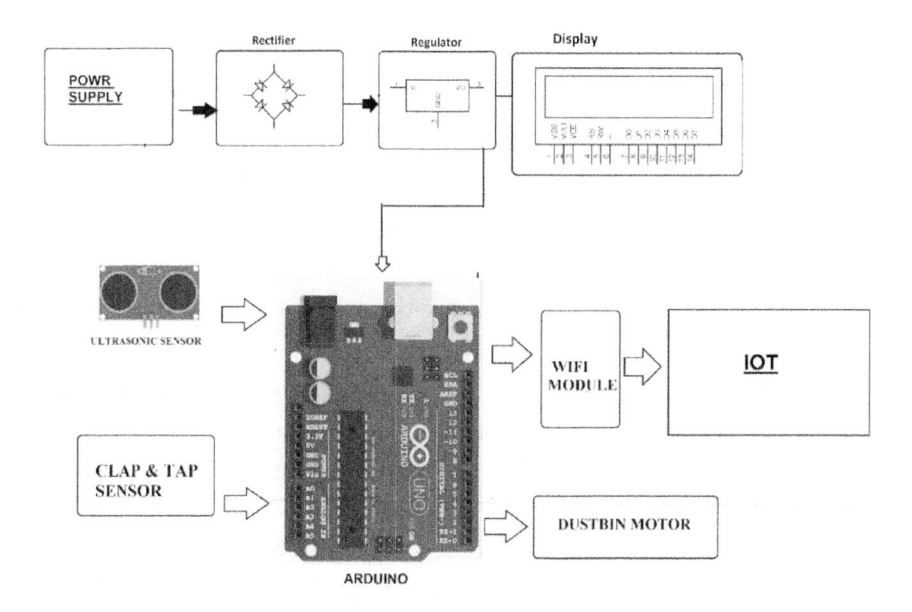

Upon receiving the signal, the dustbin opens its hatch and closes it. Additionally, the dustbin has an ultrasonic level sensor that continuously looks for level changes and triggers an automatic alarm if garbage is expected to fill the bin. A smart circuitry inside the dustbin sends data to the garbage collector over the internet, so he can empty it, if necessary. Web development of the IoT system is carried out using IoT Gecko. In offices, homes and public places this bin is of great use for garbage disposal. Therefore, garbage can be automatically cleaned with the help of an automated smart dustbin.

Hardware Specifications

- Arduino Uno
- Ultrasonic Sensor
- Mic Sensor
- Wi-Fi Module
- Resistors
- Capacitors
- Transistors
- Cables and Connectors
- Diodes
- PCB and Breadboards
- LED
- Transformer/Adapter
- Push Buttons
- Switch
- IC
- IC Sockets

- Bin Frame
- Mounts & Joints
- Supporting Frame
- Arduino Compiler
- Programming Language: C
- IoT Gecko

314. IOT Solar Power Monitoring System

It is important that solar power plants are monitored for maximum voltage output. This monitoring system helps in recovering optimum power from plants by detecting problems like faulty solar panels, connections, dust accumulating on panels, and decreased production in addition to other such things. In response, we propose an IOT-based monitoring system for solar power that allows for the automated monitoring of solar power from any internet-connected device.

A 10Watt solar panel is monitored by an Arduino based system. The solar panel is continuously monitored by our system and the power output is transmitted to the IoT system via the internet. IOT Gecko is used here to send solar power parameters remotely to an IOT Gecko server. The new program also displays these parameters in a user-friendly interface so that you can alert the system manager when the output falls below certain limits. Solar plants can be monitored via the Internet from anywhere in the world, ensuring the best power output.

Block Diagram:

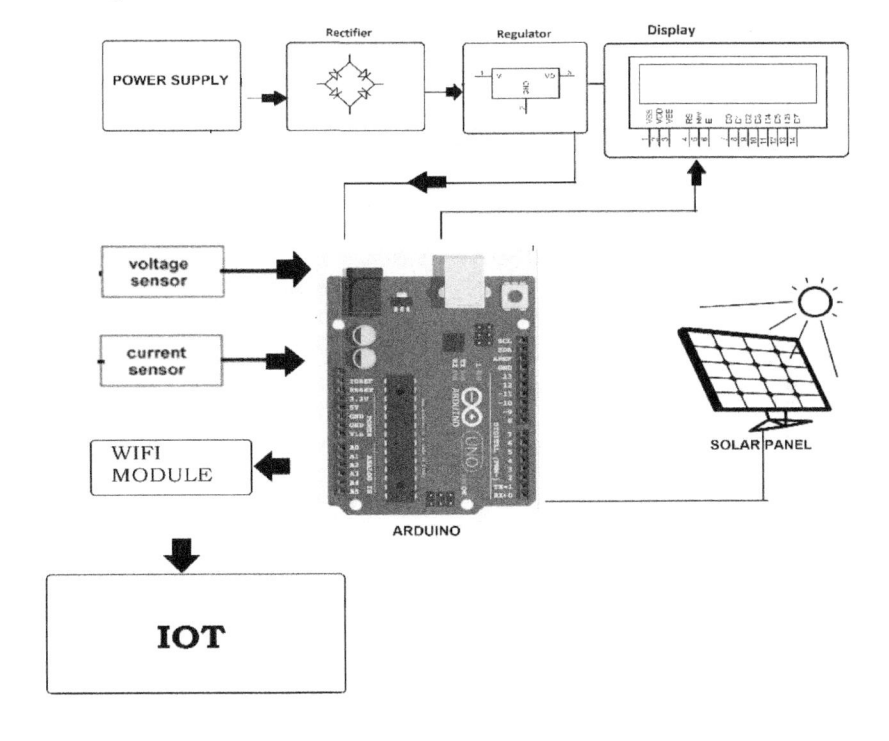

Hardware Specifications:

- Rectifier
- Regulator
- Power Supply
- LCD Display
- Voltage Sensor
- Current Sensor
- Wi-Fi Module
- Arduino Uno
- Solar Panel

Software Specifications:

- Arduino Compiler
- Programming Language: C

315. Arduino PID based DC Motor Position Control System

This motor placement control project involves the implementation of a PID Control System for an Arduino using the Derivative-Integral Formula. A PID controller with Arduino and a basic DC motor allows position control to be precise. With one shaft connected to the encoder and the other side connected to a pointer, a DC gear motor with two shafts can also be used as a power source.

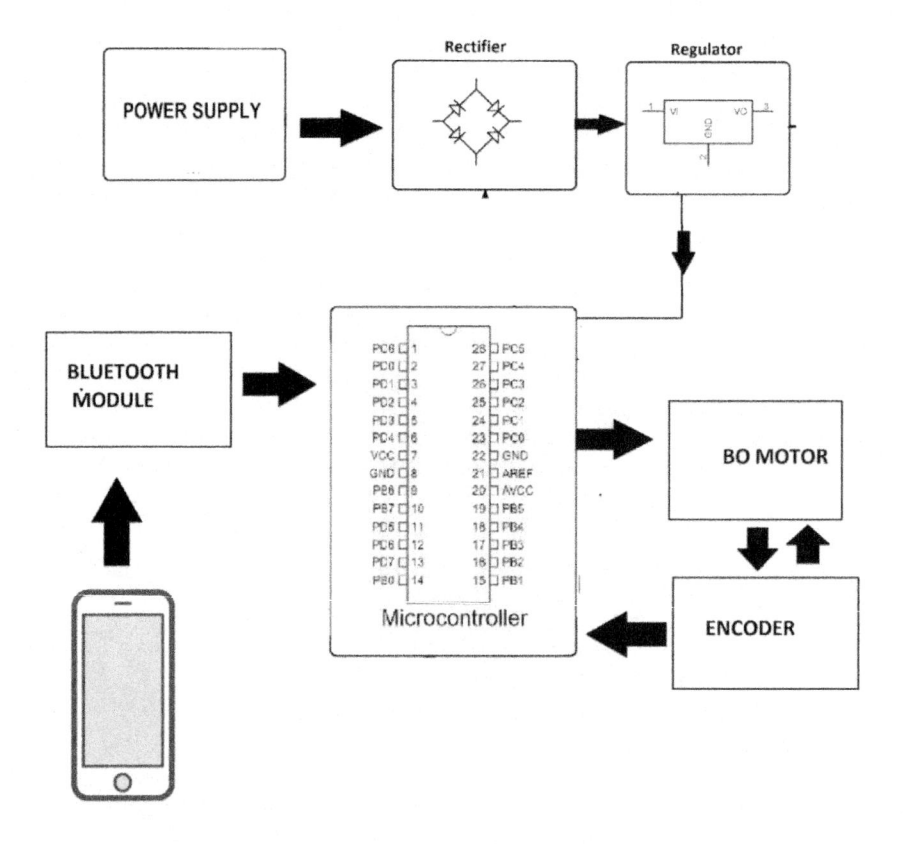

One encoder connected to Arduino interrupt pins points to the angle that is set on the protractor; one L293D motor IC is connected to our system, and an HC-05 module is used to connect it to an android device. The encoder sends a real-time input to Arduino in the meantime when sending a predefined angle set point from the robot device. If Arduino detects the encoder pulse matches a required position, it halts the DC motor at that position. A PID system controls the entire process for sleek, accurate motion.

Hardware Specifications

- Arduino
- Dc motor
- Protractor
- L293d IC
- Optical encoder
- Crystal Oscillator
- Resistors
- Capacitors
- Transistors
- Cables and Connectors
- Diodes
- PCB and Breadboards
- LED
- Transformer/Adapter
- Push Buttons
- Switch
- IC
- IC Sockets
- Arduino Compiler
- MC Programming Language: C

316. Open-Source COVID-19 Pulmonary Ventilator

The non-invasive, open-source ventilator is easy to build and is low-cost if there are no ventilators available and no patient is sedated or intubated while the patient needs to be ventilated. This project was inspired by a challenge I accepted from my former teacher and friend Serafim Pires. He presented a Spanish project to me and asked me to create a project to help fight the worldwide economic crisis.

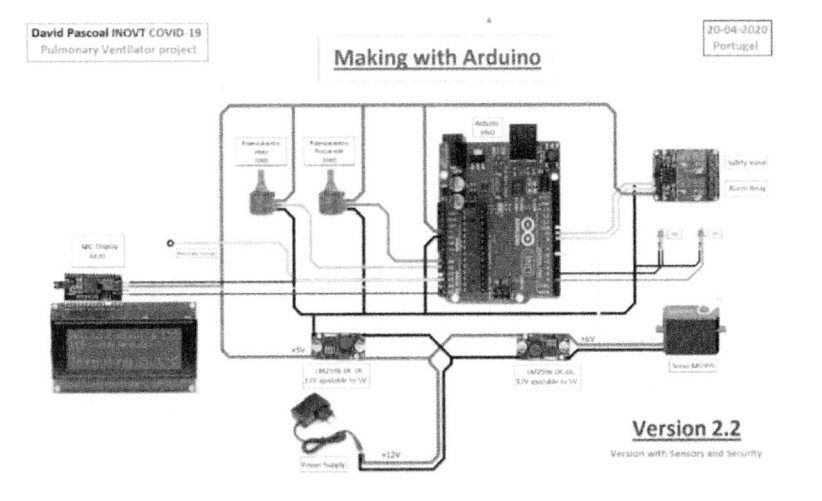

David Pascoal INOVT COVID-19
Pulmonary Ventilator project

Portugal
13-06-2020

Adaptation of the full face mask

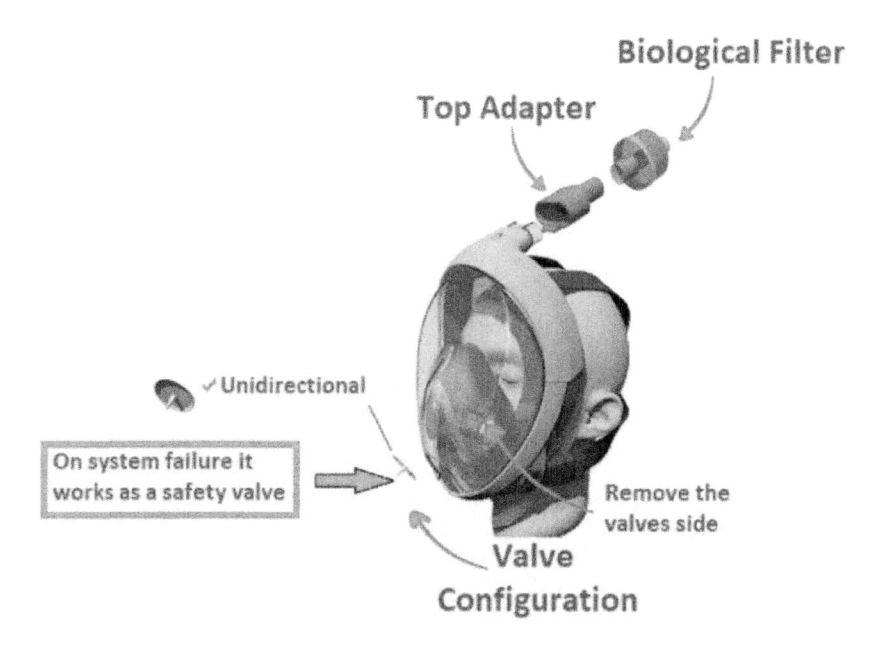

Biological Filter

Top Adapter

✓ Unidirectional

On system failure it works as a safety valve

Remove the valves side

Valve Configuration

This functional prototype was built on the basis of two existing technologies, after conducting several researches and tests on the topic of non-invasive ventilation. All tests were successful and the functional test lasted in excess of 20 days without any interruption.

Components Required

- Arduino UNO, Arduino 4 Relay Shield
- Digital Servo MG995, LM2596S Module
- 10k linear Multi-turn Potentiometer
- Digilent 60W PCIe 12V 5A Power Supply
- 5mm LED Red, 5mm LED Green, Alphanumeric LCD 20*4
- Switch button 220V
- Snorke Full Face, Solenoid Valve, 2 ways

Software Specifications

- Arduino Compiler, C Programming

Through the use of nasal and face masks, non-invasive ventilation is made possible by delivering controlled amounts of compressed air into the lungs. This helps the body fight infection and recover when the lungs are failing from disease. During the height of the coronavirus outbreak, and based on research conducted in Italy, I turned a Decathlon snorkel mask into an emergency ventilator for patients suffering from COVID-19, in order to reduce the lack of ventilators. Because of their low cost and ease of adaptation, these masks are used throughout the world.

WARNING:

- This prototype does not have official validation, nor will I accept any responsibility with it.
- Furthermore, this equipment is intended to save lives in an extreme peak situation and will be used as a last resort by trained medical personnel.
- In Portugal, the Portuguese authorities may approve the duplication of the non-profit project in bulk.

317. Arduino based Snake Robot Controlled using Android

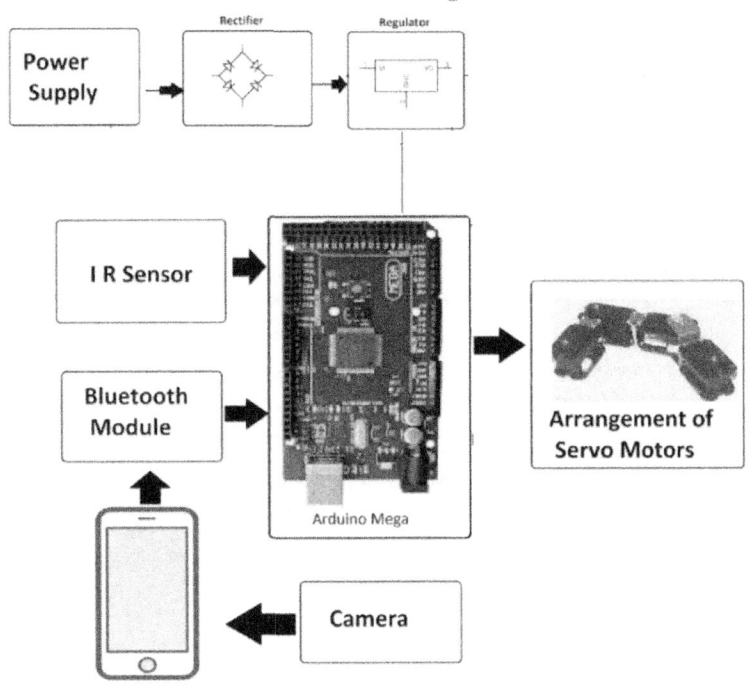

Twelve segments are driven by servo motors, all of which are joined together with metal brackets. An Arduino Mega controls the servos, which are powered by a 7.4-volt battery pack. A mobile device can control the snake via a Bluetooth app. The snake can also move autonomously. Servos and brackets can be used to make such a robot. One set of Lego wheels is built with each segment. Each segment includes a servo motor, a side bracket, a wire clip, and a C-bracket.

A Lego wheel axle had to be drilled with two screw holes so it can be connected to a C-bracket. It would be recommended to add rows and columns at the head and tail so that the Arduino and batteries can be accommodated. The brackets have two long C-braces that can connect on the side. A battery holder for 5AA batteries will power the Arduino, which is installed in the tail section of the snake. Battery packs with 7.4 volts powered the servos, which were mounted in the snake's head. An Arduino is connected to the receiver's ground on the bottom pin. A 5-volt pin from the Arduino is connected to the voltage pin.

Hardware Specifications

- Arduino mega
- IR Sensor
- Servo Motors
- Bluetooth module
- Camera
- Cables and Connectors
- PCB and Breadboards
- Push Buttons
- Switch
- IC

Software Specifications

1. Arduino Compiler
2. Programming Language: C

318. Advanced Automatic Self-Car Parking using Arduino

Both developed and developing countries facing a major car parking issue in urban areas. Several cities lack car parking areas as a result of the rapid rise of car ownership. A significant part of this imbalance results from ineffective land use planning and a miscalculation of space requirements at the start of the planning process. There are several examples of problems that arise from parking throughout the day, such as lack of parking space, high parking rates, and the congestion of traffic caused by visitors in search of a parking spot. Parallel parking is generally the worst nightmare a driver has because not only is it difficult if you do it well, but it increases the risk of other drivers hitting your parked vehicle. We developed an automated parking system to solve the above parking problems, thus enabling cars to park themselves. A self-parking car project is a self-parking car which utilizes an Arduino board, obstacle sensor, ultrasonic range finder to identify the parking distance. This robot uses a small LCD module to display various program information, a DC motor and servo motor for steering, and several algorithms for path finding.

Hardware Specifications

- Robotic Chassis
- Arduino
- Ultrasonic
- Servo Motor
- Resistors, Capacitors, Transistors
- Cables and Connectors
- Diodes, PCB and Breadboards
- LED
- Transformer/Adapter
- Push Buttons, Switch
- IC
- IC Sockets

Block Diagram

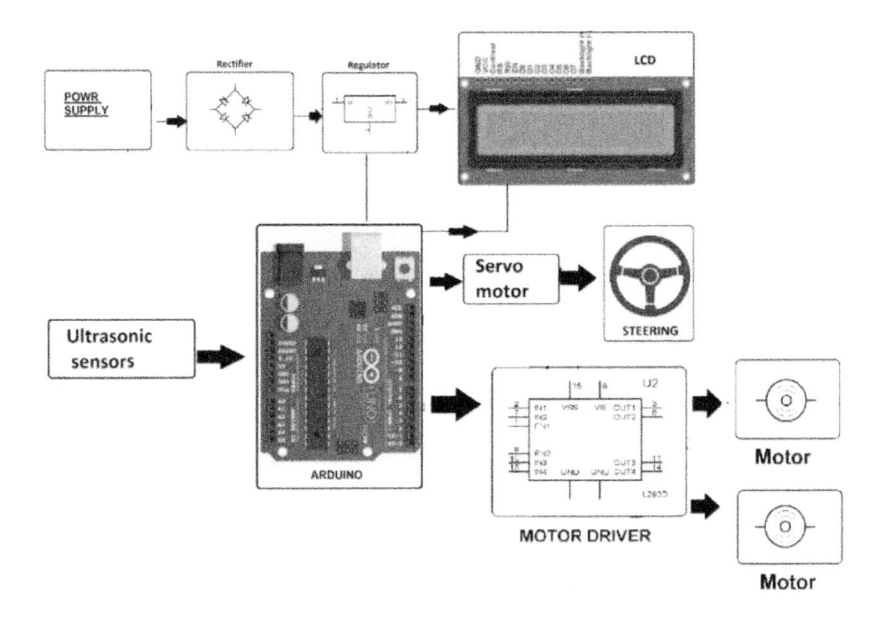

319. IoT Industry Protection System Arduino

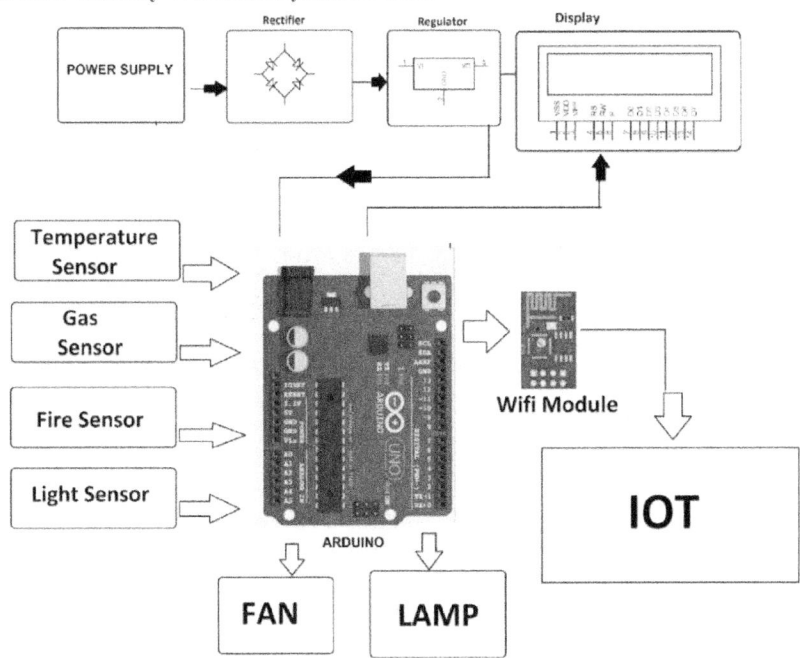

A Smart Industry Protection System, which is designed for industries to protect themselves from losses caused by incidents using the Internet of Things, protects them from unnecessary costs. Gas leaks can cause fires, which can cause massive industrial losses, as well as needing instant fire detection in case of blasts in furnaces, etc.

In addition, dim lighting may lead to improper lighting conditions in certain industries, which may result in increased accident risks. Arduino is used to make this system work. Temperature, light, and gas sensing make up the system, which works diligently to facilitate industrial accidents and loss prevention by detecting fire, gas leakage, and low lighting. Sensors for light, gas, and temperature are interfaced to Arduino devices and an LCD screen to constitute the system, which has a display and a keyboard. Sensors continuously scan data for fire, gas leak or low light exposures, record values, and then submit this information online for transmission. The internet is achieved via the Wi-Fi module. Once this data is stored, it is displayed online using IOT gecko, and the desired output can be achieved.

Hardware Specifications

- Arduino Uno
- LCD Display
- Wi-Fi Module
- LDR Sensor
- LPG CNG Gas Sensor
- Temperature Sensor
- Resistors
- Capacitors
- Transistors
- Cables and Connectors
- Diodes
- PCB and Breadboards
- LED
- Transformer/Adapter
- Push Buttons
- Switch
- IC
- IC Sockets

Software Specifications

- Arduino Compiler
- MC Programming Language: C
- IoT Gecko

320. Rotating Solar Panel Using Arduino

The energy demand in the commercial and residential markets is growing rapidly over the last few years. It leaves no other choice but to rely on renewable resources to generate usable energy as non-renewable resources are rapidly dwindling. Solar panels are another way to harness solar energy as it is the easiest and most abundant resource. Using this method, solar energy can be harnessed more efficiently.

It uses a Solar Panel attached to a rotating platform, which is powered by a motor, to charge a 12VDC Battery. An Atmega328 microcontroller attached to an Arduino Uno board, mounted on the PCB, is controlling this motor. In order to know the current position of the sun, the Rotating Solar Panel system scans from one horizon to another. This allows the greatest solar energy to be harnessed from that position. It is chosen to charge the Battery at the position that has the highest energy capacity. By aligning the Solar panel against the Sun, we can harness the most benefit from it. Thus, harnessing solar energy under this project is more efficient and thus smarter

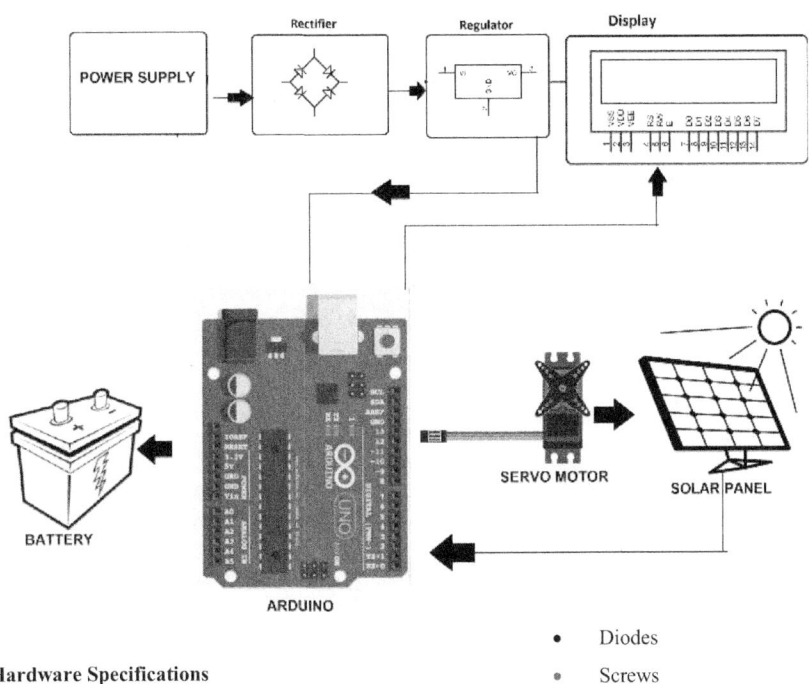

Hardware Specifications

- Arduino
- Servo Motor
- Solar Panel
- Solar Panel Mount
- Resistors
- Capacitors
- Diodes
- Screws

Software Specifications

- Arduino Compiler
- Programming Language: C

321. GPS Clock using Arduino

An exact time can be obtained from GPS synchronized clocks. Generally, clocks like this are used at railway stations, airports and other transport stands. In general, these are also used in military applications. You may have used the GPS module to create the tracking system, but did you know that the GPS module can also be used to create the Arduino Clock? Yes, the GPS module delivers time-related information, which we will use to construct an Arduino GPS Clock. As you know, UBlox NEO-M8M GPS Module sends serial data, including location. GPS also provides time, date, and latitude/longitude. We are using an Arduino board to separate the needed information from the data being sent by GPS.

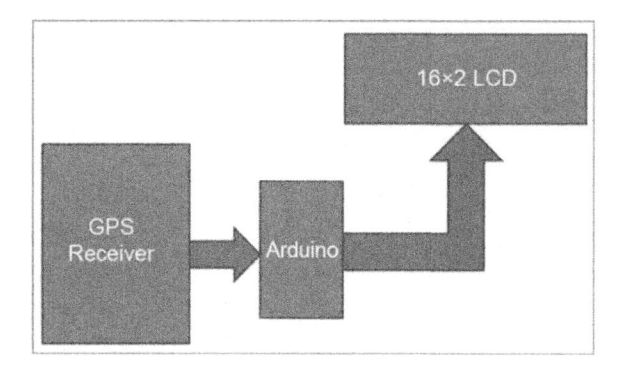

Figure: Block diagram of GPS clock using Arduino

An Arduino Uno R3 with six simple pin inputs and 14 digital inputs and outputs (I/O) pins referred to as a GPS clock has been developed. Its flash memory is 32kB, its EEPROM memory is 2kB, and its ISP flash memory is 8kB.

Circuit diagram

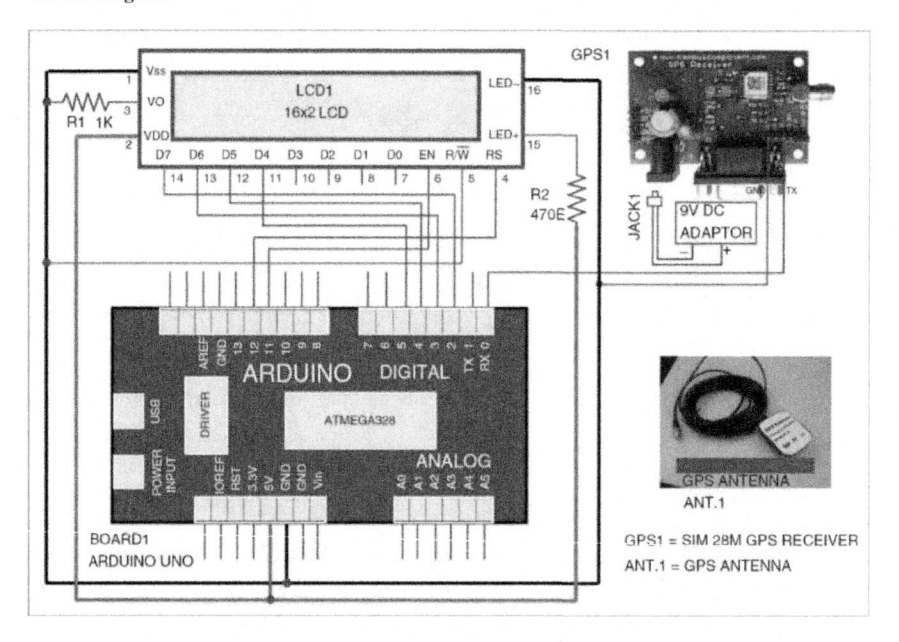

A serial connection can be made between the board and the computer via UART, SPI, and I2C.

322. Touch Free Hand sanitizer dispenser using LDR

Touch-free hand sanitizer dispenser that automatically dispenses sanitizer utilizing LDR sensors and MOSFETs to switch the motor we designed this to be in accordance with COVID 19 coronavirus. Coronavirus (COVID-19) is spreading across the world. Nearly every country is experiencing the devastating effects of the Coronavirus. A Pandemic disease has been declared by WHO and many cities are in lockdown situations. Many people's lifestyles have been changed drastically. Globally, the WHO is advising disease control officials to maintain Healthy Hand washing and Sanitation Habits, but our problem mainly involves our physical contact when we do this. The virus can be spread by touching infected bottles of alcohol or sanitizers with infected hands. We will make an automated hand sanitizer dispenser by using infrared sensors to detect a hand, and an automatic pump will pour the liquid on the hand.

Many Arduinos automated liquid dispensers are floating around the web. Nevertheless, my aim is to keep it straightforward and inexpensive, so anyone can produce it. A simple transistor or MOSFET with an IR proximity sensor would probably be the easiest solution for this purpose, and it would drastically reduce the costs as well. Because no microcontroller is present, spills are unlikely to be controlled, but the use of a smaller nozzle may reduce the flow of liquid.

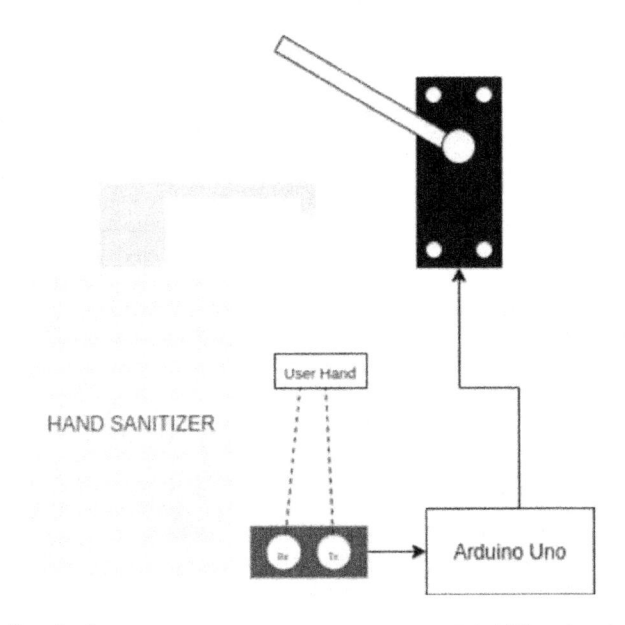

Components Required

- Arduino UNO & Genuine UNO
- Fairchild semiconductor
- Power MOSFET N-Channel
- LDR, 5 Mohm
- Fairchild semiconductor 1n4004.
- 1N4007 - High Voltage, High Current Rated Diode
- Fairchild semiconductor 1n4004.
- 1N4007 - High Voltage, High Current Rated Diode
- 09590 01
- LED (generic)
- Resistor 220 ohm
- Keystone 233 image 75px
- 9V Battery Clip

Circuit Diagram:

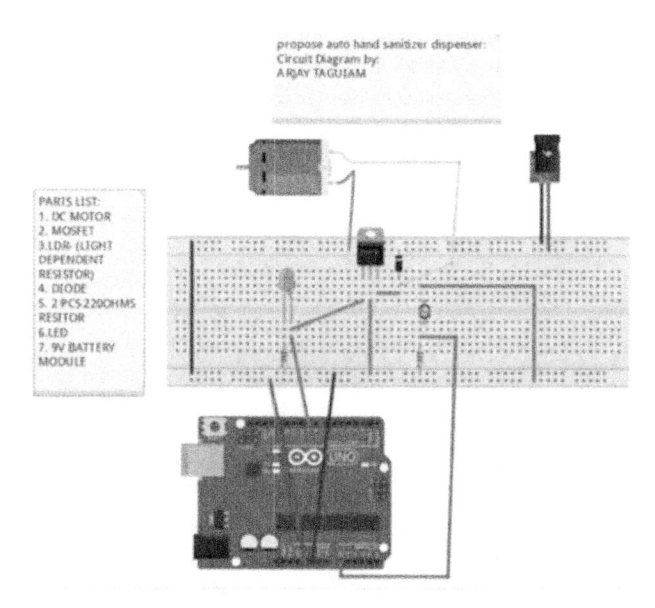

A basic working principle of579 the automatic sanitizer dispenser is that when the distance sensor detects an obstruction within its line-of-sight, it will trigger the servo to turn on the sanitizer tap. When the person's hand gets in the way of the sensor and obstructs the line of sight, the Arduino board detects the low distance and instructs the servo motor to activate the sanitizer. Here is the Circuit Diagram of a sanitizer or alcohol dispenser based on IR sensor. No microcontroller is required.

Automatic Dispenser

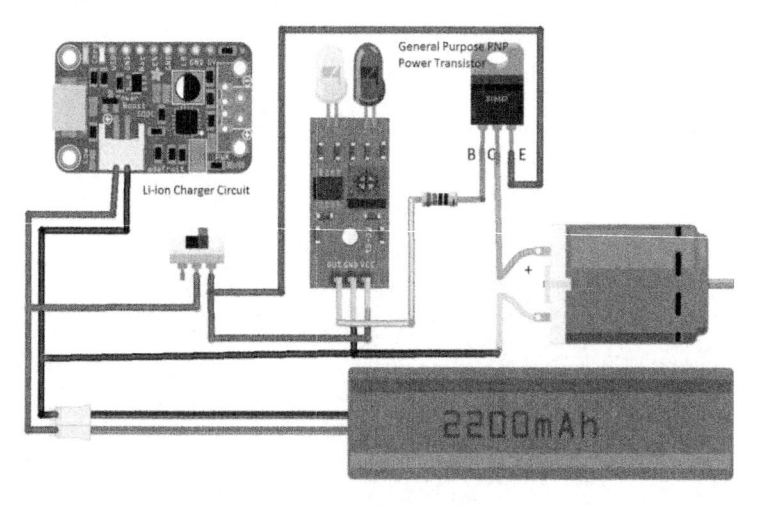

Figure: Circuit Diagram of a sanitizer or alcohol dispenser based on IR sensor

323. Line Follower Robot with Arduino

An object that is near can be detected by detecting its proximity without physical contact. Sensors that detect proximity usually emit electromagnetic fields (outside infrared light, for instance) in which changes in these fields or returns are monitored.

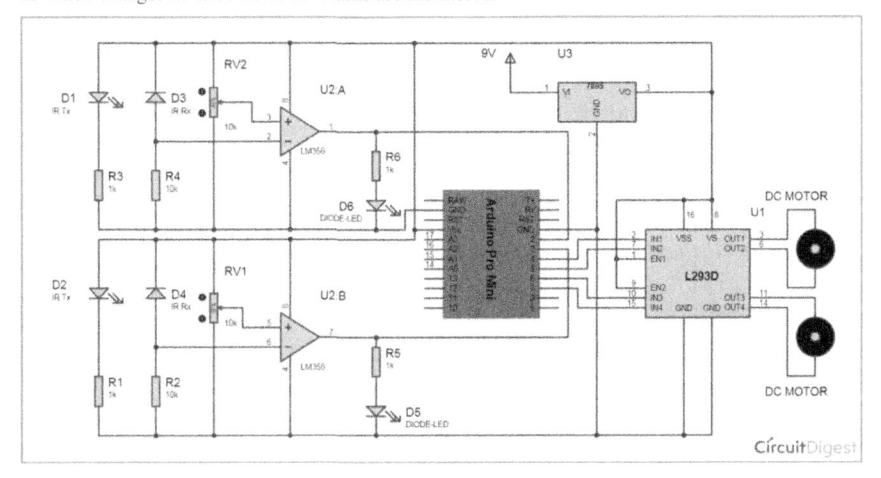

Concepts of Line Follower

Line following involves the use of light. This article discusses the behavior of light at white and black surfaces. White surfaces reflect light almost completely while black surfaces absorb it entirely. An automated line follower robot is built using this behavior of light. IR Transmitters are an essential part of this Arduino Line Follower Robot, also referred to as photodiodes. These devices are used to send and receive light. Infrared transmitters transmit infrared lights. Photodiodes generate voltage changes when infrared rays fall on white surfaces, and they reflect back. The photodiode does not receive any light or rays when infrared waves fall on a dark surface, since light is absorbed by the dark surface.

This Arduino line following robot receives 1 input as the sensor detects white surfaces while 0 input is received when it detects black surfaces.

There are three sections in the robot: A sensor section, Control section and a driver sections.

Sensor section:

In this section, you will find IR diodes, potentiometers, comparators (Op-Amps) and LEDs. The comparator's two terminals receive their references through a potentiometer, while IR sensors provide the voltage change at the comparator's second terminal. Using the comparator, then, both voltage signals are compared, producing a digital signal. Using two comparators for two sensors we have implemented a line follower circuit here. A comparator is created using the LM 358, which has two ultra-low noise Op-amps inbuilt.

Working of Line Follower Robot using Arduino

It is very fascinating to operate as a line follower. It uses a sensor that senses a black line and then transmits the signal to an Arduino board, Afterwards, the motor is driven by Arduino based on the sensors' output

Basic Components
- Arduino UNO & Genuino UNO
- SparkFun Dual H-Bridge motor drivers L298
- Proximity Sensor

324. IoT Based Home Automation controlled by smartphone

Human life is increasingly driven by the use of automation, whether it's at home or at work. Automation in the manufacturing industry is a concept that is frequently used to automate large machines and/or robots to facilitate increased production, energy, and time efficiency. In contrast, home automation affects the environment of the homeowners.

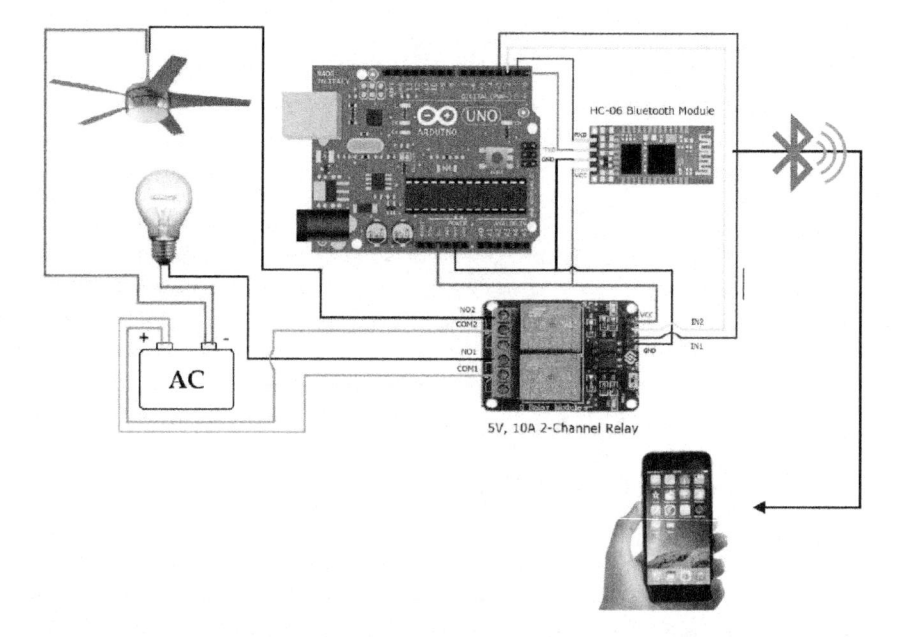

The smartphone and the internet have allowed us to do this. There are two main ways of home automation. One type is controlled by just a smartphone, while the other type involves sensors and actuators to control lighting, temperature, door locks, electronic gadgets, and electrical appliances.

Essential Components
- Arduino UNO

- 12 V Relay X 4
- HC – 05 Bluetooth Module
- Prototyping board (Bread board)
- Connecting wires
- 12 V Power supply
- Smartphone or tablet (Bluetooth enabled)

Circuit Diagram:

Modern homes require sophisticated control in the electronic appliances they have in their homes. As a result of the integration of home appliances with smart phone and tablet connectivity, the home automation field has been transformed, enabling a greater level of affordability and simplicity. In addition to the features, they have already, smart phones can be made to communicate with any other device in an ad hoc network through connectivity options like Bluetooth. As mobile phones have become more prevalent, development of mobile applications has also increased. A mobile phone commonly found in a traditional household can be connected to the electronic equipment of a smart home in a temporary network using the opportunity of automating tasks for a smart home.

The Android mobile application platform is provided by Google Inc., which is used to develop applications for Android phones and tablets. Android-based home automation system will benefit the masses as it targets a large market who uses it for their mobile devices. Android maintained its leadership position in the Worldwide Quarterly Mobile Phone Tracker, published by International Data Corporation (IDC). Using Bluetooth for home appliances and mobile phones in an ad hoc network environment, such as in your home, is an excellent solution for short-range wireless communication. A wireless technology works over 2.4 GHz frequency range up to a distance of 100 m with 1 Megabit per second, making it a secure and efficient method for controlling home automation.

325. Covid-19 Patient Monitoring Device based on LoRa using The Things UNO

Using sensors and connected networks, we have developed a patient monitoring system that can autonomously monitor patients' health conditions. The Covid-19 system was specially manufactured for patients with this condition.

The biological behavior of a patient can be gathered by several sensors. Information about biological processes then goes into the IoT cloud. By processing sensor data, the system is more intelligent, and can tell when a patient is in critical condition. Nurses and doctors receive instant alerts and hospital personnel receive push notifications. Nurses and doctors' benefit from this system because they can observe the patients remotely without having to visit them personally. Relatives of patients can also gain access to the system with limited access.

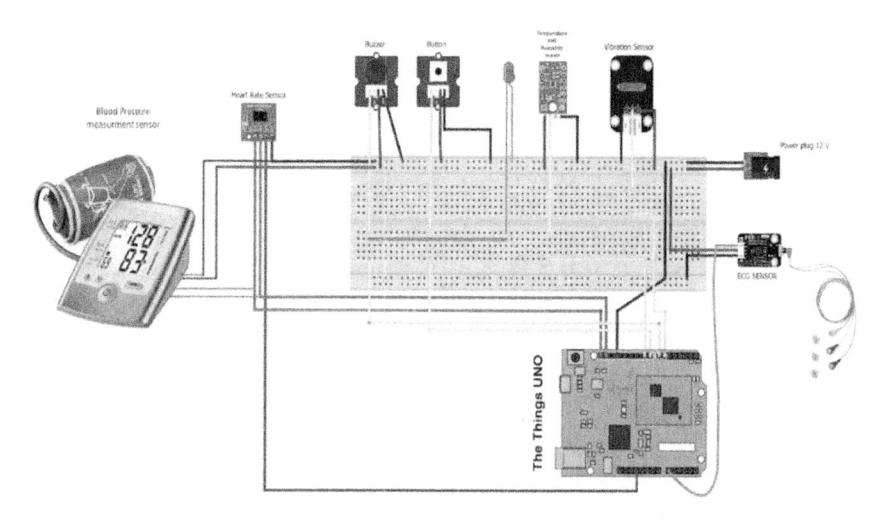

Figure: Lora network based covid-19 patient monitoring device.

Name of the components	Purpose
The Things UNO	The main controller board
Temperature and Humidity sensor	For body temperature and humidity measurement
Heart rate sensor (MAX30100)	Measure the pulse
ECG sensor	Measure the ECG data
Buzzer	Emergency Alarm
LED	Emergency indicator
Push button	call for assistance
Movement sensor	Detect the unexpected movement
Blood pressure sensor	Measure the blood pressure
360 Camera, Optional	Optionally streams the video
Others sensors	Air quality sensor and room temperature sensor are used for measuring room environment
12volt DC power supply	Power source

Sensor (notation)	Threshold	Threshold level [1]
Heart rate, h_r	$T_{heartrate}$	less than 50 and greater than 120
Temperature, $temp_{body}$	$T_{temperature}$	less than 35 and greater than 39 in Celsius
Humidity, $humidity_{body}$	$T_{humidity}$	less than 40% and greater than 55%
Movements, $move_{body}$	$T_{movements}$	Unexpected
SPO2, $spo2_{blood}$	T_{SPO2}	Under 90 %
Upper blood pressure, $upper_{blood}$	$T_{upperblood}$	less than 120 and greater than 180
Lower blood pressure, $lower_{blood}$	$T_{lowerblood}$	less than 80 and greater than 110
Push button, $button_{call}$	T_{button}	On
ECG, ecg_{heart}	T_{ECG}	N/A

This monitoring system is controlled using the Things UNO, a Lora development board. This board collects information from various health sensors (described in the Hardware Components section) that provide information about patient health parameters. Data transmission from the Things UNO to the Lora Gateway (The Things Gateway) is also handled by the Things UNO. The Lora gateway provides a connection to the Amazon Web Services IoT cloud platform. This cloud is used for managing this system.

The data has been visualized using a Mobile application. For displaying real-time sensors data, such as the present health condition of a patient, various charts and gauges have been employed. Doctors and nurses can use this application remotely to monitor patients without visiting an ICU unit.

A push notification is sent to appropriate doctors or nurses regarding the emergency situation of the patient due to the nature of intelligence, by processing the sensor data, Equation-I identifies the patient's emergency condition. During a 24-hour period, the hospital in charge personnel (ICU in charge person) continuously monitored multiple patients online via our cloud-based desktop application (shown in Figure), which makes the ICU process more efficient, Throughout the application's lifetime, all of the applications tapped into the Internet of Things and visualized the data in real-time, using visualizations such as gauges, Sparklines, and Text.

326. Open-Source Pulse Oximeter for COVID-19

This is an easy-to-make, USB-powered pulse oximeter that can be built for around $20 and features an OLED display.

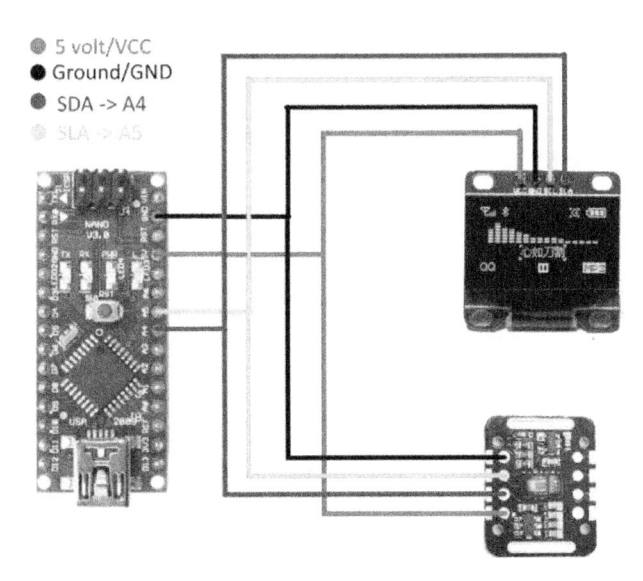

Fig.1: Open-Source Pulse Oximeter for COVID-19

The SARS-CoV-2 virus is the virus responsible for causing COVID-19, a disease which mostly attacks the respiratory system. Fever, chills, and muscle aches and pains are some of the milder symptoms, but a severe case can lead to pneumonia. A person suffering from pneumonia or even slight shortness of breath might not recognize when to seek medical attention, especially when they begin to feel even worse. Hence, I am developing this open-source pulse oximeter so that the people can be better informed about their current condition and which can assist them in getting the help they need.

Disclaimer: Using this device for accurate medical diagnosis is not recommended!

327. Touch less doorbells can operate without touching the switch.

One of the most effective ways to escape from COVID-19 is through social isolation. Staying at home is strongly recommended in the beginning days. However, we still have to make some emergency visits to certain homes. We first searched the doorbell button of a house when we arrived. Then press the doorbell button. However, in this specific situation, this doorbell button can cause the virus to spread. A virus is held on the button when someone who is not infected presses the button, and when a second person touches the button, the virus is spread. A touchless doorbell will help to eliminate this danger. Touchless doorbells are available to convert existing doorbells.

Circuit and working:

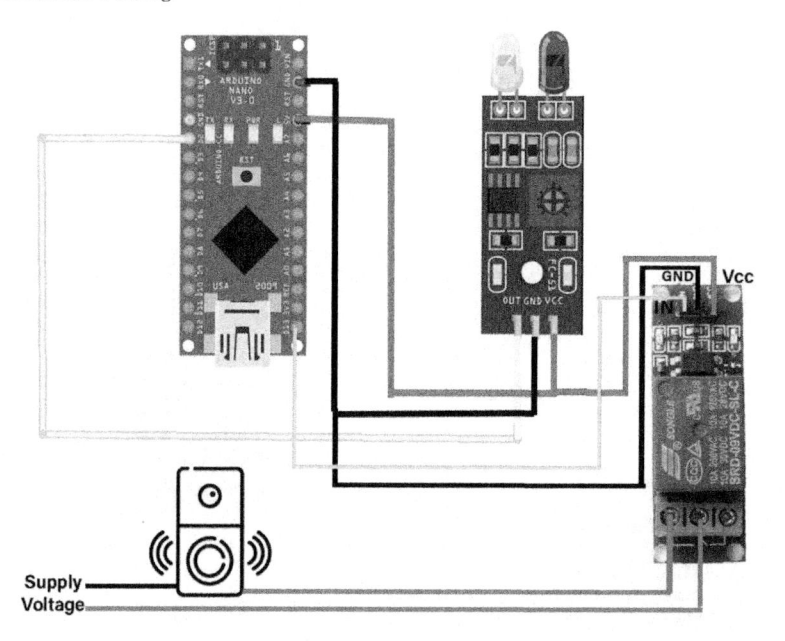

Infrared LED transmitters transmit light within the range of the infrared spectrum. Wave length of IR waves is longer than wave length of visible light. This transmitted IR light will be picked up by the photodiode receiver. The photodiode will only conduct when it is illuminated. The semiconductor is reverse biased as well. It can be shown that the current flow is directly proportional to the amount of light it receives. In this photo, we see the LM358 Operational amplifier in voltage comparator mode. Comparison is made between the voltage set by the variable resistor and the voltage set by the photodiode series resistor

1. (PSR voltage). A ground connection is maintained between the "OUTPUT" pin and the OP-AMP output.
2. The PSR Voltage has dropped below the Threshold Voltage - the output is HIGH
3. Voltage drops on PSR < Threshold Voltage - Output is LOW

4. By calibrating the variable resistor, the distance at which objects should be detected can be determined.

Essential Components

Arduino Nano R3, Relay, IR Sensor, Buzzer

Key point: In case an object is presented in front of the sensor, the sensor output will be HIGH, but if not, the sensor output will be LOW

A signal will be sent to the Arduino board when we show our hand to the IR Sensor. And Arduino drives the relay. The relay is connected to the doorbell. The bell will ring the moment the relay is activated.

328. Social Distancing Device (Safety Card)

During this pandemic, we trust everyone is following social distancing and keeping safe, which is why we made an ultrasonic sensor device at my home to do social distancing. With this device, a buzzer sounds and vibration occurs when the distance between two people is less than one meter, signaling that the distance needs to be maintained.

Required Components: Arduino UNO, Ultrasonic Sensors, Buzzer, LED

The device automatically turns off if the space is greater than 1 meter. I think this is a very interesting and useful project that you can easily build at home.

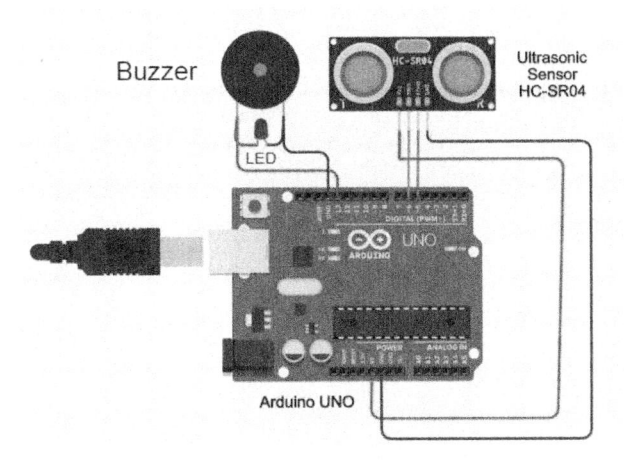

Figure: Social Distancing Device (Safety Card)

Wearable's can be manufactured by using an Arduino Nano or you can shrink their size by using an Arduino Micro. Put together the materials like Arduino Uno Arduino Nano, ultrasonic sensor, buzzer, vibrator motor, 9v battery, switch, card or cardboard box, Velcro strips, take the card and glue the face of the box with the electronics, proceed to upload the code, Once the code has been uploaded, place the Arduino inside the box, along with the switch, nine-volt battery, and circuit symbols, and do circuit connections. Attach the front portion of the box and then check that everything functions well, now decorate the box and stick it with Velcro.

329. Automatic Faucet (Touchless) for COVID-19 Using Arduino

Wash your hands comfortably and avoid getting coronavirus disease. Don't touch the surface of the faucet after you wash them. A disease caused by the severe acute respiratory syndrome coronavirus 2 is known as coronavirus disease (COVID-19), also known as the coronavirus severe acute respiratory syndrome (SARS-Cov-2). More than a million people worldwide have been affected by COVID-19 and hundreds of thousands of people have lost their lives as a result.

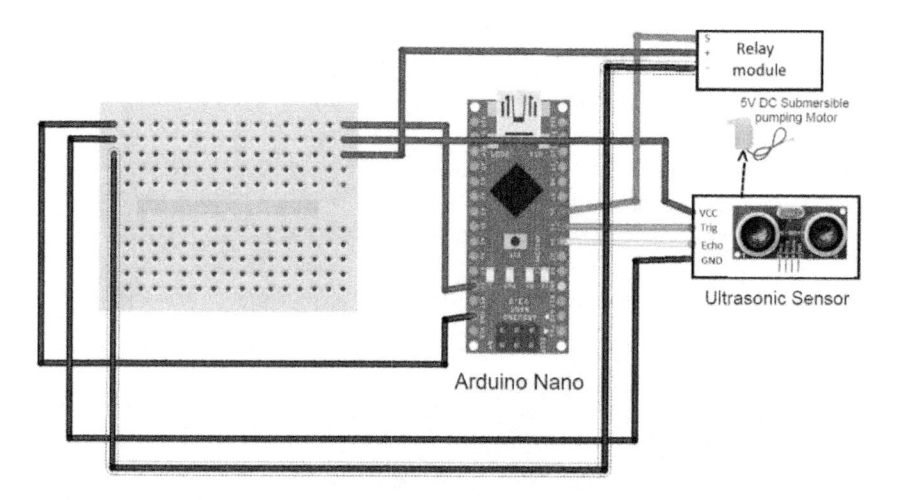

Fig.1: Circuit Diagram of Automatic Faucet (Touchless) for COVID-19 Using Arduino

People are affected by this disease in different ways. Many people develop mild to moderate illnesses that require no hospitalization or special treatment, while others develop severe illnesses that ultimately lead to death. A person infected with this virus can incubate for an average of 5-6 days, but it can also incubate for 2 weeks. It could be contagious during this period, even though the person may not be experiencing any symptoms. If the person doesn't take any precautions, he will be a virus carrier and will spread the illness easily.

Components Required

- Arduino Nano R3
- Solderless Breadboard Half Size
- Jumper wires (generic)

- Ultrasonic Sensor - HC-SR04 (Generic)
- Submersible water pump - 5V
- Relay Module (Generic)

Coronavirus has been spreading rapidly around the world and will continue to spread. The second wave of the coronavirus is still affecting some countries while others are still in lockdown and still aren't seeing any recovery.

330. Automatic Hand Sensing Water and Soap Tank with Tap

This automatic tank for water and soap prevents the spread of coronaviruses in public places while, at the same time, prohibiting touch-ups of the faucet. In March 2020, the World Health Organization (WHO) declared a global pandemic of a novel coronavirus outbreak that caused a respiratory illness that was first discovered in Wuhan, China. The World Health Organization estimates that the number of confirmed HIV cases and deaths is over 3 million by the end of April 2020.

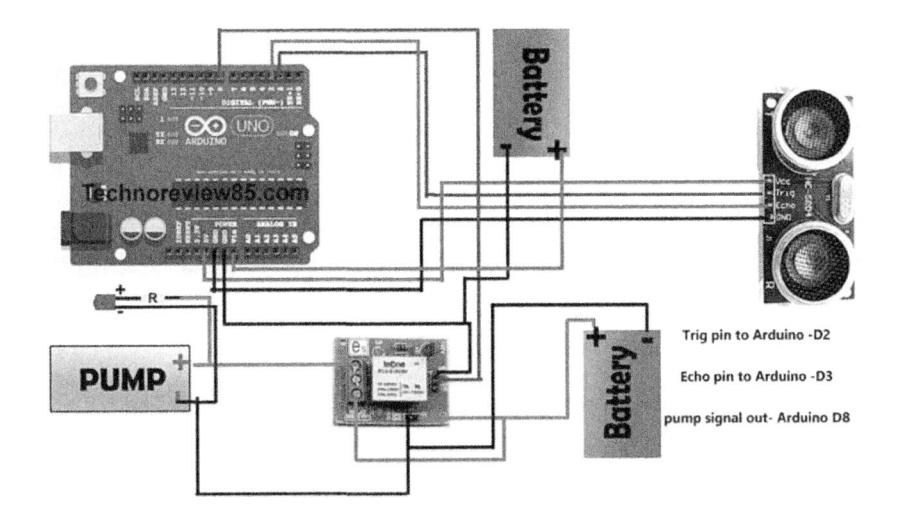

Trig pin to Arduino -D2
Echo pin to Arduino -D3
pump signal out- Arduino D8

People can contract the disease directly from an infected person or indirectly through touching surfaces contaminated with viral droplets. Since specific treatments for the outbreak or a vaccine are not yet available, it is crucial to prevent the outbreak from spreading from person to person and affecting people's lives, health, livelihoods, and the healthcare systems we are all relying on.

Infected people spread the COVID-19 virus mainly by sneezing or coughing droplets of saliva. It is important to properly wash your hands in order to prevent COVID-19. By providing safe water, sanitation and basic hygiene measures during hand washing, as well as prohibiting retouching of the faucet to avoid decontaminating the water source or tap and posing further risks to another user, this can be achieved.

Materials:

The major Materials used in this project are:

1. Transparent Pipes
2. Jumper Wire
3. 9 volts Battery
4. Ohms Resistor
5. Arduino Uno
6. Hc-sr04 Ultrasonic sensor
7. Relay Module
8. DC Powered Pump
9. Water Tank
10. Soap Container
11. Metal tank stand
12. Solar Power Source
13. Ceramic Zinc
14. Outlet Pipes

People who obtain automatic hand sensing water and soap dispensers will not only be able to wash their hands with ease, but will also be afforded proper hygiene facilities, however limiting the time in which people are in contact with the tap would prevent recontamination.

331. DIY GPS Speedometer using Arduino and OLED

Speedometers are used to determine how fast a vehicle is traveling. We previously built Analog speedometer and digital speedometer using the IR and hall sensors respectively.

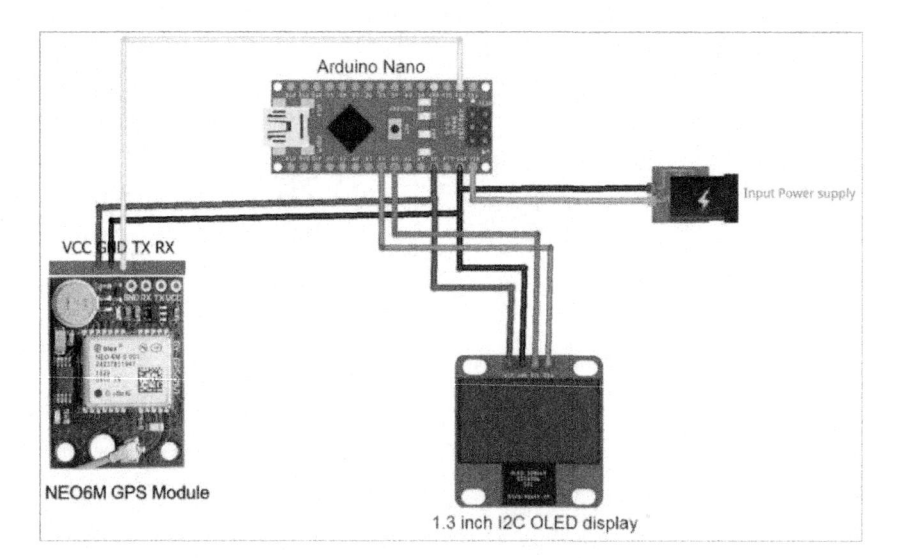

The vehicle speed will be measured using a GPS today. It is more accurate to use a GPS speedometer than a conventional speedometer because it can constantly locate the vehicle and can calculate the speed. Smartphones and vehicles incorporate GPS technology to provide navigation and traffic alerts.

Materials Used

- Arduino Nano
- NEO6M GPS Module
- 1.3-inch I2C OLED display
- Breadboard
- Connecting Jumpers

332. Automatic Bottle Filling System using Arduino

The industry which uses automatic bottle filling machines the most is the beverage & soft drink industry. With a conveyor belt being used, these machines are an economical and efficient way to fill bottles. Automation of bottle filling machines is mostly done using PLCs, but an Arduino can also be put to work in this regard. IR or ultrasonic sensors can be programmed to automatically detect the bottle and stop the conveyor belt for a short period of time thereby permitting the bottler to fill the bottle. Continue to move the belt where you stopped when the next bottle was detected.

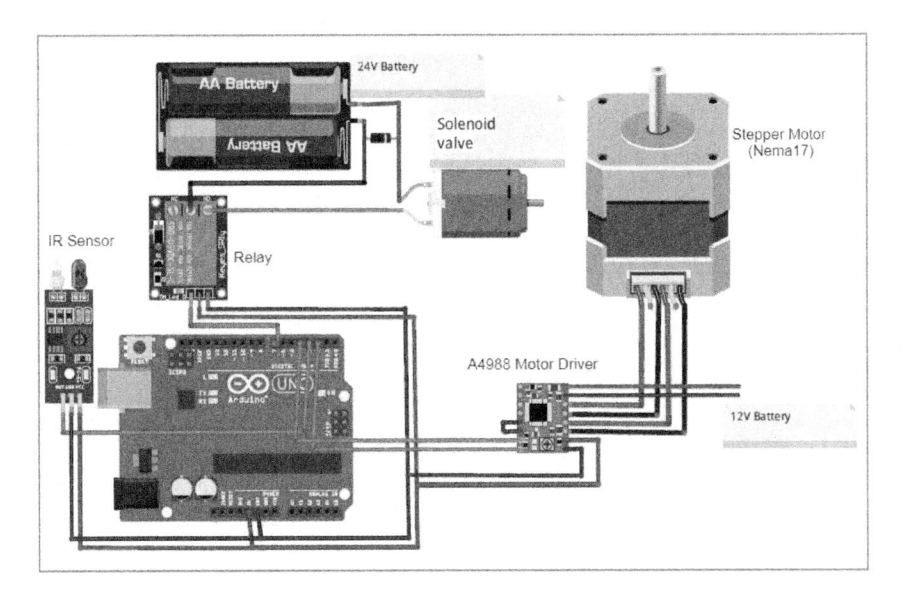

With an Arduino Uno, conveyor belt, solenoid valve, infrared sensor, and stepper motor, we are going to build a prototype for an Automatic Bottle Filling Machine. An electronic stepper motor controls the speed of the belt conveyor. Once the IR sensor is detecting the bottle, the stepper motor will continue driving the belt. As an external trigger, we used the IR sensor. A solid-state relay switches on the solenoid valve whenever the IR sensor goes high. For bottle filling, the code already describes a delay that is necessary. In that case the Arduino will continually power on the solenoid valve and turn off the stepper motor. A solenoid valve turns off the filling after a fixed amount of time, enabling the conveyor to move to the next bottle.

Components Required

- Arduino Uno
- Stepper Motor (Nema17)
- Relay
- Solenoid Valve, IR Sensor
- A4988 Motor Driver
- Battery

Previously, we used Arduino with an IR Sensor and Solenoid valve, so you can learn more about the basics of interfacing Arduino with these components by visiting the links.

333. Control a Solenoid Valve with Arduino

In many process automation systems, solenoids play an important role in actuating the components. In addition to solenoid valves, there are solenoid plungers which produce linear motion and can be used to open and close water or gas pipelines. In most homes and offices, we are all familiar with ding-dong doorbells, which use solenoid technology. Upon being energized with AC power, a small rod will be moved up and down by the Doorbell's plunger-type solenoid coil. A rod attached to the solenoid will strike metal plates connected to each side, producing the soothing sound. It's also used as a starter for vehicles or in sprinkler systems and RO systems.

Components Required

- Arduino UNO
- Solenoid Valve, IRF540 MOSFET
- Pushbutton - 2 nos.
- Resistor (10k, 100k), Diode - 1N4007
- Breadboard, Connecting Wires

How Does a Solenoid Valve Work?

Solenoids are devices that convert electrical energy into mechanical energy. In this setup, there is a coil wrapped around conductive material, acting as an electromagnet. Electric magnets are better than natural magnets, because they can be switched on or off using a coil electrically charged. As a conductor is turned energized, a magnetic field is generated around it because the current-carrying conductor is a coil. Since a coil is a magnet, a strong magnetic field is created that magnetizes the material, thus creating linear movement.

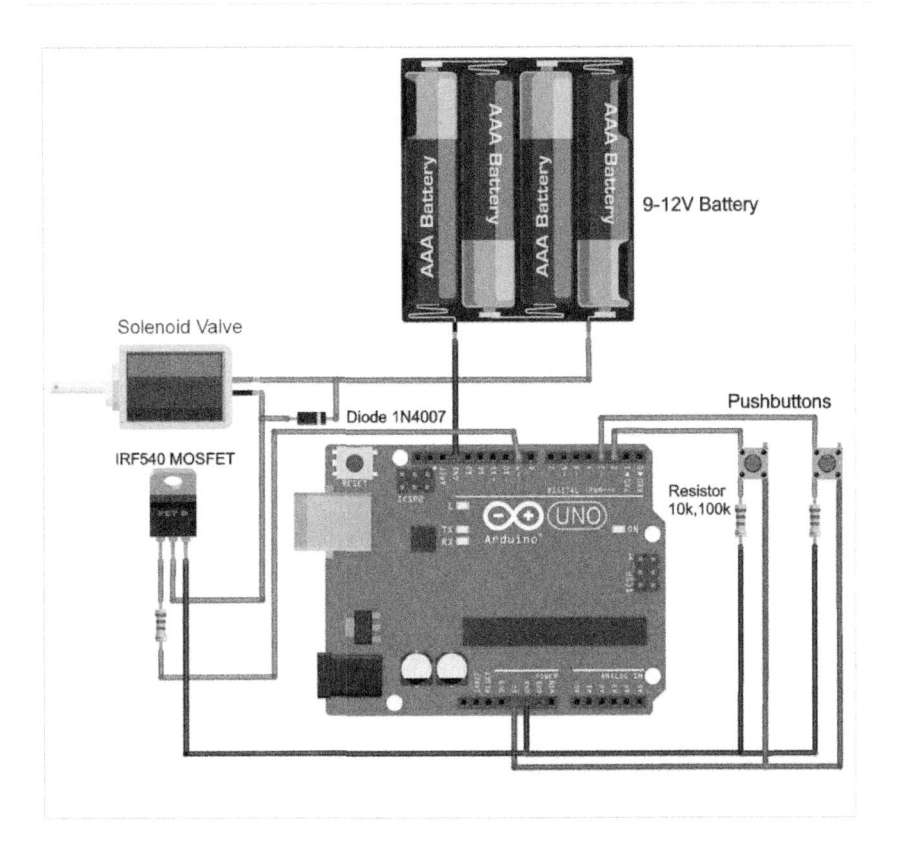

A type of relay, it operates by means of a coil which when energized pulls a conductor (piston) inside it, which then lets liquid flow through it. The spring force pushes the piston back in the previous position when the electric motor is de-energized, which again blocks the liquid flow. Therefore, it is not possible to control a Solenoid coil directly through a logic circuit during this process, as it draws large amounts of current and produces hysteresis problems. It is common to control flow of liquids with a 12V solenoid valve when building a pump. Because this particular solenoid valve draws up to 1.2A of peak or continuous current during energization, it has to be taken into consideration when designing the solenoid driver circuit.

334. An Arduino-based Gesture Controlled Air Mouse that uses Accelerometer

Did you ever wonder why we're moving toward an immersive reality? With the advent of virtual reality, mixed reality, augmented reality, etc., we are constantly finding new ways to interact with our surroundings. We are continually impressed by these fast-paced interactive technologies of new devices coming out every day. They are used in a wide variety of applications, including gaming, entertainment and interactive activities. In this tutorial, you will learn about a new sort of user interface that will replace the traditional mouse with something more interesting. It is obvious to our game geeks that Nintendo was the company to devise a way to interact with video games with a 3D interactive motion console known as a Wii. Gestures for games are sent wirelessly through the accelerometer to the system using the accelerometer. Check out their patent EP1854518B1 to find out more information about this technology. This will give you a complete understanding of how it works.

Pre-requisites

- Arduino Nano (any model)
- Accelerometer ADXL335 Module
- Bluetooth HC-05 Module
- Push buttons
- Python Installed computer

Circuit Diagram

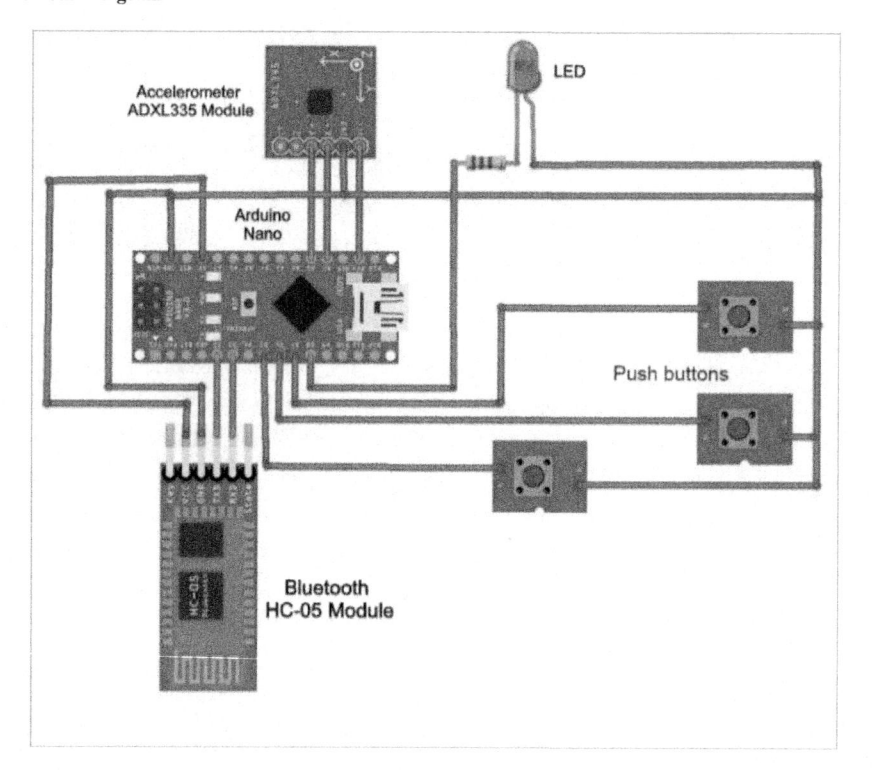

This idea inspired us to create an "Air mouse", which will make it possible to control a system just by waving it in the air, however, instead of using 3D coordinates, we will only be using 2D coordinates in order to make the computer mouse mimic actions since the mouse works in two dimensions X and Y With the Wireless 3D Air Mouse, the technical concept is very simple, through the use of an accelerometer we will measure the speed of the actions and motions of the Air Mouse along the x and y axis, with the help of the Python software drivers running on the computer, we will control the mouse cursor and perform certain actions based on the values of the accelerometer.

335. Arduino Whistle Detector Switch using Sound Sensor

In my childhood I was fascinated by a music car that got triggered when you clapped your hands, and as I grew up, I wondered if we can use the same thing to control lights and fans in a home. My fan and light switches could be operated with a simple handclap rather than walking up to them. As this circuit constantly responds to any loud noise, like a loud radio or my neighbour's lawn mower, it would often malfunction. However, clap switches can also be fun to build.

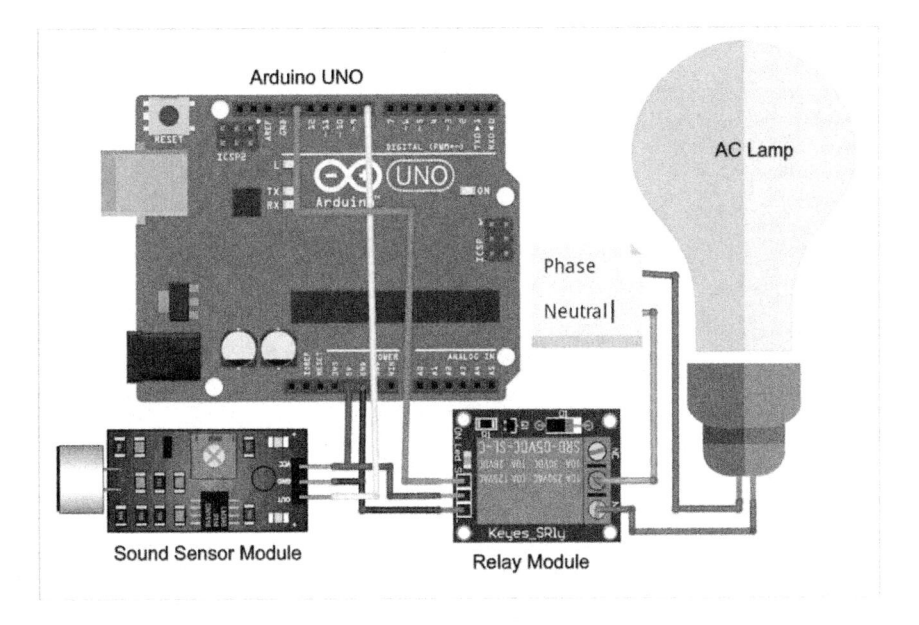

That is when I discovered this whistle detecting circuit, which detects a whistle. Like speech or music, a whistle has a specific frequency that is constant for a particular duration, thus can be distinguished from other sounds. We'll see how to identify whistles by using a Sound Sensor and an Arduino for this tutorial, and we will use an AC lamp and a relay to control it when a whistle is heard. As we proceed, we will learn how sound signals are received by devices such as microphones and learn how to use Arduino to measure frequency. This sounds interesting, so let's get started with our home automation project using Arduino.

Materials Required

- Arduino UNO
- Sound Sensor Module
- Relay Module

- AC Lamp
- Connecting Wires
- Breadboard

Sound Sensor Working

As we know, the microphone is the part of the sound sensor that converts acoustical vibrations, called sound waves, into electrical energy. An electronic signal is created on an output pin when the microphone's diaphragm vibrates to sound waves in the environment. A microcontroller like Arduino cannot directly process these signals since they are very small in magnitude (mV). The output from the microphone is by default analog, so it is a sine wave of variable frequency, but electronic microcontroller are digital devices and hence can handle square waves better.

336. Obstacle Avoiding Robot using Arduino

A simple Robot Obstacle avoidance project is designed here. The field of robotics is fast growing and has a lot to offer. Technology advances are increasing the applications of robotics since it is a branch of engineering. Mobile Robots are becoming more and more complex, and the number of mobile robots and their applications are increasing rapidly. The navigation techniques used for mobile robots include path planning, self-localization and map interpretation. Robot-type vehicles known as Obstacle Avoiding Robots are designed to avoid collisions with unexpected obstacles. We have designed an obstacle avoidance robot in this project. Ultrasonic range finders are the key to avoid collisions with this Arduino-based robot.

Ultrasonic Sensor

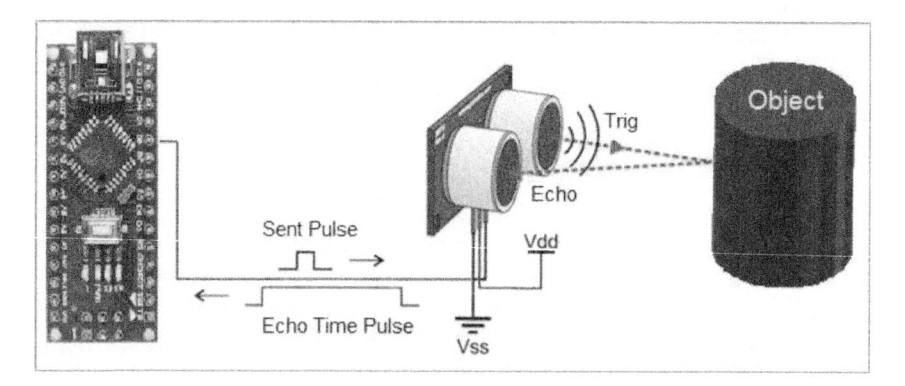

You need to know how the ultrasonic sensor works before tackling the project, read chapter 1 for more information on ultrasonic sensors. Ultrasonic Ranger - uses ultrasonic waves to measure distance between sensors and objects based on the time between generating the ultrasonic wave and receiving its echoes. There are four pins on the HC-SR04 Ultrasonic sensor: Vcc+, Trigger, Echo, and Ground.

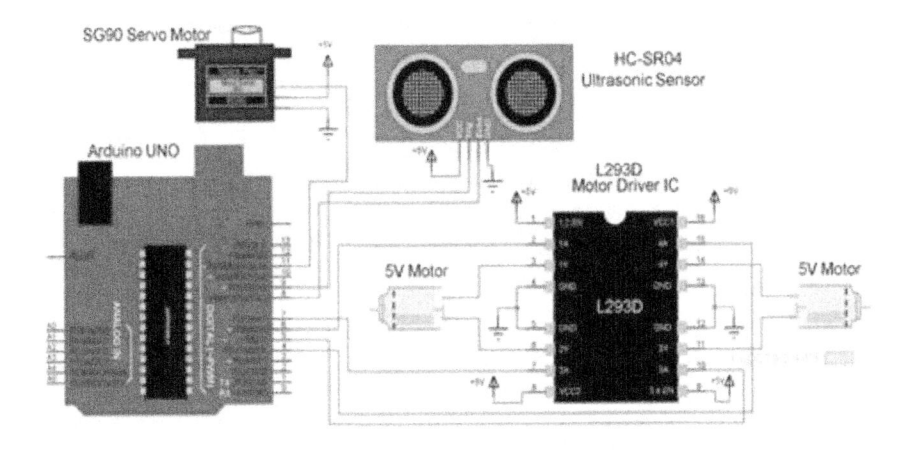

Hardware Required

- Arduino Uno
- Ultrasonic Range Finder Sensor – HC - SR04
- Motor Driver IC – L293D
- Servo Motor (Tower Pro SG90)
- Geared Motors x 2
- Robot Chassis
- Power Supply
- Battery Connector
- Battery Holde

337. Speed, Distance and Angle Measurement for Mobile Robots using Arduino and LM393 Sensor (H206)

Throughout human history, robots have gradually made our lives easier. Starship has already deployed six wheeled robotic food delivery vehicles on the UK's roads, smartly avoiding motorists to reach their destinations. All mobile robots that navigate within the environment need to be aware of their own position and orientation with regard to the environment.

A variety of technologies can be used to accomplish this, including GPS, RF Triangulation, Accelerometers, and Gyroscopes. All of the techniques have their own advantages, so each is unique in its own way. Here we will be reading speed and distance from the Arduino microcontroller using the simple and readily-available LM393 speed sensor. By using these parameters, the robot will be able to gain an understanding of its current status in the real world and, as a result, navigate safely.

Required Components

- Arduino Nano
- 16x2 LCD module
- L298N H-Bridge Motor Driver
- Analog Joystick, H206 Sensor
- LM393 Speed Sensor

Circuit Diagram

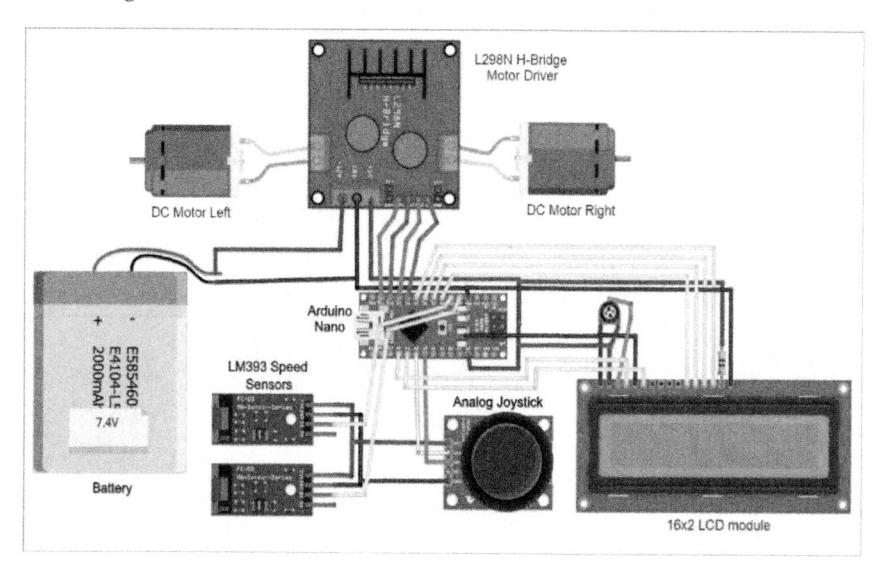

Here is the circuit diagram for this speed and distance sensing robot. A L298N H-Bridge Motor Driver module drives two DC motors powered by the Arduino Nano as the Robot's brain. Both the Joystick and the H206 speed sensors are used for controlling speed, direction, and angle of the bot, while the Joystick and second speed sensor are used to measure distance, speed, and angle of the bot respectively. Displayed in the LCD 16x2 module are the measured values. LCD contrast can be adjusted by the potentiometer connected to it, and the resistor is used to limit the current flowing through the LC's backlight. This 7.4V Lithium cell is used to power the whole circuit. Its 12.0V pin is connected to the Motor driver module. Once the motor driver module is connected to the Arduino, +5V is regulated by the voltage regulator, which powers the LCD, Sensors, and Joystick. Using Ariana's digital pins 8,9,10, and 11, the Motor can be controlled. In order to control the motor's speed, PWM signals must be provided to the positive terminal of the motor. We have a PWM capable pin which corresponds with pin 9 and an analog pin A2, which is read from the Joystick.

When a gap in the grid plate is detected by the H206 sensor, it will generate a trigger. Both the triggers (output pins) from both boards are connected to External interrupt pins 2 and 3 of the Arduino board since they should not always be read accurately to calculate the correct speed and distance. I assembled my bot like the following, with the circuit board mounted on the chassis and the speed sensor installed as explained, after the connections were made, it looked like this.

338. Build a Smart Watch by Interfacing OLED Display with Android Phone using Arduino

A 162 Dot matrix LCD display is probably familiar to most of us, but we are also familiar with the 16x2 LCD display used as some kind of information display in our everyday lives. However, there are many limitations in what these LCD displays can do. The basic information from the Android smartphone will be displayed using OLED in this tutorial like the time, date, network strength, and battery level. Here is a collection of simple tools and pictures that will allow you to build a simple but powerful Arduino based Smartwatch, which will also act as a watch for incoming calls and messages on your OLED display.

In this case, we are using an android phone application to fetch data from the OLED Display and then Bluetooth Module and Arduino Pro Mini are used to send this data to the OLED Display. Bluetooth modules with Arduino work well for sending data to Android smartphones. It is also possible to use the Bluetooth module HC-05 instead of HC-06.

Circuit Diagram:

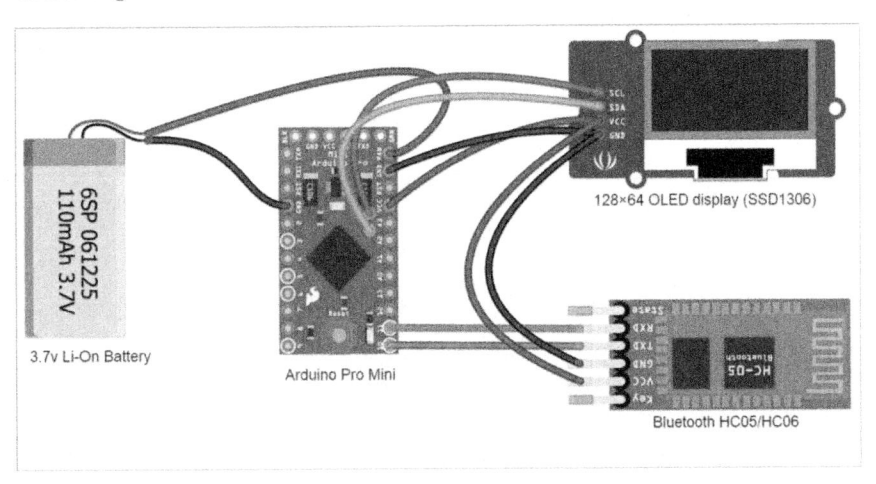

The Arduino is connected to the mobile phone by means of a String, which fetches data from the mobile phone and is sent to the Arduino. When Arduino receives the string of bytes, it decodes it into a string of temporary variables that will be displayed on an OLED display. In OLED displays, some graphics have been created that help with displaying values

Hardware Required

- 128×64 OLED display Module (SSD1306)
- Arduino (we have used Arduino Pro Mini. But we can use any Arduino Board)
- Bluetooth HC05/HC06
- Connecting Wires
- 3.7v Li-On Battery

JumperOur board for controlling all the operations is an Arduino Mini. One of the reasons why Arduino pro mini should be chosen is that it uses 3.3v power supply. The HC-06 OLED can be operated at 3.3v as well. Therefore, all of these modules can be operated from a single 3.7v Li-on. Li-on batteries are the most compact and lightweight type of battery. Wearable devices benefit greatly from this technology. This project also includes a wearable smartwatch that connects to a smartphone. Several questions arise regarding the power supply. All the modules are working on 3.3v but the 3.7V Li-ion battery is causing damage to the modules. The solution to this problem we used was to apply 3.7 volts of battery power to a raw pin of Arduino pro mini that could change that voltage into 3.3 volts.

339. Arduino Bluetooth with MATLAB for Wireless Communication

This protocol is the most popular solution for wireless communication in embedded systems due to its simplicity and ability to support short range communication. Besides being used to transfer data between two devices but also to control them wirelessly, Bluetooth is also used for controlling certain devices. Bluetooth is built into almost every electronic gadget nowadays, so securing Bluetooth control in your embedded app is a wise choice.

Circuit Diagram

The following schematics are needed to communicate between MATLAB and Arduino using Bluetooth.

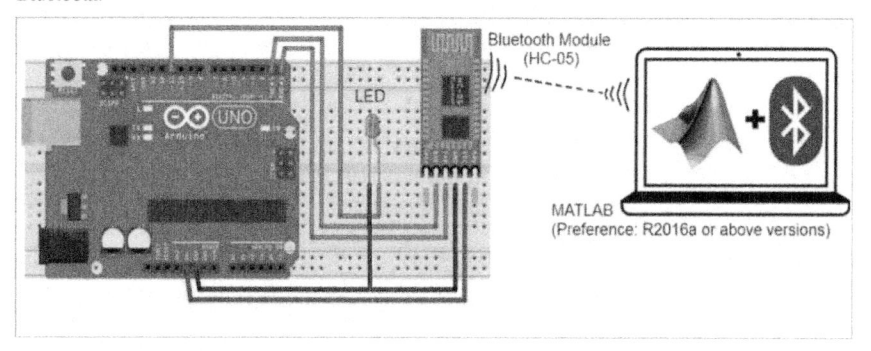

Using Bluetooth in MATLAB and wireless communication, we will learn how to accomplish this in this tutorial. On one side, we will use Bluetooth in MATLAB, and on the other, we will use HC-05 with Arduino. MATLAB and Arduino can communicate via Bluetooth in two ways, one via the command window and the other via MATLAB GUI. Both of these methods use the same Arduino code.

Components Required

- MATLAB installed Laptop (Preference: R2016a or above versions)
- Arduino UNO, Bluetooth Module (HC-05)
- LED (any color)
- Resistor (330 ohm), Jumper Wires

The Arduino UNO needs to be uploaded with the given code and then MATLAB Command Window should be used to start programming.

```
#include <SoftwareSerial.h>
int TxD;
int RxD;
int data;
SoftwareSerial bluetooth(TxD, RxD);
void setup() {
  Serial.begin(9600);
  bluetooth.begin(9600);
}
void loop()
{
if(bluetooth.available() > 0)
{
data = bluetooth.read();
Serial.print(data);
Serial.print("\n");
if(data == '1')
{
digitalWrite(11, HIGH);
}
else if(data == '0')
{
digitalWrite(11, LOW);
}}}
```

Then, copy and paste the below MATLAB code in the Command window for Bluetooth communication between MATLAB and Arduino.

To demonstrate Bluetooth Communication using MATLAB GUI, we will develop three graphical buttons in MATLAB to turn on, off, and flash the LED linked to Arduino. After clicking on the graphical buttons, the data will be transmitted from MATLAB to HC-05 using Bluetooth. The LED is controlled by Arduino according to the data that is received through the Bluetooth transmission of MATLAB data to HC-05. Arduino contains the code necessary to accomplish this. The code for the Arduino will remain the same as it was before; the only change will be that, whereas before we were sending the data '1' and '0' through the command window of MATLAB, the same data will now be transmitted on clicking on three graphical buttons.

```
instrhwinfo('Bluetooth','HC-05');
bt = Bluetooth('HC-05', 1);
fopen(bt);
```

```
Command Window
>> instrhwinfo('Bluetooth','HC-05');
bt = Bluetooth('HC-05', 1);
fopen(bt);
fx >>
```

In the below code, command *fprintf(bt,'0')* is used to turn OFF the LED by sending '0' to the Arduino. Now, if you want to turn ON the LED just send '1' instead of '0' using the below command.

```
fprintf(bt,'1');
```

```
Command Window
>> instrhwinfo('Bluetooth','HC-05');
bt = Bluetooth('HC-05', 1);
fopen(bt);
>> fprintf(bt,'1');
>> fprintf(bt,'0');
fx >>
```

To check the information about the available hardware, use below command

```
instrhwinfo('type','Name of device');
```

To open the bluetooth port, below command in used

```
fopen(bt);
```

Check the video below to understand the complete process of **Sending Data from MATLAB to Arduino using Bluetooth**.

Arduino can be programmed in MATLAB. For MATLAB to run on Arduino target hardware, you need to install a support package from add-ons. The Bluetooth modules HC-05 or HC-06 can be used. For more information about Bluetooth communication through MATLAB, please refer to the below link

- https://www.mathworks.com/help/instrument/reading-and-writing-data-over-the-bluetooth-interface.html

340. Smartphone Controlled Arduino Mood Light with Alarm

It's quite impressive how this LED strip works, I've recently purchased the NeoPixel LED strip. We control each LED individually on the board by using a driver IC which delivers a vibrant spectrum of colours. As an obsessive colour freak, I was intrigued by the tiny LED's changing colours during night times, which is why I realized that I could build a similar project for myself.

Circuit Diagram

The complete circuit diagram for this Bluetooth Controlled Arduino Mood Lamp Project is given below.

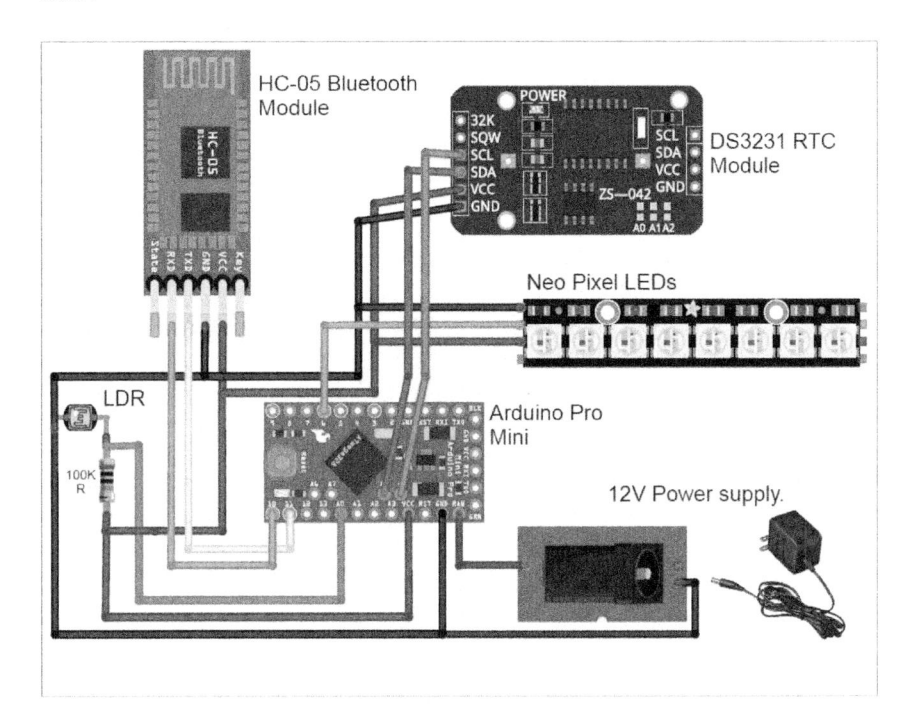

This LED light should not only be able to change colour, but should have a rational reasoning behind it as well. A professor of Color Science and Technology at University of Leeds by the name of Stephen Westland, I discovered that article. In response to colored light, humans exhibit a physical and psychological response based on the color. There have been many studies conducted on this process under the name of chronotherapy and the equipment used to accomplish this has been called a Mood Lamp. Hence, I decided to venture deeper into chronotherapy and build a lamp with color changing capabilities that also can be controlled by a phone. Also, I enhanced the setting by adding a daytime dimming screen and an option to set a bright orange wake-up light (sunshine), This alarm allows you to sleep by putting the LEDs into sleep mode in a mild purple (night sky) color. Isn't that cool? Let's get building.

Materials Required

- Enough science we are supposed to be working with electronics, so let's gather the required components.
- NeoPixel LEDs
- Arduino, DS3231 RTC module

- HC-05 Bluetooth Module
- LDR, 100K resistor, 12V Power supply.
- Chronotherapy - Mood Lamp

Many of these DIY mood lamps offer nothing more than switch on and off at random, without any purpose behind it. A mood lamp should have a minimum brightness, be gradually colored, and have progressively varying intensities. After a bit of research, I discovered that such a lamp should have a minimum brightness. Below is a table which compares how each color affects mental and physical level.

341. Interfacing nRF24L01 with Arduino: Controlling Servo Motor

Wireless communication is becoming increasingly ubiquitous, with more machines/devices communicating on cloud-based platforms such as Internet of things (IoT), Industry 4.0, Machine to Machine communication, etc. Bluetooth Low Energy (BLE 4.0), Zigbee, ESP43 Wi-Fi modules, 433MHz RF modules, Lora, nRF etc. are some of the wireless communication systems that engineers use, and their choice of medium is determined by the type of application. Most popular among all is the nRF24L01, which is a radio frequency-based interconnect system. In addition, modules of this type operate on 2.4GHz bands with baud rates between 250Kbps and 2Mbps and have been legal throughout the world.

They claim that they can also transmit and receive 100-meter distances with proper antennas. So, what will this tutorial teach you? This tutorial will give you a deeper understanding of how to successfully interface this module with microcontroller platforms such as Arduino. This module also provides solutions to some of the problems that may occur when using it.

Getting to know the nRF24L01 RF Module

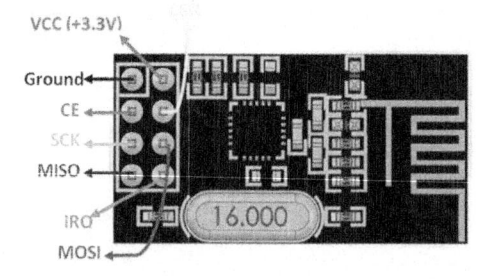

A transceiver, the nRF24L01 modules are capable of communicating in both directions but are half-duplex, which means they can only send or receive data at the same time. It uses the Nordic semiconductors nRF24L01 chip, which is responsible for transmit and receive functions. SPI is the protocol the IC uses, so it can communicate with any microcontroller. Because Arduino has libraries readily available, it gets a lot easier.

Transmitter side

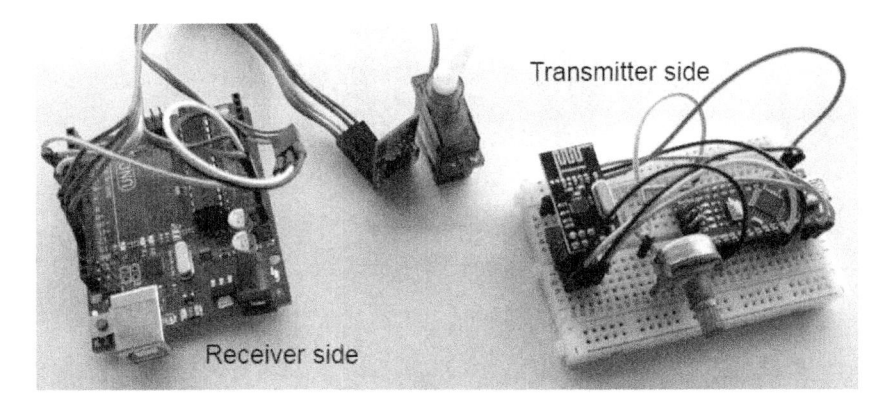

Receiver side

Here is a list of pinouts of nRF24L01 modules. Operating at voltage levels from 1.9V to 3.6V (typically 3.3V), the module consumes only 12mA during normal operation, making it battery efficient and therefore allowing the module to run on coin cells. The pins are tolerant of 5V despite the operating voltage of 3.3V, and so can be directly connected to 5V microcontrollers like Arduino.

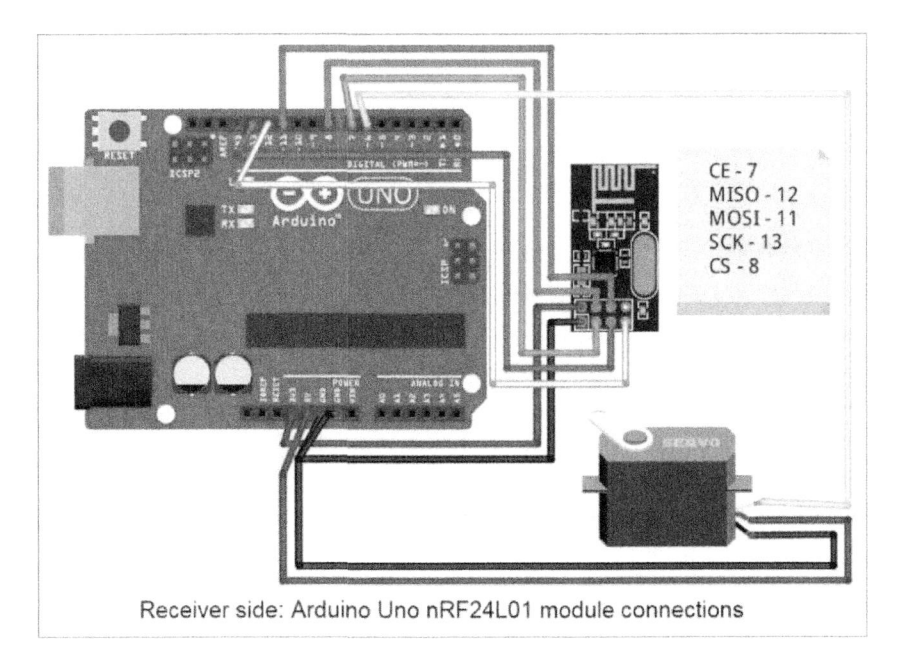

CE - 7
MISO - 12
MOSI - 11
SCK - 13
CS - 8

Receiver side: Arduino Uno nRF24L01 module connections

Using these modules has another advantage in that each one comes with 6 Pipelines. Specifically, each module can communicate with 6 other modules each time data is transmitted and received. As a result, the module can be used to create star networks and mesh networks in IoT applications. Their address range is wide as well. 125 unique identifiers are capable of being used in a closed area, so if you have 125 of them you can't interfere with each other.

```
CE - 7
MISO - 12
MOSI - 11
SCK - 13
CS - 8
```

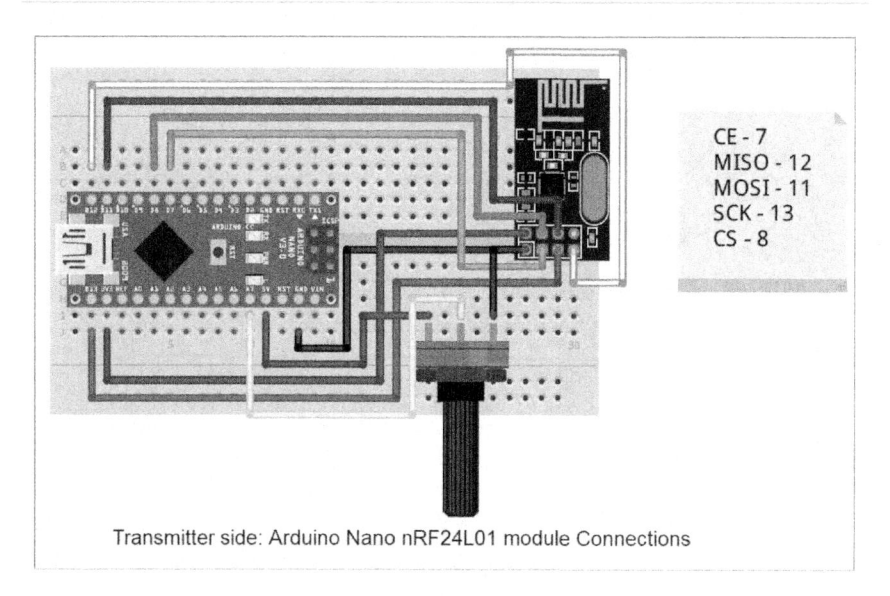

Transmitter side: Arduino Nano nRF24L01 module Connections

342. Build your own self-balancing robot with Arduino

After being inspired by and wanting to create something similar to the Segway self-balancing scooter models, I built the RYNO motor. I decided to build a self-balancing Arduino robot based on my thinking. Therefore, I can learn about the PID algorithms while understanding the underlying theory behind these scooters.

Circuit Diagram

I realized that building this bot was quite difficult the first time I did it. With so many options to choose from, including motor selection and PID tuning, it is only natural to be confused. There are a number of variables to consider, including the CoG, the battery type, the battery location, wheel grips, motor drivers, etc. However, let me reveal to you that once you learn how to do it, you will see for yourself that it's not as difficult as it sounds. In this tutorial we'll discuss the path I took in order to build the self-balancing robot. This may be the first time you are getting started with bots or maybe you have been frustrated for a long time and are now looking for help. Having arrived here, you will feel like you have arrived at your final destination.

Required Components

- Arduino UNO
- Geared DC motors (Yellow colored) - 2Nos
- L298N Motor Driver Module
- MPU6050, A pair of wheels
- 7.4V Li-ion Battery, Connecting wires
- 3D Printed Body

Circuit Diagram:

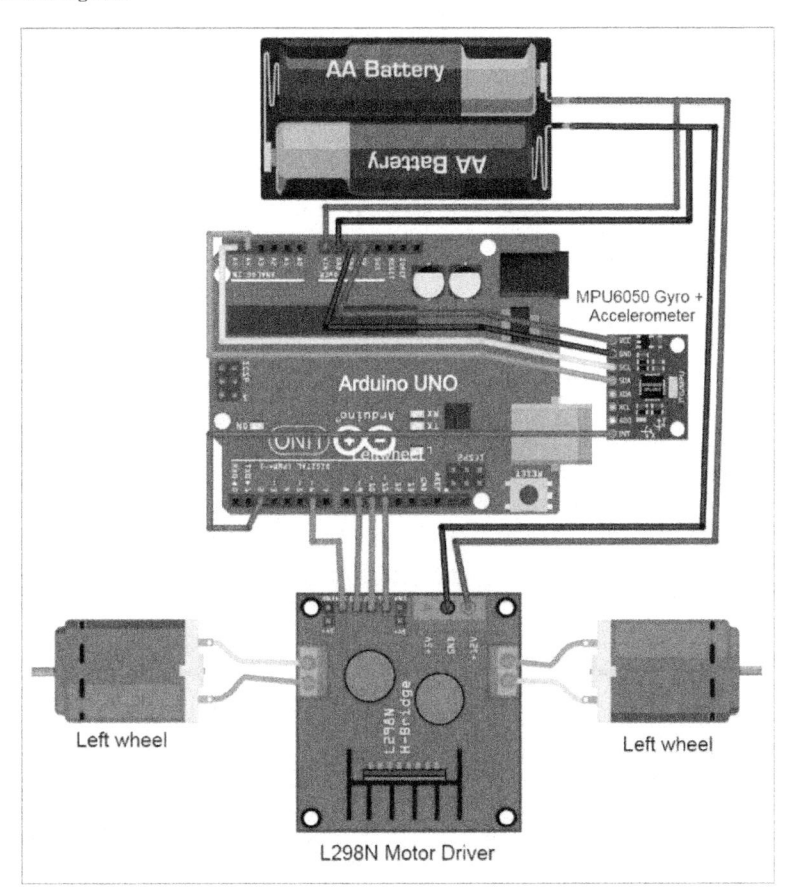

343. Automatic Water Dispenser using Arduino

Water covers about 71% of the earth, but sadly there is only drinking water in 2.5% of it. In 2025 we can expect perennial water shortages due to increased population, pollution, and climate change. We waste a lot of drinking water each year due to human negligence. On one hand, there are a number of minor disputes among nations and states regarding water sharing rivers, and on the other, there are national disputes among states regarding water sharing rivers.

One gallon of water is enough for an average human to live two days if your tap drips a drop of water every second. This might not seem like a big amount of water at first, but it would take you about five hours to waste one gallon of water. This problem can be solved through technology improvement. The answer always lies in technology development. The water consumption rate can be drastically decreased by replacing all manual taps with a smart faucet that automatically opens and closes without requiring us to touch the handle. Therefore, we will build a Solenoid Valves and Arduino based Automated Water Dispenser that can automatically dispense water to a glass when it is placed near it.

Materials Required

- Solenoid Valve, Arduino Uno (any version)
- HCSR04 – Ultrasonic Sensor
- IRF540 MOSFET, 1k and 10k Resistor
- Breadboard, Connecting Wires

Working Concept

The water dispenser works by dispensing water at the touch of a button. A HCSR04 Ultrasonic Sensor will be used to verify that no glass is placed in front of the dispenser. A solenoid valve will control the flow of water. Which generates electricity to operate when activated, and de-energizes when not in use. The solenoid will be turned on and will wait until the object is removed. We will write an Arduino program that checks if anything is placed near the tap, on the other hand if it is, the solenoid will be turned off, when you remove the object, the solenoid will turn off the water supply automatically.

Circuit Diagram

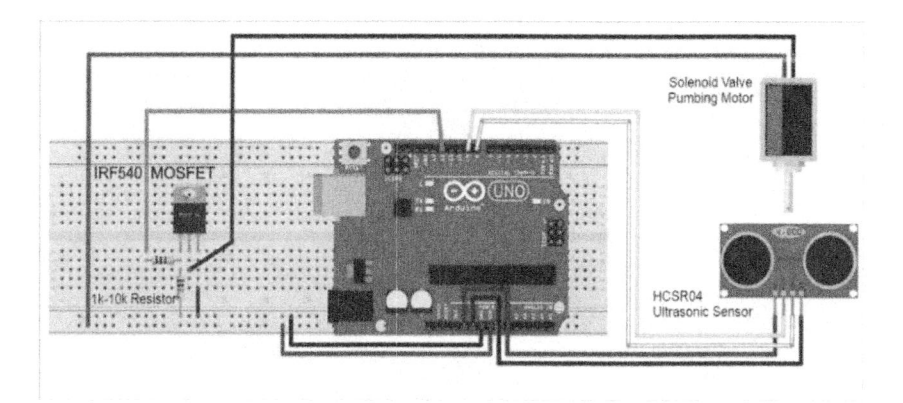

An electromechanical solenoid containing a 12-volt battery is used in this project. It has a continuous current capacity of 700mA. In other words, when the valve is on, it uses about 700mA to keep it on. Hence, a Solenoid driver circuit needs a switching driver to operate an Arduino board that operates with 5V and hence requires a 5V switching driver circuit.

IRF540N is the MOSFET used in this project and its switching device is referred to as a N-Channel MOSFET with the Gate, Source and Drain pins originating from pin 1. According to the circuit diagram, the Arduino's Vin pin powers the solenoid's positive terminal. As the Arduino will be powered by a 12V adapter and its Vin pin is 12V, the control of the solenoid can be performed. Several connections are made between the negative terminal and the ground via the MOSFET's Source and Drain pins. Only if the MOSFET is switched on will the solenoid be powered. Turning the MOSFET on or off is done with the gate pin. A gate pin grounded to ground will remain off while a gate voltage applied to it will turn it on. The gate pin of the MOSFET is pulled down to ground by a 10k resistor when no power is applied to it. A 1K resistor limits the current flowing to Arduino pin 12, which controls the MOSFET.

The Ultrasonic Sensor is powered by connecting a power supply to the Arduino's +5V and ground pins. To pins 8 and 9, respectively, are connected the trigger and echo pins. The Ultrasonic sensor can then be programmed to make use of the Arduino in order to detect objects and turn on the MOSFET if one is detected. My circuit was somewhat like this below after connecting all the wires. The whole circuit is simple and can be easily built on a breadboard.

344. Interfacing Flame Sensor with Arduino to Build a Fire Alarm System

Generally, a flame detector is a sensor that is designed to detect and respond if there is a fire present or if it is rapidly spreading. An alarm would sound, a fuel line would be deactivated, a fire suppression system would activate, and so on. Responses to a detected flame vary based on the installation. Flame detection methods differ. They include: Infrared detectors, UV/IR detectors, near-IR arrays, infrared thermal cameras, Ultraviolet detectors, and many others.

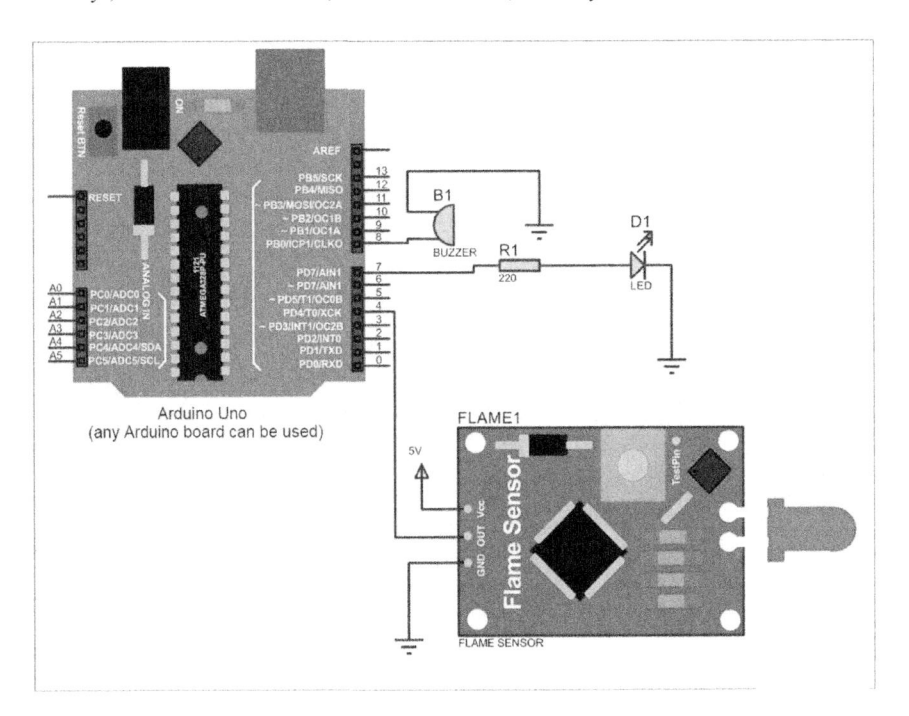

An infrared light sensor is utilized to detect the amount of Infra-red light emitted by the fire. An Op-Amp is then used to assess if any change has occurred in voltage across the IR Receiver, so that if there has been a fire the output pin (DO) will read 0V (LOW), and otherwise it will read 5V (HIGH).

In this project, we are using an IR flame sensor. This sensor utilizes a silicon phototransistor, the YG1006, which is extremely sensitive and has a high speed. Detectors for infrared light can detect wavelengths between 700nm and 1000nm, and have detection angles of 60° or greater. An integrated circuit which contains a photodiode, a resistor, a capacitor, a potentiometer, and an LM393 comparator is called a flame sensor. By adjusting the onboard potentiometer, the sensitivity can be adjusted. With a digital output, the working voltage ranges from 3.3v to 5v DC. On the output, logic high indicates the presence of fire or flame. On the output, logic low indicates the absence of fire or flame.

Applications of flame sensors

- Hydrogen stations
- Combustion monitors for burners
- Oil and gas pipelines
- Automotive manufacturing facilities
- Nuclear facilities
- Aircraft hangars
- Turbine enclosures

Components Required

- Arduino Uno (any Arduino board can be used)
- Flame sensor
- LED, Buzzer
- Resistor, Jumper wires

345. IoT Based Electricity Energy Meter using ESP12 and Arduino

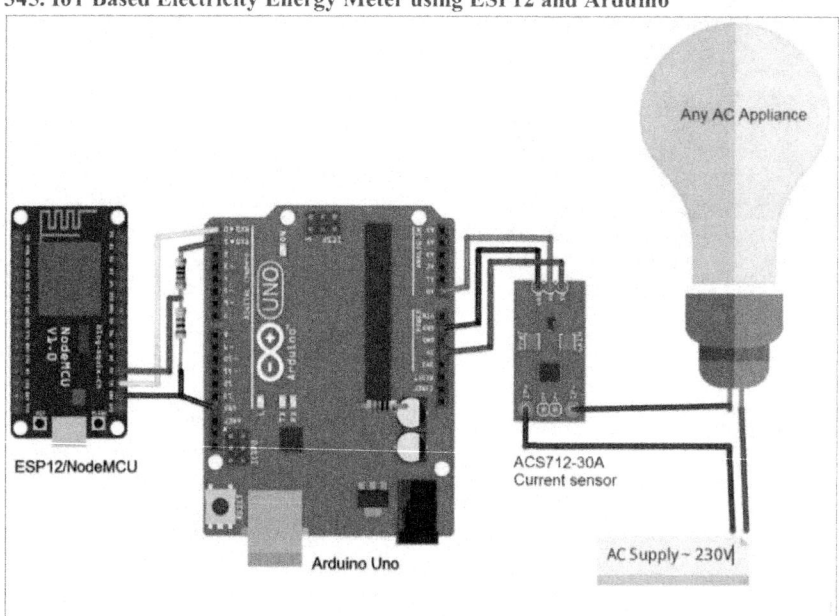

Every home in the world is equipped with electricity energy meters that measure electricity consumption. At the end of every month, many of us get worried about the high electricity bill and we have to look at the energy meter once in a while. How about being able to monitor your consumption from anywhere in the world and receive an SMS/email when the threshold value is reached

Materials Required:

- Arduino Uno, ESP12/NodeMCU
- ACS712-30Amp Current sensor, Any AC Appliance
- Male-Female Wires

An Energy Meter circuit was previously built using the GSM module which provides you with SMS notifications about your bill. Designed using the Arduino and ESP8266 Wi-Fi module, we build a Smart Electricity Energy meter that can send you an SMS/Email of the electricity bill along with real-time monitoring of your energy usage from anywhere and anytime. Our current sensor ACS712 has been used here to determine the energy consumption, and we will learn more about it shortly. In addition, we will utilize MQTT Dashboard for Android to monitor our Energy usage. Through this project, we will utilize the IFTTT platform to link our Wi-Fi with SMS and E-Mail notifications.

346. Coronavirus Sterilizer Box | Food Mask Sterilizer

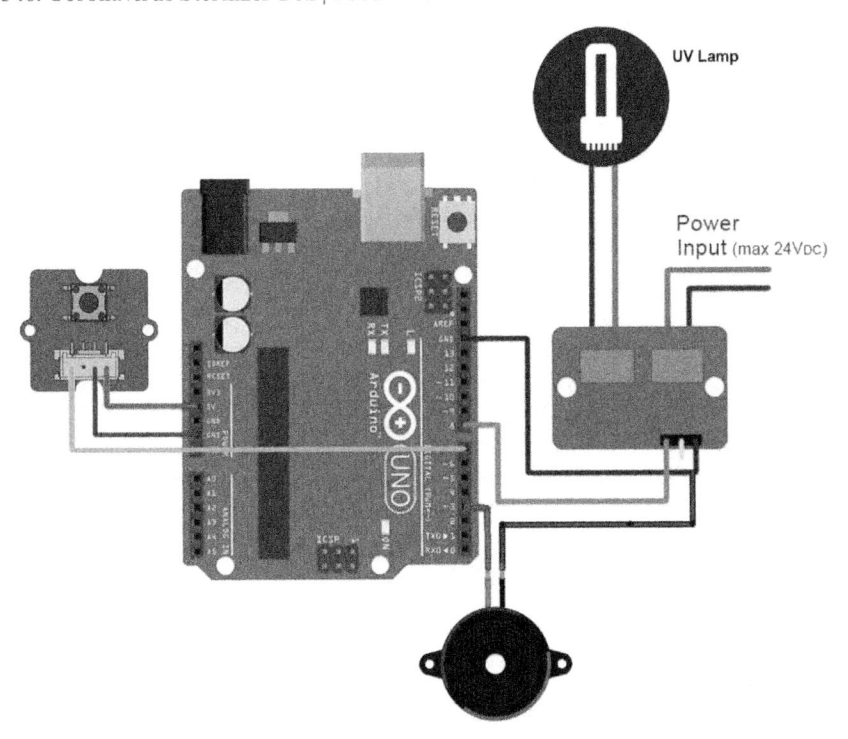

The Covid technology revolutionized 2020 for all of humanity. The way it spread rapidly forced us to wear face masks and gloves to protect our skin from everything we touched. Certainly, we can put on a mask when we are outside but what do we do if we bring something home from the store or trade with someone else? The fact that patients and employees exchange files and paperwork with doctors or with each other, cannot be sanitized by applying sanitizers to these outside items.

Circuit Diagram

Using an Arduino-powered system powered by a smart electronics chip, we solve this huge problem. To solve the problem, we designed an ultraviolet-sanitizing box that has a 60-degree angle. 8 uv tubes are employed by the system in order to achieve this task. All viruses have been killed by UV C in a matter of seconds:

Components

- Arduino Uno
- LCD Display, Buzzer
- Lid Sensor, UV-C Tubes, Buttons
- Metal Mesh, LED's

- IC's, Resistors
- Capacitors, Diodes, Transistors, Transformer
- Base frame, Supporting Frame
- Mounts and Joints
- Screws and Bolts

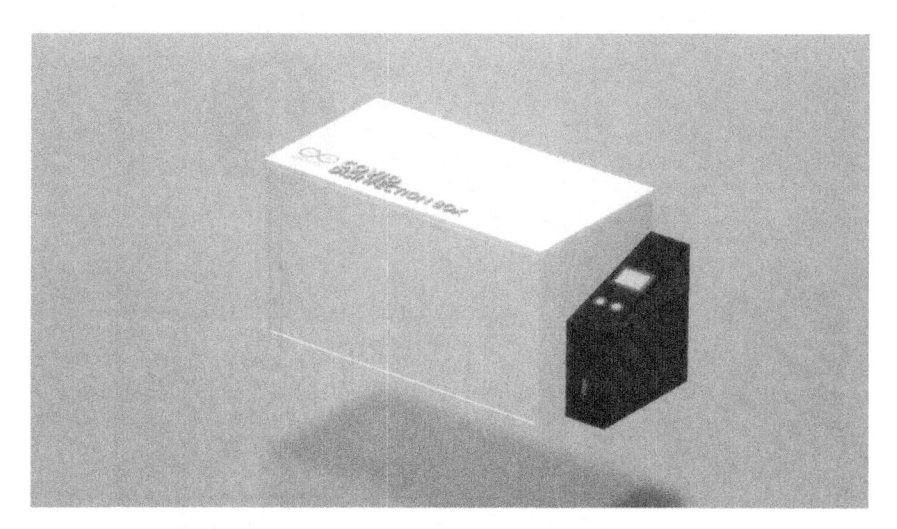

The Arduino COVID Disinfection box has the following Key Aspects

- A 360-degree approach to disinfection
- All Coronaviruses can be deactivated.
- Shutdown and alerts based on timers
- Sterilizes masks, packaged foods, electronics etc.
- Specifying the duration of the sterilization
- Shut off automatically
- Easy To Use
- We use no water and no chemicals | Environmentally friendly

Sterilization starts when the start button on the Arduino controller is pressed and the controller takes user inputs for time setting. This device shuts off automatically after it reaches the sterilization temperature. Another shutoff system also prevents lipids from being opened by users while sterilization is taking place.

Advantages

- No Chemicals or Water Sterilization
- Sterilization at 360 degrees
- Demonstrated to deactivate Bacteria and Viruses
- Sterilization time that can be adjusted
- Untimely opening will result in automatic safety shut off.

Disadvantages

- Since it is not large enough to sterilize large objects, it is not suitable.
- It does not have a battery and is not suitable for car use.

347. Play the Space Race Game using the Arduino and Nokia 5110 Graphic Display

Developing with Arduino is fun, and programming has been that way for years. Everyone out here has used the language they are learning or practicing to develop some kind of game. They have fun while practicing programming in this way. My interest in Arduino has been on the rise ever since I was introduced to it and wanted to do something cool with it. When I saw how smooth an interface could be using a Graphical LCD like the Nokia 5110 along with Arduino, the idea of developing a game came to me. So, you might also enjoy developing your game since it was an effective way to practice programming skills. As a result, in this project, you will learn how to build an entertaining game employing the Arduino microcontroller and the graphics LCD display.

Circuit Diagram:

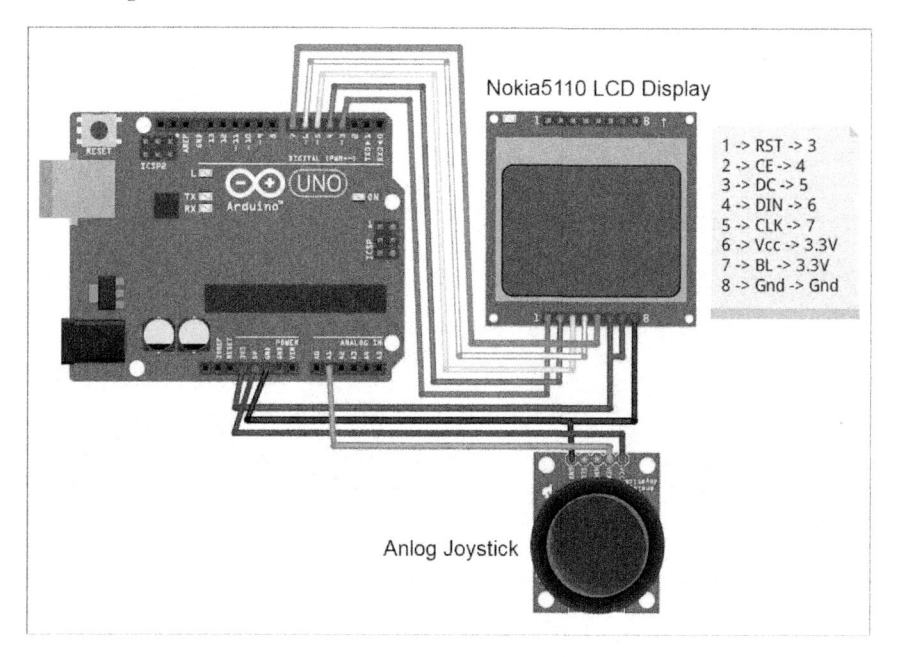

Nokia5110 LCD Display

1 -> RST -> 3
2 -> CE -> 4
3 -> DC -> 5
4 -> DIN -> 6
5 -> CLK -> 7
6 -> Vcc -> 3.3V
7 -> BL -> 3.3V
8 -> Gnd -> Gnd

Anlog Joystick

This time around we are going to try a new version of the Snake game using Arduino, which we have named Future Race, in which the player needs to keep their vessel safe from enemy ships by using a joystick.

Game Plan:

The method by which your game would work is very important before we start. A graphics LCD and joystick was the hardware that I chose for my hardware selection. You have likely selected the same option in this tutorial as well. We have had to plan our entire game inside the 84 * 48-pixel dimensions of the Nokia 5110 display because it doesn't have a lot of free space. The Nokia 5110 LCD has been reprogrammed to be used with the Arduino, as well as the Nokia Joystick for Arduino. It will be difficult to arrange the gaming area within this space, as well as the scoreboard area, which displays things like score and things. In order to update your screen with the pixel locations, it is very important to know where the pixel locations are.

It's very easy to make this game with Arduino; we just need to interface the Nokia 5110 LCD module, and the Joystick to Arduino. Our game characters need to be decided after we determine the appearance of the game screen. I have designed a game based on a spaceship over a planet and an enemy spacecraft disguised as a planet. So, I used the Nokia LCD's bitmap feature and my spaceship and enemies were displayed on the screen. A spacecraft will race against the alien spaceships, and have the capability to change lanes to avoid a contact with them. It should never be possible for an alien to occupy more than two tracks at a time and the player should always be able to drive over a free track. Our goal is to finish the Hardware and complete the Programming once the ideas are concluded.

348. Interfacing Tilt Sensor with Arduino

When a tilt sensor is activated, it determines whether the object is upright or tilted, and outputs high or low based on its orientation. In essence, it consists of a mercury ball which moves inside to create a circuit. Therefore, the tilt sensor is able to either turn on or off the circuit according to the orientation. We are interfacing an Arduino UNO with a Mercury switch / tilt sensor. We are controlling a LED and buzzer based on the tilt sensor's output. The alarm will be triggered upon tilting the sensor. This tilt sensor circuit also demonstrates the workings of tilt sensors.

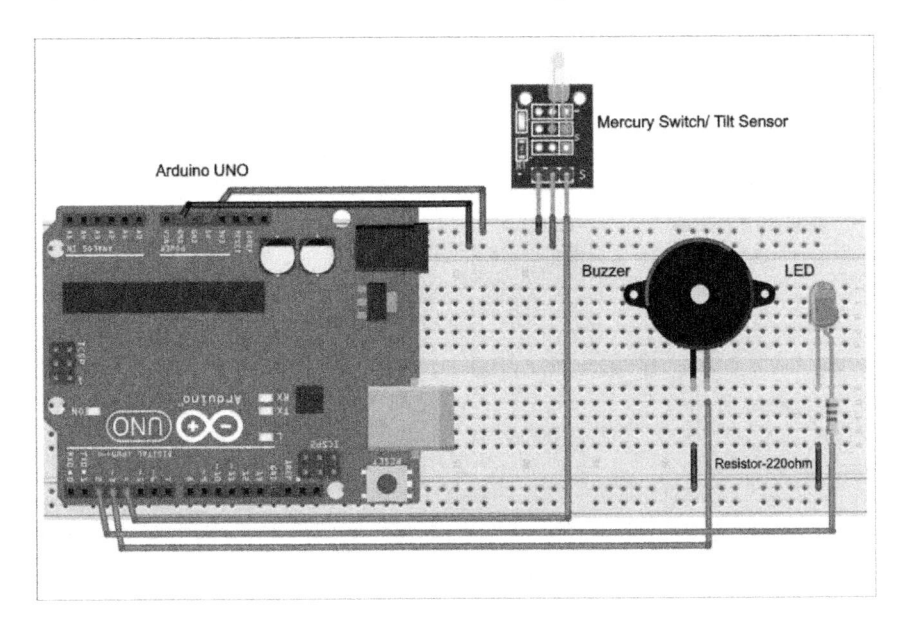

Material Required

- Mercury Switch/ Tilt Sensor
- Arduino UNO
- Buzzer, LED
- Resistor - 220 ohm
- Breadboard
- Connecting wires

An Arduino interface for tilt sensors.

A schematic of the easy connection between the sensor and Arduino follows below. Sensor pin VCC is connected to Arduino terminal 5V and sensor pin GND is connected to ground. The DO pin can be connected to any digital pin of the Arduino board. The Arduino needs 5v dc power to be able to operate the Tilt sensor. The 5V supply and Tilt sensor output is obtained by wires connected to pins 3 and 4 of Arduino. In order to prevent an overcurrent, the LED is connected with the Arduino UNO PIN 2 with a 230-ohm resistor. In addition, the buzzer is connected directly to Arduino UNO PIN 3.

Designed to measure tilt angle, this Mercury switch-based tilt sensor gives high on its output pin. 5V is required to power this device. It consists of input, ground, and output terminals. It is composed of a glass tube containing a liquid mercury ball and two electrodes. As the mercury ball is inclined a certain way, it closes and opens the circuit. Here is how the module works and is organized internally:

Working of Tilt Sensor

CASE 1: NOT TILTED

Initially, when it is in NOT tilted position as shown in the image below, it gives LOW output because of the liquid mercury complete the circuit by connecting the two electrodes. When the output is LOW on-board LED remain ON.

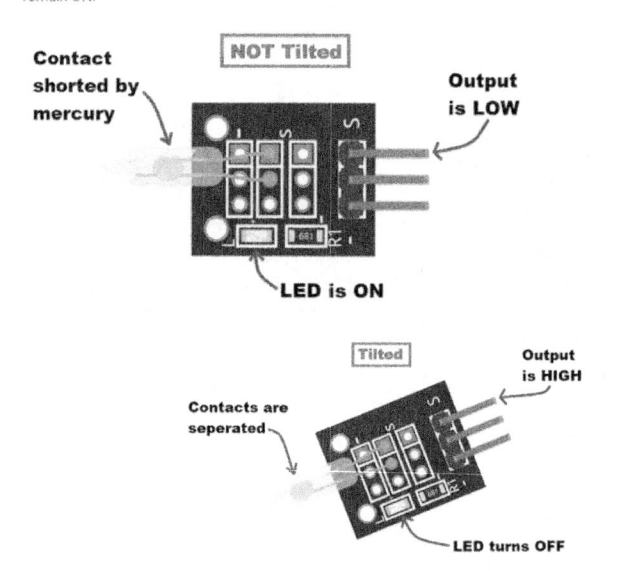

349. Bluetooth Controlled Servo Motor using Arduino

Robots are controlled precisely by using a Servo motor. Using an Arduino UNO and an Android device via Bluetooth, we will show you how to control a Servo motor via Bluetooth connection. Previously we controlled servo via Arduino, this time we will use Bluetooth Module HC-6 to control Servo wirelessly.

Material Required

- Arduino UNO,
- HC-05 or HC-06 Bluetooth module
- Servo Motor,
- Roboremo App from PlayStore
- Breadboard,
- Connecting wire
- HC-06 Bluetooth Module

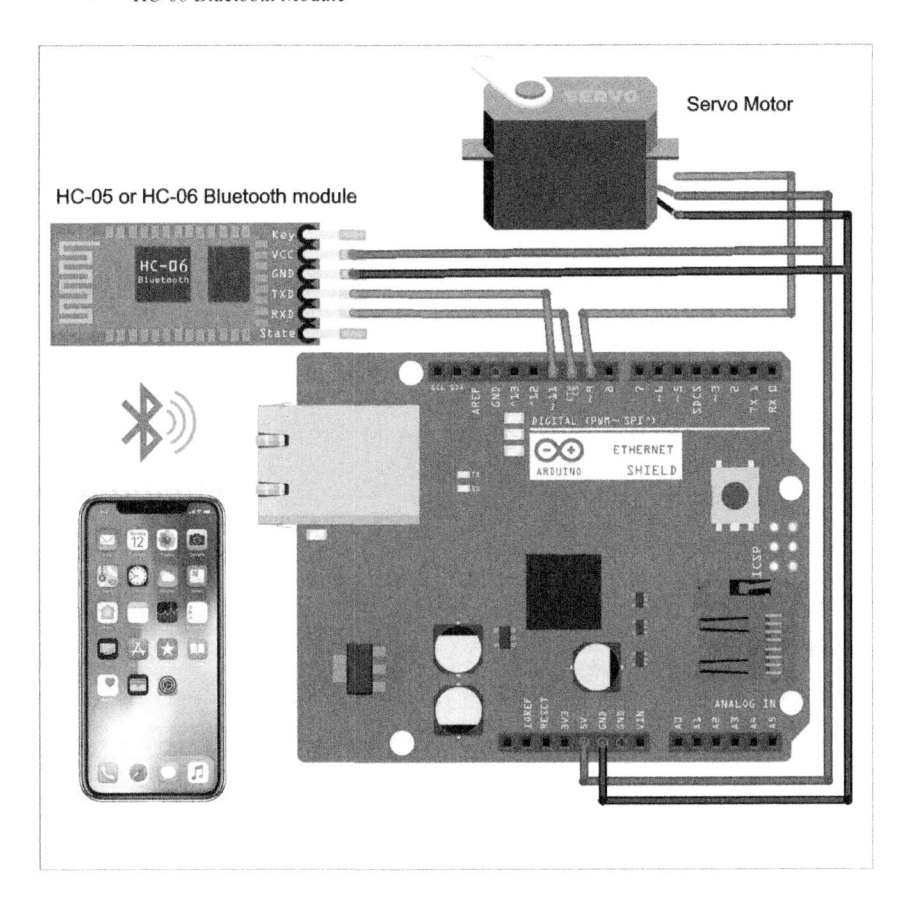

How it works

Sending data packets to the Bluetooth module is done by the android app. These data packets are then transmitted via serial communication to the Arduino Uno. A servo motor is controlled by Arduino Uno based on the value of the data packet. The flowchart below illustrates how this works.

| 500 Electronic Project Ideas for Inventors

| Android App sends Data Packets | ⇒ | Bluetooth module recieves these packets | ⇒ | Arduino sends signals to servo as per the value of data packets. | ⇒ | Servo Motor (9g) |

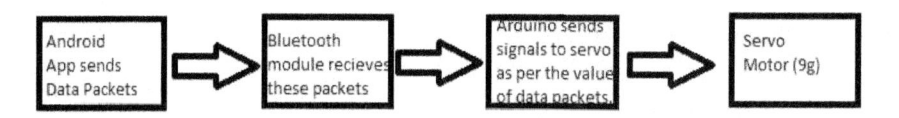

Bluetooth can operate in the following two modes:

1. Command Mode
2. Operating Mode

Command Mode will allow us to change the Bluetooth properties, like the name of the Bluetooth signal as well as the password, baud rate and range of features. This is the mode in which the PIC Microcontroller can transfer and receive data with a Bluetooth module. Thus, the Operating Mode will be our only discussion in this tutorial. We will use the default settings for the Command mode. All Bluetooth modules will use a default baud rate of 9600. The Device Name is HC-05 and the password is 0000 or 1234.

350. Controlling Multiple Servo Motors with Arduino

Arduino is great for controlling one or two Servos, but how do we control more than one Servo? We will be demonstrating how multiple servo motors can be controlled using Arduino. When connected to Arduino supply pins multiple Servo Motors seem to be simple, but they won't work correctly because there is not enough current to drive all the motors. Therefore, you need to supply each motor with a separate power supply, whether from some adapters (5v 2A) or from good quality batteries (9v).

Circuit Diagram

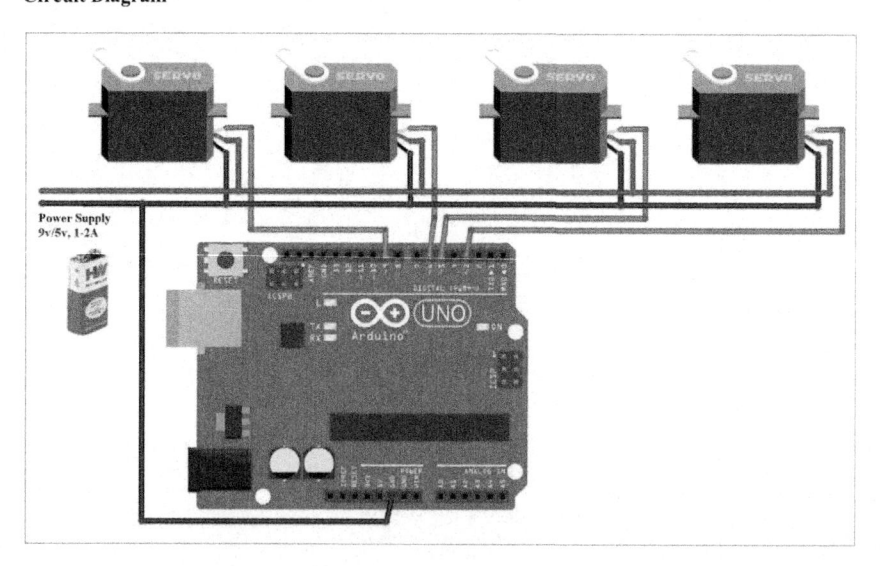

Material Required

- Arduino UNO
- Servo Motor
- Power Supply
- Breadboard
- Connecting Wires

While integrating more than two servos into one Arduino, we all experience current difficulties. There is only one solution to this problem, which is to connect an external power source rated appropriately (that is, I used a 9V supply that was rated for 2A). Powering small Servos can be accomplished by wiring in an external power supply through adapters, RPSs (Regulated Power Supplies), or good quality 9-volt batteries.

All of your servos can be controlled by a single Arduino, but they all require their own power source. It is necessary to establish a connection between the gnd pins on the Arduino and the gnd pins on the servo power supply. In case there was any confusion, I have been successfully controlling and powering all four Servos with the help of an Arduino up until this point. When using a laptop, you can also connect your USB port to power your Servos. The Arduino ground must be shorted to the external supply ground in order to use the external supply.

351. Arduino Based Countdown Timer

A timer refers to a type of clock that measures time intervals. Counting upwards from zero is one type of timer when measuring the elapsed time is called a stopwatch. It also has a second option, generally termed a Countdown Timer, which counts down based on the time duration provided by the user.

The following tutorial will provide you with detailed instructions in order to make a Countdown Timer using Arduino. Our implementation does not use the Real Time Clock (RTC) for time synchronization. With the help of the Keypad and 16x2 LCD, the time duration is provided by the user. The buzzer will be used to alert the user according to Zero on the timer.

Material Required

- Arduino UNO
- LCD 16*2
- 4*4 matrix keypad
- Buzzer
- Pushbutton
- Potentiometer (10k)
- Resistor (10k, 100 ohm)
- Connecting wires

Circuit Diagram

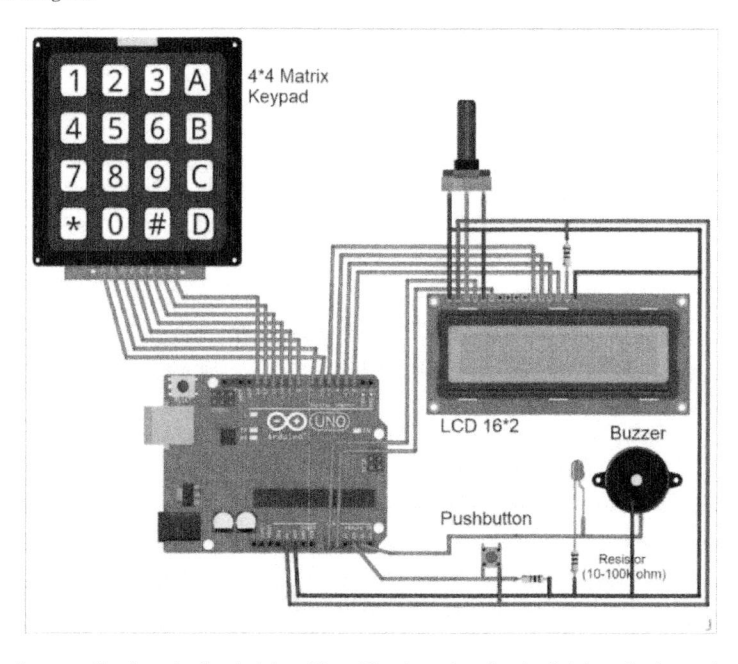

The main controller here is the Arduino Uno. The time duration is fed into the keypad and the countdown is displayed on a 16*2 LCD. To start the time, the pushbutton is pressed. Here is an Arduino tutorial on how to connect a 4x4 keypad with an LCD and a 16x2 LCD with Arduino.

352. Automatic Pet Feeder using Arduino

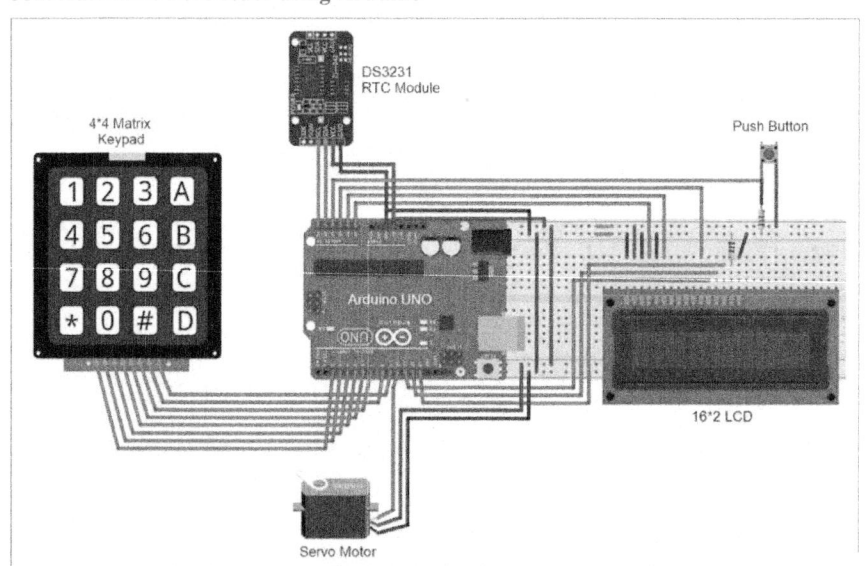

A Pet Feeder based on Arduino can automatically deliver food to your pet on schedule. Your pet should be fed on time and date set by the DS3231 Real Time Clock module. The device drops or fills the food bowl depending on your pet's eating schedule, so set the time accordingly. DS3231 RTC Module and Arduino UNO are used in this circuit to display the time on a 16*2 LCD. Also, a servo motor is used to provide the food by rotating the containers and a 4*4 matrix keypad should be used to set the feeding time.

Depending on the quantity you want to serve your pet, you can set the rotation angle and duration of dish opening. Aside from the size of your pet, the amount of food you ought to give him also depends on whether he is a cat, a dog, or a bird.

Material Required

- Arduino UNO, 4*4 Matrix Keypad
- 16*2 LCD, Push Button
- Servo Motor, Resistor
- Connecting Wires, Breadboard

We have used RTC (Real Time Clock) Module for time and date acquisition in this Arduino based Cat Feeder. With the help of the 16x2 LCD, we made the Stepper control Pet's eating time by using a four-by-four matrix keypad. When the user sets the time, the Servo motor rotates the container and drops the food on the determined date and time. In the video on the end, you can see complete working of the LCD. Date and Time can be displayed on the LCD.

DS3231 RTC Module

RTC (Real Time Clock) module DS3231 works with the DS3231 microcontroller. Many of the Electronics projects rely on it to keep track of the date and time. When the main power is removed from the module or if the MCU has undergone a hard reset, the module will maintain the date and time using the coin cell battery power supply. This module will always keep track of the date and time once the date and time have been set. In our circuit, we are using the DS3231 to make the pet's owner set the feeding schedule, like an alarm, to the pet's daily food requirements. The clock opens the container gate when the timer reaches the set time and drops the food into the Pet's bowl.

Note: You can also use the RTC IC DS1307 to read the time if you use this module for the first time.

353. Arduino Based AC Home Appliances controlling with thermistor and relay

If you were sitting in a cold room and you wanted your heater to be automatically turned on, then that might be possible. Usually, when a room temperature increases, appliances are turned on for some time and then off, making the project useful for controlling home appliances according to the temperature. With Arduino, we control our home air conditioning systems based on the temperature. A Thermistor was used to measure temperature in this case. The Thermistor was interfaced with Arduino and the temperature was displayed on LCD.

Circuit Diagram

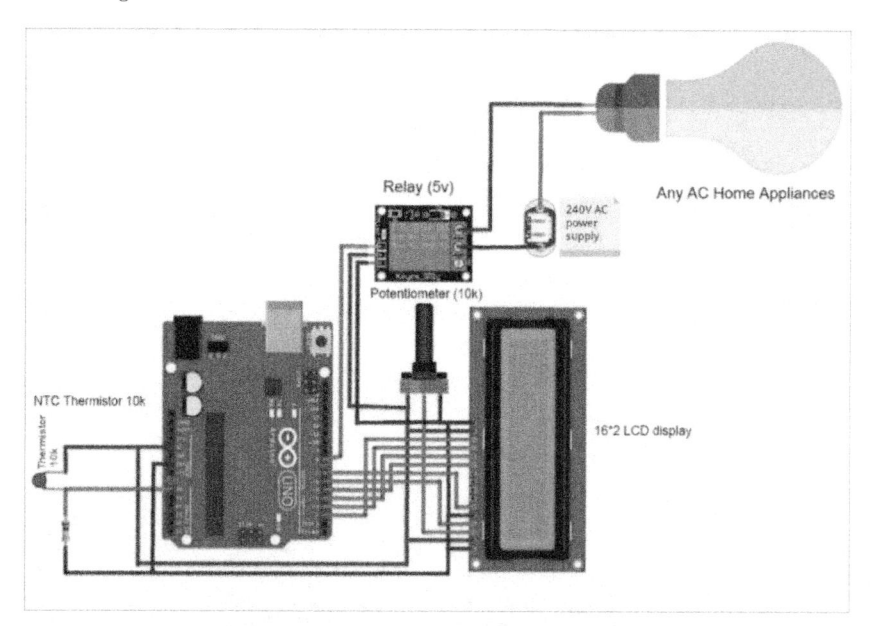

We will use the Arduino temperature-controlled system in this tutorial to be able to control an AC appliance using a Relay. Displayed on the 16*2 LCD display with connection to the circuit are the current temperature and appliance status. A variety of components are used in this Home Automation System, including an Arduino board, LCD display, relay and thermistor. Basically, the whole system works by using a relay and a thermistor; as the temperature rises, the relay will turn on and as the temperature drops below the threshold, the relay will turn off.

Material Required

- Arduino UNO, Relay (5v),
- 16*2 LCD display
- Light Bulb (CFL)
- NTC thermistor 10k, Connecting wires
- Resistors (1k and 10k ohms), Potentiometer (10k)

Relays will also enable and disable the appliance attached to them. CFL bulbs are used as AC appliances in this system. The Temperature based Home Automation System contains components such as an Arduino board, LCD display, thermistor, Relay, and Relay. The relay and the thermistor are key elements in this system. As the temperature increases so does the relay. If it drops below the preset temperature the relay is turned off. The Relay will also control the home appliance connected with it. For this example, an AC appliance is connected as a CFL bulb. The Arduino board is programmed to trigger the entire triggering procedure and set the temperature value. On the LCD screen, we can see the temperature at every half second, as well as the status of the appliances. The Arduino Servo Motor library takes care of all electronic properties of the servo, so you just need to enter this angle and there is a function servo1.write(angle); which will rotate the servo to the desired angle.

Thermistor

Thermistor is the key component in this circuit, which is responsible for detecting temperature rise. Temperature-sensitive resistors measure resistance by changing according to temperature, called thermocouples. We are being tested with a NTC thermistor.

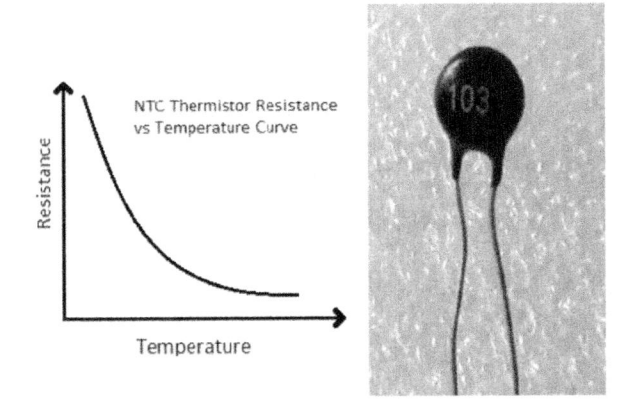

Both types of thermistors have negative temperature coefficients and positive temperature coefficients. The resistance of an NTC thermistor increases with rising temperature, while the resistance of a PTC thermistor increases with rising temperature.

354. DIY Arduino Inclinometer using MPU6050

This accelerometer and gyroscope are an integrated circuit (IC) that measures three axes of movement at the same time. Besides the temperature sensor and DCM, the module contains additional features for complex tasks. In the production of self-balancing robots and other remote devices, the MPU6050 is often used. We will build an Inclinometer or Spirit Leveler using the MPU6050 and learn how to use it.

A digital inclinometer or a spirit bubble inclinometer is used to measure inclination, but they can also be used as inclinometers to level a surface. A Digital Inclinometer is being developed in this project and it can be monitored by an Android application. By using a mobile phone for displaying the data from the MPU6050, we can do so without having to look at the hardware; this could be very useful when the MPU6050 is mounted on a drone or inaccessible place.

Materials Required:

- Arduino Pro-mini (5V)
- MPU6050 Gyro Sensor
- HC-05 or HC-06 Bluetooth module
- FTDI board, Breadboard
- Connecting wires, Smart Phone

An image of the circuit diagram is shown below for this Arduino Tilt Sensor Project. A breadboard can be used to build the circuit with just three components.

Circuit Diagram

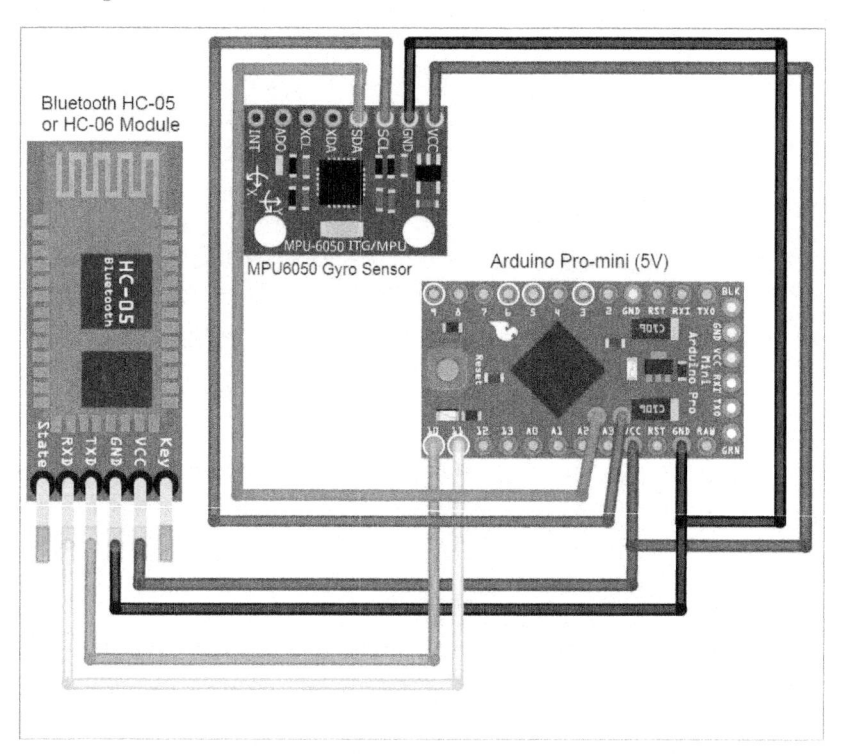

The following circuit diagram shows how Arduino Inclinometer uses MPU6050. As the breadboard moves, you can observe these values changing as the zeros become values. You should check your connections if they change, otherwise make sure your connection is correct. Look at the three values Pitches, Roll and Yaw as you tilt your sensor and note how they vary according to your tilt.

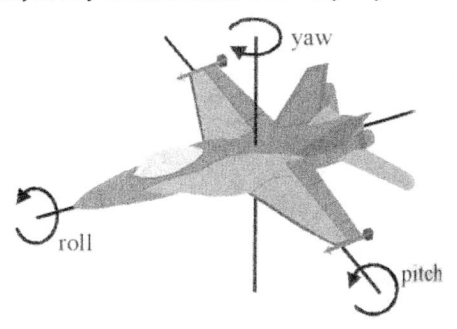

Once the Arduino has been reset, you can take a look at its values in one direction and you will be able to recognize which value changes.

355. Smart Blind Stick using Arduino

Were you ever talked about the famous American rock climber Hugh Herr? He has broken the limitations of his disabilities; Technology can help people with disabilities live a normal life. He is a strong believer in this. A TED talk given by Herr said "There is no such thing as a disabled person. a person can never be broken". Buildings and technology within our society are broken, disenfranchised, and lacking. It is not necessary for us to accept our limitations.

Through technological innovation, we can transform disability". Then, and now, he lived his life by these words, now using prosthetic legs and claiming to live a normal life. Therefore, technology can indeed neutralize human disability; to this end, let us take advantage of the power of Arduino and some simple sensors to create a Blind man's stick that can perform more than just serve blind persons.

Materials Required:

- Arduino Nano (Any version will work)
- Ultrasonic Sensor HC-SR04
- LDR, Buzzer and LED,
- 7805, Push button
- 433MHz RF transmitter and receiver
- Resistors, Capacitors
- Perf board, Soldering Kit, 9V batteries

An Ultrasonic sensor will be used on this Smart stick to assess distance from any obstacle. A light detection radio receiver to facilitate remote locating of the man's stick and a wireless RF remote control. Through a Buzzer, the blind person will get all feedback. Of course, you can swap Buzzer out for a vibrating motor, and do even more, just by putting your own creativity into it.

Circuit Diagram:

Two circuits are required to build this Smart Blind Stick. A large portion of the electrical power source will go into the main circuit, which will be mounted on the blind man's stick. This small RF transmitter circuit is intended to locate the main circuit board. The circuit diagram on the main board is shown below:

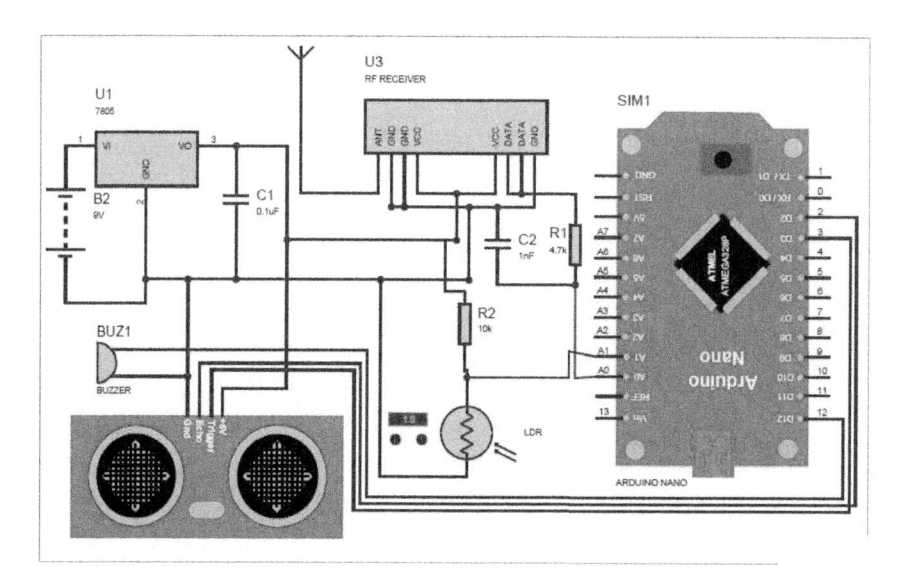

As you can see, all the sensors are controlled by an Arduino Nano. With a 9V battery plugged into the board, a Voltage regulator of 7805 rate voltage to +5V. Powered by 5V, the ultrasonic device is connected to the trigger and echo pins on the Arduino Nano.

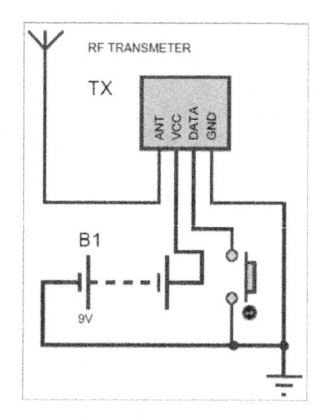

Figure: RF Remote Transmitter Circuit

Taking advantage of a LDR that will expose the ground through a resistor valued at 10K, Arduino ADC pin A1 detects any difference in voltage across pin A1, which can be used to measure the distance between the plugged-in electrode and the ground. In the output circuit, Pin 12 is connected to the buzzer which reads the signal from pin A0 of the ADC. An RF transmitter transmits the missing alert to the smart stick, which receives it via a receiver circuit. A small PCB sub-assembly houses the RF transmitter module.

356. Home Automation Using Arduino with Bluetooth Control

A human's life can be enhanced through technology today. As technology evolves, we live in progressively more advanced times. Automation used to be a science fiction story before, but is not so today. Our home can be awesome if we combine the latest technology with it. A home automation system can be created with the Arduino Uno and Windows 10 to allow home devices to operate on their own automatically.

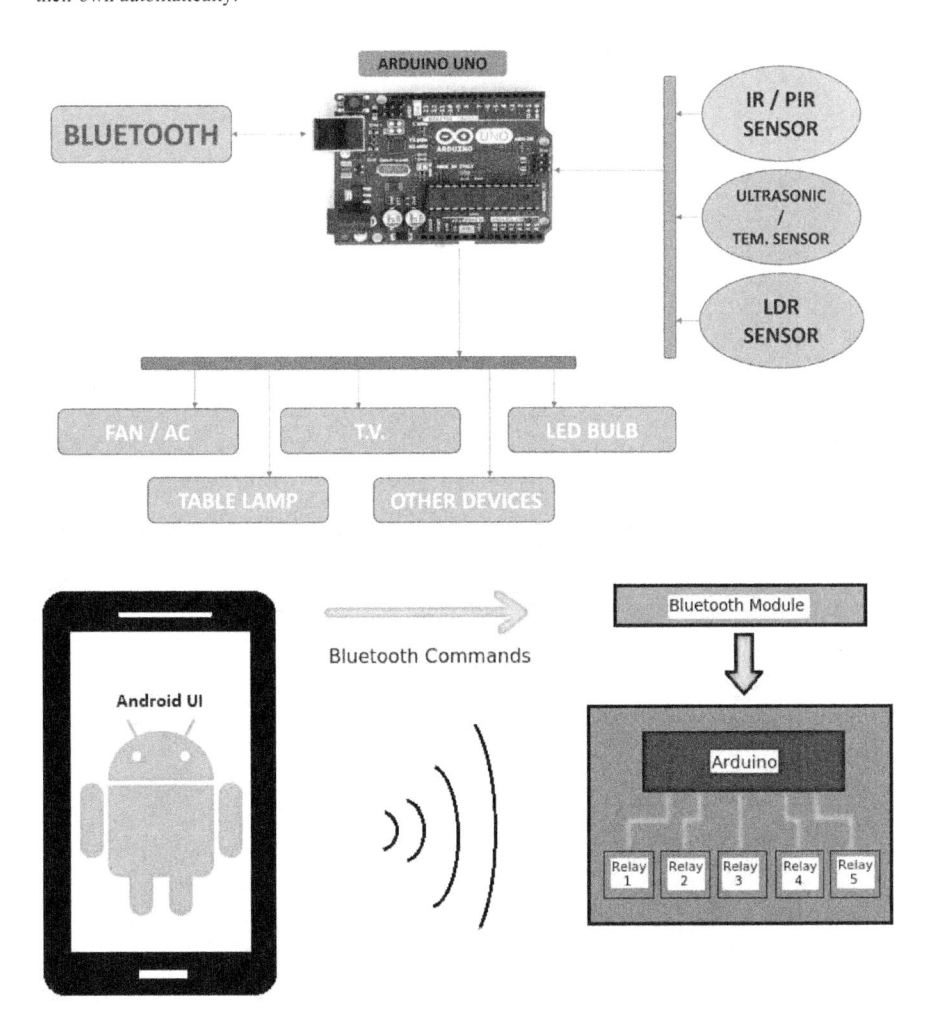

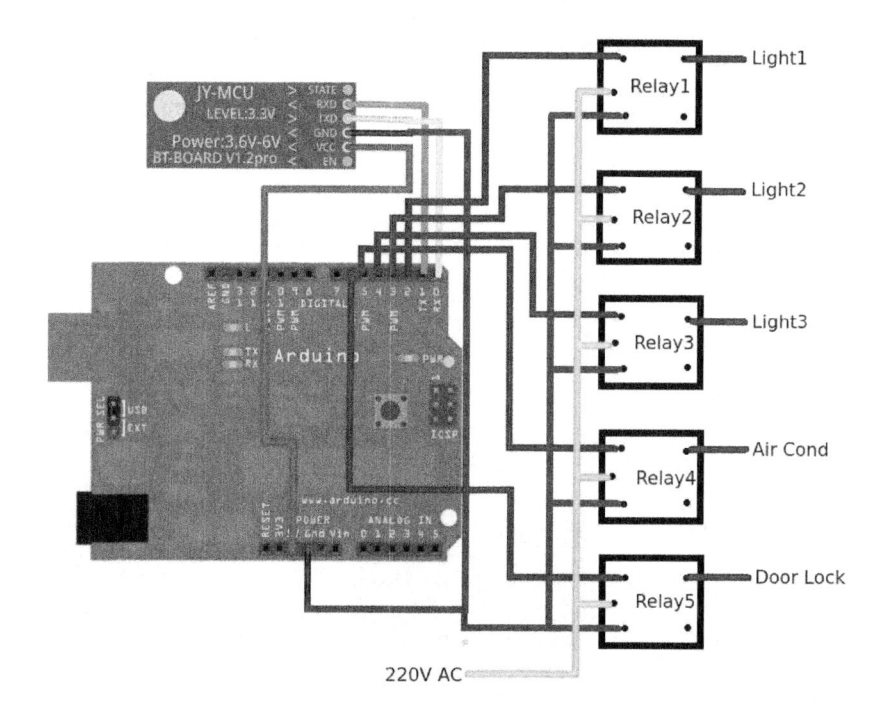

Home automation can do more than ensure the safety of your family and provide easy control of temperature, lighting, and music. Automating your home theater allows you to have perfectly balanced sound and perfect lighting that can be adapted to the time of day or the film you are watching

357. Control your Computer with Hand Gestures using Arduino

Recent market research indicates that the number of wireless computer keyboards is rapidly increasing. We can control certain functions of our computer and/or laptop by using a motion that we call Leap Motion. These laptops are rather expensive, but they are cool. Let us use the Power of Arduino and Python to provide Gesture Control for our Laptop/Computer. A VLC media player will be controlled based on the position of our hand using Ultrasonic sensors. When you are familiar with this project, however, you are able to do anything just by changing a few lines of code and manipulate your favourite application in your favourite way.

Pre-requisites:

Python has already been used in several Arduino projects before. Assuming you've already set up Python and its serial library, and have executed a few basic projects like blinking LEDs, this is what you can expect. Don't be alarmed, here's a tutorial you can use to learn how to program your led directly using Arduino and Python. Please ensure Python and the Python Serial library are installed before moving on.

Concept behind the project:

The project's concept is extremely straightforward. Our approach is to place two Ultrasonic sensors on our monitor and use Arduino to read the distance between it and our hand. We will then do what the distance reading tells us, based on it. pyautogui is an open-source Python library that allows us to perform actions on our computer. A USB connection is used to connect Arduino to the computer and send commands. Python is run on the computer and Python will read the read data and subsequently perform an action based on the read data.

Circuit Diagram:

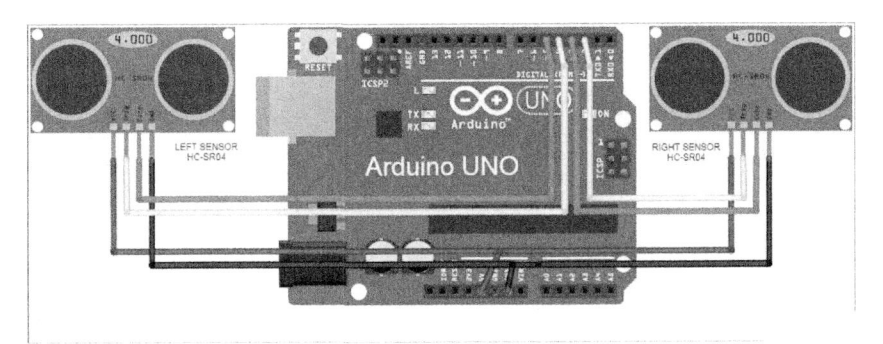

The Arduino will connect to the two Ultrasonic sensors and control the computer with hand gestures. Knowing that US sensors require 5 Volts of Voltage, Arduino's on-board voltage regulators are powering them. Besides being connected to the PC/Laptop for powering the board and serial communication, the Arduino can also be used as a computer control system. The connection should look like this once it's finished. You can use your own creativity to stick it to your monitor but I used double sided tape to do so. We can then begin the programming process after securing the device in place. For example, we can program five actions to be controlled by gestures by reading the distance value and adjusting our program accordingly.

Action 1: In VLC player, the video should Play/Pause when both hands are placed in proximity to the sensor.

Action 2: A video should Fast Forward one step when the right hand is held up in front of the sensor at a certain distance.

Action 3: At a particular distance from the sensor, the video should Rewind one step when the left hand is placed before it.

Action 4: When your right hand is placed up in front of the camera at a certain distance, and then your hand is moved towards the camera, and you move away, the video should fast forward and rewind.

Action 5: After placing the left hand close to the sensor, it should increase the video volume and when it is moved away from the sensor, it should decrease the volume.

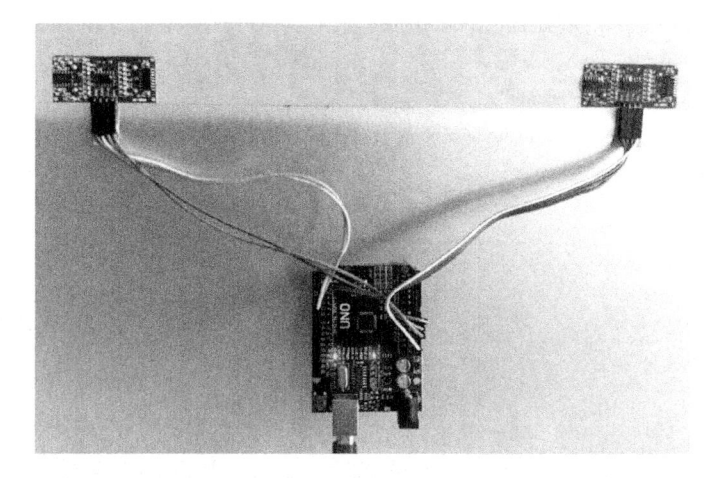

Here, we will see how the program is built to accomplish the above actions. So, just like any other program, we define the I/O pins first as you can see above. Digital pin 2, 3, 4, and 5 are used to power the two US sensors which are powered by +5V pins. The Arduino's trigger pins are input pins, and the Arduino's echo pins are output pins. Serial communication takes place with a baud rate of 9600 for the Arduino and Python programs.

358. Floor Cleaning Robot using Ultrasonic Sensor with an Arduino

Floor cleaning machines don't do anything new, but they all share the same problem. There are currently no Robots for House cleaning that aren't too expensive for what they do. So today, we are making an Automatic Robotic Cleaning Machine. Its cost will be a small fraction of the one on the market. In the event the Robot detects obstacles it can continue progressing, avoiding obstructions, until the entire room has been cleaned. The floor is cleaned with the help of a small brush on the side of the machine.

Component Required:

- Arduino UNO R3, Ultrasonic Sensor.
- Arduino Motor Driver shield.
- Wheel Drive Robot Chassis.
- Computer to Program the Arduino.
- Battery for the Motors.
- A Power Bank to Power the Arduino
- A Shoe Brush.
- A Scotch Brite Scrub Pad.

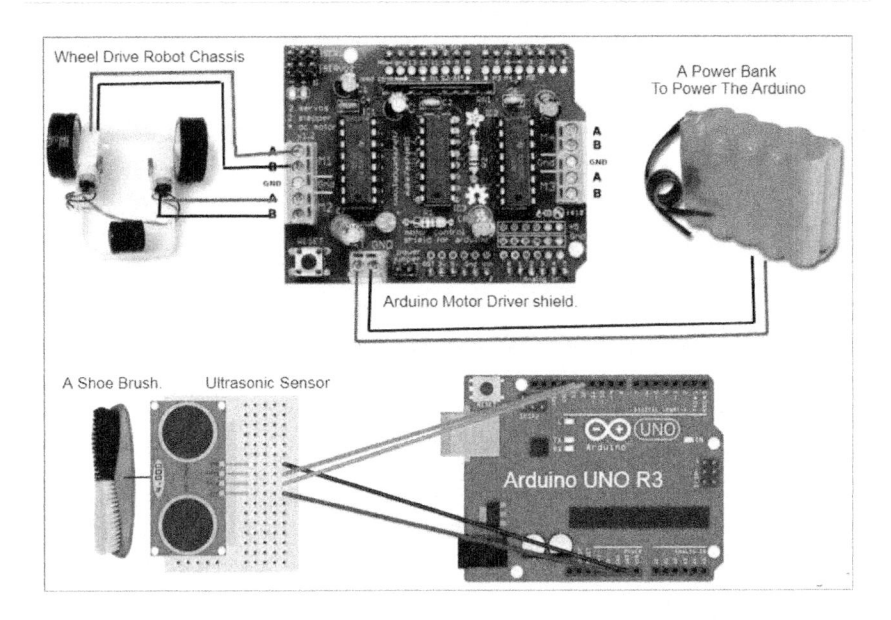

Note: A four-stranded wire can be used instead of batteries. That is what we did. If it is not something you plan to use in the real world every day, it is a feasible solution even though it is not elegant or practical. Confirm that the cable's length is sufficient.

359. Controlling a Stepper Motor using Potentiometer with Arduino

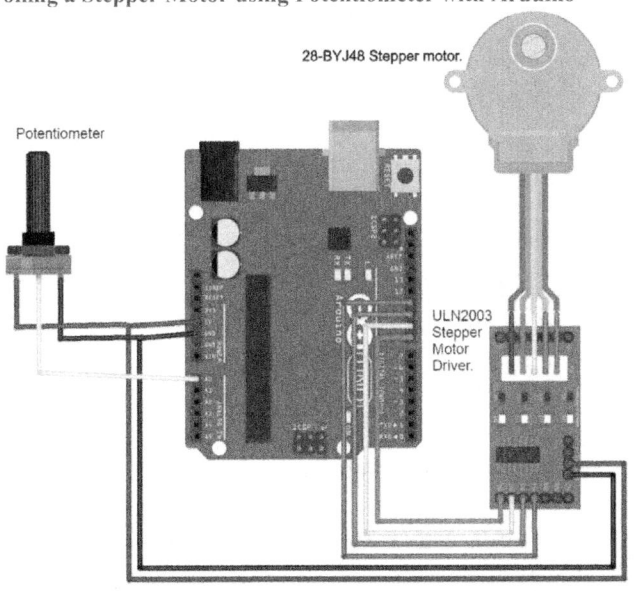

As stepper motors become increasingly popular in electronic products, they will continue to hold a stronghold in the electronics market. Everywhere, stepper motors are used as actuators for easy control ranging from surveillance cameras to complex CNC machines and robots. This tutorial demonstrates the 28-BYJ48 stepper motor, commonly available and cheaply. We'll also explore the ULN2003 stepper module that enables us to interface the 28-BYJ48 with an Arduino controller. The previous project we completed was Interface Stepper Motor with Arduino, which allowed you to control the rotation angle of the stepper motor via the serial monitor. In this project, we are going to rotate a stepper motor using an Arduino and a potentiometer. The potentiometer will rotate clockwise when you turn it clockwise and anticlockwise when you turn it the other way.

A five-lead unipolar coil arrangement can be seen on the motor. Each coil requires an individual sequence of energizing. A + 5V supply will be applied to the red wires while the rest of the wires will be pulled to ground for triggering the respective coils. These coils are energized in a particular sequence using a microcontroller such as Arduino. Now that I think about it - I have no idea why this motor is called 28-BYJ48. It doesn't appear that this motor has any technical reason for its title; perhaps we should delve further into it. In the picture below we have taken some important technical data from the data sheet of this motor.

Here is an example that shows using the Arduino Stepper Library to control a stepper motor using a potentiometer (or other sensor) on analog input 0. With either unipolar or bipolar motors, the stepper can be controlled via digital pins 8, 9, 10, and 11. If you are using a unipolar stepper, you will connect the Arduino board to a U2004 Darlington array; if you are using a bipolar motor, you will connect it to a SN754410NE H-bridge.

Rated voltage :	5VDC
Number of Phase	4
Speed Variation Ratio	1/64
Stride Angle	5.625°/64
Frequency	100Hz
DC resistance	50Ω±7%(25°C)
Idle In-traction Frequency	> 600Hz
Idle Out-traction Frequency	> 1000Hz
In-traction Torque	>34.3mN.m(120Hz)
Self-positioning Torque	>34.3mN.m
Friction torque	600-1200 gf.cm
Pull in torque	300 gf.cm
Insulated resistance	>10MΩ(500V)
Insulated electricity power	600VAC/1mA/1s
Insulation grade	A
Rise in Temperature	<40K(120Hz)
Noise	<35dB(120Hz,No load,10cm)
Model	28BYJ-48 – 5V

What are the reasons for the need for Driver Modules for Stepper Motors? A driver module is usually required to operate most stepper motors. We cannot drive the motor with the controller module (In our case Arduino) because the controller's I/O pins won't deliver enough current. So, we will use the ULN 2003 stepper motor driver as an external module. Different types of driver modules are used for different types of motors so the rating will change accordingly. For any driver module, the primary function will be to source/sink sufficient current for the motor to operate. Using Potentiometer and

Arduino to control a stepper motor is shown in the circuit diagram above. The ULN2003 driver module and the 28BYJ-48 Stepper motor were used. A driver module connected to the Arduino board's digital pin 8 is used to energize the four coils on the stepper motor. The Arduino 5V pin is used to power the stepper motor. We will rotate the Stepper motor based on the values of a potentiometer connected to A0. If you are connecting some load to the stepper motor, then power the driver with an external power supply. The motor has been used on the +5V rail of the Arduino Board for demonstration purposes. Make sure the Arduino is connected to the ground of the Driver module as well.

360. Arduino Based 3-Way Traffic Light Controller

An Arduino based 3-Way Traffic Light Controller demonstrates the working of traffic lights which we see around us. This is a simple, yet useful project to help you understand the way traffic lights work. Here we are demonstrating a simpler version of traffic lights that are used in three-sided or way traffic signals. Now let us move on to the project...

Components Required:

- 3 Red LED Lights, 3 Green LED Lights
- 3 Yellow LED Lights, 3n 220ohm Resistors, Breadboard
- Male To Male Connectors, Arduino Uno with Ide Cable

Circuit Diagram

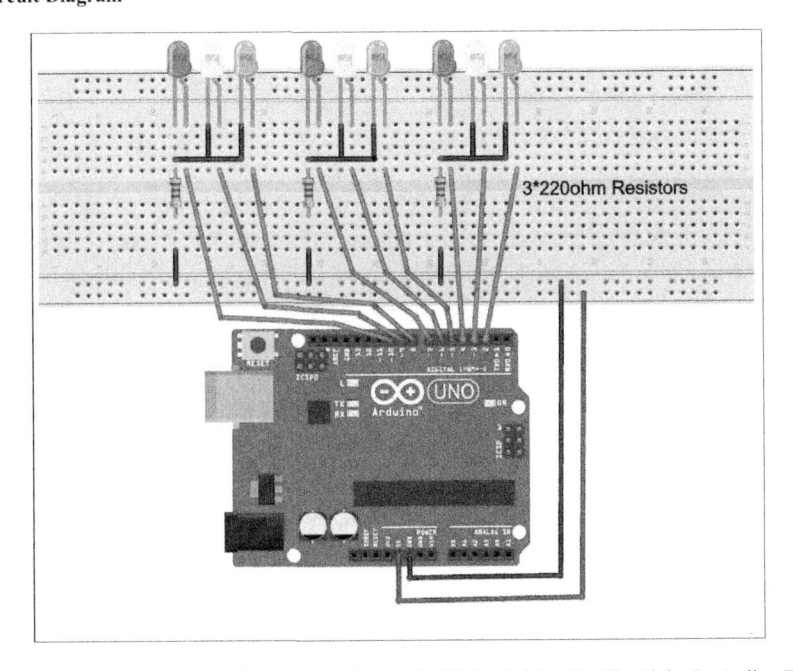

The following circuit diagram demonstrates how to build the Arduino Traffic Light Controller. This project uses an Arduino to drive traffic lights. A breadboard can easily be used to build it, as detailed in the below steps:

1. Red, Green, and Yellow LEDs should be connected in the breadboard.
2. Attach the 220ohm resistor in series with the negative terminal of the LEDs.
3. Attach your connectors in accordance with their instructions.
4. Make sure that the corresponding pins (2, 3, 4...10) on the Arduino Uno are connected to the other end.
5. The breadboard can be powered using Arduino's 5v and GND pins.

361. Simple Arduino Audio Player and Amplifier with LM386

Adding music or sounds will always make our project look a lot cooler and more appealing. The best way to add sound effects to your project is by investing in an extra SD card module and a normal speaker, especially if you are using an Arduino and you have lots of open pins. Using your Arduino Board, I have created a simple Music Player/Sound Maker. Thank you to the Arduino community for developing some libraries for you to build this quickly and easily. The IC LM386 was also used here to increase the volume and cancel out the noise.

Hardware Required:

- Arduino UNO
- SD Card Reader module
- SD card
- LM386 Audio Amplifier
- 10uf Capacitor (2 Nos)
- 100uf Capacitor (2 Nos)
- 1K, 10K Resistor
- Push buttons (2 Nos)
- Breadboard
- Connecting Wires

The goal of this project is to play the .wav music files stored on the SD card. Through the LM386 Audio amplifier, we will play these .wav files on a speaker using an Arduino programmed to read them. This project cannot technically be called an Arduino Mp3 Player since this code can only read .wav files, but you still get to hear the music. Alternatively, consider an Arduino music player without SD card if you are looking for a simple alternative

Before moving forward with the rest of the process, there are a few things you need to keep in mind. The first thing that you need to do is make sure that the audio and music files are in WAVE format, which means that their file extension should be.wav. This is due to the fact that the Arduino board is only capable of supporting PCM audio in WAVE file format (.wav). You can use any audio converter program or website to complete the conversion process. In this instance, I make use of the website known as ONLINE-CONVERT.com. It allows numerous files like archives, audio, documents, etc.

many websites provide instruments for the conversion of audio files into WAV files on your computer.

Circuit Diagram

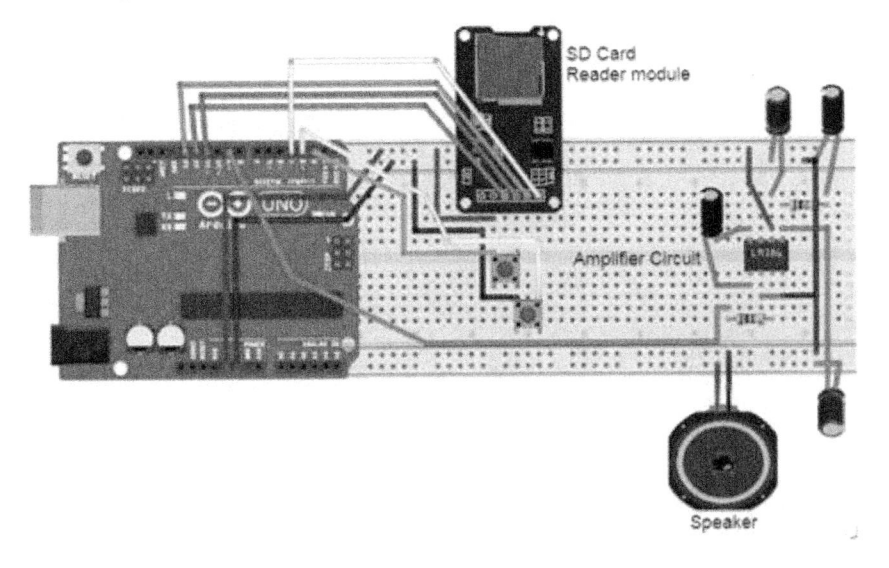

Prepare your audio files with your WAV files:

Audio files in .wav format are necessary for playing sounds from SD cards using Arduino. Due to the inherent limitations of the Arduino Board, only wav format audio can be played. An Arduino mp3 player can be made using many different mp3 shields which are available for Arduino. Alternatively,

Following the steps below will help you convert any audio file to wav format:

Step 1: Go to "https://audio.online-convert.com/convert-to-wav" website.

Step 2: The following format is supported by Arduino. You can experiment with different settings later, however, the following settings were experimentally the best in quality.

Bit Resolution	8 Bit
Sampling Rate	16000 Hz
Audio Channel	Mono
PCM format	PCM unsigned 8-bit

Step 3: In the website, check the "choose file" box and select the file you will be converting. Then feed the settings into the field. Once done, your conversion should look like the image below.

Step 4: Afterwards, click on "Convert File" and your Audio will be converted into .WAV format. Once the conversion is completed, the .WAV file will be downloaded.

Step 5: Once you've formatted your SD card, save the audio .wav file onto it. Before adding this file, make sure it is formatted correctly. Keep in mind the name of your audio file as well. You can do the same with your four audio files, selecting any one of them and saving them as 1, 2, 3, and 4. As shown below, I have converted four songs created by a particular digital music converter to wav, mp3, mp4, and mp3 audio formats.

We interface the Arduino with a SD card reader module since we haven't been able to interface our Arduino directly with some of our audio files. With the SD card and Arduino, communication is accomplished using SPI. Thus, the Module is connected to the Arduino's SPI pins as shown above in the diagram. The following table provides further details.

Arduino	SD card module
+5V	Vcc
Gnd	Gnd
Pin 12	MISO (Master in Slave out)
Pin 11	MOSI (Master Out Slave In)
Pin 13	SCK (Synchronous Clock)
Pin 4	CS (Chip Select)

After the SD card is read, the Arduino will be capable of playing the music on pin number 9. On pin 9, the audio signals generated by the Arduino are not loud enough to be audible much. Therefore, LM386 Low voltage Audio amplifier IC is used to amplify it. Amplifiers like the one shown above have Gains as high as 200 and the 5V pin of the Arduino is powering the Vdd pin. If you want to adjust the volume, the voltage applied to this pin can be increased/decreased. There are 200 watts of gain amplification in this device, which is ideal for use in low power circuits for LM386. Additional buttons are also connected to pins 2 and 3 on the Arduino. The switches can play/pause the music and play/skip the next track of a song respectively. The buttons below were used only to demonstrate the song's capabilities; you can play the song whenever desired.

362. Arduino based Bluetooth Biped Bob (Walking & Dancing Robot)

Welcome back to another site where we will build a small robot that can walk and dance. This project encompasses the use of Arduino and open-source software to allow the creation of 2D robotic devices (robots). This robot takes commands from an Android Mobile Phone to walk and dance following predefined actions at the end of the project. You can also control the movement of your robot by using the Serial monitor to control the position and motion of the servo motors. The program is included at the end of the tutorial. This project will be more interesting and cooler if it uses a 3D printer. In the absence of such an item, you may use an online service or take advantage of cardboard to recreate the same thing.

The following are the materials required for building this robot:

- Arduino Nano
- Servo SG90 – 4Nos
- Male berg sticks
- HC-05/HC-06 Bluetooth module
- 3D printer

Hardware and Schematics:

This Arduino biped robot controlled using a mobile phone has a simple circuit; the complete schematic is shown below.

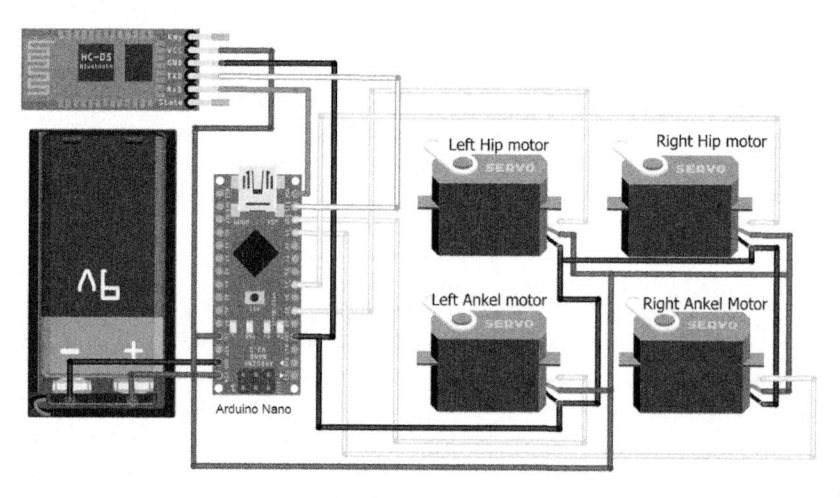

This 3D printed robotics creation requires the bare minimum number of electronic components to assemble it so that project costs are kept at a minimum. This project is purely experimental and has no immediate real-time applications so far. The connections I have made above were made on a perf board. You need to make sure that the circuit also fits into the head of the robot. The result should look something like below once your Perf board is ready.

Assembling the robot:

We can assemble the robot using the completed hardware and 3D printed parts. Make sure the motors are placed at the angles below so that the program works flawlessly before you fix them.

Motor Number	Motor place	Motor position
1	Left Hip motor	110
2	Right Hip motor	100
4	Right Ankle Motor	90
5	Right Hip motor	80

A program given at the end of the tutorial can be used to set these angles. Once you have connected the Arduino to the serial monitor and uploaded the program, enter the following commands in it. (**Note:** Baud rate is 57600). You should see the following in your Serial Monitor after all your motors have been placed.

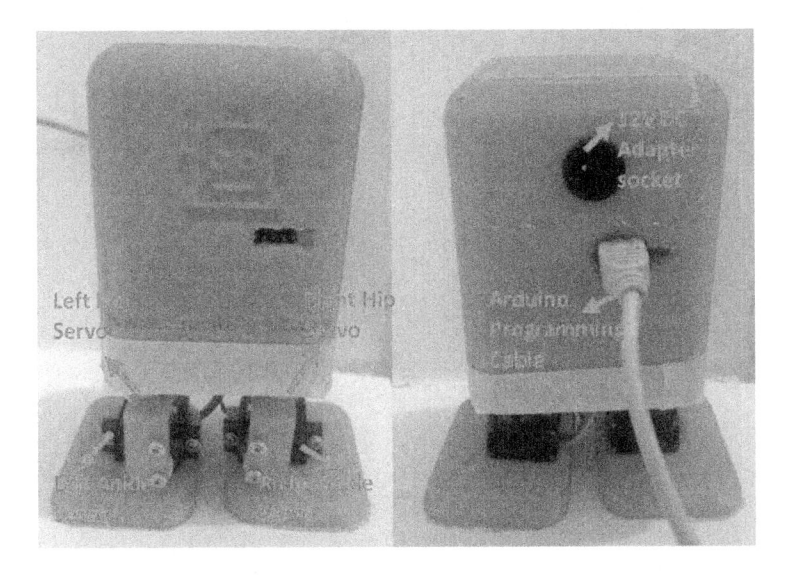

After the motors are set to the corresponding angles, they should be mounted like in the following figure.

363. Arduino Radar System Processing with Ultrasonic Sensor

The aim of this project is to demonstrate the power of an Arduino and Arduino software is used to develop a Surveillance device that transmits information via Bluetooth to an Android application. Safety and Security have always been a top priority for us. Installing a surveillance camera with night vision, tilting, and panning options will burn a big hole in our pocket. Thus, we should create an affordable device with similar functions but without video.

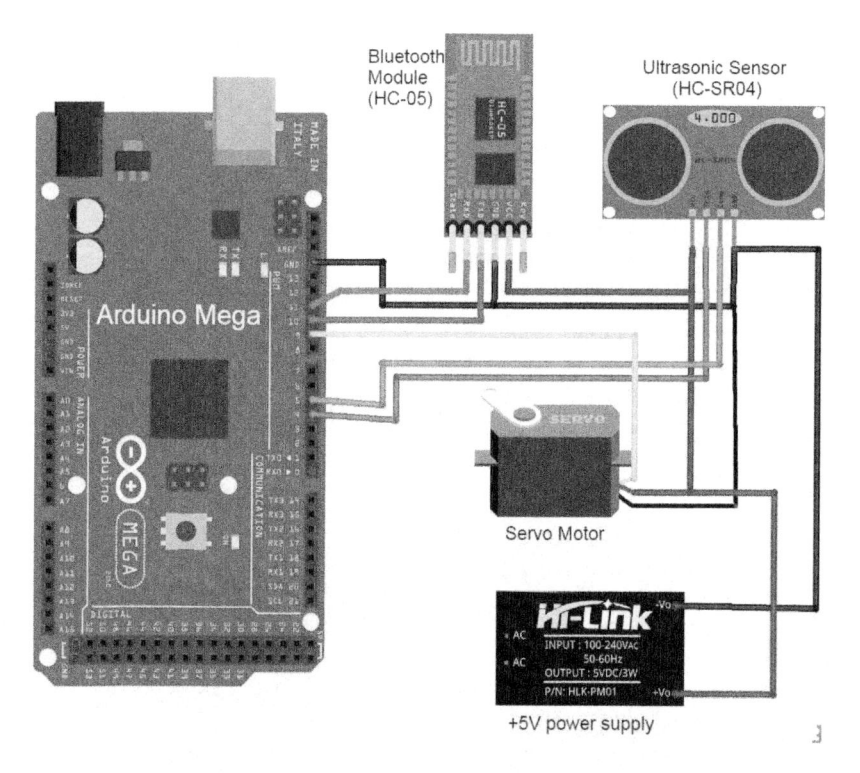

Using an ultrasonic sensor, it detects objects and can therefore be used during the night as well. We are also mounting the ultrasonic sensor on a servo motor, which is able to scan an area automatically or manually, depending on whether it is set up for automated rotation or whether it is manually rotated via our Mobile app, our ultrasonic sensors are focused in our preferred direction, allowing us to detect objects nearby. With the US sensor, we will be able to broadcast selected information to our smart phone, similar to a sonar or a radar.

COMPONENTS REQUIRED:

- A +5V power supply (I am using my Arduino (another) board for power supply)
- Arduino Mega (You can use anything from pro mini to Yun)
- Servo Motor (any rating)
- Bluetooth Module (HC-05)

- UltraSonic Sensor (HC-SR04)
- Breadboard (not mandatory)
- Connecting wires
- Android mobile
- Computer for programming
- Arduino Software
- Android SDK
- Processing Android (To create mobile application)

Let's start building the hardware once we have the materials ready. For ease of understanding, I have divided the tutorial into two parts: the Arduino Part and the Processing Part. The tutorial's code is complete and can be used directly by anyone who has no experience with Processing. The servo motor, Bluetooth module, Ultrasonic sensor and other components all play an important role in the project. As such, beginners should start with some tutorials that cover these components before returning to this one. Here are our various projects involving Servo Motors, Bluetooth Modules, and Ultrasonic Sensors.

In this project, the Arduino is not responsible for powering any components as the servo motor, Bluetooth module and US sensor all draw a great deal of current that the Arduino cannot supply. Therefore, any external +5V power supply is highly recommended. I have used two Arduino boards to share the components when I did not have access to an external +5V supply. A second Arduino mega board (red colour) has been used to power the Servos. I have also mounted the Bluetooth module HC-05 and the ultrasonic sensor HC-SR04 on the Mega. CAUTION: The voltage regulator on the Arduino will be damaged if all these modules are powered with one Arduino board. To mount the sensor, I used a piece of junk I had in my junk box, along with double-sided tape. To do the same, you can think of your own idea. Markets also offer servo holders for use with the same type of device.

Android Mobile Application for Ultrasonic Radar:

To install the same application used in this tutorial if you do not want to make your own application, you can follow the steps below.

1. This APK app is compatible with Android version 4.4.2 and above (Kitkat and above). The APK file can be downloaded directly from the link below.

https://circuitdigest.com/sites/default/files/Radar-zealotbt_1.0_apk_file.zip

(App Name: Zelobt)

2. APK files should be transferred from your computer to your phone.

3. Make sure your Android settings allow you to install applications from unknown sources.

4. Install the application.

If the app was successfully installed, you will see "**Zelobt**" installed as shown below.

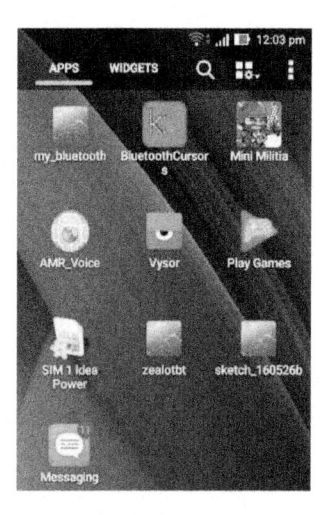

Figure. Zealotbt App

At this point, both the hardware and software are ready. Your Bluetooth module should be powered on and paired with your mobile device. You should see your Bluetooth module (HC-05) automatically get connected to your phone once you have paired them. Open the "Zelobt" application we just installed and wait a second.

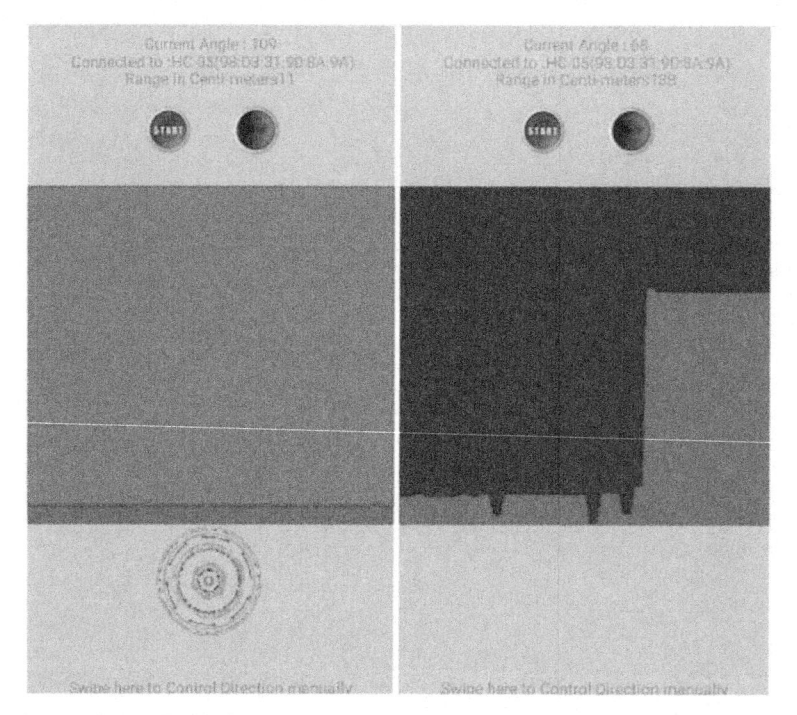

Figure. Radar data in the zealotbt mobile App

The top of the screen shows you that it is connected to: Device name (hardware address). Moreover, it shows the angle of the servomotor and the distance between the US sensors. On the red background, a blue graph is also plotted based on the measured distance. In the blue region, the closer the object, the taller it becomes. A mobile app can be used to control your servo motor, as we mentioned earlier. You can stop the servo from sweeping automatically by clicking the stop button. On the bottom of the screen, you will find a circular wheel that will rotate clockwise or anticlockwise when swiped. If you swipe this wheel, your servo motor will also turn in the same direction.

364. Heart Beat Monitoring over Internet using Arduino and ThingSpeak

The objective of this project is to create a Heart Beat Detection System using Arduino which will detect the heartbeat with the help of a pulse sensor and display the results of the measurement on an LCD connected to the Arduino board. Heart Beats can also be monitored remotely from anywhere in the world via the internet, using the Wi-Fi module ESP8266. With ThingSpeak, you can display the data online, access it when you want and where you want.

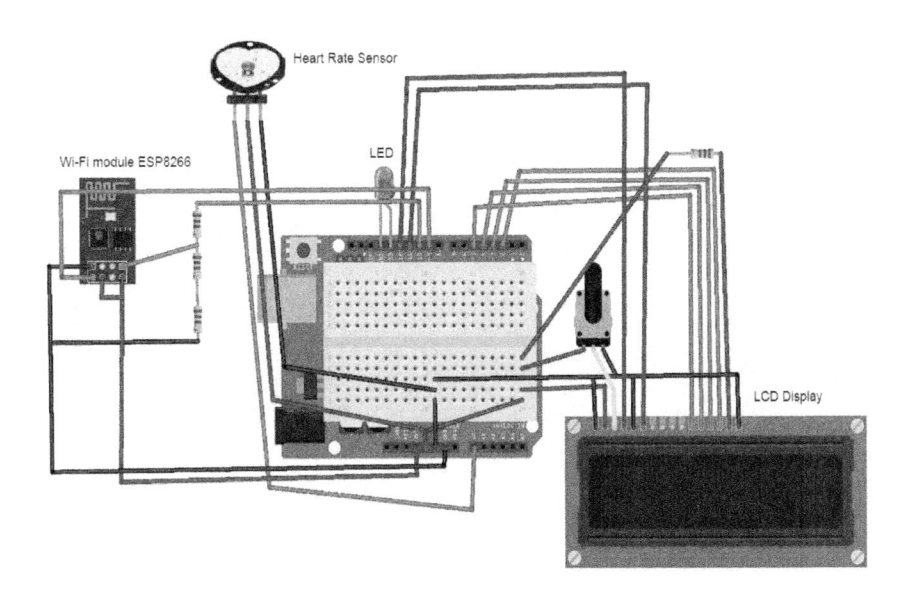

Previously, we developed a simple Heartbeat Monitor without displaying results online. Our project uses ThingSpeak for remote monitoring, so it falls under the Internet of Things category.

Components Required:

- Pulse sensor
- Wi-Fi module ESP8266
- Arduino Uno
- LCD
- Bread Board

- 10k potentiometer
- 1k resistors
- 220-ohm resistors
- LED
- Connecting wires

Circuit Explanation:

In order to use ESP8266, we need to first connect it to the Arduino. The ESP8266 runs on 3.3V and is not able to function properly if you give it 5V from the Arduino. You need to connect the VCC and CH_PD pins of the Arduino to 3.3V. When we connect ESP8266 directly to the Arduino, its RX pin will not transmit when it operates at 3.3V. Thus, we will need to create a voltage divider so that the 5V can be converted into 3.3V. You can do this by increasing the resistance of three resistors as we did in the circuit. The ESP8266's TX pin is connected to pin 9 of Arduino while its RX pin is connected to pin 10 through the resistors. Your projects can connect to the internet and Wi-Fi using an ESP8266 Wi-Fi module. Your projects become very powerful with this device, which is very cheap. In the IOT platform, it is among the most prominent devices because it can be used with any microcontroller. Check out this article to learn how you can use an ESP8266 with Arduino.

Once the Pulse Sensor is connected to the Arduino, it is ready to be used. I love how easy it is to connect the pulse sensor. An Arduino is connected to the pulse sensor through the ground pin and 5V pin. The signal pin of the pulse sensor is connected to the A0 pin.

ThingSpeak Setup:

In regard to IoT projects, ThingSpeak offers a very useful tool. Our systems can be monitored and controlled using ThingSpeak's channel and webpages, which make monitoring and controlling our systems possible. ThingSpeak collects data, analyses it, and develops a response. We have previously used ThingSpeak in Raspberry Pi weather station projects and Arduino weather station projects. The following is a short description of the IoT Heart Beat Monitoring project as it is implemented using ThingSpeak. Creating an account on ThingSpeak.com is the first step, followed by signing in and clicking on Get Started. Go to channels once you've created an account to create a channel. Put the name of the Channel and the Fields on the paper. You should also check the box below for the Make Public option before saving the channel. You have now successfully created your new channel.

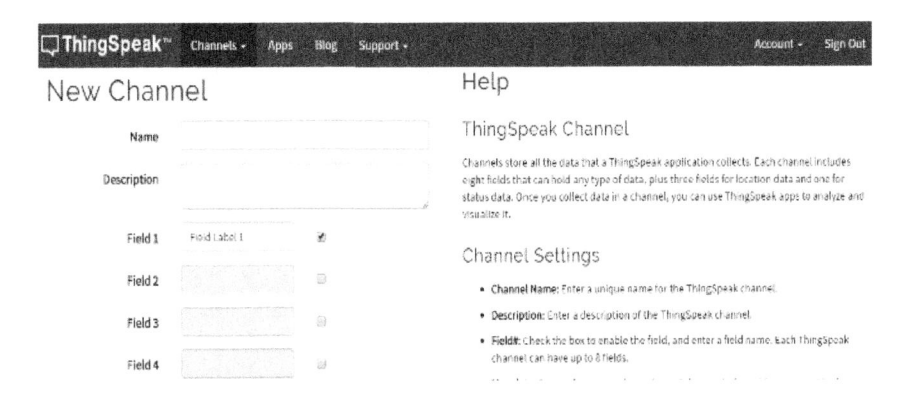

Once you have copied your API key, go to API Keys. It's needed in the code. Examine the full code at the end.

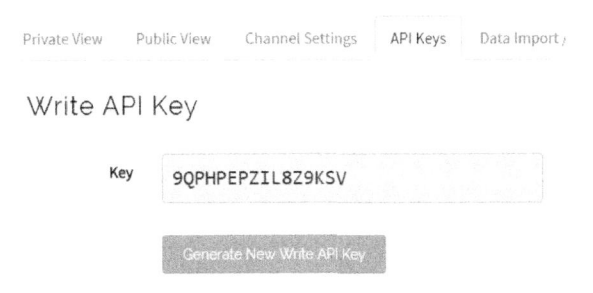

Working Explanation:

The pulse sensor must first be attached to any organ of the body where it can readily detect the pulse, like a finger, as shown in the video. As the heart pumps blood into the body, the Pulse Sensor will measure the change in volume of blood. In the same way, the change in blood volume affects the brightness of the light that filters through the organ. Once this change is observed, the Arduino will convert it to heart beats per minute (BPM). Additionally, the LED connected to pin 13 will flash in response to the Heart Beat.

An Arduino will communicate with the ESP8266 through ThingSpeak, which will send data to it. With the help of the ESP8266, you can get the data from the sensor and transfer it online via the network of your router. These readings will appear in a graph format and can be accessed from anywhere with a web browser. In addition to the BPM, the LCD connected will also display it.

Features of ThingSpeak

With ThingSpeak you can aggregate, visualize, and analyze live data streams on a cloud environment. Some of the features that ThingSpeak offers include:

- Connect ThingSpeak to a variety of IoT devices easily using popular protocols.
- Get real-time data from your sensors.
- On-demand access to aggregated data from third-parties.
- Analyze IoT data using the powerful MATLAB programming language.
- Automatically run your IoT analytics in response to schedules and events.
- IoT prototypes can be built without launching servers or creating web-based applications.
- You can use Twilio® or Twitter® to communicate using data that is automatically processed.

365. IoT based Air Pollution Monitoring System using Arduino

This project involves the development of an Internet of Things (IoT) based service that monitors air pollution over the Internet through a web server, and it will sound a warning if sufficient amounts of hazardous gasses like carbon dioxide, smoke, alcohol, benzene and NH3 enter the air at a certain level. On the LCD and on the webpage, we will be able to read the air quality in PPM, so we can easily monitor it. This time the air quality sensor was the MQ135 sensor instead of the MQ6 sensor, which we used previously for making LPG detectors. The MQ135 sensor can detect most harmful gases without affecting their amount.

You can monitor pollution levels in this IOT project using your computer or mobile device no matter where you are. In addition to installing this system anywhere, we can also set up some device that will turn on the exhaust fan or send SMS/email notifications to the user when pollution reaches some level.

Required Components:

- MQ135 Gas sensor
- Arduino Uno
- Wi-Fi module ESP8266
- 16X2 LCD, Breadboard
- 10K potentiometer
- 1K ohm resistors
- 220-ohm resistor, Buzzer

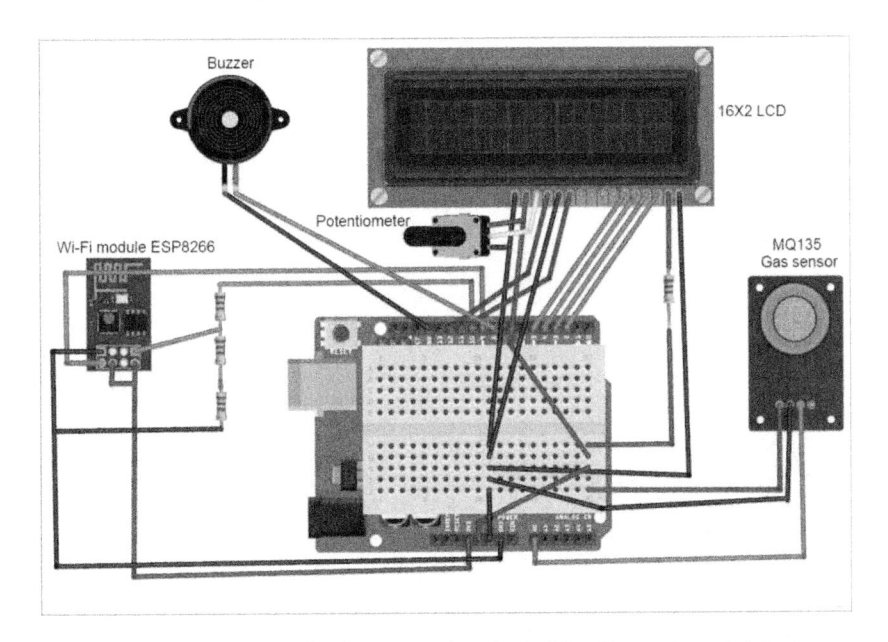

In order to use ESP8266, we need to first connect it to the Arduino. You can't use Arduino to power an ESP8266 as it runs on 3.3V, but if you give it 5V, then it won't function properly and might even be damaged. 3.3V should be connected to VCC and CH_PD on Arduino. When connected directly to the Arduino, the RX pin of the ESP8266 works with 3.3V voltage and thus cannot communicate with the Arduino. Thus, we will need to create a voltage divider so that the 5V can be converted into 3.3V. You can do this by increasing the resistance of three resistors as we did in the circuit. By connecting the ESP8266's TX pin to Arduino pin 10, and its RX pin to Arduino pin 9, you can make the ESP8266 transmit data.

366. IOT Based Dumpster Monitoring using Arduino & ESP8266

We will build an Internet of Things (IOT) based garbage can monitoring system in this DIY that will let us know when the trash can is full or empty by monitoring the webserver, enabling you to control the trash can from anywhere. Aside from being highly useful, it can also be placed on trash cans in public areas and private homes. Within this Internet of Things project, ultrasonic sensors are used to detect trash can content. The Ultrasonic Sensor is mounted on top of the trash can and measures how far the trash is from the sensor. Based on the size of the trash can, a threshold value can be set for the distance between the trash and the sensor. When the distance is less than this threshold value, the trash can will be full of garbage, and a message "Basket Full" will be printed on the webpage. If the distance is greater than this threshold value, however, the container will be empty. The threshold value has been set to 5 cm in our program code. A Wi-Fi module called ESP8266 is used to communicate between the Arduino and the webserver. On a local web server, we demonstrated our Garbage Monitoring System.

Components Required:

- Arduino Uno (you can use any other)
- ESP8266 Wi-Fi module
- HC-SR04 Ultrasonic sensor
- 1K Resistors
- Breadboard
- Connecting wires

Circuit Diagram and Explanation:

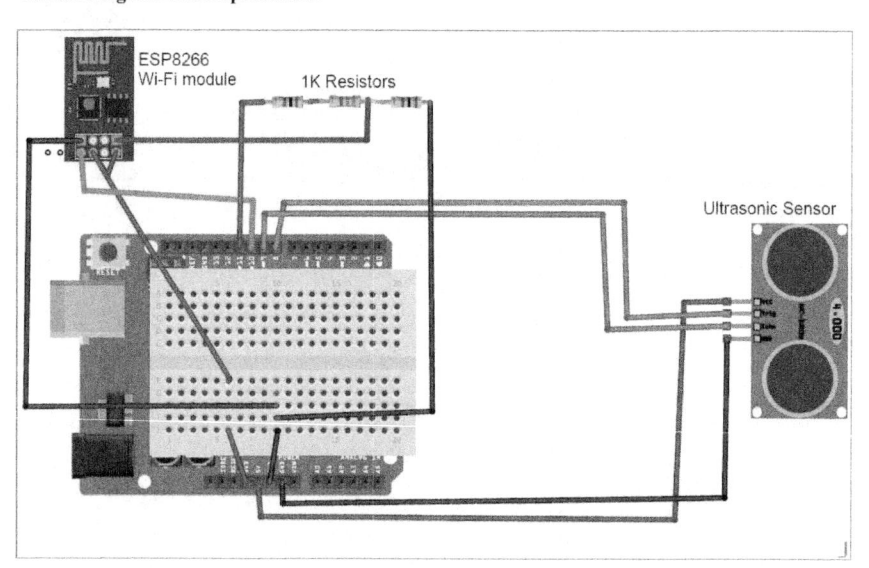

In order to use ESP8266, we need to first connect it to the Arduino. You cannot power the ESP8266 with 5V from an Arduino or it won't function properly and may get damaged. It runs on 3.3V. 3.3V should be connected to VCC and CH_PD on Arduino. When connected directly to the Arduino, the RX pin of the ESP8266 works with 3.3V voltage and thus cannot communicate with the Arduino. A voltage divider will be needed, so we will be using three 1-k resistors in series. In the circuit diagram below, you will see the RX is connected to pin 11 of the Arduino, along with the TX, as well as the TX of the Arduino is connected to pin 10. HC-SR04 ultrasonic sensor now needs to be connected to the Arduino. Ultrasonic sensors connect to Arduino boards in a very simple manner. An ultrasonic sensor's VCC and ground should be connected to the Arduino's 5V and ground. You should connect pin 8 and pin 9 of the Arduino to the TRIG and ECHO pins of the ultrasonic sensor, respectively.

367. Arduino based Vehicle Tracker using GPS and GSM

We are taking GPS one step further in this project by tracking the vehicle using GPS and GSM. With a few hardware and software changes, this system could also be used for Accident Detection Alerts, Soldier Tracking, and many more. Vehicle tracking consists of tracking the location of the vehicle using GPS coordinates (latitude and longitude). GPS coordinates represent the location of a point. A system like this is a great choice for outdoor applications.

In order to track Cabs/Taxis, stolen cars, school/college buses, etc., this type of Vehicle Tracking System Project is widely used.

Components Required:

- Arduino
- GSM Module
- GPS Module
- 16x2 LCD
- Power Supply
- Connecting Wires
- 10 K POT

GPS Module and Its Working:

With the Global Positioning System, you can find the position of any place on the planet in exact UTC time (Universal Coordinated Time). GPS stands for Global Positioning System. Vehicle tracking systems are primarily built with a GPS module as the key component. Using this device, satellite coordinates are received every second from the earth, as well as time and date. Tracking position data is sent in real time via GPS module, and all the information is formatted in NMEA (see the screenshot below). We only need one sentence in the NMEA format, which consists of several sentences. Starting from $GPGGA, this sentence contains coordinates, times, and other relevant information. The GPS fix data is known as the GPGGA, or Global Positioning System Fix Data. Read more about GPS data and its string here. Counting the commas in $GPGGA allows us to extract the coordinates for this string. Consider how Latitude and Longitude can be found from a string stored in a $GPGGA array; after two commas Latitude can be found and after four commas Longitude can be found. Latitudes and longitudes from the array can now be inserted into other arrays.

Circuit Explanation:

A Vehicle Tracking System uses a simple circuit where TX pin number 10 of the Arduino is connected directly to Tx pin number 9 of the GPS module. By using the Software Serial Library here, pins 10 and 11 have been enabled for serial communication and have been set to Rx and Tx respectively, leaving the RX pin of the GPS Module open.

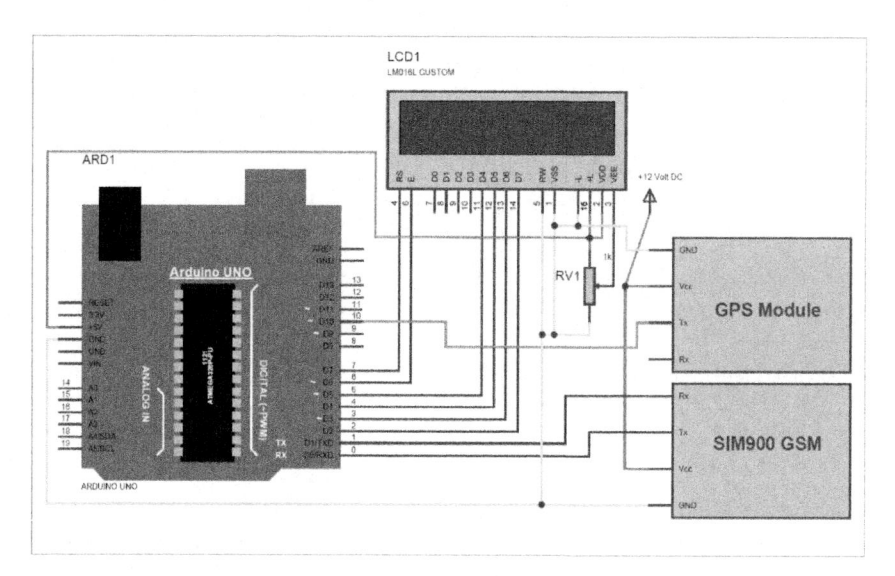

By default, serial communication is limited to Pin 0 and 1 of the Arduino board. However, the SoftwareSerial library enables serial communication on almost any pin of the board. In order to power the GPS Module, a 12-volt power supply is used. Rx and Tx pins of the GSM module are wired directly to pins Rx and Tx of the Arduino. A 12v power supply is also required for GSM modules. Pin numbers 5, 4, 3, and 2 of the Arduino are connected to the LCD's data pins D4, D5, D6 and D7. The RW pin of the LCD is connected directly to ground. Command pins EN and RS are connected to pins 2 and 3 of Arduino. The LCD can also be controlled by the potentiometer by setting the contrast or brightness.

Working Explanation:

This project uses an Arduino board along with a GPS module and GSM module to control the whole process. GSM module is used for sending GPS coordinates via SMS to the user. GPS receiver detects the location of the vehicle. Furthermore, a 16x2 LCD allows status messages and coordinates to be displayed. This module was integrated with the GPS module SKG13BL and the GSM module SIM900A.

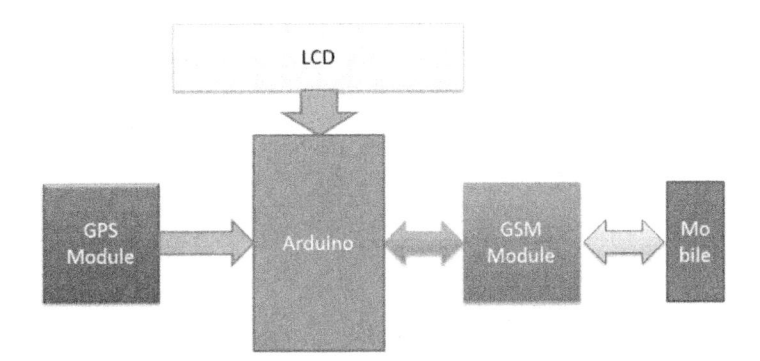

Creating our hardware is the final step in programming our vehicle. The hardware is now ready to be mounted on our vehicle and powered up. Our vehicle will receive a message stating "Track Vehicle" as soon as we send it to the tracking device. In projects like these, we have incorporated prefixes (#) and suffixes (*) to indicate the beginning and the end of the string. Similarly, we did the following in these projects: Wireless Notice Board and Home Automation with GSM The sending GSM module receives the message data, which it sends to Arduino, which reads it and extracts the main message. It will be compared with a predefined message in Arduino. Once any match occurs, the Arduino reads the coordinates by reading GPS module data and extracting $GPGGA String, which it sends via the GSM module to the user. The location of the vehicle is specified in this message.

368. Snake Game on 8x8 Matrix using Arduino

From the beginning of the mobile phones, Snake has been incredibly popular. Black and white mobile phones were initially available and became extremely popular. Cell phones have changed this game dramatically as well, which has led to a range of graphical and colorful versions being available now. Besides Snake, DIY electronics projects are also popular among students and hobbyists. Keeping it simple and yet providing all its functions is what we will be demonstrating today in the Arduino Snake Game.

Components Required:

- Arduino UNO
- 8x8 LED Dot Matrix Display

- Shift Register 74HC595
- 16x2 LCD, POT 1K
- Push Buttons
- Connecting wires
- Bread Board, Power Supply

Working Explanation:

A complex game like this requires a lot of work. Fortunately, we have simplified it in this tutorial for you. Our LCD screen displays the snake and its food dot, the 8x8 red colour Dot matrix displays the score, 5 push buttons to initiate the game and an Arduino UNO controls the entire process.

Circuit Diagram

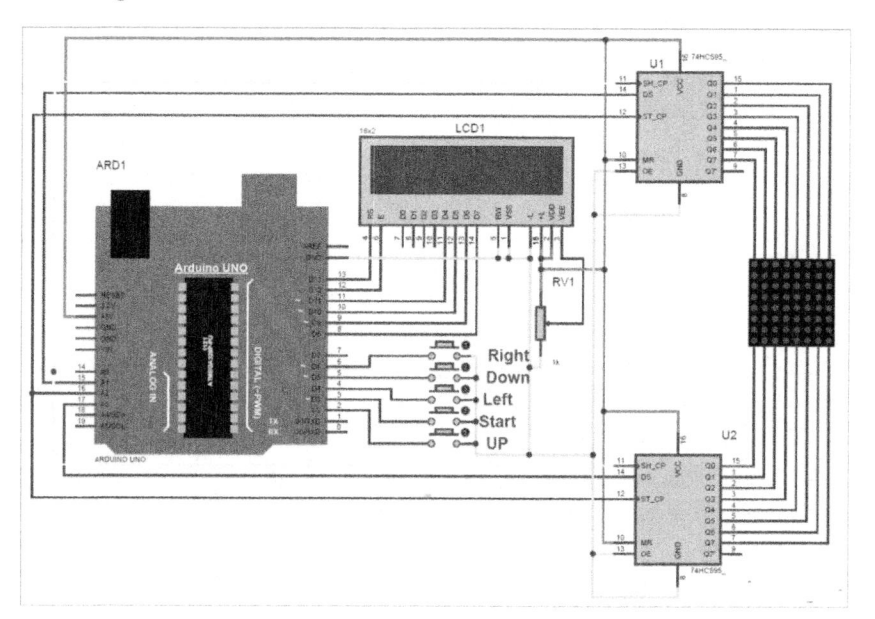

Project Snake's circuit is rather complicated. Shift Register 74HC595 is used in this example to connect the dot matrix display. For driving the columns, one shift register is used while the other drives the rows. On Arduino, pin number 14 and 16 are connected directly to the control pins for both registers, the Column Shift Register (SH, ST). Pins 15 and 17 of Arduino are directly connected to the DS pins of column shift registers and rows shift registers, respectively. Game controls are connected at pins 3, 4, 6, 2 and 5. Left and right directions are linked at pins 3, 4 and 6, up and down directions are connected at pin 5. Our hardware also includes an LCD that displays scores. There is a direct connection between pin 13 and pin 12 for RS and EN. A ground wire runs directly from the RW pin to pin 11 of the Arduino, and data lines from d4-d7 are connected to pins 8, 9, and 10 of the board. Circuit diagrams are used to show the rest of the connections.

After we power the circuit up, the LCD displays a "Press Start to Play" message and a welcome message. A second dot matrix display appears with two snakes and a single food dot, and the score is displayed as zero on the LCD.

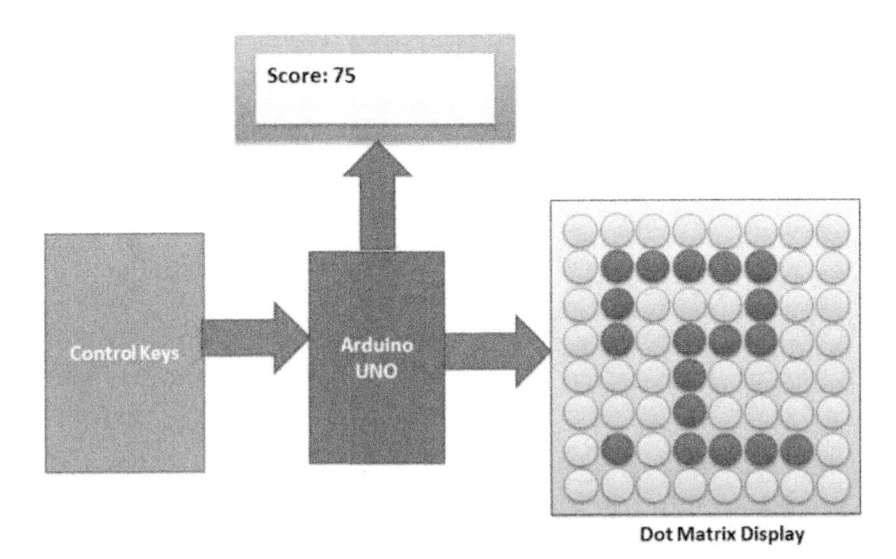

Dot Matrix Display

The game can now be started by pressing the middle button and the snake automatically moves upwards. In order to move the snake, the user must press the direction keys around the middle button. In this case we have used five push buttons (left, right, up, down, and start) respectively. Once the snake reaches the food dot or eats the food, its score will increase by five points and its length will increase by a dot (LED) every time, making it faster than before. The game would end whenever a snake struck any wall, reached the end of the LED matrix, or reached the end of the game mat. Once the user presses the start key, the game must be started again.

369. Prepaid Energy Meter using GSM and Arduino

Electricity Energy Meters with prepaid balances are a solid idea because they allow you to refill them just like we do on our mobile phones. This project involves the use of Arduino and GSM modules to create an automated system. This system allows electricity balances to be recharged through SMS messages. If the system is unbalanced or low, it can stop the power supply to the house. And this system will automatically send users updates on their mobile devices, such as low balance alerts, cut off alerts, resume alerts, and recharge alerts, depending on the energy meter readings.

Working explanation:

Using an Arduino and the power meter's calibration LED, we have interfaced the energy meter with Arduino. Our CAL LED only needs to be connected to an Arduino by means of an Optocoupler IC.

Components used:

- Arduino
- GSM Module
- 16x2 LCD
- Analogue Electricity Energy Meter
- Optocoupler 4n35
- Resistors
- POT
- Connecting wires
- Bulb and holder
- SIM card
- Power supply
- Mobile Phone

As soon as the system comes on, it reads previous rupee values from EEPROM and restores them into variables; it compares the value with the predefined one and decides whether to proceed. Use relays to switch on the electricity of a house or office if the available amount is more than 15 rupees, for example. Once the balance falls below 15 rupees, Arduino sends a SMS to the phone of the user informing them of the low balance and asking them to recharge soon. The Arduino turns off the electricity supply of the home when the balance falls beneath 5 rupees and sends a SMS alerting the user of a 'Light Cut' and requesting him to recharge soon. Here you can find information on the GSM module and AT commands used to send and receive messages.

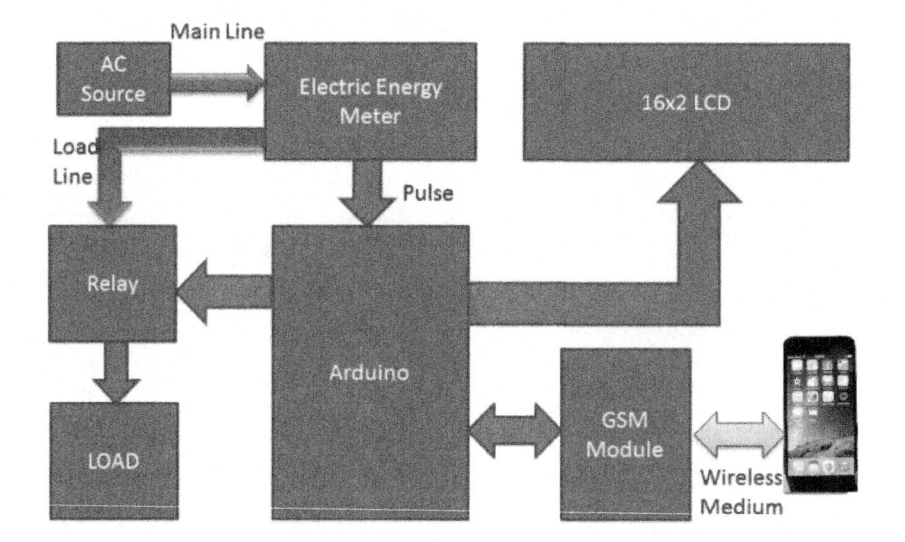

Now that our system can be recharged through our cell phones, we can do so simply by sending a SMS to it. We would send #45* if we want to recharge 45 dollars, here the # and * indicate the sum to be recharged. Upon receiving this message, the system extracts the recharge amount to update the system balance. In the video at the end, you will see how the system turns on the electricity in the house or office.

Circuit Description:

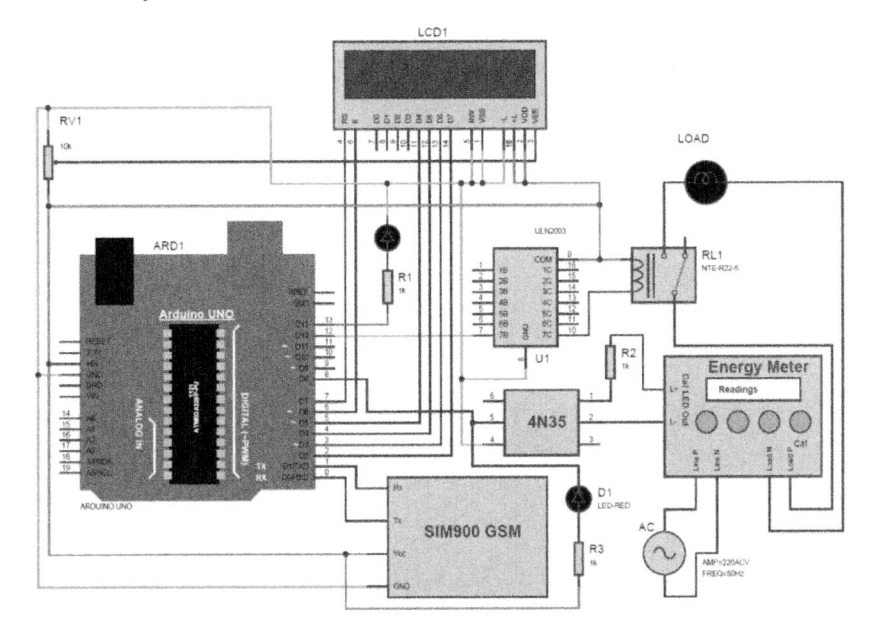

A schematic of the project circuit connections is shown; an Arduino UNO processor has been used to process all the components. Unit status and remaining balance are displayed on a liquid crystal display. LCD pins RS, EN, D4, D5, D6, D7 are connected to Arduino digital pin number 7, 6, 5, 4, 3, 2. GSM pins Rx and Tx are directly connected to Arduino pins Rx and Tx respectively. An adapter powered by 12 volts is used to power a GSM module. On pin 12 of Arduino, a relay is used to switch electricity through the ULN2003 relay driver.

How to Connect Energy Meter with Arduino:

An Analogue Electricity Energy Meter must first be purchased by the user. Once it has been opened, the user can locate the terminals of Pulse LEDs and Cal LEDs (cathode and anode). Once you have soldered two wires in each terminal, remove them from the meter and close it. Then tighten the screws. Now the LED's anode terminal must be connected to pin number 1 of the optocoupler. On an optocoupler, pin number two is connected to the cathode terminal, and pin number four should be directly connected to ground. In the pin number 5 of the optocoupler is an LED and Pull-up resistor. A similar terminal should be connected to Arduino pin 8 as well. The aim of this project is to gather information about the usage of energy by a specific consumer or user, using a wireless communications system (not requiring a visit to the consumer's home or office), and the system is called AMR (Automatic Meter Reading). AMR would involve remotely accumulating meter readings at a location through a wireless telecommunications system, instead of individuals visiting the site physically and taking notes.

370. Clap Switch using Arduino

Using ARDUINO UNO as a platform, we will prototype a Clapper circuit using the ADC concept. In order to detect sound and generate a response, we are going to use a MIC and Uno. Clap ON Clap OFF switches the device by using clapping sound, however it does not work in reverse. The 555 Timer IC was previously used to build the Clap ON Clap OFF switch and Clap ON Clap OFF switch. In case of clapping, there is a peak signal in the microphone that is much higher than normal. That signal is sent to the amplifier via a high-pass filter. ADC converts this high voltage signal into a number using an amplification circuit. In other words, the ADC reading for the UNO will peak at that time. The project is described in detail below. When the peak is detected, the LED on the board will be toggled.

MIC is a transducer that detects sound, converting audio energy into electrical energy. Hence sound is represented as a changing voltage with this sensor. A device like this is used mostly for recording or detecting sounds. All mobile phones and laptops have this type of transducer. A typical MIC looks like this A typical MIC appears like, A typical MIC can be found in all mobile phones and laptops.

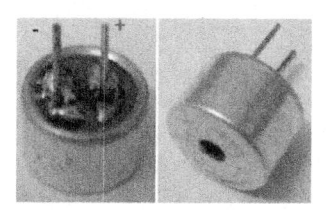

A condenser mic's polarity can be determined in several ways.

There are two terminals for MIC, one is positive, the other is negative. Using a multimeter, it is possible to check the polarity of the microphone. The positive probe of the Multimeter (the meter must be in DIODE TESTING mode) must be connected to the positive terminal of the MIC, and the negative probe must be connected to the negative terminal of the MIC. On the Multimeter, if the positive (MIC) terminal is at the negative terminal, it indicates the positive (MIC) terminal is at the negative terminal. The negative terminal has two or three soldering lines connected to the metal case. The positive terminal is similar, with one soldering line connected to the metal case. You can also check whether the connection between its metal case and the negative terminal is intact using continuity tester, in order to discover the negative terminal.

Components Required:

- **Hardware:** ARDUINO UNO, power supply (5v), a condenser mic (explained above)
- 2N3904 NPN transistor,
- 100nF capacitors (2 pieces), one 100uF capacitor,
- 1K Ω resistor, 1MΩ resistor, 15KΩ resistor (2 pieces), one LED,

- And Breadboard & Connecting wires.
- **Software:** Arduino IDE

In this figure, you can see the circuit diagram for the clapper project. There are four main parts to the process, namely: Filtration, Amplification, Analog-Digital conversion, and programming to control the LED.

Circuit Diagram

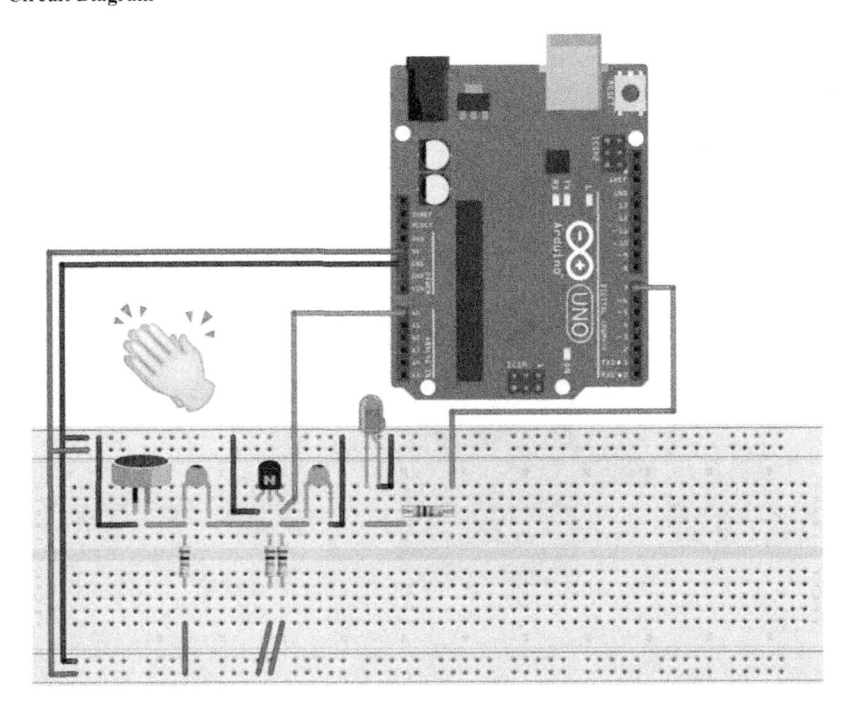

When sound is detected, the microphone can convert it into voltage that is proportional to the level of sound. We would assign a higher value to a higher sound, and a lower value to a lower sound. In order to filter this signal, it is first fed into a High Pass Filter. This filtered value is then fed into the transistor to produce an amplified output, which is delivered through the collector. ADC0 of the UNO receives this collector signal, which is then converted to a digital signal. The LED is programmed to turn on with every increment of ADC channel A0, coupled into PIN 7 of PORTD.

371. Bluetooth Controlled Toy Car using Arduino

Following the development of a few popular robotic projects like line following robots, edge avoiding robots, DTMF robots, gesture-controlled robots, etc. This project involves developing an Android-based application that uses Bluetooth technology to control a robot car.

Components Required:

- Arduino UNO, DC Motors
- Bluetooth module HC-05
- Motor Driver L293D
- 9 Volt Battery and 6-volt battery
- Battery Connector, Toy Car

Android-based mobile devices can control Bluetooth-enabled cars instead of buttons, gestures, and other traditional control methods. In this case, I could control the car with an Android phone by simply touching buttons forward, backward or left and right. Here, the Android phone transmits data to the car's Bluetooth module, which acts as the receiver. In order to move in the required direction like forward and reverse, turning left and right, and stopping, an Android phone will transmit commands via Bluetooth to the car.

Working Explanation

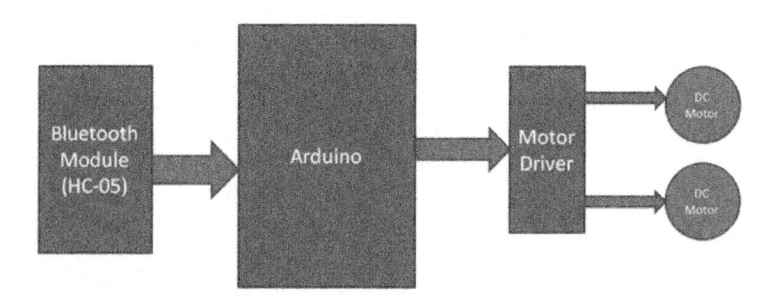

A toy car was used in this project to demonstrate the process. This RF toy car has steering features that can be moved left and right. This car's RF circuit was replaced by an Arduino circuit we purchased after buying it. Both the front and rear of this car are equipped with DC motors. An engine on the front of the car gives the car direction, such as turning left or right. Rear-wheel drive energy is used to propel the car forwards and backwards. The Arduino UNO is used to control the entire system, with Bluetooth modules receiving directions from Android phones. In the app, we can tap the left button to cause the car to move in the left direction. In addition, it continues moving left until the next command is given. When the front motors are operating in this condition, the tires will turn in the left direction, whereas the rear motors will run in the forward direction.

If we touch the right button on the Bluetooth controller app, the car will start moving in the right direction and it will continue to move until the next command is received. Under this condition, the motors on the front side move the wheels in the right direction, while the motors on the back move the vehicle forward. We can stop the car by touching the stop button. A Bluetooth-controlled car circuit diagram is shown in the above figure. To drive the car, a motor driver is connected to Arduino. Arduino's digital pin numbers 12, 11, 10 and 9 correspond to the motor driver's input pins 2, 7 and 15. This circuit uses two DC motors and a 6-volt battery to power the motors.

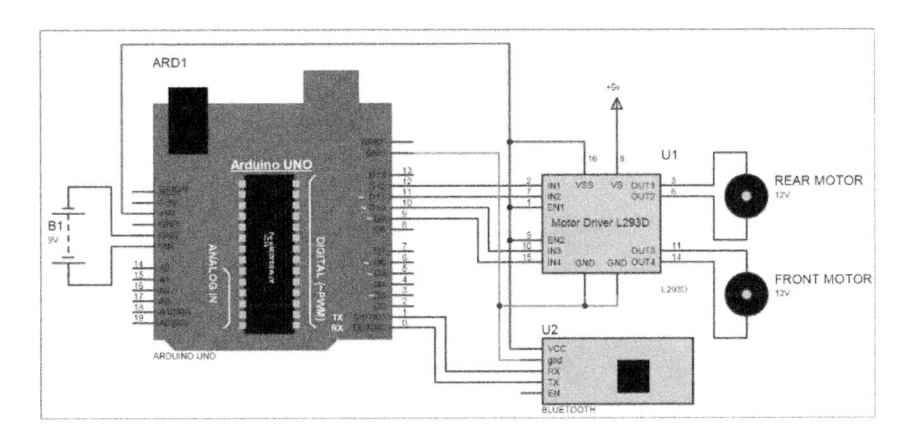

One motor is connected to the output pins 3 and 6 of the motor driver, while the other motor is connected to pins 11 and 14. RX and TX pins of the Bluetooth module are connected directly to the TX and Rx of the Arduino. The Bluetooth module's VCC and GND pins are connected to Arduino's +5 volt and GND. An Arduino's Vin pins are powered by a 9V battery.

372. Automatic Water Level Indicator and Controller using Arduino

We will use ultrasonic sensors to measure the water level in this Arduino based automatic water level controller and indicator project. A fundamental principle of ultrasonic distance measurements is based on ECHO. After impacting on an obstacle, sound waves travel back to their source as ECHO. We must then only calculate the sound's outgoing and returning times after striking the obstacle, in other words, their traveling time. In addition, we can calculate the distance and get a result from it. The water pump is automatically turned on when the tank water level drops, so that we can use this concept in our water controller project. For a simplified version of this project, you can check out this simple water level indicator circuit.

Components Required:

- Arduino Uno
- Ultrasonic sensor Module
- 16x2 LCD, Relay 6 Volt
- ULN2003, 7806, PVT
- Copper wire
- 9-volt battery or 12 Volt Adaptor
- Connecting wires

Working of Automatic Water Level Controller

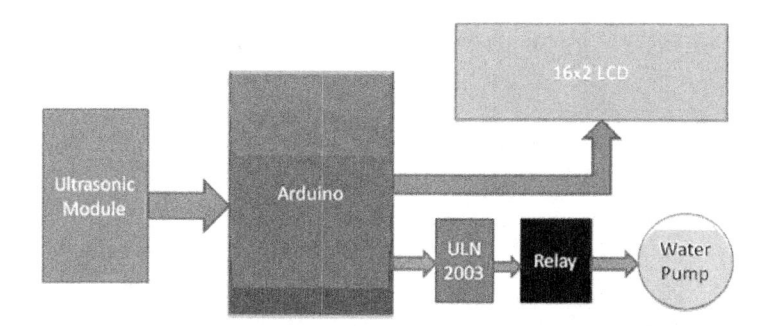

Working on this project is very straightforward. This module uses ultrasonic waves to generate sound waves which are reflected by the water in the tank and senses the sound waves as ECHO. In order to send a signal to the ultrasonic sensor module, we have to trigger Arduino to transmit it, and then we will wait for the ECHO to arrive. Arduino measures the time between triggering and receiving ECHOs. Using the following formula, we can calculate the distance using the sound speed of 340 m/s:

Distance= (travel time/2) * speed of sound

Sound travels at an average speed of 340m per second.

We can compute distances from a sensor to the surface of the water using these methods. In order to determine water level, we first need to determine depth.

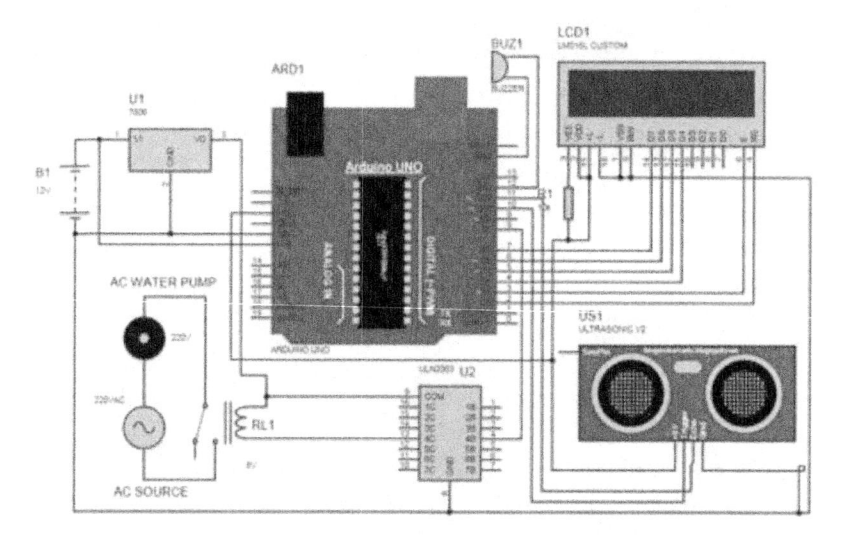

Effort must now be put into determining the total length of the water tank. By subtracting the actual distance from the tank from the total length, we can calculate the water level by subtracting the distance from the tank's length. In the next step we'll need to calculate the distance between the water level and the LCD display. We can then convert the distance into a percentage. Below is a block diagram that explains the complete workings of the water level indicator.

An Ultrasonic sensor module was placed on top of the water tank in the circuit to demonstrate the mechanism. In this sensor module, a distance will be measured between the sensor module and the water surface, and the distance will be displayed on an LCD screen as "Water Surface in Tank is:". In other words, we are displaying an empty space of volume or distance for water rather than its level. As a result of this feature, this system can be applied to any water tank. An Arduino will run a relay to turn the water pump ON when empty water levels reach about 30 cm. A green LED will begin glowing if the water level falls below the cut-off level and the motor is turned on. If the empty space is within 12 cm of the Arduino, the relay is turned off and the display shows "Tanks are full" and "Motors are off". During this time, the relay status LED will also turn off and the buzzer will beep.

373. Tachometer using Arduino

RPM is calculated by the Tachometer, which counts rotations per minute. The mechanical tachometer and the digital tachometer are separate types of instruments. Using an IR sensor module for object detection, we will design a digital tachometer based on an Arduino board. The infrared (IR) transmitter emits IR rays, which are reflected to the IR receiver, and the IR module outputs a pulse which is detected during the start of the program by the Arduino controller. There are 5 seconds of continuous counting.

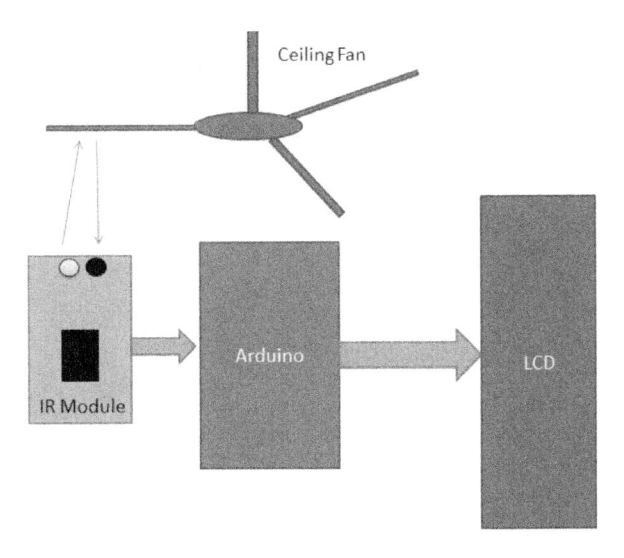

Following the given formula, the Arduino calculates the revolutions per minute for a minute.

RPM= Count x 12 for single object rotating body.

This project is demonstrated using ceiling fans. Therefore, we have made the following changes:

RPM=count x 12 / objects

Were

Object = number of blades in fan.

Required Components

- Arduino
- IR sensor Module
- 16x2 LCD
- Push button
- Bread board
- 9-volt battery
- Connecting wires

Circuit Diagram

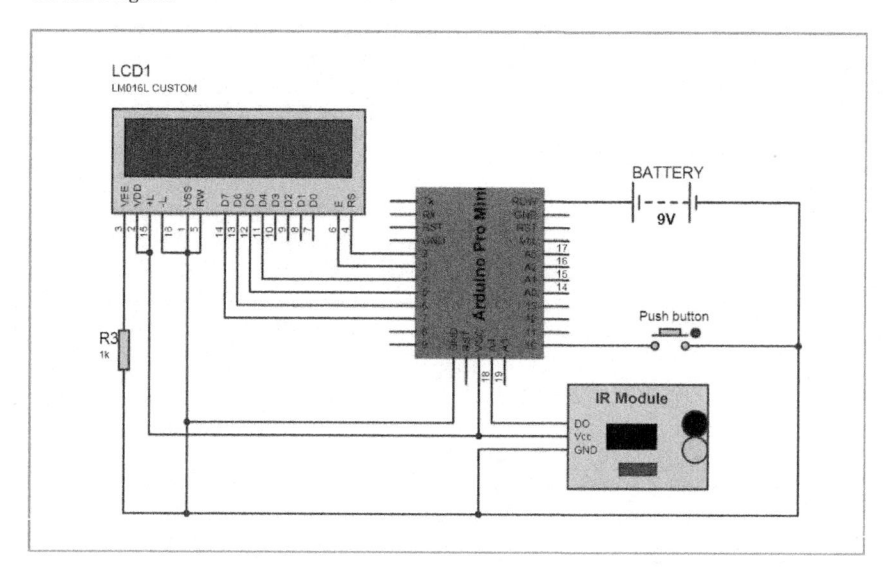

The Arduino Pro Mini is connected to an IR sensor, a buzzer and a LCD in the tachometer circuit. IR sensor module generates the reading pulse that Arduino controls. IR sensor module calculates RPM from detected objects and displays the value on the LCD. Sensors use infrared light to detect objects. Inbuilt potentiometers on the IR module allow us to adjust the sensitivity of the sensor. They are designed to detect or receive infrared rays and are composed of an IR transmitter and a photodiode. This Line Follower Robot explains that infrared rays are transmitted through IR transmitters, and they reflect back when they hit any surface. An Arduino receives the output of the photodiode through a comparator, which compares the photodiode's output voltage to a reference voltage. The Arduino Pro Mini is connected to an IR sensor, a buzzer and a LCD in the tachometer circuit. Arduino controls the entire process, including reading the pulses generated by the IR sensor module based on object detection, calculating RPM, and sending RPM values to the LCD. It uses an inbuilt potentiometer to set the sensitivity of this sensor. We can set the sensitivity of this sensor by the inbuilt potentiometer on the IR module. They are designed to detect or receive infrared rays and are composed of an IR transmitter and a photodiode. This Line Follower Robot explains that infrared rays are transmitted through IR transmitters, and they reflect back when they hit any surface. An Arduino receives the output of the photodiode through a comparator, which compares the photodiode's output voltage to a reference voltage.

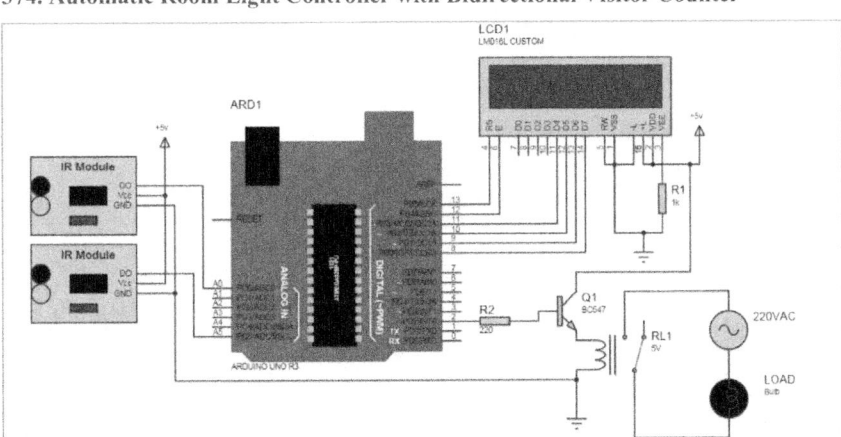

374. Automatic Room Light Controller with Bidirectional Visitor Counter

We can often find visitor counters in stadiums, malls, offices, and class rooms. What do they do with the lights when nobody is present? How do they turn them on or off? Using Arduino Uno, we have developed an automatic light control project that also tallied visitors in the room in a bidirectional manner. There is a lot of fun and learning to be had with the project for hobbyists and students.

Components

- Arduino UNO
- Relay (5v), Resisters
- IR Sensor module
- 16x2 LCD display
- Bread Board
- Connecting Wires
- Led, BC547 Transistor

This project uses Arduino to interface sensors, motors, and other components with a visitor counter.

This counter can count people in both directions. You can use this circuit to count who enters a building, mall, home, or office. By incrementing the count when a person exits the hall, the number of persons left the hall can be counted. Sensors as well as gates of parking areas are examples of other public places where sensors can be used. And it depends on where they are placed in the mall/hall. Sensors, controllers, counter displays and gates comprise the four parts of this project. An interruption would be detected by the sensor, and its input would be used by the controller to increase or decrease the counter based on whether a person entered or exited. A 16x2 LCD screen displays the counting through the controller. We set a delay for the other sensor so that it won't work if the IR sensor is interrupted by an object present in the room.

A visitor counter circuit consists of the following sections: sensors, controls, displays, and drivers.

Sensor section: Two IR sensor modules with LED's, potentiometers, comparators (Op-Amps), and IR diodes were used in this chapter. Infrared sensors sense objects and cause a change in voltage at the comparator's second terminal. A potentiometer is used to measure voltage at the comparator's first terminal. This is then compared with the output voltage by the comparator and a digital signal resulted. Here in this circuit, two sensors are being compared using two comparators. A comparison is performed with LM358. LM358 has two Op-amps built into it.

Control Section: Throughout the process of this visitor counter project, Arduino UNO is being utilized. A digital pin 14 and a digital pin 19 are connected to the outputs of comparators. A relay driver circuit drives a relay by sending commands to Arduino from the LED control circuit. You can learn how to operate a relay with Arduino with this tutorial if you are having problems working with relay.

Display section: A 16x2 LCD screen is provided in the display section. During this time the number of people will be counted and the lighting status will be shown. Suitable relay driver/section: The relay driver section consists of a BC547 transistor and a 5-volt relay used to control the light bulb. Since the Arduino will not supply enough current and voltage to operate the relay, a transistor is used instead. Our relay driver circuit was added to provide enough voltage and current to work the relay. This transistor drives a relay connected to the Arduino and turns the light on/off accordingly.

The pins 14 (A0) and 19 (A5) of the Arduino digital board interface directly with the outputs of IR sensors. At digital pin 2, there is a Relay driver transistor. A four-bit LCD is connected. In the LCD, the clock and data pins are connected directly to pins 13 and 12. A direct connection is made also between the Arduino D11-D8 and LCD pin D4-D7 respectively. Additionally, this project can be equipped with a GSM modem so that the operation can be controlled remotely by SMS. This would allow the equipment to know the status and to be adjusted accordingly. Using simple microcontrollers to design an automatic room controller is the subject of this project. Therefore, it would be wise to implement this type of home automation system for energy savings.

375. Electronic Voting Machine using Arduino

Despite the fact that we have covered some other electronic voting machines here previously using RF and AVR microcontrollers, we are quite familiar with voting machines. A voting machine made using an Arduino controller has been created in this project.

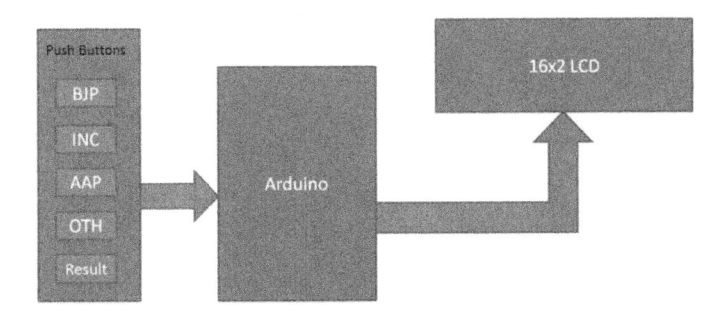

Components

- Arduino Uno, 16x2 LCD
- Push button, Bread board
- Power, Connecting wires

Circuit Diagram and Working Explanation

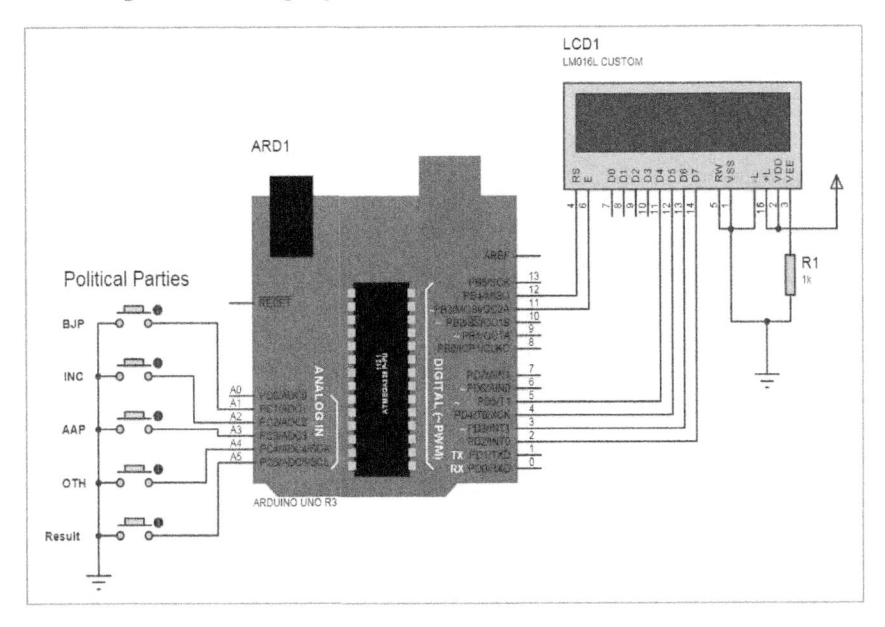

Four different candidates have been assigned four push buttons in this project. Despite the ability to increase the number of candidates, we have decided to limit it to four candidates to better understand.

Every time a voter presses one of four buttons, the voting value will be incremented by one. The results will be displayed after the voting is complete. In order to show the total number of votes for each candidate, Arduino calculates the total votes and displays them on the LCD display. Arduino, push buttons and LCD are used in this project, which is a fairly simple circuit. Arduino controls all the processes, including reading the buttons, incrementing votes, creating results, and displaying them on the LCD display. There are five buttons here, the first one showing BJP, the second showing INC, the third showing AAP, the fourth representing OTH, and the last button is for calculating and displaying results. A5-A5 on the Arduino have direct connections to pin 15-19 of the buttons. Four-bit mode is used to connect a 16x2 LCD to the Arduino. On the Arduino board, pins 12, GND, and 11 serve as control pins for RS, RW, and En, and pins 5, 4, 3, and 2 serve as data pins.

376. Humidity and Temperature Measurement using Arduino

Temperature and humidity are widely used as indicators of environmental conditions. This project will use an Arduino board to measure ambient temperature and humidity, displaying the results on a 16x2 LCD. This Celsius scale thermometer and percentage scale humidity measure project uses DHT11 temperature and humidity measurements in combination with Arduino undo. Another project that I worked on previously involved designing a digital thermometer based on the temperature sensor LM35. There are three main parts to this project: in one, a humidity and temperature sensor (DHT11) detects the humidity and temperature. Second, it determines the temperature and humidity according to DHT sensor data and converts them into percentages and Celsius values. The third component is an LCD display that displays humidity and temperature.

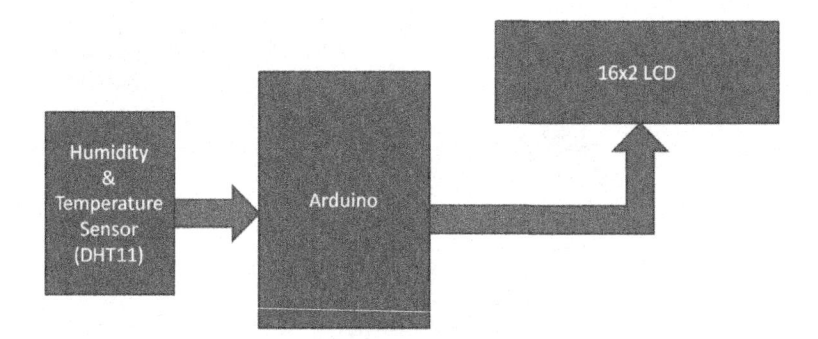

Serial communication is used in working on this project. An Arduino first sends a start signal to the DHT module, which then transmits temperature and humidity data back to the Arduino. Two humidity and temperature measurements are collected and extracted by Arduino and sent to a 16x2 LCD. The DHT11 sensor module was used in this project. An integrated humidifier and thermometer with a calibrated digital output are featured on the module. The DHT11 sensor module provides a digital output signal that is calibrated to measure humidity and temperature. DHT11's high accuracy and long-term stability guarantee high reliability and long-term stability. Incorporated into the 4-pin single row package is an eight-bit microcontroller that has an 8-bit resistive humidity and temperature measurement component that is cost-effective, has a fast response time, and is available in a 4-pin single row package. Single wire serial communication is the method used by the DHT11 module. A pulse train of data is transmitted by this module over a specific period of time. An initialization command with a time delay is required before data can be sent to the Arduino. And the total processing time is four milliseconds. In this process, there are 40 bits of data transmitted, and the format is as follows: Eight-bit integral RH + eight-bit decimal RH + eight-bit integral T + eight-bit decimal T + an eight-bit checksum.

Circuit Diagram and Explanation

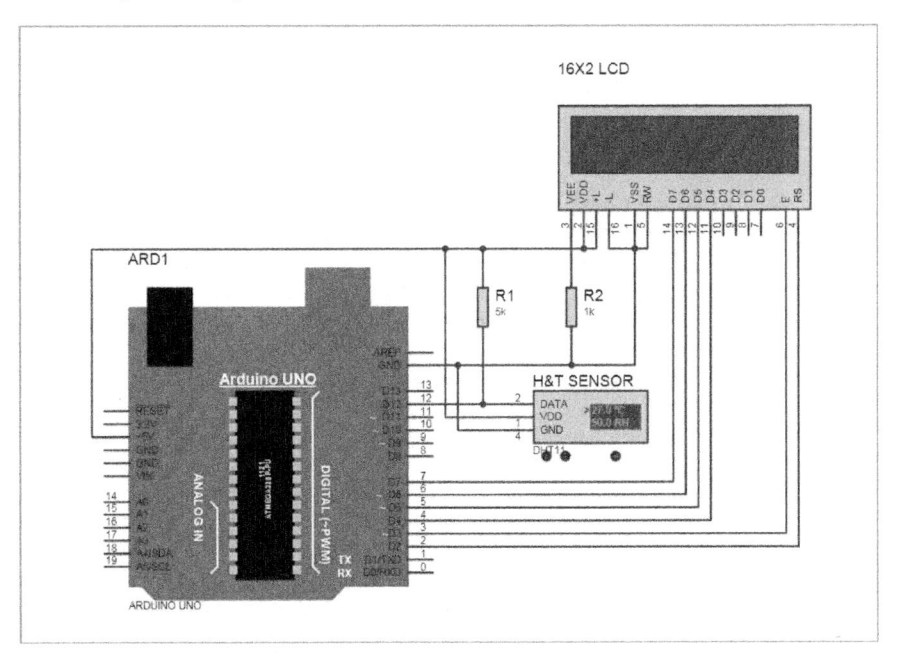

The temperature and humidity are displayed on the LCD directly connected to the Arduino in 4-bit mode. There are five LCD pins connected to the Arduino digital pins 2, 3, 4, 5, and 6. Additionally, RS, EN, D4, D5, and D7 are located in the LCD. Additionally, a 5k pull-up resistor is connected to digital pin 12 of the Arduino to connect the DHT11 sensor module.

377. Automatic Door Opener using Arduino

Automatic door openers are typically found in shopping malls and commercial buildings. Whenever someone approaches the entrance, the door is opened and then closed after a while. PIR sensors, Radar sensors, Laser sensors, Infrared sensors, etc. are all examples of systems that can be created according to various technologies. The same concept was tried in this project with an Arduino based on a PIR sensor. The door is opened or closed using a PIR sensor that detects the infrared radiation emitted from the human body. An infrared sensor detects changes in the infrared energy when a person approaches the door and responds by opening the door when a person approach. This signal is passed on to Arduino Uno, which subsequently activates the door.

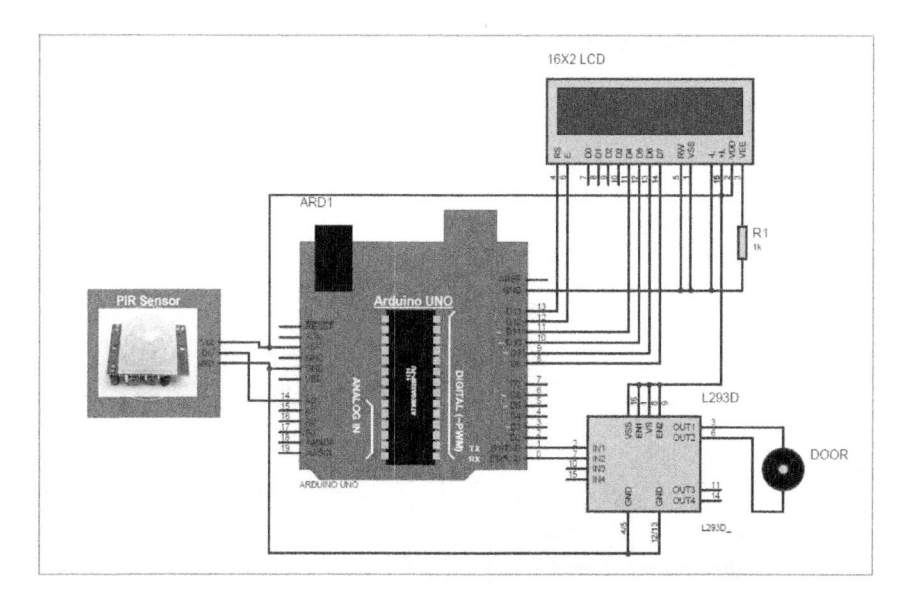

Component Required:

- Arduino UNO
- 16x2 LCD
- PIR Sensor
- Connecting wires
- Bread board
- 1 k resistor
- Power supply
- Motor driver
- CD case (DVD Trolley)
- PIR Sensor

It detects any change in heat and makes sure the output PIN is HIGH whenever it detects any change. IR motion sensors are sometimes called Pyroelectric ones. It is worth noting that all objects emit some infrared rays when heated. In addition, our bodies produce infrared because they are warm. Detecting small variations in infrared is possible with PIR sensors. Infrared is produced by friction between an object and the air, so when it passes through the sensor's range, it is caught by PIR.

Figure shows the Pyroelectric sensor (rectangular crystal behind the plastic cap) that is at the heart of the PIR sensor. A mong the components used to create PIR sensors were BISS0001, a micropower PIR motion sensor IC, resistors, and capacitors. Input signal from sensor BISS0001 is processed by BISS0001 IC to result in a HIGH or LOW output pin. Pyroelectric sensors are divided in half, so that they sense the same level of infrared no matter how fast the motion is. PIRs begin reacting when somebody enters the first half of the room, and the output pin goes high once the infrared level is larger in one half than the other. There are multiple Fresnel lenses inside a plastic cap covering the pyroelectric sensor. As a result, the lens covers a wide range so that the sensor can cover as much area as possible.

The previous circuit diagram illustrates the connections needed to build an Arduino-based door opener. An infrared sensor, based on the PIR principle, is used to detect human motion, which has three terminals: Vcc, GND, and Dout. Located at pin 14 (A0) of Arduino Uno, Doubt is directly connected to it. An LCD with a resolution of 16x2 is used to display the status. The LCD's RS and EN pins are connected to the Arduino's digital pin numbers 13, 12, and 8 while its D0-D7 pins are directly connected to digital pins 11, 10, 9, 8, respectively. The Arduino pins 0 and 1 are connected to the L293D motor driver to open and close the gate. In this circuit we are using a motor to move the gate.

378. LPG Gas Leakage Detector using Arduino

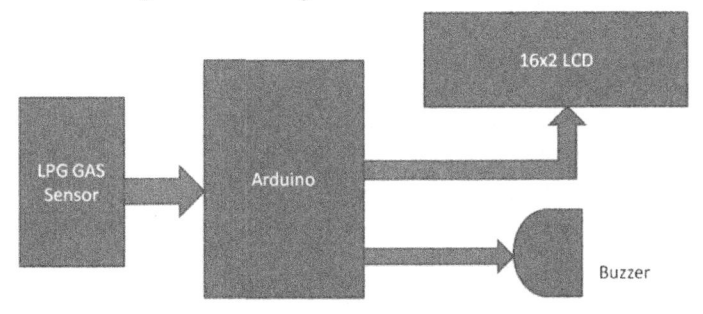

Leaking LPG may be a disaster even though it is an essential need of every household. LPG leak detection is done by various products that detect the leakage and prevent any mishappening. An LPG gas detection alarm is developed using an Arduino. The circuit buzzes if there is a leak in gas and that triggers a signal from the system. This system is extremely easy to construct, and anyone with programming and electronics skills can do it. The LPG gas was detected using an LPG sensor module. Whenever LPG leaks, it sends a HIGH pulse to the DO pin, which Arduino stays in continuous contact with. A 16x2 LCD displays the message "LPG Gas Leakage Alert" when Arduino detects a HIGH pulse from LPG Gas module. The gas detector module beeps repeatedly until no gas is detected in the environment, leading to the buzzer being activated. The LCD displays a "No LPG Gas Leakage" message when the LPG gas detector outputs a LOW signal to the Arduino.

Components Required

- Arduino Pro Mini, LPG Gas sensor Module
- Buzzer, BC 547 Transistor
- 16x2 LCD, 1K resistor, Bread board
- 9-volt battery, Connecting wires

Circuit and Description

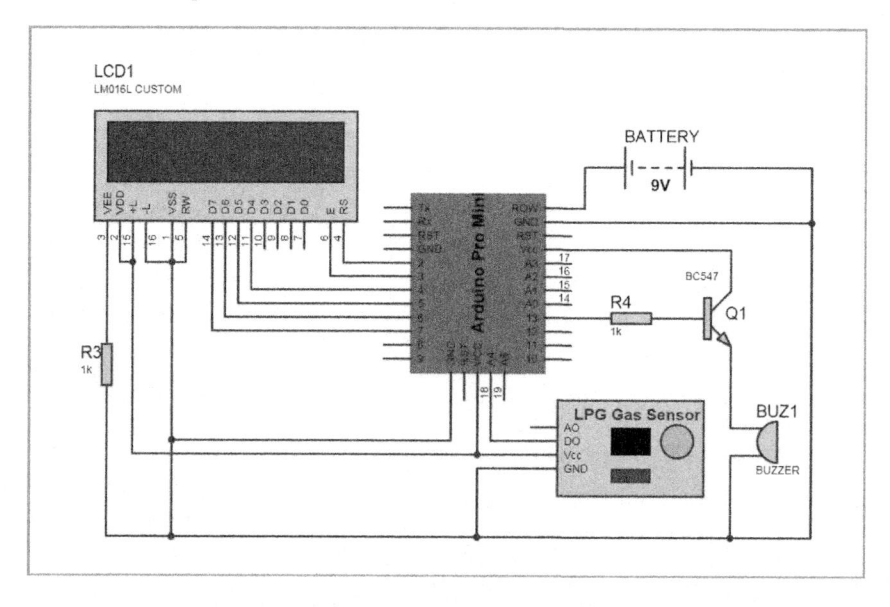

As you can see from the schematic diagram above, it has an Arduino board, an LPG GAS Sensor Module, a buzzer, and a 16x2 LCD module. This system is controlled by Arduino, which reads the gas sensor's output, sends messages to the LCD and activates the buzzer. This sensor module has an inbuilt potentiometer that we can use to set its sensitivity. DO, pin 18 (A4), of the LPG sensor module is connected directly to Arduino's Vcc and GND pins. A MQ3 sensor detects LPG gas in the LPG gas sensor module. In order for the heater to heat up, the MQ3 sensor will require some electrical power, and it may take up to fifteen minutes for it to get ready for detecting LPG gas.

An analog to digital converter circuit converts the output of MQ3 to a digital signal. Four-bit mode is used to connect a 16x2 LCD to Arduino. Arduino pin 2 and GND are directly connected to pins RS, RW and En while pins 4, 5, 6 and 7 of the Arduino are connected to the data pins. Through a binary NPN BC547 transistor coupled to pin 13 of the Arduino, a buzzer is connected.

379. IR Controlled DC Motor using Arduino

Students and hobbyists are increasingly using Arduino Microcontrollers in recent years. Since Arduino is easy to use and has a smooth learning curve, everyone uses it to make any project. Many Arduino projects are also available on our website, from basic interface modules to more complex robotic modules. IR Sensor, Relay Module and DC Motor are three basic components that we will use in our project today. Arduino will be used to control an electric motor with the IR sensor. In this case the IR sensor reads the output from the IR sensor and makes the relay high when it detects an object in front of it. As IR Sensor detects any objects in front of it, DC Motor will also be ON if relay is connected to it.

Required Components:

- Arduino UNO
- 5V-relay module
- DC motor
- IR sensor module
- Breadboard
- Connecting wires

Circuit Explanation:

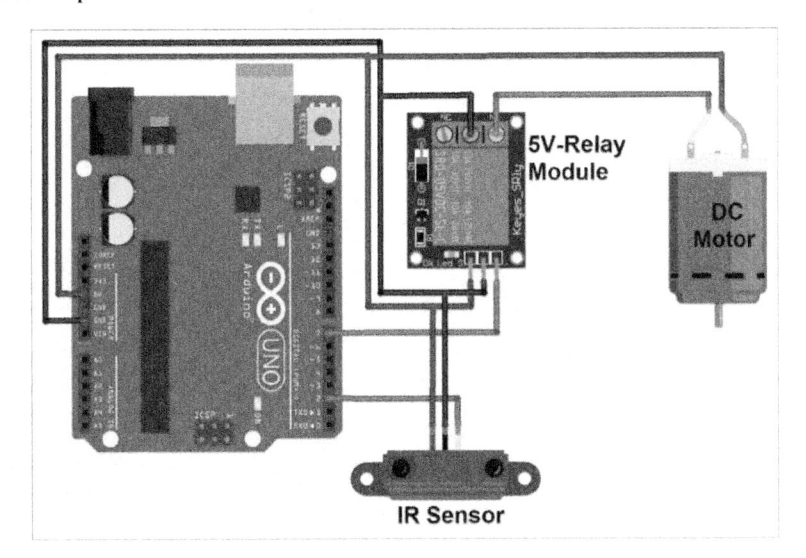

It is a simple circuit for controlling this DC Motor with an IR Sensor using Arduino. On the circuit, the output of the IR sensor module is connected to pin 2 of the Arduino and the input of the relay module to pin 7. The relay is additionally connected to a DC Motor. Project management is straightforward in this case. The IR sensor will turn on the output pin whenever it detects a movement in front of it. Arduino reads the IR sensor's output pin, so pin 7 goes high to activate the relay module, and Arduino reads pin 1 as well. Upon activation of the relay, the DC motor will begin to rotate. The output of the IR sensor remains low when nothing is in front of it, and the DC motor remains in off state, as well. With the potentiometer mounted on the module itself, the IR sensor's sensitivity can be adjusted. The sensitivity of the sensor simply means the distance at which the object can be detected.

380. DC Motor Speed Control using Arduino and Potentiometer

Electric motors are the most commonly used motors in robotics and electronics projects. Various methods exist for controlling DC motor speed, but in this project, we are using PWM to control DC motor speed. This project uses a potentiometer for controlling the speed of the DC motor. By rotating the knob of the potentiometer, the speed can be modified.

Pulse Width Modulation:

The PWM technique, also known as pulse width modulation, is used to control voltage or power. Applying 5 volts to a motor will cause it to move at a certain speed, and if we reduce a charging voltage by 2 volts, or applying 3 volts to the motor, the motor speed will also decrease. In the project, PWM is used to control the voltage using this concept.

Working and Circuit Diagram

We use a 100K ohm potentiometer to change the duty cycle of the PWM signal in this circuit to control the speed of the DC motor. Connecting the 100K ohm potentiometer to the Arduino UNO's A0 analog input pin, and connecting the DC motor to the PWM pin 12 of the Arduino, creates a simple controller to oscillate a DC motor. Using an Arduino program only requires that the voltage be read from analog pin A0. By using the potentiometer, the voltage can be varied at the analog pin. A duty cycle is then adjusted as necessary after doing some calculations.

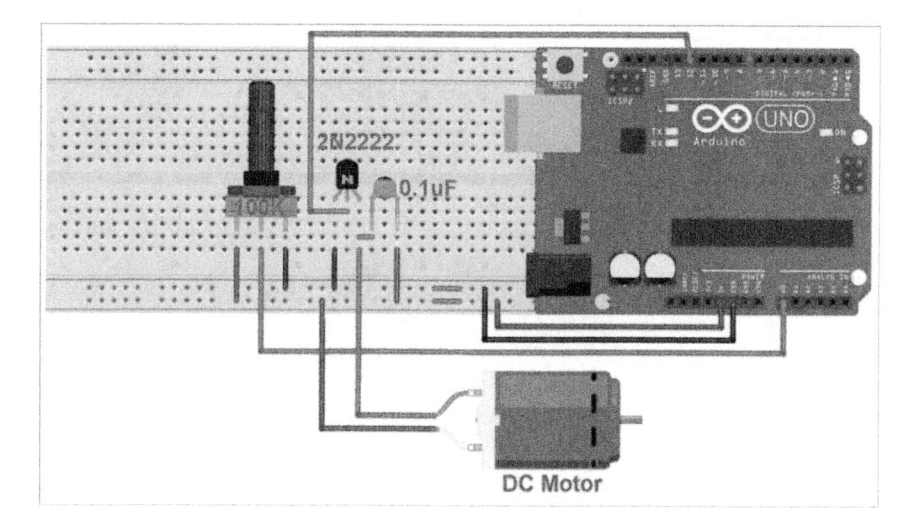

Material Required

- Arduino UNO
- DC motor, Transistor 2N2222
- Potentiometer 100k ohm
- Capacitor 0.1uF, Breadboard, Jumping Wires

The HIGH time, for example, will be 768ms (256-1024) and the LOW time, 256ms when we input 256 values to the analog input. It is only implied that 75% of frequency oscillation is being observed by our eyes. High frequency oscillation is almost imperceptible to our eyes, which makes the motor appear to run continuously with 75% of speed. This is how the speed can be controlled using a potentiometer.

381. DIY Smart Vacuum Cleaning Robot using Arduino

Robotic vacuum cleaner with four wheels that intelligently avoids obstacles while also vacuuming the floor. Robot Roomba, which appears in the image below, inspired the idea of this vacuum cleaner. I think we already know what our Automatic Vacuum Cleaner Robot is, but now we need to make it real. Thus, let's find the best position for our execution to begin. The first step in building our concept robot would be to determine the following:

Required Components:

- Microcontroller type, Sensors required
- Motors required, Robot chassis material
- Battery capacity
- Wooden sheets for chassis
- IR and US sensors
- Vacuum cleaner which runs on DC current
- Arduino Uno, 12V 20Ah battery
- Motor driver IC (L293D)
- Working tools, Connecting wires

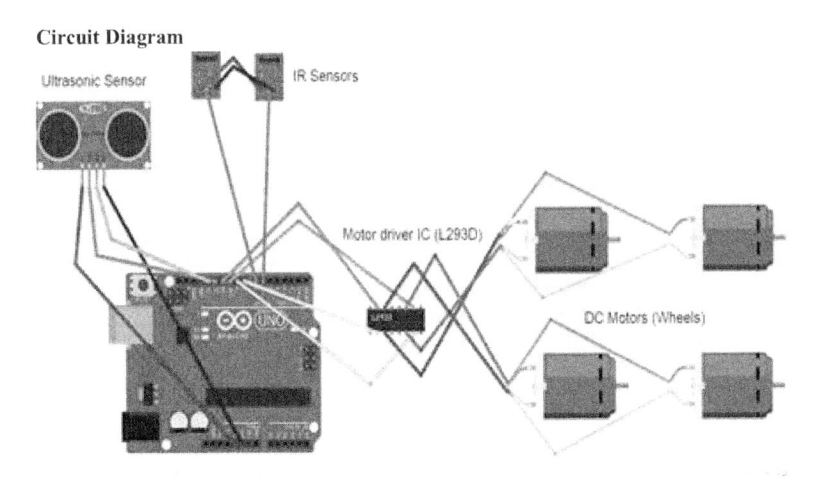

Circuit Diagram

The most critical component of robotic placement is the vacuum cleaner. The vacuum must be tilted at the angle in the photo so that it can provide effective vacuum action. Vacuum cleaners are not controlled by Arduino; they are powered on directly when the robot is powered on. We found that the wooden work was the most exhausting part of building our robot. For the sensors and vacuum cleaner, we will have to drill holes in the wood and carve them. Whenever you have the motor and the motor driver in place, it is recommended to Test Ride your robot before connecting the sensors.

Passionate energy for learning and working: Once you've verified that everything works properly, attach the sensors to Arduino using the circuit diagram provided at the end. On top and on both sides of the robot, I have added an Ultrasonic sensor and two IR sensors. On the L293D, the heat sink is fitted to prevent the IC from overheating.

There are also some extra parts you can add: We used a BLDC fan to create the vacuum and had it enclosed in a box. Those who are on a tight budget can adopt this strategy. Similarly, this approach looks good but isn't efficient. Detailed code for this robot vacuum cleaner is located below. Your robot can get started once you've connected the Arduino and loaded the program. These comments explain how a program works. The following video will show you the robot in action. Additionally, I plan to completely 3D-print the parts in the next version. It will also have some cool features and complex algorithms for covering the entire carpet area and will also be easy to use and compact. You can look forward to future updates.

382. Robot Car controlled by a mobile phone using a G-Sensor and Arduino

In this project, we will use a mobile phone's G sensor to control a robot car and you will be able to tilt the phone to control it. The G-Sensor Controlled Robot will also be controlled by Arduino and the RemoteXY app. A smartphone app called RemoteXY is used to control the Robot from the Smartphone. It will also be possible to control Robot by both tilting the phone and the joystick when the interface has a joystick.

Gravity sensors or Gravity sensors are acceleration sensors in smartphones that allow them to know the screen orientation. As the Gravitational force moves in X, Y, Z directions, the accelerometer senses how the screen rotates in accordance with it. In modern mobiles, the orientation of the screen is determined by a Gyroscope sensor that is a lot more sensitive and accurate. The Robot car in this project will follow the angle that the phone is tilted in. For example, if we tilt the phone forward then it will move forward. The car will then move backward when the tilt is lowered. A G sensor is used in most car games in Mobile, too, so the car moves accordingly.

Required Components:

- Two-wheel robot car chassis
- Arduino UNO
- L298N Motor Controller
- HC-06 Bluetooth module (HC-05 will work too)
- Power supply or Cells, Connecting wires

Creating Interface for Robot using RemoteXY app:

- You will need to access the following link in order to create the interface for controlling the robot car with RemoteXy.
- http://remotexy.com/en/editor/ the webpage will look like this

Circuit Diagram

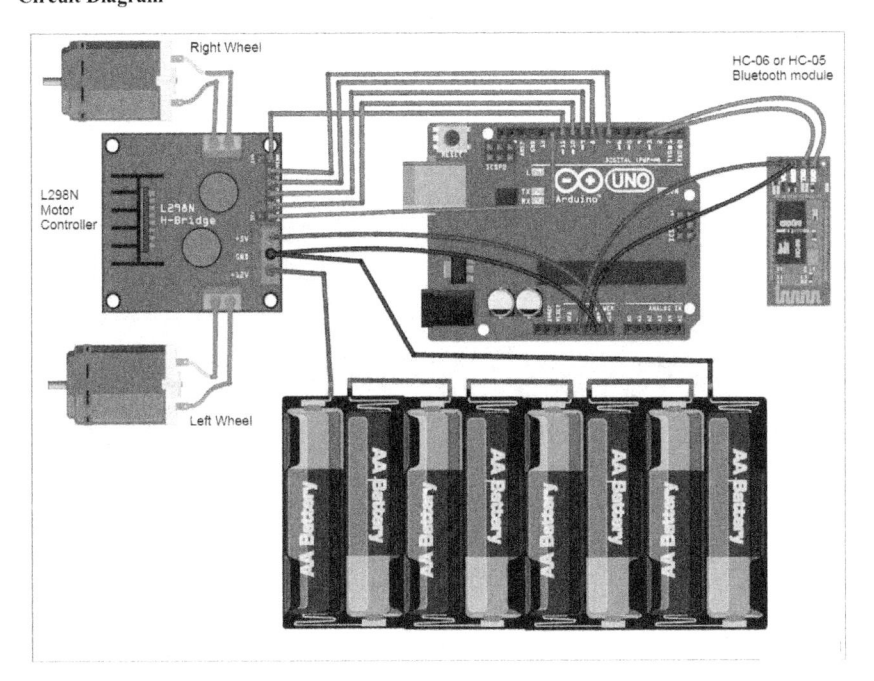

Put the switch button and joystick into the mobile interface from the left side of the screen. By pressing the button, you'll turn on Pin 13 on the Arduino, which is internally connected to the car, and by pressing the joystick, you'll control it. After you have placed the switch and joystick, you will see the following webpage.

After that, we will need to connect the G sensor enable/disable button to the joystick, to enable us to move the Robot Car with the phone by tilting it left, right, up and down. The G sensor can be turned on and off using that button, and the car can be controlled using the joystick when the G sensor is disabled.

Click the joystick you placed on the interface to place the G sensor enable/disable button. You will see the properties section on the left, and at the end, there is an option to place the G sensor button near the joystick, so you can place it wherever you like. This is what the webpage will look like after this.

You should then click the "Get source code" button and save it on your computer. Download the library from here http://remotexy.com/en/library/ and save it into the Arduino library directory. Make sure that the downloaded code is error-free by compiling it. These are not the actual Robot code, but rather the code that helps to utilize the Arduino with the App. Download the app from here http://remotexy.com/en/download/ or. You can download RemoteXY from the Google Play Store on your Android smartphone.

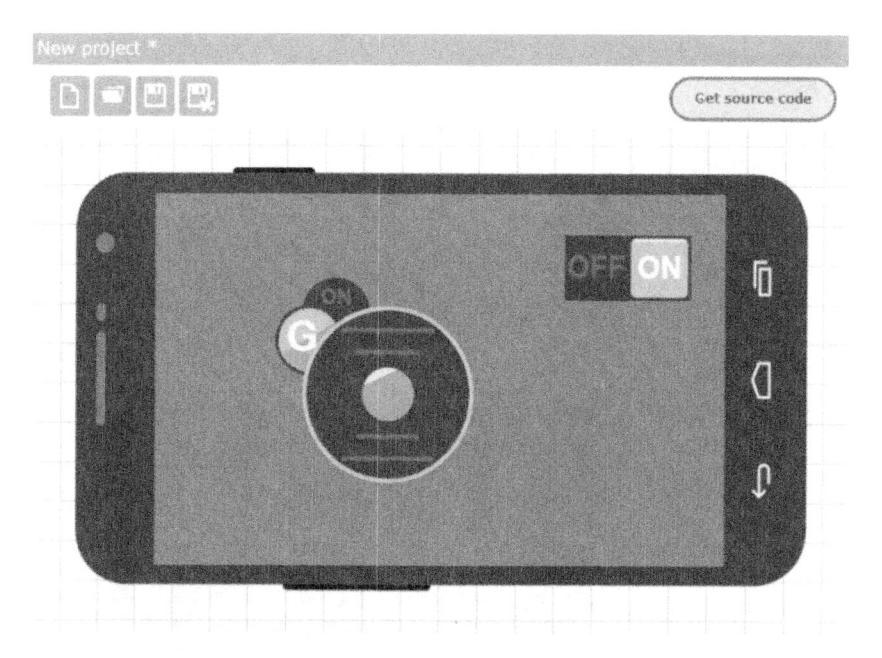

Circuit Explanation:

The L298N motor controller needs to be interfaced with the Arduino first. Motor controller pin ENA and ENB should be connected to Arduino pins 12 and 11, respectively. The motor can be controlled with these two pins via PWM. Our car can be sped up or slowed down using these pins. To connect the IN1, IN2, IN3 and IN4 to the Arduino pins 10, 9 and 8, follow the steps below. The motors will rotate both clockwise and anticlockwise with these pins. Battery positive and negative should be connected to 12V and ground on the motor controller to power the motor. To connect the Arduino to the motor controller, connect the motor controller's 5V and ground lines to the Arduino Vin and ground lines.

If you have an HC-06, you can connect it to the Arduino, but if you have an HC-05, it will work too. The 5V and ground of the Bluetooth module should be connected to the Arduino's 5V and ground. After that, connect the RX pin of the Bluetooth Module to pin 3 of the Arduino and the TX pin to pin 2 of the Bluetooth Module. You can also learn about Bluetooth controlling a toy car with an Arduino using Bluetooth Controlled Toy Car.

383. Weight measurement using Arduino, the HX711 Module, and a load cell

The Load Cell and Weight Sensor HX711 will be connected to the Arduino to measure weight. The weight of an item is displayed by an automated weight machine in many shops. Users place the item on the platform and the machine displays the weight. Therefore, there is a weighing machine that has been built with Arduino and Load cells, which is capable of weighing up to 40kg. Further increase of the limit is possible by using more capable load cells.

Required Components:

- Arduino Uno
- Load cell (40kg)
- HX711 Load cell Amplifier Module
- 16x2 LCD
- Connecting wires
- USB cable
- Breadboard
- Nut bolts, Frame and base

Load Cell and HX711 Weight Sensor Module:

Load cells are electronic transducers that create an electrical signal from force or pressure. Indirectly related to the force applied is the magnitude of the electrical output. During application of pressure, strain gauges in load cells deform. When the strain gauge is deformed, the effective resistance changes, so electrical signals are generated. Four strain gauges are typically used as part of a Wheatstone bridge to form a load cell. These load cells can weigh up to 40kg, and come in different ranges like 5kg, 10kg, 100kg and even more. Here, we are using a Load cell that can weigh up to 40kg.

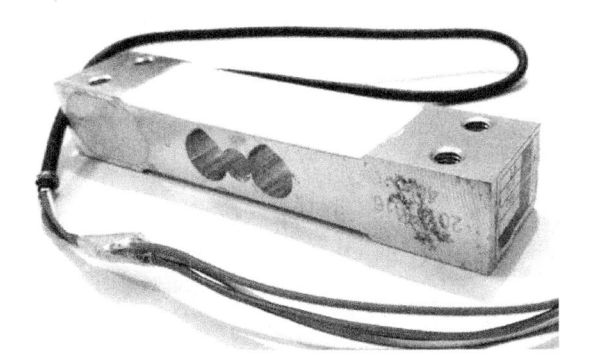

Because the load cell has an output voltage in the range of a few millivolts, these signals require further amplification, hence the HX711 Weighing Sensor is required. An HX711 chip is used in the HX711 Weighing Sensor Module, which is an A/D converter with 24 high-precision channels. Two analog input channels can be programmed in the HX711 for gains up to 128 by programming them. So the HX711 module outputs the low electrical signal from the load cells which is then amplified and digitally converted before being fed to Arduino to calculate weight.

Circuit Explanation:

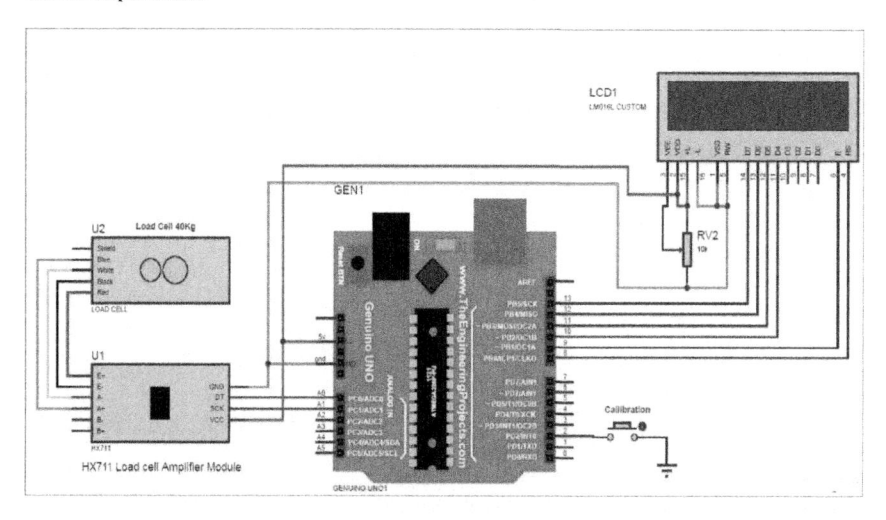

Schematic and connections are provided below for this project. Pin numbers 8, 9, 10, 11, 12 and 13 of Arduino are connected to 16x2 LCD pins RS, EN, d4, d5, d6, and d7 respectively. Arduino's A0 and A1 pins are connected directly to the DT and SCK pins of HX711 Module. Circuit diagrams of the HX711 module and load cell connections have also been explained earlier.

Working Explanation:

It is easy to execute this Arduino weight measurement project. In advance of discussing details, this system needs to be calibrated for accurate weight measurements. A calibration process starts automatically when the system is powered up. In case the user wishes to calibrate it manually, he or she can use the push button. The following code describes how we created a void calibrate() function for calibration purposes.

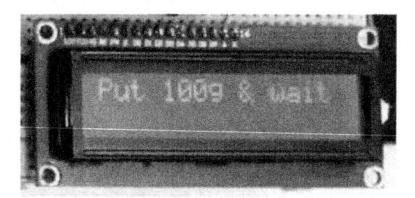

When calibrating the load cell, wait for the LCD display to appear once 100 grams of load is placed on the cell as demonstrated in the picture below. You should put the 100g weight over the load cell when the LCD reads "put 100g" and then wait for it to appear on the LCD. Upon completion of the calibration procedure, the process will take a few seconds. Any weight can be put over the load cell after calibration (maximum 40kg) and the calculated value will appear over the LCD in grams.

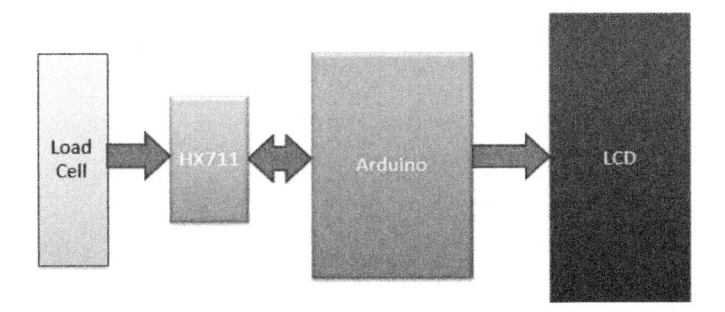

Arduino was used to manage the entire process in this project. HX711 Load Amplifier Module receives the weight signal from the load cell in the form of an electrical analog voltage. Using the HX711 ADC, a 24bit input signal is amplified, and then digitally converted into a 24-bit output, which is then fed to the Arduino. After the data from HX711 has been calculated by Arduino, the weight values are converted into grams and displayed on LCD. The calibration of the system is accomplished by pushing a button. In order to facilitate the process, we created an Arduino program.

384. Automated Plant Irrigation System Using Arduino with Message Alerts

It is always difficult to leave our plants unattended if we are going out of town for a few days. Our plants require regular watering. We are making an Arduino-based Automated Plant Irrigation System, which sends out messages to you as well as automatically provides water to plants. Water is provided to the plant through a water pump if the soil moisture level drops due to a low moisture level detected by the soil moisture sensor. Once the system detects sufficient soil moisture, it automatically turns off the water pump. An update on the status of the water pump and soil moisture is sent to the user whenever the water pump is turned on or off via the GSM module. Farms, gardens, homes, etc, would benefit all from this system. No human intervention is required because this system is completely automated.

Required Components for Arduino Plant Watering System Project

- Arduino Uno
- GSM Module
- Transistor BC547 (2)
- Connecting wires
- 16x2 LCD (optional)
- Power supply 12v 1A
- Relay 12v
- Water cooler pump
- Soil Moisture Sensor
- Resistors (1k, 10k)
- Variable Resistor (10k, 100k)
- Terminal connector
- Voltage Regulator IC LM317

GSM Module:

Using the SIM800 GSM module, we have used it here. Customers as well as hobbyists can easily embed the SIM800's quad-band GSM/GPRS module. GSM/GPRS 850/900/1800/1900MHz performance for voice, SMS, data is provided by SIM900 GSM Module, while SIM800 GSM Module employs an industry-standard interface. Slim and compact, the SIM800 GSM Module follows a modern design aesthetic.

Quad - band GSM/GPRS module in small size.

- GPRS Enabled
- TTL Output

Circuit Explanation:

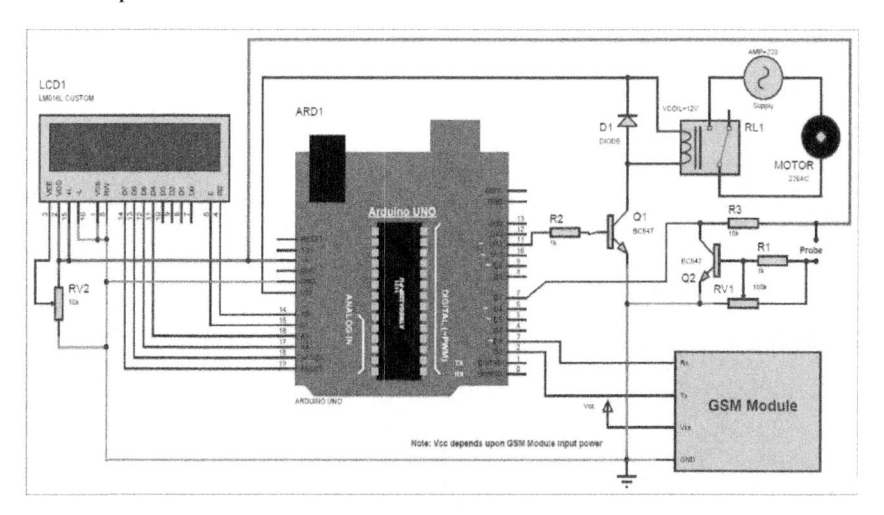

The homemade soil moisture sensor probe used in this system allows us to measure moisture levels in the soil. As shown in the following image, a copper clad board was cut and etched to make the probe. There is a direct connection between the probe and Vcc, and the other probe terminal is attached to BC547's base. Sensor sensitivity is adjusted using a potentiometer connected to the base of the transistor.

This Automatic Plant Watering System is controlled entirely by an Arduino. Directly connected to the digital pin D7 of Arduino is the soil sensor circuit's output. The sensor circuit uses a LED, which indicates whether moisture is present in the soil by its ON state and whether it is not present by its OFF state.

An SMS can be sent to the user using the GSM module. As you can see, here we are using a GSM SIM800 card, which is a device that directly gives and takes TTL signals (to be used by any GSM module). The SIM800 GSM module is powered by the LM317 voltage regulator. It is recommended to read the data sheet of the LM317 before using it, since voltage rating is very sensitive to that. 3.8 to 4.2 volts is its operating voltage rating (please use 3.8 volts when operating it). A SIM900 TTL Module should be used with 5V, and a SIM900 Module should be used with 12v in the DC Jack slot on the board. The 220VAC small water pump is controlled by a 12V Relay. An Arduino digital pin 11 is connected to the BC547 transistor, which drives the relay. Status and messages are also displayed on an optional LCD. Several LCD pins are directly connected to the Arduino, including RS, EN, and D4-D7, which are directly connected with the Arduino on pins 16, 17, 18 and 19. In this case, Arduino uses the LCD library built into the board to drive a 4-bit LCD display.

Working Explanation:

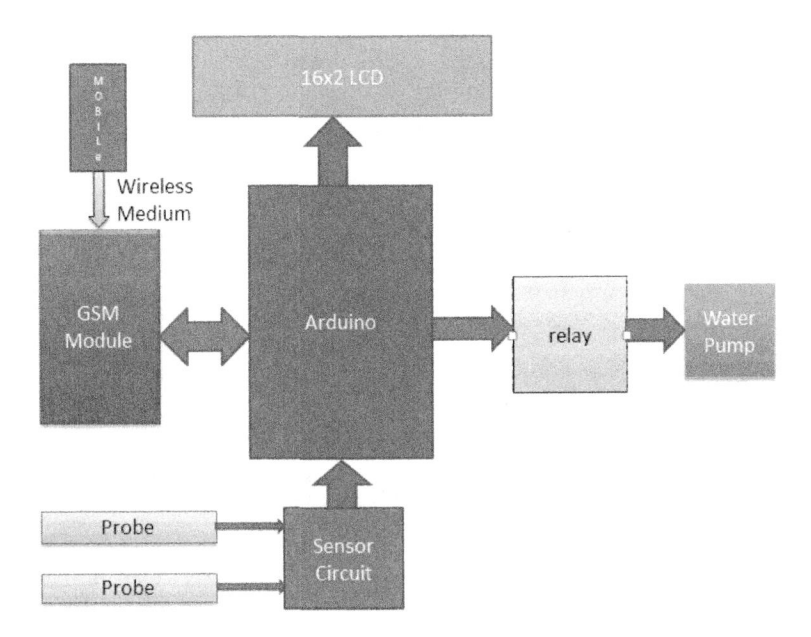

A Plant Irrigation System like this is pretty straightforward in terms of how it works. The first thing to point out is that the system is completely automated, and it does not require any manual labor. A GSM module sends alert messages to the cell phone of the user based on the Arduino's handling of the entire process. The Irrigation System with Arduino and a Soil Moisture Sensor is shown on the following block diagram: When soil contains moisture, conduction occurs between the two probes of the Soil Moisture sensor, resulting in transistor Q2 remaining in the triggered/on state, and Arduino Pin D7 remaining low. The Arduino sends the SMS message "Soil Moisture is Normal" when it detects the LOW signal at D7. The water pump is still in an off state because the motor switched off.

Transistor Q2 will become off if no moisture is present in soil, and Pin D7 will become high. A message is then sent to the user stating that "Low Soil Moisture has been detected" and Arduino turns on the water motor. Motor turned ON". The soil will automatically turn off the motor when the moisture content is sufficient.

385. Making calls and sending messages with Arduino and GSM modules

Microcontrollers sometimes have difficulty communicating with the GSM Module, specifically for functions such as SMS, calls, and texting. With the help of the Arduino, we will build a simple mobile phone. The GSM Module involved in this project can make and receive calls, as well as send and receive SMS, and the Arduino phone also has a Microphone and Speaker so you can talk over it. Besides interfacing with the GSM Module, the project will be able to run any phone's basic functions by using all the necessary code on the Arduino.

Components Required:

- Arduino Uno
- GSM Module SIM900
- 16x2 LCD, 4x4 Keypad
- Breadboard or PCB
- Connecting jumper wire
- Power supply
- Speaker, MIC
- SIM Car

Working Explanation:

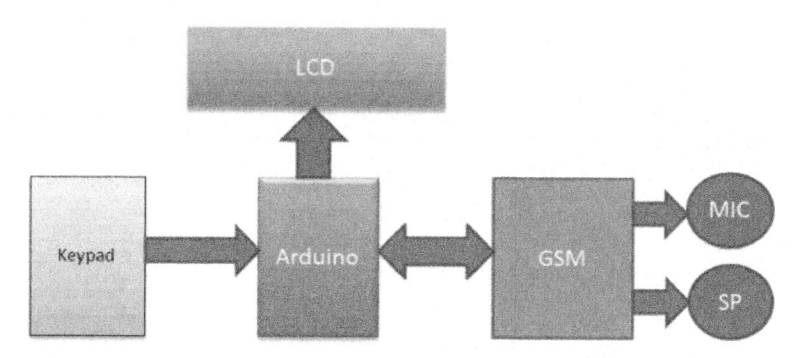

Arduino Uno is used to control all the features of this Arduino Mobile Phone Project, as well as to interface all the components. An alphanumeric keypad is used to make all kinds of inputs, such as entering mobile numbers, typing messages, making and receiving calls, and sending and receiving SMS. The GSM Module communicates with the network in order to make and receive calls and messages. As well as ICs and speakers, a 16x2 LCD shows messages, instructions, alerts, and a MIC picks up the voice call and ring sound.

Using the same keypad for both numbers and alphabets, alphanumeric is a way to enter data. Check out the Code in Code section below for the Arduino code to accept the 4x4 keypad interface as well.

Circuit Diagram and Explanation:

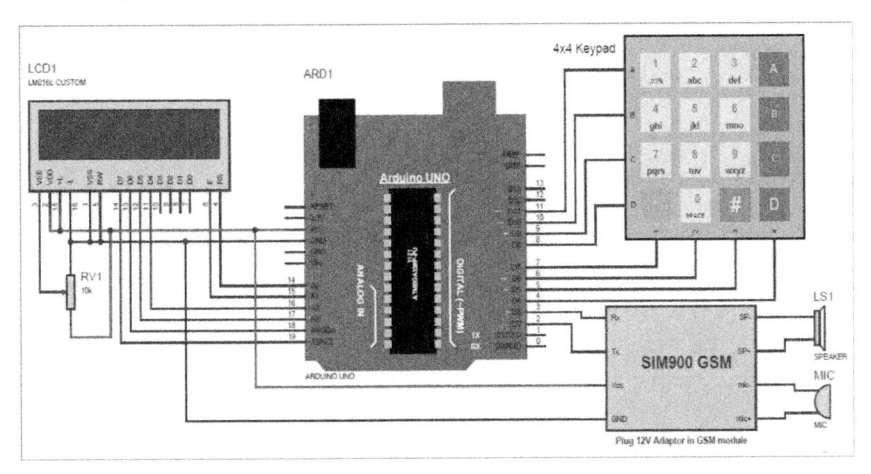

Diagram showing how GSM SIM900 and Arduino can be interconnected is shown above. Pin 14 is connected to pins 15, 16, 17, 18, 19 of Arduino, while pins EN, D4, D5 and D6 are connected to pin 16 of the LCD. (Ground of Arduino and GSM are connected.) The RX and TX pins of the GSM module are directly connected to Arduino's pins D3 and D2. Row pins R1, R2, R3, R4 of the 4x4 keypad are directly connected to pins 11, 10, 9, 8 of Arduino, and column pins C1, C2, C3 are directly connected to pins 7, 6, 5, 4 of Arduino. The MIC pins of the GSM Module are directly connected to mic+ and mic-, and the speaker pins are directly connected to pin SP+ and SP- for the GSM Module.

386. Fingerprint Based Biometric Attendance System using Arduino

The presence of students in an office or school is typically marked by attendance systems. The attendance management system has greatly improved over the years, from marking attendance manually in attendance registers to using high-tech applications and biometric systems. Other electronic attendance system projects that we covered in previous projects included RFID, AVR, and 8051 microcontrollers. We used an Arduino to take attendance records and use a fingerprint module to take attendance data. The system will become more secure for users if it uses a fingerprint sensor. A biometric attendance system based on fingerprints is described in the next section of the project.

Required Components

- Arduino -1
- Finger print module -1
- Push Button   - 4
- LEDs -1
- 1K Resistor -2
- 2.2K resistor -1
- Power

- Connecting wires
- Box
- Buzzer -1
- 16x2 LCD -1
- Bread Board -1
- RTC Module -1

Block Diagram:

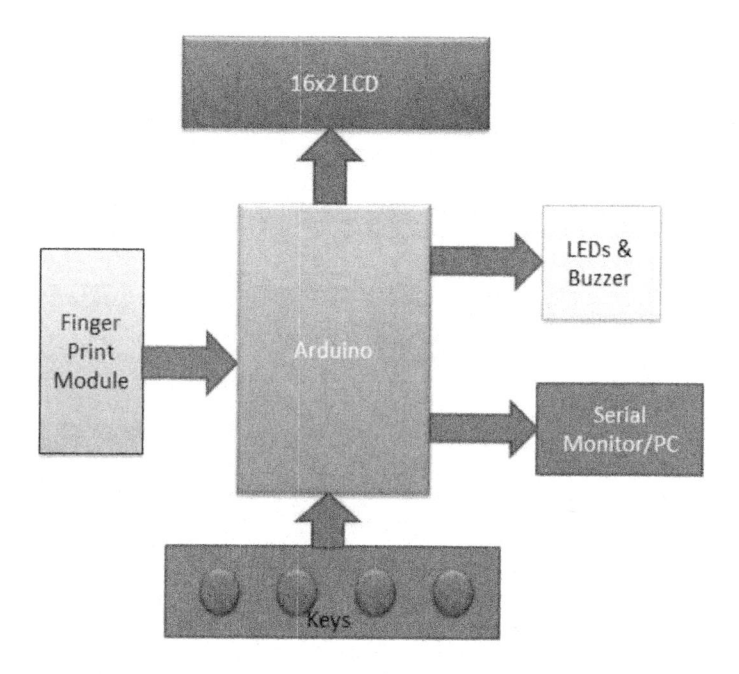

A fingerprint attendance system circuit includes a Fingerprint Sensor module that operates by recognizing a person's fingerprint. This allows the system to authenticate the person or employee. Four push buttons are being used here: Up/Down, Delete, Enroll. A key has three functions: ENROLL, DEL, and END. A new user is enrolled into the system by using the ENROLL key. The user must then press the ENROLL key when enrolling a new finger. LCD will then ask the user for an ID and where the fingerprint image is to be stored. In this case, the user may press ENROLL again to return to the first step if he or she does not want to go any further. The ENROLL key behaves as the Back key this time. Enrollment and backtracking are both possible with the ENROLL key. As well as downloading attendance data over the serial monitor, the enroll key is also used for enrollment. A similar double function is provided by DEL/OK, since the user must select finger ID using UP and DOWN after enrolling a new finger. After pressing the DEL/OK key (this time it acts like OK), the user can proceed to selecting the ID. Deleting data from the EEPROM of Arduino is accomplished by the Del key.

The fingerprint sensor module captures an image of the finger's print and uses that to create an equivalent template. It then saves the template as selected by Arduino into its memory. A fingerprint image is captured, the fingerprint is converted into templates, and the finger is stored as an ID using Arduino. Our fingerprint module has a yellow LED that we have added to indicate that it is ready to take a fingerprint image. Additionally, buzzers are used to indicate various things. This system is controlled by an Arduino; it controls the entire thing.

Circuit Diagram

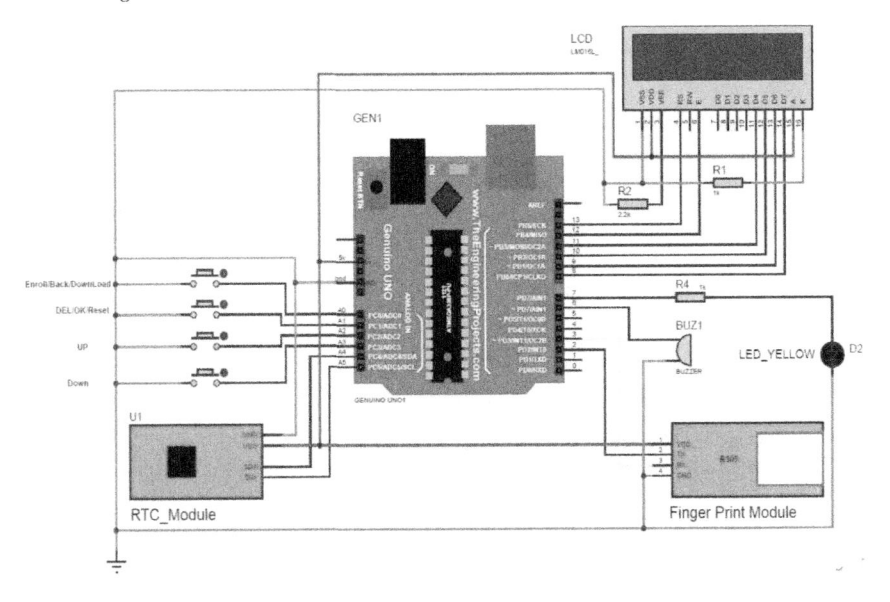

In the above diagram, we can see that this fingerprint-based attendance system project involves a fairly simple circuit. An Arduino controller controls the entire process, push buttons for enrolment, deletion, selection of IDs, and attendance, LEDs for indication, and an LCD to display the message to users. It can be seen in the circuit diagram that the push button is directly connected to pin A0 (ENROL), With respect to ground, pins A1(DEL), A2(UP), and A3(DOWN) of an Arduino are connected to one k resistor, and pin D7 of the Arduino is connected to the yellow LED via a 1k resistor. Arduino's Serial pins D2 and D3 (Software Serial) are directly connected to the RX and TX of the fingerprint module.

Power is supplied by a 5v supply to an Arduino board with a fingerprint module. The pin A5 is also connected to a buzzer. In this example, a 16x2 LCD is configured in 4-bit mode, with D4, D5, D6 and D7 connected directly to digital pins D13, D12, D11, D10, D9, and D8 of the Arduino.

Working of fingerprint-based attendance systems

It is a relatively simple project, working with fingerprint attendance systems. With the help of push buttons, the user first needs to register his or her fingerprints. The user must press ENROLL to save the fingerprint in memory by ID name and then the LCD will ask for the user's ID name. As a consequence, an ID must now be entered by using the UP/DOWN keys. Press the OK button (DEL) once the user has selected the ID. In order to use the fingerprint module, LCD now asks the user to place their finger there. As a result, the user must now place his finger over the fingerprint module in order for it to generate an image. Once you take your finger off the fingerprint module, the LCD will ask you to repeat the process. As soon as the user places his finger again, the module takes an image and converts it into a template which is stored by selecting the ID into the fingerprint module's memory. By now, the user has been registered and can use the fingerprint module to feed attendance. The system will allow all of the users to enrol in the same way. The user now needs to press the DEL key if he/she wants to delete the stored fingerprint or ID. LCD will then prompt you for the ID to be deleted after pressing the delete key. After selecting an ID, the user must press the OK key (the same key as DEL). When the fingerprint is successfully deleted, the LCD will let you know.

How Attendance works:

After placing his finger over the fingerprint module, the user will see that the fingerprint device is captured and that it is searching if any recorded ID is associated with these fingerprints. Upon detecting a fingerprint ID, the LCD will display Attendance registered, while a buzzer will burst out once and the LED will turn off until a new input can be entered. We have also implemented a Time and Date module with the fingerprint module. The system keeps track of the time and date continuously. In this context, Arduino automatically grabs time and date from a user's finger when they place it over the fingerprint and saves them in the allocated memory slot. This system allows us to create five user spaces for 30 days. When the RESET key in Arduino is pressed and once the program is enrolled, the key will download information from the Arduino EEPROM over the serial monitor.

Memory Management:

Our Arduino UNO comes equipped with 1023 byte of memory of which 1018 byte are available for data storage, and we have taken five days' attendance data from the Arduino UNO. In addition to recording time and date, every attendance recording will be 7-bytes in size. As a result, total memory requirements are 5*30*7=1050 so here we need more 32 bytes However, if four users are used, then we need 4*30*7=840. Here, we have taken the memories of five users to demonstrate this project. The 5th user's attendance records will be unable to be stored since we will not be able to store 32 bytes of data each time. Change some lines in the code to make it work with four users. **Fingerprint sensors are commonly used in mobile, lock and unlock, on a mobile display, on a phone screen, in security systems, attendance systems, and security locks.**

387. Generating Tones by Tapping Fingers using Arduino

We will use Arduino to build an entertainment system in this project. Creating random sounds out of the pen or table is an ingrained habit for all of us. There's no doubt that doing this at least once is not considered good manners, but we're all used to doing it. Therefore, I decided to take it one step further by using Arduino's tone-playing abilities. With this project you will be able to tap on anything conductive and generate tones, like playing the Piano on your palm, and create your own rhythms.

Components required:

- Arduino Pro Mini
- Piezo Speaker
- Flex Sensor
- Finger Gloves
- 10K Resistors
- BC547 Transistors
- 9V Battery

Circuit Diagram and Explanation:

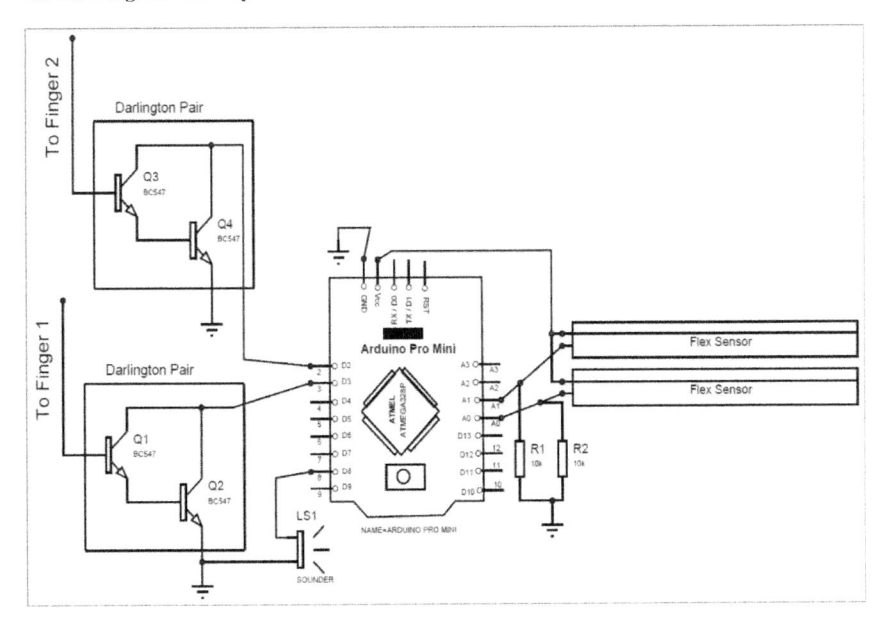

During the development of this project, we used a total of four sensors: two flex sensors and two Darlington pairs functioning as a touch sensor. Secondly, we have attached two 10k resistors R1 and R2 to the Flex sensor as pull-down resistors. Using one finger to generate three distinct tones based on how much it has bent, the Flex sensor here makes use of this technology. This means that two fingers can generate six different sounds.

Darlington Pair:

It is important that we understand what Darlington is and how it relates to our project before we move forward. The Darlington pair can be defined as two bipolar transistors connected in a way that in case of current amplification by the first transistor, the current is amplified further by the second transistor. The following image shows a pair of Darlington:

This circuit consists of two BC547 transistors whose collectors are tied to their respective collectors and whose emitters are connected to the bases of their respective transistors, as shown above. A small signal applied to the first transistor base will bias the second transistor base, which means that the circuit functions as an amplifier.

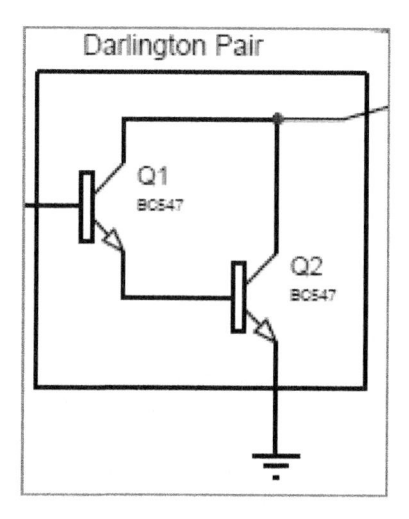

Our body serves as a ground for the second transistor so the transistor becomes biased whenever our bodies touch the base of the transistor. In order to make this project a success, we utilized this knowledge to build the touch sensor. Two Arduino interrupt pins, numbers 2 and 3, are pulled high using internal pull-up resistors. After the Darlington switch closes, these pins will be grounded. The interrupt will be triggered every time the wire touches the base of the transistor (1st transistor) on the Arduino. I added a flex sensor that alters the tone according to how much the finger is bent, so I can generate more tones using fewer fingers. I have programmed the system to produce three different tones for each finger depending on how far it is bent (flex sensor). Having access to more tones at your fingertips is possible by increasing the number.

In order for the board to fit easily in my palms, I made it on a perf board, however you can also build it on a breadboard. It is important that you touch the ground of the circuit during the course. You should have something similar to this once you've soldered everything As displayed above, I have secured the Darlington pair wires and the flex sensor with two finger gloves. While playing your tone, you can come up with a better idea (if possible) to secure the earphones in place.

Working: You can then mount them on your fingers once the hardware is ready. Ensure that you are touching the ground at some point on the circuit. You should now be able to hear the tone by touching any conductive material or your body. The taps can be played at different intervals, at different positions, to create your own melody.

388. The Arduino and Thingsboard are used to create a biometric attendance system based on IoT

With this project, we are aiming to create a smart, efficient and engaging attendance system by integrating IoT into the boring one. The vast majority of modern attendance systems store data on a micro-SD card, which must be accessed by computer software via a PC. This project is about building a biometric attendance system using Arduino that can scan a fingerprint. If a touch is successful, the data is sent to ThingsBoard via the ESP8266 wireless module and logged there. It is capable of displaying all of this information on the ThingsBoard dashboard, allowing authorities to easily view and analyse information while not requiring direct physical access to the hardware.

However, it is also possible to build the conventional Attendance system without involving the Internet of Things, and the Fingerprint sensor can be further used in further biometric applications such as Voting Machines and Security System.

Preparing the ESP8266-01

This project will use both the AT command mode and the programming mode of the ESP8266. The ESP8266 module can be powered with a LM317 regulator and the Tx and Rx pins can be hooked up to the FTDI board as shown below.

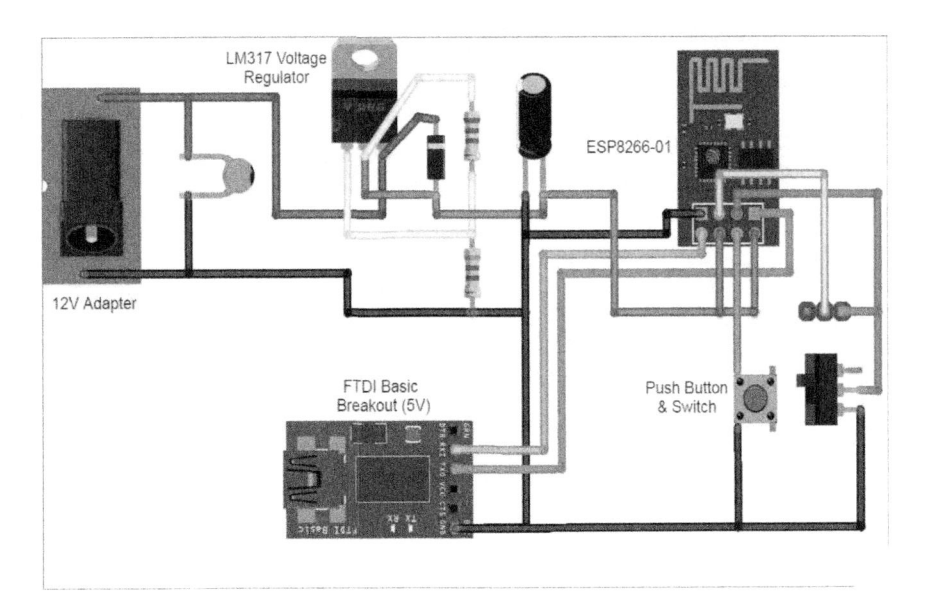

Hardware Required

- Arduino UNO
- 16x2 LCD Display

- Arduino Wi-Fi Shield
- ESP8266-01
- GT511C3 Fingerprint sensor (FPS)
- 12V Adapter

In AT command mode, a toggle switch toggles the ESP8266 into AT commands, and in programming mode, a push button can reset the module. Every time you upload code to the ESP8266, it must be reset.

Circuit diagram

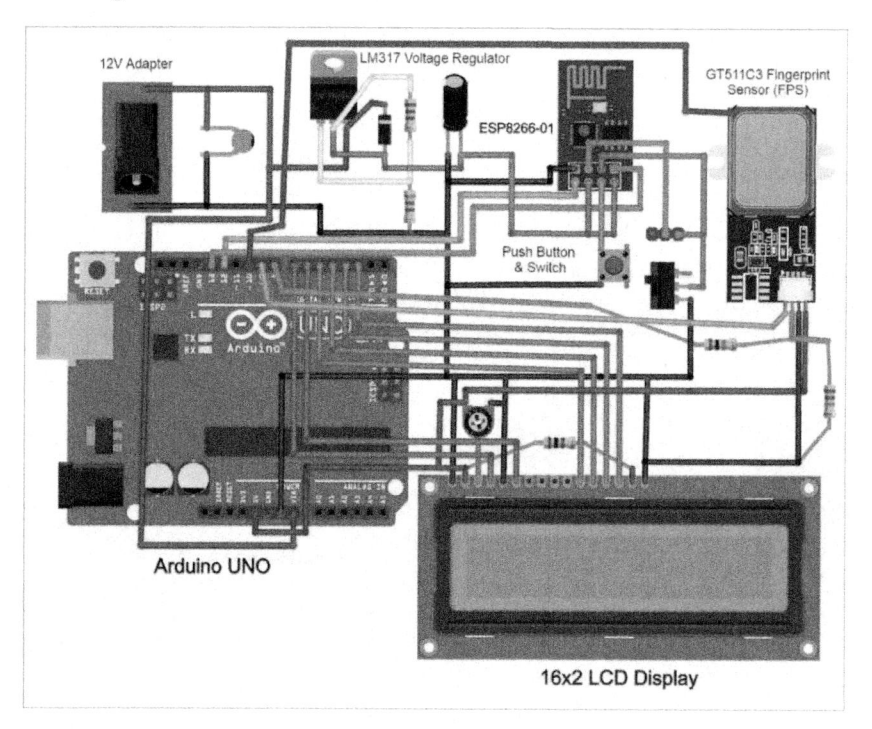

389. Real Time Face Detection and Tracking Robot using Arduino

Have you ever wanted to develop a robot while tracking your face by just using an Arduino and without having to learn programming languages such as OpenCV, Visual Basic and C#? If you're interested in learning more about how we can implement face detection with Arduino and Android, read on. In this project, you'll move the mobile camera with the help of servos to follow you around. Using an Android Mobile Phone offers the advantage of not having to invest in a camera module, and all of the image detection is done on the phone itself, so you do not need an Arduino-based computer for this. Our Bluetooth Module communicates wirelessly with mobile devices using the Arduino Uno.

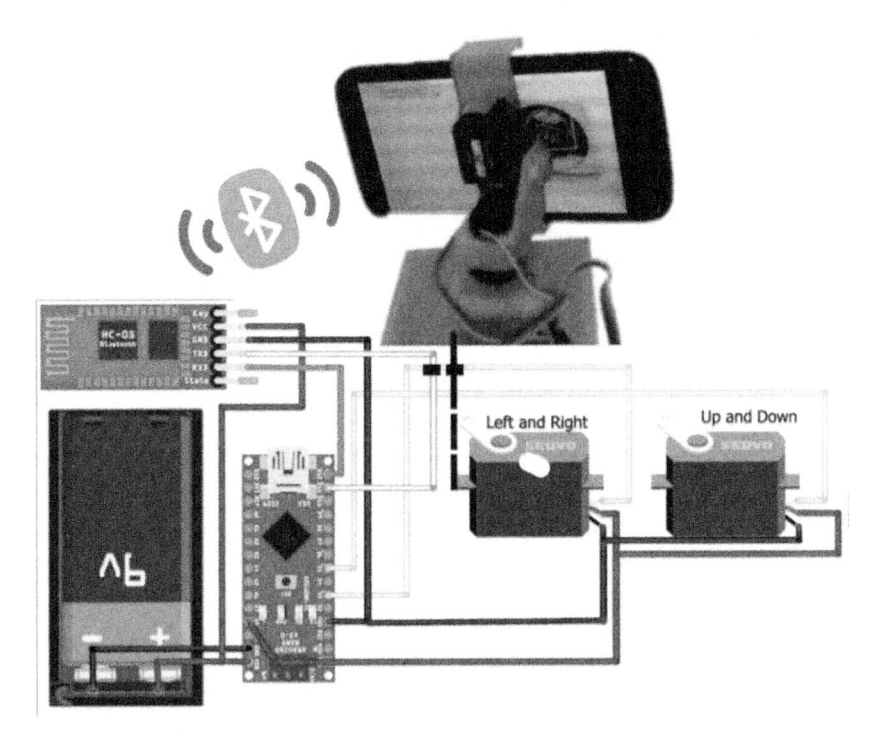

Processing Android was used to create the Android application used in this project. You can either download the APK file (see below for details) or install the application directly from the Processing Android website. Alternatively, you may use the Processing Code provided in the Tutorial to create a more interesting Android Application. Our previous Processing projects will give you more information about Processing. After following this, you will be able to build a Mini Tilt and Span Robotic Arm with the capability of tracking and moving along with your face. This (with further advancement) can be used to record vlog videos or even take selfies using the front and rear cameras on your mobile phones. The exact centre of the screen is positioned exactly where your face is. Therefore! How does that sound? The Demo Video at the end of this article shows how it works. Let's figure out how to do it...

This project has been designed to be as simple as possible, anyone with a basic understanding of hardware or programming can use these guidelines to make it work in no time. However, once you make it, I suggest you dig deep into the code so you can get a real understanding of how it works.

Materials Required:

- Arduino Nano
- Servo motor SG90 – 2Nos
- Android Phone with decent camera

- HC-05/HC-06 Bluetooth Module
- Computer for programming
- 3D printer (optional), 9V Battery

Two servo motors are used in the circuit, one for moving the mobile phone left and right, and another for tilting it up and down. As a result, the Arduino Nano will instruct the servo to move in the direction requested by the Bluetooth module (HC-05). All of the circuit components are powered by a 9V battery.

Working:

Then, once our hardware, code, and Android application are ready, we can start the building process. Simply power your Arduino and open the android application. The HC-05 Bluetooth module (must be called HC-05) will automatically connect to the application when it detects a face. In order to use our mobile holder, simply place the phone in the holder and sit back. As your phone's servo motors move, you should notice your face cantered on the display. Your mobile phone will now follow your movements when you move within the camera's range.

390. Arduino Touch Screen Calculator using TFT LCD

Building projects with Arduino is always easy and makes them more appealing. Arduino libraries and shields made it really easy to program an LCD screen with touch screen capability. We will use a 2.4" Arduino TFT LCD screen in this project to build an Arduino Touch Screen calculator that can perform basic arithmetic operations like Addition, Subtraction, Division and Multiplication.

Materials Required:

- Arduino Uno
- 2.4" TFT LCD display Shield
- 9V Battery.

Getting to know the TFT LCD Screen Module:

2.4" TFT LCD Modules come in many varieties, so it is important to understand how they work before we proceed with the project. Here is what the pinouts look like for a 2.4" TFT LCD screen. These boards will perfectly fit into any Arduino Mega or Uno board, as can be seen from the photo above. In the table below are some small classifications of these pins. There are four basic classifications for pins, namely LCD Command Pins, LCD Data Pins, SD Card Pins and Power Pins. We do not need to understand the details of how these pins work since the Arduino library will handle them. Using the Arduino program, we can display the images on the TFT LCD screen by loading them into an SD card, and this image can be displayed on the SD card slot at the bottom of the module.

TFT Pinout:

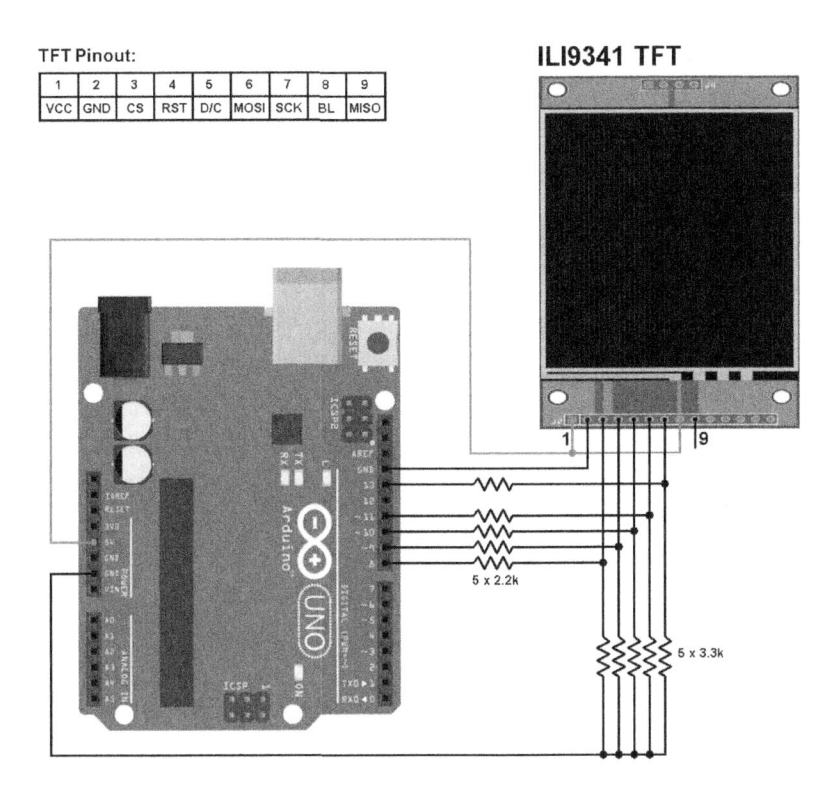

1	2	3	4	5	6	7	8	9
VCC	GND	CS	RST	D/C	MOSI	SCK	BL	MISO

ILI9341 TFT

The Interface IC is another important aspect of the kit. There are many types of TFT modules on the market, from the original Adafruit TFT LCD module to inexpensive Chinese knockoffs. If your Adafruit shield works perfectly on a Chinese breakout board, it may not work the same way on your Adafruit shield. As a result, it's essential to know what kind of LCD display you have in your hand. It is the vendor's responsibility to provide this information. If you are using a cheap clone like mine, you are almost certain to be using the ili9341 driver IC.

Calibrating the TFT LCD Screen for Touch Screen:

It is essential for you to calibrate your TFT LCD module if you intend to use the touch screen functionality. If you touch one place on an LCD screen but the TFT responds at another, for instance, you might not be able to use it. You need to calibrate each board separately because the results won't be the same for each of them. Using the calibration example program (which comes with the library) or a serial monitor is the best way to calibrate.

TFT LCD Connections with Arduino:

Arduino shields that have TFT LCD screens are ideal. The LCD screen will slide perfectly into the Arduino Uno once you push it directly on top of it. Cover the TFT LCD screen's Programming terminal with a small piece of insulation tape as a matter of safety. Here is an example of how the LCD will look when it is assembled on the UNO.

How to Program

To get our Arduino calculator code to work, we're using the SPFD5408 library. With this library, our LCD TFT Module can readily work with Adafruit's LCD TFT Module.

Note: It is vital that you have this library installed in your Arduino IDE for the program to compile correctly.

391. Arduino Motion Detector using PIR Sensor

Many projects have always required motion detection or movement tracking. With the help of the PIR Sensor, it has become easy to detect movement from people or animals. This project will demonstrate how a PIR sensor can be connected to a microcontroller like an Arduino. The Arduino will be interfaced with a PIR module so that whenever movement is detected, a buzzer and LED will beep. In order to build this project, you'll need the following components.

Materials Required:

- PIR Sensor Module
- Arduino UNO (any version)
- LED, Buzzer
- Breadboard
- Connecting Wires
- 330-ohm resistor

PIR sensor: Infrared passive sensors are referred to as PIR sensors. Using this sensor, humans and animals can be detected without incurring high costs. In addition to the pyroelectric crystal, the sensor also has a Fresnel lens to enhance the range of the sensor. A pyroelectric crystal is used to track heat signatures of living organisms (humans, animals). As shown below, we can also set the sensor's working by adjusting the options provided by the PIR sensor modules.

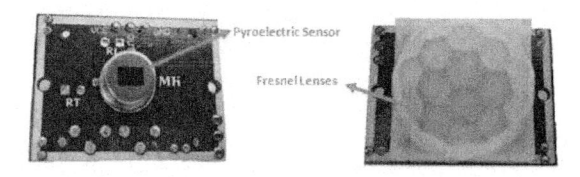

Sensor sensitivity and trigger time of the sensor may be controlled using the two potentiometers (orange color). It is essentially the Dout pin that is present between the Vcc and GND pins of the sensor. A 3.3V power supply may also be used, but the module operates on 3.3V. Additionally, there is a trigger pin setup on the left side of the module that can be used to make it work in two different ways. In one mode, the "H" key is pressed and in the other, the "I" key is pressed. A person will be detected within range when the Dout output pin goes high (3.3V) and will go low at a certain time (time is controlled by potentiometer). It does not matter if the person remains inside the range or has left the area, the output pin will stay high. In our project, our module is being used in the "H" mode. As long as a person remains within the limits of the sensor range, the output pin Dout will go high (3.3V).

Note: Depending on your PIR sensor vendor, potentiometers and pins may be positioned differently. Pinouts can be determined by following the Silk screen

Circuit Diagram and Explanation:

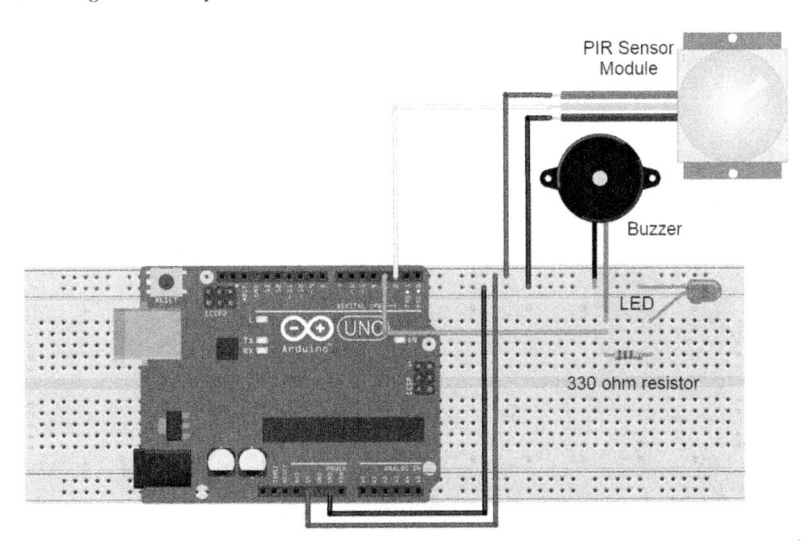

The following image shows the schematic for the Arduino PIR motion detector project by combining it with an LED/Buzzer and a PIR unit. Our 5V rail of the Arduino is used to power the PIR sensor. On the Arduino, the second digital pin is connected to the output pin of the PIR Sensor. The Arduino INPUT pin will be connected to this pin. A buzzer and LED are then connected to the 3rd pin of the Arduino. In this case, the output pin of the Arduino will be used. An output at the 3rd pin will be triggered by an input at the 2nd pin if an Arduino is programmed to do so.

392. Interfacing Hall Effect Sensor with Arduino

The use of sensors has always been crucial to any project. It is these that create the digital/variable data that is used by electronics to process the real-time environmental data. The market offers many types of sensors, and you can choose one that suits your needs. We will use an Arduino to create a project that uses Hall sensors, also known as Hall effect sensors. Using the magnet sensor, you can determine the magnet's pole as well as detect its magnetic field. Magnets are detected for what reason? Perhaps you'd like to know. Actually, there are a number of applications that use Hall Effect sensors, but most of us probably are unaware of them. The speed of rotating machines or bicycles can be measured using this sensor. BLDC motors also utilize this sensor to track the Rotor Magnet position and immediately activate the Stator coils accordingly. Let's learn how to add another tool to our arsenal by learning how to interface Hall effect sensors with Arduino. Some projects involving Hall sensors are as follows: As part of this tutorial, we will utilize Arduino interrupts to detect a magnet near a Hall sensor and flash an LED. We will use interrupts in our tutorial as well, since Hall sensors will typically only be used with interrupts due to their applications that require high reading and executing speeds.

Materials Required:

- Hall Effect Sensor (any digital version)
- Arduino (Any version)
- 10k ohm and 1K ohm Resistor
- LED, Connecting Wires

Circuit Diagram and Explanation:

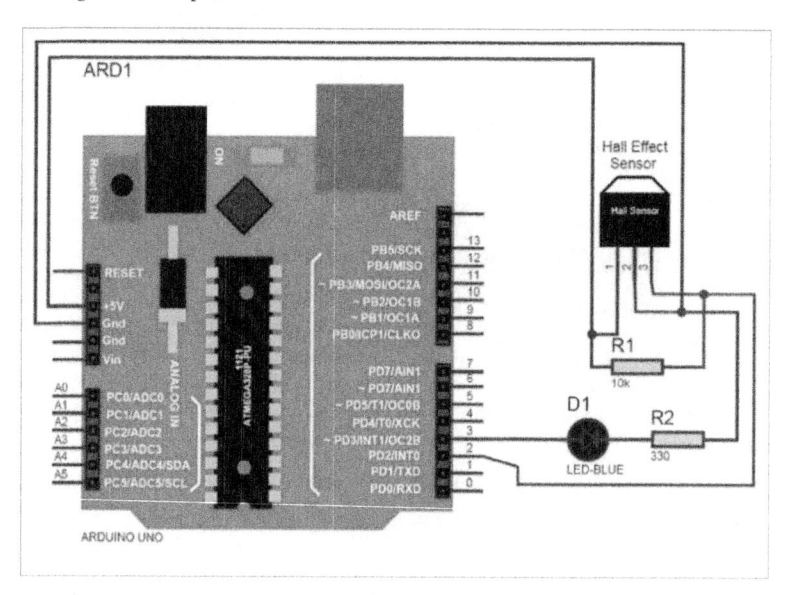

This circuit diagram demonstrates how to connect a Hall sensor to an Arduino. This Arduino circuit diagram for a Hall Effect sensor is quite straightforward as you can see. It is when we try to figure out hall sensor pin numbers that we usually make mistakes. When positioned face-on, the Vcc and Ground pins are on the left, followed by the Signal pin. As we mentioned earlier, we will use interrupts, which is why pin 2 of the Arduino is connected to the output pin of the Hall sensor. Magnets are detected by switching on the Pin, which is connected to an LED.

Hall Effect Sensors:

Several things should be kept in mind before we dive into the connections for Hall Effect sensors. Digital Hall sensors and analog Hall sensors are the two main types of Hall sensors. As opposed to the digital Hall sensor, which detects whether a magnet is present or not (0 or 1). The analog Hall sensor, on the other hand, can detect the strength of or the distance from the magnet, based on its output. Because these are the most common digital Hall sensors, this project will only focus on them. By its name, Hall Effect sensors work on the basis of the "Hall Effect". Having carried out this experiment the law states that when current flows perpendicular to the direction that the current is flowing, a voltage can be measured at the angle at which the current flows. It will be possible for the hall sensor to detect magnets around it using this technique. I'm done with theory, let's move onto hardware.

Arduino Hall Effect Sensor Working:

You can now upload the Arduino code once you have created your code and hardware. A 9V battery has been used to power the entire setup. Power can be provided by any preferred source. You will see the LED light up as soon as you place the magnet close to the sensor and it will turn off as soon as you remove the magnet.

Note: Hall sensors are pole sensitive, which means that each side of the sensor can either detect only the North Pole or only the South Pole, only one side of the sensor can detect both poles. Thus, if you bring the north sensing surface close to the south pole, the LED will not glow.

Throughout the sensor, we bring the magnet near to it, causing it to change its state when we do that. An interrupt pin is triggered by this change, which then calls the toggle function, which changes the variable "state" from 0 to 1. In this way, the LED will illuminate. Once the magnet is moved away from the sensor, the sensor output will again change. We notice that this changes again by using an interrupt statement and thus the variable "state" is set to zero. The LED will darken if the switch is turned off. Each time you approach the sensor with a magnet, the same happens.

393. Automatic Call answering Machine using Arduino and GSM Module

Across the world, we are all reliant on mobile communications as our primary way to communicate. Although we have all encountered situations when we were unable to answer our calls, they were either important personal calls or business calls that changed our lives. Due to the fact that you could not answer the call at that time, you might have missed that opportunity. A number of Arduino and GSM modules will be used in this project in order to create an Automatic Call Answering Machine to solve the problem. If you are switching to a new number or going on a long pilgrimage, remember this next time, or just record your voice stating your reason for absence while on vacation or enjoying a well-deserved break and Your recordings will be automatically played to all the people on this machine when you call them. Your business numbers can also be used for answering customer calls during off-hours. Sounds interesting right? So let us build it.

Materials Required:

- Arduino Uno
- GSM module - Flyscale SIM 900
- ISD 1820 Voice Module
- 12V adapter to power GSM module
- 9V battery to power Arduino
- Connecting wires

Before we actually proceed into the project, let us get familiar with the GSM module and ISD 1820 Voice Module

Fly Scale SIM900 GSM Module:

GSM modules are extremely useful to us in our project, especially when we want to access the system remotely. In such a module, phone calls could be placed and received, SMS messages could be sent and received, and GPRS Internet connections could be made, for example. An USB-to-RS232 adapter is included with GSM modules, which can be plugged directly into a computer or to a microcontroller using either the TX or RX pins. A microphone or a speaker can also be connected to other pins besides MIC+ and MIC-. Power for this module can be provided by a 12V adapter through a regular DC barrel jack. Upon inserting your SIM card into the module, the LED should automatically become active. During the next minute or two, you will see a red (or other color) LED flashing every 3 seconds. Therefore, a connection was established between the Module and the SIM card. Getting started is easy once you have connected the module to a phone or microcontroller.

ISD1820 Voice module:

With the ISD 1820 Voice module, you can easily integrate voice announcements into your Projects. A 10 second audio clip can be recorded in this module, and you can play it when needed. An example of the device is shown below. The module comes with a microphone and a speaker (8-ohm, 0.5W).

Powered by berg sticks on the left, the modules operate on +5V. Rec, PlayE, and PlayL are the three buttons on the bottom of the screen. Click the corresponding button. Recording your voice is as simple as pressing the Rec button. By selecting PlayE, you will be able to play it. You can play the voice by pressing and holding the PlayL button. On the left, we see pins that are used to interface with a microcontroller. ESP8266/Arduino pins, which can handle 3V-5V, can act as direct drivers. In our project, we control the PLAYE pin on the Arduino module with the D8 pin.

Circuit Diagram and Explanation:

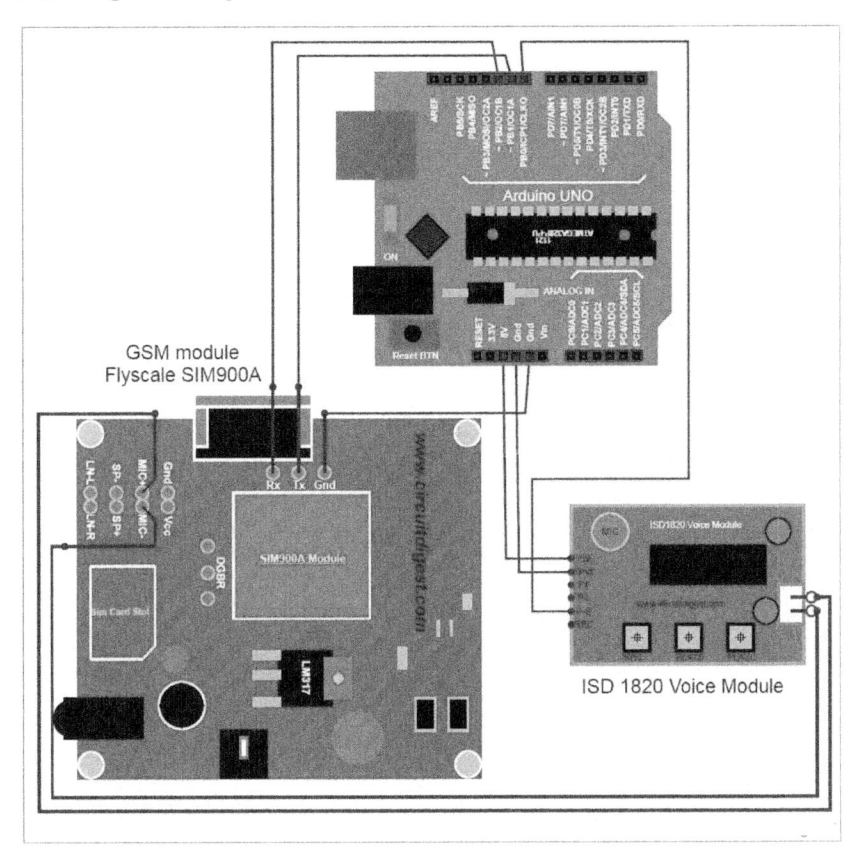

The previous circuit diagram describes in detail this automatic voice call answering machine project. There are a lot of simple connections to be seen. The GSM module is powered by a 12V 1A adapter, while the Arduino runs on a 5V battery. We can play back our voice recordings on the voice module whenever we press the rec button on the voice module and then push P-E. In this case, the microphone of the GSM module will be used to capture the audio. Unlike the GSM module, the voice module has a microphone pin that is connected to the speaker pin on the GSM module.

In order for the Arduino and the GSM module to communicate, you must use a serial connection. A chain is connected between the Arduino's X and Y pins. In this way, Arduino will have the ability to communicate with the GSM module. The Arduino requests that the GSM module answer a call when it is received by the GSM module. The Arduino demonstrates that the call is active by turning pin 8 high for 200ms (connected to pins P-E on the voice module).

394. Smart Blind Stick using Arduino

Has Hugh Herr ever caught your attention? His disability has not held him back; he is a staunch believer in the ability of technology to provide the same opportunities for disabled people as the able-bodied. Herr said in a TED talk that human do not have disabilities. There is no such thing as a broken person. We have a broken and disabled built environment and technology. Through technological innovation, we the people do not have to accept our limitations and can transfer them. In truth, he was living his life by these words, today he wears prosthetics to walk and claims a normal existence. Therefore, technology does indeed have the ability to neutralize human disability. With this in mind, let's build a blind man's stick that can do more than just assist the visually impaired. Its ultrasonic sensor will detect the distance from an obstacle, its LDR will detect lighting conditions, and its RF remote will allow the blind man to find his stick remotely. Using a Buzzer, all feedback will be given to the blind man. The buzzer can certainly be replaced with a vibrating motor and you can advance a lot further using creativity.

Circuit Diagram:

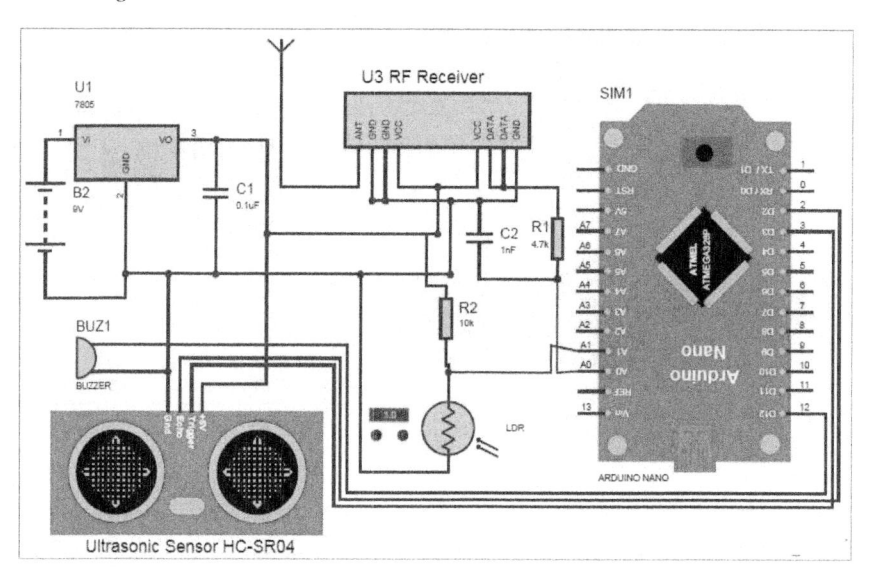

Materials Required:

- Arduino Nano (Any version will work)
- Ultrasonic Sensor HC-SR04
- LDR
- Buzzer and LED
- 7805
- 433MHz RF transmitter and receiver
- Resistors
- Capacitors
- Push button
- Perf board
- Soldering Kit
- 9V batteries

Two separate circuits are required for this Arduino Smart Blind Stick Project. It consists of the main circuit, which is mounted on the stick by the blind man. There will also be a small relay for locating the main circuit, which is an RF transmitter. As seen in the following circuit diagram, the main board consists of: As can be seen, all the sensors are controlled by an Arduino Nano. With a 7805-voltage regulator, a 9V battery provides power to the entire board. Powered by 5 volts, the Ultrasonic sensor's trigger and echo pins are connected to the Arduino nano's pins 3 and 2 as shown in above diagram. LDRs are connected to resistors of 10K, creating a potential divider, whose differential voltages are read using Arduino ADC pin A1. Reading the RF receiver signal from A0 is done using the ADC pin. Pin 12 of the board is connected to the buzzer, which provides the board's output.

The goal of this project was to develop an ultrasonic blind walking stick that utilized Arduino. There are 30 million permanently blind people and 285 billion people with some form of vision impairment according to WHO. Below is a circuit diagram of the RF remote. The workings of this system are also discussed.

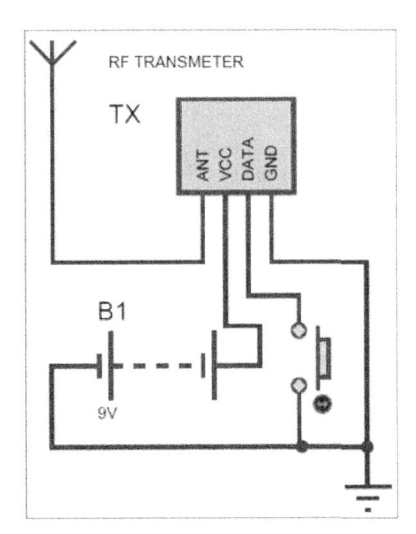

Figure: RF transmitter Circuit

As soon as you notice them, you will realize that they are unable to walk without the assistance of someone else. They require assistance to get to their destination. As a result, they have more challenges to face in everyday life. Blind sticks allow people to walk with greater confidence. In this stick, the object in front of the person is detected and a response is given either through vibrations or commands. This allows the individual to walk fearlessly. We can help them overcome their difficulties with this device. I have modified a remote-control circuit to make it function using a small hack. It usually takes two MCUs or an encoder and decoder to use this 433 MHz module. We rely on the receiver and transmitter to detect if any signals are being transmitted in our application. Thus, ground or Vcc is connected to the Data pin of the transmitter.

An RC filter is used to pass the data pin from the receiver to the Arduino. The example below illustrates this. Now, the Receiver continually outputs the same ADC value whenever the button is pressed. It is impossible to observe this repetition without pressing the button. To detect whether the button has been pressed, we write an Arduino program that checks for repeated values. Therefore, that is how Blind people can track their sticks. You can learn about how RF transmitters and receivers work by visiting this page. All connections were soldered using a perf board so that it would be intact with the stick. A breadboard can also be used to make them.

395. Arduino Metal Detector

Metal Detectors are used in various places to detect metals that are dangerous, such as airports, shopping malls, cinemas, etc. We have previously made a very simple Metal detector without a microcontroller; now we incorporate an Arduino to make the Metal Detector. The coil and capacitor that will be used in this project will be responsible for detecting metals. We have built this metal detector project using an Arduino Nano. All electronics lovers should find this project very interesting. A very loud buzzer will start beeping the moment the metal detector detects it.

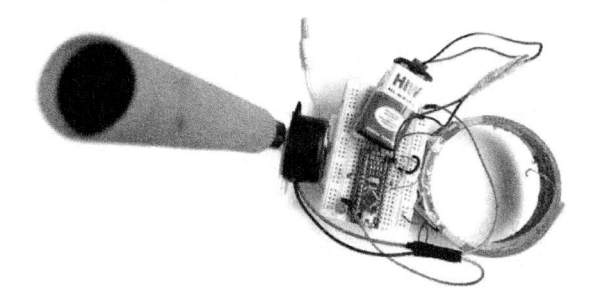

Required Components:

- Arduino (any)
- Coil
- 10nF capacitor
- Buzzer
- The 1k resistor
- 330-ohm resistor
- LED
- 1N4148 diode
- Breadboard or PCB
- Connecting jumper wire
- 9v Battery

Working Concept:

Any time there is current passing through the coil, a magnetic field is generated around it. Magnetic fields generate electric fields when they are changed. The Faraday Law states that due to the Electric field, a voltage is generated across the coil that opposes the change in magnetic field. The result is an increase in current, causing the voltage to oppose the change in magnetic field. The Inductance value is measured in Henrys using the following formula

$L = (\mu_o * N^2 * A) / l$

Were,

L- Inductance in Henries

μo- Permeability, its $4\pi*10^{-7}$ for Air

N- Number of turns

A- Inner Core Area (πr^2) in m2

l- Length of the Coil in meters

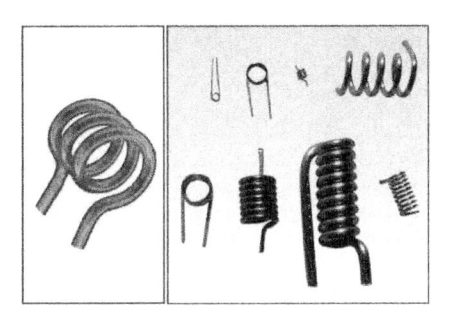

Any metal that comes into contact with the coil causes the coil to change its inductance. The type of metal determines the change in inductance. Non-magnetic metals have a smaller magnetic moment, while iron has a greater magnetic moment.

Inductance value changes drastically depending on the core of the coil. As you can see in the figure below, the inductors have an air-cored core, so there is no solid core in these inductors. This is basically just a lot of coils lying around. No matter what the inductor generates, its magnetic field flows in nothing or air. Inductors with low inductances are used in these applications.

When the values of a few microHenry are required, these inductors are used. If your value is greater than a few milliHenry, then this is not the right choice. An inductor with ferrite core can be seen in the following image. Despite the large inductance values of these Ferrite Core inductors. In this case, the coil wound within the inductor is air cored, so when a metal object is brought to the coil it serves as a core.

Circuit Diagram:

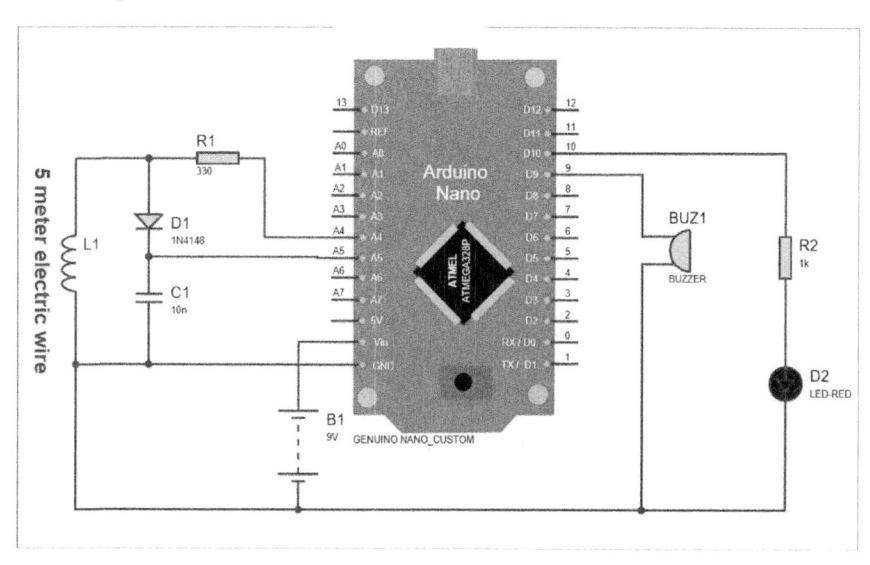

Inductance of the coil changes or increases greatly when this metal acts as a core. By adding a metal piece to the LC circuit, a significant amount of inductance is introduced, altering the overall impedance or reactants of the circuit. In this Arduino Metal Detector Project, we need to figure out how to detect metals by measuring the inductance of the coil. Thus, we did this by using the LR circuits (Resistor-Inductor Circuits) that we described previously. A coil with about 20 turns has been used here with a winding of around 10 cm in diameter. A tape roll has been used to wind up wire on, and the wire has been wound around it.

Metal Detectors are used to detect land mines, detect weapons such as knives and guns at airport security checkpoints, conduct geophysical prospecting, archaeology and treasure hunting. The whole Metal Detector Project has been controlled by an Arduino Nano. Indicators such as LEDs and buzzers are used for metal detection. For the detection of metals, coils and capacitors are used. Reduced voltage is also achieved by using a signal diode. Additionally, a resistor is used to limit the current flowing through the Arduino pin.

Working Explanation:

The metal detector on this Arduino is a little challenging to use. The high pass filter of the LR is fed the block wave or pulse generated by the Arduino. Therefore, each transition will be marked by short spikes caused by the coil. Having a coil with a high inductance result in pulses with a shorter pulse length. We can measure the inductance of a coil with the help of these shock pulses. Due to the very short duration of the spikes (approx. 0.5 microseconds), it is very difficult to successfully measure inductance with that.

The capacitor we used instead of this, is charged by rising impulses or spikes. The capacitor was charged with a few pulses so that A5 on the Arduino can read its voltage. This capacitor's voltage is read using ADC by Arduino. Cap Pin was made the output pin as soon as the voltage was read and set to low, quickly discharging the capacitor.

About 200 microseconds are required to complete this process. The measurement was repeated and the resulting average was taken to produce the best results. Using that method, we can calculate Coil's approximate inductance. Using the result, we transfer the data to LED and buzzers to see whether metal is present. The complete code that follows this article will help you understand how the program works. At the end of this project, you will find the full Arduino code. We will use two Arduino pins in this project, one to generate block waves that will be fed into the coil, and the second to read voltage from the capacitor. We have also connected LEDs and buzzers to two other Arduino pins besides these two.

396. Arduino Based Fire Fighting Robot

It is estimated that between 2010 and 2014 more than 1.2 lakh deaths were caused by fire accidents in India, according to the National Crime Records Bureau (NCRB). While there are many precautionary measures taken to prevent Fire accidents, they do sometimes occur as a result of natural disasters or human error. Human resources will be used to extinguish a fire in the event of a fire breakout so that people can be rescued. It is very much possible that humans can be replaced with robots for fighting fires thanks to the advancements in technology, especially robotics. Firefighters would be more effective, and they would also avoid endangering people on the job. The goal of this tutorial is to create an Arduino Fire Fighting Robot, which will detect fire and start the water pump automatically

This project will teach you how to make a simple robot that can pump out water around a fire and move towards it to put it out. Once you understand the following robotic basics, it will be easier for you to build more complex robots. Okay, let's start at the beginning.

Material Required:

- Arduino UNO
- Fire sensor or Flame sensor (3 Nos)
- Servo Motor (SG90), L293D motor Driver module
- Mini DC Submersible Pump, Small Breadboard
- Robot chassis with motors (2) and wheels (2) (any type)
- A small can, Connecting wires

Working Concept of Fire Fighting Robot:

The Arduino is the main processor of the project. For detecting fire, we are using the Fire sensor module (flame sensor) as shown below. It's possible to see that these sensors are equipped with an IR receiver (Photodiode), which is used to detect fire. This is impossible, how is it possible? An infrared receiver on the sensor module will be able to pick up this light from the fire when it burns. Using an Op-Amp, we then monitor the voltage change across the IR Receiver, so if there is a fire, our output pin (DO) will be 0V (low) and if there is none, our output pin will be 5V (high). To determine in which direction the fire is burning, we use three sensors located in three directions on the robot. By using the L293D module, we can drive our motors close to the fire once we know the direction in which the fire is coming from. We must use water to put out a fire when near one. The water is carried in a small container, a 5V pump is also tucked inside, and the entire device sits atop a servo motor so that the water can be sprayed from any direction. Now that we have the connections, let's move on as a profession, firefighting is an important one, but it is very dangerous.

Robots are designed for that reason, so that they can find a fire before it rages out of control. The system could be used to reduce victims' injury risks by working with fire fighters. Firefighting Robots are compact and portable emergency response robots that assist firefighters, especially in highly dangerous environments where people cannot enter, in fighting high-rise fires

Circuit Diagram:

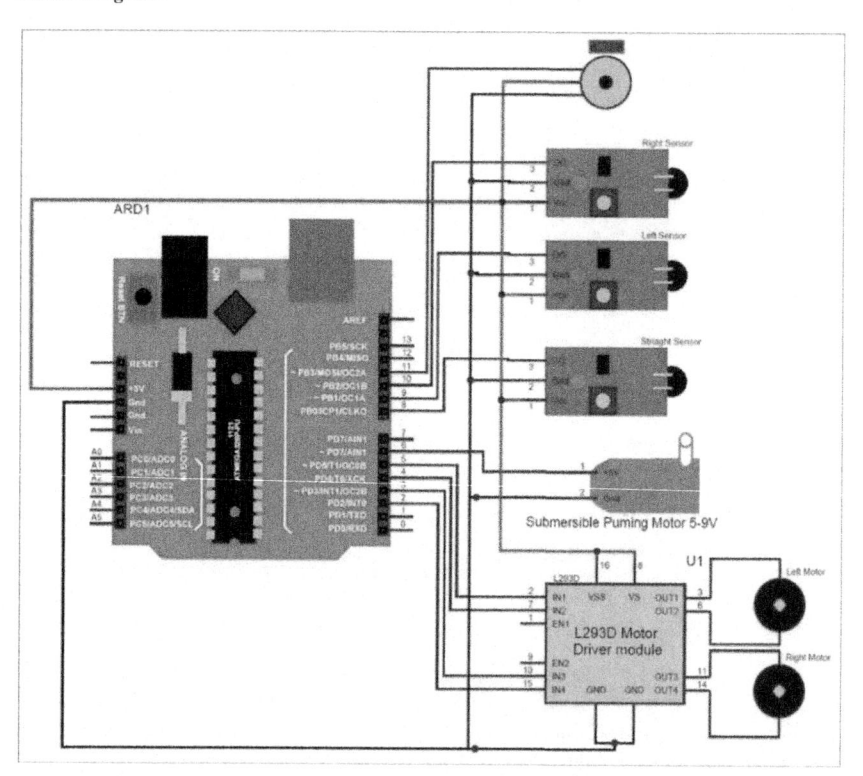

It is either possible to connect each connection to upload the bot to verify its functionality or you can assemble the bot completely. After that, connect the devices. It's very simple to make both connections and you should have no trouble doing it.

The container that I am using might not be the same for the robotic chassis that you are using. Set up the pumping system then according to your own creativity. It will, however, continue to be the same code. My pump is housed in a small aluminum can (cool drinks can) that I filled with water and set it in place. In order to control the direction of the water, I assembled the whole can on top of a servo motor. I built my robot in a way that looks like this. Servo fins were made by hot gluing the bottom of the container to the servo motor, and servo motors were attached to the chassis by nuts and bolts. The container just needs to be placed on top of the motor, then the pump inside it can be triggered to force water outside via the tube. Using the servo to rotate the whole container, the water can be directed in any direction.

397. Interfacing Joystick with Arduino

When we hear the word Joystick, we immediately think of the game controller. The same applies here, and you can use it for gaming as well. The technology is not only useful for gaming, but also in DIY electronic projects. X and Y planes of this joystick are controlled with separate potentiometers. Through the potentiometer, it can receive voltage and send it to Arduino as a number. The number changes as we move the joystick shaft (which actually is the potentiometer pointer).

Circuit Diagram

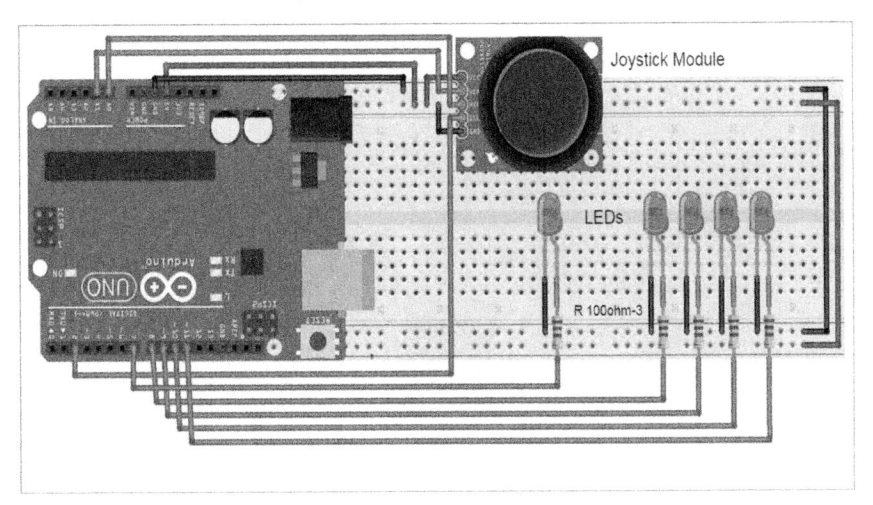

We simply control four LEDs via the Joystick in this Circuit by interfacing it with the Arduino. A motorized joystick is a device used to move the shaft of the joystick. 4 LEDs indicate the direction of motion of the shaft. Besides the joystick, it possesses a push button you can use to control other functions, or you can leave it untouched. The joystick also has a LED, which, when pressed, turns on as soon as the joystick button is pressed.

Material Required

- Arduino UNO
- Joystick Module
- LEDs-5
- Resistor: 100ohm-3
- Connecting wires
- Breadboard

Joystick Module

Joysticks come in a variety of shapes and sizes. In the following figure you can see a typical joystick module. The joystick module provides Analog Outputs, which change voltage with the direction the joystick is pointed. Through interpreting these voltage changes with a microcontroller, we can determine the direction of the movement. Joysticks have been interfaced with the AVR and Raspberry Pi in the past. See how there are two axes on this joystick module. The X-axis is horizontal, and the Y-axis is vertical. A potentiometer or potentiometer is mounted on each axis of JOYSTICK. As the pots are driven out, the midpoints are represented as Rx and Ry. These pots get their points based on Rx and Ry. Rx and Ry work together as a voltage divider when the Joystick is in standby.

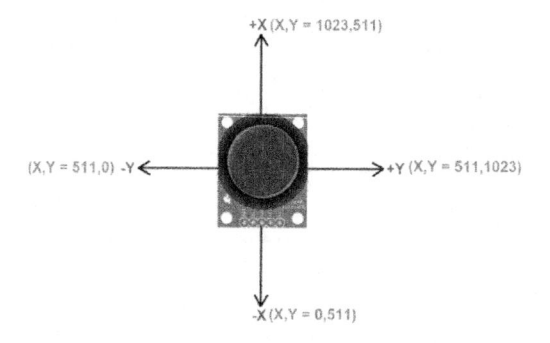

Whenever a joystick is moved horizontally, a change in voltage occurs at the Rx pin. A similar change in voltage occurs at the Ry pin when the pin is moved along the vertical axis. Therefore, we have four directions of joystick output connected to two ADCs. A voltage difference between the pins occurs when the stick is moved in either direction. Using this Joystick module, we are going to connect it to the Arduino UNO, which comes with an inbuilt ADC (Analog to Digital Converter) mechanism

Controlling LEDs using Joystick with Arduino

The Arduino code has been uploaded to it, and the components have been connected according to the circuit diagram. Now we can control the LEDs using the joystick. The four LEDs can be turned ON in each direction according to the Joystick shaft movement. One potentiometer is used for movement along the X axis, and another for movement along the Y axis, in the Joystick. The potentiometers are powered by 5 volts from the Arduino. We can see here how driving the joystick will change the voltage value and the analog values corresponding to Analog pins A0 and A1. We are reading the analog values of the X and Y axes from the Arduino and turning ON the LEDs in accordance with the joystick movement.

398. Arduino RFID Door Lock

The RFID door lock mechanism can be seen in many hotels and other places that don't require a key to unlock the doors. The key is provided to you, and by simply inserting it into the RFID Reader box, you will hear a Beep and see a blink of LEDs, unlocking the lock. Any door can be fitted with this RFID door lock, which can be easily made and installed at home. It is simply a door lock that operates when the door is turned on by a voltage (typically 12 volts).

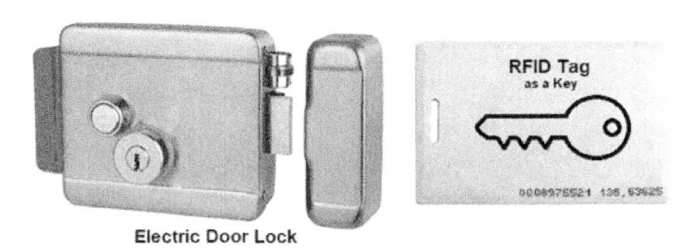

Electric Door Lock

The relay in this project is used to control the Electric Door Lock. RFID tags will be used as keys, so the Arduino and the relay will be used to trigger the lock. You will be alerted about the wrong card if you place it near the RFID reader. Start by reading about RFIDs working and how it can be interfaced with an Arduino.

Material Required:

- Arduino UNO
- EM-18 Reader Module with Tags
- Relay 5v, LED
- Buzzer, Connecting wire, Resistors

EM-18 RFID Reader:

A radio frequency identification system uses radio waves to identify objects. It is possible to read the RFID card number by using a RFID reader, which embeds a unique ID in the RFID card. The EM-18 RFID reader operates at 125 KHz and can be powered with a 5V power supply. While it comes with a built-in antenna, it also comes with an on-chip antenna. Besides Weigand output, it provides serial output as well. There is a range of approximately 8-12cm. Data and stop bits for serial communication are 8 bytes, 9600bps. There are many applications for wireless RF identification, for example

RFID Based Attendance System,

1. Security systems,
2. Voting machines,
3. E-toll road pricing

RFID Reader Module   RFID Tags

In ASCII format, EM-18 RFID readers provide 12-digit output. In a card number, the first 10 digits are the number of the card; the final two are the result of XORing the number of the card. To check for errors, two digits are added to the end.

Working of Arduino Based RFID Door Lock

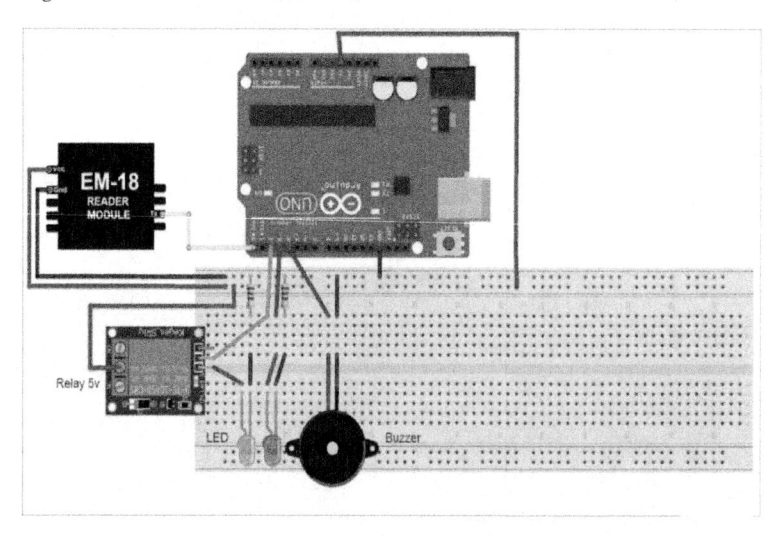

RFID systems are composed of two components: RFID tags and RFID readers. Integrated circuits are used to store data, and antennas are used to transmit the data to RFID readers. RFID tags consist of an integrated circuit and an antenna. RFID tags are powered by RF signals whenever they are in range of RFID readers. The tags transmit data serially when they are powered by RF signals. Afterwards, the RFID reader reads the data and transmits it to the Arduino microcontroller. Following that, different tasks are performed in accordance with the microcontroller's code. The value of the RFID tag has already been saved in the code of our circuit.

As a result, the relay gets activated when that tag gets within the range. To demonstrate the power of a relay, a LED has been connected. However, you can replace the LED with an Electric Door Lock to guarantee the lock will open whenever a relay is activated. If another RFID card is scanned, the buzzer will start beeping because it's the wrong RFID tag. For this reason, the door lock system relies on the fact that an RFID tag is required for it to open. It is possible to adjust the delay in codes for when the relay itself gets deactivated after 5 seconds and when the door is closed after 5 seconds.

399. An introduction to Brushless DC Motors (BLDC) and how to control them on an Arduino

We have always enjoyed building things and getting them to work as we wanted. It would definitely be an anxiety pump for hobbyists and hardware tinkerers to build something that could fly. Absolutely! Among the various aircraft I refer to are gliders, helicopters, planes, and primarily multi-copter aircraft. Due to the community support available online today, it has become very easy to build one on your own. The BLDC motor is a feature common to all things that fly, but what is it? What is the purpose of it in order to fly? How does it differ from other software? Is there a way you can interface your motor with your controller and buy the right motor? What are ESCs and why do we need them? These are just a few questions you can get answered in this tutorial. The main purpose of this tutorial is to control the speed of a 2112/13T sensorless BLDC outrunner motor (commonly found in drones), using an ESC (Electronic Speed Controller).

Materials Required

- A2212/13T BLDC Motor
- ESC (20A)
- Power Source (12V 20A)
- Arduino
- Potentiometer
- Understanding BLDC Motors

BLDC motors operate smoothly, which makes them common in ceiling fans and electric vehicles. BLDC motors, on the other hand, are equipped with three wires, thus forming three phases. Hold on... what!!??

BLDC motors still operate by using pulsed waves, though they are classified as DC motors. The DC voltage from the battery is converted into pulses by the electronic speed controller (ESC), and the motor receives the pulses via its three wires. Current can only enter and leave the motor through two phases at a given time, so that one phase powers the motor and the other phase delivers power. When the motor is in this position, the coil inside is energized, and therefore the magnet on the rotor aligns itself with the energized coil. A motor is rotated by energizing the next two wires and then turning them by the ESC. In order to function, the coils need to be energized according to their order of energization and their speed depends on its speed. The remainder of this article will discuss ESC in greater detail.

Why do Drones and other multi-copters use BLDC Motors?

Drones come in many forms, from quadcopters to helicopters and gliders, all of which have one thing in common: hardware. What are BLDC motors? Why do they exist? Compared to DC Motors, what is the purpose of using BLDC motors, which are a bit more expensive?

One important reason for this is the very high torque generated by these motors, which is important for gaining thrust or losing thrust rapidly in order to launch or land a drone. As well as having these motors as out runners, they are offered as out runners as well, enhancing their thrust. Our drone will stay steady in mid-air due to the smooth vibration less operation of the BLDC motor. BLDC motors have a powerful to light weight ratio. In order for drones to perform well, they need motors that are both powerful (high speed and torque) and lightweight. In order to match the performance and torque of a BLDC motor with a DC motor, the motor would have to be twice as heavy.

Arduino BLDC Motor Control Circuit Diagram

BLDC motors can be easily connected to Arduinos via a straightforward interface. A minimum 12V and 5A source of power is required for the ESC. My RPS has been used in this tutorial, but you can also use a Li-Po battery because it's just as powerful. Connect the three phase wires of the ESC to the three phase wires of the motors, there is no specific order in which the wires should be connected, you can connect them in any order.

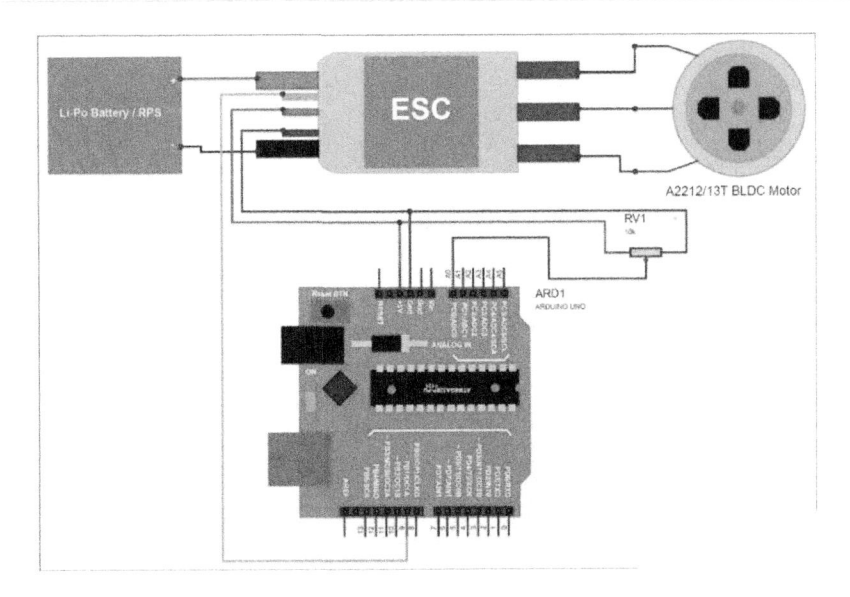

Controlling BLDC Motor with Arduino

The circuit diagram should be followed and the code uploaded to the Arduino. The ESC should be powered up. Make sure you mount the BLDC motors securely, or they will jump around when they are rotating. In the event that the throttle signal is not within the threshold limits within a few seconds after you start your setup, your ESC will make a welcoming tone and keep beeping. The beeping tone will stop when the POT from zero is gradually increased. The motor will start spinning more slowly as you increase the PWM signal beyond the lower threshold value. Once the voltage reaches the upper threshold limit, the motor will stop due to a lack of power. The more voltage you provide, the more speed you will see. Once you have completed the process, you can repeat it.

400. Automatic Medicine Reminder Using Arduino

It is always our intention to keep our dear ones in good shape when it comes to health. The question is, what will happen when people get ill and forget to take their medications. Right? We'd be concerned, wouldn't we? Hospital patients need to be reminded to take their medicine on time, but it's difficult to do so because there are so many of them. To remind people to take their medicine on time, the traditional methods require human effort. Machines can perform that task in the digital era and we aren't bound by those rules. Besides doctors in hospitals and patients at home, Smart Medicine Reminder has many other uses.

When it comes to reminding, there can be many ways to remind it:

- Show it on a display
- Send notification on email or Phone
- Using mobile apps

- Buzz alarm
- Using Bluetooth/ Wi-Fi
- Get a call
- Remind for next medicine time while reminding current time

Arduino Medicine Reminder Circuit Diagram and Connections

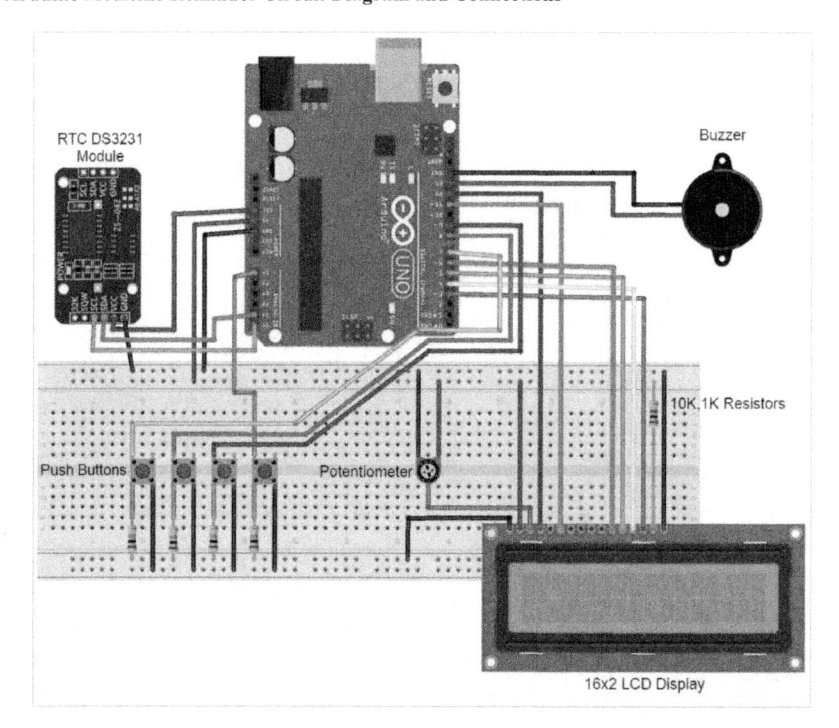

Depending on the circumstances, we can combine different methods. Our Medicine Reminder has a simple Arduino circuit that reminds us to take our medicines 1 or 2 or 3 times a day. Push buttons allow you to select a time slot. Additionally, it displays the current date and time. This article will be further extended into an IoT project, where a notification will be sent by email or SMS to users. Patient Monitoring Systems can also be used to send medication reminders. Remembering to take medications may seem like an unnecessary task, but we cannot emphasize its importance enough. Most seniors are prescribed medication to manage illness due to increased visits to the doctor and hospital. Many of the patients find it difficult to remember to take their medication and not just because of their hectic schedules. Some patients have trouble adhering to their medication regimen because of memory loss.

Our loved one's health and quality of life depend on making sure they follow their doctor's orders for prescribed medications. One small change can make all the difference in the effectiveness of a health care plan.

We value patient's health and ensure he or she takes their medication properly. To facilitate this, this project and caregivers are standing by to encourage your loved one to take their medications on time.

Components Required

- Arduino Uno (We can use other Arduino boards also, like Promini, Nano)
- RTC DS3231 module
- 16x2 LCD Display
- Buzzer
- LED (any color)
- Breadboard
- Push Buttons
- 10K Potentiometer
- 10K, 1K Resistors
- Jumper Wire

Working of Automatic Medicine Reminder System

In order to power the Pill Reminder Alarm, 5V is required. When Circuit Digest is first launched, it displays the welcome message "Welcome to Circuit Digest". Three screens are displayed on the LCD screen at a time. As soon as the screen loads, a message appears saying "Stay Healthy, Get Well Soon". On the next screen, a help screen is displayed that instructs you how to choose a time-slot to remember (once, twice, or three times per day). A time slot can be configured in the program in accordance with the user's preference. The duration has now been reduced to three, which are 8am, 2pm, and 8pm.

There are three modes of dividing up the time slots. When the user presses the first push button, the user is instructed to take medicine once a day at 8am. When the user presses the second push button in mode 2, the system selects to take medicine twice daily at 8am and 8pm. In Mode 3, when the third push button is pressed, the user will take their medication three times daily at 8am, 2pm and 8pm.

Additionally, the buzzer can be snoozed for a period of ten minutes (not included in this project). Push buttons allow the user to choose desired slots, and a RTC is used to determine the current time. The buzzer starts to buzz when the time matches the selected time slot. Users have the option of stopping the buzzer by pressing the STOP button. Similarly, the next reminder is sent by the same method.

Summary

It is assumed that you have already tried out more than 100 Arduino Project Ideas and found them to be helpful in constructing your ideal project, and that this chapter will therefore be of great interest to you. The next sessions will conclude with the top 100 NodeMCU ESP8266 and ESP32 project ideas.

CHAPTER 4: TOP 100 NODEMCU-ESP8266, ESP32 PROJECT IDEAS

Introduction

The ESP8266 is a microcontroller manufactured by Espressif Systems. It is a self-contained WiFi networking solution that can act as a bridge between an existing microcontroller and WiFi, and it is also able to execute programs that do not require any other external components. It can monitor and control objects from anywhere in the world for less than $3, making it ideal for any Internet of Things project. The ESP8266 is a chip that has a high level of integration and was designed to meet the requirements of the new connected world. It has a complete and self-contained Wi-Fi networking solution, so it can either host the application or take over all Wi-Fi networking tasks from another application processor.

https://espressif.com/en/products/esp8266/.

http://mcuoneclipse.com/2014/10/15/cheap-and-simpl...

https://scargill.wordpress.com/?s=esp826

MQTT is a way for machines to talk to each other over the "Internet of Things" (M2M). It was made to be a very small publish/subscribe messaging transport. It can be used to connect to remote locations when a small amount of code is needed and/or when network bandwidth is limited. It has been used, for example, in sensors that talk to a broker over a satellite link, in healthcare providers' occasional dial-up connections, and in a variety of home automation and small device situations.

http://mqtt.org/

http://en.wikipedia.org/wiki/MQTT

Contents:

- Pinout and description
- Power Requirement
- Various Peripherals and I/O
- On-Board buttons and LED
- Development Platforms
- Applications of ESP8266

1. Pinout and description

The NodeMCU ESP8266 has 30 total pins, 17 of which are GPIO pins. GPIO is an abbreviation for General Purpose Input Output. There are 9 digital pins, numbered D0 through D8, and only one analog pin, A0, which is a 10-bit ADC. The D0 pin is not capable of performing any additional functions and can only be used to read or write data. When the EN pin is pulled HIGH, the ESP8266 chip is given permission to operate. The chip uses the least amount of power when it is pulled LOW. The CP2102 serves as the USB to TTL converter for the board, while the board itself features a 2.4 GHz antenna that extends the network's range. The development board is outfitted with an ESP-12E module that contains an ESP8266 chip. This chip has a Tensilica Xtensa® 32-bit LX106 RISC microprocessor and can have its clock frequency adjusted anywhere from 80 to 160 MHz. It also supports real-time operating systems.

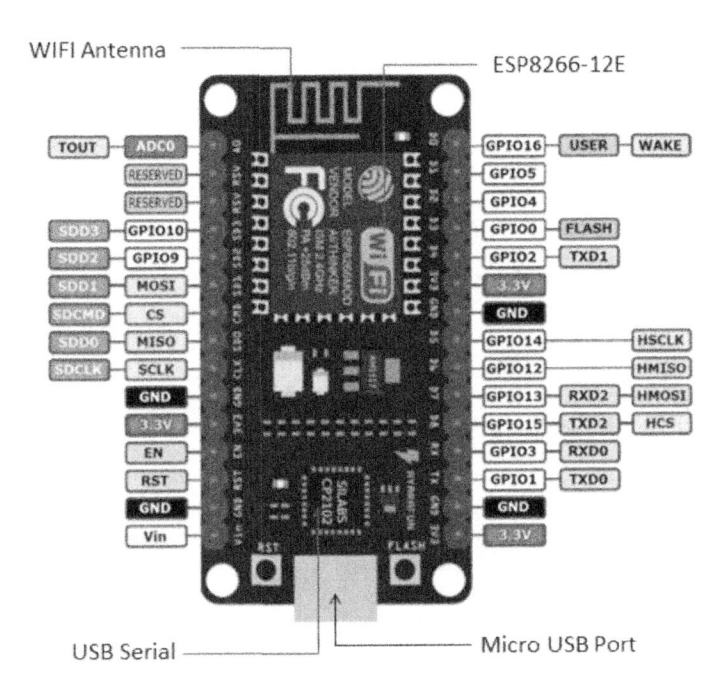

There's also 128 KB of RAM and 4MB of Flash memory (for application and data storage), which is more than adequate to handle the long strings that make up web pages, JSON/XML data, and everything else we throw at IoT devices these days. Because the ESP8266 contains an integrated 802.11b/g/n HT40 Wi-Fi transceiver, it is not only able to connect to a WiFi network and communicate with the Internet, but it can also set up a network of its own, enabling other devices to connect directly to it. This makes it possible for the ESP8266 to interact with other devices. Because of this, the ESP8266 NodeMCU has an even wider range of applications.

2. Power Requirement

Since the ESP8266 can work with voltages between 3V and 3.6V, the board comes with an LDO (low dropout) voltage regulator to keep the voltage at 3.3V. It is able to give up to 600mA in a dependable manner. It also has four GND pins in addition to its three 3v3 pins. The onboard Micro-B USB port serves as the source of the power supply. If you want, you can utilize the VIN pin to directly give power to the ESP8266 if you have access to a voltage source that is regulated at 5V. In addition to this, it calls for an Operating Current of 80mA and a Sleep Mode Current of 20 A.

3. Various Peripherals and I/O

The ESP8266 can communicate using the UART, I2C, and SPI protocols. It also has 4 PWM channels that can be used to control motor speed, LED brightness, etc. In addition, the UART protocol makes use of two separate channels. The Analog-to-Digital Converter, often known as A0, is capable of controlling any analog device. The SPI protocol makes use of a pin known as the CMD for its Chip select functionality.

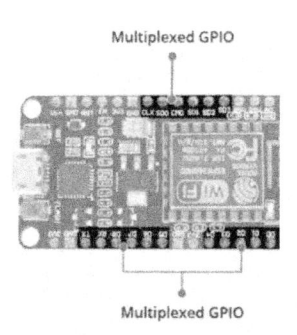

4. On-Board buttons and LED

The ESP8266 features two onboard buttons as well as an onboard LED that links to the D0 PIN. They are called FLASH and RST.

- **FLASH pin** – It is to download new programs to the board
- **RST pin** – It is to reset the ESP8266 chip

5. Development Platforms

The Arduino IDE and the ESPlorer IDE are two of the most prominent platforms. Other development platforms include Espruino, which is a JavaScript SDK and firmware that closely emulates Node.js, and Mongoose OS, which is an operating system for Internet of Things devices. Both of these platforms can be fitted to program the ESP8266.

6. Applications of ESP8266

The NodeMCU ESP8266 is essentially a WIFI module combined with a Microcontroller, making it an extremely helpful gadget in the field of IoT. The best illustration of this is seen in the fact that it has **17 GPIO** pins. The ESP8266 finds its primary application in the home automation industry, which is seeing significant growth at the moment. This is as a result of the little amount of power that it consumes while in Sleep mode. In addition to this, some further instances of where the ESP8266 could be used are as follows:

- Developing a web server with ESP8266
- Taking control of DHT11 with the help of NodeMCU
- ESP8266-based weather station that makes use of BMP280

- OTA programming
- ESP8266 NTP server for retrieving the current time.

How are ESP32 and ESP8266 different from each other?

	ESP8266	ESP32
MCU	Xtensa Single-core 32-bit L106	Xtensa Dual-Core 32-bit LX6 with 600 DMIPS
802.11 b/g/n Wi-Fi	HT20	HT40
Bluetooth	X	Bluetooth 4.2 and BLE
Typical Frequency	80 MHz	160 MHz
SRAM	X	✓
Flash	X	✓
GPIO	17	34
Hardware /Software PWM	None / 8 channels	None / 16 channels
SPI/I2C/I2S/UART	2/1/2/2	4/2/2/2
ADC	10-bit	12-bit
CAN	X	✓
Ethernet MAC Interface	X	✓
Touch Sensor	X	✓
Temperature Sensor	X	✓(old versions)
Hall effect sensor	X	✓
Working Temperature	-40°C to 125°C	-40°C to 125°C
Price	$ ($3 - $6)	$$ ($6 - $12)
Where to buy	Best ESP8266 Wi-Fi Development Boards	ESP32 Development Boards Review and Comparison

Should your projects use the ESP32 or the ESP8266? In this lesson, we will evaluate both the ESP32 and the ESP8266, discussing the benefits and drawbacks associated with each board. Both the ESP32 and ESP8266 are inexpensive Wi-Fi modules that are excellent choices for Do It Yourself (DIY) projects involving the Internet of Things (IoT) and Home Automation. Both chips contain a processor that is 32 bits in size. In comparison, the ESP8266 is a single-core processor that operates at 80MHz, while the ESP32 is a dual-core CPU that ranges from 160MHz to 240MHz. These modules come equipped with general-purpose input/output (GPIO) pins that support a variety of communication protocols, including SPI, I2C, UART, ADC, DAC, and PWM. The fact that these boards come with built-in support for wireless networking is the most notable feature that differentiates them from similar microcontrollers, such as the Arduino. This translates to the fact that you can easily control and monitor devices remotely via Wi-Fi or Bluetooth (in the case of ESP32), and that you can do so at a very low cost.

Alternately, if you don't need to use its wireless capabilities, you can control inputs and outputs with the ESP32/ESP8266 in the same way that you would with an Arduino if you don't need to use its wireless capabilities. Nevertheless, you need to keep in mind that the logic voltage used by the ESP32 and ESP8266 is 3.3V, whereas the logic voltage used by the Arduino is 5V.

Following in the footsteps of the ESP8266 is the ESP32. It supports Bluetooth 4.2 as well as Bluetooth low energy and comes with an additional CPU core, faster Wi-Fi, and additional GPIOs. Additionally, the ESP32 comes with touch-sensitive pins that can be used to wake up the ESP32 from deep sleep. It also comes with a built-in hall effect sensor and a built-in temperature sensor (although more recent versions of the ESP32 do not come with a built-in temperature sensor anymore).

Although the ESP32 costs a little bit more, it is still a very affordable option. The ESP32 can cost anywhere between $6 and $12, whereas the ESP8266 can cost between $4 and $6 (although the exact price you pay will depend on where you buy it and which model you select).

It is not simple or practical to use bare ESP32 or ESP8266 chips, especially when testing or prototyping, because of their complexity. The ESP32 and ESP8266 development boards are the ones you should look to use the majority of the time. These boards come equipped with all of the necessary circuitry to power the chip, connect it to your computer, include a circuit to make it simple to upload code, pins to connect to peripherals, built-in power and control LEDs, as well as other helpful features.

What is Thingspeak?

ThingSpeak is an open-source application and application programming interface (API) that enables users to store and retrieve data from "Internet of Things" devices by utilizing HTTP either over the Internet or a local area network (LAN). With ThingSpeak, you can make apps that log data from sensors, apps that track where things are, and a social network of things that sends out status updates. https://thingspeak.com/

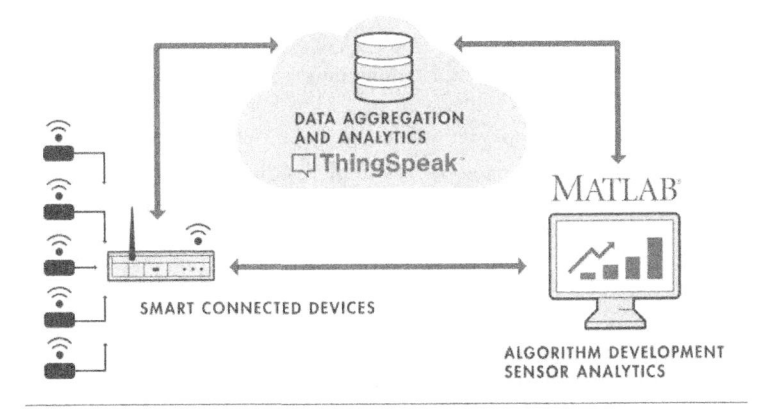

ThingSpeak Key Features

ThingSpeak enables the collection, visualization, and analysis of live data streams in the cloud. ThingSpeak's primary capabilities include the ability to:

1. You can easily set up devices to use popular IoT protocols to send data to ThingSpeak.
2. Real-time visualization of your sensor data.
3. Gather data from third-party sources on demand.
4. Make sense of your Internet of Things data by using MATLAB's analytical power.
5. Automatically execute your IoT analytics based on schedules or events.
6. Build prototypes of IoT systems without having to set up servers or write web software.
7. Automate the processing of your data and the communication with the help of third-party services such as Twilio and Twitter.

100 ESP8266, ESP32 PROJECTS

401. ESP 8266 Wifi Controlled Home Automation

ESP8266 is a great way to get started with WiFi and Internet of Things. It's also inexpensive and can be used to make cool projects that connect to the Internet. Learn how to use it to create a simple IOT project. The ESP8266 WiFi Module is a self-contained SOC that comes with an integrated TCP/IP protocol stack.

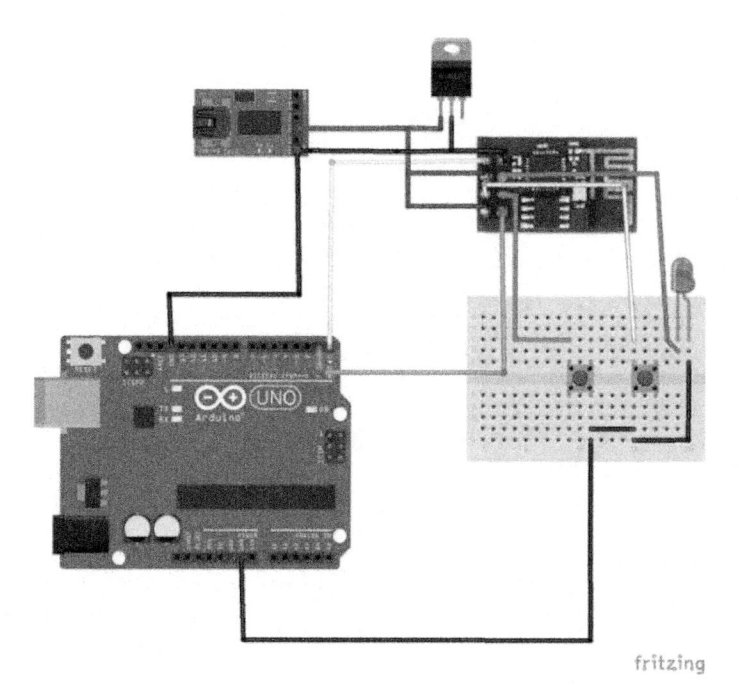

Required Components

- Esp 8266
- Perfboard
- Ams1117
- FTDI
- Arduino UNO
- Headers
- Breadboard

It can provide access to your WiFi network for any microcontroller you choose to use. The ESP8266 can either run an application on its own or take over all Wi-Fi networking tasks from another application processor. Each ESP8266 module comes pre-programmed with an AT command set firmware. This means that you can just connect it to your Arduino device and get about the same WiFi capabilities as a WiFi Shield (and that's right out of the box)! The ESP8266 module is a board that is very cheap and has a very large and growing community.

402. World Wide Web Control via ESP8266

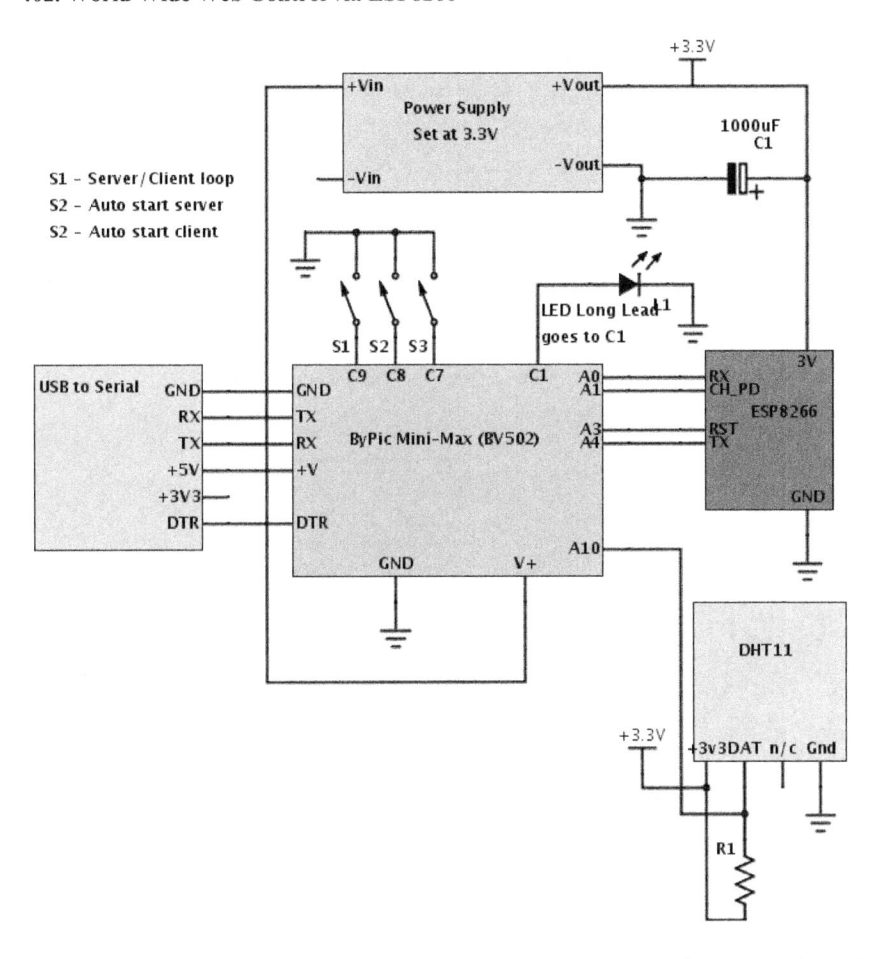

Any browser on any device (tablet, PC, phone) can be used to control things and get information, and low-power mode can be used to send information from faraway places, such as to check on water levels, wildlife, etc. The project has a temperature and humidity sensor that can be read and an LED that can be turned on and off to show what can be done. It's not meant to be a finished product, but rather an example of what can be done. It is hoped that the reader will be inspired to make their own project for something more specific. The circuit is built on a solderless breadboard, and all of the power for the circuit comes from the USB to serial adapter. The separate power supply, which MUST be set to 3.3V, gets its power from the USB, which goes into the +V pin of the BV502. The ESP8266 gets its power from this separate supply. Even though the Mini-Max (BV502) has a 3.3V regulator built in, it can't give the ESP8266 enough power (it can but only just). Experiments have shown that there MUST be a large capacitor (about 1000uF) across the power supply to the ESP8266. Without it, the results are not reliable. The port and pin names are written on the Mini-Max circuit board. For example, C0 is port C pin 0. If the pins on the Mini Max are soldered facing up, you can build the circuit without a breadboard. The stuff

There are three switches shown, but these are just wire jumpers on the breadboard and not real switches. The program's flow is controlled by these switches, which are read by the firmware. The free IDE can also be used to control the programme through the serial interface. For this project, you don't need to connect any of the switches. For now, leave lines C7 to C9 unconnected. If you are new to ByPic, you should first connect the BV502 and USB to serial, and then start the IDE to see how the system works. ByPic is an interactive Rapid Application Development system that works well with the Wi-Fi controller ESP8266.

403. Arduino + ESP8266: How to Make a DIY World Clock and Weather Bot

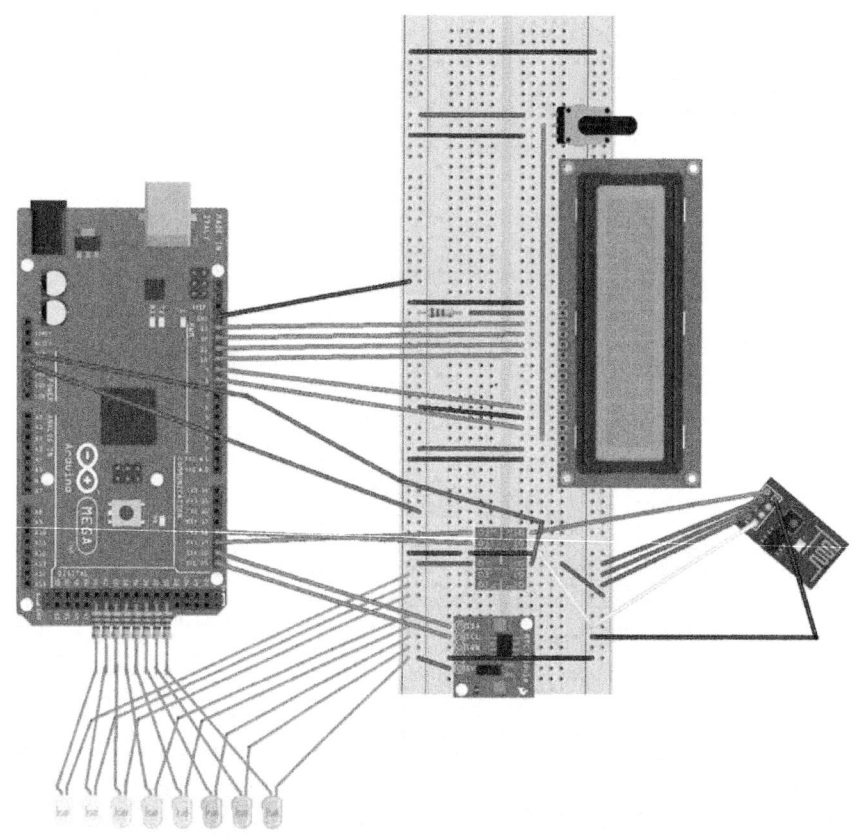

This project will teach you how to build a global clock weather bot with an Arduino Mega and an ESP8266 (AT mode). First, you need to cut holes for the LCD screen and the potentiometer. The LCD screen should show the mouth, and the potentiometer should show the nose, so just leave the right amount of space. On a piece of paper, you should also cut these holes. You should draw three raindrops on a piece of paper to represent the three blue LEDs, a sun to represent the yellow LED, a cloud to represent the white LED, and one or two lightning bolts to represent the other white LED. You can decorate the eyes and nose however you like, but don't forget to put green LEDs behind the eyes. If your box isn't clear, you'll need to cut holes behind each of your drawings to let the light from the LEDs shine through. When you are done, you can put the papers on a box by taping them on

Since LCD has many pins and it would take much space to describe how to connect LCD to Arduino, I will not describe it but you can reference to the picture above. Same goes for the LEDs. Since Arduino Mega runs on 5V and ESP8266 runs on 3.3V, you can't (or at least shouldn't) connect them directly. To lower the voltage for ESP8266, you should use Logic Level Converter (LLC), as in this project, or another method.

Components Required:

- Arduino Mega
- ESP8266
- Logic Level Converter
- 3 x blue LED
- 2 x green LED
- 2 x white LED
- yellow LED -
- DS3231 RTC (kinda optional but makes things much easier)
- 16X2 LCD screen (or bigger)
- wires, wires, wires

- 9 x 220 ohm resistor
- 10K ohm potentiometer
- a box (plastic transparent box in this case)
- white papers
- adhesive tape
- scalpel
- crayons
- Open Weather Map Api Key
- TimezoneDB Api Key

404. Emergency Button for 7$, Arduino, WIFI and ESP8266

There are many projects going on right now that aim to help older people stay more independent. Since my friend's grandmother is 95 years old and still lives alone in her apartment, I wanted to see what I could do to help her stay there as long as possible. One of the hardest things for old people to do when walking is to keep their balance. It's getting harder and harder to move around, and if they fall down, it's hard to get back up. My family calls my grandmother every night to see if something like this is going on. There are some emergency buttons that can help, but the call center usually costs a lot each month. Also, most older people don't like wearing grey plastic gadgets around their necks. Most apartments today have 10 or 20 WiFi connections that are easy to reach. So I decided to build a device that could connect to WiFi and make it easy to talk to the outside world. Since many older people don't have WiFi, I decided to ask the neighbors if I could use theirs. Most of them were happy to help and even asked to have their email address added to the list so they could help quickly in an emergency.

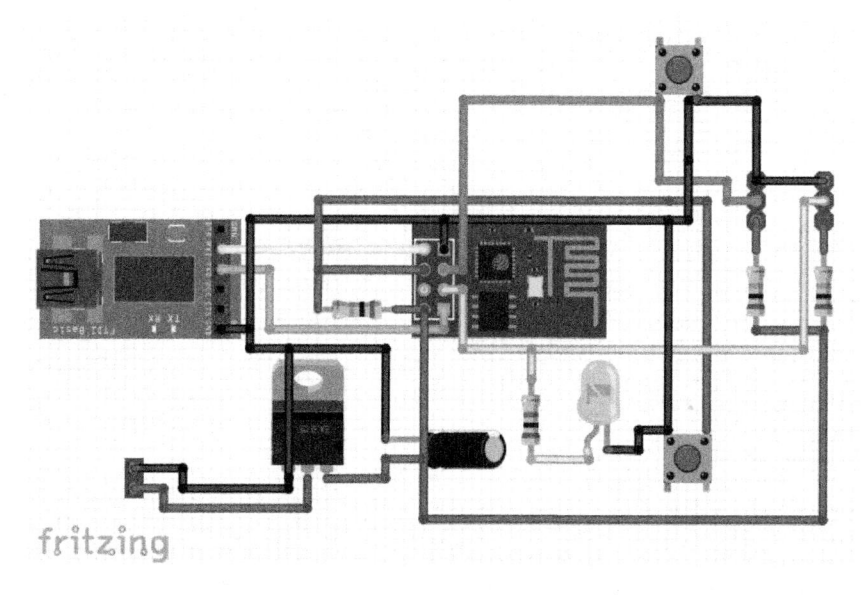

If someone is in trouble, all they have to do is push a simple button on the device. An LED lights up to show that the emergency is on. The device will connect to the internet box and send an email to an email address that has already been entered. As many people (like me) get email on their phones and check them 30 times a day, the emergency can be handled quickly by calling the person or knocking on the door. The device has to be simple to put on and work with. So, it was important to drastically cut down on the size and weight. So, the device is 2.5 x 1.5 x 1 cm (1 x 0.6 x 0.4 inches) in size, which makes it easy to add to bracelets, necklaces, or clothing. The button is big because some older people shake a lot, which makes it easier to press.

A very important question is also the price. The ESP8266 board is a great choice if you want to make a WiFi-connected device right now. It only costs a few dollars and has its own 32-bit microcontroller that can be changed (The Atmega328p on the Arduino UNO is only 8bits). This device is now used by a large group of people, and it recently became compatible with the Arduino IDE. Most people can afford to pay about $8 for the electronics, and you just have to add a battery of your choice. If you want to make your own design, it's easy to find and change any of the parts. A standard 110 mA battery should last more than 2 months, a CR2032 battery should last 5 months, and a 1000 mA battery should last 2 years. I used the cheapest and smallest parts I could find for this project. The ESP8266 is a great tool for any microcontroller/WiFi application today. Plus, it now works with Arduino! As I was adding code to the ESP8266 100 times a day, I decided to also make a programmer that was easy to use. If you don't want to change a lot of things in the code, you don't really need it.

Components Required:

- ESP8266, which costs $7, comes with WiFi and an antenna for WiFi.
- Button: $0.50. Pick the one that fits your model best. I used a "pushbutton 12*12*8cm."
- LED 0.35$
- The price of a resistor is $0.02 (I used a flat top 3mm bought on Ebay)
- Battery is $6.95 In a bigger device, you could use a CR2032 to make it last 2.3 times longer, or this bigger battery to make it last 9 times longer.
- Account SMTP2GO is free. It sends mail from the device to your mail account (the Free version is limited to 20 mails a day).
- FREE Email Address. With this code, I used a GMAIL one.
- You'll need a few wires and the following to programme your device:
- FTDI basic board and a 3.3V (minimum 500 mA) power supply or
- A chip can be taken out of the UNO-style Arduino board, but not the SMD version.
- In the second part of this Instructable, I show how to make a programmer that can be counted on. If you want to build it, you need to add the following parts:
- 3.3V regulator
- Male Header (angle or straight)
- Capacitor 10uF Female Header and a few resistors
- Connector for battery on proto board

405. DIY Arduino Wi-Fi Shield with ESP8266 for Home Automation That Can Be Controlled by Voice

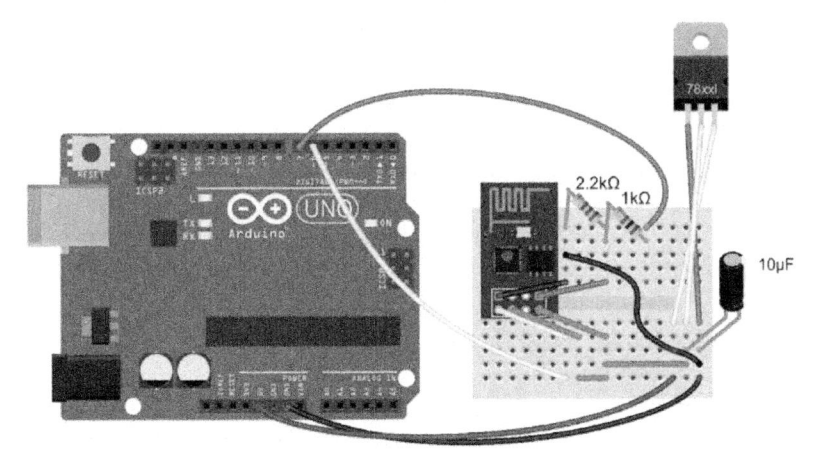

Figure: Arduino Uno Wi-Fi Shield Using ESP8266

People and machines can interact and talk to each other in many ways. Computers have a monitor, keyboard, and mouse, while smart phones have touch screens, gesture controls, and more. Augmented reality and virtual reality are also on the way. But the most basic way people talk to each other is by using their voices. We can both listen and speak, and if machines could do the same, it would be very easy to talk to them. Smart watches and smart speakers like Google Home, Amazon Echo, and others are slowly making their way into our homes, and it's getting easier and easier to talk to machines. So, in this tutorial, we'll learn how to use the Google assistant, which you can call up from your phone, Google home, or smart watch, to control the lights and fans with just your voice. We're not too lazy to turn loads on and off with switches, but at the end of the day, we just love being able to tell things to do things with our voices. We're using an Arduino UNO as the microcontroller and an ESP8266 module to connect to the internet. Really, there are a lot of other ways to do this. You can use more powerful processors like Raspberry Pi or SOCs with built-in Wi-Fi like the ESP12E or ESP32. But I chose the old-school Arduino and ESP8266 board to keep costs low and not over-engineer anything.

For this project, we will also use the ESP8266, which fits nicely on top of the Arduino UNO board, to build an Arduino WiFi shield. This shield can be used to programme the ESP8266 using AT commands or the Arduino IDE. It also has the option to connect an FTDI module directly to the ESP8266, which lets the ESP8266 be programmed as a stand-alone device without Arduino and also flash a new firmware into the ESP8266 module if needed. So, the shield can be used in a lot of other creative Arduino projects that need to connect to the internet.

We've used the IFTTT services to talk to the Google assistant on our phones. These services set up the assistant to listen for a certain command and open a link if it hears that command. Now, as you may already know, the ESP8266 can only read information from the internet through API calls, so we need a platform that can give us this API option. This is where ThingSpeak comes in. Basically, when you tell Google Assistant something with your voice, it changes the value of a field in our ThingSpeak channel to match. While the ESP8266 checks the value of this field every so often with API calls and sends this information to Arduino through serial communication. The Arduino then does what needs to be done, like turning on and off a relay, based on the value it got.

406. ESP8266 Weather Station with Arduino

When I first heard about the ESP8266 in last year, I had no idea what to do with it. I'm now really interested in how easy it is to connect an Arduino to the Internet. Like others, I started by putting together a weather station on a breadboard and sending data to thingspeak.com. This leads to a very small stripboard layout and the software package that is needed.

Why stripboard rather than etch a PCB?

A lot of good PCB layouts are out there. But I want to make something simple enough that anyone can do it at home. It's easy to get a soldering iron, stripboards, and parts. But it's often hard to etch a PCB. So, my task was to make the design on a stripboard that was the same size and worked as well as a PCB that had been etched.

What functions does it have?

- Measure the temperature, the humidity, the pressure, and the amount of light.
- All the information to Thingspeak
- Arduino can turn on and off ESP8266 to save energy and add more digital IOs.
- Works on 5 to 12 V and much more.

Why do you need ESP8266 and Arduino?

Each one has its own good points. When it comes to sensors, the Arduino is perfect. And the ESP8266 is great for connecting to the Internet. And if I want to save energy, I can turn off the Internet.

And best of all: The weather functions are just one part of the app. Because the Arduino has the following sensor pinouts, you can use this board to measure almost anything: 2x analogue, 6x digital I2C connections for WIRE, 1x reset, 1x click

Circuit Diagram:

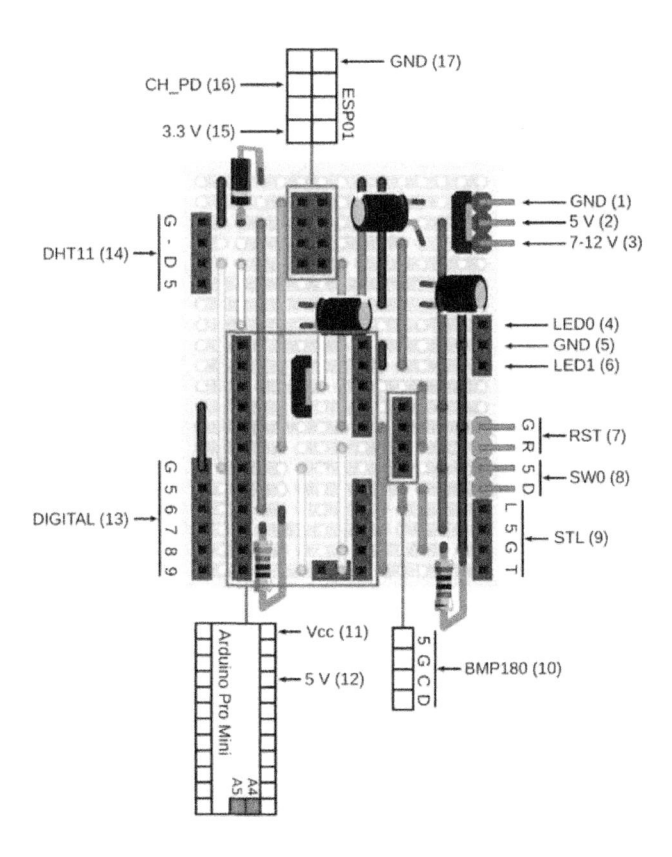

You need the following parts to build this weather station:

- 1x stripboard 20x15 vertical
- 1x ESP01
- 1x Arduino Mini Clone (this layout Arduino Mini, I chose the clone version because the position of A4 and A5 is more stripboard friendly compared to the official one)
- DHT11 or DHT22
- 1 shield for a BMP180 (the one with 4 pins)
- 1x sensor shield for temperature and light
- Capacitors:
- 3x 100 µF (*)
- Z-diode
- 1x 3.3 V
- Resistor: 1x 10 kΩ
- 1x 360 Ω
- 1x 150 Ω (for red LED) (for red LED)
- AMS 1117
- 1x 3.3 V
- 1x 5 V (*)
- LEDs: 1x red
- 1 green or one with two colors
- Headers:
- 1x 12x1 female
- 1x 6x1 female
- 2x 5x1 female
- 5x 4x1 female
- 1x 3x1 female
- 1x 2x1 female
- 3x 3x1 male (*)
- 1x 4x1 male
- 1x 4x1 male 90°
- 1x 3x1 male 90°

407. ESP8266 WiFi Temperature Logger

This is a very simple demo that shows how the ESP8266 and Arduino can be used with a digital temperature sensor to update a remote server (https://thingspeak.com/). The Internet of Things (r)evolution is going through some very exciting times. Prices are going down, and people in the Maker community are excited to make the next generation of connected devices. The next set-up could be done for less than $20. This uses "expensive" components you can buy off the shelf, like Arduino, but you could programme your own MCU with UART support and make it cheaper.

Components Required:

- 1 x ESP8266
- 1 x Arduino Pro Mini 328 - 3.3V/8MHz
- 1 x DS18B20 Digital Temperature Sensor
- 1 x 4.7k resistor
- 1 power source (3.3V up to 12V, I used a 9v)

This is when things started to get hard. I spent a lot of time trying out different ways of setting things up. Note that the ESP8266 comes in two different forms. The first one has the LEDs right next to the pins on the board. The LEDs are by the antenna on the newer second one. The second one is mine. When I loaded V0.922, which let me change the baud rate to 9600, I got the best results. To load this firmware, do these things. http://www.electrodragon.com/w/Wi07c#Firmware uploading tool Using a USB-to-TTL cable and a terminal like CoolTerm is the best way to test a connection. To change the baud rate, use this command: AT+CIOBAUD=9600

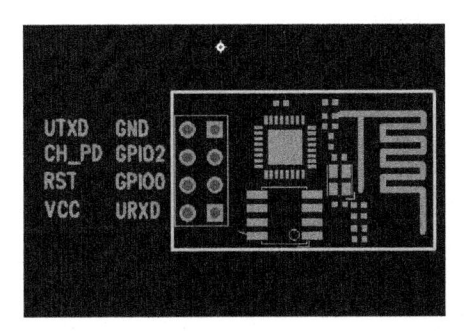

These are the pin connections I used to connect ESP8266 to USB-to-TTL. I powered the ESP8266 with the Arduino's 3.3v vcc, which is regulated. I know that the maximum output of Arduino vcc 3.3v is 150 mA and that the maximum output of ESP8266 is 240 mA. But at the time, I didn't have any other 3.3v that was stable. The ESP8266 is usually used at 70 mA. When you are uploading new firmware, don't forget to connect GPIO0 to GND. After that, take it out to get back to normal.

408. Send Sensor Data (DHT11 & BMP180) to ThingSpeak with an Arduino, Using Cable or WiFi (ESP8266) or Use ESP8266 Alone

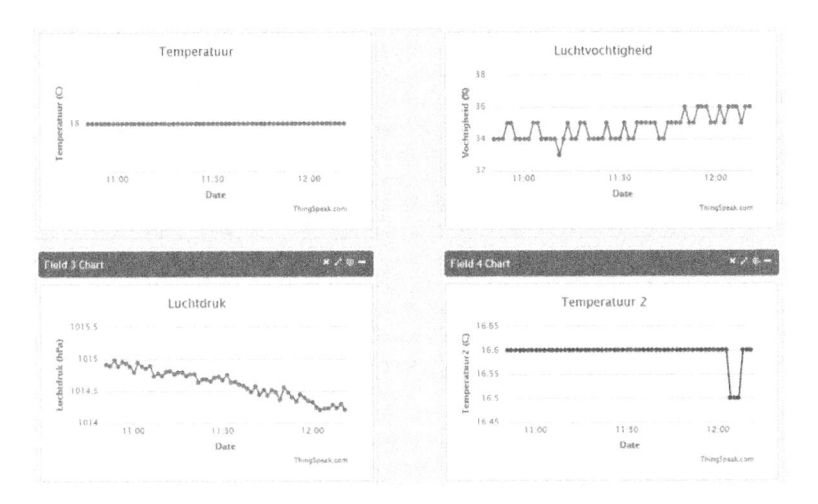

Despite how simple it is to set up, Thingspeak has a few obstacles that I encountered and, based on responses, other users appear to be having trouble with as well. This project works well, but I probably wouldn't use an Arduino and an ESP8266 together again because it is too much. The Arduino can do a lot of things that the ESP8266 can also do, especially if all it needs to do is gather data from sensors. Check out the "ESP8266 only" step of this project or one of my more recent ones, like this one.

Component Required:

- Arduino
- WS5100 EthernetShield
- Thingspeak Account
- Sensors (such as DHT11 and BMP180)
- Internet connection
- For the WiFi:
- ESP8266-01 (or other ESP8266

I wanted to use my Arduino to gather weather data and put it on a website in the form of nice graphs so I could keep an eye on it from far away. I could just use an Arduino as a web server, but if I want to do more than send numbers to a webpage, the Arduino will soon run out of memory. There are services that can take your data and make it public: Pachube, which changed its name to Xively, is well-known, but there is a waiting list for their free accounts right now. I really do have a Pachube account that became a Xively account, but I never got any results from it.

There aren't many other sites like Xively:

- o http://2lemetry.com
- o http://exosite.com
- o https://www.carriots.com
- o https://www.grovestreams.com
- o https://thingspeak.com
- o http://openenergymonitor.org

I chose "Thingspeak." I won't go into too much detail about the signup process or creating a channel or anything else because it's pretty straightforward and easy to understand. After you sign up, you basically make a channel and add fields to it where sensors will send their data later.

Under the API tab, you'll find an API that you'll need to add to your programme later. I talk about a simple connection with an ethernet cable and a wireless connection.

Step 1: Hurdles and Solutions

Hurdle 1

Where do I look for a programme? Wouldn't it be nice to have an example that works? In the upper right corner, there is a "Support" button that takes you to "Tutorials."

You'll find these things under "Tutorials":

"Updating a ThingSpeak Channel with an Arduino and an Ethernet Shield"

Sounds great, so you download that programme to your IDE, add the API key, and then compile it.

Darn, it doesn't compile, and you try to fix it (which is certainly possible) until you don't know what's needed.

The programme still seems to think that everyone will use the 022 or 023 IDE.

There is a link to a Github page, but that will give you a programme to tweet, which is not what you want, at least not right now.

Solution

You won't find a better place to start than this:

https://github.com/iobridge/ThingSpeak-Arduino-Exa...

That programme reads the A0 port and sends that information to "Field1" in your datastream.

Ok, so you try that. You put a variable resistor like an LDR or NTC on port A0, add your API to the programme, and run it.

That works fine, but I didn't just want to read a value from an analogue port. I also had a DHT11 Moisture Temperature sensor and a BMP180 Pressure & Temperature sensor. I didn't think it would be too hard.

Hurdle 2

I made sure the Thingspeak programme had the right libraries, added the objects, and read the sensors into a variable.

Thingspeak, on the other hand, wants you to send strings, but the sensors return floats.

Most variables can be turned into strings with the simple "string" function, but floats are a bit trickier. You have to use the "dtostrf" command, which I think stands for "double-to-string-function," to work with floats.

When I tried to find information about that function on the Internet, I quickly found endless discussions about how "stupid" it was, and people who asked questions were often told, "Why would you need that? Serial.print will do that for you." Yeah, that's true, but I don't want to print. I need it because Thingspeak wants it.

Solution

To use the dtostrf command, you must set up a buffer space where the string will be stored. It works like this:

It's important to have bufferspace. I got it to work with "7" or even "5", but when I added a second sensor that needs this function, my datastream would crash or I would get the strangest results. I also thought I could switch between using the same bufferspace for each sensor, but that didn't work either, so now I have a bufferspace for each sensor.

Now, I'm no expert in C, so if there's a better way to do this, I'd love to hear it, but this is what I did and it worked.

Hurdle3

Once I knew how to convert strings, I was able to add the data to the datastream.

The Thingspeak example programme only shows this for one field, but it's easy to see that you need to add the strings and use the right number of plus signs and ampersands.

Solution

So, let's say there are 4 different fields:

updateThingSpeak("field1="+temp+"&field2="+humid+"&field3="+pres+"&field4="+temp2

Step 2: Connecting with Ethernetshield

The BMP180 is like the BMP085, but it has been updated. The BMP180 can use the libraries from the BMP085.

There are two versions of the library for AdaFruit. I chose version 1 because it was easier to use. Version 2 also needs the "Sensor" library to be installed.

In the code, I also include an extra float:'m,' which gives the pressure in "mmHg." Since I haven't used it yet, there is no string conversion and it isn't added to the datastream, but adding it should now be as easy as adding 1 and 1.

Step 3: Connect to Thingspeak Using an ESP8266 WiFi Module

The internet connection shown before was made with a cable. But the Arduino can be connected to a cheap WiFi module called the ESP 8266.

A cheap WiFi module is the ESP8266. Remember that it requires 3.3 Volt. Some models, on the other hand, say they can work with 5 Volts. I added two circuits that can be used to change the voltage.

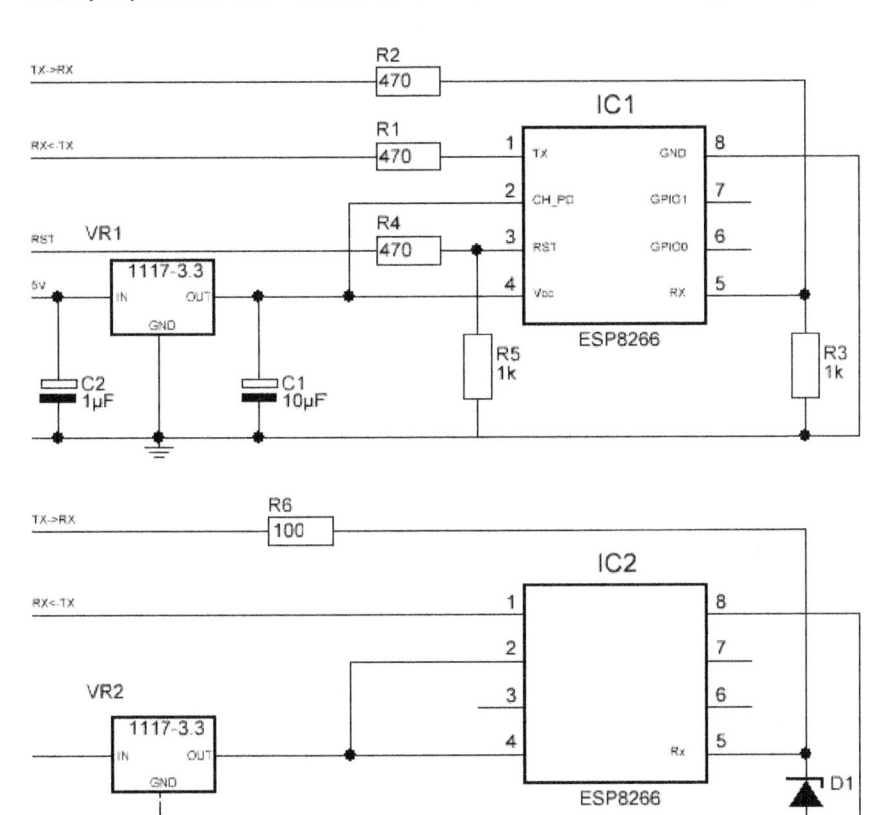

The ESP8266 needs its own source of 3.3 V because the current from the Arduino just isn't enough. There is a lot of information about how to connect the module, but I want to talk about the software needed to connect to Thingspeak.

For debugging, the SoftSerial library is added. Once the programme is running, it is not really necessary. All of the print statements to the Software serial port are the same.

I use 3 analogue values instead of an example with the BMP108 and DHT11 because that would require libraries. This makes it easier to understand how to use the ESP8266. Once you know that, adding other sensors is easy. Just make sure that everything is turned into a string.

Step 4: Using Just the ESP8266

If you see gibberish on this page, it's because there was a full step here, but our thought it would be better to replace it with gibberish, even though I had already saved it. I will try it again. In the last step, I used an ESP8266 to send to Thingspeak the sensor values that an Arduino read. But you can also do it without using an Arduino at all. I connected a DHT11 to pin 2 of my ESP8266-01 and ran the following programme. I can't take all the credit for the programme. I think Jeroen Beemster came up with it first.

This idea is not about how to programme the ESP8266. There are plenty of other places to learn that. If you already have a USB-TTL module at 3.3 Volt, you don't have to worry about voltages. If all you have is a 5V USB-TTL module, you can still use it, but you have to put a voltage divider between the module's Tx and the ESP8266's Rx. Never ever give the ESP8266 5 volts.

Note: The Adafruit DHT library has a mistake that usually doesn't show up when using an Arduino, but can when using an ESP8266, especially if the programme is a bit larger. Changes you need to make to the cpp file are shown in the picture. It's mostly just comments on two lines.

409. An Inexpensive IoT Enabler Using ESP8266

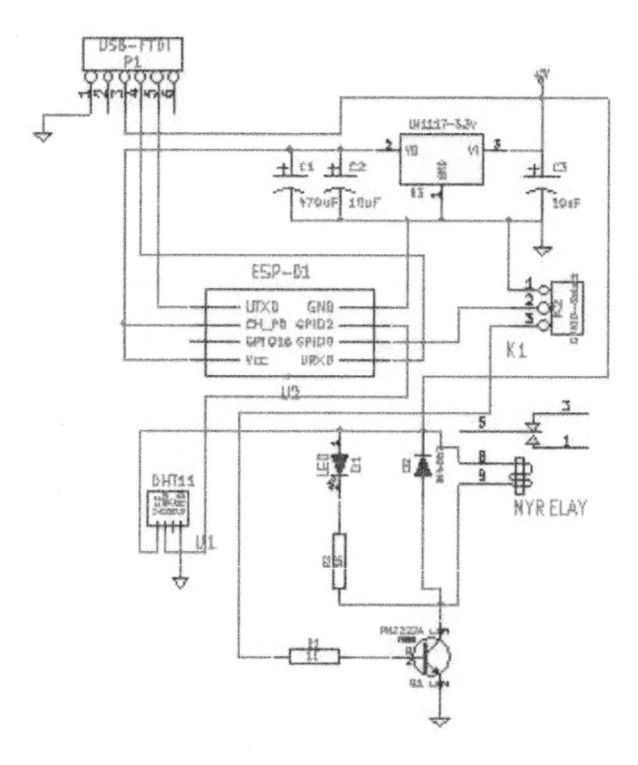

The Internet of Things, or IoT for short, focuses on the various ways in which devices can be connected to a network so that they can communicate with one another and send and receive data and commands. This is arguably the most important aspect of the IoT itself. There are already technologies like

Bluetooth, WiFi, NFC, etc. that can help with last-mile connectivity, but most of them are hard to set up and often need extra hardware like a local control server or appliance.

In this project, I show you how to build and set up a simple standalone board that, with the help of a wifi network, can send environmental data to the internet and receive commands to turn on/off a switch. This project, which costs less than 10 or 15 US dollars to build, does the following:

- Send readings from a DHT11 sensor about temperature and humidity to a MQTT broker and a Thingspeak channel.
- Listen for MQTT messages and use ESP8266 to turn on or off a relay.
- Check a Thingspeak channel every so often, and if the field has changed, turn on or off a relay.

In simple terms, this is a temperature/humidity sensor and relay that can connect to the internet. The relay can be used to turn on or off any AC appliance that is connected to it. The parts needed cost less than 15 US dollars, and you don't need any other hardware. All of the software used is open source, and to connect the board to the internet, only free online services were used. It is assumed that there is a WiFi network that the board can connect to that works. Even though there are already a few projects that show how to set up an ESP8266 module to send data to the internet, I have not seen a solution that also lets you use the internet to control a device that is connected to the ESP8266 module. This project shows how simple it is to do that.

Component Required

1.	ESP8266 ESP-01 module	7.	2 x 10 uF Tantalum capacitor
2.	DHT11 or DHT22 sensor	8.	2 x 1K resistor
3.	5v relay	9.	1 LED
4.	LM1117 3.3v LDO voltage regulator	10.	male breakaway headers
		11.	PCB Board
5.	N2222A transistor	12.	FTDI USB to TTL adapter cable
6.	1 x 470 uF capacitor		

You can order the ESP8266 module, the DHT11/22 sensor, and the USB to TTL cable from ebay.com. Buying all three should not cost more than $9. The rest of the parts can be bought for a few dollars at any store that sells electronics. The ESP8266 module should be able to connect to a working WiFi access point and know how to log in to it. The circuit is pretty simple. The board can be powered either by the USB cable alone or by a 5v DC supply plugged into the power socket at the top. A LM1117-3.3 LDO is used to give the ESP8266 module the 3.3v it needs. On the GPIO2 pin of the ESP8266, there is a DHT11 sensor. Depending on where the jumper is, the GPIO0 pin can be linked to either the ground or the relay. The temperature and humidity readings, as well as the relay state, are sent from the board to: MQTT broker: Any MQTT broker can be used (http://mqtt.org/). In my case, I used www.cloudmqtt.com, which has a free online broker. You'd need to make both an account and an instance. And put the details of the instance in the user config.h file. Using an online broker is helpful because you can access it from any network. If a local broker is used, like one that runs on a Raspberry Pi, it may be hard to connect to it from outside the home network because ISPs usually block incoming ports. Thingspeak.com: On Thingspeak.com, you need to set up an account. Also, a channel with these three channels needs to be made and the details of that channel need to be entered in user main.c:

- field1 - relay state
- field2 - temperature
- field3 - humidity

Even though, in its current state, the board needs to be connected to both cloudmqtt and Thingspeak for all functions to work, the code can be easily changed to use only one of these options.

410. Firebase: Control ESP8266 NodeMCU GPIOs from Anywhere

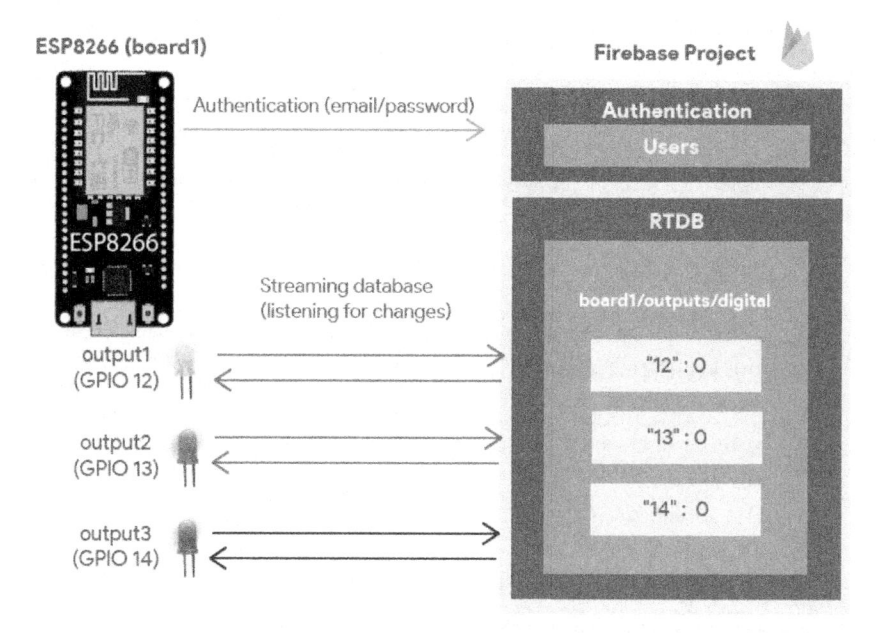

In this project, you will get an idea of how to use Firebase to control the GPIOs of an ESP8266 from anywhere. We'll make nodes in the Firebase Realtime Database to store the current GPIO states. The ESP8266 updates its GPIOs whenever there is a change in the database nodes. You can change the states of the GPIOs by writing to the database or by making a web app.

What's Firebase?

Google's Firebase is a platform for making mobile apps that helps you build, improve, and grow your app. Firebase gives you free services like hosting, authentication, and a real-time database that let you build a full-featured web app to control and monitor the ESP32 and ESP8266 boards that would be much harder and more time-consuming to build and set up on your own. The Firebase Realtime Database enables the development of rich, collaborative applications by providing direct, secure database access from client-side code. Data is stored locally, and real-time events continue to fire even when the application is not connected to the internet, providing the end user with a responsive experience.

The diagram gives a high-level look at the project overview

1. To access the database, the ESP8266 verifies itself as a user with an email address and a password (this user must be added to the Firebase authentication methods);
2. Database rules are used to protect the database. We'll add the following rule: only authenticated users can access the database. The database has several nodes that store the ESP8266 GPIO states.
3. We'll control three GPIOs as an example (12, 13, and 14). You can add or take away nodes to change how many GPIOs you can control.
4. The ESP8266 will watch the GPIOs database nodes for changes. When something changes, it will change the GPIO states to match.
5. You can change the GPIO states manually on the database using the Firebase console, or you can make a web page that anyone can access with buttons to control the GPIOs and show the current GPIO states.

Here are the main things you need to do to finish this project:

- Make a project in Firebase
- Set Authentication Methods
- Get the Project API Key
- Set up the database for real time
- Set up Database Security Rules
- Organizing your Database Nodes
- ESP8266: Monitoring for Any Changes to the Database (control GPIOs)

411. ESP8266 NodeMCU with Load Cell and HX711 Amplifier (Digital Scale)

In this project, you'll learn how to use a load cell and the HX711 amplifier to make a scale with the ESP8266 NodeMCU. First, you'll learn how to connect the scale's load cell, HX711 amplifier, and ESP8266. Then, we'll show you how to set up the scale and give you an easy way to figure out how much something weighs. Later, we'll also add a screen to show the measurements and a button to "tare" the scale.

Load Cell:

A load cell turns the force on it into an electrical signal that can be used to measure the force. The electrical signal changes in a way that depends on how much force is used. Load cells come in different kinds, such as strain gauges, pneumatic, and hydraulic. We'll talk about strain gauge load cells in this lesson. Strain gauge load cells have a metal bar with strain gauges attached to it (under the white glue in the picture above). A strain gauge is an electrical sensor that measures how hard something is being pulled or stretched. When an outside force is put on an object, the resistance of the strain gauges changes, which causes the shape of the object to change (in this case, the metal bar). Since the change in resistance is proportional to the load, we can figure out how much something weighs. Most load cells have four strain gauges connected in a Wheatstone bridge (shown below), which lets us measure resistance accurately. To learn more about how strain gauges work, click here.

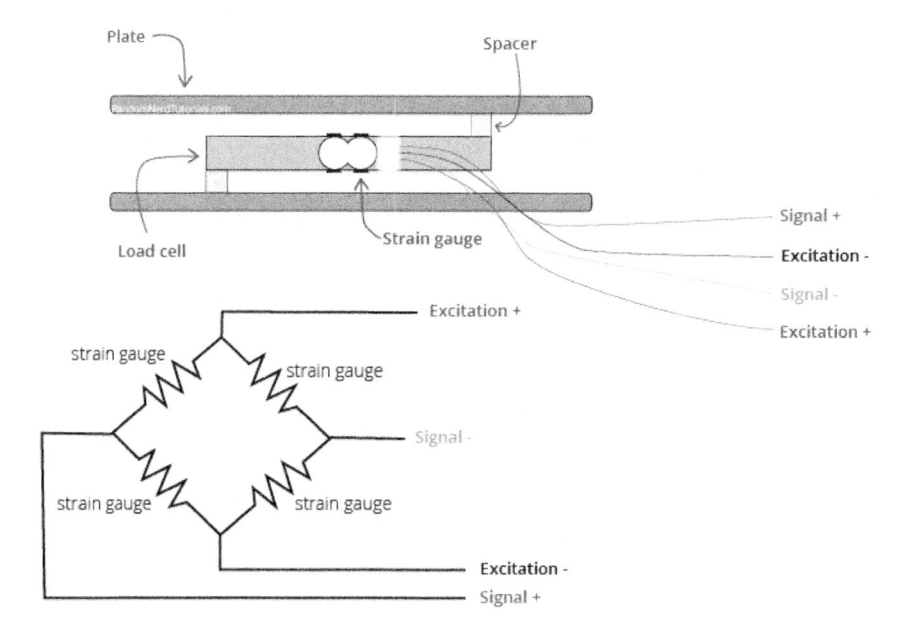

The wires coming from the load cell usually have the following colours:

- Red: VCC (E+)
- Black: GND (E-)
- White: Output – (A-)
- Green: Output + (A+)

Applications

- Load cells that use strain gauge technology have a wide range of potential applications. For instance:

- Check to see if the weight of something changes over time;
- Find out how much something weighs;
- Find out if something is there;
- Estimate a container's liquid level; and etc.

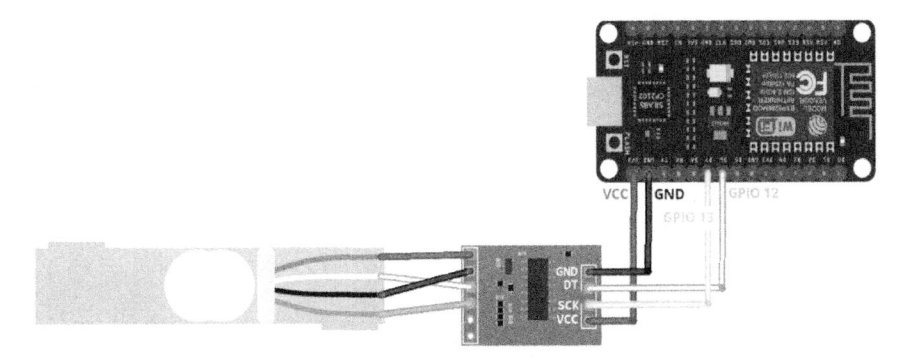

Figure: Circuits Diagram of ESP8266 NodeMCU with Load Cell and HX711 Amplifier (Digital Scale)

We need an amplifier because the changes in strain when we weigh things are so small. Most of the time, a HX711 amplifier is sold with the load cell we use. So, that will be the amp we use.

Amplifier HX711

The HX711 amplifier is a breakout board that makes it easy to measure weight with load cells. On one side, you connect the load cell wires, and on the other, you connect the microcontroller. Two-wire interface is how the HX711 talks to the microcontroller (Clock and Data). To connect to the ESP8266, you need to solder header pins to the GND, DT, SCK, and VCC pins. I soldered the wires from the load cell directly to pins E+, E-, A-, and A+. The wires on the load cell were very thin and fragile, so you had to be careful when soldering not to break them. There are several different libraries that can be used with the HX711 amplifier to get measurements from a load cell. We'll use the Bodge HX711 library. It works with the Arduino, the ESP32, and the ESP8266. The two-wire interface is how the HX711 amplifier talks to other devices. You can hook it up to any of the GPIOs on the microcontroller you choose. We are connecting the data pin (DT) to GPIO 12 (D6) and the clock pin (CLK) to GPIO 13. (D7). You can use any pins that work (check the ESP8266 pinout guide).

412. ESP8266 NodeMCU with TDS Sensor (Water Quality Sensor)

In this project, you'll learn how to use an ESP8266 NodeMCU board with a TDS meter (Total Dissolved Solids). A TDS metre shows the number of salts, minerals, and metals that have dissolved in a solution. You can use this parameter to get an idea of how good the water is and to compare water from different sources. One of the main uses of a TDS metre is to check the quality of the water in an aquarium.

We'll use the TDS metre from keystudio to show you a simple way to use Arduino IDE to measure TDS in ppm units.

Introducing the TDS Meter

A TDS metre counts the number of salts, minerals, and metals that have been dissolved in water. As the number of dissolved solids in water goes up, so does the conductivity of the water. This lets us figure out the total amount of dissolved solids in parts per million (mg/L). Even though this is a good way to check on the quality of the water, keep in mind that it doesn't measure contaminants. So, you can't just look at this sign to figure out if the water is safe to drink or not. A TDS metre can be used to check the quality of water in pools, aquariums, fish tanks, hydroponics, water purifiers, and many other places.

Features and Specifications

This guide talks about the keystudio TDS Meter V1.0. Here are the settings for the sensor:

TDS Meter:

- Input Voltage: DC 3.3 ~ 5.5V
- Output Voltage: 0 ~ 2.3V
- Current at Work: 3 to 6mA
- TDS Measurement Range: 0 ~ 1000ppm
- Accuracy of TDS measurement: 10% F.S. (25 °C)
- Module Interface: XH2.54-3P
- Electrode Interface: XH2.54-2P

TDS Probe:

- Number of Needle: 2
- Length as a whole: 60cm
- Connection Interface: XH2.54-2P
- White: Color
- Waterproof Probe

Interfacing the TDS Meter with the ESP8266

The TDS meter sends out an analogue signal that can be measured with the analogue pin A0 on the ESP8266. The TDS Meter, which is based on the ESP8266 microcontroller, connects to a Wi-Fi network and continuously transmits data to the Blynk application. You are able to obtain a report concerning the quality of your water around the clock. Therefore, the TDS sensor is the most effective sensor for monitoring water quality. Connect the sensor as shown in the following table.

TDS Sensor	ESP8266
GND	GND
VCC	3.3V
DATA	A0

If the probe is not in the water, it will show a value close to 0. Check the TDS of a solution by putting the probe on it. You can use tap water and salt to see if the values go up. The TDS value of the tap water in my house is about 100ppm, which is a good value for drinking water. I also tested tea and found that the TDS value went up to about 230ppm, which seems like a good number. Lastly, I measured the TDS value of bottled water and found that it was about 25ppm.

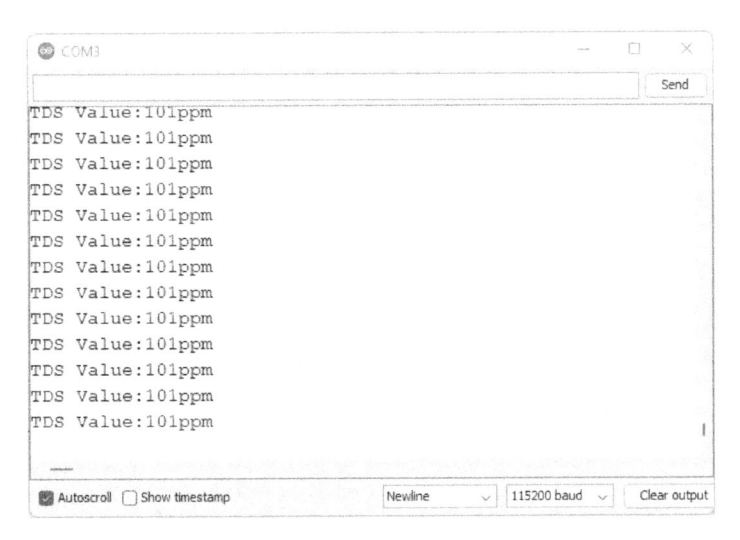

```
TDS Value:101ppm
TDS Value:101ppm
TDS Value:101ppm
TDS Value:101ppm
TDS Value:101ppm
TDS Value:101ppm
TDS Value:101ppm
TDS Value:101ppm
TDS Value:101ppm
TDS Value:101ppm
TDS Value:101ppm
TDS Value:101ppm
TDS Value:101ppm
```

A TDS meter can figure out how many solids are in a solution as a whole. It can be used to tell how good the water is and to describe what it is like. The meter tells you how much TDS there is in parts per million (ppm) or mg/L. The TDS value can be used for many things, but it can't be used alone to tell if water is safe to drink or not. A great way to use this kind of sensor is to check the quality of the water in an aquarium. You can keep an eye on your fish tank with this sensor and a waterproof DS18B20 temperature sensor.

413. ESP8266 NodeMCU: K-Type Thermocouple with MAX6675 Amplifier

In this project, you'll learn how to use a K-Type Thermocouple, a MAX6675 amplifier, and an ESP8266 NodeMCU board to read temperature. A K-type thermocouple is a type of temperature sensor that can measure temperatures from 200 to 1260oC (326 to 2300oF).

Thermocouple:

A thermocouple is a tool for figuring out how hot or cold something is. It is made up of two different metal wires that are joined together to make a junction. When the junction is heated or cooled, a small voltage is created in the thermocouple's electrical circuit. This voltage can be measured, and it shows how hot or cold the junction is. Thermocouples are used in everything from home appliances to industrial processes to making electric power to monitoring and controlling furnaces to making food and drinks to sensors in cars and aeroplane engines to rockets, satellites, and spacecraft.

What is a thermocouple K-type?

A thermocouple is a device made up of two different electrical wires that meet at an electrical junction called a "thermal junction." The temperature change at the junction causes a small but measurable change in voltage at the reference junction, which can be used to figure out the temperature. Different metals can be used to make a thermocouple. The voltage range, cost, and sensitivity will all depend on the metals that are used. The different types of thermocouples are B, E, J, N, K, R, T, and S. These letters stand for the standard metal combinations that make up each type.

The k-type thermocouple is the subject of our lesson. A k-type thermocouple has conductors made of chrome and alumel, and its temperature range is usually from 200 to 1260oC (328 to 2300oF).

Amplifier MAX6675

We need a thermocouple amplifier to get the temperature from the thermocouple. The temperature that comes out of the thermocouple amplifier depends on the voltage read at the reference junction. The difference in temperature between the reference junction and the thermal junction affects the voltage at the reference junction. So, we need to know how hot or cold the reference junction is.

The MAX6675 thermocouple comes with a temperature sensor that measures the temperature at the reference junction (cold-compensation reference) and amplifies the small voltage at the reference junction so that we can read it with our microcontrollers. The MAX6675 amplifier talks to a microcontroller using the SPI protocol, and the data is sent out at a resolution of 12 bits.

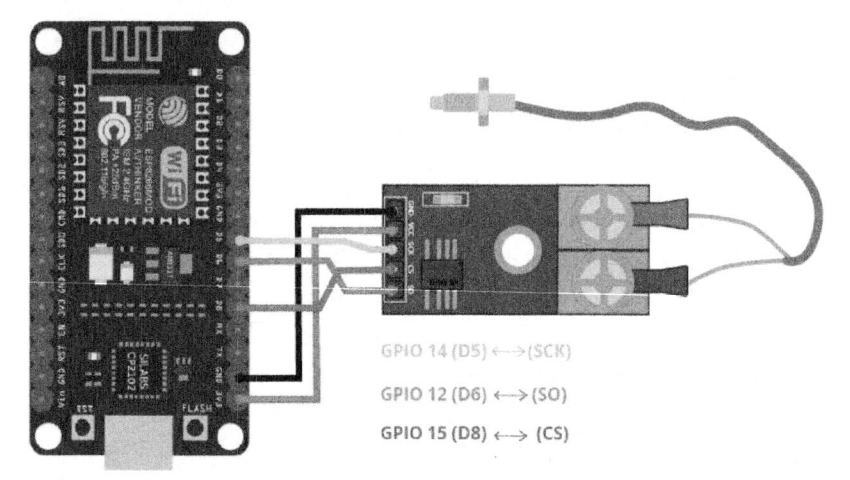

Schematic of ESP8266 with K-type thermocouple and MAX6675 amplifier

GPIO 14 (D5) ←→ (SCK)
GPIO 12 (D6) ←→ (SO)
GPIO 15 (D8) ←→ (CS)

Most of the time, you can get a k-type thermocouple and a MAX6675 amplifier in the same package. Here is a list of the most important features of the MAX6675. Please look at the MAX6675 datasheet for a more in-depth explanation.

- Direct digital conversion of k-type thermocouple output
- Cold-junction compensation
- Simple SPI-compatible serial interface
- Operating voltage range: 3.0 to 5.5V
- -20oC to 85oC is the range of operating temperatures.
- Temperatures can be read up to 1024oC (1875oF) and are accurate to 0.25oC.

K-Type Thermocouple Interface with MAX6675 Amplifier. As we've already talked about, the MAX6675 talks to a microcontroller using the SPI protocol.

Parts Required

- K-type thermocouple amplifier MAX6675
- ESP8266 NodeMCU (read Best ESP8266 development boards)
- Switcheroos (female-to-female)

414. ESP8266 NodeMCU with BH1750 Ambient Light Sensor

The BH1750 is a light sensor with 16 bits. In this project, you'll learn how to use ESP8266 NodeMCU board and the BH1750 ambient light sensor. I2C communication protocol is used for the sensor to talk to a microcontroller. You'll learn how to connect the sensor to the ESP8266 NodeMCU board, install the necessary libraries, and use a simple sketch to show the sensor readings in the Serial Monitor.

Introducing BH1750 Ambient Light Sensor

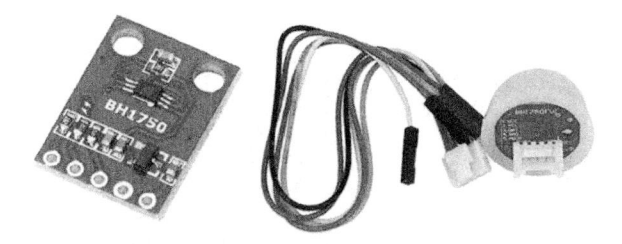

The BH1750 is a 16-bit light sensor that talks to other devices using the I2C protocol. It gives measurements of brightness in lux (SI derived unit of illuminance). It can measure anywhere between 1 lux and 65535 lux. The sensor may come in more than one type of breakout board. Look at the photos below. Both pictures show what a BH1750 sensor looks like.

BH1750 Features

Here is a list of the things that the BH1750 sensor can do. Check out the BH1750 sensor datasheet for more information.

- Interface for I2C bus
- Spectral responsibility is about the same as how the human eye reacts.
- Illuminance to digital converter
- Range: 1 – 65535 lux
- Low current by turning off the power
- 50Hz / 60Hz Getting rid of light noise
- You can choose from two different I2C slave-addresses.
- Small changes in measurements (+/- 20%)
- Very little is changed by infrared.
- Supports continuous measurement mode
- Supports one-time measurement mode

Schematic - ESP8266 with BH1750

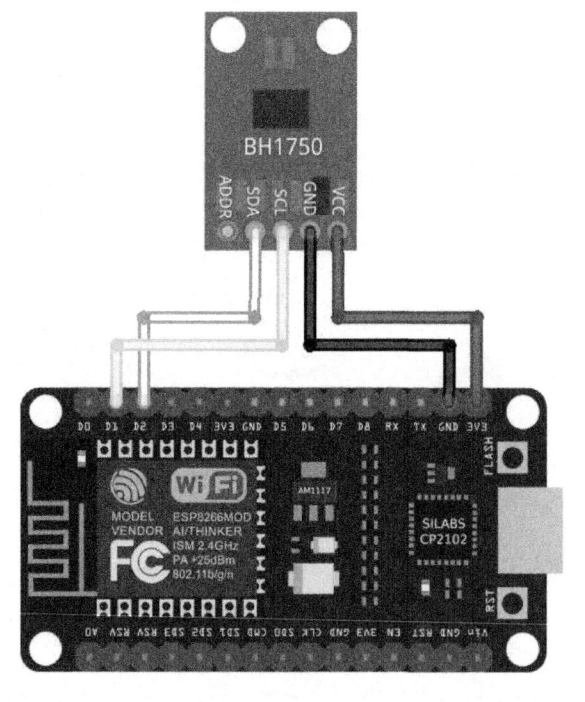

Measurement Modes

The sensor can be used in two different ways to take measurements: continuously or just once. There are three different resolution modes for each mode. In continuous measurement mode, the sensor measures the amount of light in the room all the time. In one-time measurement mode, the sensor only measures the amount of light in the room once, and then it shuts off.

Applications

- The BH1750 is an ambient light sensor, making it suitable for a variety of applications. For example: to detect if it is day or night;
- to change the brightness of an LED or turn it on or off based on the amount of light around it;
- to adjust the LCDs and the brightness of the screen;
- to find out if an LED is turned on. etc.

Parts Required

- BH1750 ambient light sensor ESP8266 (read Best ESP8266 development boards)
- Breadboard (optional) (optional)
- Jumper wires (optional)

Open the Serial Monitor at a baud rate of 9600 after the code has been successfully uploaded.

```
COM4                                                           —    □    ×
                                                                      Send
Light: 1417.50 lx
Light: 219.17 lx
Light: 99.17 lx
Light: 102.50 lx
Light: 106.67 lx
Light: 109.17 lx
Light: 109.17 lx
Light: 110.00 lx
Light: 366.67 lx
Light: 72.50 lx
Light: 2.50 lx
Light: 2.50 lx
Light: 2.50 lx

Autoscroll  Show timestamp      Newline    9600 baud    Clear output
```

The Serial Monitor should show new measurements of brightness.

415. ESP8266 NodeMCU Web Server: Display Sensor Readings in Gauges

Learn how to use the ESP8266 NodeMCU to build a web server that can show sensor readings in gauges. As an example, we'll show the temperature and humidity from a BME280 sensor using both a linear and a radial gauge. You can change the project easily to plot any other kind of data. We'll use the canvas-gauges JavaScript library to make the gauges.

Project Overview

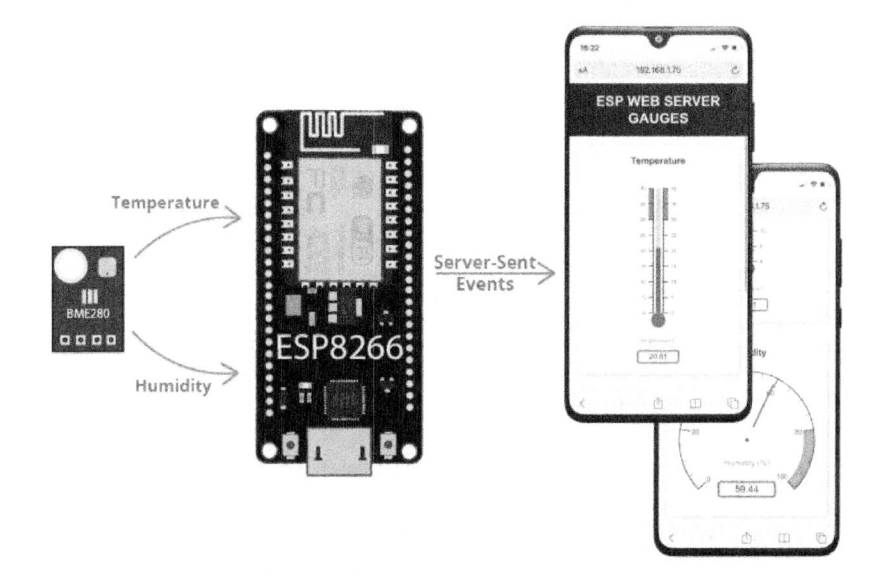

With the ESP8266, this project will make a web server that shows temperature and humidity readings from a BME280 sensor. To show the temperature, we'll make a linear gauge that looks like a thermometer, and to show the humidity, we'll make a radial gauge.

Parts Required

- ESP8266 (read Best ESP8266 development boards)
- BME280 Sensor
- Jumper wires
- Breadboard

PREREQUISITES

Before you move on with the project, make sure you've done everything in this section.

1. Add the ESP8266 board to the Arduino IDE

We'll use Arduino IDE to programme the ESP8266. So, the ESP8266 add-on must be in place. If you haven't already, read the next lesson:

2. Filesystem Uploader Plugin

We'll use a plug-in for Arduino IDE called LittleFS Filesystem Uploader to send the HTML, CSS, and JavaScript files to the ESP8266 filesystem (called LittleFS). To install the filesystem uploader plugin, follow the steps in the next guide:

3. Installing Libraries

You need to install the following libraries to build this project:

- Adafruit BME280 (Arduino Library Manager) (Arduino Library Manager)
- Adafruit Sensor library (Arduino Library Manager)
- Version 0.1.0 of the Arduino JSON library (Arduino Library Manager)
- ESPAsyncWebServer (.zip folder); ESPAsyncTCP (.zip folder).

The Arduino Library Manager can be used to set up the first three libraries. Go to Sketch > Include Library > Manage Libraries, then look for the names of the libraries.

Since you can't install the ESPAsyncWebServer and ESPAsynTCP libraries through the Arduino Library Manager, you have to copy the library files to the Arduino Installation Libraries folder. You can also download the libraries'.zip folders. Then, in your Arduino IDE, go to Sketch > Include Library > Add.zip Library and choose the libraries you just downloaded.

Schematic Diagram

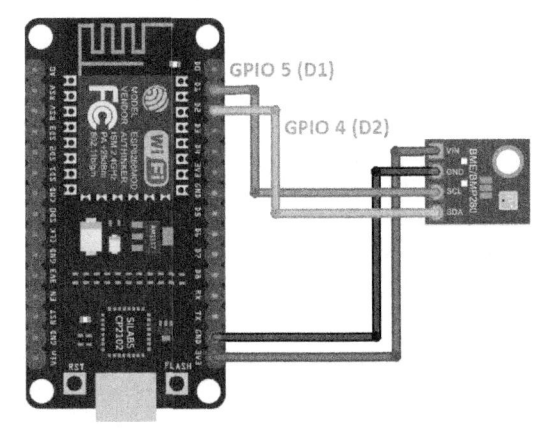

From a BME280 sensor, we'll send the temperature and humidity. With the BME280 sensor module, we will use I2C communication. Wire the sensor to the ESP8266's SCL (GPIO 5) and SDA (GPIO 4) pins by default.

416. ESP8266 NodeMCU Door Status Monitor with Telegram Notifications

In this project, you'll use an ESP8266 NodeMCU board and a magnetic reed switch to keep track of the state of a door. When the door is opened or closed, you'll get a message in your Telegram account. As long as your smartphone can connect to the internet, you'll get a message no matter where you are. The Arduino IDE will be used to programme the ESP8266 board.

Project Overview

In this project, we'll make a Telegram Bot that will send you messages when a door's status changes. We'll use a magnetic contact switch to tell when something has changed.

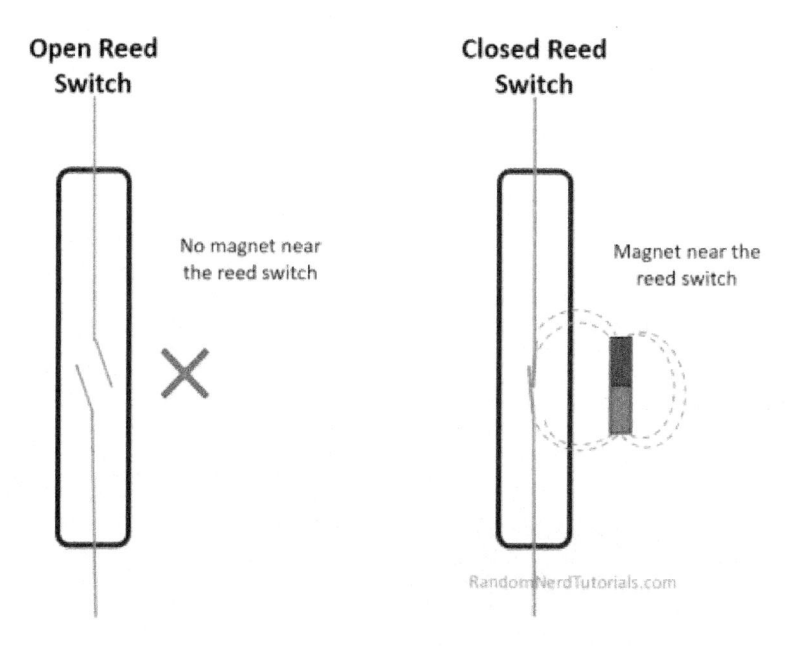

A magnetic contact switch is essentially a reed switch that has been encased in a plastic casing so that it can be easily applied to a door, a window, or a drawer to determine whether or not it is open or closed. This allows the switch to detect whether or not the object being monitored is open or closed.

When a magnet is close to a switch, the circuit is closed, which means the door is shut. When the door is open and the magnet is far from the switch, the circuit is open. Look at the image below. We can connect the reed switch to a GPIO on the ESP8266 to see when its state changes.

Telegram: An Introduction

Telegram Messenger is an instant messaging (IM) and voice over IP (VoIP) service that runs in the cloud. You can easily put it on your computer or phone (Android or iPhone) (PC, Mac, and Linux). There are no ads and it is free. You can make bots that you can talk to using Telegram. "Bots are applications made by third parties that run inside Telegram. Users can talk to bots by sending them messages, commands, and inline requests. You use HTTPS requests to Telegram Bot API to control your bots. The Telegram bot will talk to the ESP8266, which will then send messages to your Telegram account. When the state of the door changes, you'll get a message on your phone (as long as you have access to the internet).

Creating a Telegram Bot

1. Download and install Telegram from Google Play or the App Store.
2. To make a Telegram Bot, open Telegram and follow the steps below. First, look for "botfather" and click on BotFather, as shown below. Or, go to this link on your phone: t.me/botfather.
3. The next window should pop up, asking you to click the "Start" button.
4. To make your bot, type /newbot and follow the instructions. Name it and give it a user name. Door Sensor is the name of mine, and ESPDoorSensorBot is my username.
5. You will get a message including a link to access the bot and the bot token if your bot is successfully created. This message will be sent to you if your bot was successfully formed. Save the bot token because you'll need it so that the ESP8266 can talk to the bot.

Sending a Message to the Bot

Before it is able to send you messages, your Telegram Bot must first receive a message from your Telegram account that it can read and respond to.

1. Go back to the chats tab and type the username of your bot into the search field.
2. Choose your bot to start talking.
3. Click the link that says "Start."

Universal Telegram Bot Library

To talk to the Telegram bot, we'll use Brian Lough's Universal Telegram Bot Library, which gives a simple way to talk to the Telegram Bot API.

Follow the steps below to put the latest version of the library in place.

- ➤ To get the Universal Arduino Telegram Bot library, click here.
- ➤ Click Sketch > Include Library > Add.ZIP Library...
- ➤ Add the library that you just got.

Important: Don't use the Arduino Library Manager to install the library because it might install an old version.

Parts Required

- ESP8266: Best ESP8266 Development Boards
- 1 The Magnetic Reed Switch
- 1× 10kΩ resistor
- 1 Jumper wires for a breadboard

Schematic of ESP8266 with Reed Switch

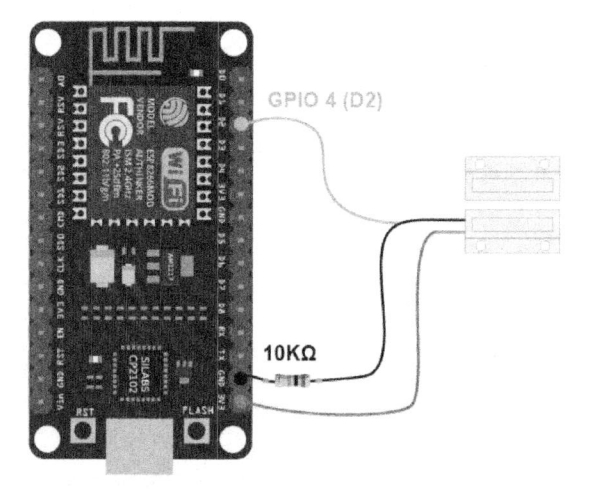

We connected the reed switch to GPIO 4 (D2), but you can connect it to any GPIO that works.

417. ESP8266 NodeMCU Web Server: Control Stepper Motor (WebSocket)

In this Project, you will learn how to use the ESP8266 NodeMCU board to make a web server that shows a web page to control a stepper motor. On the website, you can type in the number of steps and choose whether to go clockwise or counterclockwise. It also shows whether the motor is running or has stopped. The WebSocket protocol is what makes it possible for the client and the server to talk to each other. All clients are kept up to date on the current state of the motor.

1) Parts Required

- 28BYJ-48 Stepper Motor + ULN2003 Motor Driver
- ESP8266 (read Best ESP8266 Development Boards) (read Best ESP8266 Development Boards)
- Connecting wires
- 5V Source of Power

2) Add-on for Arduino IDE and ESP8266 Boards

We'll use Arduino IDE to programme the ESP8266, so you'll need to have the ESP8266 add-on installed.

3) Filesystem Uploader Plugin

We'll use a plugin for Arduino IDE called LittleFS Filesystem uploader to send the HTML, CSS, and JavaScript files we need to build this project to the ESP8266 filesystem (LittleFS).

4) Libraries

There are different ways to use a microcontroller to control a stepper motor. We'll use the AccelStepper library on the ESP8266 to control the stepper motor. This library makes it easy to move the motor a certain number of steps, set its speed, acceleration, and do a lot more. There is a lot of good documentation about how to use the library's methods. Check it out here. Use the steps below to add the library to your Arduino IDE.

Click on **Sketch > Include Library > Manage Libraries...**

Look for the word "accelstepper."

Mike McCauley's AccelStepper library should be put in place. Version 1.61.0 is what we're using.

You need to install the following libraries to build the web server:

- ESPAsyncWebServer (.zip folder) (.zip folder)
- ESPAsyncTCP (.zip folder) (.zip folder)

You can't use the Arduino Library Manager to install the ESPAsyncWebServer and ESPAsynTCP libraries. To get the library files, you need to click on the links above. Then, in your Arduino IDE, go to Sketch > Include Library > Add.zip Library and choose the libraries you just downloaded.

5) Schematic Diagram

The stepper motor is connected to the ESP8266 in the way shown in the following diagram.

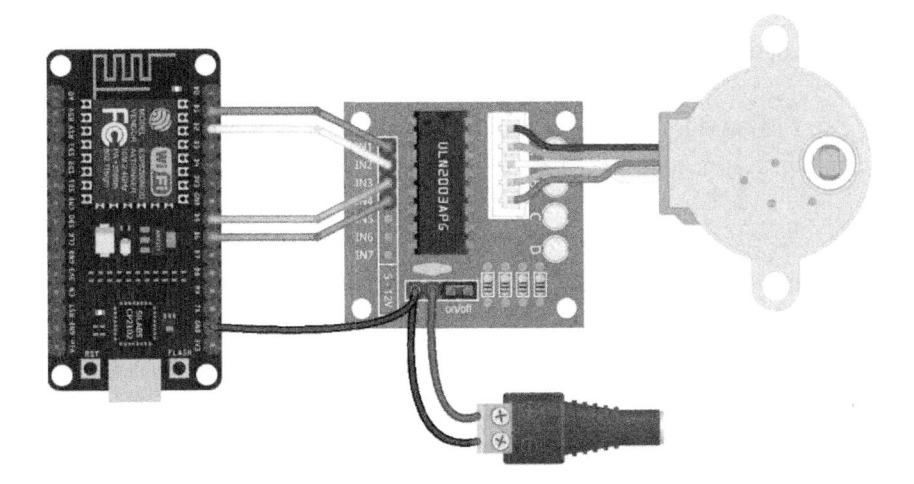

Note: You should use an outside 5V power supply to power the ULN2003 motor driver.

Project Overview

The web page you'll make for this project is shown in the picture below. On the web page, there is a form where you can choose how many steps you want the motor to move and whether you want it to move clockwise or counterclockwise.

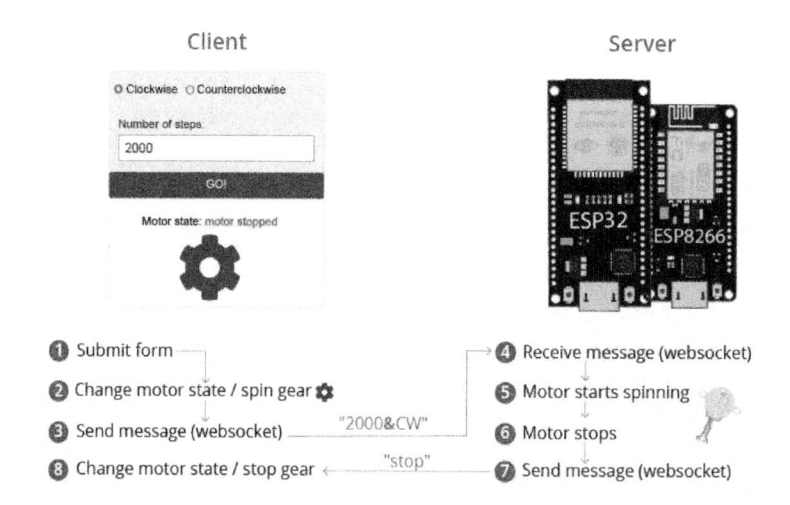

It also shows what the motor is doing, whether it is spinning or not. Additionally, there's a gear icon that spins as long as the motor is spinning. The gear spins clockwise or counterclockwise direction accordingly to the chosen direction.

- WebSocket protocol is how the server and the client talk to each other.
- When you click on the GO! button, it calls a JavaScript function that sends a message using the WebSocket protocol with all the information: steps and directions **(3)**. This is how the message is set up:
- So, if you put in 2000 steps and the direction clockwise, it will send this message:
- At the same time, it will change the motor state on the web page, and the gear will start spinning in the proper direction **(2)**.
- The server then gets the message **(4)** and turns the motor in the right way **(5)**.
- When the motor stops turning **(6)**, the ESP will send a message to the client(s), also using the WebSocket protocol, telling them that the motor has stopped **(7)**.
- When this request comes in, the client(s) update the web page with the motor state **(8)**.

418.ESP8266 NodeMCU with HC-SR04 Ultrasonic Sensor with Arduino IDE

This project shows how to use the Arduino core to connect the HC-SR04 ultrasonic sensor to the ESP8266 NodeMCU board. Sonar is used by the ultrasonic sensor to figure out how far away an object is. We'll show you how to connect the sensor to the ESP8266 and give you a few examples sketches for using the HC-SR04 to figure out how far away something is.

The HC-SR04 Ultrasonic Sensor:

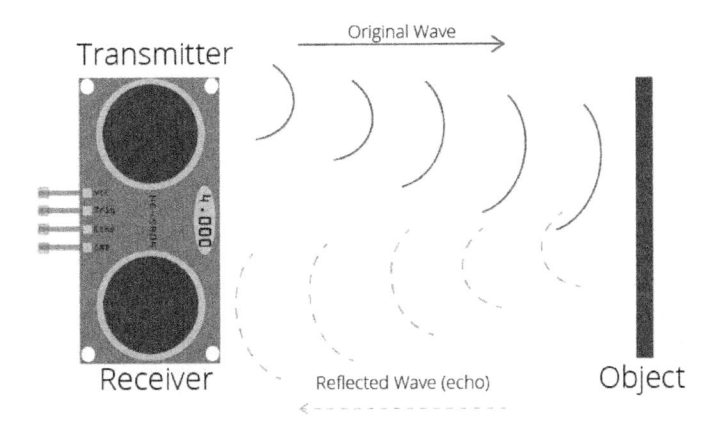

Sonar is used by the HC-SR04 ultrasonic sensor to figure out how far away an object is. This sensor has a reading range of 2 centimeters to 400 centimeters (0.8 inches to 157 inches) and an accuracy of 0.3 centimeters (0.1 inches), making it suitable for the majority of projects undertaken by hobbyists. Also, this module has ultrasonic transmitter and receiver modules built in.

Parts Needed

Here's a list with the parts required to complete this example:

- HC-SR04 Ultrasonic Sensor
- ESP8266 (read Best ESP8266 development boards)
- 0.96-inch I2C OLED Display SSD1306
- Breadboard
- Jumper wires

Schematic Diagram: ESP8266 with HC-SR04 and OLED Display

After the uploading process is complete, start the Serial Monitor with a baud rate of 115200. If you restart the board by pressing the RST button that's built into it, the Serial Monitor will begin to display the distance between the board and the nearest object. Something like what is in the picture.

Code to Display the ESP8266 Distance Sensor (HC-SR04) on an OLED Display, to follow along with this example, you will need to ensure that the Adafruit SSD1306 and Adafruit GFX libraries are installed on your computer. Through the Arduino Library Manager, you'll be able to install these libraries on your computer.

419. ESP8266 NodeMCU with BMP388 Barometric/Altimeter Sensor (Arduino IDE)

The BMP388 is a miniature absolute barometric pressure sensor that is renowned for its accuracy. Because of the accuracy it provides, it is frequently utilized in applications using drones for the purpose of estimating height. It can also be used for GPS, navigation indoors and outdoors, and other things. In this tutorial, you'll learn how to use the BMP388 pressure sensor with the Arduino board. We'll show you a wiring diagram and an example of code.

Introducing BMP388 Barometric Sensor

Absolute pressure and temperature are both precisely measured using the BMP388 absolute barometric pressure sensor, which is precise, low-power, and low-noise. Because pressure changes with height, we can also figure out how high we are with a lot of accuracy. Because of this, this sensor is useful for drones and navigation. It can also be used for other things:

It can also be used for other things:

- vertical velocity calculation;
- internet of things;
- weather forecast and
- weather stations;
- health care applications;
- fitness applications;

Part Required:

- BMP388 sensor module
- ESP8266 (read Best ESP8266 development boards)
- Breadboard
- Jumper wires

Schematic for ESP8266 NodeMCU with BMP388

The I2C and SPI communication protocols can both be used by the BMP388 when it comes to exchanging data.

ESP8266 with BMP388 using I2C

To connect the BMP388 to the ESP8266 via making use of the I2C pins by default, refer to the following schematic design.

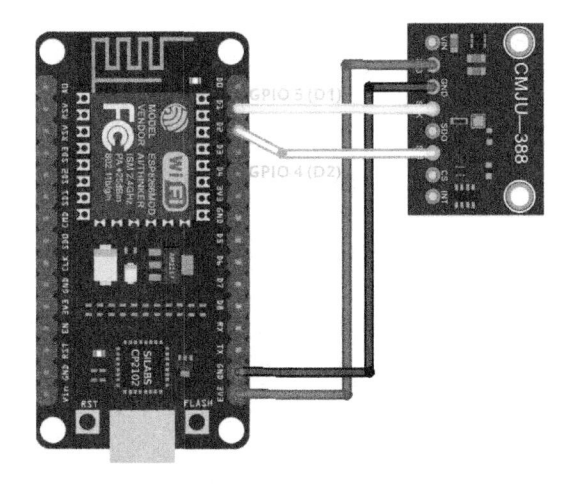

ESP8266 with BMP388 using SPI

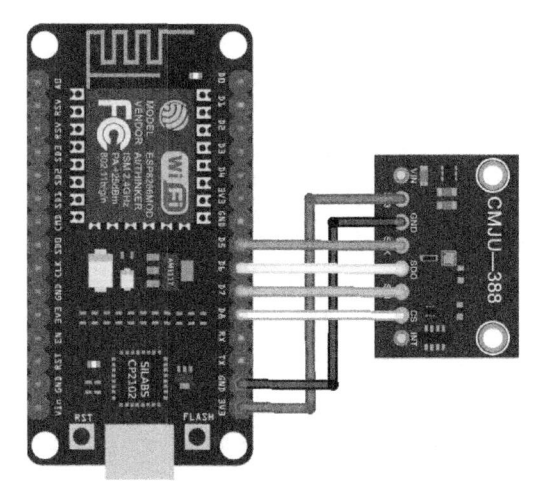

The SPI communication protocol is an alternative option to consider. In such case, refer to the subsequent schematic diagram for instructions on how to connect the BMP388 to the ESP8266 utilizing the SPI pins that are standard.

420.Multiple Sliders for ESP8266 NodeMCU Web Server (WebSocket): Control the brightness of LEDs (PWM)

This project demonstrates how to construct a web server with the ESP8266 NodeMCU board by displaying a web page that contains many sliders on the screen. The brightness of many LEDs can be adjusted by adjusting the duty cycle of several PWM signals, which is controlled by the sliders. This idea can be used to drive DC motors or other actuators that require a PWM signal in place of the LEDs if you so want. The WebSocket protocol is utilized in order to facilitate communication between the clients and the ESP8266. In addition, each time there is a modification, the values of the sliders on all of the clients are updated concurrently.

Project Overview

The web page we'll make for this project is shown in the picture below:

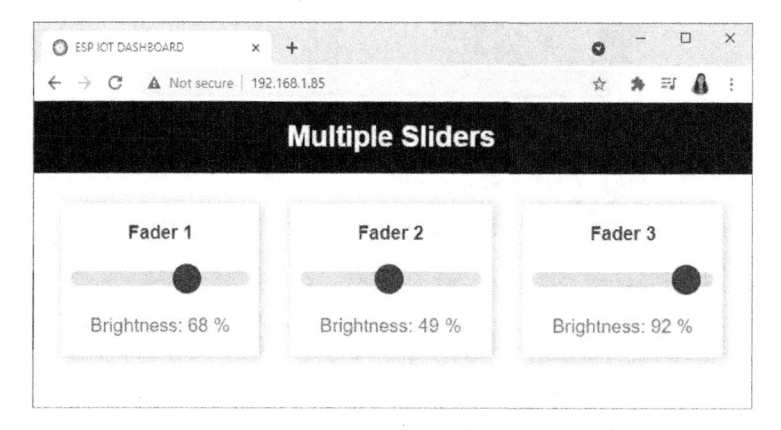

The web page has three cards: Fader 1, Fader 2, and Fader 3. Each card has its own paragraph with the title of the card. On each card is a range slider that you can use in order to adjust the amount of light emitted by the appropriate LED; Another paragraph on each card displays the brightness of the LEDs at the moment (shown as a percentage); If you have more than one tab open in your web browser (or more than one device), all of those tabs and devices will update almost instantly whenever there is a change. When you move the slider to a new position, all clients are notified of the change.

How Does It Work?

A web server that the ESP maintains and operates presents a webpage that has three sliders;

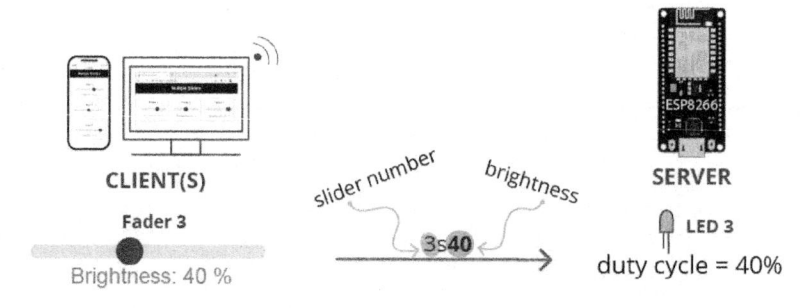

When you move a slider to a new location, the client will send the number of the slider as well as the value of the slider to the server via the WebSocket protocol. For instance, if you moved slider number 3 to position number 40, it would communicate that setting to the server with the message "3s40."

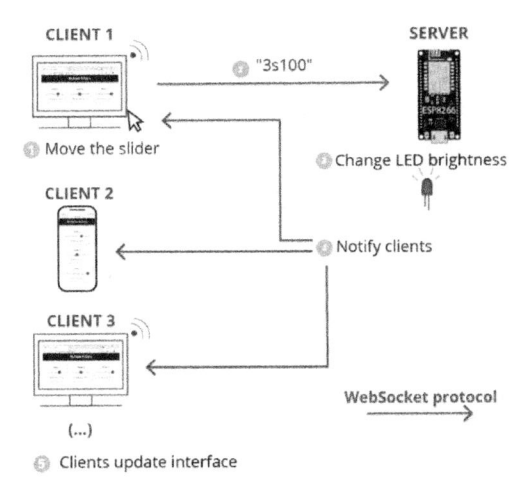

Figure: WebSocket communication between clients and ESP server showing slider interaction, LED brightness change, and client notification.

- Move the slider
- Change LED brightness
- Notify clients
- Clients update interface
- WebSocket protocol

The server (ESP) will make the necessary adjustments to the PWM duty cycle after it has received the slider number and the related value. Additionally, it alerts all the other clients with the new current slider settings, which enables us to have all clients changed in what feels like almost no time at all. To adjust the level of brightness emitted by the LED, the ESP8266 sends out a PWM signal complete with the appropriate duty cycle. If the LED has a duty cycle of 0%, then it is completely dark; if it has a duty cycle of 50%, then it is only half lighted; and if it has a duty cycle of 100%, then it is completely lit; It will send a message to the ESP8266 (also using the WebSocket protocol) with the message getValues whenever you open a new web browser window. This is when a new client connects. When the ESP8266 receives this message, it will communicate the values of the slider at the current position. When you start a new tab, the values will always be displayed as they have been most recently changed in this manner.

Schematic Diagram

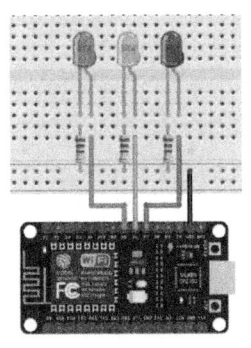

Figure: Three LEDs connected to an ESP8266 on a breadboard.

Connect three LEDs to the ESP8266 using the wires. GPIOs 12 (D6), 13 (D7), and 14 are being utilized here (D5). You are free to use any other GPIOs that are appropriate.

1) Parts Required

- ESP8266 NodeMCU Board - read our Best ESP8266 Wi-Fi Development Board - Buying Guide for more information.
- 3x LEDs
- 3 resistors each rated at 220 ohms
- Jumper wires for the breadboard

You can test this project with just one LED, the results of which can be seen in the Serial Monitor, or you can use other actuators whose operation requires a PWM signal. You do not need three LEDs to test this project.

2) Add-on for the Arduino Software and ESP8266 Boards

Using Arduino IDE, we'll write some code for the ESP8266. Therefore, the ESP8266 add-on component needs to be installed on your system.

3) Filesystem Uploader Plugin

We will be utilizing a plugin for the Arduino IDE known as the LittleFS Filesystem uploader in order to upload the HTML, CSS, and JavaScript files required to construct this project into the flash memory (LittleFS) of the ESP8266.

4) Libraries

You will need to install the following libraries in order to construct this project: ESPAsyncWebServer (a.zip folder), ESPAsyncTCP, and the Arduino JSON library by Arduino version 0.1.0 can be found in the Arduino Library Manager (.zip folder). Utilizing the Arduino Library Manager, you will be able to install the initial library. Find the library's name by going to Sketch > Include Library > Manage Libraries and searching for it there. Because the ESPAsynTCP and ESPAsynWebServer libraries cannot be installed through the Arduino Library Manager, you will need to manually copy the library files into the Arduino Installation Libraries folder. These libraries are not accessible for installation through the Arduino Library Manager. You can also select the libraries you just downloaded by going to the Sketch menu in your Arduino IDE, selecting the Include Library option, and then clicking the Add.zip Library button.

421. ESP8266 NodeMCU Plot Sensor Readings in Charts (Multiple Series)

This project demonstrates how to construct a web server using the ESP8266 NodeMCU board so that sensor readings can be plotted in charts that contain more than one series. On the same chart, we will plot the readings from four different DS18B20 temperature sensors so that you can see how this works. You are free to modify the project in order to plot any other data you require. We will be utilizing the Highcharts JavaScript library in order to create the charts.

Project Overview

This project will build a web server with the ESP8266 NodeMCU board. The web server will display temperature readings from four DS18B20 temperature sensors on the same chart; the chart will have multiple series. The chart can show a maximum of forty data points for each series, and it refreshes itself with fresh data every thirty seconds. These values can be changed in your code as you see fit.

DS18B20 Temperature Sensor

The DS18B20 temperature sensor is a digital temperature sensor that only requires a single wire to operate. This indicates that only a single data line is necessary in order to successfully communicate with your microcontroller.

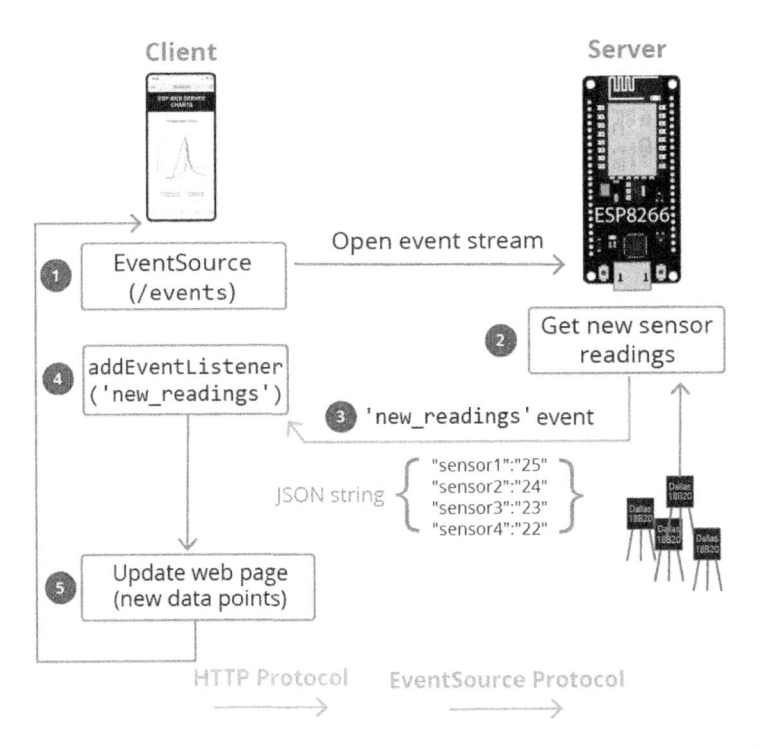

Because every sensor has its own one-of-a-kind 64-bit serial number, it is possible to connect multiple sensors to the same GPIO, which is exactly what we're going to do in this tutorial. Find out more information regarding the DS18B20 temperature sensor here:

Server-Sent Events

Server-Sent Events are used to ensure that the readings on the website are kept up to date automatically (SSE). Those files that have been saved on the filesystem (LittleFS)

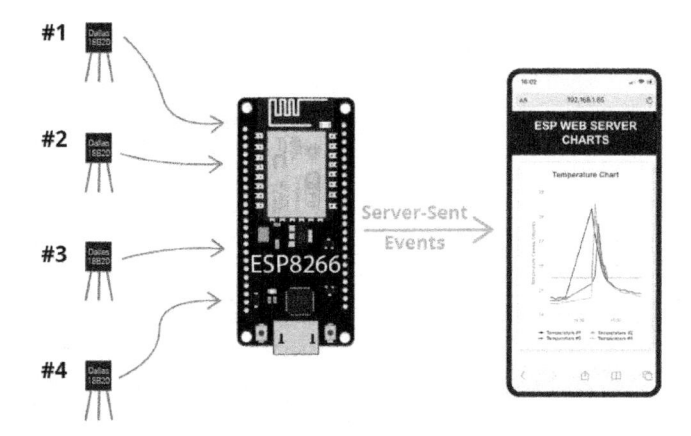

The HTML, CSS, and JavaScript files that will be used to construct the web page will be saved on the filesystem of the board. This will ensure that our project is better organized and easier to comprehend (LittleFS).

Components Required:

- ESP8266 (read Best ESP8266 development boards)
- 4x DS18B20 temperature sensor (one or multiple sensors), in a water-resistant version
- 4.7k Ohm resistor
- A set of jumper wires
- Breadboard

Prerequisites

Before moving on with the project, you must first ensure that all of the prerequisites outlined in this section have been met.

1. Open the Arduino Software and Install the ESP8266 Board

Using Arduino IDE, we'll write some code for the ESP8266. Therefore, the ESP8266 add-on component needs to be installed on your system.

2. Filesystem Uploader Plugin

We will be utilising a plugin for the Arduino IDE known as the LittleFS Filesystem uploader in order to upload the HTML, CSS, and JavaScript files to the flash memory (LittleFS) of the ESP8266.

3. Installing Libraries

You will need to install the following libraries in order to construct this project: OneWire, which was developed by Paul Stoffregen and is managed by the Arduino Library Manager; DallasTemperature, which is also managed by the Arduino Library Manager; the Arduino JSON library, which was developed by Arduino and is currently at version 0.1.0 (Arduino Library Manager)

ESPAsyncWebServer (.zip folder); ESPAsyncTCP (.zip folder). You will need to use the Arduino Library Manager in order to install the first two libraries. Find the library's name by going to Sketch > Include Library > Manage Libraries and searching for it there. Due to the fact that the ESPAsyncWebServer and AsynTCP libraries cannot be installed using the Arduino Library Manager, it is necessary for you to manually copy the library files into the folder designated for Arduino Installation Libraries. Alternately, you can select the libraries you just downloaded by going to the Sketch menu in your Arduino IDE, selecting the Include Library option, and then clicking the Add.zip Library button.

Schematic Diagram

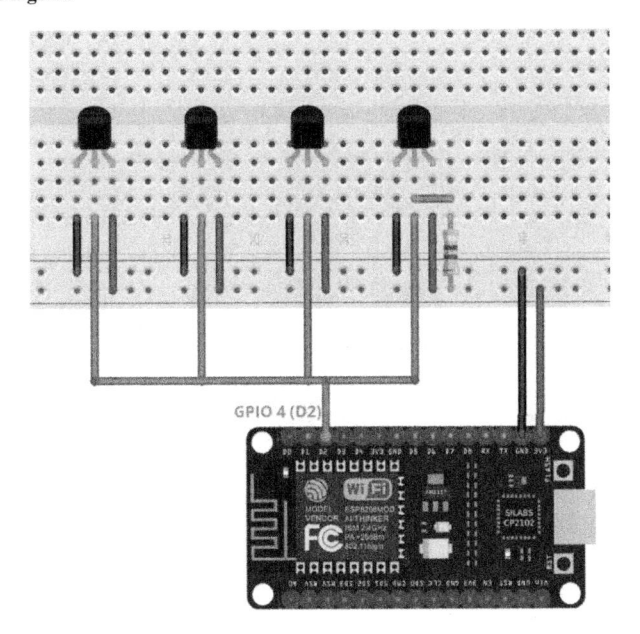

422. ESP8266 NodeMCU Integrated with MPU-6050 Accelerometer, Gyroscope, and Temperature Sensor (Arduino)

Through the completion of this project, you will acquire the knowledge necessary to interface the ESP8266 NodeMCU with the MPU-6050 accelerometer and gyroscope module. The MPU-6050 Inertial Measurement Unit (IMU) is a sensor that contains a 3-axis accelerometer as well as a 3-axis gyroscope. Both the accelerometer and the gyroscope measure the rotational velocity of the object. The accelerometer measures the acceleration due to gravity. In addition to that, this module is also capable of measuring temperature. This sensor works wonderfully for determining the orientation of an object that is in motion. The MPU-6050 is a module that contains a three-axis accelerometer as well as a three-axis gyroscope. The rotational velocity (in radians per second) along the X, Y, and Z axes is what the gyroscope measures. This can be thought of as the change in angular position over time (roll, pitch and yaw). Because of this, we are able to figure out the orientation of an object.

Acceleration, or the rate at which the object's velocity is changing, is what the accelerometer measures. It is able to detect dynamic forces such as vibrations or movement in addition to static forces such as gravity (9.8m/s2). Acceleration along the X, Y, and Z axes can be measured with the MPU-6050. In a perfect world, the acceleration along the Z axis of a static object should be the same as the gravitational force, while the acceleration along the X and Y axes should be zero. Calculating the roll and pitch angles using trigonometry is possible if the values from the accelerometer are used. The yaw, on the other hand, cannot be determined by calculation.

Installing Libraries

Readings can be obtained from the sensor in a variety of different ways. We'll be utilizing the Adafruit MPU6050 library throughout this tutorial. In addition, the Adafruit Unified Sensor library and the Adafruit Bus IO Library need to be installed on your computer before you can use this library.

Launch the Arduino Integrated Development Environment (IDE), then navigate to **Sketch > Include Library > Manage Libraries**. It ought to start up the Library Manager.

Install the library by entering "adafruit mpu6050" into the search box on your computer.

The next step is to conduct a search for "Adafruit Unified Sensor." To locate the library and to install it, scroll all the way to the bottom.

Finally, look for "Adafruit Bus IO" in the search engine, and then install it.

Readings Obtained From the MPU-6050 Sensors, Including Accelerometer, Gyroscope, and Temperature

- Readings of acceleration (x, y, and z), angular velocity (x, y, and z), and temperature will be discussed in this section so that you can learn how to retrieve them from the MPU-6050 sensor.
- Readings from the Accelerometer, Gyroscope, and Temperature Sensors on the MPU-6050 Can Be Obtained Using the Following Code:
- This sensor is demonstrated in a number of different ways within the Adafruit library. In this part of the tutorial, we are going to look at the fundamental example that prints the readings from the sensors in the Serial Monitor.

- Navigate to the basic readings folder by going to File > Examples > Adafruit MPU6050. It is expected that the following code will load.
- It obtains the temperature as well as the angular velocity (gyroscope) on the x, y, and z axes, as well as the acceleration on the x, y, and z axes.

Parts Required

The following components are necessary to finish off this example:

- MPU-6050 Accelerometer Gyroscope
- ESP8266 (read Best ESP8266 development boards)
- Jumper wires for the 0.96-inch I2C OLED Display SSD1306 on the Breadboard
- Jumper wires

ESP8266 NodeMCU Schematic Diagram, Including MPU-6050 and OLED Display

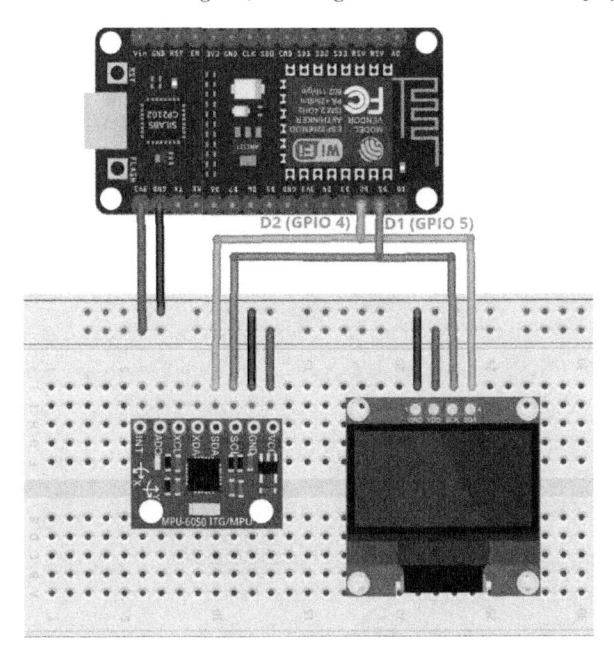

Connect each of the components using the wiring shown in the schematic diagram below. We are able to connect the OLED display and the MPU-6050 sensors to the same I2C bus because these two types of components use different I2C addresses (same pins on the ESP8266). Display the MPU-6050 sensor readings on the OLED display with this code. In order to follow along with this example, you will need to ensure that the Adafruit SSD1306 library is properly installed. The Arduino Library Manager allows for the installation of this library on your device. You can get the SSD1306 library from Adafruit by going to **Sketch > Library > Manage Libraries**, searching for "SSD1306," and then installing it.

423. The ESP8266 NodeMCU can be used to get Epoch/Unix Time (Arduino)

Using the Arduino IDE and the ESP826 NodeMCU board, this project demonstrates how to obtain the epoch time or the unix time. Obtaining the time of the epoch can be helpful for a variety of applications, including timestamping your readings, providing files with names that are distinct, and more. Because we will be requesting the current epoch time from an NTP server, the ESP8266 board really has to be connected to the internet.

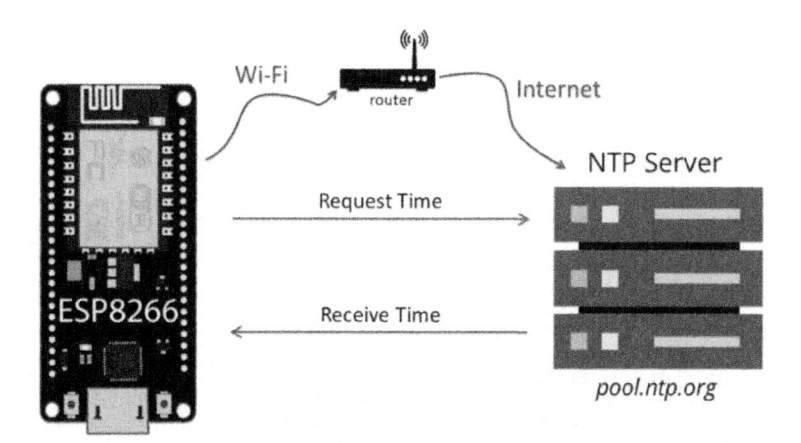

What is epoch time?

The number of seconds that have transpired since the Unix epoch, which occurred at 00:00:00 Coordinated Universal Time (UTC) on January 1, 1970, minus any leap seconds, is referred to as the Epoch time or the UNIX time. It monitors the passage of time in seconds and does all calculations beginning at the same instant in time regardless of the time zone. It is frequently utilized in a variety of file formats and operating system configurations.

What is the purpose of the epoch time in Unix?

Unix time is represented by a single signed number that increases by one every second. Because of this, it is far simpler for computers to both store and operate than other date systems. After that, interpreter programmes can turn information into a format that is readable by humans. The time 00:00:00 Coordinated Universal Time (UTC) on January 1, 1970 is the Unix epoch.

Why is it necessary for us to use epoch time?

In order for the computer to keep track of the time, it has to know when to begin. When computers keep track of time, one common standard they use is called the Epoch time. The time from the Epoch, which is measured in seconds, begins at 00:00:00 on Thursday, January 1, 1970 UTC. Therefore, the number of seconds that have passed since Epoch zero is the Epoch time.NTP (Network Time Protocol) (Network Time Protocol)

Network Time Protocol is a networking protocol that allows computer systems to synchronize their clocks with one another. NTP is an abbreviation for "Network Time Protocol." In other words, it is utilized to synchronize the clock times of computers that are connected to a network.

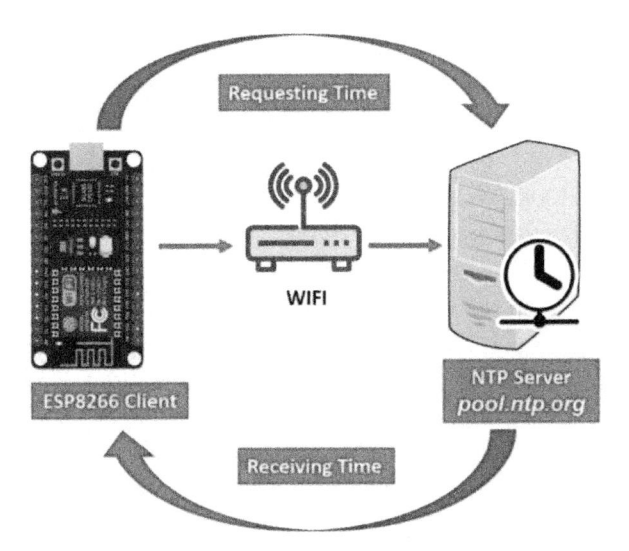

NTP is a protocol that is commonly utilized to synchronize the clocks on many computers to a reference network. It synchronizes the epoch time of all networked devices to the Coordinated Universal Time (UTC) with an accuracy of around 50 milliseconds over the wide-area network (WAN) and less than 5 milliseconds over the local area network (LAN).

In order to retrieve epoch time from an ESP8266 NodeMCU connected to an NTP server, the server will function according to the client-server model. We will make use of the NTP server located at pool.ntp.org, which is readily available to people all over the world. As the client, our ESP8266 NodeMCU development board will connect to the NTP server over UDP on port 123. This connection will take place on our behalf. The server, which is located at pool.ntp.org, will be able to connect with the client by using this port. When the connection has been successfully established, the ESP8266 NodeMCU board will then submit a request to the server. After the NTP has been given the request, it will send out a time stamp that includes all of the relevant information regarding the time.

Bringing in the NTP Client and the Wi-Fi Libraries

In order to retrieve the time at the current epoch using ESP8266, we will be including the following three libraries. Because we want our ESP8266 board to connect with the local network, we need to make sure that we have the ESP8266WiFi.h library. Because of this, we will have access to the NTP server via the internet, making this library an absolute necessity. Second, the NTPClient.h library will take care of the synchronization of the NTP server, which is necessary in order to retrieve the current time from the NTP server.

424.ESP8266 NodeMCU MQTT - Publish Temperature, Humidity, Pressure, and Gas Readings from a BME680 Sensor (Arduino IDE)

Learn how to use the ESP8266 NodeMCU and MQTT to send temperature, humidity, pressure, and gas air quality readings from a BME680 sensor to any platform that supports MQTT or any MQTT client. As an illustration, we will send the readings from the sensors to the Node-RED Dashboard MQTT is gaining popularity as an Industrial IoT (Internet of Things) data protocol. It is a lightweight method that is both efficient and secure, and it was developed for the purpose of connecting remote devices to a central server. On the other hand, IoT installations are becoming larger and more complicated, and there is a rising demand for OT/IT connectivity.

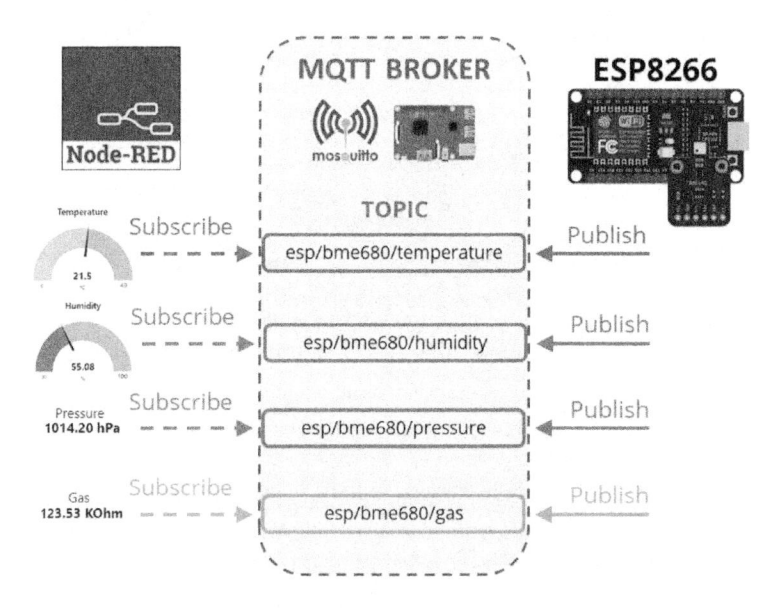

The MQTT protocol is therefore good for the environment and simple to implement for a large number of devices. Keeping devices connected despite having unreliable networks: Even when connections between devices are unstable, MQTT in the Internet of Things is able to leverage QoS levels to assure that messages will be delivered to their intended recipients.

Components Required

- The ESP8266 is a required component (read Best ESP8266 development boards)
- Raspberry Pi board equipped with BME680 and ESP8266 Guide (read Best Raspberry Pi Starter Kits)
- Class 10 MicroSD Card with 16 GB of Storage
- Supply of Electricity for the Raspberry Pi (5V 2.5A)
- A set of jumper wires, Breadboard

Project Overview

A high-level overview of the project that we are going to construct can be seen in the accompanying diagram. Readings from the sensor are requested by the ESP8266, which is connected to the BME680. Node-RED is subscribed to those topics; Node-RED receives the sensor readings and then displays them on gauges and text fields; The temperature readings are published in the esp/bme680/temperature topic; the humidity readings are published in the esp/bme680/humidity topic; the pressure readings are published in the esp/bme680/pressure topic; and the gas readings are published in the esp/bme680/gas topic. You have the ability to receive the readings in any other platform that supports MQTT, and you may treat the readings in any manner that you see fit.

BME680 Integrated Environmental Sensor

The BME680 is a digital sensor that measures gas, humidity, pressure, and temperature all in one package, and it does so use sensing methods that have been tried and tested. The BMP180, BMP280, and BME280 are all predecessors to the BME680, which is an updated and superior version of those models. The BME680 has a gas sensor that can detect a wide variety of volatile organic chemicals, which allows it to monitor the quality of the air within. High linearity and accuracy are both characteristics of the sensor. The BME680 was designed specifically for mobile applications and wearables, two categories of electronics in which the criteria for size and power consumption are particularly stringent. The gas sensor contained within the BME680 is capable of detecting a wide variety of gases, including volatile organic compounds, in order to carry out air quality measurements (VOC).

Schematic Diagram

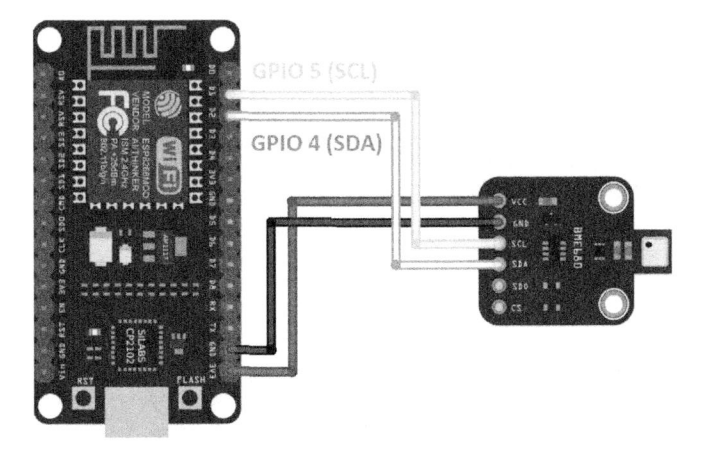

Connect the BME680 to the ESP8266 in the manner depicted in the accompanying block diagram of the schematic. The SDA pin should be connected to GPIO 4, and the SCL pin should be linked to GPIO 5.

425. Web Server for ESP32/ESP8266 Relay Modules Built with the Arduino IDE (1, 2, 4, 8, 16 Channels)

ESP8266 NodeMCU Web Server that has the capability of controlling any relay module. We will develop an ESP32/ESP8266 Web Server that is responsive on mobile devices and that can be accessible from any device in your local network that has a browser installed on it. A relay is a switch that is electrically operated, and just like any other switch, it can be switched on or off to either allow electricity to pass through it or prevent it from doing so. It is possible to control it with low voltages, such as the 3.3V that is supplied by the ESP GPIOs, and it also gives us the ability to manage high voltages, such as 12V, 24V, or the mains voltage (230V in Europe and 120V in the US).

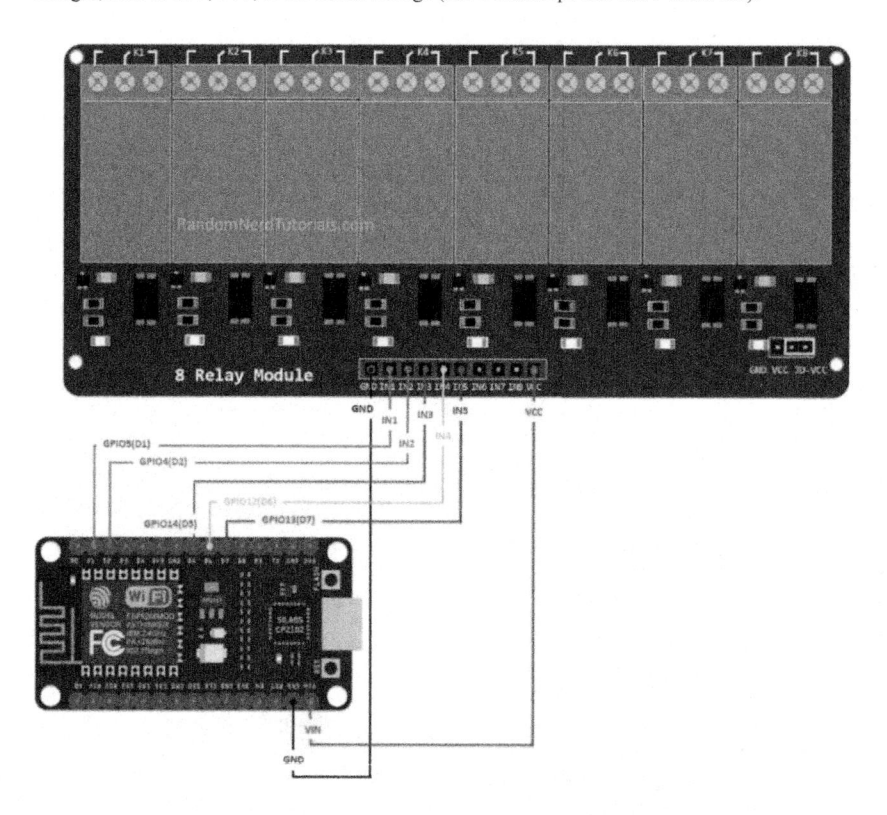

There is a variety of relay hardware available, each with its own distinct number of channels. There are relay modules available with a single channel, two channels, four channels, eight channels, and even sixteen channels. The number of outputs you can control is directly proportional to the number of channels available.

Control Relay Modules with Multiple Channels using an ESP32/ESP8266 Web Server

You can control an unlimited number of relays via web server with the help of this code for the web server, regardless of whether the relays are set up to be generally opened or normally closed. To determine the number of relays you wish to manage and the assignment of their pins, all that is required is a simple update to a few lines of code.

Wiring a Relay Module to an ESP Board

The goal of this example requires us to operate five different relay channels. Following the instructions provided in the following set of schematic diagrams, connect the ESP boards to the relay module.

Caution: the majority of relay projects involve working with mains voltage. Inappropriate use may result in severe harm. In the event that you are not familiar with mains voltage, you should seek the assistance of someone who is. Make sure that everything is removed from mains voltage before you begin programming the ESP or wiring your circuit. You also have the option of controlling 12V appliances with a power source that operates at 12V.

426. DHT Temperature and Humidity Readings Displayed on the on-board ESP8266's OLED Display

Discover how to read the temperature and humidity from a DHT11 or DHT22 sensor on an SSD1306 OLED display by utilizing an ESP32 or ESP8266 along with the Arduino IDE. The objective of illustrating how one may develop a physical user interface for their boards is the primary motivation behind combining the OLED display with either the ESP32 or the ESP8266. In this project, we will make use of an I2C SSD1306 128x64 OLED display. A DHT22 temperature and humidity sensor will be utilized in order to obtain accurate readings of the surrounding environment (you can also use DHT11).

Components Required:

- 0.96 inches of OLED screen real estate
- ESP32 or ESP8266; both are available (read ESP32 vs ESP8266)
- Temperature and humidity sensors of the DHT22 or DHT11 kind
- Breadboard
- 10k Ohm resistor
- A set of jumper wires

Schematic

Because the OLED display that we are communicates using the I2C communication protocol, you will need to connect it to the I2C pins on either the ESP32 or the ESP8266. The following pins will be assigned as the default I2C connections if you are using an ESP8266: GPIO 5 (D1): SCL GPIO 4 (D2): SDA

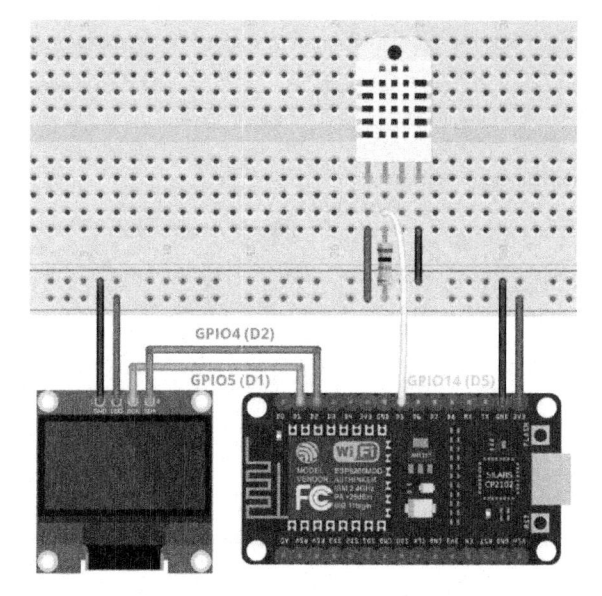

If instead you are working with an ESP8266, go to the following diagram. In this example, the data pin of the DHT sensor is connected to GPIO 14, but you are free to use any other GPIO that is appropriate.

427. Temperature and Humidity Web Server for ESP8266 DHT11/DHT22 with Arduino IDE

In this project, you'll use the Arduino IDE to turn an ESP8266 into a stand-alone web server that shows the temperature and humidity with a DHT11 or DHT22 sensor. You will be able to access the web server that you will develop using any device that is connected to your local network and has a browser installed on it. Throughout the lesson of this project, we will demonstrate how to construct two distinct web servers, namely:

Web Server #1 is an asynchronous web server that automatically refreshes the website with the current temperature and humidity readings without the need for the user to do so. This web server also uses custom CSS to decorate the page.

Web Server #2 is a basic HTTP web server that updates a raw HTML page with the most recent sensor values whenever the page is refreshed. This web server only supports HTTP.

Component Required:

- ESP8266 development board (read ESP8266 development boards comparison)
- Breadboard with either a DHT22 or DHT11 Temperature and Humidity Sensor and a 4.7k Ohm Resistor

- A set of jumper wires

ESP8266 with DHT11/DHT22 Schematic Diagram

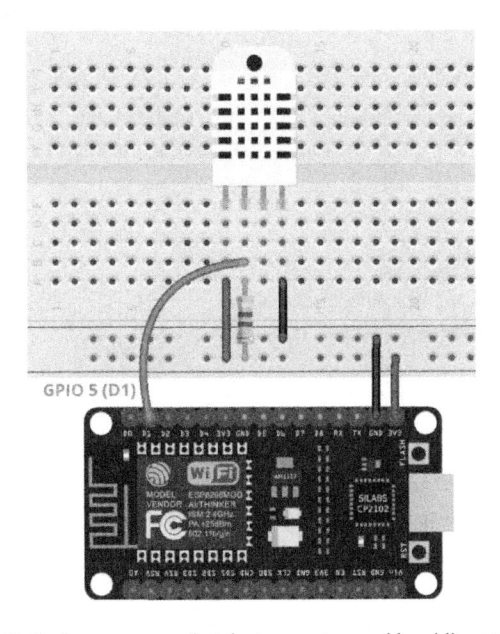

Before continuing with the lesson, ensure that the temperature and humidity sensor, either a DHT11 or DHT22, is connected to the ESP8266 in the manner depicted in the following schematic diagram.

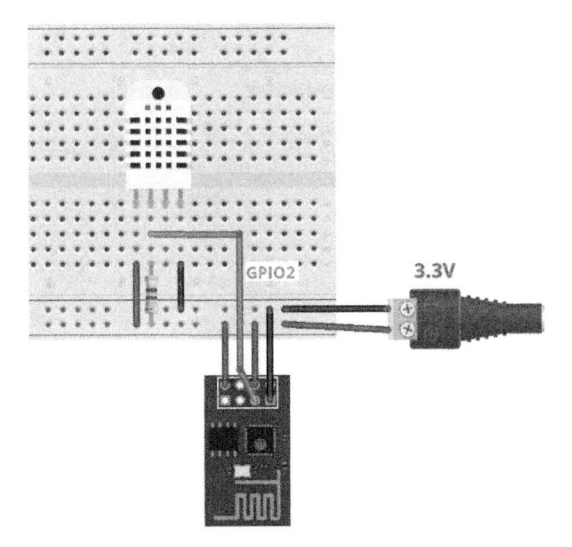

In this demonstration, the DHT data pin is connected to GPIO5 (D1); however, you are free to use any other GPIO that is appropriate. Learn more about the GPIOs available on the ESP8266 by consulting our ESP8266 GPIO Reference Guide. As the previous diagram demonstrates, the GPIO 2 pin on an ESP-01 is the one that is best suited for making the connection to the data pin on a DHT sensor.

428. ESP8266 ADC – Read Analog Values with Arduino IDE, MicroPython and Lua

Both the ESP8266-12E and the ESP8266-07 have an easy-to-reach ADC pin. This means that the analogue signals can be read by these ESP8266 boards. In this tutorial, we'll show you how to use Arduino IDE, MicroPython, or Lua firmware to read analogue data from an ESP8266.

This project is divided in three sections:

1. ESP8266 Analog Read with Arduino IDE
2. ESP8266 Analog Read with MicroPython
3. ESP8266 Analog Read with Lua/NodeMCU

Part Required:

- ESP8266-12E NodeMCU Kit is suggested (read ESP8266 development boards comparison)
- 1x ESP8266-07 chip or 1x ESP8266-12E chip plus an FTDI programmer is another option.
- If you are using a bare chip, you will need a 100 Ohm resistor.
- If you're using a bare chip, you'll need a 220 Ohm resistor. 1k Ohm Potentiometer
- Breadboard, Switchers

Schematic Diagram

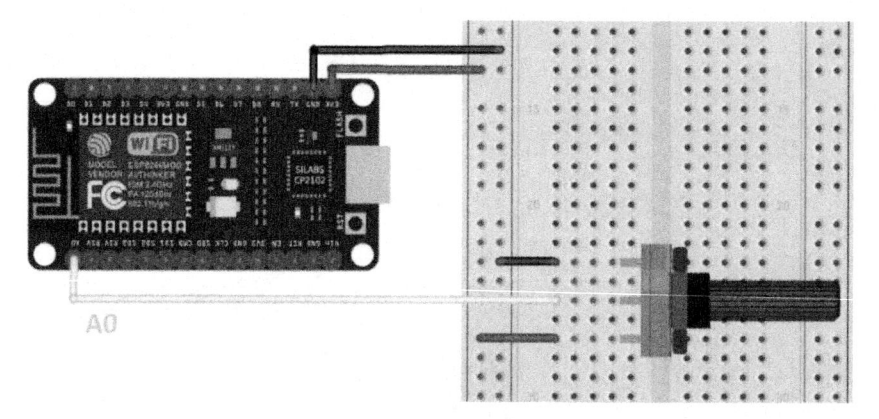

If you are using an ESP8266 chip with an input voltage range of 0V to 1V, the input voltage on the A0 pin can't be more than 1V. So, you need a voltage divider circuit.

Wrapping Up

In this guide, we showed you how to use the ESP8266 analogue pin to read analogue values (A0). One important thing to remember is that the ESP8266 analogue input range is either 0-1V or 0-3.3V, depending on whether you're using a bare chip or a development board No matter what, you should never go over the maximum voltage that is recommended. When you need a wider range of input voltage, you might want to add a voltage divider circuit.

429. Hack a PIR Motion Sensor with an ESP8266

With an ESP8266, you can hack a PIR motion sensor. In this project, we'll connect an ESP8266 to a commercial (mains-powered) motion sensor and change it so that it logs data whenever it detects motion. MQTT, a type of communication protocol, will be used to send the data to Node-RED. The HLK-PM03 AC/DC converter will be used to power the ESP8266 through the motion sensor phase out. The HLK-PM03 AC/DC converter will be used to power the ESP8266 through the motion sensor phase out wire. Motion sensors, also called motion detectors, are electronic devices that can detect and measure movement. Most home and business security systems use motion sensors, but you can also find them in phones, paper towel dispensers, game consoles, and virtual reality systems.

What You Need:

- PIR 220V Motion Sensor (or 110V PIR Motion Sensor)
- ESP8266-01 – See Best ESP8266 Wi-Fi Development Boards
- ESP8266-01 Serial Adapter (to upload code to the ESP8266)
- Hi-Link HLK-PM03 (to convert mains voltage to DC 3.3V)
- Slow Blow Small Protoboard Fuse (200mA)
- 47 uF electrolytic capacitor
- Pi Raspberry (to host Node-RED and MQTT broker)

There are three parts to this project:

- Putting the wires together
- Putting the ESP8266 code on paper and uploading it
- Setting up the flow in Node-RED

We'll hack a commercial motion sensor that has enough room for an ESP-01 and an HLK-PM03 AC/DC converter module. We got our motion sensor at a store near us for $5. When the PIR sensor sees movement, power flows through the red-hot wire and can be used to turn on a lamp or other device. Your motion sensor should have a similar wiring diagram on the top or in the instructions. In our project, the motion sensor output load is the HLK-PM03 AC/DC converter module, which will power the ESP8266. The HLK-PM03 module AC/DC converter takes either 110VAC or 220VAC and turns it into 3.3V. This makes it perfect to use mains voltage to power the ESP8266.

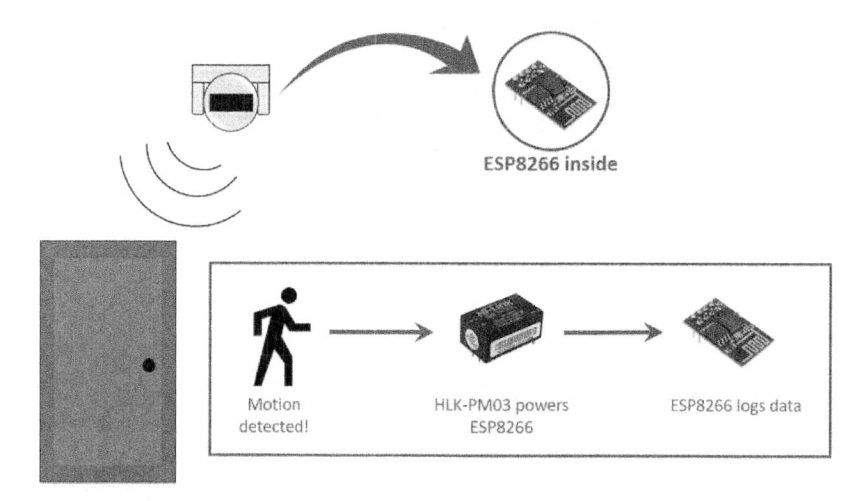

Motion detected! → HLK-PM03 powers ESP8266 → ESP8266 logs data

In short, power is sent to the ESP8266 when motion is detected. As long as the motion sensor is set off, the ESP8266 can run tasks. You might need to change how long the sensor stays on so that the ESP8266 has enough time to do its job. There should be a knob on the sensor to change the time (and another to adjust luminosity). In our example, when the ESP8266 is turned on, it runs a sketch that sends information to Node-RED via MQTT to record the date and time that motion was detected. You don't have to send information to Node-RED. You can do other things, like: Log information into a Google Spreadsheet, send an email when motion is detected, and get alerts on your phone. IFTTT makes it easy to do all of these things.

Schematic Diagram:

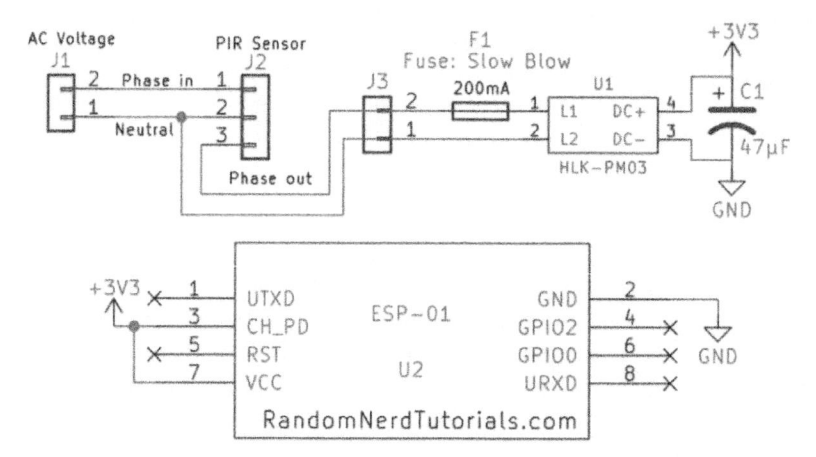

Take off the lid of your PIR sensor. Inside, it should have three wires: one for phase in, one for phase out, and one for neutral. Do these things: Wire the motion sensor with the phase-in (brown) and neutral (blue) wires.

Wire the HLK-PM03 input with the neutral (blue) and phase out (red) wires. It is recommended to add a slow-blow fuse and a capacitor to the output right before the HKL-PM03 converter.

Note: We recommend using this protection circuit if you are using an ESP8266 that is always on with the HLK-PM03. The 3.3V and GND come out of the HLK-PM03. These are hooked up to the VCC and GND pins of the ESP8266 to power it. To save space, we put the HLK-PM03 and ESP8266 circuit on a small protoboard. We've also added some header pins so the ESP8266-01 can be put in place. So, every time you need to upload new code, you can just plug and unplug the board.

430.Power ESP8266 with Mains Voltage using Hi-Link HLK-PM03 Converter

Using the Hi-Link HLK-PM03 converter, you will learn how to power the ESP8266 (or ESP32) using mains voltage in this tutorial. We'll demonstrate how the ESP8266-01 may be used to control a relay by connecting it to a web server. Both the ESP32 and ESP8266 are inexpensive Wi-Fi modules that are excellent choices for Do It Yourself (DIY) projects involving the Internet of Things (IoT) and Home Automation. When combined with a relay, an ESP32 or ESP8266 gives you the ability to remotely control any AC electronic appliance using Wi-Fi (using a web-server, for example). Finding an appropriate power source for the ESP32/ESP8266 that has a tiny form factor at the same time is one of the most challenging aspects of these projects (in a final application, you do not want to power the relay and the ESP32/ESP8266 using two different power supplies). Using the AC/DC converter Hi-Link HLK-PM03 to draw power from the mains voltage in order to power the ESP8266 or ESP32 is one method (or HLK-PM01 model).

Introducing the Hi-Link HLK-PM03/01 Converter Modules

The Hi-Link HLK-PM03 is an AC/DC converter module that is quite compact. The HLK-PM03 AC/DC converter can obtain 3.3V from either 110VAC or 220VAC, depending on which input voltage is used. Because of this, it is ideal for use in modest applications that require a supply of 3.3V from the mains voltage. The HLK-PM01 is another option for achieving the desired output of 5V. We will be utilizing the HLK-PM03 to supply 3.3V to the VCC pin of the ESP8266-01 so that it can be powered by the mains voltage. If you need to power other ESP8266 models or an ESP32 through the VIN pin with 5V, you can use the HLK-PM01 model, which has a 5V output and functions in a manner that is analogous to that of the original.

Safety Warning: This project deals with voltage that comes from the mains. Check that you have a firm grasp on the activities at hand. Please make sure you give the safety warning a thorough read.

Powering the ESP8266 with AC using the Hi-Link HLK-PM03 module

You don't need any additional circuitry to utilize the HLK-PM03; simply connect it straight to the VCC pin on the ESP8266. Having said that, I don't think that's a good idea. It is highly recommended that a protective circuit consisting of a thermal fuse and fast-acting fuses be added. When powering an ESP8266 device, it is a good idea to add capacitors to the output of the HLK-PM03 so that the voltage peaks are smoothed out and unexpected resets or unstable behavior are avoided. In addition, we have installed a varistor across the mains input in order to shield the circuit from sudden spikes in voltage.

Parts Required

- HLK-PM03\ESP8266-01\ESP8266-01
- Programmer Relay Module Compatible with FTDI Serial Adapter (3.3V)
- Terminal blocks consisting of two connections each
- Terminal blocks consisting of three connectors each
- Electrolytic capacitor 10uF
- Electrolytic capacitor 22uF
- Slowly Blow the Fuse (200mA)
- Thermal Fuse (72 degrees Celsius)
- Blow the Fuse Quickly (630mA)
- Varistor Stripboard (Prototyping Circuit Board)
- Connecting Wires

Circuit Diagram:

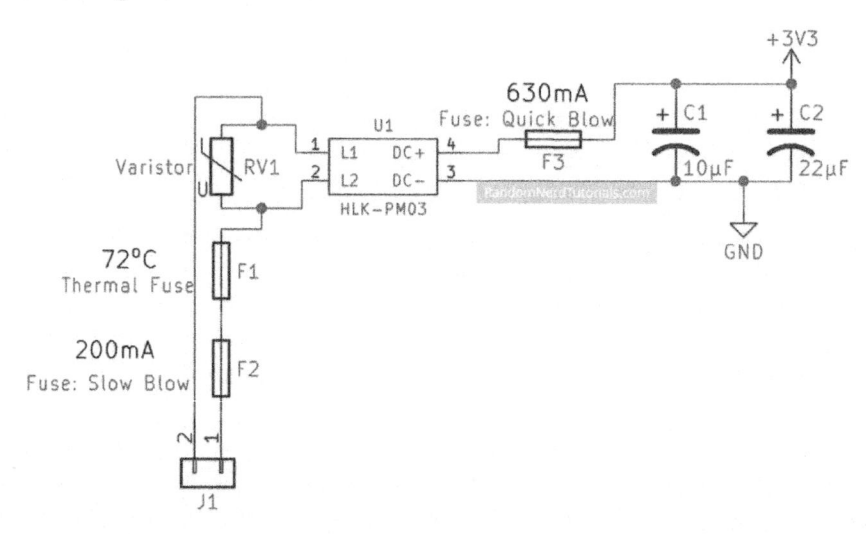

The connection to the mains voltage should be made at the J1 terminal block. The ESP8266-01 will be powered by the 3.3V and GND that are provided after the capacitors. A three-terminal block that is attached to the ESP8266 has also been added in order to gain access to 3.3V, GND, and GPIO 2 in order to operate an output (relay module).

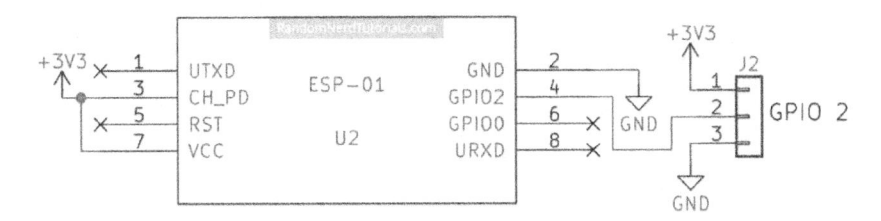

Due to the fact that we are working with 3.3V, you should make use of a 3.3V relay module similar to this one.

431. ESP8266 Multisensor Shield with Node-RED

You will learn how to integrate the ESP8266 Multisensor Shield for the Wemos D1 Mini board with Node-RED by following the instructions in this project. An MQTT connection will be established between the ESP8266 Multisensor Shield and a Raspberry Pi that is running Node-RED and the Mosquito MQTT broker.

Project Overview

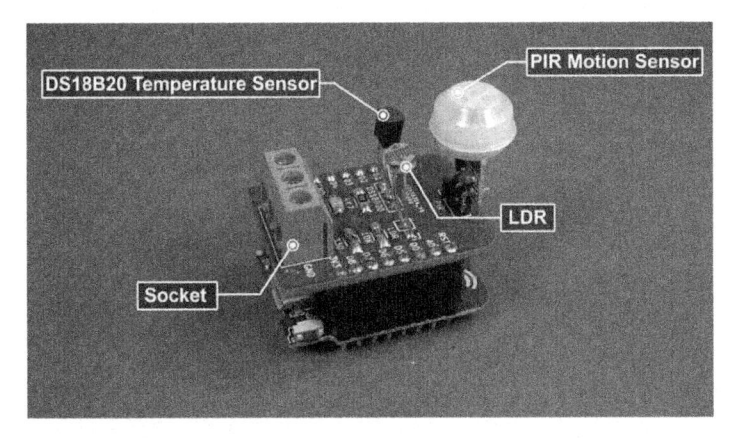

The Multisensor shield includes a temperature sensor, a motion sensor, a light-dependent resistor (LDR), and a three-pin socket that can be used to connect any output, such as a relay module, for instance. On the same Raspberry Pi that is running Node-RED, we will use the Mosquito broker that has been installed. The broker is in charge of receiving all messages, sorting through those messages to determine who might be interested in them, and then publishing those messages to all of the clients who have subscribed to receive them. Read "What is MQTT and How It Works" if you're interested in finding out more about MQTT. The application known as Node-RED will publish messages (such as "on" or "off") in the topic known as esp8266/output. That particular topic has been subscribed to by the ESP8266. Therefore, it turns the output on or off based on whether it receives the message that contains "on" or "off." The temperature is published on the esp8266/temperature topic, and the luminosity is published on the esp8266/ldr topic. The Node-RED application has been set to receive updates regarding those topics. Therefore, it obtains readings of the temperature and luminosity, which are shown on a chart and a gauge respectively. In the following picture, you can see an overview of what we'll do in this project.

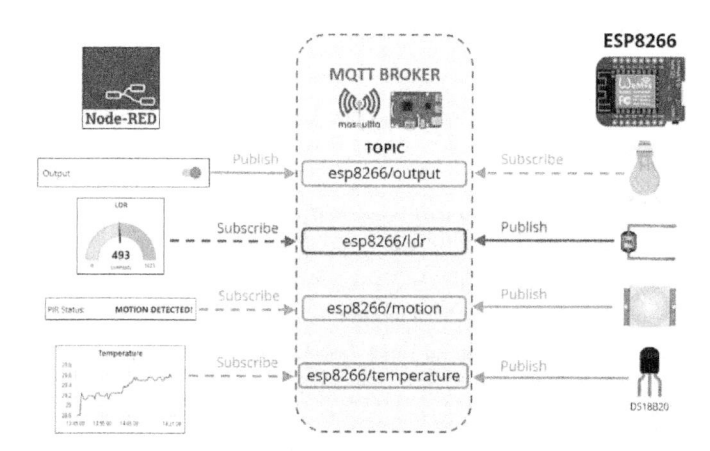

Node-RED has indicated that it is interested in the topic esp8266/motion. When motion is detected, the ESP8266 sends out a publication on that subject. In the text widget of the Node-RED Dashboard, the message "MOTION DETECTED!" appears whenever motion is detected and remains there for ten seconds.

Circuit Diagram;

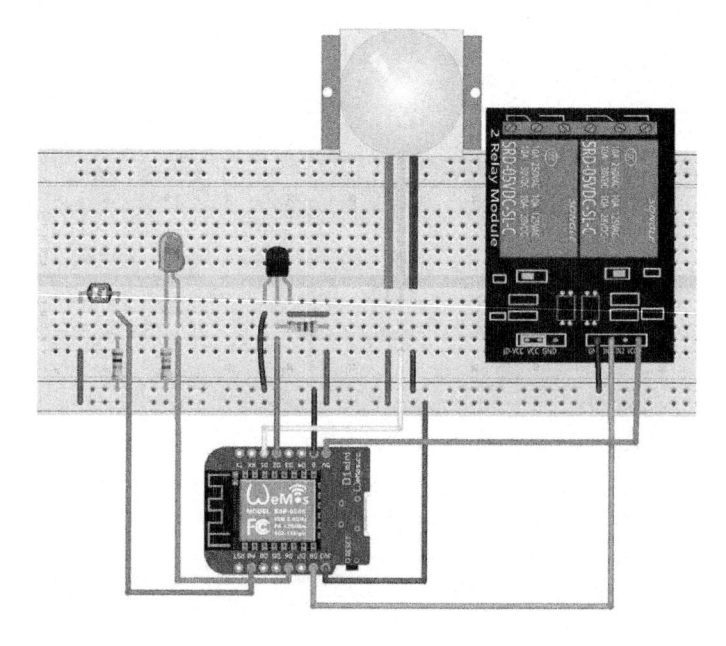

Part required:

- ESP8266 Wemos D1 Mini - see Highest-Rated ESP8266 Wi-Fi Development Board
- 1x 5mm LED, 1x 330 Ohm resistor
- 1x DS18B20 temperature sensor 1x mini-PIR motion sensor 1x light dependent resistor
- 2x 10k Ohm resistor, 1x relay module
- Breadboard, A set of jumper wires

432. Control Sonoff Basic Switch with ESP Easy Firmware and Node-RED

You will gain the knowledge necessary to install the ESP Easy firmware on a Sonoff basic smart switch and utilize the Node-RED dashboard to exercise control over the switch. On a Raspberry Pi, the Node-RED software is currently being executed. The ESP Easy firmware can also be coupled with various other home automation platforms, such as Home Assistant, Domoticz, openHAB, and so on.

Project Overview

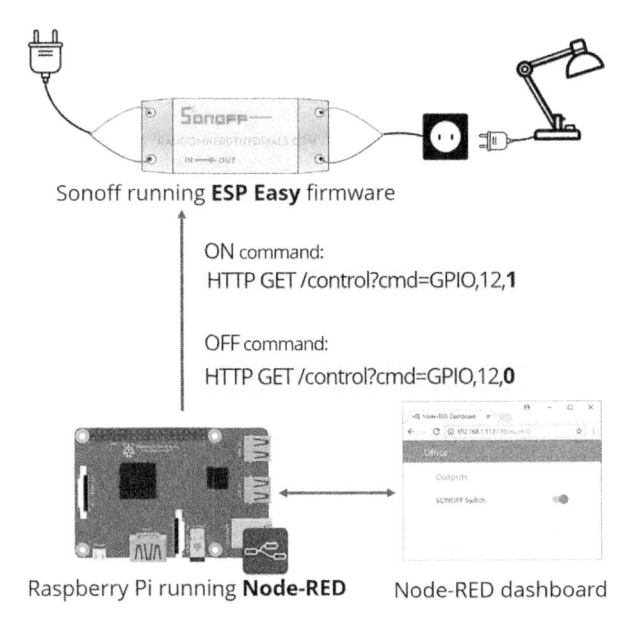

Figure: Diagram provides a high-level summary of how the project is carried out.

You will finish this project with a Sonoff that is loaded with the ESP Easy firmware and that can be controlled with HTTP GET requests made from a web browser or another Wi-Fi capable device. You will be able to integrate it with the majority of home automation platforms once you have finished configuring it on the ESPEasy web interface. In this demonstration, I'll be utilizing Node-RED, which will be running on a Raspberry Pi. After connecting a few nodes, you will see a button appear on your Node-RED dashboard that you can use to control the Sonoff.

Parts Required

- Smart Switch by Sonoff with Wi-Fi
- FTDI Programmer
- Read more about the Raspberry Pi Board in our article about the Best Raspberry Pi Starter Kits.
- MicroSD Card with at least 8 GB and Class 10 speed
- Raspberry Pi Power Supply (Raspberry Pi) (5V 2.5A)

About the Sonoff

The Sonoff is a device that is connected in series with your home's power lines and gives you the ability to control any electronic appliance from a distance. The app platform known as Sonoff eWeLink is compatible with several different brands of smart devices, including Sonoff products. It enables connections between various types of smart hardware and integrates popular smart speakers like Google Home and Amazon Alexa. To explain it more simply, a Sonoff device consists of an ESP8266 chip that is connected to a relay. Through this method, you will be able to connect to the Sonoff via Wi-Fi in order to operate the relay. The eWeLink app can be used to take control of the Sonoff thanks to the firmware that comes preinstalled on the device. On the other hand, the majority of people favor flashing the Sonoff device with their own personal firmware, just as we will do in this project.

Safety warning:

During the process of uploading a new firmware, you need to ensure that your Sonoff is unplugged from the voltage supplied by the mains. When a Sonoff component is linked to mains voltage, you should never touch it in any way. After everything has been removed and disconnected, the plastic box enclosure of the Sonoff should be opened.

Sonoff Pinout

The Sonoff is supposed to be hacked, and you can see plainly that some connections were left out, so that you can solder some pins and install a modified firmware. The figure below depicts the pinout.

Preparing 3.3V FTDI Programmer

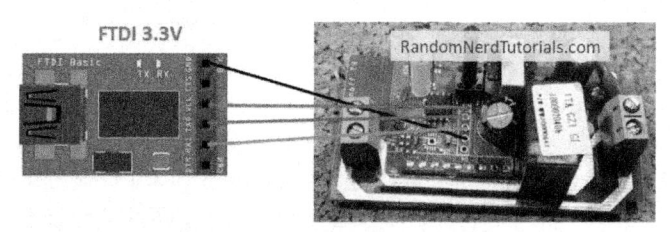

You need an FTDI programmer to upload a new firmware to your Sonoff.

Warning: downloading a custom firmware is irrevocable and you'll no longer be able to utilize the app eWeLink. I've placed a toggle switch in the power line, so that I can easily turn the Sonoff on and off to flash a new firmware without having to remove the FTDI programmer. I used hot glue to connect the ends of the wires together. This stops you from establishing erroneous connections between the FTDI programmer and the Sonoff in the future.

433.ESP8266 Daily Task - Publish Temperature Readings to ThingSpeak

In this project, you will learn how to build a sensor node with an ESP8266 that publishes the temperature and humidity levels once a day to a free service called Thing Speak. The data can be accessed from anywhere via the internet.

Parts required

- ESP8266: for more information, see the article titled Best ESP8266 Wi-Fi Development Boards
- Resistor with a 4.7k Ohm value for the DHT11 temperature and humidity sensor
- Breadboard, A set of jumper wires

Circuit Diagram

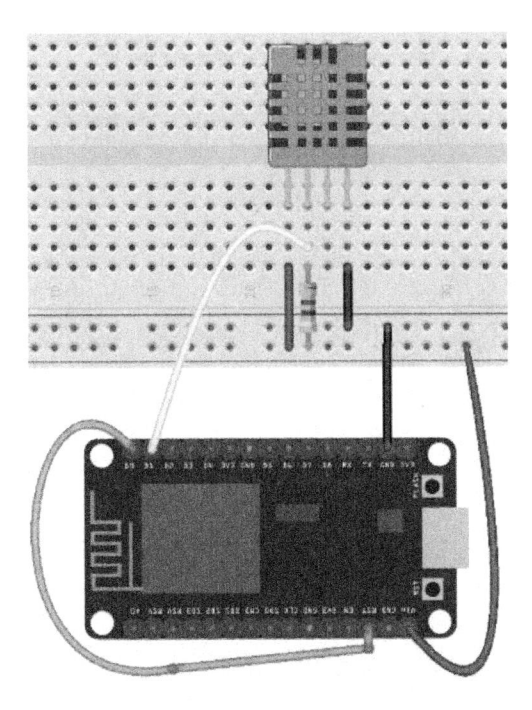

Simply follow the diagram below to put together the circuit, which is extremely simple and does not require any soldering.

434. Touchscreen user interface for Node-RED provided by the Nextion Display with ESP8266.

This project will show you how to build a touchscreen user interface for Node-RED using the Nextion display and the ESP8266 so that you can control the electronic appliances in your home. The goal of this project is to give you the ability to control your home automation system through the Nextion display so that you don't have to use your smartphone or computer to access the Node-RED user interface. This will be accomplished while ensuring that the Node-RED Dashboard is always kept up to date.

Project Overview

In order to exert control over four distinct outlets, you will construct a physical Node-RED interface using the Nextion display. Take a look at the chart that can be found below:

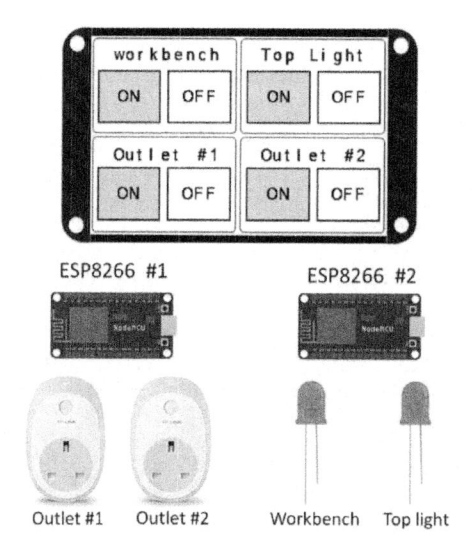

The user interface of Nextion controls four distinct outputs. Through the use of a 433 MHz transmitter.

ESP8266 #1 controls Outlet #1 as well as Outlet #2. Additionally, this ESP8266 is connected to the display of the Nextion.

ESP8266 #2 is in charge of controlling two LEDs that are referred to as the Workbench and Top light. The plan is for you to replace these two LEDs with other outputs that are of use to you, such as a relay or a SONOFF smart switch, for example.

When you use the Nextion display to turn an output ON, the state of the corresponding node in the Node-RED Dashboard is automatically updated to reflect the change. The Node-RED Dashboard also gives you the ability to exercise control over all of these outputs.

How does the project work?

The steps involved in completing this project are outlined in the figure below.

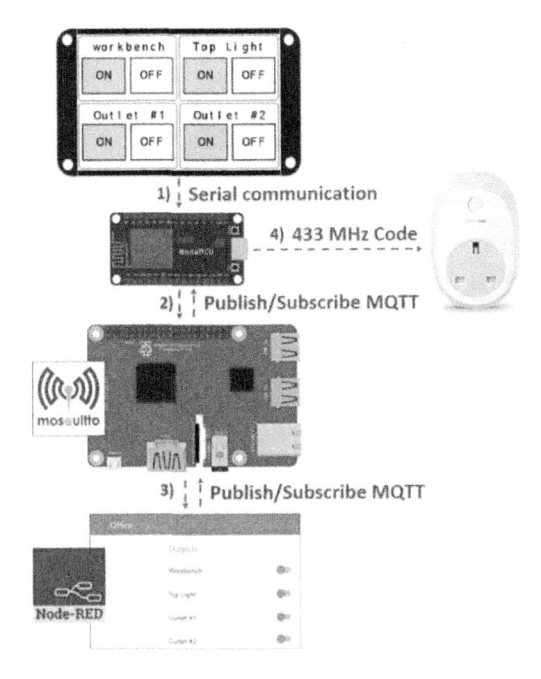

1) When you tap the Outlet #1 ON button, the Nextion display communicates with the ESP8266 via serial communication to let it know that the button was tapped. This lets the ESP8266 know which button was tapped.

2) The ESP and Raspberry Pi are able to communicate with one another thanks to a communication protocol known as MQTT. The Raspberry Pi is equipped with the Mosquitto broker, which is responsible for collecting all of the MQTT messages and distributing them to the various devices that have subscribed to a specific topic. The ESP will publish a "true" message on the topic office/outlet1/buttonState whenever you tap the ON button that is displayed on the Nextion display.

3) The Node-RED message will be sent to the Outlet #1 button, which is subscribed to this topic and will therefore receive it. When it receives that message, it toggles the state of the button that corresponds to it so that it is ON. When something like this takes place, the Node-RED sends out a message with the subject office/outlet1.

4) Since the ESP is subscribed to this topic, it is aware that Node-RED has switched the button state to the ON position. In order to activate Outlet #1, the ESP8266 initiates the transmission of a signal at 433 MHz via a 433 MHz transmitter. Controlling any of the other outputs follows the same procedure as described here.

Parts Required

- The Raspberry Pi
- 2x ESP8266 – read more about them on the Best ESP8266 Wi-Fi Development Board 3.2" Nextion display basic model – read more about it here. Purchasing advice for the Nextion display
- MicroSD card 433 MHz RF Plugs that are operated by remote control
- Transmitter and receiver operating at 433 MHz
- 2x LEDs
- 2 resistors with 330 Ohm each
- Jumper wires for the breadboard

Developing the User Interface for the Nextion

The user interface for the 3.2-inch Nextion basic model has been developed by our team. If you are utilizing a Nextion display of a different size, you will need to make some modifications to the user interface in order for it to function properly for your particular model. In order to do this, you will need to edit the.HMI file and produce a new.TFT file.

ESP8266 #1 – Schematics

By following the schematic in the following figure, you can successfully connect an ESP8266 to a Nextion display as well as a 433MHz transmitter. It is necessary to connect the RX pin of the ESP8266 to the TX pin of the Nextion display.

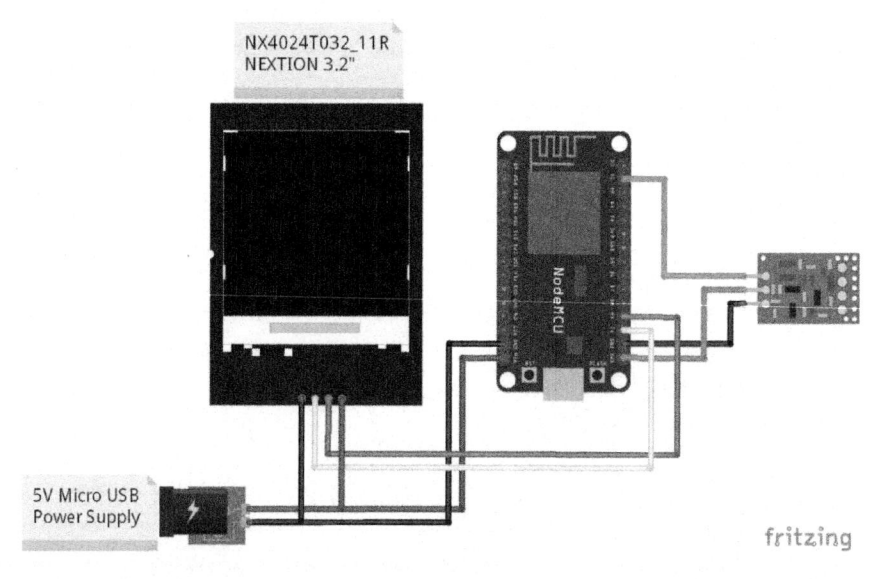

It is necessary to connect the RX pin of the Nextion display to the TX pin of the ESP8266.The data pin of the 433 MHz transmitter is connected to the GPIO 5 of the ESP8266 (D1 pin). A connection has been made between the VCC pin of the 433 MHz transmitter and the 3V3 pin of the ESP8266.

ESP8266 #2 – Schematics

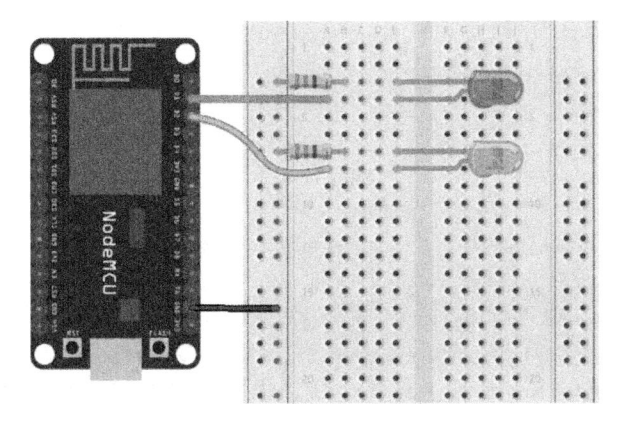

Connect two LEDs to the second ESP8266 in the manner denoted by the accompanying circuit diagram.

- GPIO 5 is the point of connection for the red LED (D1 pin).
- GPIO 4 is the point of connection for the green LED (D2 pin).

Decoding the RF Signals from the Sockets

Decoding the signals that cause the sockets to turn on and off is necessary if you want to use 433 MHz signals to control the sockets in your device.

Code: Each individual ESP8266 device requires a unique sketch to be uploaded by the user. Be sure to make the necessary adjustments to each code in order to get them to function properly for you.

Note: Before you upload any code, check that the Nextion library for the Arduino IDE is properly configured to work with the ESP8266. This must be done before you upload any code.

435. ESP8266 Voltage Regulator (LiPo and Li-ion Batteries)

Following the steps in this project will allow you to construct a voltage regulator for the ESP8266 that is compatible with both LiPo and Li-ion batteries.

ESP8266 power consumption

It is common knowledge that the ESP8266 consumes an excessive amount of power when working in a Wi-Fi environment. It has a power consumption range of 50 mA to 170 mA. Therefore, the use of a battery with it is not ideal for a great number of applications.

If you want to avoid having to worry about power consumption or charging batteries, the best option is to make use of an adapter for power that is connected to the mains voltage.

ESP8266 powered by lithium polymer and lithium-ion batteries

Nevertheless, using the ESP8266 with rechargeable LiPo batteries is an excellent solution for certain ESP8266 projects, such as those that make use of Deep Sleep or do not require a continuous Wi-Fi connection. Because it has fewer components already on board, the ESP-01 variant is the board that is recommended for use with applications that are powered by a battery.

Because it contains fewer components on board, the ESP-01 variant is the board that is recommended for use with applications that are powered by a battery. Because they contain additional components such as resistors, capacitors, chips, and so on, boards such as the ESP-12 NodeMCU consume a greater amount of power. we'll show you how to power the ESP8266 with LiPo batteries because those are the most readily available type of batteries to use in this scenario. This project is not about the various kinds of batteries, and I won't even attempt to explain how LiPo battery's function. I'll just provide you with the information you need to finish the circuit that has been presented...

LiPo/Li-ion batteries fully charged

When fully charged, LiPo/Li-ion batteries can be recharged using the appropriate charger, and when they do so, they produce an output voltage of approximately 4.2V. It is recommended that the ESP be run at 3.3V, but the device is capable of functioning with voltages ranging from 3V to 3.6V.

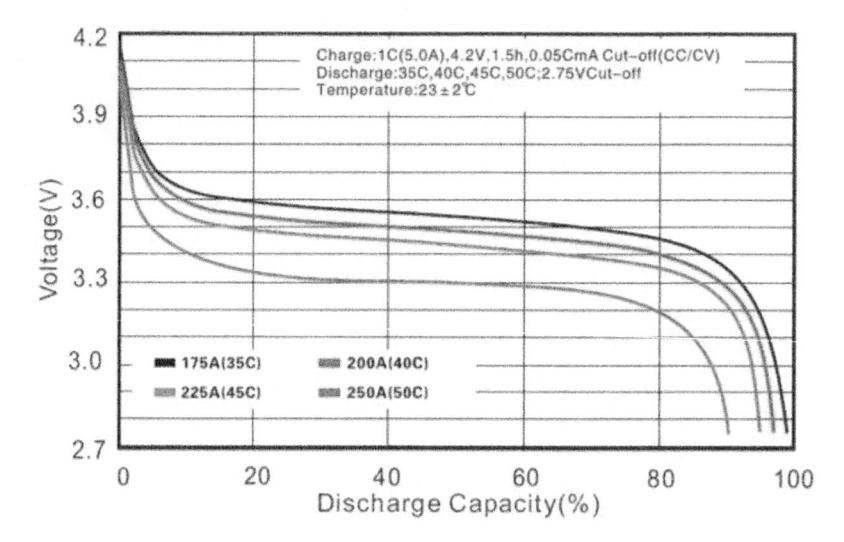

Because of this, you won't be able to connect a LiPo battery straight to an ESP8266; instead, you'll need to use a voltage regulator.

Typical Linear Voltage Regulator

It is not a good idea to attempt to reduce the voltage from 4.2V to 3.3V by utilizing a standard linear voltage regulator. For instance, because it has a high cutoff voltage, your voltage regulator will stop functioning if the battery discharges all the way down to 3.7 volts.

Low-dropout or LDO Regulator

When working with batteries, you need to use a low-dropout regulator, also known as an LDO regulator, so that you can drop the voltage in an efficient manner. This type of regulator can also regulate the output voltage. A low dropout voltage ensures that the device will continue to function normally even if the battery is only producing 3.4V of power. It is important to keep in mind that the LiPo battery should never be completely drained because doing so can cause damage to the battery or reduce the battery's lifespan. After doing some research on LDOs, I found a few other organizations that are viable options. The MCP1700-3302E is without a doubt one of the finest LDOs that I've come across.

It is not very big, and it resembles a transistor in appearance. There is also another viable option available, which is the HT7333-A. Other good alternatives include any LDO that has parameters that are comparable to those detailed in the datasheet that can be found below. Your LDO ought to have comparable specifications with regard to the following:

- Output voltage (3.3V)
- Quiescent current (~1.6uA)
- Output current (~250mA)
- Low-dropout voltage (~178mV)
- ESP8266 Circuit with a Li-ion Battery and LDO

Part Required:

- Lithium-ion or lithium-polymer batteries, along with a battery holder
- Low-dropout regulator, also known as an LDO (MCP1700-3302E)
- 1000uF electrolytic capacitor 100nF ceramic capacitor
- Pushbutton 10k Ohm resistor
- For more information on the ESP-01, see Best ESP8266 Wi-Fi Development Boards.
- Breadboard
- A set of jumper wires

If you want to design your own voltage regulator circuit, take a look at the diagram that is provided in the next page. The pushbutton is connected to the RESET pin of the ESP-01. Although it is not required for this specific project, it will come in handy for a future project.

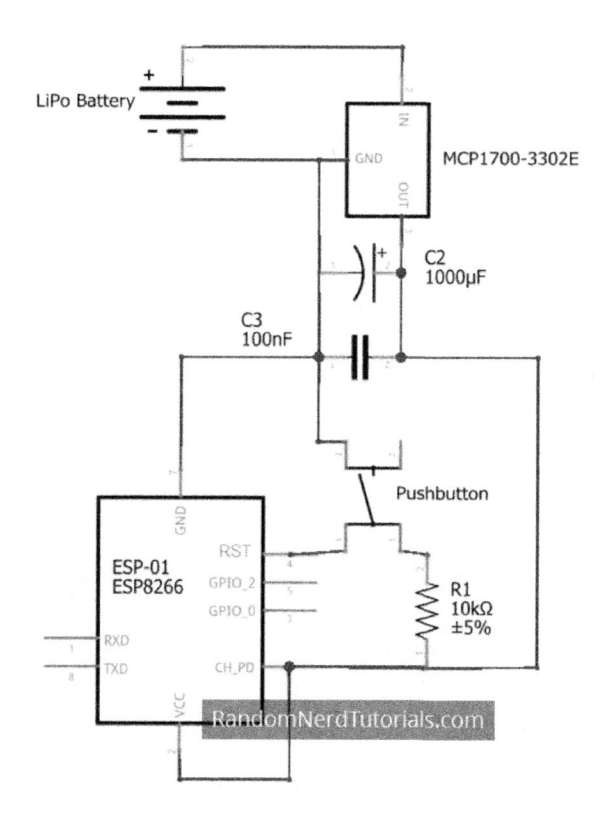

Here's the final circuit:

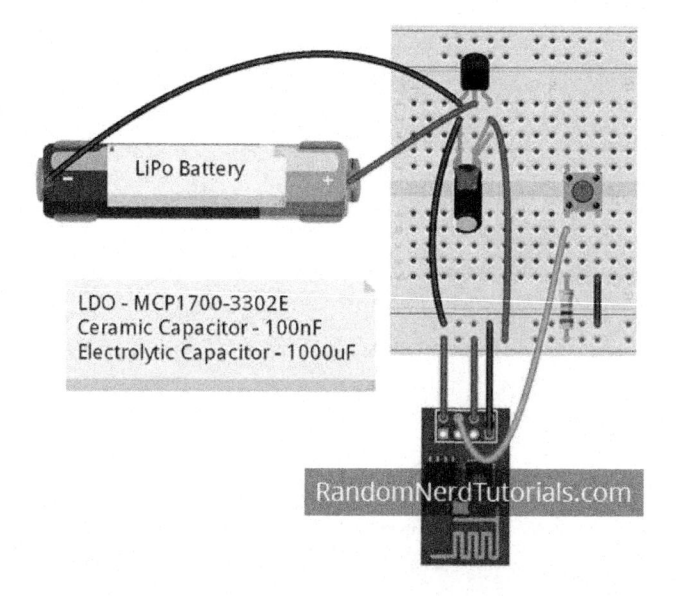

About the capacitors

In order to smooth out the peaks in the voltage, the LDOs should each have a ceramic capacitor and an electrolytic capacitor connected in parallel to GND and Vout. Your ESP8266 will not experience any unexpected resets or unstable behavior as a result of the capacitors.

Testing

Let's give the circuit some power and see how it performs. When you measure the Vin voltage of the LiPo battery with a multimeter, you can see that it outputs approximately 4.2V because the battery is currently at its maximum capacity and has been fully charged.

Let's start by putting the probe of the multimeter on the Vout. At this time, the multimeter is measuring somewhere around 3.3V, which is the voltage that is recommended to supply power to the ESP8266.

Voltage regulator

The following is an example of a popular design for the voltage regulator on the ESP8266: After soldering the capacitors to the LDO, you will have created a voltage regulator that has a small form factor and can be easily incorporated into your projects as a result. I have high hopes that this guide was helpful. This idea will be of great assistance in the operation of future projects.

To that end, we offer our sincere appreciation for your participation in this endeavor. After gaining an understanding of the concepts presented in this project, you will be able to modify it to control virtually anything you want by utilizing the Nextion display interface and Node-RED.

436. ESP8266 Weather Forecaster

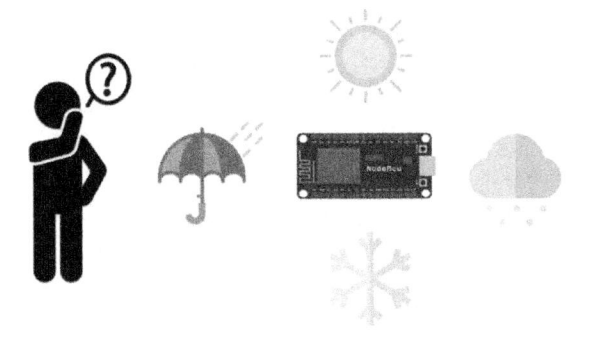

In this project, we will use an ESP8266 to create a weather forecaster. It's the worst feeling in the world to be caught outside on a wet day without an umbrella, and I think most people feel the same way. When we are in a hurry, we frequently fail to pay attention to the weather widget that is displayed on our smartphones, even though it provides a forecast of the upcoming weather.

Therefore, it would be convenient to have a device that hangs on the back of the door and serves as a constant reminder to take an umbrella with you whenever you leave the house on a day that there is a chance that it will rain. The weather forecast is displayed on this device in the form of a change in the color of the LED. This piece of technology wouldn't have an alarm or a screen; instead, it would just have a few LEDs that were designed to blend in with their surroundings.

Parts required

- ESP8266 12-E, more information can be found at Best ESP8266 Wi-Fi Development Boards.
- 4x LEDs (different colors to represent different weather conditions)
- 4 resistors in total (220 or 330 ohms should do the trick)
- Breadboard
- A set of jumper wires

Open Weather Map

1. Because the Open Weather Map API is the foundation of this project, it is necessary to register on their platform and acquire an API key before we can begin the process of putting together the schematics and writing the code for this project.
2. The free plan offered by OpenWeatherMap gives you access to all of the features you require for this demonstration. You will need an API key, also known as the APIID, in order to use the API. To get an APIID:
3. Launch your web browser, and navigate to the **OpenWeatherMap** website.
4. Click the button labeled **"Sign up"** to set up your free account.
5. After you've successfully created an account, you'll be taken to a dashboard that features a number of different tabs.
6. Go to the tab labeled **API Keys**, and copy your individual Key.
7. You will need this one-of-a-kind key in order to retrieve information from the website. Make a copy of this key and save it to your computer; you'll need it in a little while.
8. Enter the following URL into your browser's address bar, making sure to replace the information in the sections denoted by curly brackets with the appropriate details for the location you've selected as well as your own unique API key:
9. Replace "your city" with the city whose data you require, "your country code" with the country code corresponding to that city, and "your API key" with the one-of-a-kind API key we discovered for you earlier. For instance, our Application Programming Interface (API) URL for the city of Porto in Portugal, after replacing the details with their own, would be:
10. Please take note that additional information regarding the use of the API to get weather information can be found here. If you paste your URL into your browser, it should give

you a bunch of information that is relevant to the weather forecast for where you are. Be sure to guard your one-of-a-kind API key, and we will then proceed to the section on the code.

Circuit Diagram:

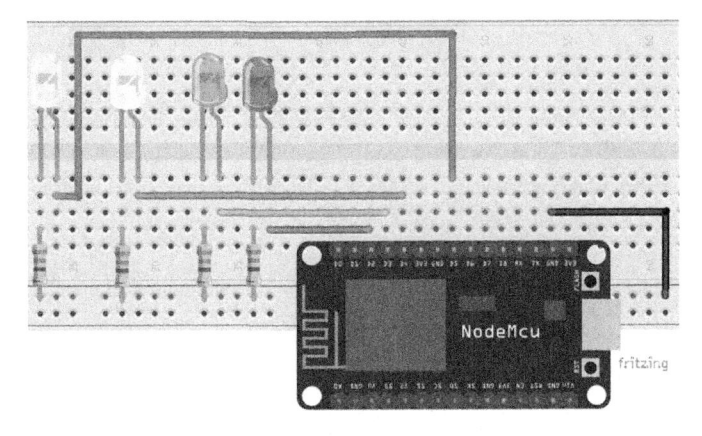

Demonstration

The following figure provides an explanation of what each LED represents: Rain on Day 2, a clear sky on Day 3, snow on Day 4, hail on Day 5, and hail on Day 5:

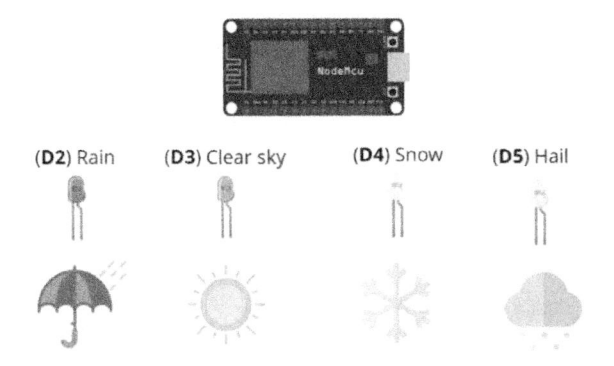

(**D2**) Rain   (**D3**) Clear sky   (**D4**) Snow   (**D5**) Hail

These days, the purpose of computing is to develop a natural user interface that will allow people to communicate with their computers. We are looking to technologies such as AR (Augmented Reality) to help make this a reality, but the fact of the matter is that we are all growing tired of having to constantly look at our phones and computers for even the smallest piece of information. I believe that it would be extremely helpful to have this project hung up around the house in a location where it would be easy to check the weather before leaving the house. In order to take this project one step further, I might incorporate a battery backup into a 3D-printed enclosure and give it a more streamlined appearance. When that is complete, you can count on me to let you know about it.

437. ESP8266 Publishing DHT22 Readings with MQTT to Raspberry Pi

This project uses a Raspberry Pi to display temperature and humidity readings from a DHT22 sensor. Using the MQTT protocol, it is also possible to control two outputs coming from an ESP8266. Flask is the name of the Python microframework that you will be utilizing in order to create the web server. The following is an overview of the system at a high level:

Initial Configuration of a Raspberry Pi

Before you proceed with reading about this project, check to see that the Raspbian operating system has been installed on your Raspberry Pi.

You can install Raspbian and finish the basic setup by reading my guide to getting started with the Raspberry Pi.

Run and install Mosquitto broker

The MQTT protocol is going to be used to facilitate communication between the Raspberry Pi and the ESP8266. In addition to having Mosquitto broker installed on your system, you are required to have Mosquitto broker actively running in the background:

Installing Flask

For the purpose of transforming the Raspberry Pi into a web server, we will be utilizing a Python microframework known as Flask.

You will need to have pip already installed in order to install Flask. To bring your Raspberry Pi up to date and install pip, run the following commands:

pi@raspberrypi ~ $ sudo apt-get update

pi@raspberrypi ~ $ sudo apt-get upgrade

pi@raspberrypi:$ sudo apt-get install python-flask git-core python-pip git-core

The next step is to install Flask and Paho MQTT by utilizing pip:

pi@raspberrypi ~ $ sudo pip install flask

pi@raspberrypi ~ $ sudo pip install paho-mqtt

pi@raspberrypi ~ $ sudo pip install paho-mqtt

Installing SocketIO

This project makes use of SocketIO, which makes it possible for you to create a web page in Python Flask that can be updated asynchronously by the Python Flask application you've created. This indicates that you do not need to refresh the web page in order to view the most recent readings because they are immediately updated. Installing the Flask SocketIO Python package is going to be your next step.

pi@raspberrypi ~ $ sudo pip install flask-socketio

Developing the Script Using Python

The essential script for our application can be found here. When these buttons are pressed, it publishes a MQTT message to the ESP8266 and then sets up the web server. In addition, it has readings for temperature and humidity MQTT topics subscribed to so that it can receive them.

Programming an ESP8266 module

Installing the PubSubClient library on the Raspberry Pi is necessary in order for the ESP8266 to communicate with the web server running on the Pi. This library offers a client that is capable of carrying out straightforward publish/subscribe messaging with a server that supports MQTT (basically allows your ESP8266 to talk with Python web server).

Installing the PubSubClient library

1. To download the PubSubClient library, please click on the link provided. Within your Downloads folder, there should be a folder labeled ".zip."
2. After you have unzipped the .zip folder, you should see a folder named pubsubclient-master.
3. Rename your folder's current name, which is pubsubclient-master, to just pubsubclient.
4. Transfer the pubsubclient folder to the libraries folder that was created during the installation of your Arduino IDE.

The library includes several different examples in the form of sketches. Within the Arduino IDE software, navigate to the File menu and select Examples before selecting PubSubClient.

Installing the DHT sensor library

Your ESP8266 or Arduino boards will be able to read temperature and humidity with ease when you use the DHT sensor library. This library makes it simple to use any DHT sensor.

1. To download the DHT sensor library, click on the link provided. Within your Downloads folder, there should be a folder labeled ".zip."
2. After you have unzipped the.zip folder, you should have access to the DHT-sensor-library-master folder.
3. Rename your folder's current name, which is DHT-sensor-library-master, to simply DHT.
4. Transfer the DHT folder to the libraries folder that was created during the installation of your Arduino IDE.
5. Next, reopen the Arduino IDE on your computer.

Uploading sketch

At this point, you are ready to finish the process by uploading the complete sketch to your ESP8266 (replace with your SSID, password, and RPi IP address):

Circuit Diagram:

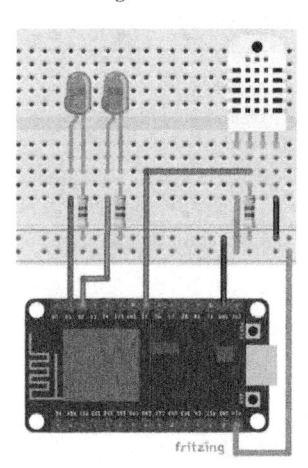

It is imperative that you use the Vin pin that outputs 5V on your ESP8266, as the DHT sensor works correctly only when it receives the required 5V voltage to do so.

Part Required:

- 1x ESP8266 12E – read Best ESP8266 Wi-Fi Development Boards
- 1x DHT22 Sensor, 1x 4700 Ohm Resistor
- 2x 470 Ohm resistors, 2x LEDs

Activating and Starting the Web Server

In order to initiate the launch of your Raspberry Pi web server, navigate to the folder that houses the app.py file:

pi@raspberrypi:/web-server/templates $ cd.. ;;;;;;;;;;;;;;;;;;;;;;;;;;

Execute the following command after that:

pi@raspberrypi:/web-server $ sudo python app.py and then press enter.

Your web server needs to start up right away on port 8181!

438. ESP8266-Based Do-It-Yourself Wi-Fi RGB LED Mood Light for $10 (Step by Step)

This project shows how to make a mood light. You can change the color of your light from a distance by utilizing an ESP8266 and a browser on your smartphone or any other device that you have access to. $10 DIY Wi-Fi RGB LED Mood Light is the name of this particular project.

Parts Required

- 1x ESP8266-12E - for more information, read the article on the best ESP8266 Wi-Fi development boards.
- 1x RGB LED Strip
- 1x Power Supply Device with 12V Output Capable of Reducing Voltage to 5V Alternative: LM7805 with Heat Sink
- 3x NPN Transistors 2N2222 or equivalent
- 3x 1k Ohm Resistors
- 1 set of jumper wires for the breadboard
- Lamp for the Table With an Appearance of Mood Light

Flashing Your ESP with NodeMCU

In this tutorial, we will be utilizing the firmware that comes with the NodeMCU board. It is necessary for you to flash your ESP with the NodeMCU firmware.

Downloading ESPlorer IDE

To send commands to your ESP8266, it is highly recommended that you make use of the ESPlorer IDE, which is a program that was developed by 4refr0nt.

- Download ESPlorer Unzip that folder.
- Navigate to the primary file folder.
- Run "ESPlorer.jar" file
- Launch the IDE for ESPlorer.

624

Circuit Diagram:

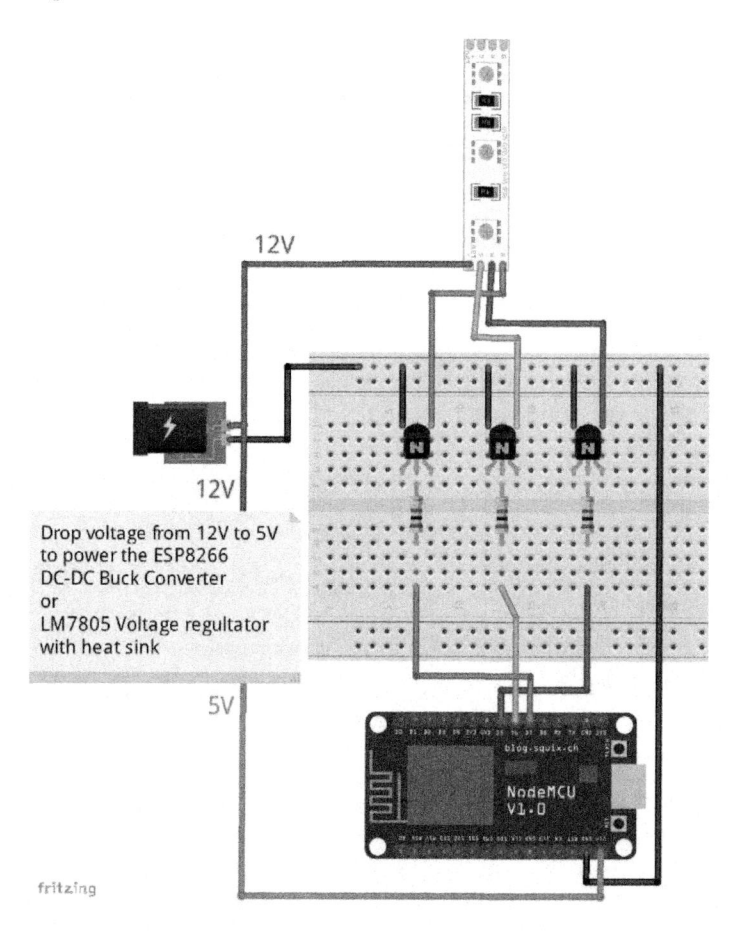

Drop voltage from 12V to 5V to power the ESP8266
DC-DC Buck Converter
or
LM7805 Voltage regultator with heat sink

Uploading Code

You should see a window that looks like the one in the figure before this one; in order to upload a Lua file, follow these instructions:

1. Connect your ESP8266-12E that has built-in programmer to your computer
2. Choose the port for your ESP8266-12E.
3. Use the Open/Close button.
4. Choose the tab labeled "NodeMCU+MicroPtyhon."
5. Make a new file and save it with the name init.lua.
6. Click the Save to ESP button.

Everything that requires your attention or requires you to make a change is highlighted in red box.

Your ESP IP Address

When you restart your ESP8266, the IP address of the ESP will be printed in the serial monitor. Keep that IP address in mind for when you need it in the future.

Opening Your Web Server

Open up any browser on your computer, and type in the IP address of the ESP8266. This is what you need to keep in mind: When you click on the input field, a small window containing a color picker will open. To choose the color for your RGB LED strip, simply drag your finger or the mouse in the appropriate direction:

Finally, select the desired hue by clicking the "**Change Color**" button: Now, your living room is the perfect place to put your mood lighting:

An application of the ESP8266 board in the real world is demonstrated by this project. If you don't have an RGB LED strip but you still want to try out this project, you can learn how to change the color of an RGB LED using an ESP8266 by reading this blog post, which is titled "ESP8266 RGB Color Picker."

439. The ESP8266 is controlled by an Android app (MIT App Inventor)

In this project, you will create an Android app that allows you to control the ESP8266 GPIOs using the MIT App Inventor software.

Flashing Your ESP with NodeMCU

In this lesion, we will be utilizing the firmware that comes with the NodeMCU board. It is necessary for you to flash your ESP with the NodeMCU firmware.

Downloading ESPlorer IDE

- ➤ To send commands to your ESP8266, it is highly recommended that you make use of the ESPlorer IDE, which is a program that was developed by 4refr0nt.
- ➤ Download ESPlorer and then extract the compressed folder.
- ➤ Navigate to the primary file folder.
- ➤ Run "ESPlorer.jar" file

Launch the IDE for ESPlorer.

- You should see a window that looks like the one in the figure before this one; in order to upload a Lua file, follow these instructions:
- Establish a connection between your computer and the FTDI programmer.
- Choose the appropriate port for your FTDI programmer.
- Use the Open/Close button.
- Choose the tab labeled "NodeMCU+MicroPtyhon."
- Make a new file and save it with the name init.lua.
- Click the Save to ESP button.
- Everything that requires your attention or requires you to make a change is highlighted in red box.

Parts Required:

- 1x ESP8266 (for more information, see the Best ESP8266 Wi-Fi Development Boards).
- 1x FTDI programmer
- 2x LEDs, 2x 220Ω Resistors
- 1x Breadboard, 1x Android Phone

Circuit Diagram:

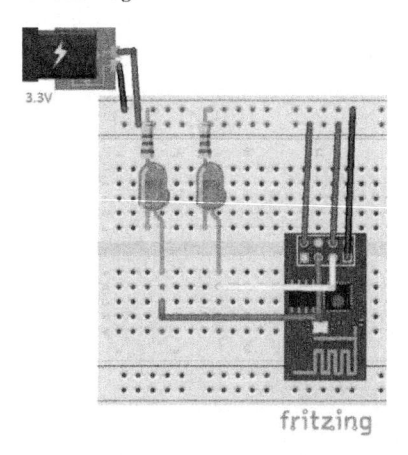

Now, construct the circuit that controls two LEDs by following these schematics in the appropriate order.

Here's how to edit the ESP8266 Controller app:

- Download the .aia file
- Unzip the folder
- Visit the MIT App Inventor website.

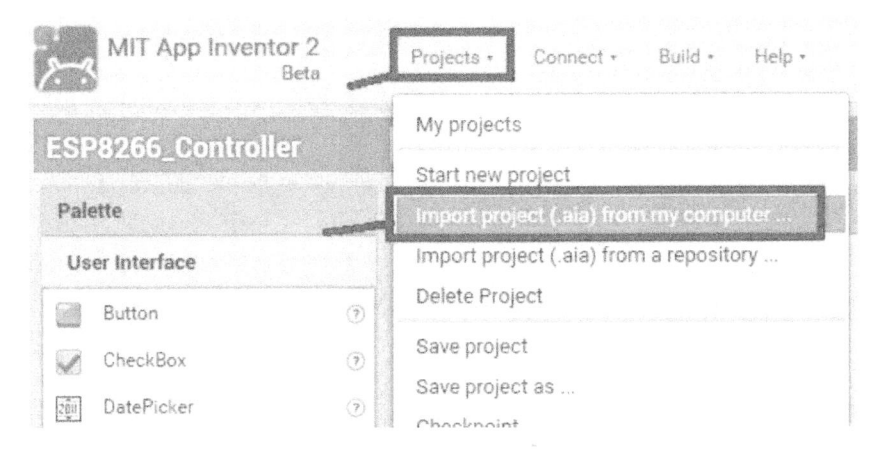

To create an application, select the "**Create Apps**" button located in the upper right corner.

Navigate to the "**Projects**" tab, then select "Import project (.aia)" from the drop-down menu.

The IP Address of Your ESP

- When you restart your ESP8266, the IP address of the ESP will be printed in the serial monitor. Keep that IP address in mind for when you need it in the future.
- Regarding myself, the IP address of the ESP is 192.168.1.95. Please read this troubleshooting guide if you are having issues viewing your IP address.
- Creating an Android App with MIT App Inventor MIT App Inventor is a drag-and-drop software that enables you to create an Android app that is simple but fully functional in under an hour.
- You will be able to make changes to the app and observe the construction process once you have imported the .aia file

Designer

You can make changes to the way the app appears by using the designer tab. Please feel free to make any changes you see fit to the text, the colors, the buttons, or the features.

Blocks

Within the blocks section of your app, you will be able to specify what each button does as well as add logic to your application.

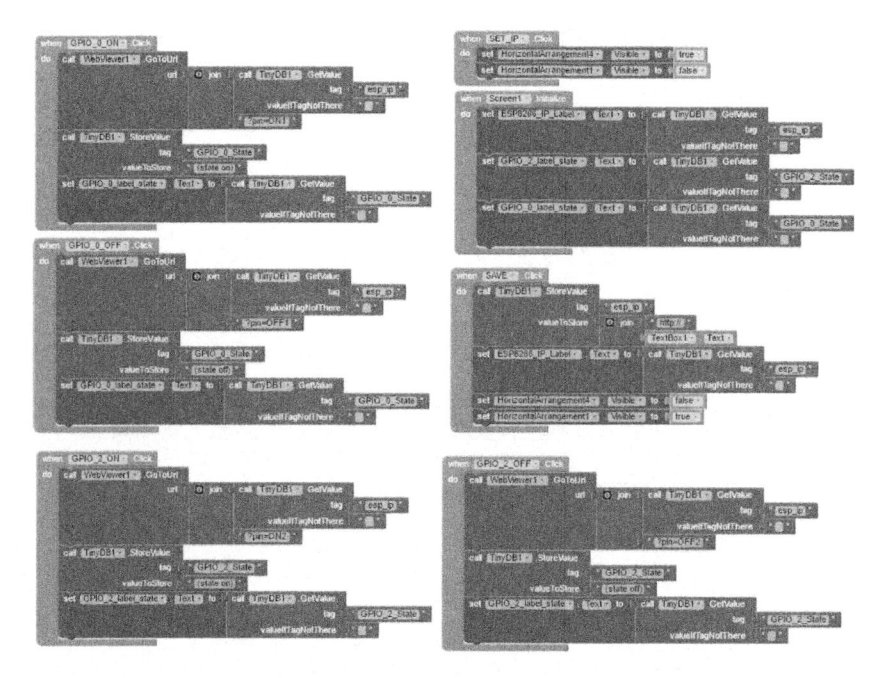

When you are finished making changes to the app, you can click the "Build" app tab to generate the .apk file and then install it on your Android device. Uploading the application that is linked below is the first step that I would recommend you take to ensure that everything operates as it should (later you can edit the app).

Installing the Android App

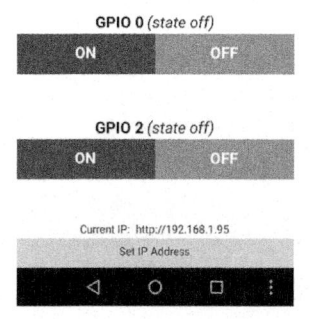

To successfully install the default app that we've created, please follow these instructions:

- Please visit this link in order to download the .apk file.
- Do not compress the folder.
- Transfer the .apk file to the Android device you are using.
- Start the installation process by opening the .apk file.
- When you first open the ESP8266 Controller app, it will appear as shown in the following image.
- The configuration couldn't be simpler. Your IP address can be changed by selecting the "Set IP Address" button at the bottom of the screen and entering it (in my case 192.168.1.95). You're all set!

Developing It Further

An Android application can be easily integrated with the ESP8266, as demonstrated by this simple example, which can be found here. You are free to adapt this example to fit your needs. You could add multiple screens to the application in order to control additional ESPs or add buttons in order to manage additional GPIOs.

440. ESP8266 - Wireless Weather Station with Data Logging to Excel

In this project, we're going to connect two ESPs wirelessly and send data from three sensors to an Excel spreadsheet. This tutorial walks you through the process of installing a wireless weather station in your own home, complete with data logging capabilities.

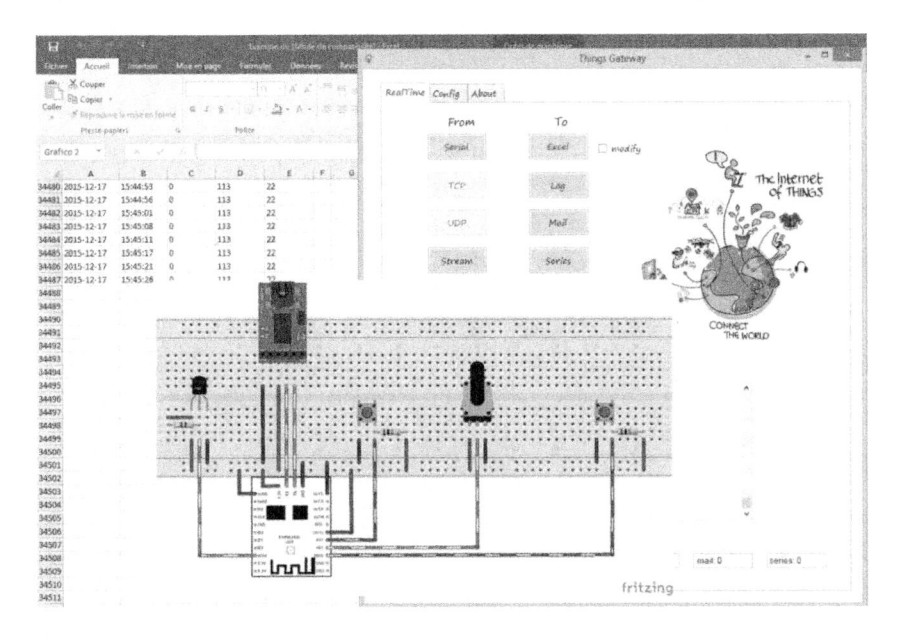

Parts Required:

- Recommended: 2x ESP-12E Read More About the Best ESP8266 Wi-Fi Development Boards.
- Alternative: 2x ESP-201 + 1x FTDI Programmer
- 1x DS18B20, 1x Breadboard, 3x Pushbutton
- 3x 10k Ohm Resistor 1x 4700 Ohm One ten-thousand-ohm resistor Potentiometer

DS18B20 - One Wire Digital Temperature Sensor

We will be measuring temperatures with a DS18B20 one-wire digital temperature sensor for the duration of this project. First, let's learn how to wire up our temperature sensor so that we can move on to the programming portion of this lesson. Because the DS18B20 can be powered by voltages ranging from 3.0V to 5.5V, you simply need to connect its GND pin to the ESP8266's GND pin, and its VDD pin needs to be connected to the ESP8266's 3.3V. After that, connect the DQ pin on the ESP8266 to the IO04 port. Pulling the DQ pin up to 3.3V requires the use of a pullup resistor with a value of 4K7 ohms.

Reading ADC Value

We are going to read an analog value using the ADC pin, and when people talk about the ADC pin on the ESP, you will frequently hear the following different terms used interchangeably: ADC (Analog-to-digital Converter) TOUT Pin6 A0 Analog Pin 0 These are all different names for the same pin on the ESP8266, which will be discussed in more detail in the following section (read this article for more information on the ADC pin). When TOUT is connected to an external circuit, the input voltage range is 0 to 1.0 V, and the current precision of TOUT (Pin6) is 10 bits.

Getting Access to the Analog Pin on the ESP8266, it is very simple to gain access to the ADC when using an ESP-201; all you need to do is connect a jumper wire to the pin that is highlighted in the figure below.

Client Circuit

If you want to build your own ESP client, follow this circuit, and if you are going to use an ESP-12, you can look at the schematics. Simulating wind speed requires the use of the right switch, which is connected to IO05. The program does not currently provide a real-time wind speed reading; rather, it provides an average wind speed over the previous 5 seconds. Therefore, it makes no difference how quickly you press the right switch. If you press the right switch five times within the sampling period of five seconds... That translates into an average of once per second, which is 1.492 miles per hour (or 2.4 kilometers per hour). Since we are using the integer version of NodeMCU at the moment, you will get 1.492 miles per hour and 24 kilometers per hour. You will need to do some division in the spreadsheet by both 100 and 10, in that order.

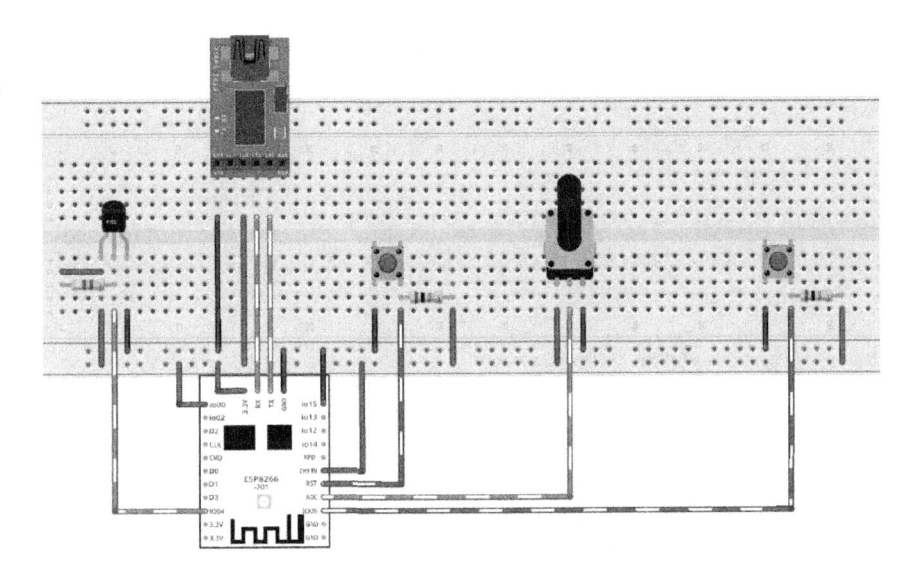

Final Server Circuit

If you want to make your own ESP server, follow this circuit, and if you're going to use an ESP-12, you can look at the schematics.

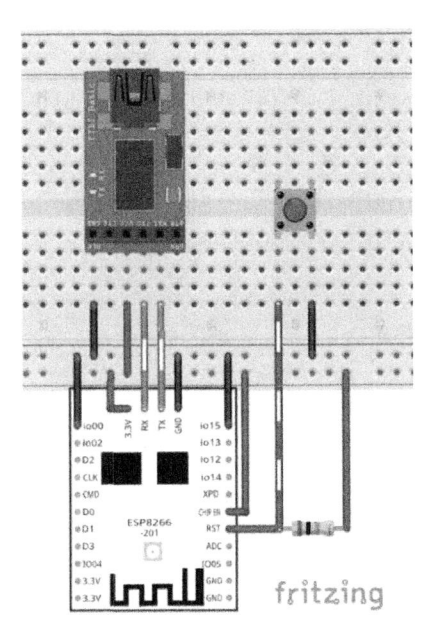

We have designed a unique printed circuit board (PCB) for the ESP-201. This is what you need to know: yellow wire: can either connect GPIO 0 to GND to flash NodeMCU firmware or to VCC to save scripts to ESP. GPIO 15 is hard wired to GND Reset button and has a 10K Ohm pull-up resistor connected to the RST pin (in order to reset the board, the RST pin has to be pulled down and then has to go back up)

Putting Everything Together

Turn on both of your ESPs and check to see that the ESP server is still able to establish serial communication with your computer. This is necessary because you will need to post the data to your Excel spreadsheet using this method.

Downloading and Installing Things Gateway

In order to write data on an Excel spreadsheet and display self-updating real-time charts, the team working on this project is going to use a piece of software developed by Roberto Valgolio called Things Gateway. Things Gateway is a program for personal computers that connects a microcontroller and gives it the ability to do the following things:

- Get data from Excel files
- Write Excel files
- Write CSV log files
- Send emails (when certain conditions are met)
- Please display the values and charts (charts are Excel independent)

Things Additionally, Gateway can connect to a GPS and display the following information on the screen:

- Information on speed, heading, and altitude, as well as other navigational elements
- Tracks updated in real time on Google Maps

Visit the website for Things Gateway, download the program, and then install it on your device. The Beta4 version of the software was used to test this project.

Launching Your Things Gateway Application

Getting the application for Things Gateway up and running is a very simple process. Open the application by navigating to Your Computer > Documents > ThingsGateway on your computer.

Then follow these instructions:

- Navigate to the Configuration tab.
- Choose the Serial COM Port that is connected to your ESP8266 Server.
- Adjust the speed of the serial port to 9600.

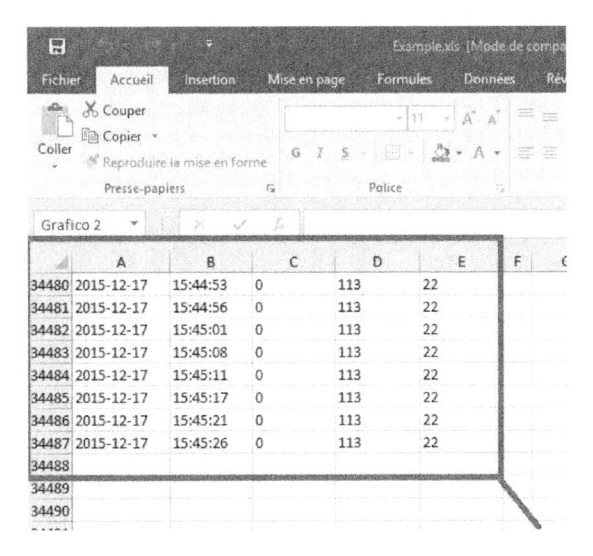

	A	B	C	D	E
34480	2015-12-17	15:44:53	0	113	22
34481	2015-12-17	15:44:56	0	113	22
34482	2015-12-17	15:45:01	0	113	22
34483	2015-12-17	15:45:08	0	113	22
34484	2015-12-17	15:45:11	0	113	22
34485	2015-12-17	15:45:17	0	113	22
34486	2015-12-17	15:45:21	0	113	22
34487	2015-12-17	15:45:26	0	113	22
34488					
34489					
34490					

Your information is being saved in the sample spreadsheet created in Excel that is located in that folder.

Future Improvements

Throughout the course of this guide, we built a fundamental system that makes use of two ESP8266 microcontrollers and the Things Gateway application to transmit data wirelessly to an Excel spreadsheet.

- You'll be able to experiment with additional features offered by Things Gateway, such as the following when you use this example:
- Graph that automatically updates itself in real time (not Excel dependent)
- Data logging in .csv files
- Sending of an automatic email when predetermined conditions are satisfied (such as a wind speed greater than a certain number of kilometers per hour, a temperature lower than a certain value, etc.) warning of problems with the heating system or the freezer not being sufficiently cold etc.)
- Logging of data in stream.txt, which can be replayed at speeds of 0.1, 0.5, 1, 5, 10, 60, or 3600 times faster than the original.
- Additionally, GPS tracks can be streamed on Google Maps at the actual speed or at 0.1, 0.5, 5, 10, 60, or 3600 times the normal speed.

Further development could also include:

- Other sensors like bmp180 barometric pressure sensor
- Photosensitive cells

- Detector of humidity
- Any additional digital or analog sensor that employs multiplexing

It is important to note that Things Gateway was developed for use with Arduino boards; however, as long as you send the appropriate string via serial, you can use it with any board you like.

441. How to Make Two ESP8266 Talk

The goal of this project is for you to get two ESP8266 boards to communicate with one another.

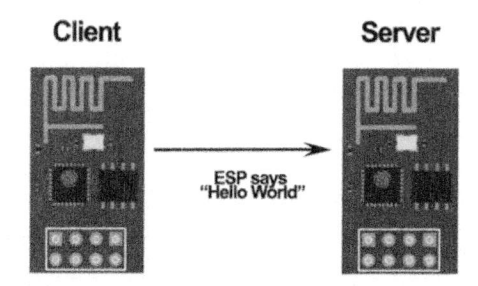

How does it operate? You are going to configure one ESP to function as an Access Point (Server), and the other ESP is going to be configured to function as a Station (Client). After that, they will set up a wireless communication, and the Client will communicate with the Server by sending a message that says "Hello World!"

Parts List

- ESP8266-01
- FTDI programmer

Schematics (3.3V FTDI Programmer) (3.3V FTDI Programmer)

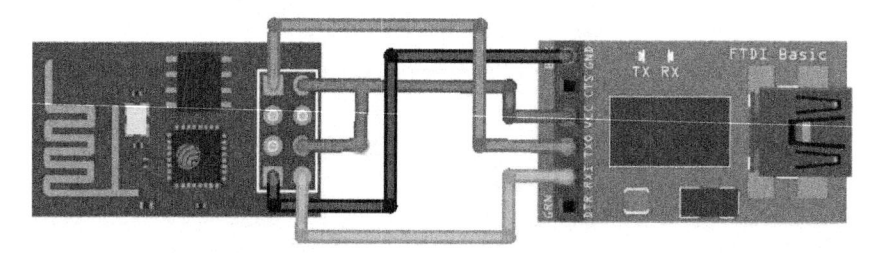

The blueprints for this project don't require much interpretation at all. In order to upload some code, all that is required of you is to establish a serial communication link between your FTDI programmer and your ESP8266. (The diagrams for the Client and the Server are presented below in identical fashion.)

Downloading ESPlorer

- It is strongly suggested that you make use of the ESPlorer program that was developed by 4refr0nt in order to generate and store Lua files into your ESP8266.
- Get ESPlorer by clicking here.
- Do not compress that folder.
- Navigate to the primary file folder.
- Run ESPlorer.jar. Because it is a JAVA program, you will need to ensure that you have JAVA installed on your computer.
- Activate the ESPlorer.

442. ESP8266 in conjunction with Node-RED and MQTT

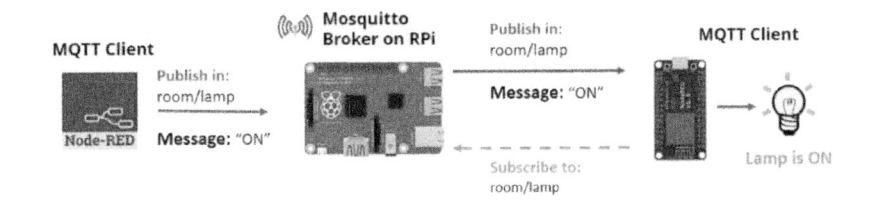

In this post, we are going to demonstrate how you can use Node-RED to control the outputs of an ESP8266 and display sensor data received from an ESP8266. The Node-RED software is being run on a Raspberry Pi, and the MQTT communication protocol is being used to facilitate communication between the ESP8266 and the Node-RED software.

Node-RED and Node-RED Dashboard

It is necessary for your Raspberry Pi to have Node-RED and the Node-RED Dashboard application installed. The blog posts listed below are helpful if you are just getting started with Node-RED and the Node-RED dashboard:

MQTT Protocol

Using MQTT, we are going to connect an ESP8266 to a Raspberry Pi that is operating the Node-RED software. The goal of this tutorial is to enable two-way communication between the two devices.

MQTT is an acronym that stands for **"MQ Telemetry Transport."** It describes a system that allows clients to both send and receive messages, and it is a nice, lightweight publish and subscribe system. It is a straightforward messaging protocol that was developed for devices that have limited resources and bandwidth. Therefore, it is an ideal option for use in applications related to the Internet of Things.

Installing Mosquitto Broker

In MQTT, it is the broker's primary responsibility to receive all messages, filter those messages, determine which clients are interested in a particular message, and then publish that message to all clients who have subscribed to it.

There are many different brokers available for your use. In this guide, we are going to make use of the Mosquitto Broker, which requires the Raspberry Pi to have the appropriate software installed.

Testing

Execute the following command to determine whether or not the Mosquitto broker was successfully installed:

pi@raspberry:~ $ mosquitto -v

This gives you the version of Mosquitto that is currently installed and operating on your Raspberry Pi. It should be 1.4 or higher at the very least.

Note that in addition to returning the Mosquitto version that is currently installed, the Mosquitto command attempts to initialize Mosquitto once more. Due to the fact that Mosquitto is already active, an error message is displayed. If you see a message like this, you don't need to worry because Mosquitto has been installed and is running correctly.

Establishing a MQTT communication with Node-RED

Using the Node-RED nodes, we are going to set up a communication using the MQTT protocol in this section.

Dashboard Layout

The first thing you need to do is design the layout for the dashboard. In this demonstration, we will make use of a button to control an ESP8266 output, as well as a chart and a gauge to display the temperature and humidity readings collected by the DHT11 sensor.

Choose the Layout tab from the drop-down menu that appears after you click the dashboard tab in the Node-RED window's upper right corner. Make a tab labeled Room, and then inside that tab, make two groups: one labeled Lamp, and the other labeled Sensor.

Building the Circuit

The components and diagrams of the circuit that you will need to construct for this project are presented in the following sections.

Parts required

To construct the circuit, you will need the following components

- Raspberry Pi
- ESP8266 (ESP-12E Nodemcu)
- DHT11 temperature and humidity sensor Breadboard resistors 330 ohm and 4700-ohm LED resistor

Circuit Diagram

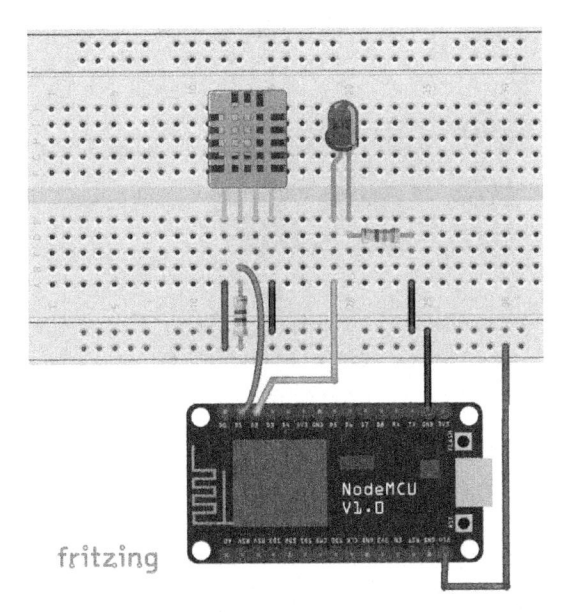

443.Blynk Controlled Automatic Pet Feeder with Timer

If you leave a pet at your house and no one is there to feed it at the appropriate times while you are away, your pet may go hungry. Therefore, we have decided to construct an Internet of Things-based Pet Feeder that is uncomplicated, effective, and cost-effective. With the help of this automatic pet feeder, you will be able to provide food for your pet through the Blynk Mobile App or Web Dashboard no matter where you are in the world. The only thing you need to do is press a button or set a timer, and this machine will take care of the rest of the work for you. For the purpose of this undertaking, we are making use of a NodeMCU

ESP8266 as the primary controller, a Servo motor to control the pet feeder, and NTP servers to acquire the most up-to-date time.

Components Required

- NodeMCU ESP8266
- Servo Motor

Automatic Pet Feeder Circuit Diagram

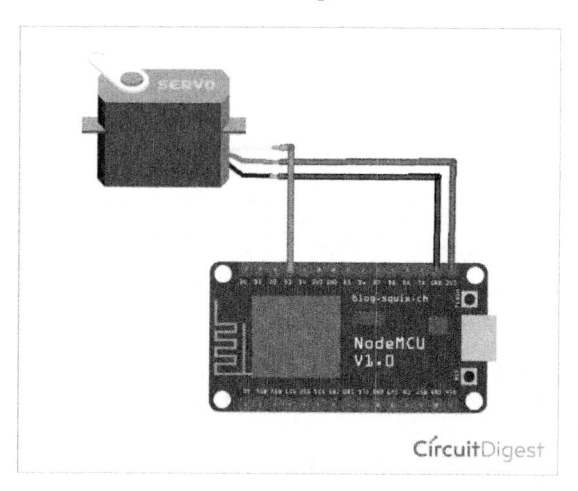

Because we are only connecting a Servo Motor with NodeMCU, the connections are extremely straightforward. The Servo's Vcc (Red Wire) and GND (Brown Wire) pins are connected to the 3.3V and GND pins of the NodeMCU, respectively. Meanwhile, the signal pin (Yellow Wire) of the Servo is connected to the D3 pin of the NodeMCU.

Configuring Blynk for Pet Feeder

Blynk is an all-inclusive collection of software that can be used for rapidly prototyping, deploying, and remotely managing connected devices at any scale, ranging from modest Internet of Things projects to millions of connected items that are commercially available. It is possible to use it to connect the hardware to the cloud and build no-code apps for iOS, Android, and the web to analyze real-time and historical data from devices, control them remotely from anywhere in the world, receive important notifications, and perform a great deal of other tasks. For the purpose of controlling the servo motor that is connected to the pet feeder setup, we are going to use the Blynk mobile application that can be downloaded onto a mobile device.

For this, the first step is to make a brand-new account on the Blynk Cloud platform; alternatively, if you already have an account, you can use the one you have been using. After successfully logging into your account, the next step is to create a template to which you can add a number of different devices. To do this, click on the "+ New Template" button that is located in the upper right corner of the screen. We ordered a pet feeder box from Amazon in order to construct an automatic cat feeder, and then we modified the box so that it could accommodate the servo and the other electronic components. DC motor is installed inside the box, and a gear that was printed using 3D printing is attached to it. The NodeMCU board is installed at the base of the enclosure, and a 12V adapter is used to supply power to the entire configuration.

Testing the Automatic Pet Feeder

Let's put this Pet Feeder set up to the test now that everything is prepared as it was supposed to be. In order to accomplish this, I put food inside the Pet Feeder box, and then I used the Blynk app, which I had previously installed on my mobile phone, as a remote control to see if the pet feeder in our home was releasing food or not. As soon as I turned on the system, the automatic pet feeder began doling out the cat food at a rate that could be adjusted. In addition to the switch, the timer widget within the Blynk app can be used to determine when we should give the cat food.

444. Raspberry Pi Pico Web Server with ESP8266 & MicroPython

We know that the best and cheapest module on the market is ESP-01. In this tutorial, we'll do what the title says and connect a separate Wi-Fi module to the Raspberry Pi Pico. We will learn how to use Micropython and the ESP8266 library to connect and program the ESP01 module with the Raspberry Pi Pico. The ESP8266-01 module comes with default firmware that we can access through UART by sending the AT command sets. With the help of Micropython ESP8266, it's easy to connect ESP8266 to Raspberry Pi Pico and finish the HTTP get/post operation.

The Pico-ESP8266 is a WiFi expansion module for the Raspberry Pi Pico. It can be controlled by the UART AT command and supports the TCP/UDP communication protocol. It's a simple way to turn on WiFi on Raspberry Pi Pico, which can be used for all kinds of IoT communication project

Components Requires

- Raspberry pi Pico
- ESP8266-01 Module
- Bread Board
- Jumper wire
- 1k or 10k resistor (optional connect between 3.3v and ESP8266 EN pin)

For the raspberry pi pico and esp8266-01 module, you can look at the diagram below. With the red wires, the VCC pin on the ESP-01 module is connected to the 3.3V pin on the Pico board, which is Pin 36. The "CH EN" Channel Enable pin on the ESP-01 Module gets its 3.3V from the "Brown" wire. In the diagram below, the Yellow Wire is used to connect the transmitter pin of the ESP-01 module to the receiver pin of the UART0 Channel, which is Pin 2 of the Pico board. The orange wire connects the Receiver pin of the ESP-01 module to the Transmitter pin of the UART0 channel, which is pin number 1 on the Pico board. All of the ground connections on the board are made with the black wires.

Circuit Diagram & Connections

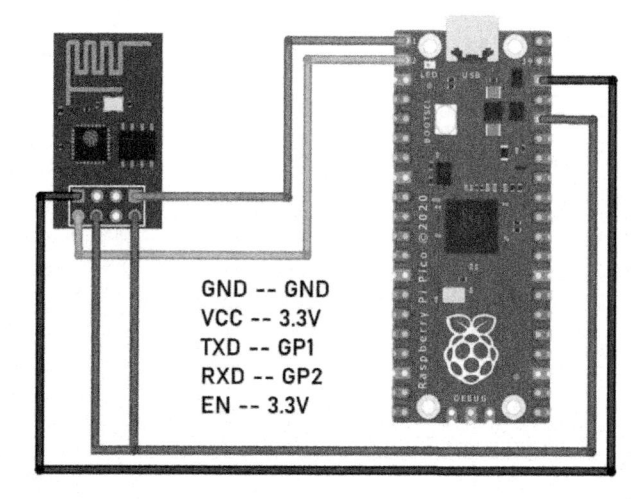

GND -- GND
VCC -- 3.3V
TXD -- GP1
RXD -- GP2
EN -- 3.3V

Now, let's look at how to connect the ESP8266 WiFi Module to the RP2040 Raspberry Pi Pico Board. Here is a simple diagram of how things connect. There are two UART built into the Raspberry Pi Pico. We will use UART-0 in this plan. In the same way, the ESP8266's baud rate is set to 115200 by default. To keep the Raspberry Pi Pico in sync with the ESP8266, we need to set up the baud rate to be the same. Connect the ESP8266's VCC and EN pins to the Pico's 3.3V pin and GND to GND. In the same way, connect the Tx and Rx pins of the ESP8266 to the Rx and Tx pins of the Pico UART-O.

445. IoT based Smart Agriculture Monitoring System

We will use IoT to build a Smart Farming System for this project. The goal of this project is to help farmers get Live Data (Temperature, Humidity, Soil Moisture, and Soil Temperature) for better monitoring of the environment, which will help them increase their overall crop yield and product quality. This smart agriculture system is powered by NodeMCU and uses IoT. It has a DHT11 sensor, a moisture sensor, a DS18B20 sensor probe, an LDR, a water pump, and a 12V led strip. When the IoT-based agriculture monitoring system starts up, it checks the soil's moisture, temperature, humidity, and temperature. It then sends this information to the IoT cloud so it can be watched in real time. If the amount of water in the soil drops below a certain level, the water pump will turn on by itself. We've already made an Automatic Plant Irrigation System that sends alerts to mobile phones but doesn't check any other parameters. Aside from this, building a Smart Agriculture Monitoring System can also be helped by a Rain alarm and a soil moisture detector circuit.

Components Required

Hardware

- NodeMCU ESP8266
- Soil Moisture Sensor
- DHT11 Sensor
- DS18B20 Waterproof Temperature Sensor Probe
- LDR
- Submersible Mini Water Pump
- 12V LED Strip
- 7805 Voltage Regulator
- 2×TIP122 Transistor
- Resistor (4.7K, 10K)
- Capacitor (0.1µF, 10 µF)

Online Services

- Adafruit IO

Smart Agriculture System Circuit Diagram

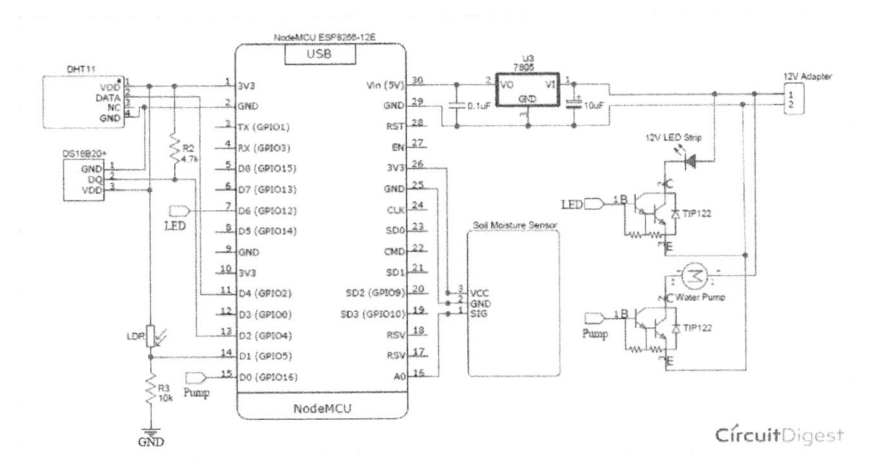

This circuit isn't very difficult. Here, we used one 12V LED Strip, a 12V water pump, a 7805-voltage regulator, and two TP122 transistors to control the LED Strip and the water pump. From the 12V adapter, the 7805 is used to get regulated 5V, and the DHT11 sensor is used to read the temperature and humidity. The DS18B20 sensor probe is used to find out the soil's temperature, and a soil moisture sensor is used to find out how wet the soil is, so that the water pump can automatically turn on and off.

Adafruit IO Setup

Adafruit IO is an open data platform that lets you collect, visualize, and analyze live data in the cloud. With Adafruit IO, you can upload, display, and monitor your data over the internet, making your project compatible with the Internet of Things. With Adafruit IO, you can control motors, read sensor data, and make cool IoT apps that run over the internet.

First, we'll add two toggle buttons blocks to manually turn ON/OFF the LED Strip and Water Pump. Then, we'll add four sliders to show the Temperature, Humidity, Soil Temperature, and Moisture Value. Finally, we'll add two graph blocks to show the Moisture and Soil Temperature Data from the last 30 days. Click on the Toggle block to add a button to the dashboard.

How to Get OpenWeatherMap's API

As we've already said, we'll also show weather forecasts on the Adafruit IO dashboard. To do this, we'll use the OpenWeatherMap API to ask for the day's weather forecast for a specific location. OpenWeatherMap offers well-known weather products that make it much easier to work with weather data. This data can be accessed through fast, reliable APIs that are in line with industry standards and work with a variety of enterprise systems. OpenWeatherMap has both paid and free plans. For this project, we will use the free plan to find out what the weather is going to be like.

Now, to get the API key, you have to sign up on their platform. To do this, first make an account, and once that's done, you'll be taken to the dashboard, as shown below. You can get information from the site by clicking on your name and then on "My API Keys." This will give you a unique API key.

Smart Agriculture System NodeMCU Programming

At the end of the document, you can find the full code for the IoT-based Agriculture Monitoring System. Here, we'll talk about some of the code's most important parts. The code uses libraries called DHT.h, OneWire, Adafruit MQTT, ArduinoJson, and DallasTemperature. You can download Adafruit MQTT.h and DHT11.h, and the rest of the libraries can be downloaded directly from the Arduino IDE library manager.

How to Read the Weather Report:

We will use the code snippets we made with ArduinoJson Assistant to read the weather forecast data from the OpenWeatherMap API. Here in the void loop, we'll only call the API after a certain amount of time has passed, so we don't go over our daily limit.

How to Read Sensor Data:

Now that we have the weather information, we will read all of the sensor information. We're using the DHT11, DS18B20, LDR, and Soil Moisture Sensor in this project. The data from the LDR and the soil moisture sensor will be used to automate the LED strip and the water pump. So, first we will check the status of the LDR, and if the LDR reading is less than 200, the LED will turn on by itself. In the same way, if the percentage of water in the soil is less than 35, the water pump will turn on. I grew seeds in a plastic tray, as shown in the picture below, to test this project.

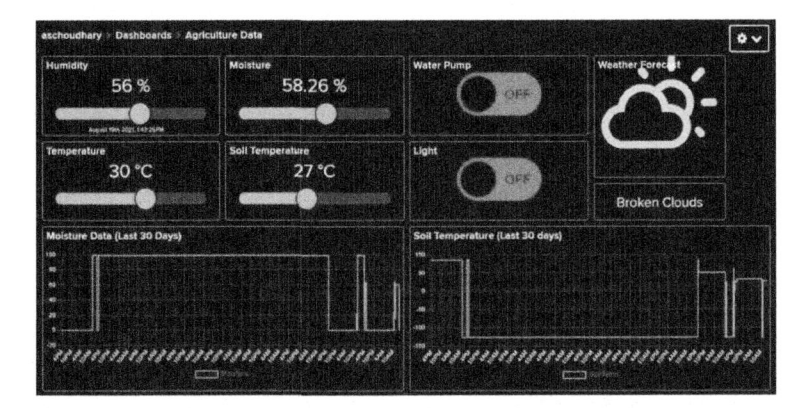

I put the hardware box next to the tray, hooked up a water pump to a bottle of water, and hooked up the power supply. After this is done, it starts to keep an eye on things like the soil's moisture, temperature, etc. All of these will be posted on the Adafruit IO dashboard. So, this is how you can set up a smart agriculture system based on the Internet of Things. You can also measure things like PH, etc., in addition to these parameters. I hope the project was fun and that you learned something from it.

446. Interfacing 5MP SPI Camera with NodeMCU ESP8266

This project is about the Arducam Mega Camera, which will be connected to the NodeMCU ESP8266 WiFi Module. Arducam just came out with a legendary camera solution that makes it easy to connect one or more cameras to any microcontroller. It is especially made for IoT devices that run on batteries, embedded machine vision, and applications that use artificial intelligence. Any microcontroller with a single standard SPI interface can work with the Arducam Mega Camera. Register settings and frame buffers do not take up any memory. The best thing about it is that it can work with both 3.3V and 5V systems.

In this project, we'll connect a 5-megapixel Arducam Mega Camera to a NodeMCU ESP8266 Board and do things like take pictures with different pixel sizes. We will also stream some videos using the ESP8266 Board by itself. Arducam has put out the Arducam Mega SDK for the Arducam Mega camera, which makes it easy to program. So, let's talk in detail about this camera and how it works.

Arducam Mega Camera

The Arducam Mega Camera is a famous camera solution that makes it easy to hook up one or more cameras to any microcontroller. It is mostly made for IoT devices that run on batteries, embedded machine vision, and applications that use artificial intelligence. Any microcontroller with a single standard SPI interface can work with Arducam Mega. If you leave out VCC and GND, you only need 4 pins (called GPIOs). Register settings and frame buffers do not take up any memory. It works well with both 3.3V and 5V systems. The camera can be used directly with Arduino, STM8/STM32, ESP8266/ESP32, MSP430, Nordic, Renesas, and other MCU systems.

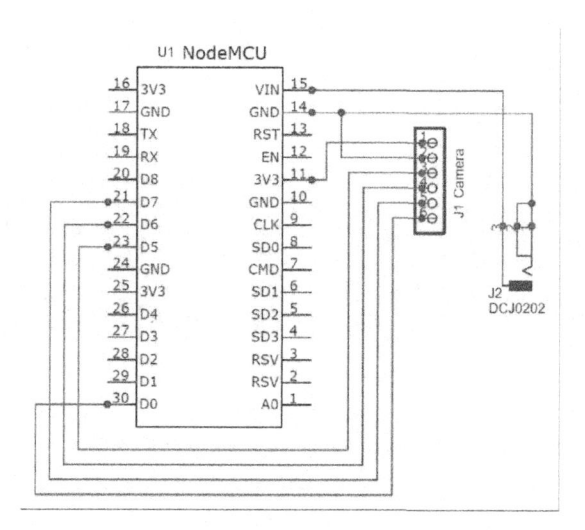

The camera was made to be used in ways that save energy. When your MCU is sleeping, you can turn off the camera completely without having to worry about loading long register settings. It does this instantly (in less than 100 ms) and on its own. There are two types of Arducam Mega cameras. One has 3MP and a fixed focus, while the other has 5MP (autofocus). It comes with a case that makes it easy to mount wherever you want.

Interfacing 5MP Arducam Mega Camera with NodeMCU ESP8266

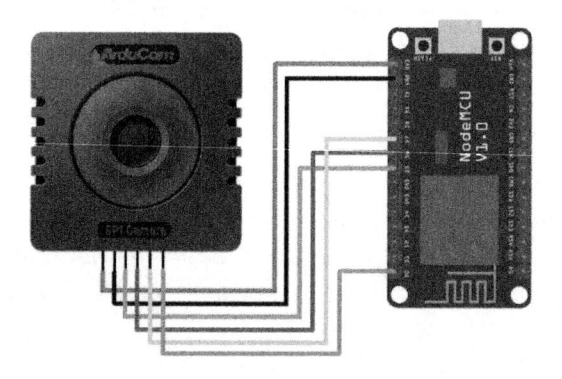

Now, let's look at how to connect an Arducam Mega SPI camera with 5 megapixels to a NodeMCU ESP8266. The picture below shows how simple the connection diagram is.

Testing the Camera

- To make sure the Camera works, we need some software that lets us see how the GUI looks. So, go to the link below to get the Arducam software.
- https://github.com/ArduCAM/Arducam Mega/releases/download/v1.0.0/ArducamMegaSetup Windows x64.exe
- Install the software on your computer after you have downloaded it.
- After installation, click open and choose the NodeMCU ESP8266 port number. The baud rate is 921600.
- Now, the Camera is linked to the software for visualizing. So, to see everything on the GUI Screen, you can move the camera in different ways.
- You can take a picture by clicking the picture button. To close the window, you can click the close button. You can also choose the size and format of the image.
- Click the Video button to start streaming video. The default resolution is 320x240. Click the "**Close**" button to stop streaming videos.
- By default, the auto exposure is turned on. If you want to use manual exposure, you need to turn off the automatic exposure function.
- By default, the auto gain is turned on. When you use manual gain, you need to turn off the automatic gain function.
- The software has controls for brightness, contrast, EV, saturation, special effects, white balance, and focus.
- On the 5MP Arducam Mega Camera, you can control the autofocus. You can turn this feature on or off. Turn on or off the continuous focus function. By default, the continuous focus function is turned off.

447. Real Time GPS Tracker using ESP8266 & Blynk with Maps

This project is about a real-time GPS tracker that uses Google Maps and NodeMCU ESP8266. Before, we made a GSM+GPS Based Vehicle Location Tracker for keeping track of where a vehicle is. We will track location with ESP8266 WiFi Module instead of GSM Module. A GPS tracker is a navigation device that is usually attached to a vehicle, asset, person, or animal. It uses the Global Positioning System (GPS) to figure out where it is and how it is moving. GPS tracking devices send out special satellite signals that are read by a receiver. The tracking device stores locations or sends them to an Internet-connected device using the cellular network or WiFi all over the world. GPS trackers talk to a network of satellites to figure out where they are. The tracker uses a method called trilateration to figure out latitude, longitude, elevation, and time. This method uses the positions of three or more satellites from the Global Navigation Satellite System (GNSS) network and how far away they are from each other. We will connect the Quectel L86 GPS Module to the NodeMCU ESP8266 Board for this project. You can use Neo-6M GPS Module or any other similar GPS Module instead of Quectel L86 GPS Module. We will find the latitude, longitude, speed, heading, and location on Map by using the TinyGPS library. We'll send all of these parameters to the Blynk App and use the map and real-time data on the Blynk Dashboard to keep an eye on things.

Part Required:

- NodeMCU ESP8266
- Quectel L86
- Jumper wire, Breadboard

Quectel L86/L80 GPS Module

The L86 is a great choice for wearable fitness devices because it is small and doesn't need much power. Its Low Power feature lets it connect to GPS while using about half as much power as normal mode does when it is in static receiving mode. With its accuracy and high sensitivity, the L86 can be used for a wide range of Internet of Things (IoT) applications, such as portable devices, automotive, personal tracking, security, and industrial PDAs. The L86 has a patch antenna on top that is 16.0mm by 16.0mm by 6.45mm and has 66 channels for acquiring and 22 channels for tracking. It finds and follows satellites as quickly as possible, even when the signal is weak inside. The module works between 2.8V and 4.3V and uses about 20mA of power on average. When it is not being used, it uses about 1mA of power.

Circuit for GPS Tracker using ESP8266

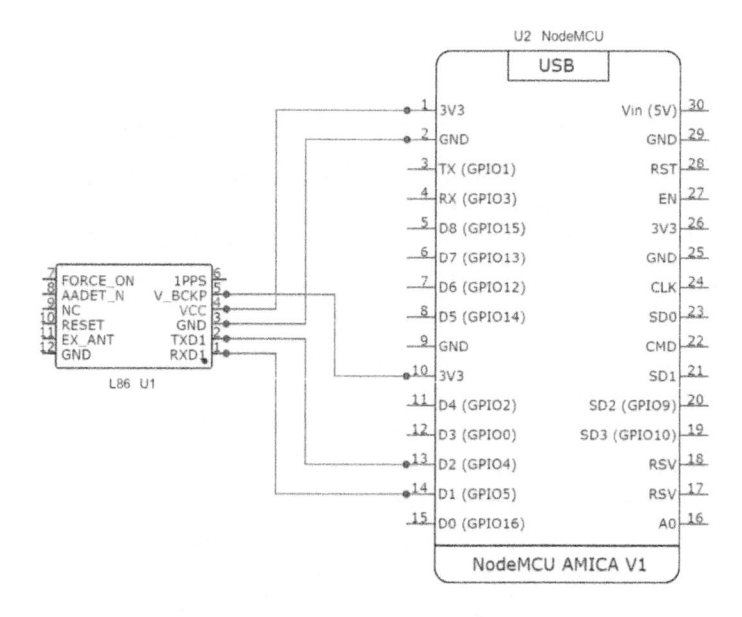

Now let's get to the project part and use ESP8266 and Blynk to make a real-time GPS tracker. The connection diagram is pretty easy to understand.

Connect the L86 GPS Module's VCC and GND pins to the 3.3V and GND pins on the NodeMCU ESP8266. Do not give more voltage than 3.3V. Connect the VCC backup (V BCKP) to the VCC power supply or an external battery. If this pin doesn't have power, it won't work Connect L86's RX and TX to the NodeMCU's D1 and D2. This is for Software Serial Serial Communication. You can connect the parts directly with a jumper wire or use a breadboard to put the circuit together. So, the hardware for the GPS Tracker that uses ESP8266 is now ready.

Setting up the Blynk app

We will use the Blynk App to keep track of GPS location and all of its values. Google Maps can also be used with the Blynk Application.

Blynk is a new platform that lets you quickly build interfaces for your hardware projects that you can control and monitor from your iOS or Android device. After you download the Blynk app, you can set up a project dashboard with widgets like buttons, sliders, graphs, and more.

Testing Real Time GPS Tracker

- Open Serial Monitor after you have uploaded the code. The ESP8266 will try to connect to the WiFi Network. Once the GPS Module is connected to the WiFi network, it will start looking for the closest satellite. Depending on your indoor and outdoor conditions, it could take a while to find your location.
- Once it gets satellite data, it will show the latitude, longitude, speed, heading, and number of satellites it is connected to.
- Now you can open your Blynk app and click the play button in the top right corner.
- The Blynk app will start showing all of the above information right away, along with the location from Google Maps.
- You can now use this GPS device to track the location of your vehicle or any other asset in real time. Always keep the tracker connected to the WiFi network.

448. WiFi Controlled Robot using ESP8266 & Android App

This project will show you how to use a NodeMCU ESP8266 Board to build a robot that can be controlled over the Internet or WiFi. There are many different kinds of robots and robotic cars, from simple ones like toys to more complex ones like robotic arms used in factories.

The Wemos D1 Chip is used as the control unit for this WiFi-controlled robot. We used the L298n Motor Driver IC Module to control the pair of motors. We can use any battery to turn on the circuit because it needs more than 5V. A MIT APP Inventor software is used to make an Android app that can be used to control the robot.

Circuit Diagram & Hardware

Here is the schematic for this WiFi-Controlled Robot Project, which was made with the software Fritzing. We will use an L298 Motor Driver IC to run the two DC motors. For this task, you can use a 200-300 RPM DC motor. Wemos D1 Board is the main control unit. It connects and controls the whole circuit and all of the equipment. We'll use a 6V DC battery or two Lithium-Ion batteries connected in series to power the circuit.

Connect the battery to the power input on the L298 Motor Driver. Connect the L298's four inputs to the ESP8266's D3, D4, D7, and D8 pins. L298 5V Pin is where you give Wemos 5V. Connect the left and right motors to the pins that come out of L298.

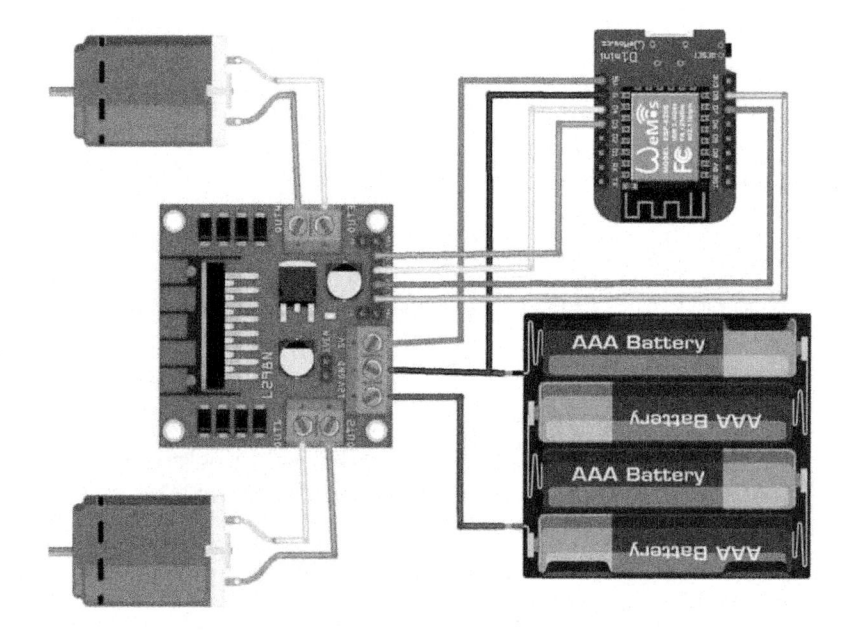

Put the base and chassis of the robot car together. I powered the circuit with two 3.7V Samsung 18650 batteries. Since the batteries are linked together in a series, the total voltage is about 8V. I have used the glass-fiber chassis that is clear. You can use metal, wood, or anything else that fits your needs. Screw all the parts together tightly and put them on the chassis. Use wheels that are strong and of good quality so that the robot can move even on rough ground.

The Android App Design

Now we need to make an Android app for ESP8266 WiFi Controlled Robot. Using MIT App Inventor is the easiest way to make an Android app. With the MIT APP Inventor, you can make apps for Android phones by using a web browser and a connected phone or an emulator. The App Inventor servers save your work and help you keep track of your projects.

With App Inventor, you can create pretty much any kind of app you can think of, including games, informational apps that use user-generated data, personal convenience apps, apps that facilitate communication, apps that use the phone's sensors, and even apps that connect to online services like Twitter.

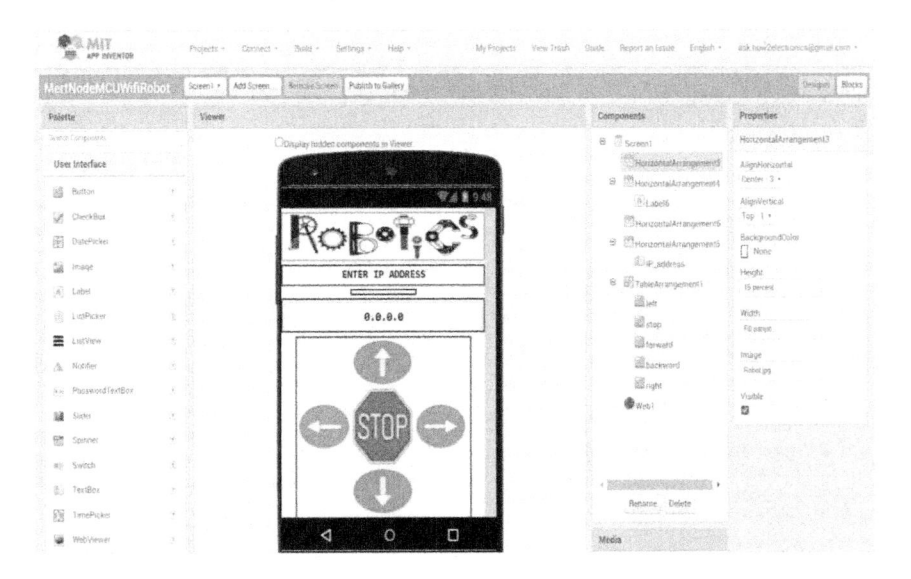

I just made a User Interface for this Robotic Project. The app has 5 sets of switches that send the 0 and 1 commands to the web server.

Testing the WiFi Controlled Robot

Once the code is uploaded, click on the serial monitor. So, once the Serial Monitor is connected to the WiFi Network, it will show the IP Address. Note this IP Address, as the Android App needs it. Open the Android app you have on your phone and type in the IP address you wrote down earlier. Now you can tell the Robot what to do. Press the Up-arrow key to move the Robot forward, and press the DOWN arrow key to move it backward. In the same way, press the Left and Right arrow keys to move the Robot left and right

449. IoT Based Smart Kitchen Automation & Monitoring with ESP8266

In this project, we will use NodeMCU ESP8266 to build an IoT-based smart kitchen with automation and monitoring. One of the most important rooms in a house is the kitchen. The safety factor is the main aspect that must be taken into account during the activity in the kitchen. Gas leaks, fires that are out of control, high temperatures, and a damp environment must be found quickly and fixed. Aside from this, it's important to be able to monitor and control things like the lights, fridge, oven, and other kitchen appliances from afar.

The main goal of this project is to use the Internet of Things to make a prototype of an IoT-based smart kitchen. The system uses multiple sensors, relays & NodeMCU ESP8266 Board. We can monitor all the sensor data on Blynk Applications. We can also send commands to kitchen appliances from the Blynk app.

Component Required:

- ESP8266, NodeMCU
- DHT11 Sensor
- MQ-135 Sensor
- HC-SR501 Sensor, Passive Infrared (PIR) Sensor
- OLED Display
- Buzzer 5V, Active Piezo Buzzer
- Relay 4 Channel Relay Board
- Jumper Wires, Male/Female Jumper Wires
- Breadboard, 830 Points Breadboard
- 1USB Cable, Micro-USB Data Cable

Basically, the IoT Smart Kitchen does the following tasks:

- Use the DHT11 sensor and the Blynk app to keep an eye on the kitchen's temperature and humidity.
- Using the MQ-135 Gas Sensor on the Blynk app, you can check the Air Quality Index (Gas).
- The 0.96-inch OLED display shows the temperature, humidity, and gas level in the kitchen.
- When the gas level gets too high, the alarm goes off and the exhaust fan turns on.
- Detects the presence or absence of a person in the Kitchen using a PIR sensor
- Sends Alarm Staus, Exhaust Fan Status & Person in Room Status to Blynk App
- The Blynk app lets users turn on and off their fridge, oven, and room lights from a distance.

System Design & Circuit Diagram

We will utilize the sensors like DHT11 Humidity Temperature Sensor, MQ-135 Gas Sensor, Passive Infrared Sensor to monitor the Indoor Air Quality Parameters. Similarly, a simple 5V buzzer can work as an alarming system. A relay is connected to an automatic exhaust fan. When the gas level goes over the threshold value, the relay turns on automatically.

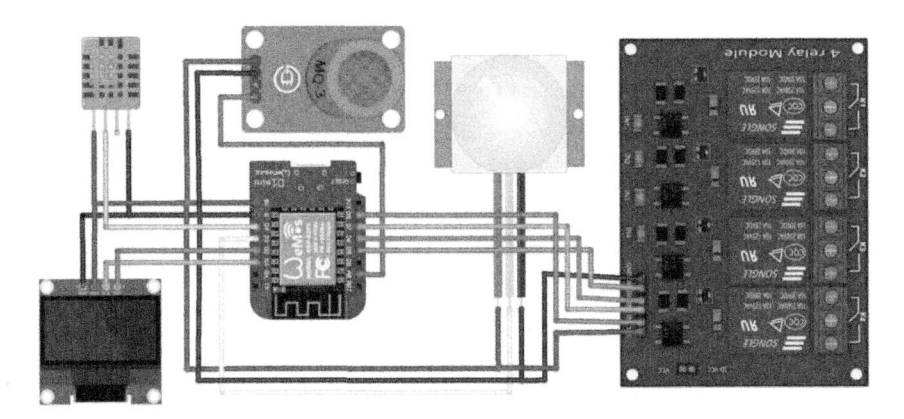

Since we are using a 4-channel relay, the other 3 relays can be connected to kitchen appliances like a mixer, refrigerator, oven, water heater, induction, etc. A simple 0.96-inch I2C OLED can show the temperature, humidity, and gas level in the room in real time. Wemos D1 Mini Board or NodeMCU ESP8266 Board are the brains and hearts of this project. You can use any board that is based on the ESP8266-12E. The ESP8266 chip connects to the WiFi network and makes a connection with the Blynk application.

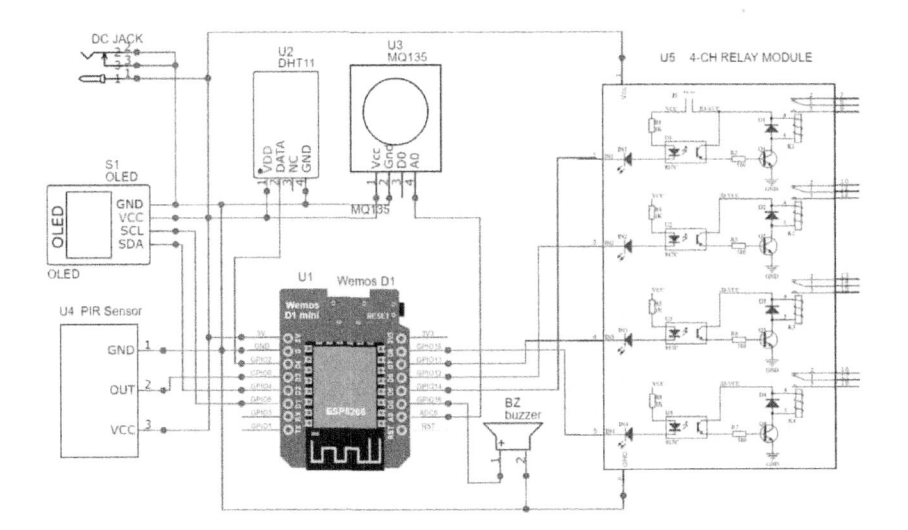

Use the diagram below as a guide and put the circuit together on a breadboard. Connect the SDA and SCL pins of the OLED display to the Wemos D2 and D1 pins. In the same way, connect the output pins of the DHT11, MQ-135, and PIR sensors to the Wemos D4, A0, and D3 pins. For the alarm system, you can connect the 5V active Buzzer to the D0 Pin of Wemos. We can use a 4 channel Relay Module to control the appliances in our homes. So using the jumper wires, connect the 4 channel relay input pin to the D5, D6, D7 & D8 of Wemos.

450. IoT Indoor Air Quality Monitoring with BME680 BSEC & ESP8266

In this project, we will use the BME680 Sensor with the ESP8266 to measure the quality of the air inside using the BSEC Library and the Blynk Application, which is an IoT Cloud platform. Earlier, we made an Indoor Environment Monitoring System with ESP32 and a TFT Color LCD Screen. We learned about the BME680 integrated Environmental Sensor in projects we did before. First, we connected BME680 to Arduino and made a simple project for a weather station. Also, we used an ESP32 and a BME680 sensor to make an IoT-based Weather Station. We used the Ubidots MQTT Platform to keep an eye on the weather. But one problem with the project was that we could only measure things like temperature, humidity, pressure, altitude, dew point, and gas resistance. We couldn't figure out the IAQ Value, which stands for the Index of Air Quality. We couldn't even get a reading for the same amount of CO2 and a percentage of VOCs.

In this article, we will use the BSEC library, which stands for Bosch Sensortec Environment Cluster and is a very advanced BME680 library. Using this library, we can find the value of Indoor Air Quality (IAQ) as well as the equivalent amount of carbon dioxide or Total Volatile Organic Compound (TVOC). The Blynk Cloud platform is used to keep an eye on data about the environment from afar. The ESP8266 connects to the WiFi network and sends data to the Blynk server on a regular basis. So, we can use the BME680 sensor with the ESP8266 to check the quality of the air inside on the Blynk app.

BME680 Integrated Environmental Sensor

Breakout Board   **BME680 Chip**

The BME680 is a digital 4-in-1 sensor that can measure gas, humidity, pressure, and temperature using tried-and-true methods. The BME680 is a better and more advanced version of the BMP180, BMP280, and BME280. The BME680's gas sensor can pick up on a wide range of volatile organic compounds to check the quality of the air inside. The sensor is very accurate and has a high degree of linearity.

The BME680 was made especially for mobile applications and wearables that need to be small and use little power. The BME680's gas sensor can pick up a wide range of gases, such as volatile organic compounds, to measure the quality of the air (VOC). Between 1.7V and 3.6V, the sensor works. This module uses between 0.29 and 0.8 uA of power when it's not being used. When it's in sleep mode, it uses between 0.15 and 1 uA. The BME680 Sensor is able to measure temperatures from -40°C to

+85°C. And the range of the humidity measurement is from 0 to 100%. The Air quality index (IAQ) can be measured from 0 to 500 PPM. The sensor's I2C Address is set to 0x76 by default, but you can change it to 0x77 by connecting SDO to GND.

Circuit: Interfacing BME680 with ESP8266

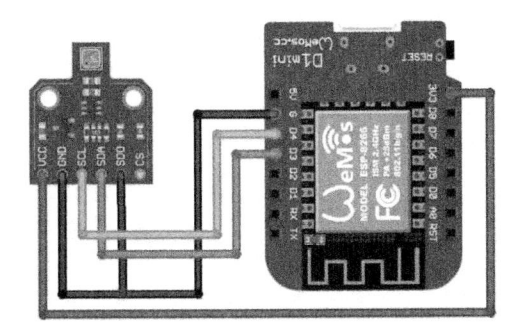

Wemos D1 Mini Board is the main thing that is used in this project. NodeMCU Board is another option. All of these boards have an ESP8266-12E Chip, which is a fast, 32-bit controller with a lot of features. The chip has a WiFi chip built in that can use a WiFi Network to send the data to the internet or a server. Here is a diagram of how to connect a BME680 sensor to a Wemos D1 Mini or ESP8266. Connect the SCL and SDA pins of the BME680 to D4 and D3 on the Wemos Board. Supply the sensor as 3.3V VCC through 3.3V Pin of Wemos Board. Link the SDO to the GND. It's important to connect the BME680's SDO pin to GND because the original code was written to use a different I2C address (0x77). If you connect the SDO pin of the BME680 sensor to Ground, you can get to this I2C address. You can test this connection on a breadboard or on a PCB Board that you made yourself. For testing the circuit, I like to use a breadboard connection.

Go to the "Tools" menu and pick "Wemos D1 Board" from the list. If you are using NodeMCU Board, you can also choose it from the list. Then, choose the COM port and click the upload button to send the code to the board. Now, you can open the Serial Monitor, The following parameters will be displayed every 3 seconds:

1. Millisecond time stamp
2. Temperature in °C (raw)
3. Pressure in millibars
4. Rough Relative Humidity in Percent
5. The gas sensor's raw data is given as a resistance value in Ohm.
6. IAQ index
7. IAQ Accuracy (begins at 0 after startup, goes to 1 after a few minutes, and reaches 3 when the sensor is calibrated).
8. Temperature in degrees Celsius

9. Relative Humidity in Percent
10. Fixed IAQ
11. CO2 equivalent (estimation of the CO2 equivalent in ppm in the environment)
12. Breath VOC equivalent output (estimates the total VOC concentration in ppm in the environment)

Once the readings from the BME680 sensor are stable, you can check the right values for IAQ, CO2, and VOC.

Indoor Air Quality Monitoring using BME680 & ESP8266 on Blynk

With the SSID and password given, the ESP8266 Board will try to connect to the wifi network. Every 3 seconds, the data from the BME680 IAQ is sent to the Blynk application. Both Serial Monitor and Blynk Application can be used to look at the data.

Every time the sensor sends some values, the data changes. Here will be the lovely gauge for pressure, temperature, humidity, IAQ, CO2, and VOC. This is how you can use BME680 and ESP8266 to check the quality of the air inside and outside. It's a nice and easy way to keep track of the air quality on Blynk Cloud.

451. IoT MQTT Based Heart Rate Monitor using ESP8266 & Arduino

In this project, we will use NodeMCU ESP8266, Arduino, and an Easy Pulse Sensor to make an IoT-based heart rate monitor. We will show the Pulse Rate or BPM value on both the OLED Display and the MQTT Dashboard. In one of our previous projects, we used an Optical Pulse Sensor with NodeMCU ESP8266 that could connect to a WiFi network and regularly send the Heart Rate (BPM) value to the Thingspeak Server.

With the free version of Thingspeak Server, you can only do certain things. The data doesn't get sent to Thingspeak Server until 15 seconds have passed. Because of this, we need to choose a method that can easily meet our needs. MQTT protocol fulfills our requirements. MQTT is a lightweight messaging protocol for networks with low bandwidth, high latency, and low reliability. Because of its features, MQTT is a great way to send a lot of sensor messages to platforms for analytics and cloud solutions.

So, Ubidots is a good MQTT platform for IoT projects. With the Ubidots platform, we can send data to the cloud from any device that can connect to the Internet. In this IoT MQTT Based Heart Rate Monitor Project, the Easy Pulse Sensor will be connected to Arduino and ESP8266. First, we will show the Pulse Rate Data on the OLED Display. The data will then be sent to the Ubidtos MQTT Cloud through the WiFi connection.

Easy Pulse Sensor

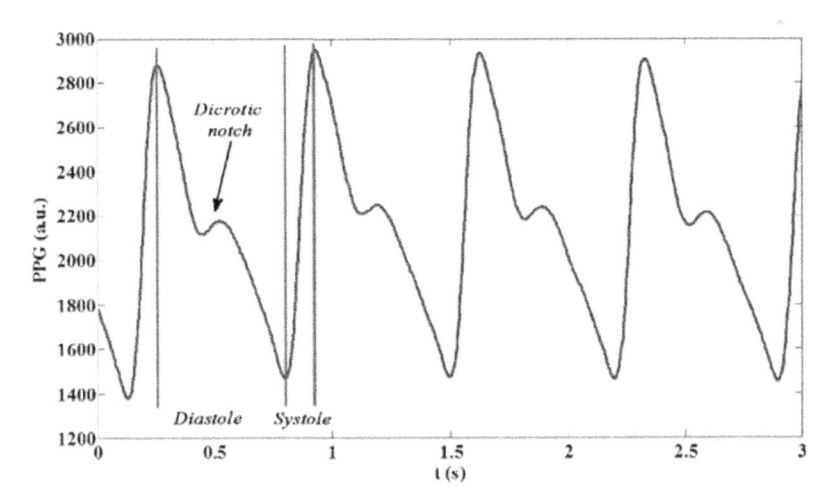

The Easy Pulse Sensor is a pulse sensor that you can make yourself. It is made for hobbyists and educational uses. It is used to explain how photoplethysmography works (PPG). PPG is a non-invasive way to find the pulse wave of the heart and blood vessels from a fingertip. A transmission mode PPG probe (HRM-2511E) is used in the Easy Pulse Sensor.

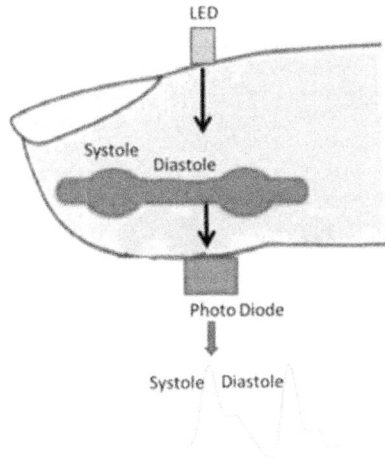

On one side of the Sensor, the finger is lit up by an infrared light source. On the other side of the sensor is a photodetector that measures small changes in the intensity of the light that gets through. These changes are caused by changes in the amount of blood in the tissue. The onboard components and instruments give an analog PPG waveform that is clean and filtered. The digital pulse output is also shown by the LED on the board. Both the analog and the digital signals match the heartbeat.

Block Diagram: IoT Based Heart Rate Monitor using ESP8266 Arduino on MQTT

Let's look at the block diagram to make things clear and easy to understand. This is a simple diagram that shows how the project will work: ESP8266, Arduino, and a pulse sensor are used to make an IoT-based heart rate monitor.

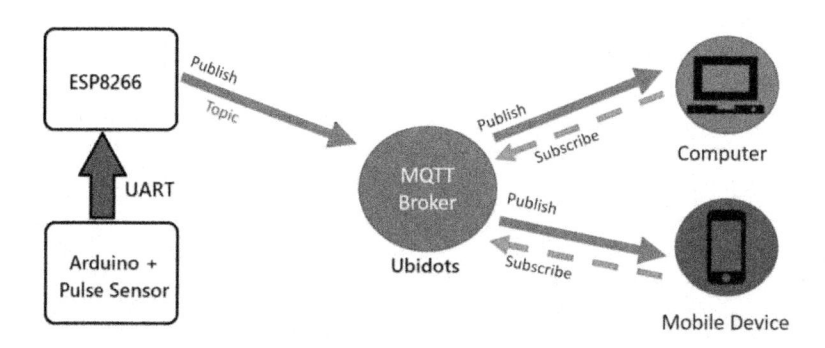

First, we hook up the Pulse Sensor to the Arduino. We send the data from Arduino to NodeMCU ESP8266 through UART Communication. We could have put the Pulse Sensor directly on the NodeMCU ESP8266 Board. But the pulse sensor doesn't seem to be working and the Serial Monitor doesn't show anything. So, it's easy to get the data from Arduino to ESP8266 by using the UART method. The NodeMCU ESP8266 can connect to the WiFi Network. The BPM topic is then uploaded or published to MQTT Cloud Called Ubidots. As a subscriber, you can use the Ubidots Dashboard to look at the published data on your computer or phone. This is how the IoT Based Heart Rate Monitor Project works in its entirety.

Part Required:

- Arduino Board or Arduino Nano
- NodeMCU Board   ESP8266
- Easy Pulse Sensor, HRM-2511-E Pulse Sensor
- OLED Display, 0.96" I2C OLED Display
- Power Supply, 5V Supply
- Connecting Wires, Breadboard

Circuit Diagram & Connections

You can turn the block diagram shown above into a circuit diagram. I make the schematics with Fritzing. Pulse Sensor ESP8266 and Arduino are easy to connect to each other. To talk between ESP8266 and Arduino, we use the Software Serial Method. Connect the TX and RX pins of the ESP8266 to the digital 7 and 8 pins of the Arduino. Connect the pulse sensor's input to the Arduino A0 Pin. In the same way, connect the Pulse Sensor's VCC and GND pins to Arduino's 5V and GND pins.

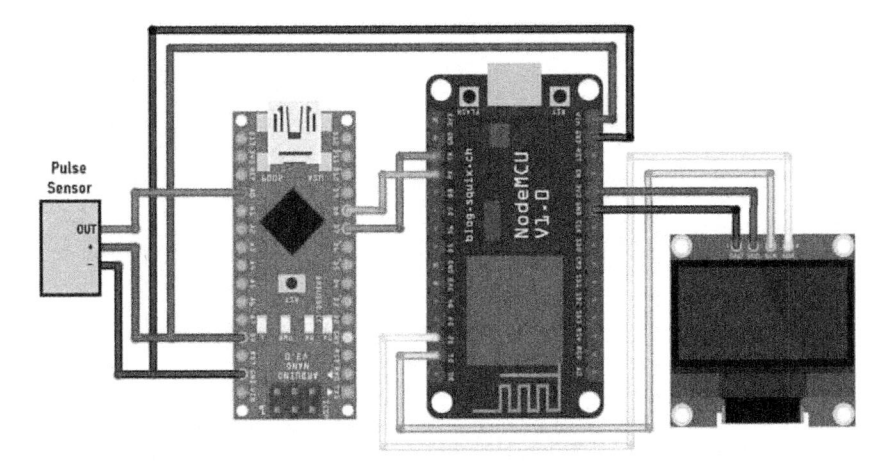

Since the OLED Display is an I2C Module, connect its I2C Pins (SDA and SCL) to D2 and D1 on the NodeMCU. Connect the OLED Display's VCC and GND pins to the 3.3V and GND pins on the ESP8266.

452. IoT Bidirectional Visitor Counter using ESPP8266 & MQTT

In this project, we'll use NodeMCU ESP8266 and Ubidots MQTT to make an IR-based, two-way visitor counter that works with IoT. You can use Ubidots Dashboard to keep track of how many people are coming to, leaving, and staying on your site from anywhere in the world. Use an Infrared or IR Sensor to keep track of how many people are coming and going. The NodeMCU ESP8266 WiFi Module automatically sends the data about visitors to Ubidot's cloud. This NodeMCU Bidirectional Visitor counter can be used to count the number of people who come into a room, office, shopping mall, or event at the front gate. The device counts how many people come in and out of the gate. It also counts how many people leave through the different gates. And finally, it figures out how many people are currently online by subtracting the number of people who left from the number of people who came in. If there is even one person in the room, the light will turn on by itself. When no one is in the room, the light automatically turns off.

In one of my past projects, I used Arduino to make the Visitor Counter Project. But this time, instead of watching the data on an OLED screen, we will send it to the cloud. We can use an ESP8266 Wifi Module, a pair of IR Sensors, an OLED Display, and a Relay Module to make this IoT Visitor Counter. Aside from these main parts, we also need a PCB Board and some passive electronic parts. The project is so simple that even a newbie can do it with no trouble.

IR Sensor as Visitor Detector

The IR sensor, which works as an obstacle detector, is the most important part of this IoT project. When the IR sensor sees a person, it counts them and adds them to the total from before.

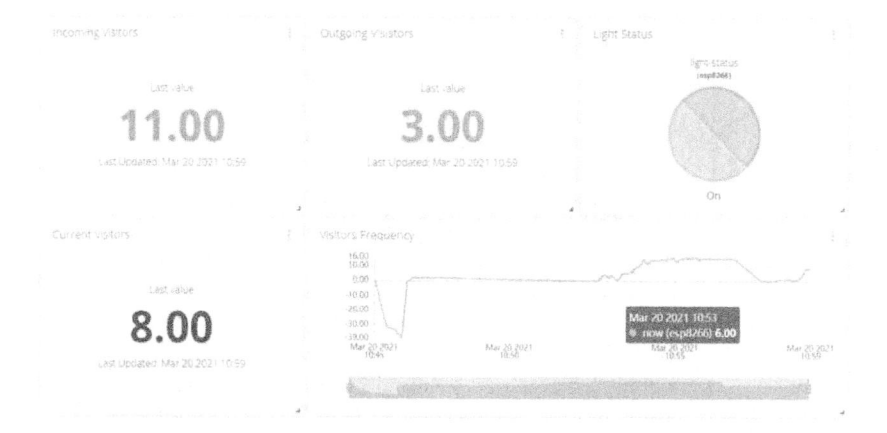

The IR Sensor module is very good at adapting to the light around it. It has a sender and a receiver for infrared. The infrared emitting tube sends out a certain frequency, which gets reflected back to the signal when it hits something. The receiver tube then gets the signal that was sent back. Opamp, a variable resistor, and an output LED are the other parts of the circuit.

The Sensor consists of the following electronics components.

1. IR LED Transmitter

IR LED emits light, in the range of Infrared frequency with a wavelength of 700nm – 1mm. IR LEDs send out light at an angle of about 20 to 60 degrees and can reach up to 5 to 10 cm away.

2. Photodiode Receiver

The photodiode is the IR receiver because it conducts electricity when light hits it. The outside of a photodiode is black, making it look like an LED.

3. LM358 Opamp

Operational Amplifier (Op-Amp) LM358 is used in the IR sensor as a voltage comparator. The comparator circuit checks the difference between the voltage set by the preset and the voltage of the photodiode's series resistor. When the voltage drop across the Photodiode's series resistor is more than the threshold voltage, the Op-output Amp's is high, and vice versa.

When the Op-Amp output is high, the LED at the output terminal lights up. This means that an object has been found.

4. Variable Resistor

In this case, the variable resistor is set. It is used to set the range of distances where the object should be picked up.

IoT Two-Way Visitor Counter Schematic or Circuit

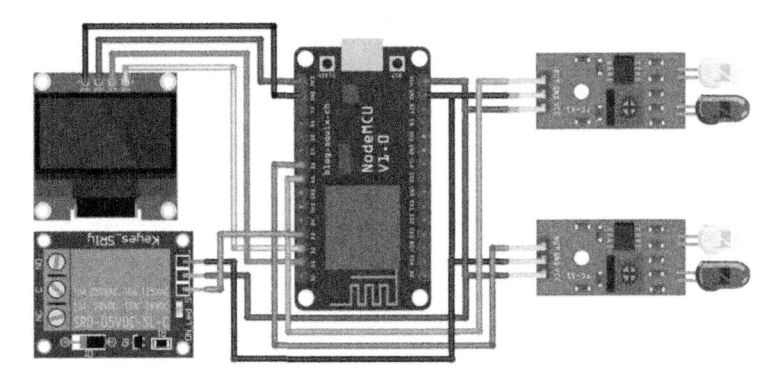

The Circuit for NodeMCU ESP8266 Bidirectional Visitor Counter is very easy to understand. Here is a diagram of how the project will work. You can use Fritzing Software to make the schematic.

Connect the 0.96" OLED Display's I2C pins (SDA and SCL) to the NodeMCU's D2 and D1 pins. Connect the IR sensors' output pins to NodeMCU's D5 and D6. One of the IR sensors will count how many people come in, and the other will count how many people leave. Connect a 5V Relay Module to NodeMCU's D4 Pin in the same way. At 5V VCC, both the IR Sensors and the Relay Module work. You can get 5V from the Vin pin on the Amica NodeMCU or the VU pin on the Lolin NodeMCU.

Schematic & PCB Designing

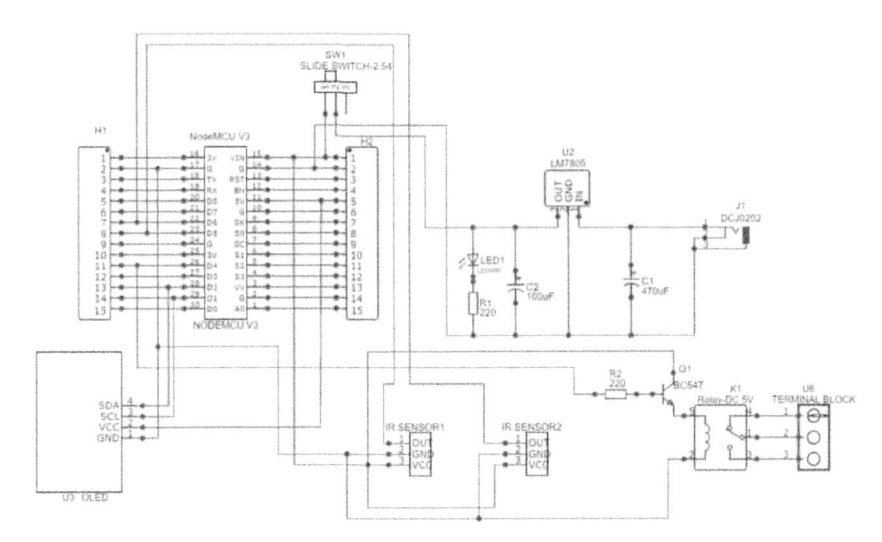

You can put this circuit together and test it on a breadboard. But if you don't want to put the circuit together on a breadboard, you can look at this schematic and make a custom PCB. For drawing the Schematic, I like to use EasyEDA, which is an online Schematic & PCB Designing Tool.

453.IoT IR Thermometer using MLX90614 & ESP8266 on Blynk

In this project, we will use MLX90614 and ESP8266 to build our own IoT-based IR thermometer and then use the Blynk app to check the temperature. This DIY Infrared Thermometer is a low-cost, non-contact thermometer that can be used to measure the temperature of the body or of very hot objects. An infrared thermometer is a tool that measures an object's infrared radiation. Infrared radiation is a type of electromagnetic radiation that is below the range of light that we can see. Using the SMBus Protocol, which is similar to the I2C Protocol, the IR Temperature Sensor MLX90614 from Melexis can be easily connected to any microcontroller. Here, we'll connect the MLX90614 sensor to the NodeMCU ESP8266 and the 0.96" OLED display.

We will connect the ESP8266 to the WiFi network and send the temperature data, including both the ambient temperature and the temperature of the object, to the Blynk Application Dashboard. Blynk is an IoT platform that lets you quickly build interfaces for controlling and monitoring your hardware sensors from your smartphone.

Components Required:

- NodeMCU (ESP8266 12E Board)
- IR Temperature Sensor (GY-906 MLX90614 IR Temperature Sensor)
- OLED Display (0.96" I2C OLED Display)
- Jumper Wires
- Breadboard

Circuit Diagram:

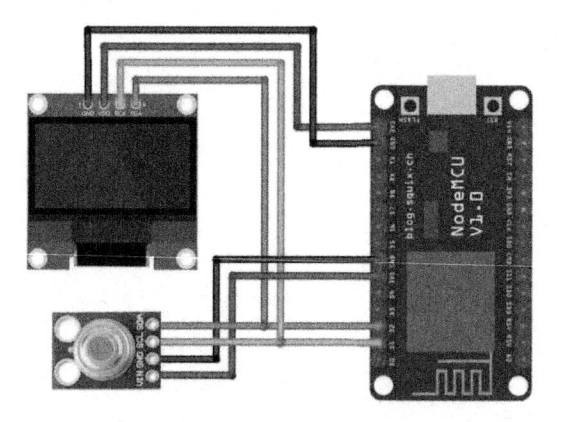

Now, let's connect the MLX90614 IR temperature sensor with the NodeMCU ESP8266 and the OLED display. Here is a picture of the connection. Connect the VCC pin of the MLX90614 and the GND pin of the OLED display to the 3.3V pin of the NodeMCU ESP8266. In the same way, connect the SDA and SCL pins of the MLX90614 to the D2 and D1 pins of the NodeMCU.

454. IoT Temperature Monitor for Industry with MAX6675 and ESP8266

In this IoT-based project, we will connect a Thermocouple Temperature Sensor MAX6675 to a NodeMCU ESP8266 Board and show the temperature data on a 0.96" OLED Display. Since the project is based on the Internet of Things (IoT), we will send the temperature data to an IoT app called Blynk. On the Blynk Application, you can check the temperature data from anywhere in the world. The device can measure temperatures between 0°C and 1024°C. So, it can be used in the industrial world. In industry, it's hard to get close to a hot object and check its temperature by hand with an Industrial Thermometer because the object is hot or the body gives off heat. So, the solution is to use some IoT devices and a Thermocouple Temperature Sensor to monitor the temperature from afar.

Here, we'll use Thermocouple MAX6675 and ESP8266 Wifi Module to make a simple project. The ESP8266 will connect to the wifi router, and since it will be connected to the internet, it will regularly send the temperature data to some online servers. The Blynk is the best and free online server.

MAX7765 K-Type Thermocouple Temperature Sensor

The Maxim MAX6675 K-Thermocouple to digital converter IC is used in this MAX6675 Module + K Type Thermocouple Sensor. It has an SPI-compatible digital serial interface that works with microcontrollers and gives accurate temperature measurements that take into account the temperature.

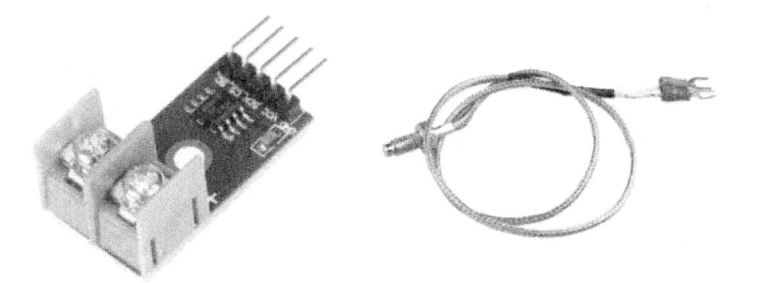

It is made up of 12 bits. This converter can measure temperatures from 0°C to +700°C with a thermocouple accuracy of 8 LSBs and a resolution of 0.25°C. It can also read temperatures as high as +1024°C. Screw terminals let you connect to the spade connectors on the thermocouples, and a 5-pin standard 0.1" header lets you connect to a microcontroller.

Part Required:

- Nodemcu ESP8266-12E Board
- OLED Display (0.96" I2C OLED Display)
- MAX6675 (Thermocouple Temperature Sensor)
- Jumper Wires
- Breadboard

Circuit Diagram:

The circuit diagram for interfacing MAX6675 Thermocouple with NodeMCU ESP8266 is given below.

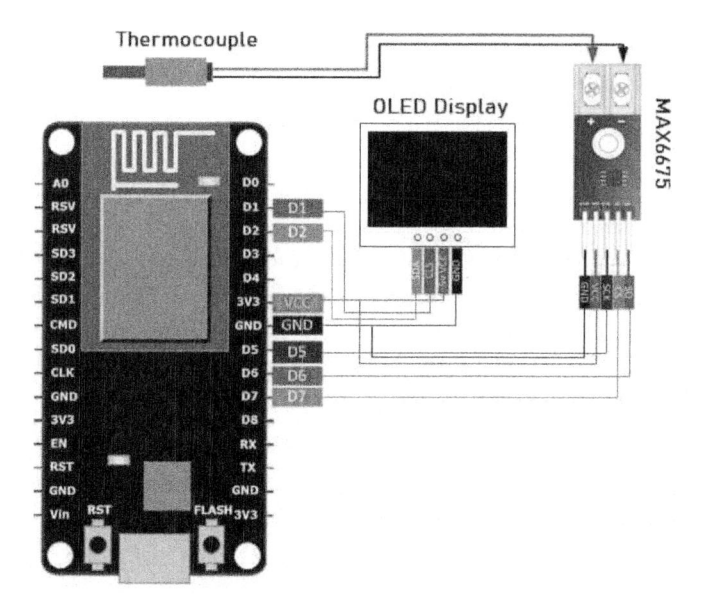

The MAX6675 board's SCK, CS, and SO pins are connected to the NodeMCU board's D5, D6, and D7 pins. The 0.96″ I2C OLED Display's SDA and SCL pins are connected to Nodemcu's D2 and D1 pins. Both the OLED Display and the MAX6675 work between 3.3V and 5V. So, hook up the VCC Pin to 3.3V and the GND Pin to GND.

IoT Industrial Temperature Monitor with MAX6675 & ESP8266

The code should be uploaded to the Nodemcu Board. After the code is uploaded, the Nodemcu will try to connect to the network using the Wifi SSID and Password that were given. The temperature can be seen in both degrees Fahrenheit and degrees Celcius on the OLED Display. You can use the Gas Lighter to heat the temperature sensor, which is called a thermocouple. The OLED will show how hot it is getting. You can also look at the Blynk Application. The data will come from the Blynk Server and be shown on the Blynk Application Dashboard.

455. IoT ESP8266 Lux Meter using BH1750 Light Sensor & Blynk

In this project, we'll make an IoT ESP8266-based Lux meter or light meter using a BH1750 ambient light sensor and watch it on Blynk. With this sensor, we can figure out how much light there is in lux units and see how bright the light is around us. This sensor can be used for a lot of different things. For example, it can be used to change the brightness of the LCD screen on a mobile phone by detecting the light around it. It can also be used to turn street lights on and off based on how dark it is. Earlier, we made Lux/light Meter using BH1750 Sensor & Arduino. On a 162 LCD Display, the amount of light in lux is shown. Now, instead of showing the data on an LCD screen, we will send it to the Blynk application. With the Blynk app, we can check on the amount of light in a certain place from anywhere in the world. We need some hardware, like the NodeMCU ESP8266 & BH1750 Ambient Light Sensor Module, to do that. This is a BH1750 light intensity sensor breakout board. It has an I2C bus interface and a digital Ambient Light Sensor IC. This IC is best for getting information about ambient light in mobile phones so that the brightness of the screen can be changed based on how bright the room is. This sensor can measure light accurately up to 65535 LUX. It uses very little electricity and uses a photodiode to detect light. 2.4V to 3.6V can be used to power the BH1750. The main part of the sensor is the BH1750FVI module, which needs 3.3V to work. So, the circuit uses a voltage regulator. I2C communication with the address 0x23 uses the pins SDA and SCL. With these pins, pullup resistors with a value of 4.7k are used.

Part Required:

- NodeMCU Board (NodeMCU ESP8266 12E Wifi Module)
- Light Sensor (BH1750 Ambient Light Sensor)
- Jumper Wires
- Breadboard

Circuit Diagram

Now, let's connect the NodeMCU ESP8266 Board to the BH1750 Ambient Light Sensor. It's not hard to see the link. Connect the I2C pins of the BH1750, which are SDA and SCL, to D1 and D2 on the NodeMCU board. Connect BH1750's 3.3V pin to NodeMCU's 3.3V pin and GND to GND. The circuit is easy to put together on a breadboard.

456. IoT Based TDS Meter using ESP8266 for Water Quality Monitoring

In this project, we'll learn how to make our own IoT-based TDS meter using NodeMCU ESP8266 and a TDS sensor for monitoring water quality. The TDS value, which stands for "Total Dissolved Solids," tells how many solids are dissolved in water. Salts, minerals, and metal ions that conduct electricity are all examples of these solids. The value of the water is also called its conductivity. Because the water conducts electricity better when there are more of these solids or ions in it.

Most TDS meters measure this conductivity in micro siemens or parts per million (ppm). This means that there are a certain number of solid particles for every million water particles. With a value of 40 ppm, there are 40 dissolved ions for every million particles. This means that the rest of the particles (=999,960) are water molecules. The TDS Sensor can measure both the EC and the TDS of a liquid (Electrical Conductivity). A device called an electrical conductivity meter (EC meter) measures how well a solution conducts electricity. It is often used in hydroponics, aquaculture, aquaponics, pisciculture, and freshwater systems to measure the amount of nutrients, salts, or impurities in the water.

To make our own IoT-based TDS meter, we'll need a TDS sensor and a waterproof temperature sensor. We need an IoT platform to send the TDS, EC, and temperature measurements. Blynk is the best free IoT platform. The ESP8266-based TDS Meter connects to wifi and continuously sends data to the Blynk app. So, you can get a report on your water quality 24 hours a day, 7 days a week. So, the TDS sensor is the best sensor for monitoring water quality. You can also add a Ph Sensor and a Turbidity Sensor to learn more about the Water Quality.

The diagram and connections for the TDS meter

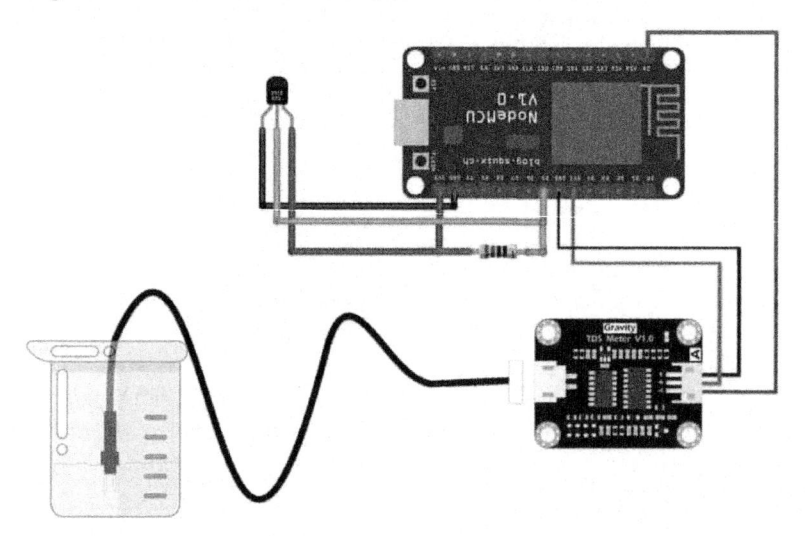

Above is a diagram of how to connect NodeMCU ESP8266 and TDS Sensor. The TDS Sensor's analog pin is hooked up to NodeMCU's A0. The VCC pin is connected to 3.3V and the GND pin is connected to GND.

Part Required:

- Nodemcu Board (NodeMCU ESP8266 Wifi Module)
- TDS Sensor
- DS18B20 Sensor (DS18B20 One-Wire Waterproof Temperature Sensor)
- Resistor   4.7K
- Jumper Wires
- Breadboard

Calibration is what the DS18B0 Temperature Sensor is used for. This is because the TDS value changes as the temperature does, so the temperature of the liquid needs to be taken into account. The DS18B20 Waterproof Temperature Sensor does this.

IoT Based TDS Meter with ESP8266 for Monitoring Water Quality

The NodeMCU tries to connect to the network after the code has been uploaded. Once it is connected to a WiFi network, it will start sending data to the Blynk Server. You can take different samples of water and check its TDS Value, EC Value, and temperature. When you add salt or any other ionic solute, the TDS value goes up quickly. On the Blynk Application Dashboard, you can keep an eye on the data.

457. IoT Smart Agriculture & Automatic Irrigation System with ESP8266

We will learn about the IoT-based smart agriculture and automatic irrigation system with Nodemcu ESP8266 in this project. Agriculture is a key part of how agricultural countries grow and change. Some problems with agriculture have always kept the country from getting better. So, the only way to solve this problem is through smart agriculture, which involves making traditional ways of farming more modern. So, the method is to use automation and Internet of Things (IoT) technologies to make farming smart. Internet of Things (IoT) makes it possible to do things like monitor and choose crops based on their growth, help with automatic irrigation decisions, and more. We suggested the ESP8266 IoT Automatic Irrigation System to make the crop more modern and productive.

This project shows how to make IoT Smart Agriculture with Automatic Irrigation System using some simple sensors that are already on the market. Capacitive Soil Moisture Sensor will be used to measure the amount of water in the soil. In the same way, we prefer the DHT11 Humidity Temperature Sensor to measure the temperature and humidity of the air.

Circuit Diagram:

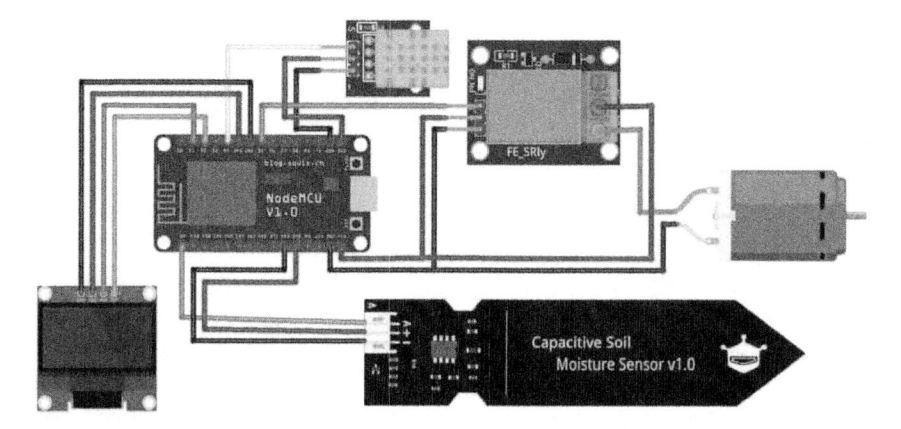

Let's look at the project plan for the IoT Smart Agriculture and Automatic Irrigation System. For most of my projects, I use Fritzing to make a schematic. All you have to do is put something in place and connect it.

We'll be able to control the Water Pump with a 5V Power relay. When the sensor sees that the soil doesn't have enough water, the motor turns on by itself. So, the field will automatically get water. When the ground gets wet, the motor stops. Through the online Thingspeak Server, you can watch all of this from anywhere in the world.

Capacitive Soil Moisture Sensor: This is an analog capacitive soil moisture sensor. It uses capacitive sensing to measure the amount of water in the soil. This means that the capacitance changes based on how much water is in the soil. You can turn the capacitance into a voltage level from 1.2V at the lowest to 3.0V at the highest. The good thing about Capacitive Soil Moisture Sensor is that they are made of a material that doesn't rust, so they last a long time.

DHT11 Temperature and Humidity Sensor: The DHT11 is a simple, very cheap digital sensor for measuring temperature and humidity. To measure the air around it, it has a capacitive humidity sensor and a thermistor. On the data pin, it spits out a digital signal. It's easy to use, but you have to time it right to get data. The only real problem with this sensor is that it only gives you new data every two seconds. So, when using the library, sensor readings can be up to two seconds old. We will use this sensor in this project to measure the temperature and humidity of the air.

DC 3-6V Mini Submersible Micro Water Pump: The DC 3-6 V Mini Micro Submersible Water Pump is a small Submersible Pump Motor that doesn't cost a lot of money. It runs on a power supply of 2.5 to 6V. It can hold up to 120 liters per hour and only uses 220 mA of electricity. Just hook up the tube pipe to the motor outlet, put it under water, and turn it on.

Part Required:

- NodeMCU ESP8266
- Soil Moisture Sensor
- OLED Display
- DHT11 Sensor
- Relay Module 5V
- DC Motor Pump 5V
- Connecting Wires
- Breadboard

458. Home Automation using Google Firebase & NodeMCU ESP8266

In this project, we will learn how to use Google Firebase and NodeMCU ESP8266 to make an IoT-based home automation project. Home Automation Project is one of the most common and well-known hobby projects you will find on the internet. When we say "Home Automation," we mean that home appliances can be controlled without a switch. Home devices are an important part of the Internet of Things ("IoT") Application when they are connected to the Internet. We've already learned how to use the Blynk Application, WebServer, Amazon Alexa, Arduino IoT cloud, and AWS IoT Core to automate our homes. But we will use Google Firebase for this project. Google Firebase is software for making apps that is backed by Google. It can be used to create, manage, and change data from any Android or iOS app, web service, IoT sensor, or hardware.

To find out more about the Google Firebase Console, type "/OFF" as "1/0" from the app. Google Firebase gets the information from the app. Now, the data from Google Firebase is updated on the NodeMCU Board through the Internet. So, the digital GPIO pins go high and low, which turns on and off the devices that are connected to the relay. We will use MIT APP Inventor to make an Android app for home automation. We will link the Android app to Google Firebase by using the Firebase Host and Authentication Key. Then, we'll send the ON/OFF command from App as 1/0.

Google Firebase gets the information from the app. Now, the data from Google Firebase is updated on the NodeMCU Board through the Internet. So, the digital GPIO pins go high and low, which turns on and off the devices that are connected to the relay.

Part Required:

- NodeMCU ESP8266
- Relay - 5V Relay
- Voltage Regulator IC - LM7805 5V
- Female DC Power Jack
- Diode - 1N4007

- Resistor - 330ohm
- NPN Transistor - BC547
- Terminal Block - 5mm
- LED - 5mm LED Any Color
- 1Female Header - 2.54mm

Circuit Diagram & Connection

Below is a diagram of how to use ESP8266 with Google Firebase to automate your home. With this circuit diagram, you can put together the circuit on a Breadboard using a 4-channel relay and a NodeMCU board.

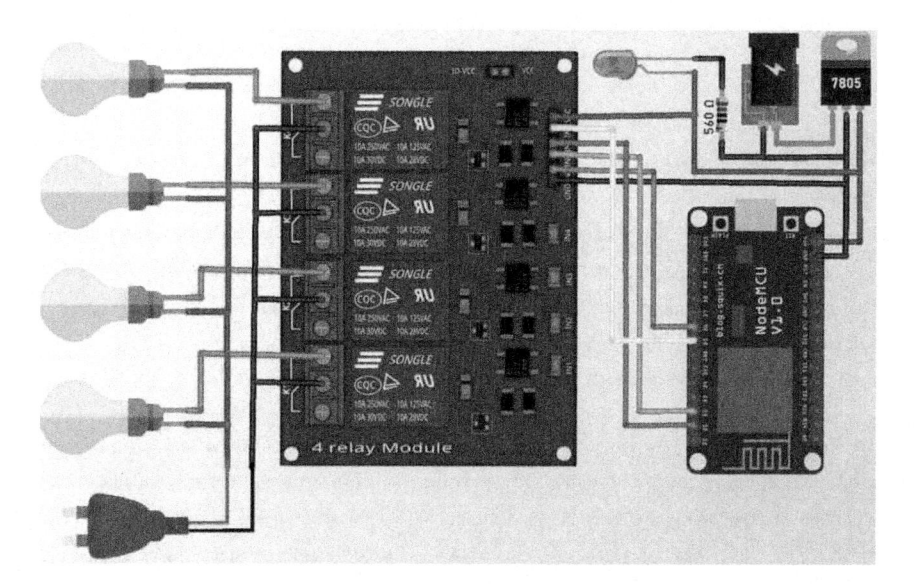

Home Automation using Google Firebase & ESP8266

The NodeMCU will connect to the WiFi Network after the code is uploaded. Using the Firebase Authentication Token & Host, both the Nodemcu Board and the Android App can connect to Google Firebase. Now you can send ON/OFF commands from your mobile app to your home appliances to turn them on or off.

459. IoT Water Flow Meter using ESP8266 & Water Flow Sensor

In this project, we'll learn how to use ESP8266 and a water flow sensor to make an IoT-based water flow meter. We will connect the NodeMCU ESP8266 Board to the YFS201 Hall Effect Water Flow Sensor. On a 0.96" OLED Display, we will show the water flow rate and the total volume. After that, we'll connect the hardware to IoT Server. Thingspeak App will be used for the IoT Server. The water flow rate and volume data can be uploaded to Thingspeak Server and viewed or monitored from anywhere in the world. A key part of City Management is the Water Management System. Managing water means giving out water only when it is really needed and not wasting any. So, it is very important to measure how fast and how much water flows. Almost nothing can be done about water without measuring these parameters. Also, it has become very important to be able to check the water's volume, flow rate, and quality from anywhere with an Internet connection. So, there is a need for an online system for water management. There are a lot of Water Flow Sensors on the market, but they cost too much to use and buy. Because of this, you need a low-cost water flow meter. So, we'll use the YFS201 Hall Effect Water Flow Sensor with the ESP8266 to make a simple IoT-based water flow meter.

Hall-Effect Water Flow Sensor YF-S201

The YF-S201 Hall-Effect Water Flow Sensor is shown in this picture. This sensor has both an inlet and an outlet, so it can be hooked up to the water line. A pinwheel inside the sensor measures how much liquid has passed through it.

Here is a magnetic Hall effect sensor built in that sends out an electrical pulse every time the motor turns.

The sensor comes with three wires:

1. Red (5-24VDC power) (5-24VDC power)
2. Black (ground)
3. Yellow (Hall effect pulse output)

By counting the pulses that come out of the sensor, you can figure out how fast the water is moving. About 2.25 milliliters are in each pulse. This Sensor is the cheapest and best, but it's not the most accurate because the flow rate/volume changes a little depending on the flow rate, the pressure of the fluid, and the way the sensor is positioned. A lot of calibration needs to be done to get more than 10% accuracy. With this Sensor, you can make a simple IoT-based water flow meter. The pulse signal is a simple square wave, which makes it easy to record and convert to liters per minute using the following formula.

Pulse rate (L/min) = pulse rate (Hz) divided by 7.5

IoT Water Flow Meter with ESP8266 and Water Flow Sensor: Now let's connect the YF-S201 Hall-Effect Water Flow Sensor to the Nodemcu ESP8266 and OLED Display. The OLED Display will show how fast water is flowing through the pipe and how much water has gone through it so far. After every 15 seconds, the same Flow Rate and Volume data can be sent to Thingspeak Server. If you want information right away, you can switch to the Blynk App. Using the MQTT Protocol is another way to improve wireless communication.

Part Required:

- NodeMCU ESP8266
- Water Flow Sensor (YF-S201 Hall-Effect Water Flow Sensor)
- OLED Display
- Connecting Wires   Jumper Wires
- Breadboard

Circuit Diagram.

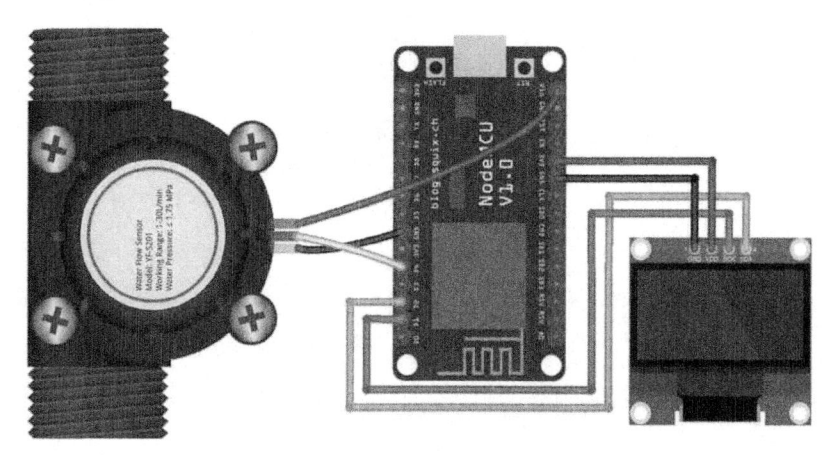

Since the Water Flow Sensor is a digital sensor, its output pin can be connected to any digital pin on the ESP8266. In my case, I linked to GPIO2, which is also called D4. The sensor works with 5V and can be hooked up to the Vin port on the ESP8266. In the same way, the I2C OLED Display SDA and SCL pins are connected to D2 and D1 on the ESP8266. Since the OLED Display works at 3.3V, it can be connected to Nodemcu's 3.3V pin.

460. IoT Decibel meter with Sound Sensor & ESP8266

In this project, we'll make an IoT-based decibel-meter with a Nodemcu ESP8266 and a sound sensor, which we'll be able to check online using the Thingspeak Server. Acoustic (sound that moves through the air) measurements are made with a sound level meter. The condenser microphone is the best type of microphone for sound level meters because it is accurate, stable, and reliable. The mic's diaphragm reacts to changes in air pressure that are caused by sound waves. Because of this, the tool is sometimes called an SPL (Sound Pressure Level) Meter. Noise pollution studies often use decibel-meters to measure different types of noise, especially industrial, environmental, mining, and airplane noise. A sound level meter's reading doesn't match up well with how loud something sounds to a person. A loudness meter is a better tool for measuring how loud something is. Specific loudness is a nonlinearity caused by compression, and it changes at certain levels and frequencies. There are also different ways to figure out how to measure these things.

In this IoT project, we will use ESP8266 and a sound sensor to make a decibelmeter. This easy DIY project can be made at home to measure loudness in decibels (dB). We will use a Nodemcu ESP8266, a sound module, and either a 16x2 LCD Display or an OLED Display for the display. The sound will be picked up by the Sound Sensor, which will then turn it into an analog voltage that the Nodemcu ESP8266 can read. The Nodemcu connects to wifi and sends the data to Thingspeak Server.

Microphone Sound Sensor

Like its name says, the microphone sound sensor can pick up sounds. It lets you know how loud something is. The sound sensor is a small board with a microphone (50Hz-10kHz) and some processing circuitry to turn sound waves into electrical signals. This electrical signal is sent to the on-board LM393 High Precision Comparator, which converts it to a digital signal and makes it available at the OUT pin. The module has a potentiometer built right in so that the OUT signal's sensitivity can be changed. With the help of a potentiometer, we can set a threshold. So that if the sound is louder than the threshold value, the module will send out LOW. Otherwise, it will send out HIGH. Besides this, there are two LEDs on the module. When the module is turned on, the Power LED will light up. When the digital output goes LOW, the Status LED will light up.

There are only three pins on the sound sensor: VCC, GND, and OUT. The power for the sensor comes from the VCC pin, which works with 3.3V to 5V. When there is no sound, the OUT pin sends out HIGH, but when there is sound, it sends out LOW.

Setting Up Thingspeak

You need to Setup Thingspeak before you can monitor sensor data on the Thingspeak Server. Go to https://thingspeak.com/ to learn how to set up the Thingspeak Server. Sign up for an account or just sign in if you already have one.

IoT Decibelmeter with ESP8266 and 16x2 I2C LCD Display

First, let's build a decibelmeter by connecting a sound sensor to a Nodemcu ESP8266 and an 16x2 I2C LCD display. Here is a picture of the connection. Connect the Sound Sensor's Analog output pin to the ESP8266. Connect the LCD display's I2C pins (SDA and SCL) to D2 and D1 on the ESP8266. Vin Pin is where you put 5V to power the LCD Display. In the same way, the 3.3V Pin should be used to power the Sound Sensor. You can also add 3 LEDs of different colors to the Nodemcu D3, D4, and D5 Pins. This LED lights up based on how loud the sound is.

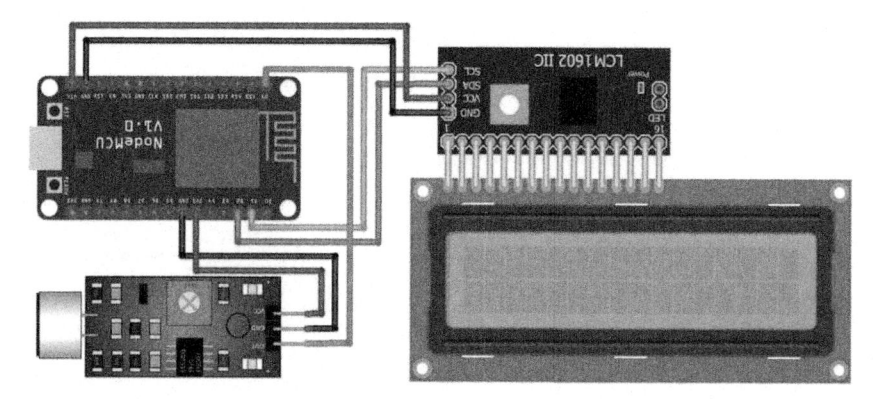

Monitor Sound dB on LCD Display & Thingspeak

The Nodemcu will try to connect to wifi after the code is uploaded. All of the process can be seen on the LCD screen. Now you can play music and see the value on the LCD Display.

You can check out the Decibelmeter dB data online right now by going to Thingspeak. Just go to the private view of Thingspeak to see the logs and graphs.

IoT Decibelmeter with ESP8266 & OLED Display

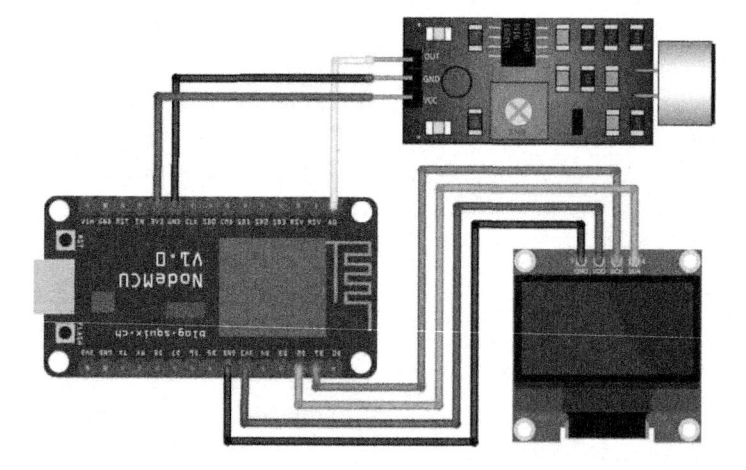

Above is a diagram of how to connect a sound sensor to a NodeMCU ESP8266 and an OLED display to make a decibelmeter. Both the sound sensor and the OLED screen need 3.3V power. So, connect their VCC and GND pins to the 3.3V and GND pins on the NodeMCU. Connect the sound sensor's output pin to the A0 pin on the Nodemcu. Connect the OLED Display's I2C pins (SDA and SCL) to the Nodemcu's D2 and D1 pins, respectively.

461. BMP180 Pressure Temperature Monitor on Thingspeak with ESP8266

In this project, we'll connect a BMP180 barometric pressure sensor to a NodeMCU ESP8266 and an OLED display. We'll show the values for Temperature, Pressure, and Altitude both in the Serial Monitor and on the 0.96" OLED Display. Then, we'll send these values to Thingspeak Server via IoT Cloud. With the Bosch BMP180 barometric pressure sensor, you can predict the weather, find out how high you are, and measure your vertical speed. It can be used to measure absolute and relative pressure, altitude, and the temperature of the environment. Barometric pressure is another name for atmospheric pressure. It is the force that the weight of the air has on the Earth. Since the air in the atmosphere has weight, gravity makes that column of air push down on the surface. The SI unit for the BMP180 is Pascals, which is how it measures pressure. We can also use hectoPascals (hPa) or millimeters of mercury to measure the pressure (mm of Hg).

In this IoT project, we will connect the BMP180 Barometric Pressure Sensor to the NodeMCU ESP8266-12E Board. We will use a 0.96" I2C OLED Display to show weather parameters like Pressure, Temperature, and Altitude. Then, we'll send these data/parameters to Thingspeak Cloud Server over the internet.

BMP180 Barometric Pressure Sensor: The BMP180 Breakout is an I2C ("Wire") interface barometric pressure sensor. The absolute pressure of the air around barometric pressure sensors is measured. This pressure changes with the weather and with how high you are. Depending on how you interpret the data, you can track changes in the weather, measure altitude, or do any other task that needs an accurate pressure reading.

How BMP180 Barometric Pressure Sensor Works?

The BMP180 is a pressure sensor called a piezoresistive sensor. Piezoresistive sensors are made up of a semiconducting material (usually silicon) that changes resistance when a mechanical force like atmospheric pressure is applied.

The BMP180 can measure both pressure and temperature. This is because the density of gases like air changes with temperature. At higher temperatures, the air is not as dense and heavy, so it puts less pressure on the sensor. At lower temperatures, the air is more dense and heavier, so it presses harder on the sensor. The sensor takes measurements of the temperature in real time to adjust the pressure readings for changes in the density of the air. First, we measure the temperature, and then we measure the pressure.

Components Required:

- NodeMCU ESP8266
- BMP180 Sensor, OLED Display
- Connecting Wires, Breadboard

Circuit Diagram:

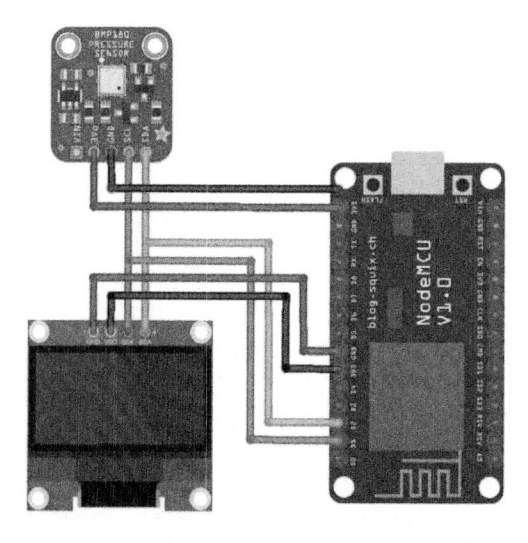

Now, let's add another OLED Display to the Circuit. The SSD1306 I2C OLED Display will be used. Connect its SDA & SCL pins to D2 & D1 of NodeMCU.

462. IoT ECG Monitoring with AD8232 ECG Sensor & ESP8266

In this project, we will learn how to make an IoT-based ECG monitoring system using an AD8232 ECG sensor and a NodeMCU ESP8266. We will use an IoT platform called Ubidots to watch the ECG waveform/Graph made by the AD8232 Sensor online.

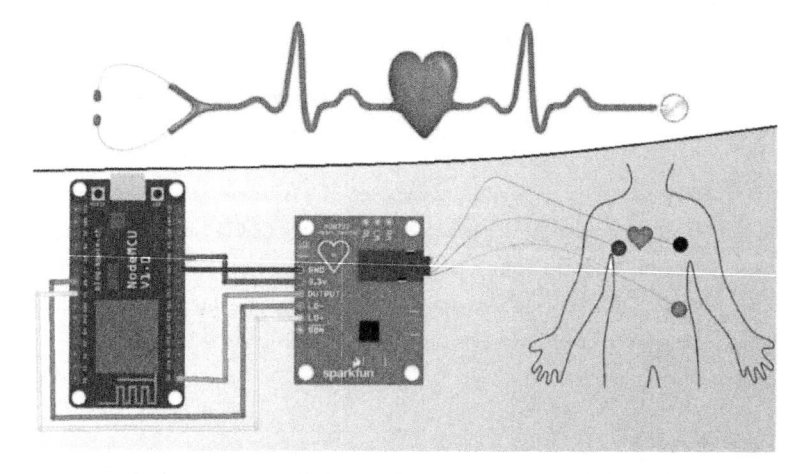

In the last few decades, heart diseases have become a big problem, and many people die because of health problems. So, heart disease is not something to be taken lightly. So, there should be a technology that can keep an eye on the patient's heart rate and how it works. By analyzing or keeping an eye on the ECG signal early on, different types of heart disease can be avoided. This is why I'm showing you this great Internet of Things (IoT) project. In this project, I'll show you how to connect the AD8232 ECG Sensor to the NodeMCU ESP8266 Board and watch the ECG Waveform on the Serial Plotter Screen. In the same way, you can send the ECG waveform over the IoT Cloud platform and monitor the signal online using a PC or a smartphone from anywhere in the world. If you want to watch your heart's activity or behavior, you don't have to stay in the hospital. You can do it online from anywhere. So, it can be said that the Patient Health Monitoring System has made progress. The IoT platform that I am gonna use here is Ubidots. Ubidots is an Internet of Things (IoT) platform that lets innovators and businesses build prototypes of IoT projects and scale them up to make them. You can send data to the cloud from any device that can connect to the Internet by using the Ubidots platform. Then, you can set up actions and alerts based on your real-time data and use visual tools to get the most out of your data.

AD8232 ECG Sensor

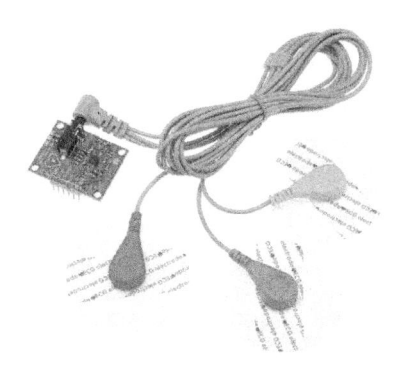

This sensor is a cheap board that is used to measure the heart's electrical activity. This electrical activity can be charted as an ECG or Electrocardiogram and output as an analog reading. ECGs can be extremely noisy, the AD8232 Single Lead Heart Rate Monitor acts as an op-amp to help obtain a clear signal from the PR and QT Intervals easily. The AD8232 is a built-in signal conditioning block that can be used to measure ECG and other biopotentials. It is made to find, boost, and filter small biopotential signals even when there is a lot of noise, like when the electrodes are moved or placed far away.

The AD8232 module breaks out nine connections from the IC that you can solder pins, wires, or other connectors to. This monitor can be used with an Arduino or other development board through its SDN, LO+, LO-, OUTPUT, 3.3V, and GND pins. There are also RA (Right Arm), LA (Left Arm), and RL (Right Leg) pins on this board so you can attach and use your own custom sensors. There is also an LED indicator light that pulses to the beat of your heart.

Note: This product is NOT a medical device, and it is not meant to be used as one or as a part of one. It is also not meant to be used to diagnose or treat any health problems.

Component Required:

- NodeMCU (ESP8266-12E Board)
- ECG Sensor (AD8232 ECG Sensor Kit)
- Data Cable (5V Micro USB Data Cable)
- USB Adapter - 5V
- Breadboard

Circuit Diagram: Interfacing AD8232 ECG Sensor with NodeMCU ESP8266

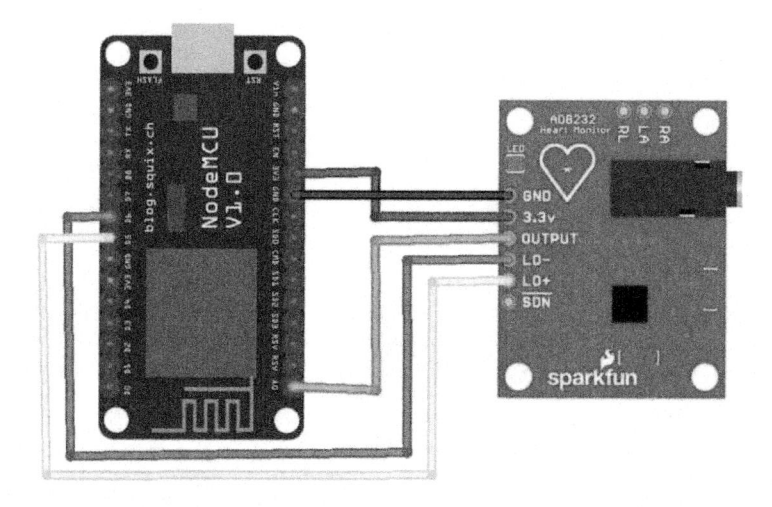

Here is a diagram of how to connect the NodeMCU ESP8266 to the AD8232 ECG sensor. AD8232 Breakout Board has six pins. SDN is not linked together. Connect the OUTPUT to Nodemcu's analog A0. Connect the LO+ and LO- wires to NodeMCU's D5 and D6. Give the AD8232 kit 3.3V VCC and connect its GND to the GND on the kit.

ECG Leads/Electrode Placement

It is recommended to snap the sensor pads on the leads before application to the body. The better the reading, the closer the pads are to the heart. The cables are color-coded to help identify proper placement.

- Red: RA (Right Arm)
- Yellow: LA (Left Arm)
- Green: RL (Right Leg)

We connected an ESP8266 with an AD8232 to a patient chest, or you can just put it in your chest, as shown in the figure picture.

463. ESP8266 and Android Home Automation with WiFi and Voice Control

This project is all about using NodeMCU and Android to automate your home with WiFi and your voice. Here, we'll do the same thing, but instead of using Alexa, we'll make our own app on an Android phone that will let us control our home devices locally (with buttons or by voice). In this project, we will control 4 different LEDs by connecting a 4-channel relay to a NodeMCU ESP8266 12E Wifi Module. Four of these LEDs look like four different home appliances. We'll send a signal to NodeMCU using an Android app with 5+5 ON+OFF buttons to control relay output. There is a unique IP Address for each NodeMCU. We'll give this IP address to the Android app to control who can use it. In the same way, NodeMCU must be connected to local Wifi.

GPIO Pins of NodeMCU are used to connect to a 4-channel relay. With the help of a 220-ohm resistor, the NodeMCU's output pin is linked to 4 different LEDs. The NodeMCU is hooked up to Local Wifi and can get signals from the Internet. The "Home Automation Control" app for Android is already on the Android Device. To control the inputs and outputs of NodeMCU, the IP address of NodeMCU is typed into the IP box of the Android App.

Circuit Diagram:

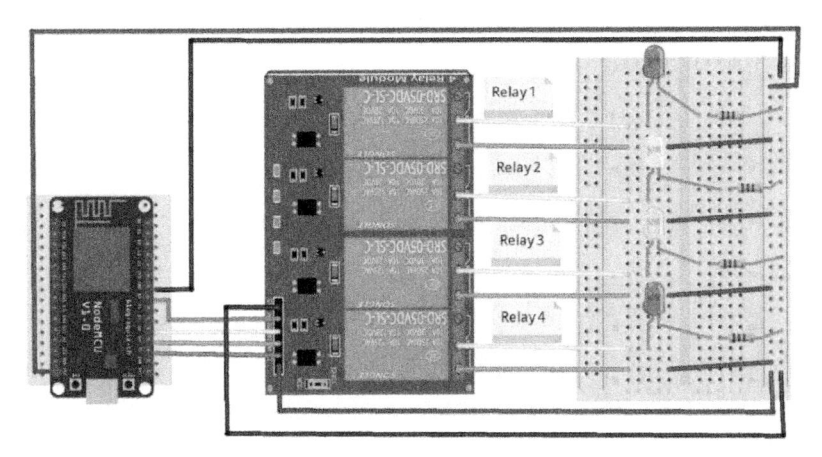

After this is set up, you can use the Android App from anywhere in the world. Your Android phone works as a remote, NodeMCU works as a receiver, and the signal is sent over the Internet We will use a 4 Channel relay module to control 4 LEDs that will act like 4 home devices.

Component Required:

- NodeMCU ESP8266, Relay Module
- LED 5mm LED of Any Color
- Resistor - 220-ohm, Power Supply
- Connecting Wires, Breadboard

Connect Relay Input Pin with NodeMCU as follows

NodeMCU GPIO Pin 1 - Relay Input Pin 1 NodeMCU GPIO Pin 3 - Relay Input Pin 2 NodeMCU GPIO Pin 12 - Relay Input Pin 3 NodeMCU GPIO Pin 14 - Relay Input Pin 4

In the same way, connect the four outputs to four different LEDs using four resistors, as shown above.

The Android app for controlling WiFi and voice:

The Android app will send a string that the code needs to figure out how to read in order to turn on each of the relays listed below

Relay1:

Turn-On: "r1on";

Turn-Off: "r1off";

Relay2:

Turn-On: "r2on";

Turn-Off: "r2off"

Relay3:

Turn-On: "r3on";

Turn-Off: "r3off"

Relay4:

Turn-On: "r4on";

Turn-Off: "r4off"

If "r1on" is sent as a command from the Android app, Relay1 must be turned on. We have also set up "group commands" to turn all devices on ("allon") or off ("all off") at the same time. In the same way, there is also an image for voice input that, when clicked, brings up a window for Google Assistant to accept voice commands.

464. ESP8266 and DS3231 Based Real Time Clock (RTC)

In this project, we will show the time and date on a 16x2 LCD Display by interfacing a Real Time Clock (RTC) Module DS3231 with a NodeMCU ESP8266 Board. We are going to learn how to interface the RTC Module DS3231 with the NodeMCU ESP8266 12E Board and the 16x2 LCD Display via the course of this project. We will utilize a DS3231 Real Time Clock (RTC) module to keep track of the exact time and date, and we will use an ESP8266 as our microcontroller. This information will be shown on a 16x2 LCD display.

DS1307 is an alternative integrated circuit to DS3231. This project is made significantly simpler because to the built-in alarm functionalities and temperature sensor that are included in the DS3231 RTC. The resolution of the temperature sensor is 0.25 and the accuracy is 3 degrees Celsius.

Component Required:

- NodeMCU ESP8266
- LCD Display - 16X2 I2C LCD Display
- RTC Module - DS3231 Real Time Clock
- Connecting Wires Jumper Wires
- Breadboard

DS3231 RTC Module

The DS3231 is a temperature-compensated crystal oscillator (TCXO) and crystal that are combined into a low-cost, incredibly accurate I2C real-time clock (RTC) that bears the model number DS3231. When the device's primary source of power is cut off, it is equipped with a battery input so that it can continue to keep precise time even without that power.

Information regarding seconds, minutes, hours, days, dates, months, and years are all stored in the RTC. For months that have fewer than 31 days, including adjustments for leap year, the date at the end of the month is automatically modified to reflect the new length of the month. The time can be displayed either in a 12-hour or 24-hour format, and there is an active-low AM/PM indicator on the clock. There are two time-of-day alarms that can be programmed, as well as a square-wave output that can be programmed. The state of VCC is monitored by a precise temperature-compensated voltage reference and comparator circuit. This allows for the detection of power failures, the provision of a reset output, and the automatic switching to the backup supply when it is required. In addition, the active-low RST pin serves as a pushbutton input that is monitored for the purpose of creating a P reset.

Key Features:

1. An extremely accurate RTC that is in charge of all aspects of timekeeping functions
2. The real-time clock keeps track of seconds, minutes, hours, the day of the month, the month itself, the day of the week, and the year, and it accounts for leap years up to the year 2100.
3. Accuracy of 2ppm between 0 and 40 degrees Celsius
4. Accuracy of 3.5ppm from -40 degrees Celsius to 85 degrees Celsius
5. Digital Temp Sensor Output: ±3°C Accuracy
6. Register for Aging Trim
7. Pushbutton RST Output and Active-Low RST Output, with Debounce Input
8. Two Alarms for Different Times of the Day
9. Programmable Square-Wave Output Signal
10. Connects to the Majority of Microcontrollers Using a Simple Serial Interface
11. Fast (400kHz) I2C Interface
12. Input for Continuous Timekeeping that Uses a Battery Backup
13. Extended runtimes with low power consumption Battery-Backup Run Time
14. 3.3V Operation
15. Temperature Ranges for Operation: Commercial (-70 degrees Celsius to +70 degrees Celsius) and Industrial (-40 degrees Celsius to +85 degrees Celsius)
16. Recognized by Underwriters Laboratories® (UL) Certification

Circuit Diagram:

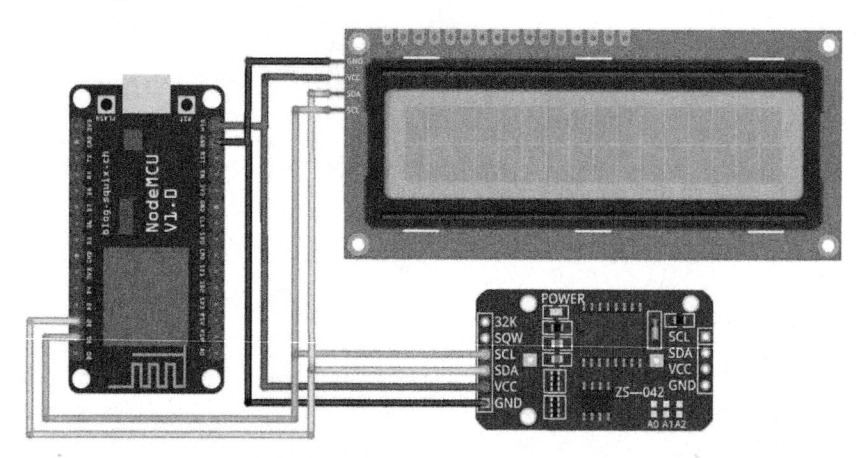

ESP8266 and DS3231 are used in this real time clock circuit.

The following is a diagram of the circuit that will be used to interface the DS3231 Module with the NodeMCU ESP8266. The relationship is not too difficult to understand. Breadboard construction is another option for putting together the circuit. I2C Modules are what both the DS3231 and the 16x2 LCD are. Therefore, the connection just requires two pins. Therefore, connect the pins labeled Serial

Data (SDA) to the NodeMCU D2 port, and link the pins labeled Serial Clock (SCL) to the NodeMCU D1 port. Vin on the NodeMCU should be used to supply 5V to the LCD and RTC Module. The DS3231 Module is also compatible with a source of 3.3V voltage. The RTC Module will begin operating as soon as the code is uploaded successfully. On a 16 by 2 LCD Display, both the time and the date will be shown. There is no need for any more settings, and there is also no requirement for any additional buttons or a switch.

465. MAX30100 Pulse Oximeter with ESP8266 on Blynk IoT App

We are going to learn how to link a MAX30100 Pulse Oximeter with a NodeMCU ESP8266 throughout the course of this project. On the Blynk application, we will do online monitoring of the blood oxygen level and heart rate. As part of this do-it-yourself Internet of Things project, we are going to attempt to build a smart health monitoring device that is capable of measuring both the percentage of oxygen in the blood (SpO2) and the user's heart rate in beats per minute (BPM) (Beat Per Minute). During their training, athletes can utilize this wearable device to keep track of their blood oxygen levels as well as their heart rate. The most impressive aspect of this project is that it allows you to connect this gadget to an Android software called Blynk, which will record and frequently update the data for SPO2 and BPM that is available on the internet. Anyone, from anywhere in the globe, is able to see the data in real time as it is being posted to the server.

Because there is data readily available online, it should be possible to use this initiative to keep an eye on a patient's health while doing so online. We are able to create our own pulse oximeter using a module that is both straightforward and affordable, in contrast to the highly pricey pulse oximeters that are already on the market. Now that we have everything out of the way, let's get into making a MAX30100 Pulse Oximeter with ESP8266.

Explanation of How the MAX30100 Pulse Oximeter and Heart-Rate Sensor Works

The component consists of two LEDs, one of which emits red light and the other of which emits infrared light. Only the infrared light is required to determine the pulse rate. When determining the amount of oxygen present in the blood, both red light and infrared light are utilized as measuring tools. Because there is a greater volume of blood being pumped by the heart, there is a corresponding rise in the percentage of oxygen-rich blood.

When the heart slows down and relaxes, the amount of blood that is oxygenated in the body also reduces. The length of time that elapses between the two changes in the amount of oxygenated blood in the body is used to calculate the pulse rate. It has been discovered that oxygenated blood absorbs more infrared light and transmits more red light, whereas deoxygenated blood absorbs more red light and transmits more infrared light. This is the primary task that the MAX30100 is designed to perform; it analyses the levels of absorption produced by both of the light sources and stores the results in a buffer from which they can be retrieved using I2C.

Component Required:

- NodeMCU ESP8266
- Pulse Oximeter Sensor MAX30100 Module
- OLED Display
- Connecting Wires
- Breadboard

Circuit Diagram:

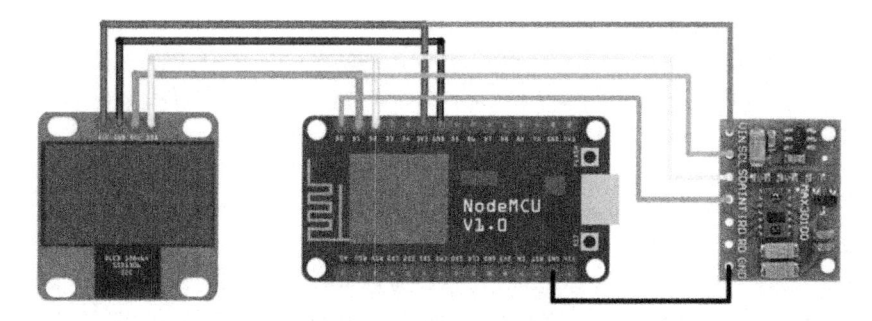

The MAX30100 Pulse Oximeter will be interfaced with NodeMCU ESP8266 and an I2C 0.96″ OLED Display at this time. The connection as well as the circuit schematic are shown below. You can put together the device in the exact same way that is depicted in the diagram below. Both the MAX30100 and the OLED Display share I2C Pins between them. Therefore, connect their SDA pins to D2 of the NodeMCU ESP8266 Board and their SCL pins to D1 of the same board. 3.3V is the minimum voltage for the power supply that OLED Display and NodeMCU require. Therefore, connect the VCC terminal of their device to the 3.3V of the NodeMCU.

MAX30100 Not Working Troubleshooting

If you bought the MAX30100 Module that is displayed below, then it is possible that it will not function properly because of a significant design flaw. The MAX30100 IC requires 1.8V for VDD, and in order to attain this value, this particular module employs the use of two regulators. There's absolutely no problem with that. However, if you take a closer look, you'll notice that the SCL and SDA pins are being pulled up to 1.8V by the 4.7k ohm resistors! Because of this, it will not function properly with microcontrollers that have higher logic levels.

1st Method

The answer is to remove the resistors from the board (which are circled in the image below), and then to attach external 4.7k ohm resistors to the INT Pin, SDA Pin, and SCL Pin respectively.

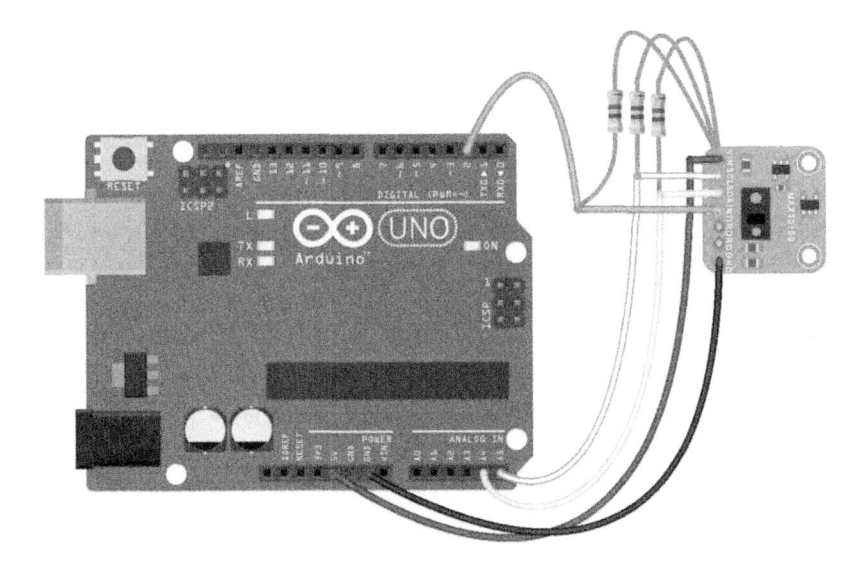

After taking off all of the 4.7K resistors, connect the INT, SDA, and SCL pins to the 4.7K Pull up resistor that is external to the circuit.

2nd Method

If you don't like the first solution to this problem, you can always try the second one, which is exactly the same thing. It is sufficient to make a jumper in the manner depicted by the yellow line and to cut the path where it is marked with a red cross. A wire that is insulated is not required for the jumper. You are able to remove a strand of tinned wire from the stranded wire. A protective mask is applied to the board, and there is no possibility of a short circuit leading to the copper pour.

466.IoT Based Air Pollution/Quality Monitoring with ESP8266

With the help of an ESP8266, a PM2.5 particulate matter sensor, a MQ-135 air quality sensor, and a BME280 barometric pressure sensor, we are going to create an Internet of Things–based air pollution and quality monitoring system. Through the use of the internet, we will keep an eye on the Air Quality on the Thinspeak Server. This is a basic prototype for an Internet of Things (IoT)

Environmental Air Pollution/Quality Monitoring System that will monitor the concentrations of the most significant air pollutants. The PMS5003 PM2.5 Particulate Matter Sensor, the MQ-135 Air Quality Sensor, and the BME280 Barometric Pressure Sensor are the three sensors that are utilized by the system. Through this Internet of Things project, you will be able to monitor the level of pollution using your computer or mobile device from any location. Plantpower's PMS5003 PM2.5 Particulate Matter Sensor is capable of determining the concentration of particles in PM1.0, PM2.5, and PM10. This MQ-135 Sensor for Air Quality measures the concentrations of gases such as CO, CO_2, SO_2, and NO_2 and provides the results in parts per million (PPM) (Part per Million). In a similar vein, the BME280 can measure the Temperature, Pressure, and Humidity of its surroundings. The environmental sensors will collect data on a variety of environmental parameters and then transmit that data to the Thingspeak server, which will then display the data online at regular intervals of 15 seconds. We are able to install this system anywhere, and it has the capability to activate certain devices, such as an exhaust fan, if the level of pollution in the air reaches a certain threshold.

A low-cost laser particle counter, the Plantower PMS5003 is one of a range of sensors made by Plantower that also includes the PMS1003, PMS3003, and PMS7003. PMS5003 is a type of digital and universal particle concentration sensor that can be used to obtain the number of particles that are suspended in the air, also known as the concentration of particles, and output them in the form of a digital interface. This can be accomplished by using the sensor to determine the number of particles that are suspended in the air. This sensor can be installed into variable instruments that measure the concentration of suspended particles in the air or other types of environmental improvement equipment in order to provide accurate data on the concentration of the substance in question at a given point in time.

Component Required:

- NodeMCU - ESP8266
- PM2.5/PM10 Sensor - PMS5003
- Air Quality Sensor - MQ-135
- Barometric Pressure Sensor - BME280 Temperature, Humidity, Pressure Sensor
- Connecting Wires
- Breadboard

Circuit Diagram: IoT Based Air Pollution/Quality Monitoring with ESP8266

The connection diagram or circuit diagram for the IoT-based air quality and pollution monitoring system can be found. There is a total of three sensors, each of which is linked to the wifi chip NodeMCU ESP8266 12E. The PMS5003 Sensor uses UART Communication to function. There are 8 pins in total, and the counting starts from the right. Connect the VCC pin of Pin1 to the Vin pin of NodeMCU, and connect the GND pin of Pin2 to GND.

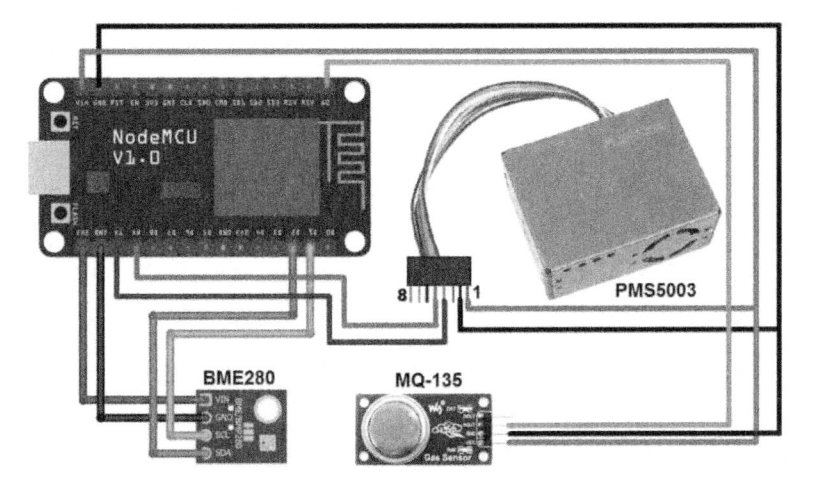

The receiving (Rx) pin is the fourth one, and it is connected to the transmitting (Tx) pin on NodeMCU. In a similar manner, the fifth pin is known as the Tx pin, and it is connected to the NodeMCU Rx pin. MQ-135 is an analog sensor. Therefore, connect its A0 pin to the corresponding A0 pin on the NodeMCU. Join the VCC pin to the Vin 5V pin, and the GND pin to the GND pin. The BME280 Sensor is also included, and it operates using the I2C Communication protocol. Therefore, connect the SCL and SDA pins of it to the I2C pins of NodeMCU, which are D1 and D2.

467. IoT Biometric Fingerprint Attendance System using NodeMCU

In this project, we will learn how to build an Internet of Things (IoT) based biometric fingerprint attendance system by utilizing NodeMCU ESP8266 12E, 0.96" OLED Display, and R305 Fingerprint Sensor. The project is called IoT Biometric Project. The fingerprint data from a number of different users will be gathered by an ESP8266 Wi-Fi Module, which will then transmit the data to a website via the internet. The enrolment of fingerprints is carried out on the server by making use of an R305 or R307 or any other compatible Fingerprint Sensor. Verification of fingerprints is carried out on the client by making use of the transmission of fingerprint templates over the network.

The database and attendance records can be accessed through the website that is written in PHP. After logging into the website, you will have access to all of the attendance records of each user, which will include their personal information as well as the times they logged in and out. Additionally, the information can be downloaded and transferred to an Excel spreadsheet. The traditional methods of authentication, such as RFID tags and authentication cards, both have a number of flaws; the biometric method of authentication is an immediate replacement for these traditional methods. Fingerprints, voices, and electrocardiogram (ECG) signals are all examples of biometrics that are completely unique to each individual and cannot be imitated. Real-time system implementations are made easier as a result of this. Biometric Attendance systems are systems that are frequently used in offices and schools to mark the presence of individuals. This project has a wide range of potential applications in settings such as schools and colleges as well as business organizations and offices where accurate and timely attendance marking is required. Therefore, the incorporation of the fingerprint sensor will result in an increased level of protection for the users of the system. Here, at Arduino Fingerprint Attendance System, you can follow the instructions for the basic level of the same project. GT511C3 is a fingerprint sensor that you can use in the event that you are looking for a better option that is also compact and lightweight.

Component Required:

- NodeMCU - ESP8266
- Fingerprint Sensor - R305/R307 Module
- OLED Display
- Connecting Wires
- Breadboard

Circuit Diagram: IOT Based Biometric Fingerprint Attendance System

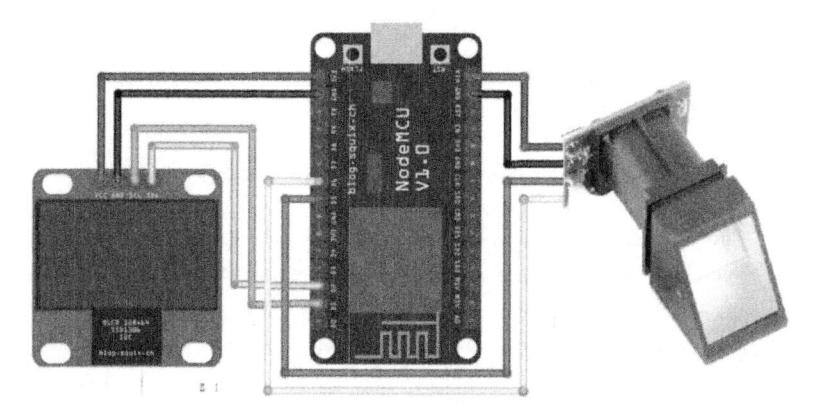

The OLED Display and Fingerprint Sensor are connected to the NodeMCU ESP8266 12E Board in the manner that is depicted in the circuit diagram that was just presented. OLED Display's SDA and SCL I2C pins are connected to NodeMCU's D2 and D1 pins, respectively, so that the display can communicate over I2C. In a similar fashion, the fingerprint sensor is connected to the UART pins labeled D5 and D6. The color of the fingerprint sensor's Tx and Rx wires might be different. In my situation, the color is yellow and blue, with yellow representing treatment and blue representing recovery. Therefore, you need to connect it by locating the appropriate color wires in order for NodeMCU to recognize the module.

The 5V that is required by the R305 fingerprint sensor is drawn from the NodeMCU's Vin pins. In my experience, applying 3.3V to the sensor caused it to become inoperable. In a similar manner, connect the Vcc pin of the OLED to the 3.3V of the NodeMCU.

468.IoT Based RFID Attendance System Using Arduino ESP8266 & Adafruit.io

In this project, you will learn how to make an Internet of Things (IoT)-based RFID attendance system using Arduino Node MCU ESP8266 Arduino and Adafruit.io Platform using the MQTT broker. Consequently, we will be utilizing an RFID MFRC522 along with an Arduino Nano and a Node MCU ESP-12E Board. The RFID cards are scanned by an Arduino and RFID scanner, and the data from those scans is then uploaded to the Adafruit IO cloud platform with the assistance of an ESP8266 Wi-Fi module. This data may be presented in the Adafruit IO dashboard and may be retrieved by the appropriate authorities to view and evaluate the attendance over the internet at any time and from any location. So, let's get started.

Component Required:

- NodeMCU - ESP8266
- Arduino Board - Arduino UNO/Nano or any other Board
- RFID Module - MFRC522 RFID SPI Module
- RFID Cards - 13.56 Mhz RFID Cards
- Connecting Wires
- Breadboard

Circuit Diagram & Connections

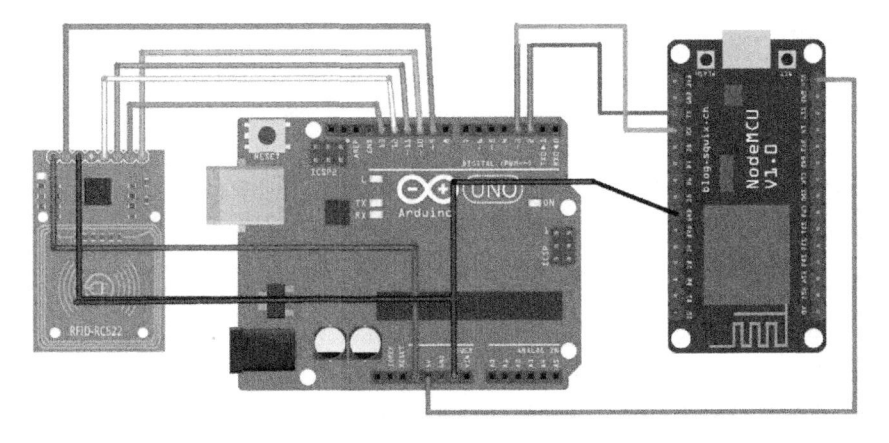

MFRC522 RFID, an Arduino Nano, and a NodeMCU Board are all that are required. Therefore, the connection diagram is precisely the same as what is depicted in the figure. Join the SDA pin of the RFID to the 10th digital pin of the Arduino. Similarly Connect the RST pin to D9 and the SCK pin to D13. Connect MOSI to D11 and MISO to D12. Connect GND to GND. Supply 3.3V power using the 3.3V Pins. There is a disconnect in the IRQ. In a similar fashion, connect the TX and RX pins on the NodeMCU to the D2 and D3 pins on the Arduino. Connecting the Node MCU's VCC and GND Pins will also allow you to supply it with power.

469. Voice Based Home Automation with NodeMCU & Alexa using fauxmoESP

The purpose of this post is to discuss voice-activated home automation using NodeMCU and Alexa with the help of fauxmoESP. The excellent open-source library and example that Xosé Perez developed served as the inspiration for this IOT Project tutorial (Tinkerman). fauxmoESP is a library for ESP8266-based devices that emulates a Belkin WeMo device. This allows you to control those devices using the WeMo protocol, in particular from Alexa-powered devices such as the Amazon Echo or the Dot. ESP is only compatible with devices that use the ESP8266 chip.

ALEXA & ECHO DOT

Alexa is capable of voice interaction, the playback of music, the creation of to-do lists, the setting of alarms, the streaming of podcasts, the playback of audiobooks, and the provision of other real-time information such as the weather and traffic. Alexa can also serve as a hub for home automation, allowing it to control a variety of smart home devices. We are going to use this project called the "Echo-Dot," which enables users to activate the device by saying a wake word, like "Alexa" or "Computer," similar to how it is done in "Star Trek!" In the realm of home automation, Alexa is capable of interacting with a variety of different devices, including SmartThings, Belkin Wemo, Philips Hue, and others.

What is Echo Dot?

The Echo Dot is a voice-activated speaker that enables users to control smart home devices, play music, make calls, get answers to questions, set alarms and timers, and much more through the use of Alexa.

Features

1. Stream songs from services such as Amazon Music, Spotify, Pandora, iHeartRadio, TuneIn, and SiriusXM.
2. Make calls or send messages to family and friends without having to physically pick up the phone, or drop in from the Alexa app on your smartphone to your Echo device.

3. Allows the user to control compatible connected devices such as lights, locks, thermostats, and more
4. If you want a louder sound, you can use the built-in speaker, or you can connect to external speakers via Bluetooth or an audio cable.

Components Required for this Project:

The following are the list of components for designing Voice Based Home Automation with NodeMCU & Alexa Echo Dot.

1. NodeMCU ESP8266-12E
2. Echo Dot (2nd Generation)
3. 2 X Mini Breadboard
4. 4-Channel Relay Module
5. Male-Female Dupont Cables
6. External 5V power supply or Battery

Block Diagram:

The following are depicted in the block diagram as being developed as part of this project:

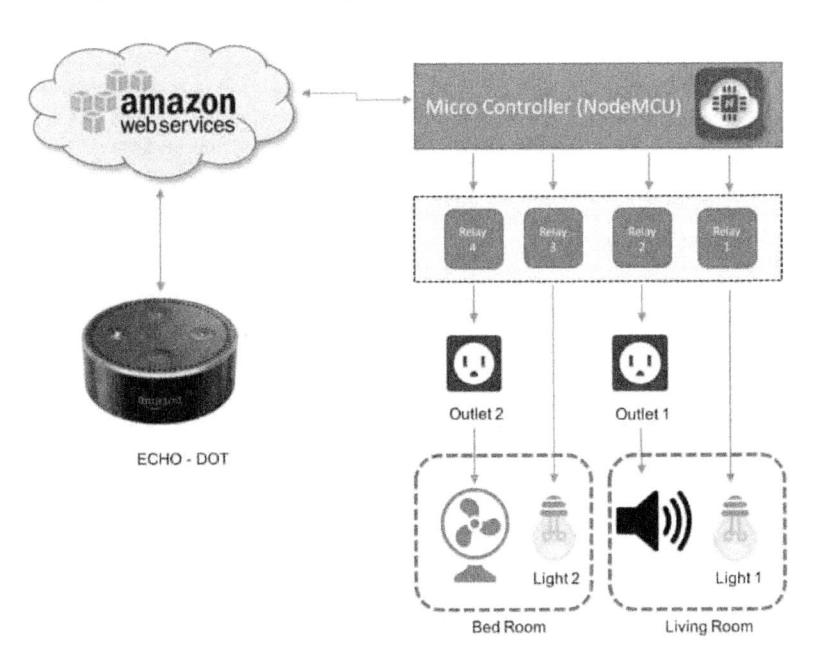

Alexa's brain is actually deployed on AWS, where she can use the same flexibility, large-scale infrastructure, and global network we built for our customers. If I am in my living room and I ask Alexa about the weather, I am setting off a complicated chain of events.

Circuit Diagram: Voice Based Home Automation with NodeMCU & Alexa

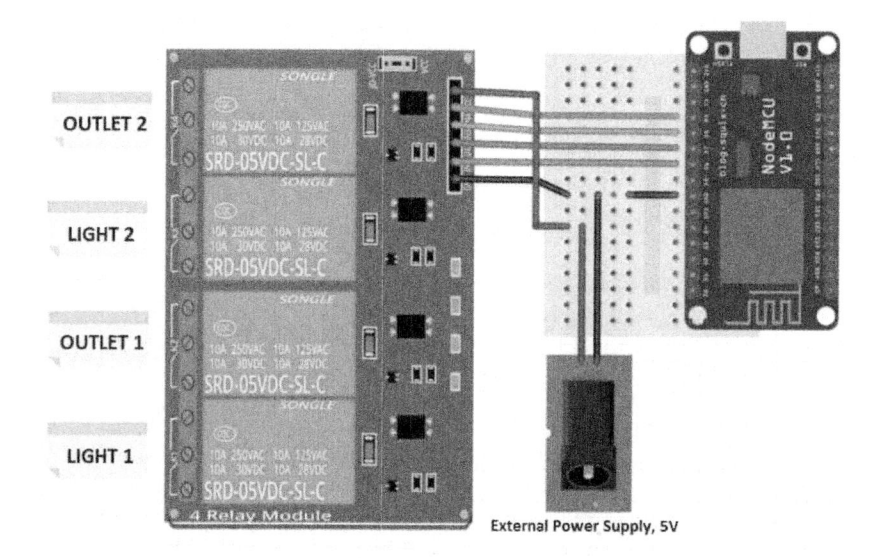

External Power Supply, 5V

In order to implement voice-activated home automation with NodeMCU and Alexa utilizing fauxmoESP, assemble the necessary hardware on your breadboard in the manner depicted in the figure.

Setup for uploading Code & Libraries:

Step 1: The first step is to configure Arduino's preferences for use with NodeMCU.

Step 2: Select Preferences from the menu that appears when you navigate to the Arduino IDE file. Then you should include the link that is shown below in the "Additional Boards Manager URLs"

Installation of the NodeMCU board is the second step.

Navigate to Board Manager and install version 2.3.0 of the NodeMCU Board. It's very important because, without it, Alexa won't be able to find devices.

Step 3: Selection of the Board

Choose the NodeMCU board that corresponds to the right model.

Step 4: Adding Libraries

It is necessary for you to download three separate libraries and then incorporate them into the Arduino IDE Library. These are the libraries:

1. The fauxmoESP Library (fake ESP)

2. ESPAsync TCP Library
3. Library for the ESPAsync Web Server

470. IoT Based Patient Health Monitoring using ESP8266 & Arduino

IoT is quickly changing the healthcare industry, which has a lot of new tech start-ups in the health field. In this project, we used ESP8266 and Arduino to make an IoT-based system for monitoring a patient's health. ThingSpeak is the IoT platform used for this project. ThingSpeak is an open-source application and application programming interface (API) for the Internet of Things (IoT). It can store and retrieve data from things using the HTTP protocol over the Internet or a Local Area Network. This Internet of Things (IoT) device could read the pulse rate and measure the temperature around it. It constantly checks the pulse rate and temperature of the area and sends updates to an IoT platform. The functions of the project are carried out by the device's Arduino Sketch, which does things like read sensor data, turn it into strings, send it to the IoT platform, and show the measured pulse rate and temperature on a character LCD.

Block Diagram:

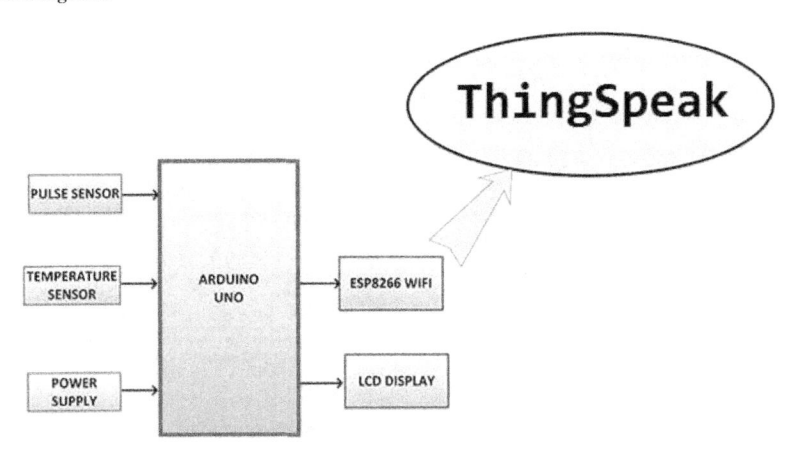

This is a simple block diagram that shows how ESP8266 and Arduino are used in the IoT Based Patient Health Monitoring System. The Pulse Sensor and the LM35 Temperature Sensor each measure the BPM and the temperature of the environment. The code is read by the Arduino, which shows it on the 16*2 LCD Display. The ESP8266 Wi-Fi module connects to Wi-Fi and sends the data to the IoT device server. Thingspeak is used as the IoT server in this case. Lastly, the data can be watched from anywhere in the world by logging into the Thingspeak channel.

Component Required:

- Arduino Nano Board
- ESP8266-01 WiFi Module

- 16x2 LCD Display
- Potentiometer 10K
- Pulse Sensor, LM35 Temperature Sensor
- 2K Resistor, 1K Resistor
- LED 5mm Any Color
- Connecting Wires 10-20
- Breadboard

Circuit Diagram & Connections:

For designing IoT Based Patient Health Monitoring System using ESP8266 & Arduino, assemble the circuit as shown in the figure below.

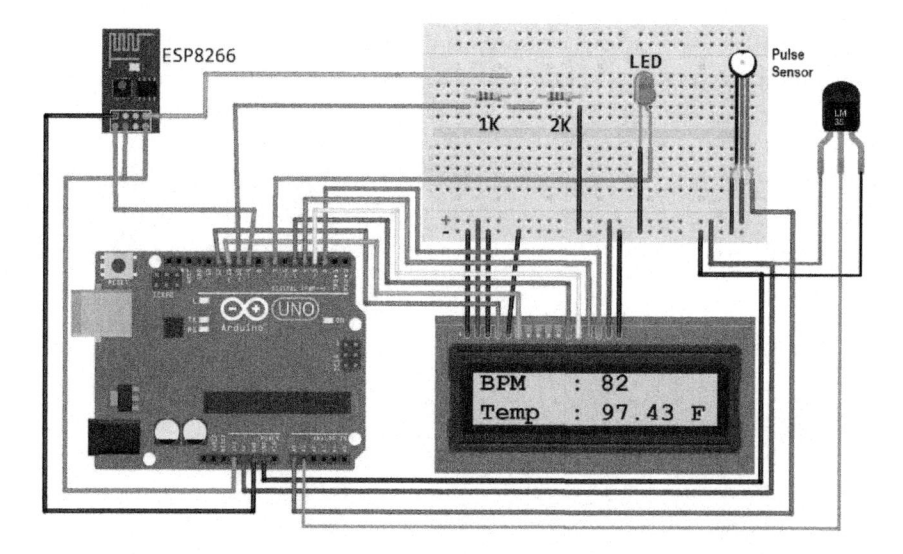

1. Connect the output pin of the Pulse Sensor to Arduino's A0 and the other two pins to VCC and GND.
2. Connect the LM35's output pin to Arduino's A1 and the other two pins to VCC and GND.
3. Using a 220-ohm resistor, connect the LED to Digital Pin 7 of Arduino.
4. Connect the LCD's Pins 1, 3, 5, and 16 to GND.
5. Connect the LCD's Pins 2, 15, and VCC.
6. Connect Pin 4,6,11,12,13,14 of LCD to Digital Pin12,11,5,4,3,2 of Arduino.

The RX pin of the ESP8266 works with 3.3V, so if we connect it directly to the Arduino, it won't be able to talk to it. So, we'll need to make a voltage divider that will turn 5V into 3.3V. To do this, connect the 2.2K resistor to the 1K resistor. This means that the resistors connect the RX pin of the ESP8266 to pin 10 of the Arduino. Connect the ESP8266's TX pin to the Arduino's pin 9.

471. Gas Level Monitoring Using ESP8266 & Gas Sensor Over the Internet

In this project, we'll learn how to monitor gas levels over the Internet using ESP8266 and the MQ135 gas sensor module. The amount of gas will be measured in percentages and sent over the internet using the thingspeak server. With this system, you can keep an eye on the data from anywhere in the world. We just need a gas/smoke/LPG sensor like MQ2/MQ3/MQ5/MQ7/MQ135 that is directly connected to Nodemcu ESP8266-12E Module. ThingSpeak is an open-source application and application programming interface (API) for the Internet of Things (IoT). It can store and retrieve data from things using the HTTP protocol over the Internet or a Local Area Network.

Component Required:

- NodeMCU - ESP8266
- Air Quality Sensor - MQ135 Air Quality Sensor
- Connecting Wires, Breadboard

MQ135 Gas/Smoke Sensor

The MQ-135 gas sensor can detect gases like ammonia, nitrogen, oxygen, alcohols, aromatic compounds, sulfide, and smoke. As a gas sensing material, the MQ-3 gas sensor has a lower conductivity to clean the air. There are polluting gases in the air, but the conductivity of the gas sensor goes up as the amount of polluting gas goes up. With the MQ-135 gas sensor, you can find smoke, benzene, steam, and other dangerous gases. It might be able to find different dangerous gases. It doesn't cost much and works well for applications that monitor air quality. The MQ135 sensor is an instruction for sending a signal. It has two outputs: one is analog and the other is TTL. The Microcontroller's IO ports can be used to access the TTL output, which is a low signal light. The analog output is a concentration, which means that as the voltage goes up, so does the concentration. This sensor lasts for a long time and stays stable.

Features

- High sensitivity to Ammonia, Sulfide, and Benze
- Stable and Long Life, Heater Voltage: 5.0V
- Detection Range: 10 – 300 ppm NH3, 10 – 1000 ppm Benzene, 10 – 300 Alcohol
- Dimensions: 18mm Diameter, 17mm High excluding pins, Pins – 6mm High
- Long life and low cost

Circuit Diagram

Connect the parts as shown in the diagram below. Connect the VCC pin of the MQ135 to the Vin pin of the NodeMCU and the GND pin to the GND pin of the NodeMCU. Connect the MQ135's analog pin A0 to the NodeMCU's analog pin A0.

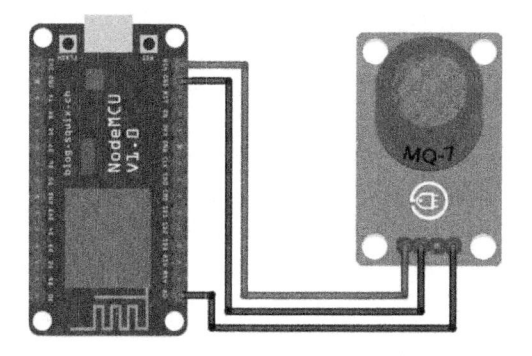

Figure: Circuit Diagram of Gas Level Monitoring Using ESP8266

472. IoT Live Weather Station Monitoring Using NodeMCU ESP8266

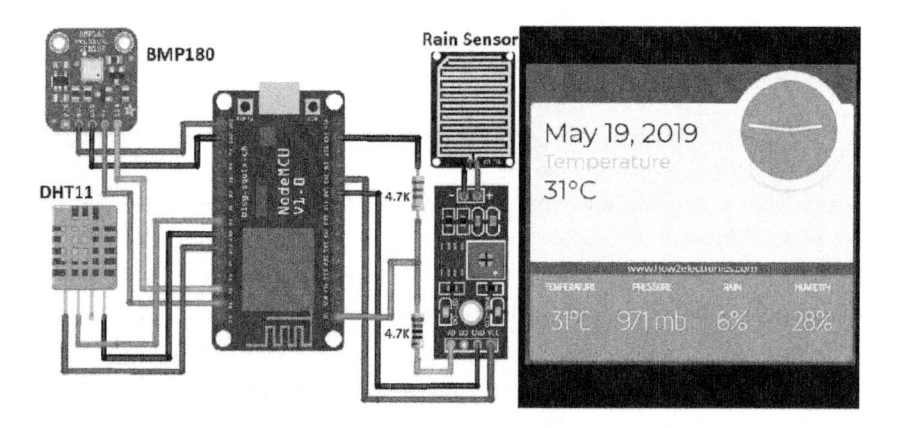

This tutorial is all about monitoring a live weather station with an ESP8266 NodemCU using Internet of Things technology. With the help of the NodeMCU ESP8266-12E Wifi Module, we will connect the DHT11 Humidity & Temperature Sensor, the BMP180 Barometric Pressure Sensor, and the FC37 Rain Sensor. The information that we gather by taking readings of the temperature, humidity, barometric pressure, and rainfall will then be uploaded to a web server.

After the code has been successfully uploaded, the IP address of NodeMCU can be located in the serial monitor. Going to any web browser of your choice and displaying the data in an appealing widget manner requires only one IP address. The project is really intriguing, and it has the potential to be put to use in secluded locations or in a refrigerator where data needs to be monitored.

What is a Weather Station?

A weather station is a device that, via the use of a variety of sensors, gathers information relating to the environment and the weather. There are two distinct categories of weather stations: those that have their own sensors and those that get their data from remote servers. The first category of weather stations is the type that has its own sensors. In this walkthrough, we will focus on the first option, which is to create our very own weather station from scratch.

A weather station's sensors might include a thermometer for obtaining temperature readings, a barometer for determining the pressure in the atmosphere, a hygrometer for determining the level of humidity in the air, a rain sensor for determining the amount of precipitation, an anemometer for determining the rate of wind, and other instruments. There are a few different names for weather stations, including weather centers, personal weather stations, professional weather stations, home weather stations, and weather forecasters.

Component Required:

- NodeMCU- ESP8266
- BMP180 Sensor- BMP180 Barometric Pressure Sensor from Bosch
- DHT11 Sensor- DHT11 Humidity Temperature Sensor
- Rain Sensor - FC-37 Rain Sensor Module
- Resistor 4.7K
- Connecting Wires
- Breadboard

Rain Sensor:

Rain sensors are used in the detection of water beyond what a humidity sensor can detect. The rain sensor is able to detect water because it is able to complete the circuits that are written on its sensor boards. When wet, the sensor board behaves like a variable resistor with a value of 100k ohms, but when dry, it has a value of 2 million ohms. In a nutshell, the amount of current that will be conducted will directly correlate to how moist the board is.

The Connections and the Circuit Diagram

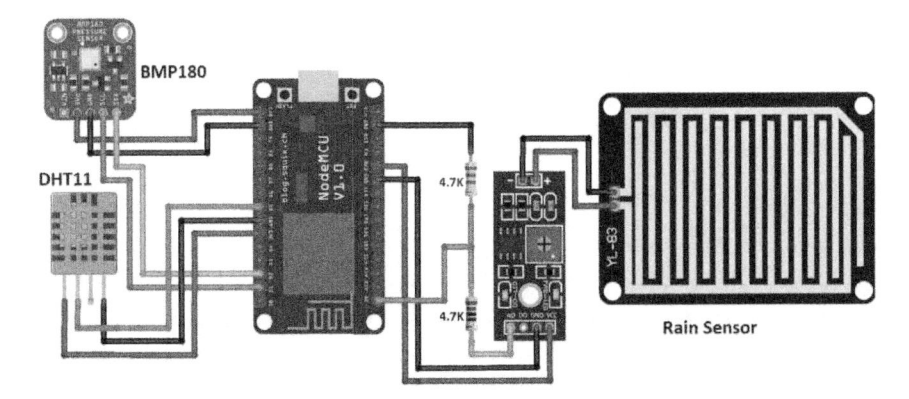

The schematic for the live weather station monitoring that may be made with NodeMCU can be found down below. Put together the circuit in the same way as is depicted in the diagram. First, put together the circuit in the same way that it is depicted in the figure, and then transfer the main ino file from your computer to the NodeMCU board. After the code is uploaded, obtain an IP address for the ESP8266 from the serial monitor. After that, make a copy of this IP address, paste it into any web browser such as Google Chrome, and press the enter key. You will receive an attractive widget containing the weather information as displayed in the following example.

473. IoT Based Analog/Digital OLED Clock using NodeMCU

In this post, we will learn how to construct an analog/digital OLED clock using NodeMCU that is based on the Internet of Things (IoT). There is no requirement for an RTC Module of any kind, such as DS1307 or DS3231. Simply the current time and date will be retrieved from a server located online. The time that was downloaded can be shown either analogically or digitally at your option.

Component Required:

- NodeMCU - ESP8266-12E Board
- OLED Display - 0.96" I2C OLED Display
- Connecting Wires
- Breadboard

0.96" I2C OLED Display:

This is a blue OLED display module with a 0.96-inch screen. SPI and IIC protocols can be used to create an interface between the display module and any microcontroller. This image has a resolution of 128 by 64 pixels. Display board, display, and a male header with 4 pins that has been pre-soldered to the board are all included in this package.

Pin 1: GND
Pin 2: 3.3V to 5V
Pin 3: SCL - Serial Clock
Pin 4: SDA - Serial Data

Organic Light-Emitting Diodes, also known as OLEDs, are a type of self-illuminating technology that are made up of a multi-layered organic film that is very thin and is sandwiched in between an anode and a cathode. OLED technology, on the other hand, does not require a backlight, in contrast to LCD. It is widely acknowledged that organic light-emitting diode (OLED) technology represents the pinnacle of innovation for the next generation of flat-panel displays and boasts a great application potential for nearly every type of display.

Circuit Diagram & Connection:

With the help of the below Circuit diagram, you can make IoT Based Analog/Digital OLED Clock using NodeMCU. Therefore, construct the circuit in the manner depicted in the image. Establish the connection between the SCL Pin of the OLED and the D1 Pin of the NodeMCU.

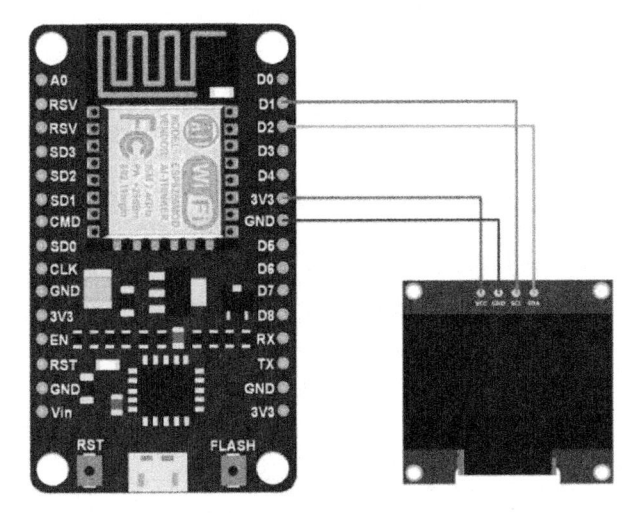

Connect the SDA pin of the OLED to the D2 pin of the NodeMCU, as well. Join the 3.3V and GND pins of the OLED to the corresponding pins on the NodeMCU.

Working of the Project:

The Arduino Integrated Development Environment (IDE) is used to construct a sketch that precisely sets the date and time by accessing an NPT time server. After that, the date and time are shown on an OLED display, and they are also visible in a web browser that is connected to the ESP8266 webserver. On this web page, you have the ability to specify variables such as the Time Zone, 24-hour clock, daylight savings time, and a variable to manage the time server update interval. AJAX is being used on the website to keep the time current without having to completely reload the page. The NTPClient library establishes a connection between the ESP8266 WiFi and a time server; the latter then provides the module with accurate time data. Network Time Protocol is abbreviated as NTP. The final one is the Arduino Time library, which translates Unix timestamps, also known as the Unix epoch, into seconds, minutes, hours, days of the week, months, and years. The number of seconds that have passed since the start of the Unix operating system on January 1, 1970 (midnight UTC/GMT) is referred to as the epoch. The time server transmits the time in the Unix epoch format, which needs to be translated. This library accomplishes all of the work necessary to convert the time. The NTPClient library is setup to obtain time information (Unix epoch) from the server time.nist.gov (GMT time), along with an offset of 1 hour (==> GMT + 1-time zone), which is equal to 3600 seconds, as specified in the RFC.

474. Guide for TCA9548A I2C Multiplexer: ESP32, ESP8266, Arduino

Using the TCA9458A 1-to-8 I2C Multiplexer, which will be covered in this guide, you will learn how to extend the I2C bus ports on the ESP32, ESP8266, and Arduino. If you wish to control many I2C devices with the same I2C address, then you can utilize this piece of hardware to do so. Multiple OLED panels are another example, as are multiple sensors such as the BME280.

Introducing the TCA9548A 1-to-8 I2C Multiplexer

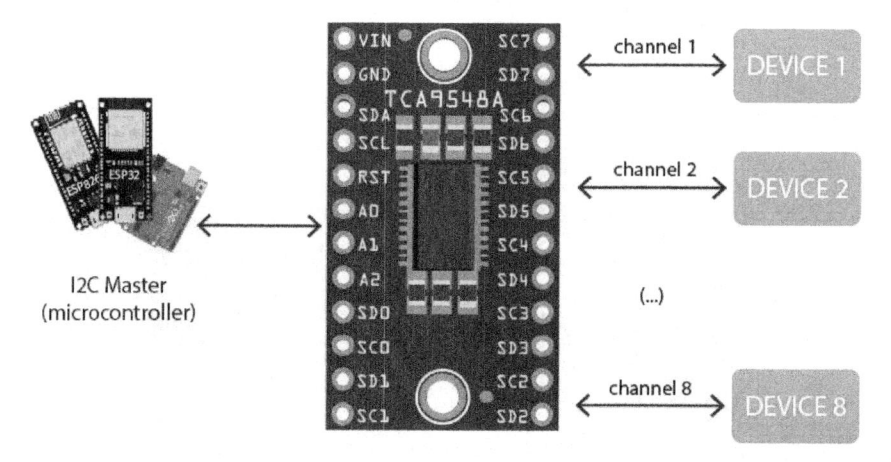

You are able to connect with several I2C devices on the same I2C bus thanks to the I2C communication protocol. The only requirement is that each device have its own distinct I2C address. If, on the other hand, you want to connect numerous I2C devices that have the same address, it will not function. Through the use of a single I2C bus and the TCA9548A I2C multiplexer, you are able to communicate with up to 8 different I2C devices. The I2C communication protocol is utilized in the exchange of information between the multiplexer and the microcontroller. After that, you may use the multiplexer to select which of the I2C buses you want to address. You only need to provide the multiplexer one byte containing the appropriate output port number in order to address a particular port on the device.

TCA9548A Multiplexer Features

Here is a list of its most important features:

- 1 to 8 translating switches capable of bidirectional flow
- a reset input with active-low threshold
- There are three address pins, allowing for up to eight TCA9548A devices to be connected to the same I2C bus.
- I2C bus-based channel selection is provided here.
- 1.65 V to 5.5 V is the range of acceptable voltage for the operating power supply. 5 V tolerant pins

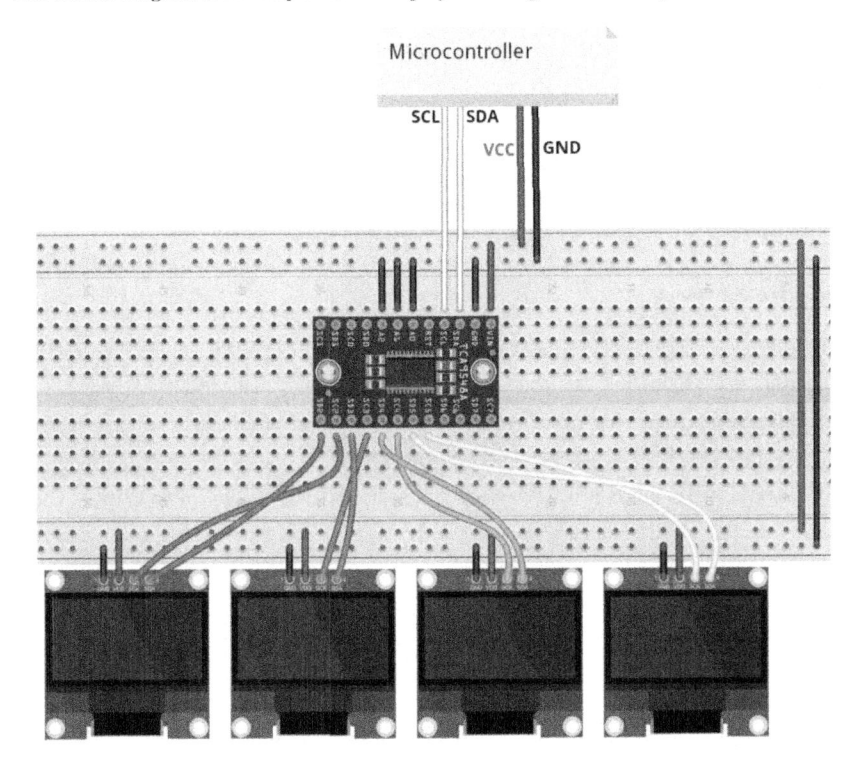

The Circuit Diagram for Multiple OLED Displays Utilizing an I2C Multiplexer

TCA9548A Multiplexer I2C Address

I2C is the communication protocol that is utilized when the TCA9548A Multiplexer has a conversation with a microcontroller. Therefore, an I2C address is required. There is room for customization in the multiplexer's address. Adjusting the settings on the A0, A1, and A2 pins allows you to choose a value in the range of 0x70 to 0x77 for the register. Therefore, it is possible to link as many as eight TCA9548A multiplexers to the same I2C bus. This would enable you to connect sixty-four devices that have the same address despite the fact that the microcontroller only has one I2C bus. If you connect A0, A1, and A2 to GND, for instance, it will cause the multiplexer's address to be set to 0x70. Connect four OLED screens in the manner outlined in the following diagram of a schematic. Buses 2, 3, 4, and 5 are being used by our group. You are free to select a different port number. In addition to that, we will connect A0, A1, and A2 to GND. The address 0x70 is selected for use by the multiplexer as a result of this.

Multiple OLED Displays with I2C Multiplexer Code

The other displays can be controlled in the same straightforward manner as a single display. Before sending the commands to write to the display, you will first need to make sure that you have selected the appropriate I2C bus.

475. ESP32: Guide for MicroSD Card Module using Arduino IDE

You will learn how to read files from and write files to the microSD card by following the instructions in this guide, which demonstrates how to use a microSD card with the ESP32. We will be making use of a microSD card module in order to interface the microSD card with the ESP32 board (SPI communication protocol). When it comes to data logging or storing files that are too large to fit in the filesystem, using a microSD card in conjunction with the ESP32 is particularly helpful (SPIFFS). The Arduino core will be utilized in the process of programming the ESP32.

MicroSD Card Module

There is more than one type of microSD card module that is compatible with the ESP32. The microSD card module that is shown in the following figure is the one that we are utilizing; it communicates via the SPI communication protocol. You are free to utilize any alternative microSD card module that possesses an SPI interface. In addition to being compatible with Arduino and ESP8266 NodeMCU boards, this microSD card module is also compatible with various types of microcontrollers.

MicroSD Card Module Pinout - SPI

The SPI communication protocol is used for communication within the microSD card module. You can connect it to the ESP32 by making use of the SPI pins that are pre-installed.

MicroSD card module	ESP32
3V3	3.3V
CS	GPIO 5
MOSI	GPIO 23
CLK	GPIO 18
MISO	GPIO 19
GND	GND

Parts Required

- ESP32 development board (read: Best ESP32 development boards)
- MicroSD Card Module, MicroSD Card
- Jumper Wires, Breadboard

Getting ready to use the microSD Card

Make sure that your microSD card is formatted using the FAT32 file system before continuing with the guide. You can format your microSD card by following the instructions that are provided below, or you can use a software application such as SD Card Formatter (compatible with Windows and Mac OS).

1. Place the microSD card inside of your computer's card reader. To format the SD card, navigate to "My Computer" and right-click on the card. Choose Format in the manner depicted in the figure below.

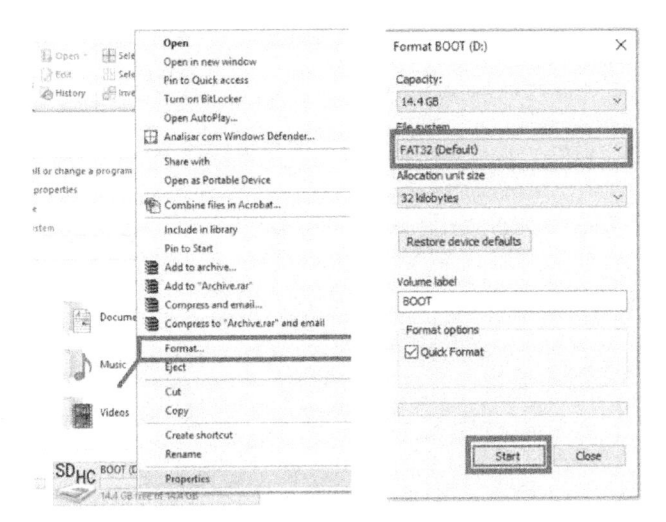

2. A new window shows up. Choose FAT32, then press Start to begin the formatting procedure, and then adhere to the directions that appear on the screen.

ESP32 Handling Files with a MicroSD Card Module

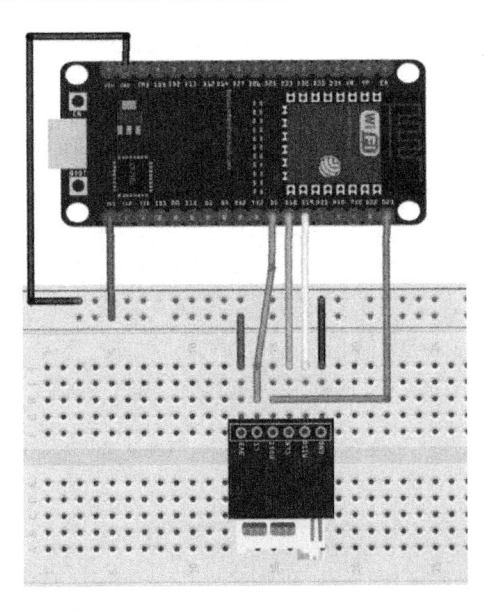

There are two distinct libraries available for use with the ESP32 (both of which are included together as part of the Arduino core for the ESP32): the SDD MMC.h library and the SD library.

If you make use of the SD library, you will be operating with the SPI controller. Using the SDD MMC library means that you are controlling the ESP32 using the SD/SDIO/MMC controlled.

476. ESP32 IoT Shield PCB with Dashboard for Outputs and Sensors

We will walk you through the process of constructing an Internet of Things shield PCB for the ESP32 as well as a web server dashboard to manage it. The shield includes a BME280 sensor, which measures temperature, humidity, and pressure; an LDR, which measures how sensitive an object is to light; a PIR motion sensor; a status LED; a pushbutton; and a terminal socket, which can be used to connect a relay module or any other type of output.

IoT Shield Features

The Internet of Things sensor shield was made specifically to be stacked atop an ESP32. Because of this, in order to use our PCB, you will need an ESP32 board that is identical to ours. The board that we're using is an ESP32 DEVKIT DOIT V1 (the model with 36 GPIOs). You can still complete this project even if you have a different model of ESP32 by building the circuit on a breadboard or modifying the PCB layout and wiring so that it is compatible with your ESP32 board. The shield includes the following components: a BME280 temperature, humidity, and pressure sensor; a light-dependent resistor (LDR); a passive infrared (PIR) motion sensor; a status on-board LED and pushbutton; a three-pin socket that provides access to GND, 5V, and a GPIO where any output can be connected; and an on-board LED that displays the current status of the shield (like a relay module for example).

ESP32 IoT Shield Pin Assignment: The pin assignments for each component of the IoT shield are detailed in the following table:

Component	ESP32 Pin Assignment
BME280	GPIO 21 (SDA), GPIO 22 (SCL)
PIR Motion Sensor	GPIO 27
Light Dependent Resistor (LDR)	GPIO 33
Pushbutton	GPIO 18
LED	GPIO 19
Additional Output	GPIO 32

Components Required:

- DOIT ESP32 DEVKIT V1 Board
- 2x 5mm LED, 2 resistors with 330 Ohm each
- 1x BME280 (4 pins) (4 pins)

- 1x mini-PIR motion sensor
- 1x light dependent resistor
- 2x 10k Ohm resistor, 1x pushbutton
- Breadboard, A set of jumper wires

Circuit Diagram:

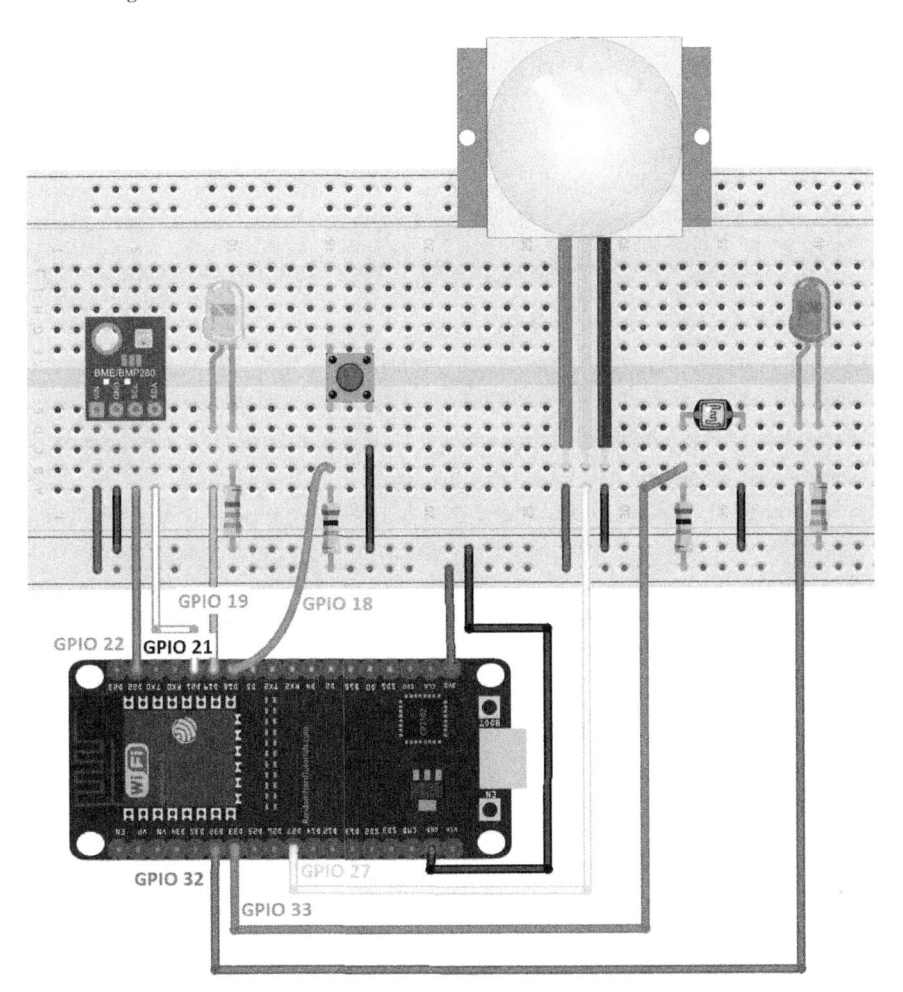

Web Server (IoT Dashboard) Features

We will construct a web server so that the shield can be controlled. You can, however, program the sensor shields however you like using any other web server, or you can integrate it with a home automation platform if you so choose.

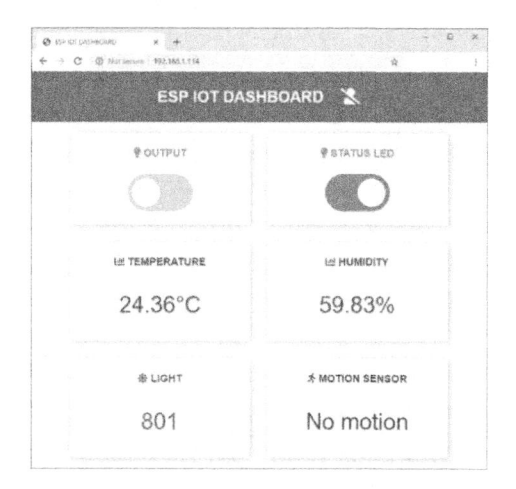

477. ESP32 LoRa Sensor Monitoring with Web Server (Long Range Communication)

In this project, you will construct a sensor monitoring system using a TTGO LoRa32 SX1276 OLED board. This board will send readings of temperature, humidity, and pressure via LoRa radio to an ESP32 LoRa receiver that you will build yourself. The most recent sensor readings can be viewed on a web server thanks to the receiver. Over the course of the past few years, a variety of communication technologies have become available for use in the interaction of Internet of Things devices. The Wi-Fi Technology and the Bluetooth Module are two of the most well-known examples. However, they have a few drawbacks, such as a restricted range, a restricted number of access points, and a high level of power consumption. Therefore, Semtech developed the LoRa technology in order to resolve all of these problems. This device can run for over a year on a single battery even though it only requires one.

As a result, in this Internet of Things project, we will design an ESP32 LoRa Web Server to enable wireless monitoring of sensor readings from distances of several kilometers. The DHT11 Sensor will be used to read the data pertaining to the relative humidity and temperature by the sender. After that, it sends the data out using a LoRa Radio. The data is obtained through the receiver module, which is the component responsible for housing the OLED Display. The receiver has the ability to link the device to a WiFi network so it can communicate with other devices. The IP Address will be displayed on the OLED screen. You can log into any web browser where the sensor reading is monitored by using the IP address.

Project Overview

The image that follows provides a high-level overview of the project that we will construct throughout the entirety of this project. Readings from the BME280 sensor are transmitted over the LoRa radio by the LoRa sender every ten seconds. The readings are obtained by the LoRa receiver, which then displays them on a web server; Accessing the web server will allow you to monitor the readings from the sensors; Depending on where both the LoRa sender and the Lora receiver are situated, there may be a distance of several hundred meters between the two devices. Therefore, if your fields or greenhouses are located some distance from your house, you can use this project to monitor sensor readings from those locations;

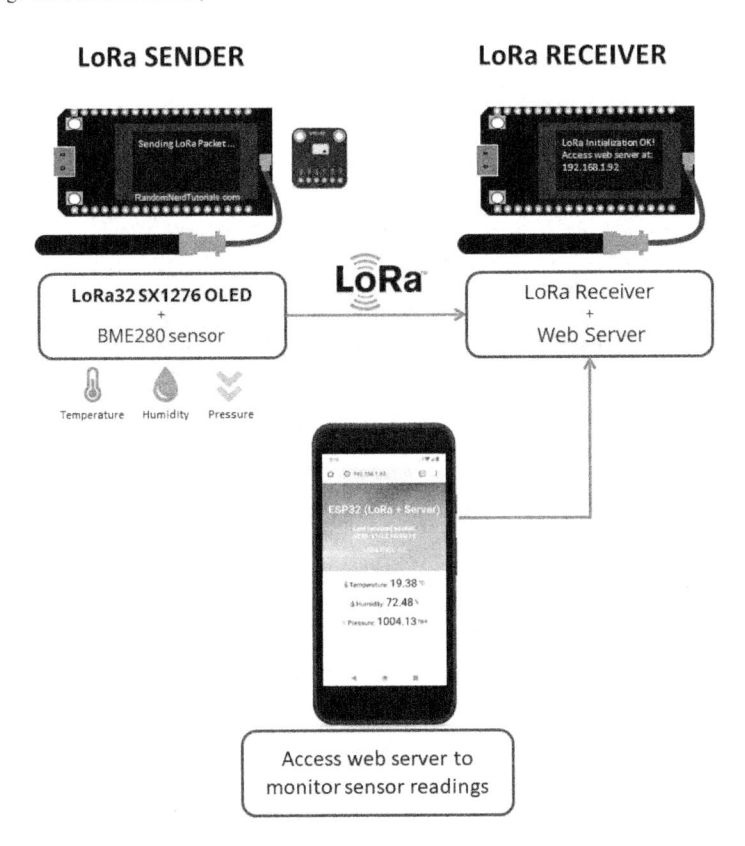

The LoRa receiver operates as an asynchronous web server, and the files for the web page are stored on the ESP32 filesystem (SPIFFS). The LoRa receiver also displays the date and time that the most recent readings were obtained. Utilizing the Network Time Protocol on the ESP32 allows us to obtain the current date and time.

Required Component:

The following elements will be utilized in the completion of this project: TTGO LoRa32 SX1276 OLED board (2x): this is a development board for the ESP32 platform that incorporates an OLED display and a LoRa chip. You can use boards that are similar to what you already have, or you can use an ESP32, LoRa chip, and OLED individually. BME280 sensor for measuring temperature, humidity, and pressure It ought to be possible for you to adapt this project so that you can use any other sensor.

LoRa Sender Circuit

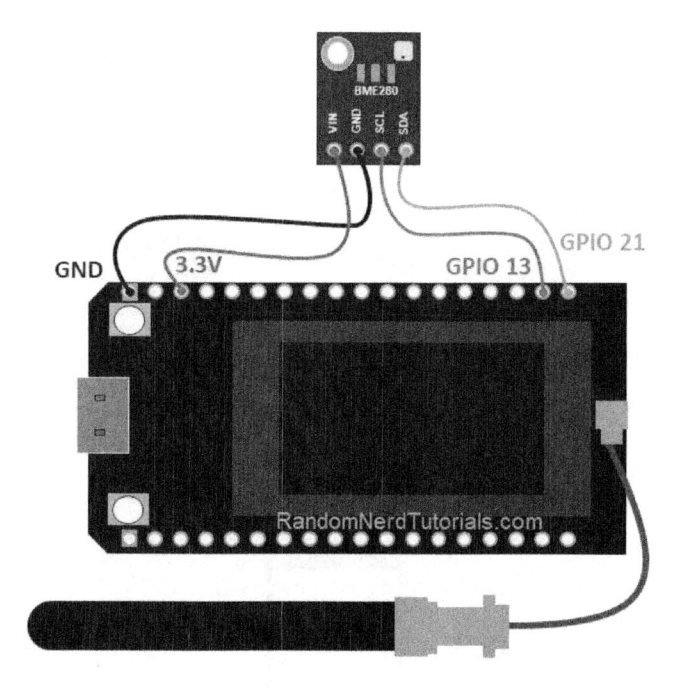

The I2C communication protocol is utilized for conversation between the BME280 and the ESP32 that we are using. Connect the sensor's wires in the manner depicted in the following wiring diagram:

BME280	ESP32
VIN	3.3 V
GND	GND
SCL	GPIO 13
SDA	GPIO 21

478. Visualize Your Sensor Readings from Anywhere in the World (ESP32/ESP8266 + MySQL + PHP)

During the course of this project, you will develop a web page that can be accessed from any location in the world and displays sensor readings in the form of a plot. In a nutshell, you will construct a client for the ESP32 or ESP8266 microcontroller family that issues a query to a PHP script in order to save sensor readings in a MySQL database. For the purpose of illustration, we will be using a BME280 sensor that is linked to an ESP board. You can either use multiple boards or modify the code that was provided so that it sends readings from a different sensor.

To create this project, you'll use these technologies:

Arduino IDE-based programming of an ESP32 or ESP8266 microcontroller Hosting server and domain name. A script written in PHP that inserts data into a MySQL database and displays it on a website. Readings will be stored in a MySQL database. A chart-generating PHP script that pulls data from a database.

1. Providing Hosted Services for Your PHP Application and MySQL Database

This project's objective is to provide you with your very own domain name and hosting account so that you can save sensor readings obtained from an ESP32 or ESP8266 microcontroller. If you access your own server domain, you will be able to visualize the readings no matter where you are in the world. An overview of the project is as follows, at a high level:

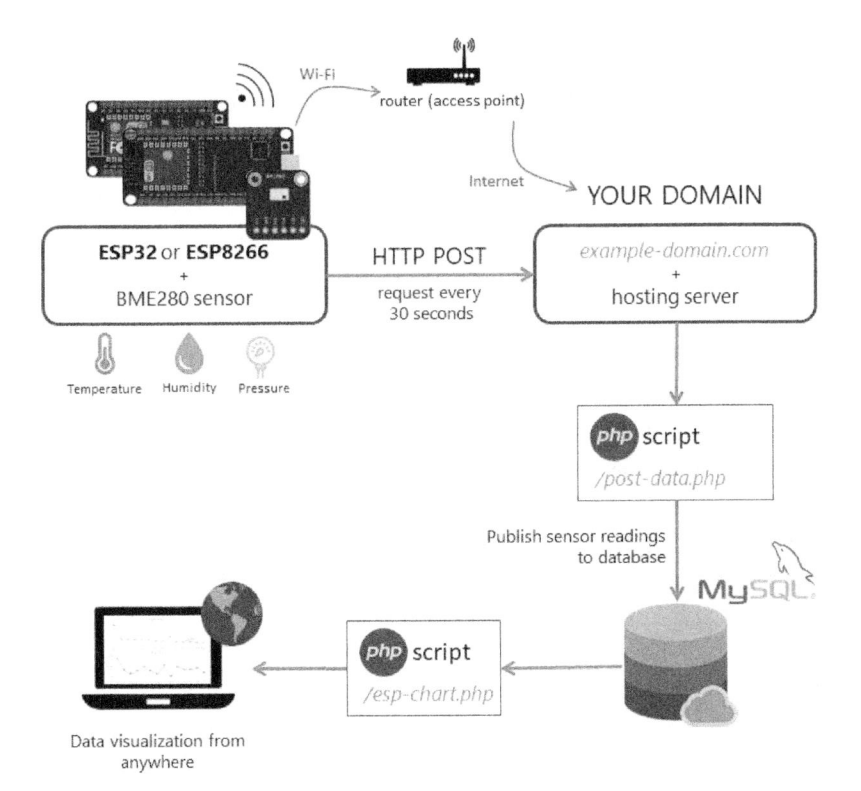

It is strongly suggested that you use one of the following hosting services, as they are able to fulfill all of the requirements for the project: When you sign up for Bluehost's 3-year plan, you'll receive a free domain name. Bluehost is known for being user-friendly and offering cPanel. It is highly recommended that you go with the option that allows unlimited websites; A Linux server hosted by Digital Ocean that can be administered via a command line. This choice is one that I would only recommend to more experienced users. Those are the two hosting services that I use and highly recommend to others, but you are free to use whichever hosting service you prefer. This project will function properly with any web hosting service so long as it provides PHP and MySQL.

2. Preparing Your MySQL Database

After registering for a hosting account and establishing a domain name, you will be able to access your cPanel or an interface that is very similar to it. After that, you will need to create your database, username, and password, as well as a SQL table, by following the subsequent steps.

479. Power ESP32/ESP8266 with Solar Panels (includes battery level monitoring)

This project demonstrates in detail how to provide power to an ESP32 development board using photovoltaic cells, a lithium-ion battery with a capacity of 18650 milliampere-hours (mAh), and a TP4056 battery charger module. The circuit that we are going to build is compatible not only with the ESP8266 but also with any other microcontroller that operates on 3.3V. When you are using solar panels to power your ESP32, it may be beneficial to make use of its deep sleep capabilities in order to save power. Our project will teach you everything you need to know about using the ESP32 to achieve a deep sleep state.

Parts Required

- ESP32 or ESP8266; both are available (read ESP32 vs ESP8266)
- 2x Mini Solar Panel (5/6V 1.2W)
- 18650 Holder for Lithium-ion and Lithium-ion batteries
- Charger for a battery (optional)
- TP4056 Charger for Lithium-Ion Batteries, Module
- Low-dropout regulators, also known as LDO regulators, are used to regulate voltage (MCP1700-3302E)
- 100uF electrolytic capacitor, 100nF ceramic capacitor
- Voltage divider for the battery monitor, which is optional:
- 27k Ohm resistor 100k Ohm resistor

ESP32 Solar Powered - Circuit Overview

The following diagram shows how the circuit to power the ESP32 with solar panels works

- When exposed to direct sunlight, the solar panels produce an output of between 5V and 6V.
- Through the TP4056 battery caharger module, the solar panels are able to supply power to the lithium battery. This module is in charge of charging the battery while also preventing it from being overcharged.
- When it is fully charged, the lithium battery has a voltage output of 4.2V.

- In order to get 3.3V from the output of the battery, you will need to use a low dropout voltage regulator circuit, such as MCP1700-3302E.
- The ESP32 will receive its power from the output of the voltage regulator, which is connected to the 3.3V pin.

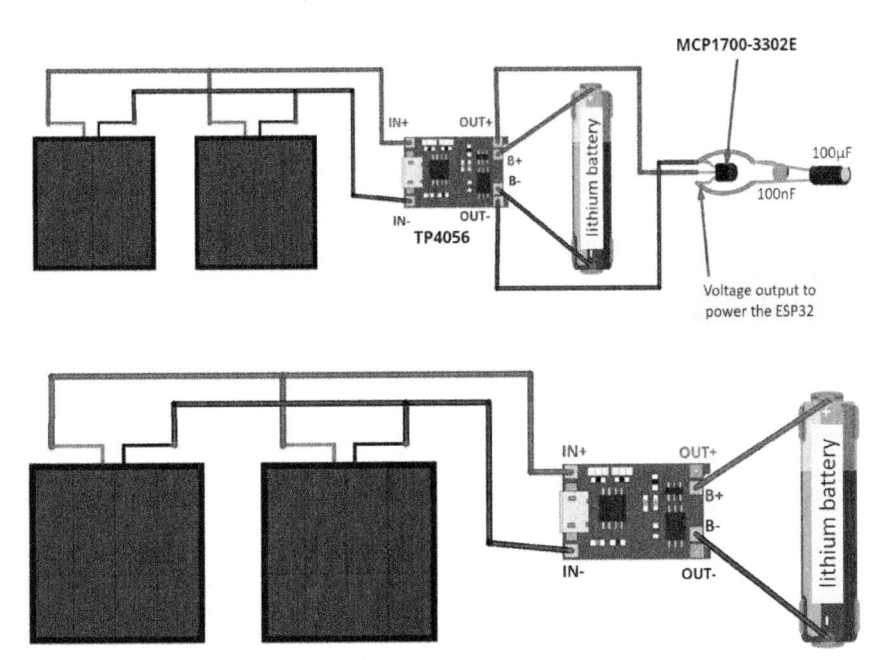

Follow the instructions on the wiring diagram that is provided below to connect the solar panels to the TP4056 lithium battery charger module. Connect the terminals that have a positive charge to the pad that is labeled IN+, and connect the terminals that have a negative charge to the pad that is labeled IN-. The positive terminal of the battery holder should then be connected to the B+ pad, and the negative terminal of the battery holder should be connected to the B- pad. The outputs of the battery are denoted by the symbols OUT+ and OUT-. When completely charged, the voltage produced by these lithium batteries can reach 4.2 volts (although they have 3.7V marked in the label). We need a voltage regulator circuit so that we can get 3.3V from the output of the battery so that we can power the ESP32 through its 3.3V pin.

Voltage Regulator

It is not a good idea to use a typical linear voltage regulator to drop the voltage from 4.2V to 3.3V because, as the battery discharges to a lower voltage, such as 3.7V, your voltage regulator would stop working because it has a high cutoff voltage. Alternatively stated, it is not a good idea to use a linear voltage regulator to drop the voltage. It is necessary to make use of a low-dropout regulator, also known as an LDO for short, which is able to regulate the output voltage in order to drop the voltage in a battery pack effectively.

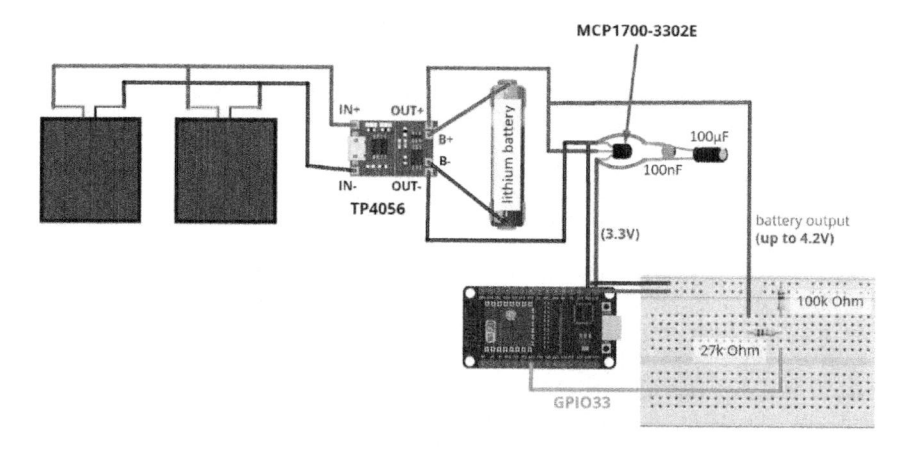

You can continue to use the same circuit even if you have switched to using an ESP8266 instead. Connect the output of the MCP1700-3302E to the 3.3V pin on the ESP8266, and connect GND to GND on both devices

480. Alexa (Echo) with ESP32 and ESP8266 - Voice Controlled Relay

You are going to learn how to give voice commands to Alexa in order to control an ESP8266 or an ESP32 device with the help of this project (Amazon Echo Dot). We'll demonstrate how to control relays by manipulating two 12V lamps that are connected to a relay module. In addition to this, we will install two 433 MHz RF wall panel switches so that the lamps can be physically controlled.

Note: this tutorial is compatible with all Echo Dot generations and with the latest fauxmoESP library (3.1.0). It is compatible with both the ESP32 and the ESP8266.

Project Overview

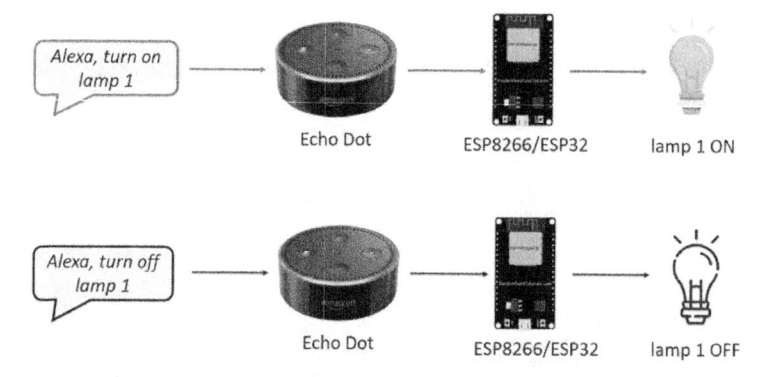

Figure: Alexa can be used to control the Lamps.

This project is compatible with the ESP8266 as well as the ESP32. Instructions are provided by us for both of the development boards. Read the rest of this section first so that you have an idea of what you will have accomplished by the time you are finished with this project. When you are finished with this project, you will be able to use voice commands to control two lamps (lamp 1 and lamp 2) that are controlled by Alexa. The diagram that follows provides a high-level overview of how the project operates to control lamp 1; it operates in a comparable manner for lamp 2.

The following are the commands that Alexa will respond to:

- "Alexa, turn on lamp 1"
- "Alexa, switch off the first lamp."
- "Alexa, turn on lamp 2"
- "Alexa, turn on lamp 2"
- When you tell Alexa to "turn on lamps," she turns on both of them.
- When you tell Alexa to turn off the lamps, both of them turn off.

A relay will be triggered by the ESP8266 or ESP32 whenever you say something like, "Alexa, turn on lamp 1." This will cause lamp 1 to turn on. When you tell Alexa to do something like "turn off lamp 1," either the ESP8266 or the ESP32 will send a signal to the relay, which will cause the lamp to turn off. Lamp 2 operates in a manner analogous to this.

Wall Switches Operating at 433 MHz Used to Regulate Lamps

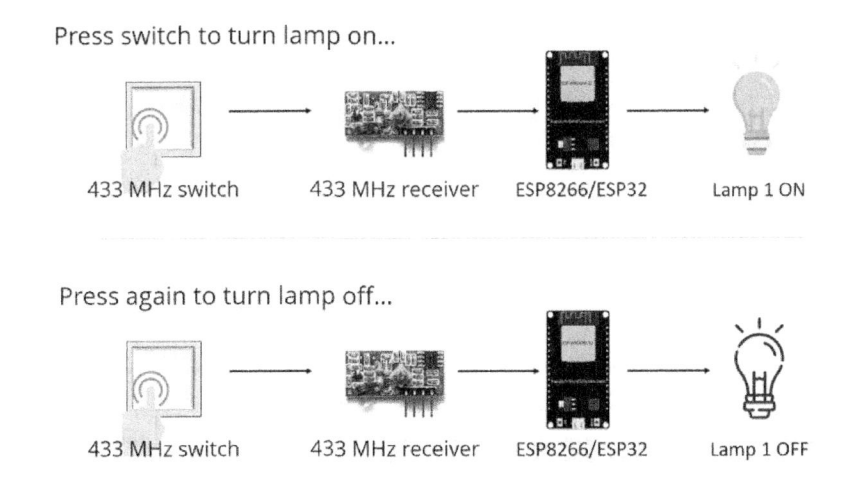

Press switch to turn lamp on...

433 MHz switch → 433 MHz receiver → ESP8266/ESP32 → Lamp 1 ON

Press again to turn lamp off...

433 MHz switch → 433 MHz receiver → ESP8266/ESP32 → Lamp 1 OFF

As part of this project, we will also install two wall switches that operate at 433 MHz in order to physically control the lamps. Each lamp will have its own separate switch. The state of the lamp is altered by the switch so that it becomes the opposite of what it was previously. For instance, if the lamp is turned off, you can turn it on by pressing the wall switch. Simply pressing the switch once more will cause it to be turned off. Take a look at the diagram that is provided below; it shows how the process works.

Parts Required

- ESP Board (you have the option of using either an ESP32 or an ESP8266):
- ESP8266: for more information, see the article titled Best ESP8266 Wi-Fi Development Boards
- ESP32, for which we make use of the ESP32 DOIT DEVKIT V1 Board, which has a total of 36 GPIOs (read ESP32 development boards comparison)
- Alexa, available on the Echo, Echo Show, and Echo Dot (read the next section for more details)
- RF Wall Panel Switch 433 MHz 433 MHz Transmitter and Receiver 12V 2A Power Adaptor
- buck converter with a step-down transformer
- Relay module 12V lamp 12V lamp holder
- Male DC barrel jack 2.1mm Stripboard or breadboard
- A set of jumper wires

You should get different values. After that, you will implement these signals into the sketch for your ESP8266 or ESP32. When the switch is pressed, it will send a signal at a frequency of 433 MHz. The receiver that is linked to the ESP is able to pick up on the presence of this signal. Because of this, the ESP is aware that the switch was pressed, and it adjusts the state of the lamp accordingly.

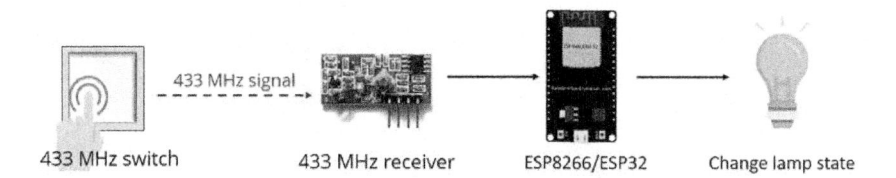

The FauxmoESP

You will need to have the FauxmoESP library installed in order to control your ESP8266 or ESP32 with Amazon Echo. By using this library, which simulates a Belkin Wemo device, you will be able to control your ESP32 or ESP8266 using the protocol developed by Belkin. After the code has been uploaded, the Echo or Echo Dot will be able to instantly recognize the device without the need for any additional expertise or services from a third party.

Circuit Diagram:

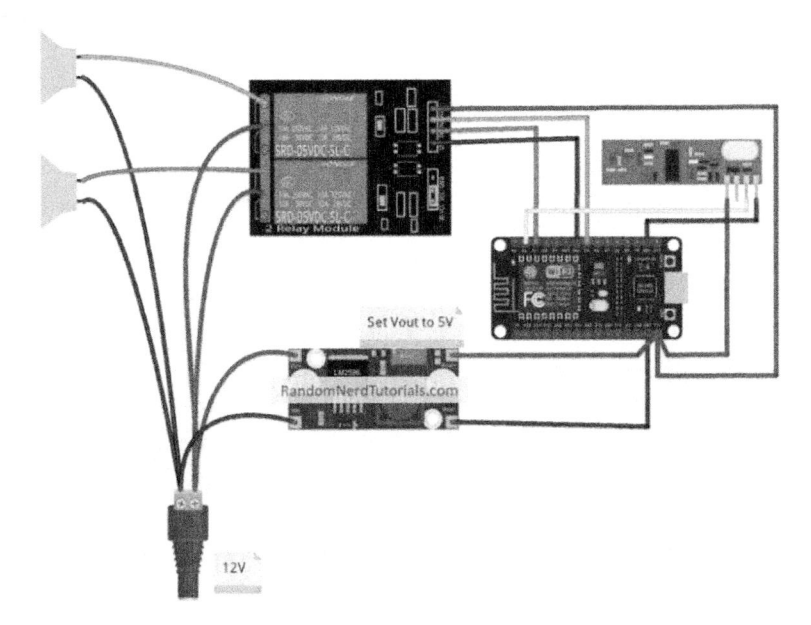

The information in the following table can serve as a reference for you if you're having trouble understanding the circuit diagram

ESP8266	Connect to
GPIO 5	433 MHz receiver data pin
GPIO 4	Relay IN1 pin
GPIO 14	Relay IN2 pin

481. How the HC-SR04 Ultrasonic Range Sensor Can Communicate with the ESP32

If you want to build a robot that can avoid obstacles but are just starting out as an electronics engineer, the first thing you need to do is educate yourself on how an obstacle avoidance system operates. Because of this, the project that we are working on is of the utmost significance. In this project, we are going to interface an HC-SR04 Ultrasonic Distance Sensor module with ESP32, and as a result, we are going to become intimately familiar with every aspect of the module. Since the module is an essential component of any obstacle avoidance or detection system, this makes this project particularly significant. As a result, during the course of this project, we will become familiar with each and every facet of the HC-SR04 Ultrasonic Distance Sensor module as well as the process of interfacing it with ESP32. The ultrasonic HC-SR04 sensor has a detection range of 13 feet and an angle of 15 degrees, making it an excellent tool for locating obstructions. Aside from that, it has a very low operating current, making it exceptionally well-suited for use in applications that are powered by batteries. In our earlier efforts, we were able to construct a wide variety of fascinating projects by combining an Arduino and a PIC microcontroller with this HC-SR04 board.

ESP32 with HC-SR04 Ultrasonic Sensor Circuit Diagram

Now that we have a comprehensive understanding of how the HC-SR04 ultrasonic module operates, we have connected all of the required wires to the ESP32 module. Next, we will write a straightforward code in order to test the module and ensure that it is functioning properly. The illustration below shows the full test schematic for connecting the ESP32 to the HC-SR04 sensor.

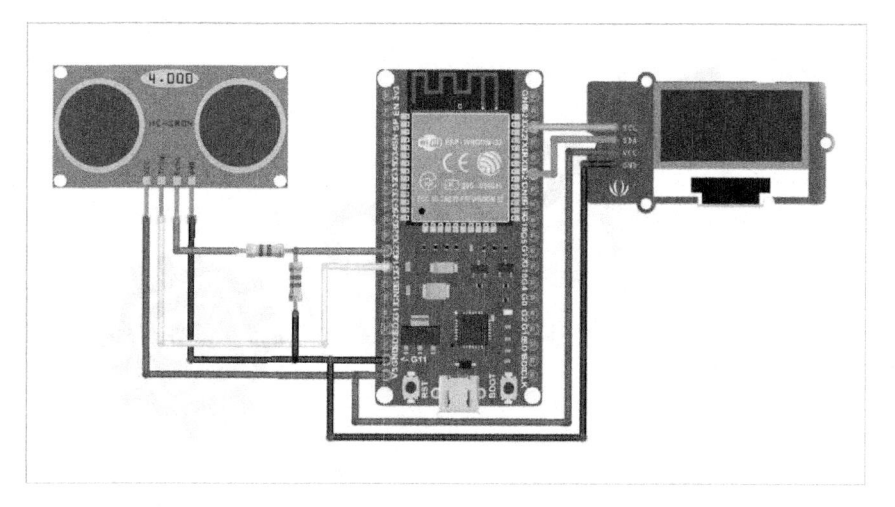

The HC-SR04 module and an OLED module have been connected to the ESP32, as can be seen in the image that is located above this one. Because the ultrasonic module requires 5V to function, the procedure for connecting it is very straightforward. Because of this, we have implemented a voltage divider in the echo pin in order to reduce the 5V output to 3.3V. For the HC-SR04 module, any digital pin of the ESP32 can be used as an I/O; however, for the OLED model, you need to use the I2C pin, which is the IO21 and IO22 of the ESP32 module. You can use any digital pin of the ESP32 for the HC-SR04 module. The procedure for making the connection is now complete; the image of the hardware that can be found below will provide you with a clearer picture of how the circuit is connected.

482. How does a Servo Motor Work and How to Interface it with ESP32?

One variety of motor known as a servo motor is one that can precisely control the motion of an appliance or machine. A DC motor, a control circuit, and a feedback mechanism are the typical components that make up this mechanism. The control circuit receives information about the position of the motor shaft from the feedback mechanism. Based on this information, the control circuit modifies the amount of power that is supplied to the motor. Because of this feedback loop, the motor shaft will always move to the precise location that has been determined by the control signal. Servo motors find use in a wide variety of applications, including robotics, CNC machines, and 3D printers, where precise control of positioning is required. Examples of these applications include: Control surfaces in aircraft, such as elevators and ailerons, are another application for these servos.

Consequently, we are going to interface an SG90 servo motor with an ESP32 in the course of this lesson, and while we are doing so, we will explain how the internal circuitry of a servo motor functions. At the very end, we will construct a hardware circuit and write a little bit of code in order to provide you with a deeper comprehension of the servo motor and how it operates.

How to Control the Angle of the Servo Motor?

In a closed-loop control system, it is the responsibility of a servo motor to control the angle of rotation. Using the feedback from the system, the servo motor then adjusts its speed and angle to achieve the result that is actively desired.

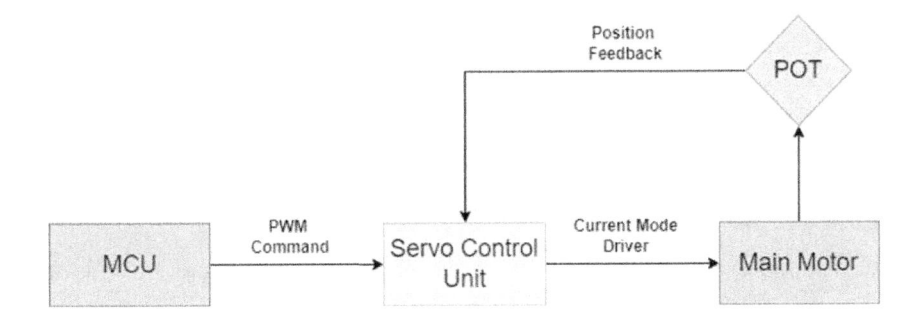

A closed-loop control system is utilized by the SG90 servo motor that we are utilizing, which allows us to control the position of the motor's shaft. It is sufficient to supply a 50-hertz pulse width modulation (PWM) signal with a variable duty cycle in order to alter the position of the motor arm.

You need to send a series of pulses to the servo motor in order to control the shaft angle of the servo motor; a conventional servo motor anticipates receiving a pulse every 20 milliseconds with a duty cycle of 50Hz. The control board that is housed inside of the servo motor is designed in such a way that the length of the pulse can be used to determine the angle of the servo shaft. If the pulse remains high for 1 millisecond, the servo angle will remain at 0 degrees throughout the duration of the event. The low pulse of 1.5 milliseconds will move the shat to an angle of 90 degrees, and a high pulse of 2 milliseconds will move the servo arm to an angle of 180 degrees. A pulse lasting between 1 and 2 milliseconds will rotate the servo shaft through 180 degrees when applied to it.

SG90 Servo Motor Parts

The S90G servo motor is a very cost-effective and low-power device that has many potential applications and can be used in a variety of contexts. The S90G servo motor's internal components are depicted in the following image.

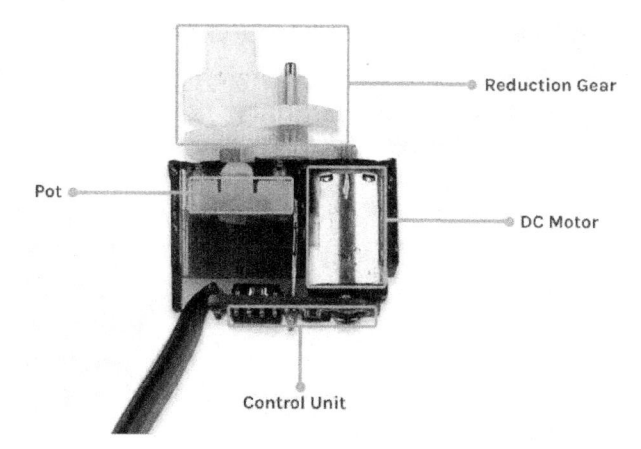

As you can see in the image that is displayed above, we disassembled a servo motor in order to show you all of the components that are contained within it. Additionally, the reduction gear, the main motor, the control board, and the potentiometer are all visible on the inside of the motor. The potentiometer is responsible for providing position feedback to the servo control unit, which then evaluates how closely the motor's current position matches the desired position.

SG90 Servo Motor Pinout

The vast majority of the servo motors you can buy off the shelf will come with three different connections. The following is a description of the pinout for the SG90 servo:

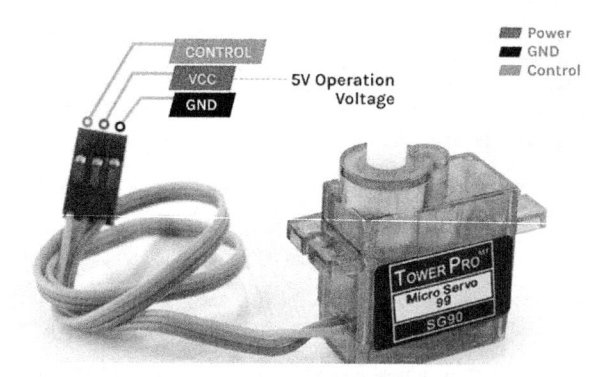

- GND - refers to a common ground that is shared by the microcontroller and the motor.
- VCC - is the voltage that the servo gets power from.
- Control - is the information that comes from a microcontroller or any other control system.

Circuit Diagram Interfacing Servo Motor with ESP32

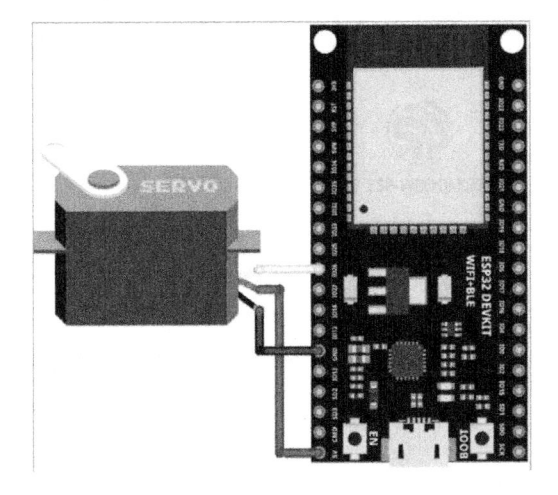

Servo Motor Connection Errors? - Here's What You Should Do

When you run your servo directly from the power supply of the ESP32, there is a possibility that the servo will behave erratically at times. Because when the motor is first started, it draws high current spikes, which can cause the microcontroller to either reset or become permanently damaged. Between the GND supply and the 5V supply, you could install a sizable electrolytic capacitor (between 470 and 1000 uF) to solve this problem.

The capacitor serves as a reservoir, and when the motor begins to draw enormous amounts of current, the capacitor can provide the current that is needed to keep the ESP from acting erratically. The capacitor has two legs: the longer one is connected to 5V, and the other one is connected to ground. The shorter of the two legs is connected to ground.

483. Designing a Smartwatch using ESP32 - Magnetometer and Gyroscope

We have looked at how to interface the BH1750 ambient light sensor and the MAX30102 Heart rate sensor for our smartwatch project in the previous part, which was titled Designing Smartwatch Using ESP32. This part was split into two parts. In this section, we will discuss the different types of sensors that can be interfaced with our smartwatch and how to do so. The HMC5883L/QMC5883L Magnetometer sensor and the MPU6050 accelerometer gyroscope sensor are going to be interfaced in this section. We are going to examine the process of interfacing each of these modules independently.

Prerequisites - Installing Necessary Libraries

The GitHub repository that is linked to at the bottom of the article contains all of the necessary library files. You will need to download them and then extract them into the library folder that is located inside of the Arduino document folder. You can skip the step before this one if you have already installed the necessary components, but you will still need to install the display library. You are required to use the TFT eSPI library that I have attached for the display. When that is finished, let's move on to the next step, which is the interface.

Required Libraries

1. TFT_eSPI Library (Modified)
2. QMC5883L Library from DFRobots

Interfacing Magnetic Sensor

The HMC5883L/QMC5883L is going to be the component that we use for the magnetic sensor. We can make use of the sensor as a digital compass given that it is able to indicate the geographic directions of north, south, east, and west. Inside of the HMC5883L/QMC5883L are three different magneto-resistive materials that are organized along the axes of x, y, and z. The strength of the earth's magnetic field has an effect on the amount of electric current that flows through these materials. Therefore, we are able to detect shifts in the magnetic field of the Earth by monitoring the change in the current that is flowing through these materials. After the change in the magnetic field has been absorbed, the values can then be sent to any embedded controller using the I2C protocol. Examples of embedded controllers include a microcontroller and a processor.

HMC5883L/QMC5883L Module Pinout

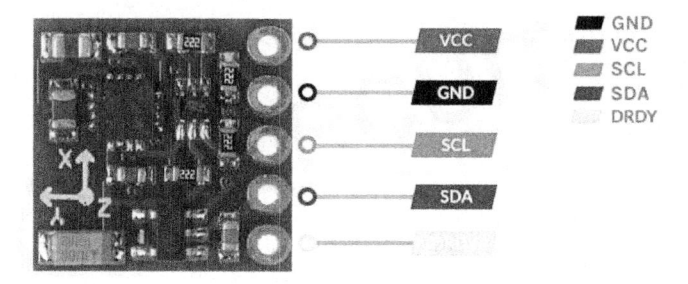

A total of five pins are present on the HMC5883L/QMC5883L module. With the exception of the VCC and Ground pins, all of the pins on this sensor module are digital. There is one optional interrupt pin in addition to the four pins that are used for interfacing with the MCU. The HMC5883L/QMC5883L module has the following pinout, as specified by the manufacturer:

- GND - Ground connection for the module. Establish a connection to the ESP32's GND pin.
- VCC - Provides power for the module. Establish a connection to the ESP32's 3.3V pin.
- SCL - stands for Serial Clock. Utilized for the purpose of providing the clock pulse for I2C Communication.

➤ SDA - Serial Data pin. Used for moving data between devices using the I2C communication protocol.

DRDY - An interrupt is generated in this pin whenever the output value of the sensor is ready to be read. By default, the module has a pull-up resistor connected to this pin. The value of the output pin is set to "0" for a period of 250 microseconds when the module's output value is ready.

Circuit to Interface a Magnetic sensor with an ESP32

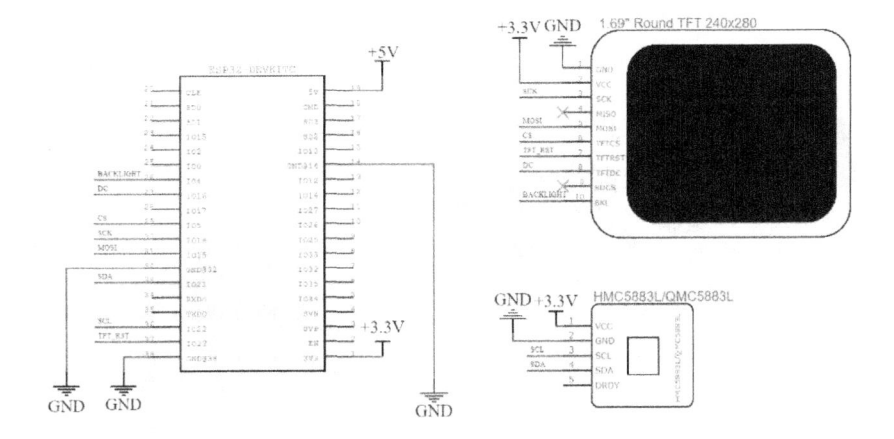

Circuit Diagram for the ESP32 and Accelerometer/Gyroscope Sensor

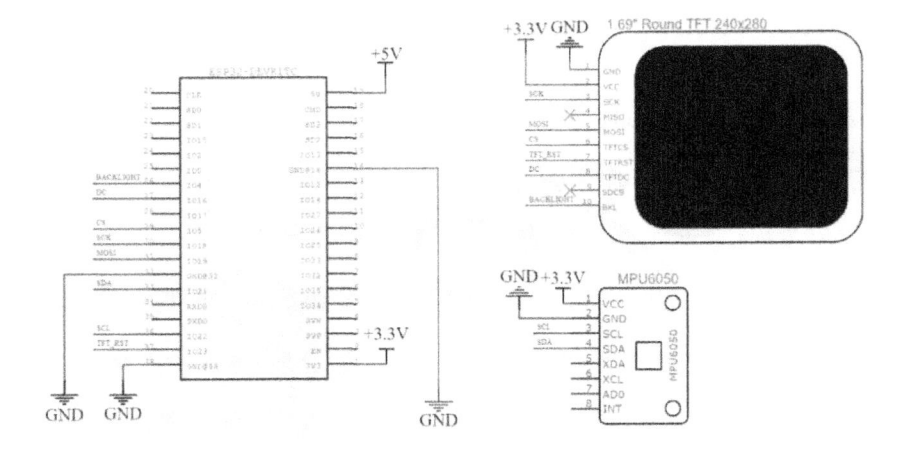

When conducting tests, ensure that the Magnetic sensor module and the display module are connected to the ESP32 Devkit in accordance with the circuit diagram that can be found below. As you did in the previous step, connect the display to the SPI bus and the sensor to the I2C line. The actual configuration can be seen here. I2C lines are used to make the connection between the ESP32 and the

QMC5883L module. Connect the accelerometer gyroscope sensor module and the display module to the ESP32 Devkit in accordance with the circuit diagram that can be found below. This will allow you to test the device. As you did in the previous step, connect the display to the SPI bus and the sensor to the I2C line. The actual configuration can be seen here. In order to have the freedom to move the module around while the test is being performed, the MPU6050 module is connected using a jumper cable.

484. How Does a NEO-6M GPS Module Work and How to Interface it with ESP32

If you are an embedded engineer who works in the electronic industry, there may come a time when you are faced with a situation in which you need to determine the position of a moving object or the altitude and velocity of a specific location. Alternatively, you may need to determine the altitude and velocity of an entire location. Given that a GPS module could prove to be rather useful in a scenario like this one, we made the decision to interface the NEO-6M GPS GSM module with the ESP32 in this post. In addition to that, we will discuss all of the benefits and drawbacks of using this device, so without further ado, let's go right into it. NEO-6M GPS Chip is capable of tracking up to 22 satellites simultaneously across 50 channels, and it achieves the industry's highest level of sensitivity, measuring in at -161 dB tracking, all while consuming only 45mA of supply current. This is a significant improvement over previous generations of GPS chips. In addition to this, the u-blox 6 positioning engine has a Time-To-First-Fix (TTFF) that is lower than 1 second.

No matter if you are connected to the internet or not, you can always use GPS on your mobile device. As long as you have a charged smartphone, you can find your way around unfamiliar or remote places. It is not necessary to have internet connectivity or cellular service for the GPS tracking feature on your phone to work properly.

NEO-6M GPS Module Pinout

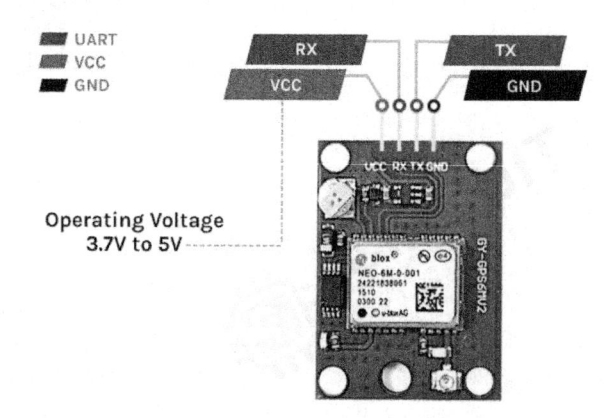

> GND, TxD, RxD, and VCC are the four pins that are included on the NEO-6M GPS

module. Communication with the microcontroller is accomplished through the use of the TxD and RxD pins.

- ➢ GND is the GPS Module's ground pin, and it should be connected to the ESP32's ground pin.
- ➢ TXD is the GPS module's transmit pin, which must be connected to the ESP32's RX pin.
- ➢ RXD is the GPS module's receive pin, which must be connected to the ESP32's TX pin.
- ➢ VCC is the GPS module's power pin, and it must be connected to the ESP32's 3.3V pin.

NEO-6M GPS Module - Parts

The NEO-6M module is a GSM module that is already configured for use and may be used in a wide variety of contexts. The following is a list of the components that make up the NEO-6M GPS module:

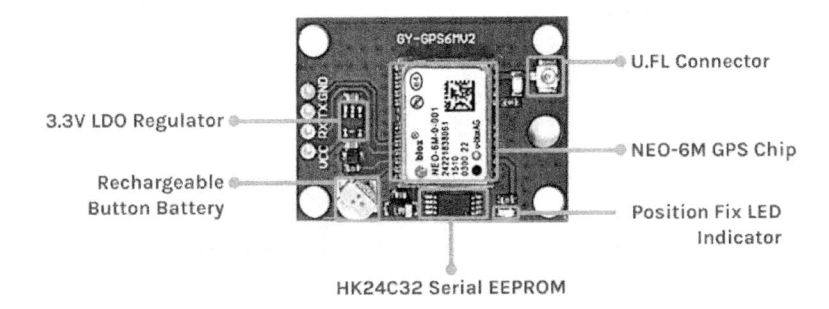

There are five primary components on the board that make up the NEO-6M GPS module. The NEO-6M GPS chip, which is located in the center of the PCB, is the first major component. Following that, we have a serial EEPROM module as well as a rechargeable battery. The clock data, the most recent position data (GNSS orbit data), and the module configuration can be retained with the use of an EEPROM and a battery, although this type of memory is not designed for long-term data storage. When the battery is removed, the GPS always starts from a cold start, which causes the initial GPS lock to take significantly more time. When electricity is supplied, the battery will begin charging itself automatically, and it can store data for up to two weeks when it is not being charged. Next, we have our LDO. The module may be powered from a 5V supply thanks to the onboard LDO, which is located on the board. At long last, we have our UFL connector, which is the point at which we will connect an external antenna in order for the GPS to function correctly.

Circuit Diagram for NEO-6M GSM Module

The circuit for the NEO-6M GPS module is really straightforward and straightforward to comprehend. The GPS module is responsible for the majority of the work in this module; nevertheless, in order for the GPS module to function, we need either two or three additional components. Following is an illustration of the NEO-6M GPS Module's comprehensive schematic.

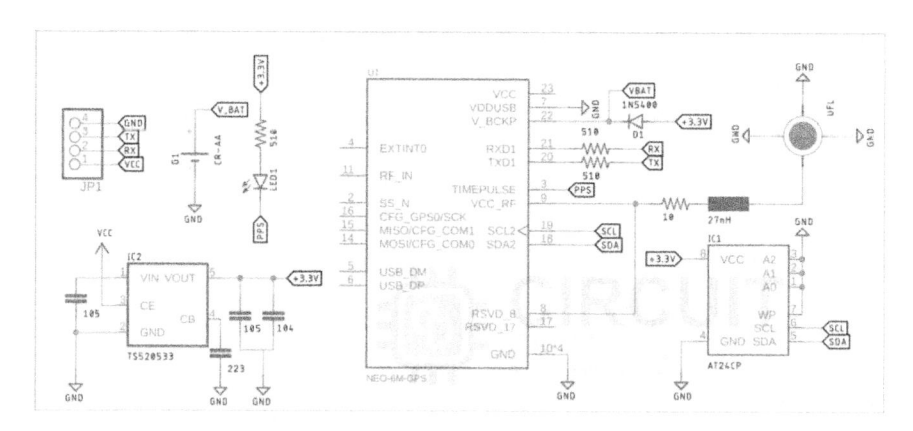

In the schematic, we can see that we have an AT24C33 EEPROM, a battery, and a 3.3V voltage regulator that is responsible for changing the input from 5V to 3.3V. Additionally, the input voltage ranges from 3.3V to 5V. EEPROM and batteries work together to help maintain data in battery-backed RAM for a limited amount of time. This RAM can keep track of the current time and location for the clock. Aside from that, we have a UFL connector that is required to have an External Patch Antenna connected to it in order for the module to function as intended.

ESP32 NEO-6M GPS Module Circuit Diagram

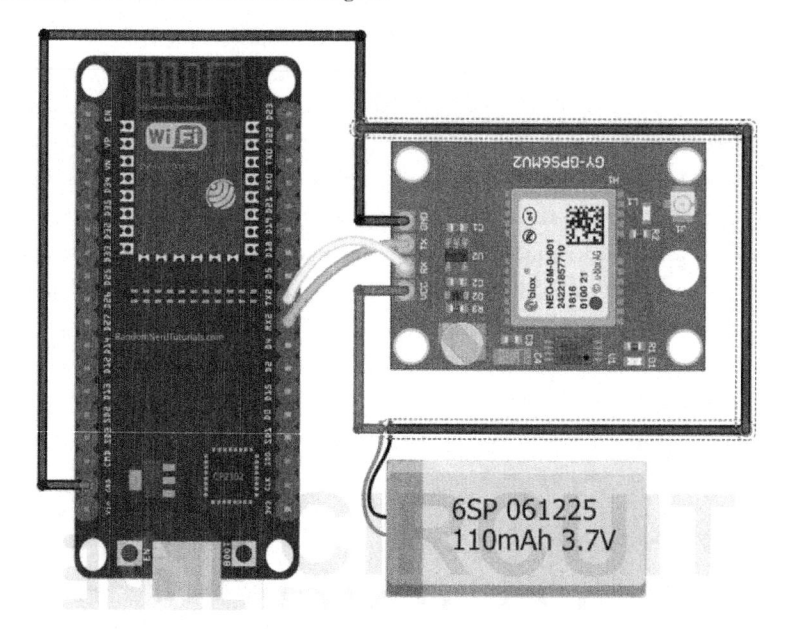

Since we now have a comprehensive knowledge of the NEO-6M GPS module, we are able to connect all of the necessary wires to the ESP32 board and examine the module to see whether or not it is operating normally or whether it is producing problems. The next step is to develop some code and decode the GPS data that we have received. Following is a schematic illustrating the connection between the NEO-6M GPS module and the ESP32.

Understanding NMEA Sentences

We can see some data flowing out of the NEO-6M GPS module if we attach it to a USB to UART converter and then open a window in which to watch the serial port on our computer. These types of statements are referred to as NMEA sentences, and NMEA is an acronym that stands for the National Marine Electronics Association. The vast majority of GPS receivers use this message format, which is standardized.

485. DIY ESP32 Oscilloscope

Every electronics expert has to have a test device called an oscilloscope in their toolbox. It can be used to view and analyze a variety of signals, and it is typically presented in the form of a two-dimensional plot in which one or more signals are plotted against time. In the process of designing and debugging electrical devices, they are utilized to view and compare waveforms, as well as determine the voltage levels, frequency, noise, and other properties of signals that are applied at the device's input as it varies over time. Because of this, oscilloscopes are a very useful instrument that should be kept on the workbench of each electronics expert or maker.

Oscilloscopes, on the other hand, may be rather expensive, with entry-level models ranging anywhere from $500 to $2,000 in price. Because of their tens of thousands of dollar price tags, more complex oscilloscopes are out of reach for consumers with more fundamental needs. But what if we were able to design one that was simpler, smaller, and less expensive to produce? This is the question that prompted us to start working on this project today.

ESP32 Oscilloscope Features

- Single-channel
- 1 million samples per second, 50 thousand at 16 bits per buffer (50ms of data at 1Msps)
- Scale from 10 microseconds per division to five milliseconds per division @ 1 msps
- In 1X mode, the maximum VPP is 3.3 volts, while in 10X mode, it is 33 volts.
- Control that is both quick and responsive, employing tactile switches.
- Frequency calculations (20hz min due to buffer size)
- Simple mean filter ON/OFF
- Maximum, minimum, typical, and highest point
- Peak voltage, The difference in time and voltage
- Analog, Digital/Data Mode
- Single TRIGGER AUTOSCALE

Components Required to build ESP32-based Oscilloscope

- 1.69-inch, 240x280 Rounded Corner TFT display included with the ESP32 Devkit (ST7789s)
- Switches that can be felt.
- SHDT and SPDT switches
- 100K resistor 10K resistor 100nF capacitor
- Clad in copper or perforated board, Soldering tools

ESP32 Oscilloscope Circuit Diagram

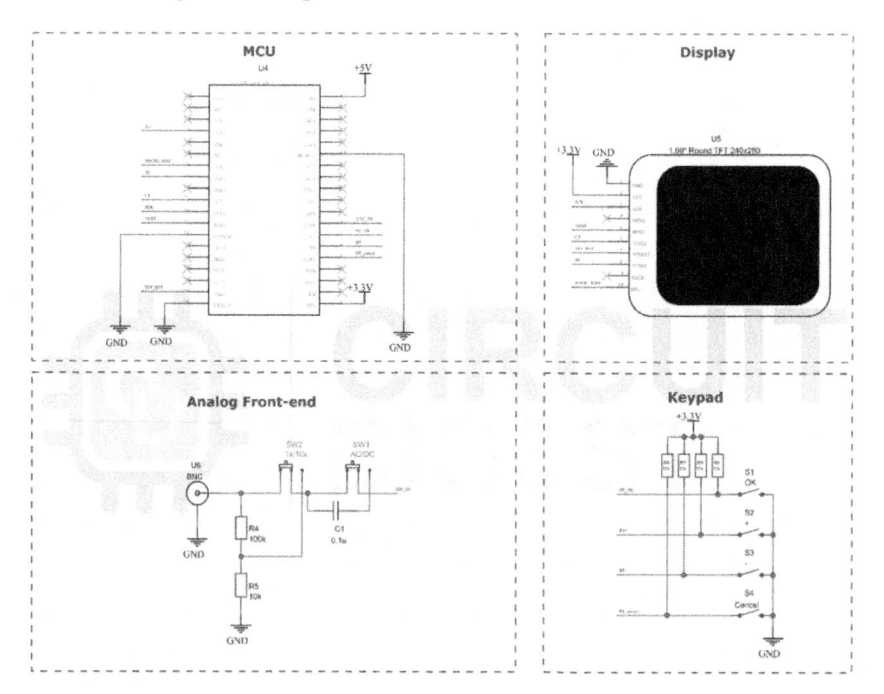

The data acquisition is controlled by an ESP32, which is employed in this application. When it comes to storing and manipulating the signals, we will be using the I2S buffer that is built right in. In this example, the 38-Pin variation is used, however other types of development modules might be used instead. We are utilizing a 1.69-inch TFT display module as our means of display. The resolution of this device is 240 by 280 pixels. The display controller that we will be using is a ST7789S, and the SPI communication protocol will be used to control it. The module also has a slot for an SD card, however we haven't utilized it yet. In a future version, we might make use of this for waveform capturing or other applications along those lines. The keypad is a pretty straightforward device. For this particular reason, tactile switches that include pullup resistors are utilized. To identify each individual key press, we are making use of the hardware interrupt. This will provide us with a keypad that is incredibly responsive.

The analog input part is a rather straightforward component. It is made comprised of two SPDT switches, one for selecting the range, and another for selecting the AC/DC coupling. A voltage divider that can be used to feed signals with a peak voltage greater than 3.3V has been introduced by our team as part of the range selection process. This divider can be used to feed the signals. Signal will be converted to a 10:1 ratio after passing through the voltage divider.

Arduino Code for Oscilloscope

Download the complete source code from the GitHub repository hosted on the Circuit Digest website. You will also find an archive with the name TFT eSPI within the GitHub repository. In order to successfully drive the display, this modified library is required. It should be extracted into the library folder for Arduino. In the event that the TFT eSPI library has already been installed, you will need to uninstall it before extracting the new version. When it is finished, go to the board manager and choose the esp32 option. After that, you need to compile the code and then upload it. Our do-it-yourself oscilloscope is now operational and ready for usage. The Oscilloscope can be powered by using the Micro USB port that is located at the bottom. This connector is used exclusively for supplying electricity.

485. DIY ESP32 Oscilloscope

Every electronics expert has to have a test device called an oscilloscope in their toolbox. It can be used to view and analyze a variety of signals, and it is typically presented in the form of a two-dimensional plot in which one or more signals are plotted against time. In the process of designing and debugging electrical devices, they are utilized to view and compare waveforms, as well as determine the voltage levels, frequency, noise, and other properties of signals that are applied at the device's input as it varies over time. Because of this, oscilloscopes are a very useful instrument that should be kept on the workbench of each electronics expert or maker.

Oscilloscopes, on the other hand, may be rather expensive, with entry-level models ranging anywhere from $500 to $2,000 in price. In addition, more sophisticated oscilloscopes might cost a few thousand dollars, putting them out of reach for consumers with less sophisticated needs. But what if we were able to design one that was simpler, smaller, and less expensive to produce? This is the question that prompted us to start working on this project today.

ESP32 Oscilloscope Features

- Single-channel
- 1 million samples per second, 50 thousand at 16 bits per buffer (50ms of data at 1Msps)
- Scale from 10 microseconds per division to five milliseconds per division @ 1 msps
- Maximum VPP 3.3V in 1X and 33V in 10X mode
- Control that is both quick and responsive, employing tactile switches.

- Frequency calculations (20hz min due to buffer size)
- Simple mean filter ON/OFF
- Max, min, average and Peak-Peak voltage
- The difference in time and voltage
- Analog, Digital/Data Mode
- Single TRIGGER AUTOSCALE

Components Required to build ESP32-based Oscilloscope

- 1.69-inch, 240x280 Rounded Corner TFT display included with the ESP32 Devkit (ST7789s)
- Switches that can be felt.
- SHDT and SPDT switches
- 100K resistor 10K resistor 100nF capacitor
- Clad in copper or perforated board
- Soldering tools

ESP32 Oscilloscope Circuit Diagram

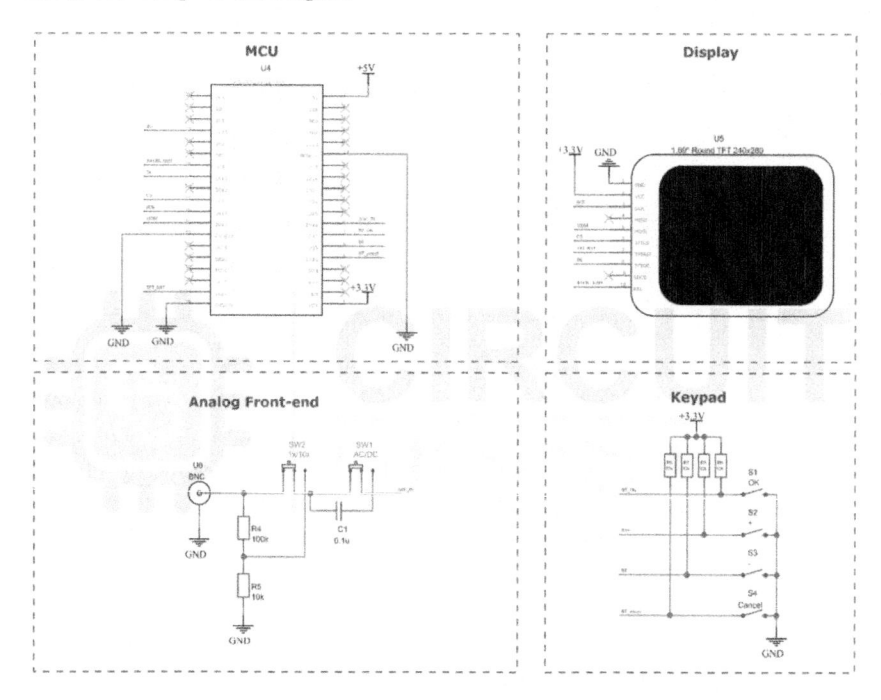

The full circuit diagram for the ESP32-based oscilloscope may be seen in the next page. The data acquisition is controlled by an ESP32, which is employed in this application. When it comes to storing and manipulating the signals, we will be using the I2S buffer that is built right in. In this example, the

38-Pin variation is used, however other types of development modules might be used instead. The data acquisition is controlled by an ESP32, which is employed in this application. When it comes to storing and manipulating the signals, we will be using the I2S buffer that is built right in. In this example, the 38-Pin variation is used, however other types of development modules might be used instead. We are utilizing a 1.69-inch TFT display module as our means of display. The resolution of this device is 240 by 280 pixels. The display controller that we will be using is a ST7789S, and the SPI communication protocol will be used to control it.

The module also has a slot for an SD card; however, we haven't utilized it yet. In a future version, we might make use of this for waveform capturing or other applications along those lines. The keypad is a pretty straightforward device. For this particular reason, tactile switches that include pullup resistors are utilized. To identify each individual key press, we are making use of the hardware interrupt. This will provide us with a keypad that is incredibly responsive. You are able to acquire knowledge regarding the ESP32 Interrupts that we discussed previously.

The analog input part is a rather straightforward component. It is made comprised of two SPDT switches, one for selecting the range, and another for selecting the AC/DC coupling. A voltage divider that can be used to feed signals with a peak voltage greater than 3.3V has been introduced by our team as part of the range selection process. This divider can be used to feed the signals. Signal will be converted to a 10:1 ratio after passing through the voltage divider.

486. Smart Wi-Fi Video Doorbell using ESP32 and Camera

The topic of security systems is currently one of the most investigated fields in today's society. In response to the growing number of security risks, many organizations are developing innovative, high-tech security products. IoT is an extra advantage in this industry that, in the event of any emergency, can immediately trigger an event, such as calling the police, fire brigade, or your neighbor. In the past, we have constructed a wide variety of security systems, such as an IoT-based door security alarm, a wireless doorbell, a video surveillance camera, and a Raspberry Pi-based visitor monitoring system. Using an ESP32 and a camera, we will construct a smart Wi-Fi doorbell in this lesson. After learning about ESP32-CAM and how to utilize it for video streaming in a previous lesson, we will now use ESP32-CAM to create a smart Wi-Fi video doorbell. This AC-powered "smart" doorbell not only alerts you to visitors at your door by playing a custom tune on your phone, but it also sends you a text message with a link to a live video feed so you can see who's there from anywhere in the globe.

Components Required

- ESP32-CAM, FTDI Programming Board
- 220V AC to 5V DC Converter
- Buzzer, Push Button, LED (2)

Circuit Diagram

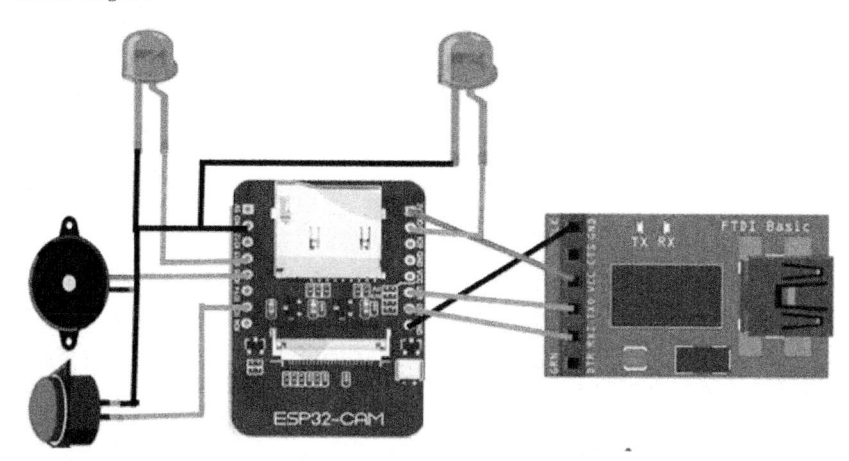

The Smart Wi-Fi Doorbell has a very straightforward circuit schematic; all you need to do is connect two LEDs, a push button, and a buzzer to the GPIO pins on the ESP32 board. When a button is pressed, the sound is produced by a buzzer that is attached to the button. The status of the network may be seen through one of the LEDs, while the status of the power supply can be seen through the other LED. If the ESP is connected to a network, the Network LED will be on, and if it is not, it will blink.

The Wi-Fi video doorbell system will look like this when it is housed in the 3D-printed casing:

IFTTT Setup for Wi-Fi Doorbell

IFTTT is a free web-based service that enables users to create chains of simple conditional statements, which are referred to as "recipes." These chains are triggered based on changes to other web services, such as Gmail, Facebook, Instagram, and Pinterest. IFTTT is a service that allows users to create chains of simple conditional statements. "If This Then That" is what "IFTTT" stands for as an abbreviation.

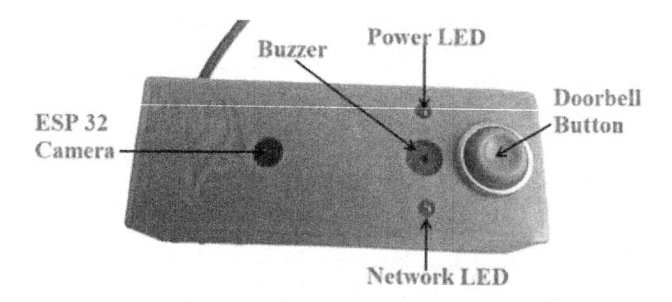

IFTTT is being used in this project to trigger the sending of an email whenever the temperature or humidity reaches a threshold that was previously determined. In the past, we utilized IFTTT in several IoT-based projects to send emails or SMS on specific occurrences, such as when there was excessive use of electricity, when there was a high pulse rate, when an intruder entered the building, etc. To begin, head over to IFTTT and either sign in with your existing credentials or create a new account if you don't already have one. Now search for "Webhooks," and when you find it, click on the Webhooks section under Services.

487. Monitoring the power output of solar panels over the Internet of Things using ESP32 and ThingSpeak

Solar energy is currently at the forefront of the field of renewable energy. This is due to the fact that harnessing the power of the sun to generate electricity is the most straightforward and economically feasible method of generating renewable energy. When it comes to solar panels, the output power of each panel needs to be carefully monitored in order to ensure that the panels are producing the most amount of power possible. Because of this, having a monitoring system that operates in real time is essential. It is also possible to use it to monitor the power output from each panel in a large solar power plant, which assists in the identification of dust buildup. In addition to this, it eliminates the possibility of any fault circumstances occurring while the machine is running. We have constructed a few projects linked to solar energy in some of our earlier posts. These projects include a solar-powered cell phone charger and a solar inverter circuit, amongst others. If you are interested in solar power and are seeking for other projects, you can look at those.

In this project, we will be constructing an IoT-based Solar Power Monitoring System by including the MPPT-based battery charging approach, which will help to reduce the amount of time needed to charge the battery while simultaneously improving the system's efficiency. In addition, in order to make the circuit more secure, we are going to take readings of the temperature of the panel, the output voltage, and the current. In the end, the cherry on top is that we are going to employ the cloud services provided by ThingSpeak so that we may monitor the output data from any location in the world. Please be aware that the MPPT Solar Charge Controller Project, which we constructed earlier, will be continued in this project. Using the ESP32 Internet of Things development board, we are going to now monitor the output voltage, current, and power of the solar panel.

How to Select the Appropriate Components for an Internet of Things-Enabled Solar Power Monitor

When a solar system is equipped with a solar monitor, it is much simpler to monitor the system and identify problems should they arise. Because of this, choosing the components to go into the system becomes a highly critical element of the design process. The list of components that we made use of may be found down below.

- ESP32 dev board, MPPT circuit (can be any solar circuit)
- A resistor that is shunted (for example 1 Ohm 1 watt - suitable for up to 1A of current)
- A battery that uses lithium (7.4v preferred).

- Connection to Wi-Fi that is active
- a sensor that can read the temperature of the solar panel
- Voltage divider circuit (see the description)
- Esp32 Dev Board:

It is essential to select the appropriate type of development board in order to enable an application to work with the Internet of Things. This board must be able to process the data coming from its analog pins and transmit it using any type of connection protocol, including Wi-Fi or to a cloud server. We went with the ESP32 because it is a microcontroller that comes packed with a lot of functions but doesn't cost that much. Additionally, it has a Wi-Fi radio already installed in it, so we can connect to the internet without much difficulty using that.

Solar Circuit:

A solar charging circuit is a circuit that receives a greater voltage from the solar panel, then transforms it down to a charging voltage in order to charge the battery in the most effective manner possible. For this project, we will be utilizing the LT3562-based MPPT Charge Controller Circuit Board that we have already developed for one of our past projects. This board was used for one of our earlier projects. But you may use any kind of solar circuit if you want to implement this Internet of Things enabling monitoring into your system. Because the circuit on this board includes Maximum Power Point Tracking (MPPT), which is advantageous for low power solar panel projects, we decided to go with it as our board of choice. Using a solar panel to charge a lithium-ion battery pack in a tiny device is an effective method.

Shunt Resistor:

Any resistor adheres to ohm's law, which states that there will be a drop in voltage proportional to the amount of current that passes through the resistor. Shunt resistors are not an exception to this rule; rather, they are utilized for the express purpose of measuring the flow of current.

Lithium Battery:

When working with solar panels, it is imperative to carefully select the lithium battery that will be used in the project. Because the microcontroller unit that is always on and continuously examines and transmits the data requires at least a hundred milliamperes of electricity for stable functioning, the battery must be replaced every few days.

Voltage Divider:

The measuring of the voltage produced by solar panels requires the utilization of a voltage divider. When selecting a voltage divider, one should look for one that can split the voltage in accordance with the microcontroller's I/O voltage input. You should select the aforementioned resistors in such a manner that the output voltage of the voltage divider does not exceed the maximum I/O voltage of the microcontroller (3.3V for ESP32). Potentiometers, on the other hand, are recommended due to the versatility they allow in terms of selecting a solar panel with either a greater or lower voltage rating and the ease with which the voltage may be set with a multimeter.

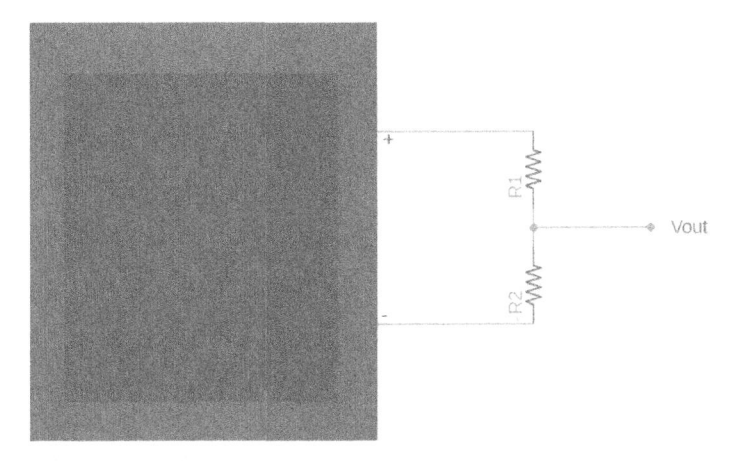

Solar Panel

In our particular setup, the MPPT board circuit includes a potentiometer that performs the function of a voltage divider. We determined that a division factor of 6V would work best for the voltage divider. We connected two multimeters, one in the input and another in the output of the pot, and fixed the value so that the output would be 3V when the input voltage was 18V. This was done because the solar panel's nominal output voltage is 18V.

Monitoring Solar Panel Temperature:

The temperature of a solar panel has a direct bearing on the amount of power that can be extracted from that panel. Why? Because as the temperature of a solar panel begins to rise, the amount of current that is produced by the panel climbs exponentially while the amount of voltage that is produced begins to decrease linearly.

According to the formula for power, which states that wattage is equal to voltage times current (W = V x A), a reduction in output voltage also results in a reduction in the solar panel's output power, and this occurs even when the flow of current is increased. The next issue that arises in our minds is, "How can we accurately estimate the temperature of the sun?" It is fairly intriguing to consider, considering that solar panels are typically subjected to a hot environment due to the fact that they are exposed to direct sunlight and for the obvious reasons.

Utilizing a temperature sensor that is designed for flat surfaces is the most accurate method for measuring the temperature of solar panels. It is also recommended to utilize a thermocouple of the K type, which should be installed directly in the solar panel.

Circuit Diagram

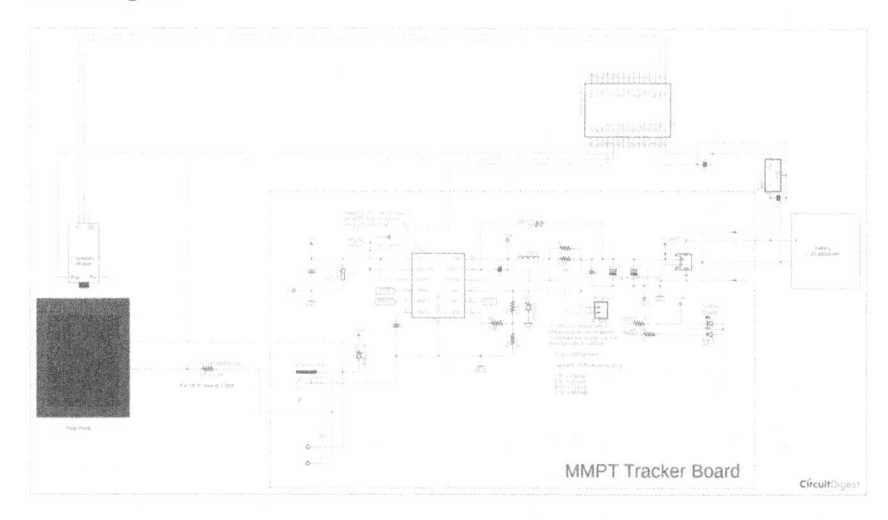

The following is an illustration of the whole circuit diagram for the Internet of Things-enabled solar power monitor. The diagram is easy to understand. Our MPPT board, which is represented by the red board with dashes on it, may be seen in this particular project.

488. Audio Player Built with ESP32 for DIY Projects

Building your own music player can be a lot of fun, and in the past, we've constructed a few DIY music players by utilizing Arduino and a specialized MP3 module. Now that we have ESP32, we are building an interesting audio player with it. If you want to add sound effects, all you have to do is attach an additional speaker to ESP32. To play audio files, we will make use of an LM386 and a speaker in conjunction with an ESP32. This program demonstrates that the ESP32 board is capable of playing audio files, despite the relatively quiet volume of the audio output.

Components Required

- ESP32, LM386 Amplifier Module
- 8-ohm Speaker, Jumper Wires

Circuit Description:

A speaker is required in order for the ESP32 to play sound. The LM386 Audio Amplifier module is what establishes the connection to the speaker. While the GPIO 25 pin of the ESP32 is linked to the IN pin of the Amplifier Module, the Vcc and GND pins of the Amplifier Module are connected to the VIN and GND pins of the ESP32. DAC stands for "digital to analog converter," and one of the two DAC pins is GPIO 25.

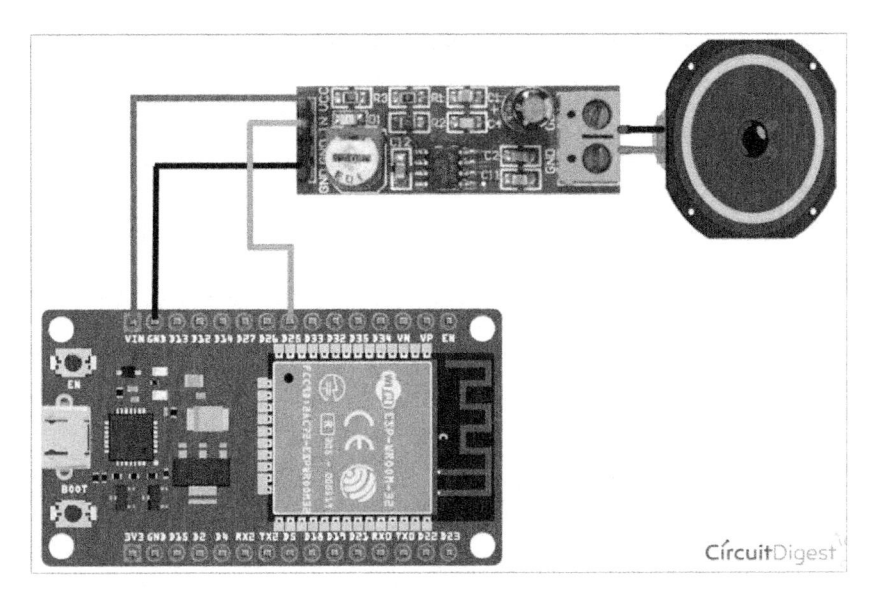

489. ESP32-CAM Face Recognition Door Lock System

Nowadays, everyone is extremely concerned about their personal security, whether it be the protection of their data or the security of their own home. These days, the use of digital door locks is extremely widespread due to the proliferation of connected devices and the general improvement of technology. The use of a physical key is not necessary to operate a digital lock; rather, the lock can be controlled by means such as RFID, fingerprints, Face ID, pins, passwords, and so on. In the past, we have built a wide variety of applications using these various technologies, including digital door locks. Through the use of ESP32-CAM, we will construct a Face ID-controlled digital door lock system in this tutorial.

The AI-Thinker ESP32-CAM module is an inexpensive development board that features a micro-SD card port in addition to a camera with a very compact form factor called an OV2640. It is equipped with a high-performance ESP32 S processor that includes built-in Wi-Fi and Bluetooth connectivity, as well as a 7-stage pipeline architecture and 2 high-performance 32-bit LX6 CPUs. In a previous article, we went over ESP32-CAM in depth and demonstrated how it might be used to construct a Wi-Fi door Video doorbell. This time around, we will make use of the ESP32-CAM to construct a Door Lock System that is based on Face Recognition by utilizing a Relay Module and a Solenoid Lock.

When a person approaches the door, the camera will take a picture of their face, check to see if it has been registered, and then either unlock the door or sound an alarm. If the face has not been registered, the camera will sound an alarm, take a picture, and send it to the number that has been registered. The operation of the system is exactly like this.

Components Required

- ESP32 CAM
- FTDI Board
- Relay Module
- Solenoid Lock
- Jumper Wires
- Solenoid Lock

The electronic-mechanical locking mechanism is utilized in the operation of a solenoid lock. This particular style of lock has a slug that has a cut that is angled and an effective mounting bracket. DC produces a magnetic field when power is applied, which moves the slug inside and retains the door in the unlocked position.

This field also prevents the door from being locked. The slug will continue to remain in its position till the power is turned off. After the electricity has been cut off, the slug will exit the building and shut the door behind it. When it is locked, it does not consume any power at all. In order to power the solenoid lock, you will need a power supply that can deliver 12 volts at a current of 500 milliamps.

Circuit Diagram

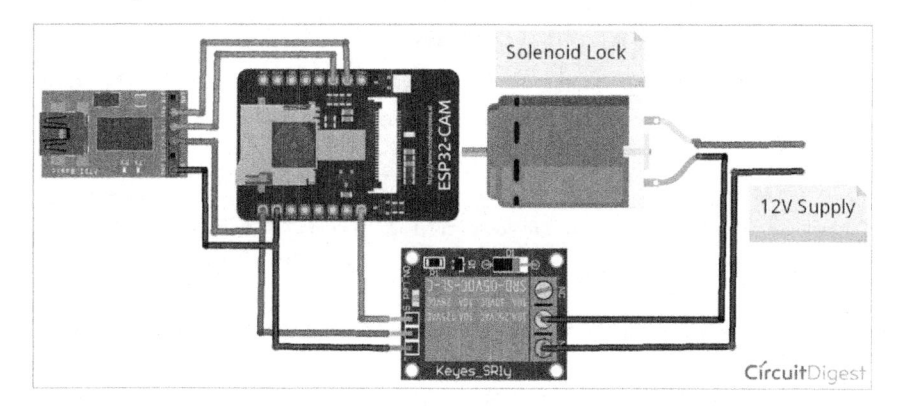

Combination of the circuit described above with an FTDI board, a relay module, and a solenoid lock. Because the ESP32-CAM doesn't have a USB connector, the FTDI board is what's used to flash the code into it. Meanwhile, the relay module is what's utilized to turn the solenoid lock on and off. The VCC and GND pins of the FTDI board and the Relay module are linked to the ESP32-CAM's Vcc and GND pins. The ESP32's RX and TX ports are connected to the FTDI board's RX and TX ports, and the IN pin of the relay module is connected to the IO4 port on the ESP32-CAM.

Important: Before you upload the code, make sure that the IO0 is connected to the ground. IO0 is responsible for determining whether or not the ESP32 is operating in flashing mode. The ESP32 enters flashing mode when the GND pin on GPIO 0 is connected to ground.

ESP32-CAM	FTDI Board
5V	VCC
GND	GND
UOR	TX
UOT	RX

ESP32-CAM	Relay Module
5V	VCC
GND	GND
IO4	IN

490. Automatic Hand Sanitizer Dispenser with COVID19

The Corona Virus, also known as Covid19, is wreaking havoc over the globe. The Corona Virus is currently affecting nearly every nation in the world. The World Health Organization (WHO) has already declared that it is a pandemic disease. As a result, several cities are currently in lockdown situations, individuals are unable to leave their houses, and thousands of people have already passed away. Numerous websites, such as Microsoft's Tracker and Esri's Covid19 Tracker, among others, are currently offering real-time data on the number of coronavirus cases.

As part of this endeavor, we are going to create an Automatic Hand Sanitizer Dispenser that has an LCD display that also provides a real-time tally of Coronavirus cases. The ESP32, an Ultrasonic Sensor, a 16x2 LCD Module, a Water Pump, and a Hand Sanitizer will all be utilized in this project. To obtain real-time information on people infected with Covid19, we are relying on Esri's API Explorer. To determine whether or not there are hands covering the opening of the hand sanitizer machine, an ultrasonic sensor is utilized. It will continually calculate the distance between itself and the sanitizer outlet and will instruct the ESP to switch on the pump to push the sanitizer out once the distance is less than 15 centimeters. The ESP32 is a Wi-Fi module that can quickly and easily connect to the internet, and it is utilized as the primary controller. In the past, we made use of it to construct a variety of IoT-based projects utilizing ESP32.

Components Required

- ESP32 Dev Module, Ultrasonic Sensor
- 16*2 LCD Display, Relay Module
- Mini DC Submersible Pump
- Hand Sanitizer

Circuit Diagram:

This Covid19 Tracker and automatic hand sanitizer dispenser machine's full circuit diagram is shown below.

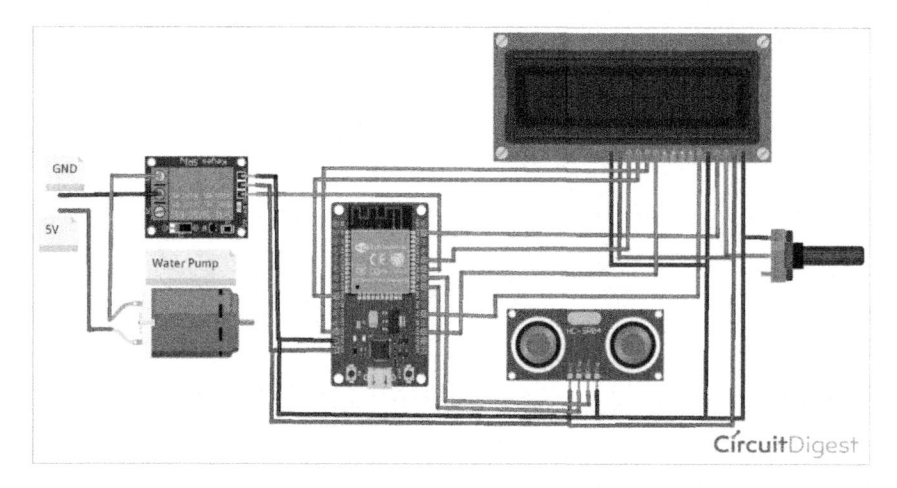

Through the use of a relay module, the ESP32 can communicate with the water pump. The relay's Vcc and GND pins are linked to the Vin and GND pins of the ESP32, while the relay's input pin is attached to the D19 pin of the ESP32. The ultrasonic sensor's Trig and Echo pins are connected to the Arduino's D5 and D18 pins, respectively.

491. Bitcoin $ Price Tracker Using ESP32 & OLED Display

In this project, we'll use an ESP32 WiFi module and an OLED display to create a Bitcoin price tracker. The device is programmable, and it has the capability of displaying the current price of Bitcoin in US dollars. The most advantageous feature of this approach is that the gadget may be configured to show the value of any cryptocurrency in any fiat currency. This is only doable if one of the exchanges provides access to an application programming interface (API). The Bitcoin Price Tracker was created with the help of the ESP32 development board, which already incorporates WiFi connectivity. It then establishes a connection to your WiFi network and, once every 15 minutes, retrieves the most recent Bitcoin price using the Coindesk API. In addition to displaying the current price of Bitcoin, the device also shows the percentage change in price from yesterday's market close. In addition to that, a red or green LED glows, depending on whether or not the price has gone up or down since it was last checked.

Component Required:

- ESP32- ESP32 WROOM WiFi Module
- LED 5mm LED Red & Green Color

- Resistor 560 ohm
- OLED Module - 0.96" SSD1306 OLED Display
- Jumper Wires, Breadboard

Hardware Setup & Circuit

The circuit for the Bitcoin Price Tracker is not a complicated one at all. For the creation of the schematic, I utilized the fritzing software.

Circuit Diagram:

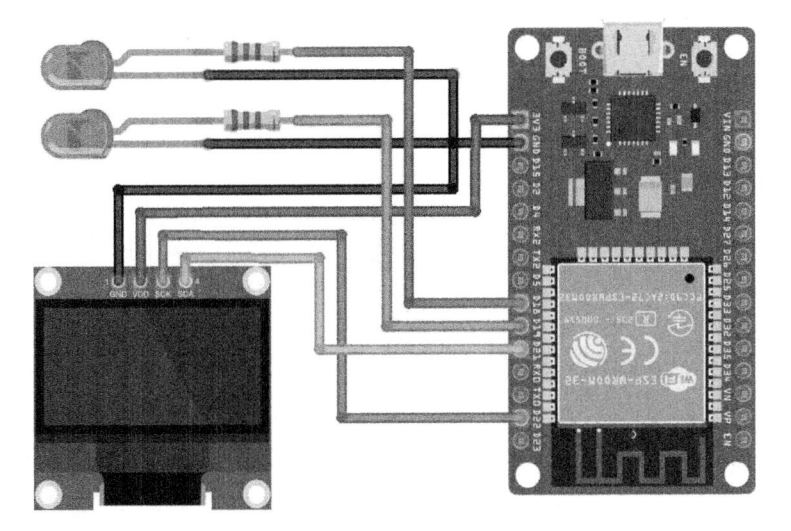

Establish a connection between the SDA and SCL pins of the OLED display and the D21 and D22 pins of the ESP32 module. Connect the red and green LEDs to terminals D18 and D19, respectively, using a resistor of 200 ohms. The 3.3V and GND pins on the ESP32 are used to supply power to the OLED Display. In this case, a red LED shows that the price of Bitcoin is decreasing, while a green LED indicates that the price of Bitcoin is increasing. In order to put together the circuit in accordance with the circuit schematic, I utilized a breadboard. For this project, you have the option of building your own personal circuit board (PCB).

492. Connecting ESP32 to Amazon AWS IoT Core using MQTT

This is an introduction to Amazon Web Services, more specifically an AWS IoT Core with ESP32 getting started lesson. The AWS Internet of Things Core is a managed cloud service that enables connected devices to interface with cloud applications and other devices in a safe and straightforward manner. In this project, we will learn how to connect the ESP32 with AWS IoT Core and publish sensor readings to AWS MQTT. ESP32 is a microcontroller that was developed by

Espressif Systems. As a demonstration, we will make use of the DHT11 Sensor and read the data pertaining to the temperature and humidity. The ESP32 will establish a connection to the neighborhood WiFi network and will then upload the data collected by the DHT11 Sensor to the AWS IoT Cloud. Not only can we submit data, but we also have the ability to receive data from the AWS Dashboard. Earlier, we used AWS IoT Core and ESP8266 to publish the data from the sensors to the AWS Dashboard.

The lesson is divided into several sections.

- Creating an account using and configuring the Amazon Web Services
- In the Arduino IDE, installing the required libraries and writing an Arduino sketch for the project are both required steps.
- The process of creating a Thing on Amazon Web Services involves generating certificates and applying policies.
- Making Adjustments to the Arduino Sketch Based on the Thing Data and Credentials
- Publish and Subscribe to Data on the Amazon Web Services Dashboard

Beginners who are interested in learning more about the Amazon AWS IoT Core for IoT Applications will benefit from this tutorial. In an earlier lesson, we discussed various Internet of Things platforms, such as Google Firebase and Arduino IoT Cloud. You are however able to construct and manage devices for use in business applications using AWS IoT Core.

Hardware Setup

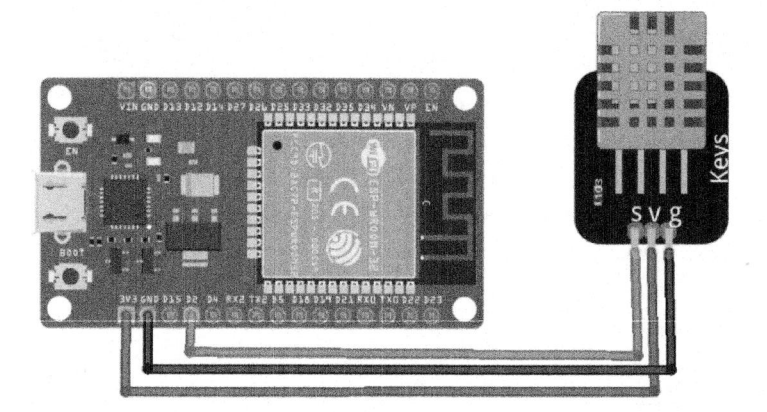

An ESP32 Wifi Module is the piece of hardware that is essential for this project. In addition, we will be utilizing a DHT11 Humidity and Temperature Sensor for the section dealing with the sensors.

Connect the DHT11 Sensor to the ESP32 Board. For the connection, you can make use of a breadboard or you can simply make use of a male-to-female connector wire.

What are Amazon AWS IoT Core?

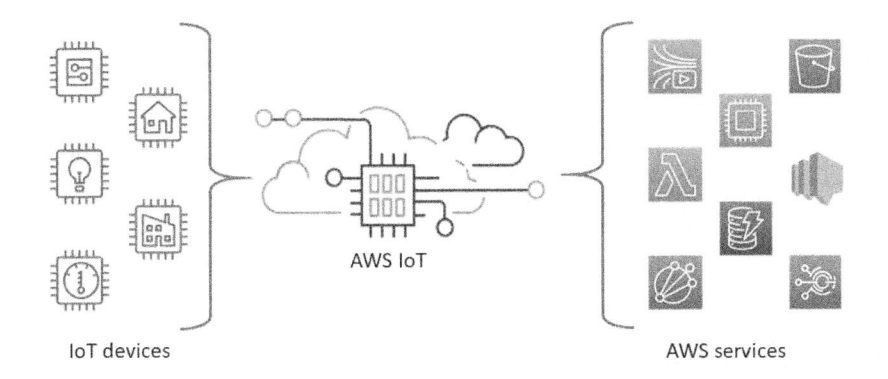

AWS provides services and solutions for the Internet of Things (IoT), which allows for the connection and management of billions of devices. Your Internet of Things devices can now communicate with other devices and AWS cloud services thanks to these cloud services. The device software that is provided by AWS IoT can assist you in integrating your Internet of Things devices with solutions that are based on AWS IoT. AWS Internet of Things will be able to link your devices to the cloud services that AWS offers if your gadgets can connect to AWS IoT.

AWS IoT gives you the ability to select the technologies that are both most suitable and up to date for your solution. AWS IoT Core supports the following protocols to assist you in the management and support of your Internet of Things devices out in the field:

- MQTT (Message Queuing and Telemetry Transport) (Message Queuing and Telemetry Transport)
- MQTT over WSS (Websockets Secure)
- HTTPS (Hypertext Transfer Protocol – Secure) (Hypertext Transfer Protocol – Secure)
- LoRaWAN (Long Range Wide Area Network) (Long Range Wide Area Network)

493. Aquarium Water Quality Monitor with TDS Sensor & ESP32

In this project, we are going to discover how to monitor the water quality of an aquarium using a TDS sensor, a temperature sensor, an ESP32 WIFI module, and a TFT LCD display. A facility in which a collection of aquatic species is displayed or studied is referred to as an aquarium. An aquarium can be a receptacle for preserving aquatic organisms in either freshwater or marine environments. In a Smart Aquarium, you need to keep an eye on the water quality and have it automatically changed and feed the fish. Monitoring the water's quality is the most difficult task because it needs the use of a variety of sensors and electrical instruments. The characteristics that make up the water quality are as follows:

pH of the water, total dissolved solids, turbidity, dissolved oxygen, temperature, and electrical conductivity, etc. But the water temperature and total dissolved solids (TDS) are the most crucial parameters to consider for aquariums and aquatic life in general, including fish. The ideal temperature range for an aquarium is between 25 and 27 degrees Celsius (76- and 80-degrees Fahrenheit). TDS levels in the water between 400 PPM and 450 PPM are recommended for the majority of freshwater fish. With the help of an ESP32 and a TDS sensor, we will create our own aquarium water quality monitoring system in this project. On the TFT Color LCD Display, we are going to display the current value of the TDS in the water as well as the temperature. You may find the information you need about the TDS Sensor and the DS18B20 Waterproof Temperature Sensor on the posts that came before this one. The utilization of a TFT LCD Display that is embedded with ESP32 is the most impressive aspect of the project. Makerfabs is responsible for the design of the individualized Display.

Part Required:

- ESP32 + LCD Display
- TDS Sensor Gravity Analog TDS Sensor
- Temperature Sensor DS18B20 Waterproof Temperature Sensor
- Resistor 4.7K
- Connecting Wires
- Breadboard

Interfacing TDS & Temperature Sensor with ESP32

Let's go on to the next step, which is to connect the Gravity TDS Sensor and the DS18B20 Temperature Sensor to the ESP32 Board. During the TDS correction process, having access to the temperature parameter is essential, which is why we have opted to make use of a temperature sensor. When the temperature rises and falls, there is a significant shift in the TDS levels.

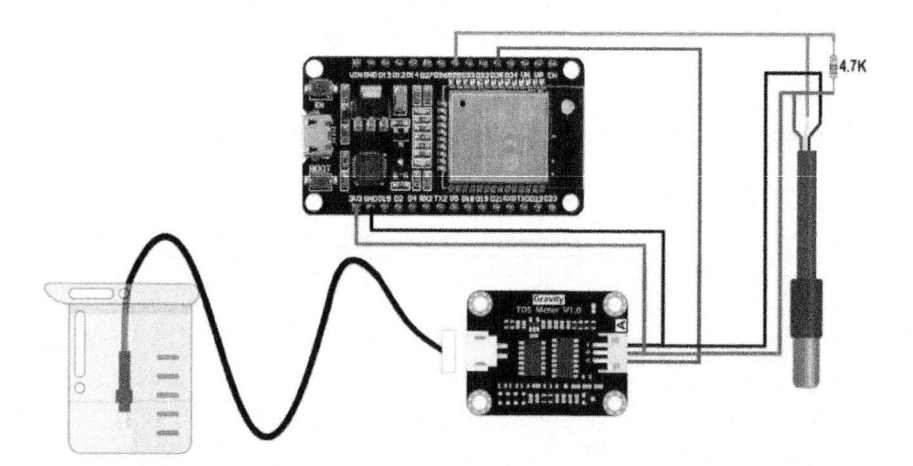

Connect the VCC and GND pins of the temperature and time sensor to the ESP32's 3.3V and GND pins, respectively. Establish a connection between the output analog pin of the TDS Sensor and the IO35 Pin of the ESP32. In a similar manner, connect the output of the DS18B20 to the IO25 Pin on the ESP32. A pull-of resistor with a value of 4.7K is required, and there must be a connection made between the DS18B20 output pin and 3.3V VCC so that the parasitic power may be supplied. You can test the system by putting the individual components together on a breadboard. Launch the Serial Monitor once the code has been successfully uploaded. Additionally, a sample of water should be dipped using the TDS and Temperature Sensor probe. Keeping an eye on the Serial Monitor, you will notice that it will start displaying the readings of both the temperature and the TDS value.

494. Monitoring the Indoor Environment Using an ESP32 and an LCD Display

This is a straightforward project for monitoring the indoor environment using an ESP32 and LCD panels. We are going to make use of an Indoor Environment Expansion board that has a DHT11 Temperature and Humidity Sensor as well as an SPG30 Air Quality Sensor installed on it. Measurements of CO2, TVOC, Temperature, and Humidity are taken with the Expansion Board. In the past, we measured the Indoor Air Quality with a BME680 Sensor. The LCD panel that is being utilized in this instance has a screen resolution of 320x480, a screen size of 3.5 inches, and an ILI9488 driver. It is essentially a TFT touch screen, but we are not going to make use of the touch function. Instead, we are going to use it as a colorful display.

The first thing that we are going to do is make a lovely gauge that will show the current temperature, humidity, and CO2 levels. The three gauges will display whether there has been an increase or reduction in the pattern of circular widgets. As a second project, we will display the CO2 and TVOC values in a graphical way. This will be similar to the first project. The value of CO2 and TVOC will cause the graphs to either rise or fall, depending on the direction of the change. Indoor Environment Monitoring is a potential use for both the ESP32 and the TFT LCD Display.

Part Required:

- ESP32+TFT Touch Display
- ESP32 3.5" TFT Touch (Capacitive) with Camera
- Indoor Environment Expansion board SPG30 + DHT11 Sensor + Buzzer
- USB Cable, Type-C USB Cable for Programming
- Connecting Wires Male to Female Jumper Wires

ESP32 3.5-Inch TFT Capacitive Touch Screen Display with Camera

This stunning touchscreen display has a 3.5-inch screen, is powered by an ESP32-WROVER chip, and incorporates an OV2640 camera with 2 million pixels. The incorporation of all of these element's results in the creation of an ideal platform for a variety of ESP32 applications and projects.

The TFT LCD driver is essentially an ILI9488, and it measures 3.5 inches in size and has a screen resolution of 320 by 480 pixels. The ILI9488 LCD communicates with the ESP32 microcontroller via SPI (Serial Peripheral Interface). Making the display smooth enough for videos may require the SPI main clock to be increased to between 60M and 80M. You are free to make use of any and all of these pins on the breakout connections while the camera is not being utilized. After that, you may utilize the ESP32 display for any Internet of Things application you want by connecting it to sensors or modules. Programming in Arduino and MicroPython are both supported by the ESP32 microcontroller.

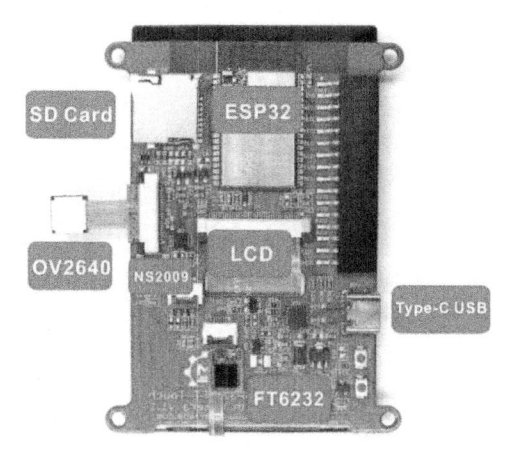

A micro-SD Card slot is provided on the board for the purpose of adding an external SD Card. Files and pictures can be saved to the SD Card and accessed whenever necessary. There is a USB Type-C port available, which functions as a USB to UART converter and can be used for ESP32 programming. You can directly upload the code to the board by connecting a Type-C data connection to it and then following the on-screen instructions.

ESP32 Touch Indoor Environment Expansion

This is an expansion board for the ESP32 3.5 Touch Screen that is designed for use in an indoor environment. Through the usage of the extension connector, the ESP32 3.5 TFT touch can also serve the purpose of a hardware expansion. As a result of the Indoor environment expansion's integration of the temperature and humidity sensor DHT11 and the SGP30 Air Quality Sensor, which allows for the detection of CO2 and TVOC, you will be able to quickly develop an indoor environment detector. In addition to that, there is a buzzer located on board that can be used for the alarm.

Indoor Environment Monitoring on Gauge

Let's utilize the ESP32 and LCD Screen to monitor the environment inside the building. The value of the temperature, humidity, and CO2 level will be displayed on the LCD Screen in the form of a Gauge. After you have connected the Type-C USB Cable to the ESP32 Board, you can then upload the code. After the code has been uploaded, you will need to press the reset button. After that, you will see three separate widgets or gauges displaying the data of the temperature, humidity, and CO2 levels.

Monitoring of CO2 and TVOC Historical Data Displayed in Graph

Let's use the ESP32 and the LCD screen to monitor the CO2 and TVOC levels once more. Graphical representations of the CO2 and TVOC levels will be shown on the LCD screen as they are measured. in order to allow you to keep track of the historical data.

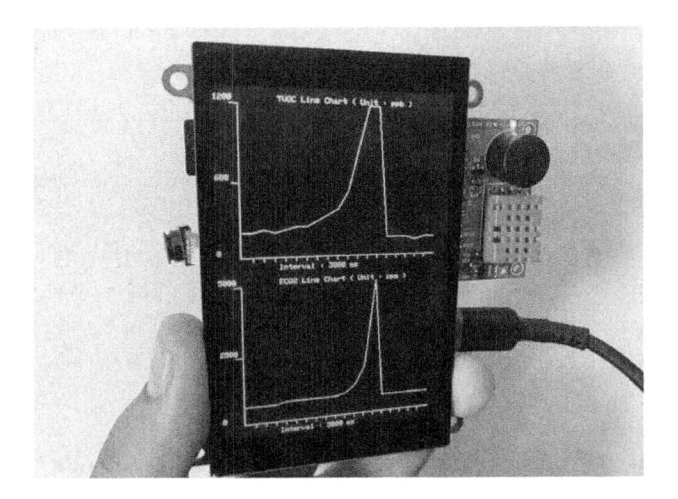

Repeat the uploading of the code. After the code has been uploaded, you will need to press the reset button. After that, graphical display data of CO2 and TVOC will be shown to you.

495. Measure Wind Speed with Anemometer on ESP32 TFT Display

With the help of an ESP32 and an anemometer, we will demonstrate how to read wind speed from a TFT display. The project makes use of an integrated board that has a 3.5-inch touchscreen display, an ESP32-WROVER Module, and an OV2640 camera that has a built-in resolution of 2 million pixels. We are going to connect the Anemometer Sensor to the analog pin on the ESP32 in an external location.

An instrument known as an anemometer is used to determine both the speed and the direction of the wind. The Adafruit anemometer is the one that serves as the sensor for this particular anemometer. The Adafruit anemometer has the ability to measure wind speeds of up to 70 meters per second, which is equivalent to 156 miles per hour. An ILI9488 TFT LCD Driver is utilized in the construction of the 3.5-inch TFT Touch Screen Display. The resolution of the screen is 320 by 480. There are a number of external pins on the ESP32 display that can be used to connect it to sensors or actuators. Because of this, they are appropriate for a wide variety of Internet of Things applications. Now that we have the Anemometer Sensor, let's take a look at how we can utilize it in conjunction with the ESP32 TFT Display to create a stunningly dynamic widget. Along with the severity of the warning, the wind speed will be displayed on the TFT Display in miles per hour.

Part Required:

- ESP32+TFT Touch Display
- ESP32 3.5" TFT Touch (Capacitive) with Camera
- Anemometer Sensor Adafruit Anemometer Cup Type
- USB Cable Type-C USB Cable for Programming
- Connecting Wires
- Male to Female Jumper Wires

Adafruit Anemometer Sensor

The Adafruit Anemometer Sensor is a Three-cup type anemometer that has the potential to measure wind speeds of up to 70 meters per second (156 miles per hour). It is made up of three different parts: the circuit module, the wind cup, and the shell. A length of wires with three cores and three connections are included with the sensor in the package. A brown wire for power, which can be anywhere from 7-24 volts DC, a third blue wire that offers measurements via analog voltage, and a black wire for power and signal ground. The analog voltage that is output will range from 0.4V (when there is no wind) up to 2.0V (when there is 32.4m/s of wind).

Connections for the Hardware of the ESP32 TFT Display with the Anemometer

Because the TFT Touch Display and ESP32-Wrover Module are already wired to the PCB on the inside, the only component we need is the Anemometer Sensor.

Connect the external 9V and GND to the VCC and GND pins of the anemometer sensor. Establish a connection between the GND terminals of the external power source and the GND terminals of the ESP32. In a similar fashion, connect the input pin for the anemometer to GPIO35 on the ESP32.

Measure Wind Speed with Anemometer on ESP32 TFT Display

Following the successful upload of the code, the TFT Display will immediately begin displaying the

Wind Speed data using an attractive widget. In order to cause the warning signal to appear on the display, you can manually rotate the cup of the anemometer at a high speed.

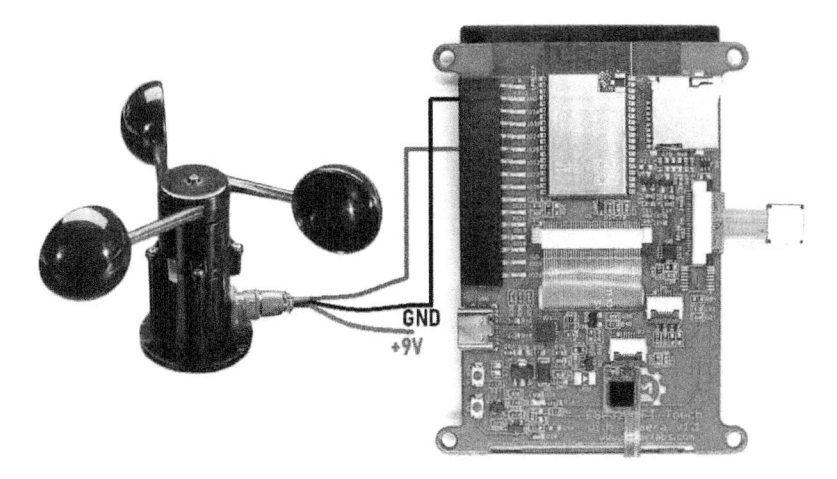

This is how you can measure the wind speed using the ESP32 and the Anemometer Sensor, and then display the wind speed value using the beautiful widget on the TFT Display.

496. UV Index Meter with ESP32 & UV Sensor ML8511

In this project, we are measuring the intensity of ultra violet light in terms of milliwatts per square centimeter by interfacing an ML8511 UV Sensor with an ESP32. Within the electromagnetic spectrum, the range of wavelengths known as ultraviolet light radiation, or UV Radiation, extends from 10 nm to 400 nm. Therefore, the GY/ML8511 sensor manufactured by lapis semiconductor is very helpful in order to get effective output in accordance with UV light. This wavelength is classified as part of the UVB-burning rays' spectrum and most of the UVA-tanning rays spectrum. The ML8511 UV sensor is able to detect light with a detection range of 280 nm to 390 nm in a more efficient manner.

The ML8511 sensor is exceptionally simple to operate. It gives forth an analog voltage that is linearly related to the UV intensity that has been measured (in mW/cm2) as its output. If your microcontroller is capable of converting analogue signals to voltage, then you will be able to determine the amount of UV present. It has a low supply current of only 300 uA, and it has a low standby current of only 0.1 a. It has a small and thin surface-mount package that measures 4.0mm x 3.7mm x 0.73mm (0.16 inches x 0.15 inches x 0.03) and is made of ceramic with 12 pins QFN.

Part Required:

- ESP32 Board ESP32 ESP-32S Development Board (ESP-WROOM-32)
- UV Sensor UV Sensor ML8511

- LCD Display 16X2 I2C LCD Display
- Connecting Wires, Jumper Wires, Breadboard

UV Sensor ML8511

The ML8511 UV sensor is an ultraviolet light sensor that is straightforward to operate. In order to perform its function, the MP8511 UV (ultraviolet) Sensor sends out an analog signal that is proportional to the amount of UV light that it detects. This breakthrough could be very useful in the creation of gadgets that alert the user of sunburn or detect the UV index in relation to the conditions of the environment. This sensor is particularly sensitive to light with a wavelength between 280 and 390 nm. This is considered to be the majority of the ultraviolet A (tanning rays) spectrum as well as a portion of the ultraviolet B (burning rays) spectrum. It gives forth an analog voltage that is linearly related to the UV intensity that has been measured (in mW/cm2) as its output. If your microcontroller is capable of converting analog signals to digital ones, then you will be able to determine the amount of UV present.

Block Diagram

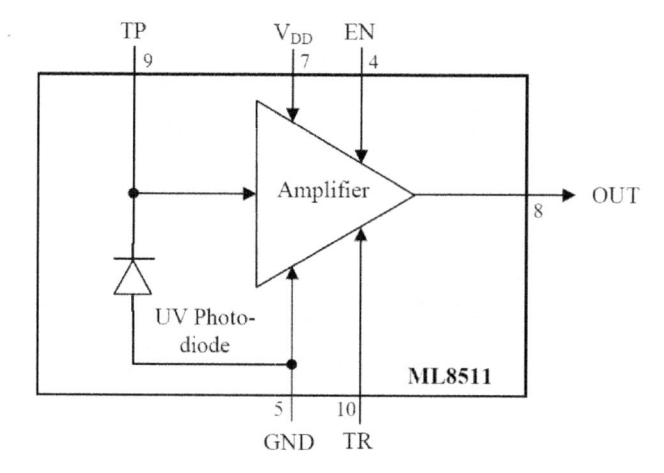

The photodiode in the UV Sensor ML8511 is sensitive to both UV-A and UV-B light. Then, it contains an embedded operational amplifier that is located internally, and this amplifier will convert photocurrent to voltage output based on how intense the UV light is. At all times, an analog voltage will be delivered as the output. The voltage output makes it simple to communicate with ADC and microcontrollers that are external to the device.

UV Index Meter with ESP32 & UV Sensor ML8511

The following is a circuit diagram that illustrates how to interface a UV sensor ML8511 with an ESP32 and an I2C LCD Display.

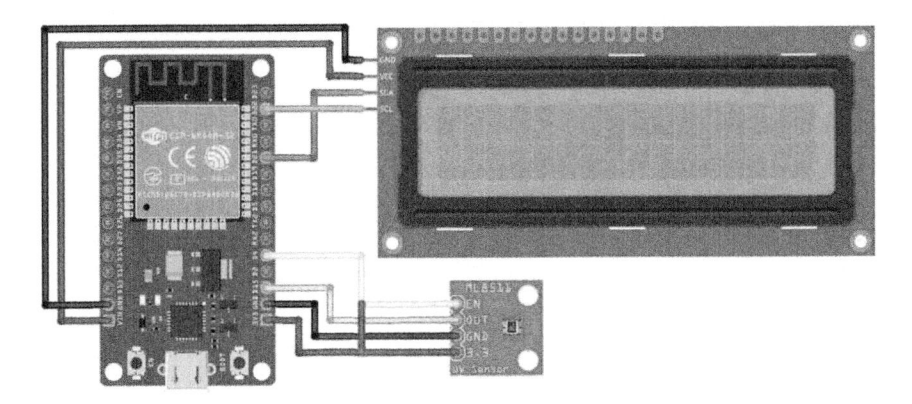

LCD receives 5V, and its ground terminal (GND) is connected to the ground terminal on ESP32. The GPIO22 and GPIO21 pins on the ESP32 are connected to the SCL and SDA pins of the I2C LCD.

Vin, 3V3, GND, OUT, and EN are the five pins that are featured on the UV Sensor. There are some modules that do not have the Vin pin, even though it is not required. The 3.3V pin of the ESP32 can be reached by connecting the EN pin and the 3V3 pin. The same 3V3 Pin that is attached to Analog pin GPIO4, which is utilized as a reference voltage, is also connected to another analog pin. The ESP32's out pin is connected to the GPIO, and the GND pin is connected to the GND pin.

This connection for ML8511 is a little bit difficult to understand. The use of VCC is necessary for any conversion from analog to digital. We are operating under the assumption that this is 5.0V; however, depending on how the board is powered, this might be as high as 5.25V or as low as 4.75V. The analog-to-digital converter (ADC) on the ESP32 is fairly imprecise as a result of this unknown timeframe. In order to resolve this issue, we make use of the extremely precise onboard 3.3V reference, which is accurate to within 1%. Therefore, we can derive a true-to-life value, regardless of what VIN is (as long as it's over 3.4V), by doing an analog-to-digital conversion on the 3.3V pin (by attaching it to GPIO4) and then comparing this result against the reading from the sensor. After the code has been uploaded, you will be able to view the UV Index when the device is exposed to sunlight.

497. IoT Based Electricity Energy Meter using ESP32 & Blynk

In this project, we will discover how to construct our very own Internet of Things–based electricity energy meter utilizing ESP32 and how to monitor data on the Blynk application. In the past, we constructed a GSM Prepaid Energy Meter. Because of the way that technology is now designed, you will need to walk into the meter reading room in order to record any readings. Therefore, keeping an eye on how much electricity you use and recording it can be a time-consuming and laborious chore. The Internet of Things is a useful tool that can help us automate this process. By automating the gathering of remote data, the Internet of Things helps users save both time and money. In recent years, the Smart Energy Meter has been the subject of a significant amount of praise all throughout the world.

Why not construct our very own Internet of Things–based electricity and energy meter?

It is necessary for us to choose the current sensor in addition to the voltage sensor so that both the current and the voltage can be measured, and therefore so that we can determine the amount of power that has been spent and the total amount of power consumed. The SCT-013 is the most accurate current sensor that can be purchased anywhere. The SCT-013 Non-Invasive AC Current Sensor Split Core Type Clamp Meter Sensor is a device that can measure AC current of up to 100 amperes and does so without causing any damage to the electrical system. In a similar vein, the AC Voltage Sensor Module ZMPT101B is the most accurate voltage sensor available. The ZMPT101B AC Voltage Sensor performs exceptionally well in situations in which an accurate measurement of AC voltage using a voltage transformer is required.

We are able to measure all of the necessary parameters for the Electricity Energy Meter by utilizing the SCT-013 Current Sensor as well as the ZMPT101B Voltage Sensor. Together, the SCT-013 Current Sensor and the ZMPT101B Voltage Sensor will be interfaced with the ESP32 Wifi Module, and then the data will be sent to the Blynk Application. The voltage, current, power, and total unit spent in kWh will all be displayed on the dashboard of the Blynk application.

Circuit Diagram & Hardware Setup

Now let's take a look at the schematic for the Internet of Things-based electricity energy meter that uses ESP32. The schematic for the circuit was created in the software known as Fritzing.

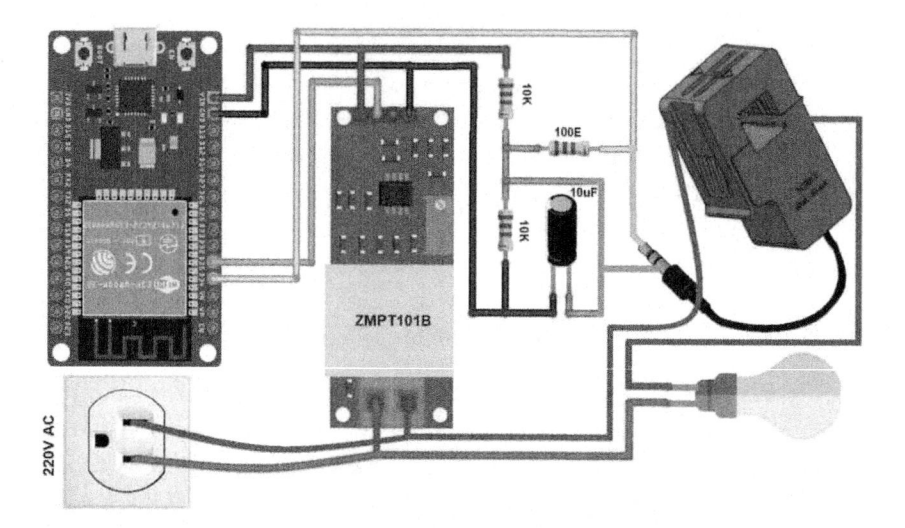

The diagram of the connections is straightforward. Both of the sensors, the SCT-013 Current Sensor and the ZMPT101B Voltage Sensor, have their VCC connections made to the ESP32's Vin supply, which is 5 volts. Both of the modules have a connection made between their GND pins and the GND pin on the ESP32. A connection has been made between the output analog pin of the ZMPT101B

Voltage Sensor and the GPIO35 of the ESP32. The output analog pin of the SCT-013 Current Sensor is also wired to the GPIO34 of the ESP32. In addition to a capacitor with a 10 uF rating, you will need two resistors with a value of 10 kilobards and a single resistor with a value of 100 ohms.

Component Required:

- ESP32 WiFi Module
- ZMPT101B AC Voltage Sensor Module
- SCT-013-030 Non-invasive AC Current Sensor
- 6x2 LCD Display
- Resistor 10K, Resistor 100ohm
- Capacitor 10uF
- Connecting Wires, and Breadboard

In addition to the circuit portion, the AC wires that need to have their current and voltage monitored are connected to the input AC Terminal of the Voltage Sensor. In a similar manner, the current sensor clip does not have any connections, and instead, a single live wire or neutral wire is inserted inside the clip component in order to complete the circuit. This is demonstrated in the diagram that was presented before. It is not necessary to complete this project with a 16x2 LCD. Because we will be watching the ESP32/SCT-013 ZMPT101B/ Energy Meter Data on the Blynk Application, there is no requirement that the LCD be connected. In the event that you want to connect the LCD, you are going to need a great deal of connections. Establish a connection between the pins 4, 6, 11, 12, 13, 14 of the LCD and the ESP32 pins D13, D12, D14, D27, D26, and D25. Additionally, connect the LCD's 1, 5, and 16 Pins to GND, and its 2, 15, and 5 Pins to VCC. To change the contrast of the LCD, connect a potentiometer with a 10K value to Pin 3 of the display.

498. IoT Based Soil Nutrient Monitoring with Arduino & ESP32

In this project, Arduino and ESP32 will be used to learn about IoT-based soil nutrient monitoring and analysis systems. Agriculture relies heavily on healthy soil. The nutrients that are provided by the soil contribute to the overall growth of a crop. A crop's production is highly influenced by a number of the soil's chemical and physical features, including its moisture level, temperature, and the proportions of nitrogen, phosphorus, and potassium present in the soil. The open-source hardware has the capability of sensing these features, and this information may then be applied in the field.

A system for the monitoring and analysis of soil nutrients, such as nitrogen, phosphorus, and potassium, is one of the things that is going to be worked on as part of this project. The farmer will be able to use this system to monitor the temperature and moisture of the soil as well as the nutrient content of the soil. All of these parameters can be monitored wirelessly on the farmer's mobile phone or on the computer system.

We are going to employ a capacitive soil moisture sensor in order to get an accurate reading of the soil's moisture content. Using the DS18B20 Waterproof Temperature Sensor, one is able to get an accurate reading of the temperature of the ground. In a similar vein, we will make use of a Soil NPK Sensor in order to determine the NPK values of the soil. The Arduino board is compatible with all of these sensors, making it simple to interface with them. We are going to monitor the data using the graphical and numerical formats provided by the Thingspeak Server. In order to transfer the data to the server, we need to be connected to a GSM or WiFi network in the field. The agricultural industry, on the other hand, does not have access to these networks. In order to find a solution to this problem, we will be utilizing the NRF2401 Wireless transceiver Module to transmit the data from the sensor Node to the Gateway.

It is possible for the data sent by the transmitter to be wirelessly relayed to the receiver from a distance of one kilometer. The ESP32 WiFi Module, which has access to the WiFi Network, is used in the construction of the receiver. The data can be sent to the Thingspeak Server if the user connects their device to this WiFi Network. Therefore, let's create an Internet of Things-based system for analyzing, monitoring, and testing the nutrient content of soil using nothing more than a wireless sensor network, Arduino, and ESP32.

Circuit Diagram:

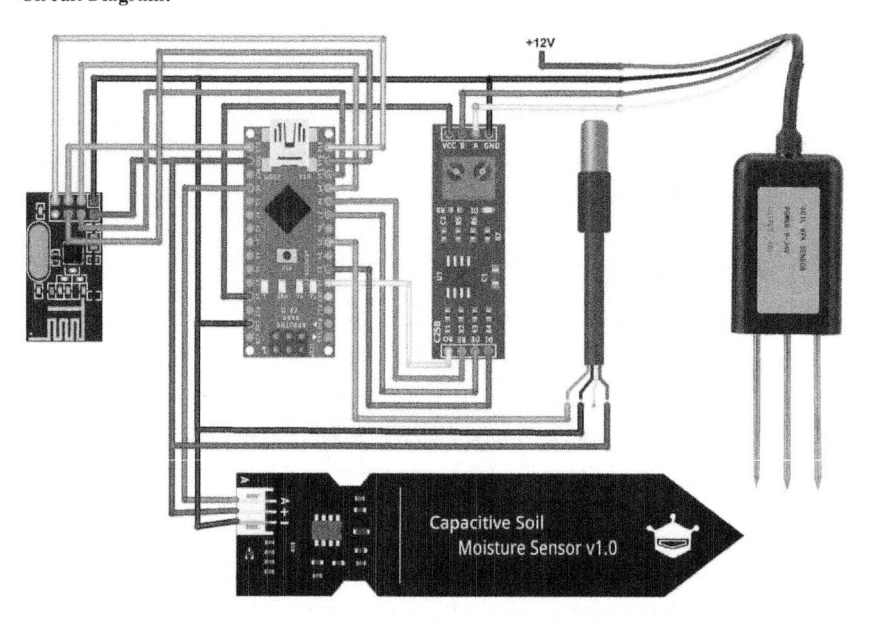

Together, the ESP32 WiFi Module and the NRF24L01 Transceiver Module made up the gateway component that we developed. The data is sent from the Sensor Node to the Gateway, and from there it is uploaded to the Thingspeak Server by the Gateway.

The circuit for the sensor node as well as its connection diagram are presented below. The Arduino Nano Board, the NRF24L01 Transceiver Module, the Soil Moisture Sensor, the DS18B20 Temperature Sensor, and the Soil NPK Sensor are the components that make up the Sensor Node.

Component Required:

- Arduino Board - Arduino Nano
- ESP32 Board ESP32 ESP-32S Development Board
- NR24L01 Module NRF24L01 PA+LNA 2.4gHz - Wireless Transceiver Module
- NPK Sensor - JXIOT Soil NPK Sensor
- Soil Moisture Sensor - Capacitive Soil Moisture Sensor v2.0
- Temperature Sensor - DS18B20 Waterproof Temperature Sensor
- Modbus Module - MAX485 Modbus
- Resistor 4.7K Resistor
- Power Supply - 9V - 12V DC Supply
- Connecting Wires, Breadboard

The connection between NRF24L01 & Arduino Nano Board is given below.

NRF24L01 **VCC** .... *3.3V of Arduino*
NRF24L01 **CSN** .... *10 of Arduino*
NRF24L01 **MOSI** .... *11 of Arduino*
NRF24L01 **GND** .... *GND of Arduino*
NRF24L01 **CE** .... *9 of Arduino*
NRF24L01 **SCK** .... *13 of Arduino*
NRF24L01 **MISO** .... *12 of Arduino*

The sensors are linked to the analog and digital pins of the Arduino board in addition to the NRF24L01 Arduino Connections. Analog pin A0 of the Arduino is connected to the capacitive soil moisture sensor's analog output. In a similar manner, the DS18B20 sensor is linked to the D5 pin on the Arduino. And the NPK Sensor is linked to the Arduino board by way of the Modbus Pin being attached to the 2,3,7,8 Pin on the Arduino. The NPK Sensor can operate on voltages between 9V and 24V. Therefore, the circuit requires an additional supply of power. The Arduino 5V/3.3V Pin is an option for powering the remaining components of the circuit.

Gateway Circuit

The Gateway is a component of both the Internet of Things-based soil nutrient content analysis and monitoring. The ESP32 WiFi Module and the NRF24L01 Wireless Transceiver Module were used in the construction of the Gateway.

The diagram below illustrates the connection that should be made between the ESP32 Board and the NRF24L01 Wireless Transceiver Module.

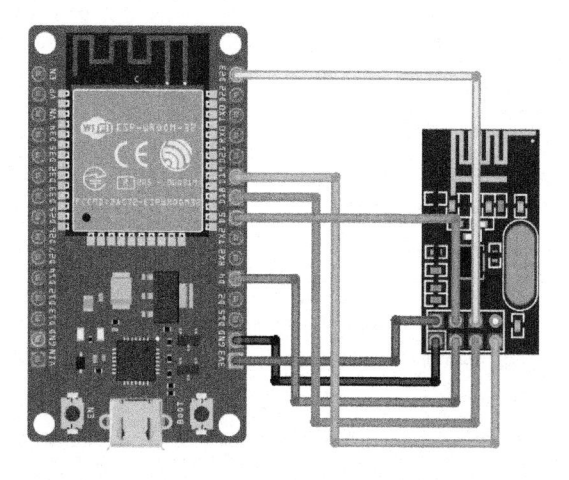

NRF24L01 **VCC**	**3.3V** of ESP32
NRF24L01 **CSN**	**D5** of ESP32
NRF24L01 **MOSI**	**D23** of ESP32
NRF24L01 **GND**	**GND** of ESP32
NRF24L01 **CE**	**D4** of ESP32
NRF24L01 **SCK**	**D18** of ESP32
NRF24L01 **MISO**	**D19** of ESP32

499. Ultrasonic Range Finder with ESP32 TFT Display & HC-SR04

In this project, we'll make an ultrasonic range finder with an ESP32 TFT display and an HC-SR04 ultrasonic sensor. The project makes use of an integrated board that has a 3.5-inch touchscreen display, an ESP32-WROVER Module, and an OV2640 camera that has a built-in resolution of 2 million pixels. We will attach the HC-SR04 Ultrasonic Sensor to the digital pins on the ESP32 board in an external connection. The HC-SR04 ultrasonic sensor uses SONAR, just like bats do, to figure out how far away something is. It provides outstanding non-contact range detection from 2 centimeters to 400 centimeters, with high precision and reliable readings in a compact that is simple to use. The range is equivalent to 1 inch to 13 feet. An ILI9488 TFT LCD Driver is utilized in the construction of the 3.5-inch TFT Touch Screen Display. The resolution of the screen is 320 by 480. There are a number of external pins on the ESP32 display that can be used to connect it to sensors or actuators. Because of this, they are appropriate for a wide variety of Internet of Things applications. Now, we'll look at how the ESP32 TFT Display and the Ultrasonic Sensor can be used to make a beautiful, dynamic analog-style display. On the TFT Touch Display of the ESP32 Ultrasonic Range Finder, the distance will be displayed in centimeters.

Part Required:

- ESP32+TFT Touch Display - ESP32 3.5" TFT Touch (Capacitive) with Camera
- Ultrasonic Sensor - HC-SR04 Ultrasonic Sensor
- USB Cable - Type-C USB Cable for Programming
- Connecting Wires - Male to Female Jumper Wires

Hardware Connections for ESP32 TFT Display with HC-SR04 Ultrasonic Sensor

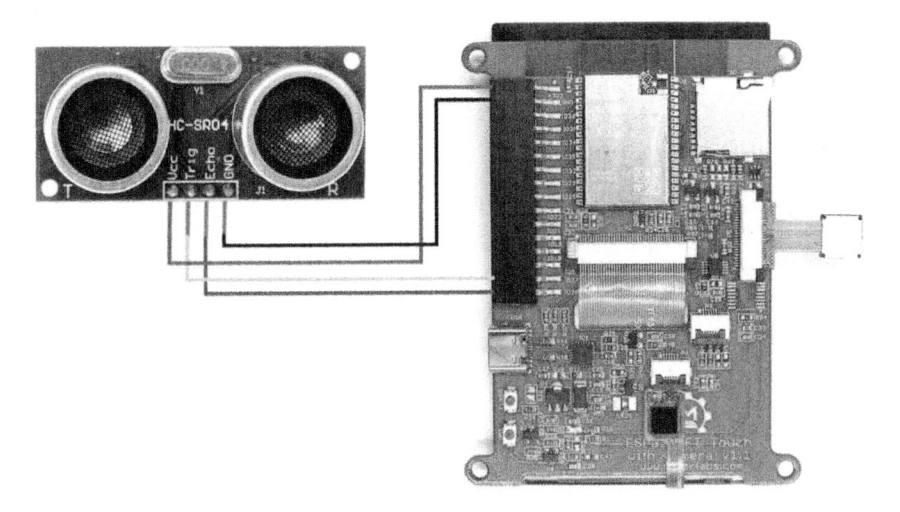

The following connections are required in order to construct an ultrasonic range finder utilizing an ESP32 and an HC-SR04 sensor. Since the TFT Touch Display and ESP32-Wrover Module are already attached on the inside of the PCB, the only connection that we need is for the Ultrasonic Sensor. Follow the instructions in the diagram below to connect the HC-SR04 Sensor. Connect the HC-SR04 Sensor's VCC and GND Pins to the ESP32's 3.3V and GND Pins respectively. In a similar fashion, connect its echo and trig pin to the GPIO26 and GPIO27 correspondingly on the ESP32.

500.DIY IoT Water pH Meter using pH Sensor & ESP32

In this project, we'll use a pH sensor and an ESP32 WiFi module to make our own DIY IoT-based water pH meter. Earlier on in this series of tutorials, we made a portable ph meter by interfacing an Arduino with a pH Sensor. On an OLED screen, the data from the pH reading was presented. This is an entirely new project, in which we will make use of a more sophisticated pH sensor than in the previous one. The pH Sensor that is being utilized in this situation is an analog type pH Sensor that provides a linear pH measurement in the range of 0 to 14 pH. We are going to transfer the data to ThingSpeak Server rather than showing it on an OLED panel. The ESP32, which is the gadget we are using, posts data to the ThingSpeak platform, which then delivers rapid visualizations of that data. When compared to Arduino, which only has a 10-bit analog to digital converter (ADC), the ESP32's 12-bit controller with built-in ADC allows for more accurate data measurement. It includes a high-resolution pH probe that takes readings of the pH of liquids at intervals that are set by the user and then sends the data to distant servers. The pH probe is equipped with a cable that is five feet long and can be submerged in any solution. The amount of time required to respond is less than one minute. We are able to implement this IoT-based pH meter in a variety of settings, including hydroponics, laboratories, and aquariums. Now that we know how to create our own pH meter, let's learn how to interface the pH sensor with the ESP32. The pH Sensor Calibration technique is additionally covered in this project.

Part Required:

- ESP32 Board ESP32 ESP-32S Development Board (ESP-WROOM-32)
- pH Sensor PH SENSOR KIT for Water
- Battery 9V Battery or DC Adapter
- Connecting Wires
- Jumper Wires, Breadboard

Precautions

PH sensor probe Because the tip of the tube is so delicate, you should avoid touching it with your hands and avoid placing it on the ground. Because of its extreme fragility, the bulb-shaped tip must always be kept in the storage solution while it is not in use.

Circuit Diagram: Testing the Interfacing between the PH Sensor and the ESP32

Let's begin with the most fundamental level of interfacing and testing of the pH sensor with the ESP32 WiFi Module before moving on to the IoT pH Meter.

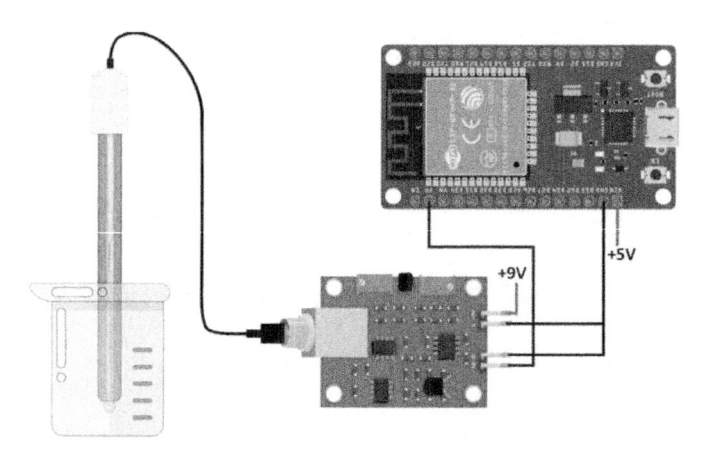

You can use either an external 9-volt battery or a 9-volt DC supply to power the pH, Sensor. Connect the A0 pin of the ESP32 to the output pin of the pH sensor signal board. The ESP32's VP pin can serve in this capacity. The output of the sensor has a range of 0.5V to 3V, which enables the sensor to be utilized with the analog pins of the ESP32.

Summary

The following chapter should be quite interesting to you, and we presume that you have practiced more than 500 Electronic Project Ideas as they will be useful to you in building your dream project. In upcoming sessions, we will cover How to troubleshoot and Fix Arduino Issues as well as how to repurpose your old Arduino.

Essential Resources

- Here are links to projects, tutorials, parts, and online communities to help you get started with electronics projects http://www.instructables.com/id/Electronic-Projects-For-Beginners/ http://www.instructables.com/id/Beginners-Electronics-Projects/
- Electronicsforu.com: https://www.electronicsforu.com/category/electronics-projects/hardware-diy
- 5 Beginner Projects That Work on The Initial Attempt: Links to videos of simple projects, with a clap on/clap off switch. http://www.buildcircuit.com/5-beginners-projects-that-work-in-the-first-attempt/
- Simple Electronics Projects for Beginners: Articles by reader comments for a project involving an FM radio transmitter, a water level indicator, and an infrared motion detector: http://www.circuitstoday.com/simple-electronics-projects-and-circuits
- Electronics Projects for Dummies: No one who has seen this magazine is a dummy, but this Dummies website has a great step-by-step group of projects for a coin toss circuit that also impart the process of designing and making electronics projects: http://www.dummies.com/how-to/consumer-electronics/electronics/Electronics-Projects.html
- Beginner Electronics Projects from Radio Shack: Actually, this is an impression article with ideas on what they offer students who want to get started with electronics projects, such as Engineer Small Notebooks, which sound interesting. http://techchannel.radioshack.com/beginner-electronics-projects-1831.html
- Electronics Projects from Makenzie: While most of their projects are not for novices, this is a great website to browse to get enthusiastic about what you might do once you finish a few

- electronics projects for beginners. On this website, there are some really well-organized projects. http://makezine.com/category/electronics/
- Ben Heck Show: Back to Basics: Important abilities include not frying your projects.http://www.youtube.com/playlist?list=PLwO8CTSLTkijrSW6DIFsQxcvjRo5fZ-y5
- Sparkfun Future Tutorials: https://learn.sparkfun.com/tutorials
- How to Solder: The ability to solder, which uses heat to fuse two soft metals, is an important skill for electronics projects. https://learn.sparkfun.com/tutorials/how-to-solder—through-hole-soldering/all
- Parts for Electronics Projects: ADAFRUIT: to concentrate on Arduino, Raspberry Pi, and Beagle Board parts and projects through lots of tutorials. https://www.adafruit.com/
- ELEMENT 14: Element 14 demonstration that features electronics projects, as well as a free online resource where you can get assistance with your projects. http://www.element14.com/community/welcome
- OCTOPARTS a search engine to discover electronic parts from various different sellers. http://www.octopart.com/
- MINIBREAD http://www.minibread.com/
- SPARK FUN https://www.sparkfun.com/
- Arduino Official Home Page https://www.arduino.cc/ Info on hardware, some libraries for sensors, the integrated development environment (IDE) and reference material on commands and structures
- www.instructables.com : Info on DIY projects prepared by others
- https://create.arduino.cc/projecthub
- https://learn.adafruit.com Adafruit makes many shields and sensors, and they have tutorials for almost everything they carry.
- http://www.arduinoclassroom.com/index.php/arduino-101 Arduino Classroom is currently doing an intro series on Arduinos.
- http://playground.arduino.cc/: The Arduino Playground is the wiki run by the Arduino Company for its products. There is a lot of helpful information on almost everything imaginable here.

- Github Arduino Page https://github.com/arduino/Arduino
- Hackster Arduino Page https://www.hackster.io/arduino
- Hackster Arduino Projects https://www.hackster.io/arduino/projects
- Circuit digest Arduino Projects https://circuitdigest.com/arduino-projects
- Electronics Hub Arduino Project Ideas https://www.electronicshub.org/arduino-project-ideas/
- Electronics for you Arduino Project Ideas https://www.electronicsforu.com/arduino-projects-ideas
- Home of Make Magazine, which has lots of Arduino projects www.makezine.com
- Arduino Projects by All about circuits dot com https://www.allaboutcircuits.com/projects/category/arduino
- Arduino Projects by How to mechatronics dot com https://howtomechatronics.com/arduino-projects/
- Arduino Official YouTube Channel https://www.youtube.com/c/Arduino/videos

Bibliography

[1]. Earl Boysen and Nacy C. Muir (2006), Electronics Projects for Dummies.

[2]. Cathleen Shamieh (2015), Electronics for Dummies

[3]. Charles Platt (2015), Make Electronics, Second Edition.

[4]. Bill Pretty (2015), Getting Started with Electronics Projects

[5]. Arduino official web site https://www.arduino.cc/

[6]. Exploring Arduino: Tools and Techniques for Engineering Wizardry; 2nd Ed; Jeremy Blum; Wiley; 512 pages; 2019;

[7]. Fritzing. http://fritzing.org/

[8]. Programming Arduino Next Steps: Going Further with Sketches, 2nd Edition, Simon Monk, McGraw-Hill Education, 2018, 320 pages

[9]. Arduino. https://www.arduino.cc

[10] J. Fraden, Handbook of Modern Sensors, Springer Verlag, Berlin, third edition, 2004.

[11]. Hackster Arduino Page https://www.hackster.io/arduino

[12]. Arduino Workshop: A Hands-On Introduction with 65 Projects; 1st Ed; John Boxall; No Starch Press; 392 pages; 2013;

[13]. Circuit digest Arduino Projects https://circuitdigest.com/arduino-projects

[14]. Arduino For Dummies; 2nd Ed; John Nussey; John Wiley & Sons; 400 pages; 2018;

[15]. Arduino Official YouTube Channel https://www.youtube.com/c/Arduino/videos

[16]. Make: Getting Started with Arduino; 3rd Ed; Massimo Banzi, Michael Shiloh; Make Community; 262 pages; 2014;

[17]. Hackster Arduino Page https://www.hackster.io/arduino

[18]. Programming Arduino: Getting Started with Sketches; 2nd Ed;

[19]. Electronics for you Arduino Project Ideas https://www.electronicsforu.com/arduino-projects-ideas

[20]. Simon Monk; McGraw-Hill Education; 192 pages; 2016

[21]. Circuit digest Arduino Projects https://circuitdigest.com/arduino-projects

[22]. Beginning C for Arduino: Learn C Programming for the Arduino; 2nd Ed; Jack Purdum; Apress; 388 pages; 2015;

[23]. Github Arduino Page https://github.com/arduino/Arduino

[24]. Electronics Hub Arduino Project Ideas https://www.electronicshub.org/arduino-project-ideas/

[25]. Hackster Arduino Projects https://www.hackster.io/arduino/projects

[26]. Arduino: A Quick Start Guide; 2nd Ed; Maik Schmidt; Pragmatic Bookshelf; Pragmatic Bookshelf; 323 pages; 2015;

[27] The home of Make Magazine, which contains many Arduino projects. www.makezine.com

[28]. Arduino Projects by How to mechatronics dot com https://howtomechatronics.com/arduino-projects/

[29]. Make: Sensors; 1st Ed; Tero Karvinen, Kimmo Karvinen, Ville Valtokari; Make Community; 400 pages; 2014;

ABOUT THE AUTHOR

Biomedical engineer and YouTuber Arsath Natheem is from India. He focuses on blockchain technology, artificial intelligence, and data science. He won the best project award for a human interaction intelligence robot and an IoT-based voice recognition robot for defense applications. His multimedia presentation "How Biomedical Engineers Save Lives," which was shown at VCET in Tamilnadu, is what he is best known for. He also presented his project at the Adhiyaman CET and won the competition. He took part in a project competition at the Madras Institute of Technology (MIT) in Chennai. He was interested in R&D in data science and content creation; he now works on Amazon as a self-publishing author and technical writer.

ONE LAST THING...

If you enjoyed this book or found it useful, I'd be very grateful if you'd post a short review on Amazon. Your support really does make a difference and I read all the reviews personally so I can get your feedback and make this book even better.

We gift you this PDF version of the book, which you can read or download through the link below or scan the QR code. https://bit.ly/3YuE4Q3

Grab an enjoyable gift and a convenient moment. I'm really looking forward to it!

Thanks again for your love & support!

Cheers,

Printed in Great Britain
by Amazon